Cyber-Physical Systems

From Theory to Practice

OTHER COMMUNICATIONS BOOKS FROM AUERBACH

Cyber-Physical Systems

From Theory to Practice

Edited by
Danda B. Rawat
Joel J.P.C. Rodrigues
Ivan Stojmenovic

CRC Press
Taylor & Francis Group
Boca Raton London New York

CRC Press is an imprint of the
Taylor & Francis Group, an **informa** business

MATLAB® is a trademark of The MathWorks, Inc. and is used with permission. The MathWorks does not warrant the accuracy of the text or exercises in this book. This book's use or discussion of MATLAB® software or related products does not constitute endorsement or sponsorship by The MathWorks of a particular pedagogical approach or particular use of the MATLAB® software

CRC Press
Taylor & Francis Group
6000 Broken Sound Parkway NW, Suite 300
Boca Raton, FL 33487-2742

First issued in paperback 2020

© 2016 by Taylor & Francis Group, LLC
CRC Press is an imprint of Taylor & Francis Group, an Informa business

No claim to original U.S. Government works

ISBN 13: 978-0-367-57542-7 (pbk)
ISBN 13: 978-1-4822-6332-9 (hbk)

Visit the Taylor & Francis Web site at
http://www.taylorandfrancis.com

and the CRC Press Web site at
http://www.crcpress.com

To Ivan Stojmenovic

Contents

Section I   Control Systems

Section II   Modeling and Design

Section III   Communications and Signal Processing

Contents

Section VII   Sensors and Applications

Section VIII   Computing Issues

Preface

Over the past decades, science and technology have gone through tremendous changes in terms of computing, communications, and control to provide a wide range of applications in all domains. This advancement provides opportunities to bridge physical components and cyber space, leading to cyber-physical systems (CPS). The notion of CPS is to use computing (sensing, analyzing, predicting, understanding), communication (interaction, intervene, interface management), and control (interoperate, evolve, evidence-based certification) to make intelligent and autonomous systems. The complexity in CPS has increased due to the integration of cyber components with physical systems. A comprehensive knowledge base of the CPS domain is required not only for researchers and practitioners but also for policy makers and system managers.

The main motivation for offering this book stems from the realization that, in spite of its importance and potential socioeconomic impact, at present, there is no comprehensive collection of research and resource of information about CPS. We believe that there is a value in bringing together a knowledge base in an integrated and cohesive manner for CPS that promises to see remarkable growth in the next few years.

This book consists of eight broad sections, each containing a number of chapters contributed by leading experts in the field. Each section covers many different areas that impact the design, modeling, and evaluation of CPS.

Section I: Control Systems
Section II: Modeling and Design
Section III: Communications and Signal Processing
Section IV: Mobility Issues
Section V: Architecture
Section VI: Security Issues
Section VII: Sensors and Applications
Section VIII: Computing Issues

This book provides state-of-the-art research results and trends related to the science of CPS, technology of CPS, and engineering of CPS from theory to practice. Some content from some of the chapters overlap. This is deliberate since it provides a benefit to the readers by providing an opportunity to get different perspectives on the same or related topics. This book has chapters on control engineering, communication, computing, signal processing, modeling and design, mobility, security, sensors, and applications that provide insight for the design, analysis, evaluation, and validation of CPS.

We expect this book to be of significant importance not only to researchers and practitioners in academia, government agencies, and industries but also for policy makers and system managers. We anticipate this book to be a valuable resource for all those working in this new and exciting area, and a *must have* for all university libraries.

This book would not have been published without the contributions of several people. First and foremost, we would like to express our warm appreciation to the authors for their willingness to share their knowledge by contributing chapters. We are indebted to the referees for their delightful work and valuable time in ensuring high quality of the offerings.

Danda B. Rawat
Georgia Southern University
Statesboro, Georgia

Joel J.P.C. Rodrigues
Instituto de Telecomunicações
University of Beira Interior
Covilhã, Portugal
University of Fortaleza (UNIFOR)
Fortaleza, Ceará, Brazil

Ivan Stojmenovic
University of Ottawa
Ottawa, Ontario, Canada

MATLAB® is a registered trademark of The MathWorks, Inc. For product information, please contact:

The MathWorks, Inc.
3 Apple Hill Drive
Natick, MA 01760-2098 USA
Tel: 508-647-7000
Fax: 508-647-7001
E-mail: info@mathworks.com
Web: www.mathworks.com

Editors

Danda B. Rawat, PhD, received his PhD in electrical and computer engineering from Old Dominion University, Norfolk, Virginia. He is an assistant professor in the Department of Electrical Engineering at Georgia Southern University (GSU), Statesboro, Georgia. His research focuses on cognitive radio networks, cyber-physical systems, software-defined networks, cybersecurity, smart grids, and vehicular/wireless ad hoc networks. He is the Director of Cybersecurity, Wireless Systems, and Networking Innovations (CWiNs) Lab (http://www.CWiNs.org) at GSU. His research is supported by the U.S. National Science Foundation and the Center for Sustainability grants. Dr. Rawat has published more than 130 scientific/technical articles. He has been serving as an editor/guest editor on over 10 international journals. He serves as the webmaster for IEEE INFOCOM 2016, student travel grant cochair of IEEE INFOCOM 2015, track chair for wireless networking and mobility of IEEE CCNC 2016, track chair for Communications Network and Protocols of IEEE AINA 2015, and many others. He has served as a program chair, general chair, and session chair for numerous international conferences and workshops, and has served as a technical program committee (TPC) member for several international conferences, including IEEE GLOBECOM, IEEE CCNC, IEEE GreenCom, IEEE AINA, IEEE ICC, IEEE WCNC, and IEEE VTC. Dr. Rawat is a senior member of IEEE and a member of ACM. Dr. Rawat has been a vice chair of the Executive Committee of the IEEE Savannah Section since 2013. He has also received the Best Paper Awards at international conferences.

Joel J.P.C. Rodrigues, PhD, is a professor in the Department of Informatics at the University of Beira Interior, Covilhã, Portugal, and senior researcher at the Instituto de Telecomunicações, Portugal. He received the Academic Title of Aggregated Professor from the University of Beira Interior, the habilitation in computer science and engineering from the University of Haute Alsace, France, a PhD in informatics engineering, an MSc from the University of Beira Interior, and a 5-year BSc (licentiate) in informatics engineering from the University of Coimbra, Portugal. His main research interests include sensor networks, e-health, e-learning, vehicular delay-tolerant networks, and mobile and ubiquitous computing. He is the leader of the NetGNA Research Group (http://netgna.it.ubi.pt), the chair of the IEEE ComSoc Technical Committee on eHealth, the past chair of the IEEE ComSoc Technical Committee on Communications Software, a Steering Committee member of the IEEE Life Sciences Technical Community, a member representative of the IEEE Communications Society on the IEEE Biometrics Council, and an officer of the IEEE 1907.1 standard. Dr. Rodrigues is the editor in chief of the *International Journal on E-Health and Medical Communications, Recent Advances on Communications and Networking Technology, Journal of Multimedia Information Systems*, and editorial board member of several other journals. He has been general chair and TPC Chair of many international conferences, including IEEE, ICC, and GLOBECOM. He is a member of many international TPCs and has participated in several international conferences. Dr. Rodrigues has authored or coauthored more than 400 papers in refereed international journals and conferences and a book and holds two patents. He has been awarded several Outstanding Leadership and Outstanding Service Awards by the IEEE Communications Society and several best paper awards. Dr. Rodrigues is a licensed professional engineer (as a senior member), member of the Internet Society, an IARIA fellow, and a senior member of ACM and IEEE. Dr. Rodrigues can be reached via e-mail at: joeljr@ieee.org.

Contributors

Baher Abdulhai
Faculty of Engineering
Department of Civil Engineering
ITS Center and Testbed
University of Toronto
Toronto, Ontario, Canada

Mainak Adhikari
Department of Computer Science and
  Engineering
IMPS College of Engineering and Technology
West Bengal, India

Daniel Alonso
Department of Information and Communication
  Technology
Centre for Integrated Emergency Management
University of Agder
Kristiansand, Norway

and

Group of Information and Communication
  Systems
Universidad de Valencia
Valencia, Spain

Cesar Asensio
Department of Information and Communication
  Technology
Centre for Integrated Emergency Management
University of Agder
Kristiansand, Norway

and

Group of Information and Communication
  Systems
Universidad de Valencia
Valencia, Spain

César Azurdia
Department of Electrical Engineering
Universidad de Chile
Santiago, Chile

Sourav Banerjee
Department of Computer Science and
  Engineering
Kalyani Government Engineering College
West Bengal, India

Baltasar Beferull-Lozano
Group of Wireless Communications
  and Embedded Systems
Department of Information and Communication
  Technology
Centre for Integrated Emergency Management
University of Agder
Grimstad, Norway

and

Group of Information and Communication
  Systems
Universidad de Valencia
Valencia, Spain

Umesh Bellur
Department of Computer Science and
  Engineering
Indian Institute of Technology Bombay
Mumbai, India

Utpal Biswas
Department of Computer Science and
  Engineering
University of Kalyani
West Bengal, India

Eugenio Celada
Department of Information and Communication
  Technology
Centre for Integrated Emergency Management
University of Agder
Kristiansand, Norway

and

Group of Information and Communication
  Systems
Universidad de Valencia
Valencia, Spain

Sandra Céspedes
Department of Electrical Engineering
Universidad de Chile
Santiago, Chile

Sanjay Chaudhary
Institute of Engineering and Technology
Ahmadabad University
Ahmadabad, India

Miguel Correia
INESC-ID
Instituto Superior Técnico
Universidade de Lisboa
Lisboa, Portugal

Siyuan Dai
Institute for Software-Integrated Systems
Vanderbilt University
Nashville, Tennessee

Mohamed El-Darieby
Software Systems Engineering
University of Regina
Regina, Saskatchewan, Canada

Mohamed Elshenawy
Faculty of Engineering
Department of Civil Engineering
ITS Center and Testbed
University of Toronto
Toronto, Ontario, Canada

Alireza Esfahani
Instituto de Telecomunicações
Campus Universitário de Santiago
Aveiro, Portugal

Claudio Estevez
Department of Electrical Engineering
Universidad de Chile
Santiago, Chile

Vijay Gupta
Department of Electrical Engineering
University of Notre Dame
Notre Dame, Indiana

Sukhendu Kar
Department of Computer Science and
  Engineering
IMPS College of Engineering and Technology
West Bengal, India

Abhirup Khanna
Department in Center of Information
  Technology
University of Petroleum and Energy Studies
Dehradun, India

Xenofon Koutsoukos
Institute for Software-Integrated Systems
Vanderbilt University
Nashville, Tennessee

Zsolt Lattmann
Institute for Software-Integrated Systems
Vanderbilt University
Nashville, Tennessee

Xue Liu
School of Computer Science
McGill University
Montréal, Québec, Cananda

Alexey Lukashin
Telematics Department
Saint Petersburg Polytechnic University
Saint Petersburg, Russia

Tie Luo
Institute for Infocomm Research (I2R)
Agency for Science, Technology and Research
  (A*STAR)
Singapore, Singapore

Wann-Jiun Ma
Department of Mechanical Engineering and
  Materials Science
Duke University
Durham, North Carolina

Naércio Magaia
INESC-ID
Instituto Superior Técnico
Universidade de Lisboa
Lisboa, Portugal

Georgios Mantas
Instituto de Telecomunicações
Campus Universitário de Santiago
Aveiro, Portugal

Vladimir Muliukha
Russian State Scientific Center for
  Robotics and Technical Cybernetics
Saint Petersburg, Russia

Alberto Nascimento
Instituto de Telecomunicações
Campus Universitário de Santiago
Aveiro, Portugal

Jose Carlos Neves
Instituto de Telecomunicações
Campus Universitário de Santiago
and
Universidade de Aveiro
Aveiro, Portugal

Sotiris Nikoletseas
Department of Computer Engineering and
  Informatics
University of Patras
and
Computer Technology Institute and
  Press–Diophantus
Patras, Greece

Mattias Nyberg
Department of Machine Design
KTH Royal Institute of Technology
Stockholm, Sweden

Athanasia Panousopoulou
Signal Processing Laboratory
Institute of Computer Science
Foundation for Research and
  Technology–Hellas
Crete, Greece

Pankesh Patel
ABB Corporate Research
Bangalore, India

Jian Pei
Department of Computer Science
Simon Fraser University
Vancouver, British Columbia, Canada

Paulo Pereira
INESC-ID
Instituto Superior Técnico
Universidade de Lisboa
Lisboa, Portugal

Daniel E. Quevedo
Department of Electrical Engineering (EIM-E)
The University of Paderborn
Paderborn, Germany

Theofanis P. Raptis
Department of Computer Engineering and
  Informatics
University of Patras
and
Computer Technology Institute and
  Press–Diophantus
Patras, Greece

Ajay Rawat
University of Petroleum and Energy Studies
Dehradun, India

Jonathan Rodriguez
Instituto de Telecomunicações
Campus Universitário de Santiago
and
Universidade de Aveiro
Aveiro, Portugal

Sandip Roy
School of Electrical Engineering and Computer
  Science
Washington State University
Pullman, Washington

Ricardo G. Sanfelice
Department of Computer Engineering
University of California, Santa Cruz
Santa Cruz, California

Lui Sha
Department of Computer Science
University of Illinois at Urbana–Champaign
Urbana, Illinois

Emad Soroush
GraphLab
Seattle, Washington

Stavros Tripakis
Aalto University
Helsinki, Finland

and

University of California
Berkeley, California

Grigorios Tsagkatakis
Signal Processing Laboratory
Institute of Computer Science
Foundation for Research and
  Technology–Hellas
Crete, Greece

Panagiotis Tsakalides
Signal Processing Laboratory
Institute of Computer Science
Foundation for Research and
  Technology–Hellas
Crete, Greece

George Tzagkarakis
Signal Processing Laboratory
Institute of Computer Science
Foundation for Research and
  Technology–Hellas
Crete, Greece

Jonas Westman
Department of Machine Design
KTH Royal Institute of Technology
Stockholm, Sweden

Kui Wu
Department of Computer Science
University of Victoria
Victoria, British Columbia, Canada

Mengran Xue
School of Electrical Engineering and Computer
  Science
Washington State University
Pullman, Washington

Du Yang
Instituto de Telecomunicações
Campus Universitário de Santiago
Aveiro, Portugal

Jianguo Yao
School of Software
Shanghai Jiao Tong University
Shanghai, Minhang, People's Republic of China

Vladimir Zaborovsky
Telematics Department
Saint Petersburg Polytechnic University
Saint Petersburg, Russia

Lichen Zhang
Faculty of Computer Science and Technology
Guangdong University of Technology
Guangzhou, Guangdong, People's
  Republic of China

Zhanyang Zhang
Department of Computer Science
College of Staten Island
City University of New York
Staten Island, New York

Guchuan Zhu
Department of Electrical Engineering
Ecole Polytechnique de Montreal
Montréal, Québec, Canada

Section I

Control Systems

1 Analysis and Design of Cyber-Physical Systems
A Hybrid Control Systems Approach

Ricardo G. Sanfelice

CONTENTS

1.1 INTRODUCTION

A *cyber-physical system* is a system that combines physical and computer or cyber components. The physical component consists of systems that exist in nature, such as biological entities as well as those developed by humans, such as transportation and energy-producing systems. This component exists, operates, and interacts with its environment in continuous or ordinary time. The computational component consists of systems and entities involved in processing, communicating, and controlling information via computational means. These include algorithms implemented in software and digital systems, interfaced to the physical components via analog-to-digital converters (ADCs) and digital-to-analog (DACs) converters and digital communication networks. They are man-made systems that operate in discrete-time or in an event-driven fashion. Evidently, the complexity of component

3

integration in cyber-physical systems stems from the fact that the computational component is distributed throughout the system, and tightly coupled with the physical component. As a consequence, cyber-physical systems are highly complex systems combining continuous and discrete dynamics.

This chapter considers cyber-physical systems that are modeled by the interconnections of by physical components, cyber components, and the systems needed to interconnect them. The physical components include the systems that exist in the physical world, for example, a process which needs to be monitored or controlled. The cyber components capture the computer algorithms in the system, for example, a communication or a control algorithm. The subsystems used to interconnect *them* include interfaces, converters, signal conditioners, and networks. Models from the theory of hybrid systems are employed to capture the mixed continuous and discrete behaviors of some of these subsystems and of the entire interconnection. More specifically, dynamical models of physical components, cyber components, and systems used to interface them have been proposed within the framework for hybrid dynamical systems [1]. In this framework, hybrid dynamical systems are the combination of differential inclusions and difference inclusions with constraints, namely, *hybrid inclusions*. Such a combination leads to dynamical models with variables that may change continuously as determined by the differential inclusions while, at times, may also change discretely according to the difference inclusions. These capabilities make hybrid inclusions very suitable for mathematical modeling of cyber-physical systems. Furthermore, the stability and robustness of hybrid inclusion models of cyber-physical systems can be systematically analyzed using a recently developed theory [1] for hybrid dynamical systems without inputs.

Motivated by the recent results on modeling, analysis, and design of hybrid inclusions, this chapter compiles a collection of methods for hybrid inclusions that are suitable for the study of the class of cyber-physical systems of interest here. A mathematical model for the physical components in terms of differential inclusions with set-based enabling conditions on the state and input is provided in Section 1.2.1. This model captures the continuous dynamics of a hybrid system with inputs, which not only extends the model [1], but is also required to be able to define an interconnection between the physical components and the other subsystems; see also the models Sanfelice [2]. Several mathematical models of cyber components are provided in Section 1.2.2 in terms of difference inclusions that are enabled when certain conditions on the state and input are satisfied. First, following the classical definitions Hopcroft and Ullman [3] and the models Girault, Lee, and Lee [4], models of finite-state machines (FSMs) in terms of difference inclusions are presented to capture the evolution of deterministic and nondeterministic FSMs (pure and with conditional structures). The model proposed for FSMs with conditional structures implements such conditions in terms of set conditions involving the state, input, and output. Next, models of digital computations and discrete-time algorithms are introduced to capture one-shot and iterative computations, as well as the algorithms obtained when discretizing the dynamics of a continuous-time system. The examples in Section 1.2.2 illustrate how these models can be used to describe operations performed by the algorithms implemented in the software of cyber-physical systems. The models of the systems used to interconnect physical and cyber components are introduced in Section 1.2.3. Due to being at the interface between physical and cyber components, these models involve both differential and difference equations. A model of the system performing the conversion of analog signals into digital signals is proposed first. This model uses a timer to trigger the sampling events and a memory state to store the samples. A similar model is proposed to capture the operation of DACs of the zero-order hold type. In addition to models of converters, a model of a digital network in which transmission of data occurs with a rate that is within a prespecified range is proposed. The model is nondeterministic because a difference inclusion is used to model all of the possible transmission rates within the allowed range. The combination of the models in Sections 1.2.1 through 1.2.3 leads to a hybrid inclusion model. Section 1.2.4 includes examples of such combinations and motivates the hybrid inclusion modeling approach for cyber-physical systems that is advocated in this chapter.

With a hybrid inclusion model for the class of cyber-physical of interest, Section 1.3 summarizes tools for analysis and design. Following the ideas Goebel, Sanfelice, and Teel [1], a notion of time and execution for these systems is introduced first in Section 1.3.1. In this notion, time is hybrid in the sense that a time instant uniquely determines the amount of time that variables have evolved continuously as well as discretely. Then, executions (or solutions) are given by a state and input pair that are parameterized by the said notion of hybrid time. Solutions are classified according to the properties of their domain of definition, which are given by sets called hybrid time domains. Conditions guaranteeing properties of their *hybrid time domains* are given at the end of Section 1.3.1. Using this notion of solution, a definition of an *invariant* for the proposed model of a cyber-physical system is proposed in Section 1.3.2. Invariants are sets of particular interest, for example, when studying reachability and safety. Sufficient conditions that are useful for determining the invariants of a system are given. Section 1.3.3 pertains to stability and attractivity, which are properties of particular interest to cyber-physical systems with a feedback topology. Following Goebel, Sanfelice, and Teel [1], a Lyapunov stability notion for sets is introduced and a sufficient condition in terms of Lyapunov functions is presented. Attractivity of a set is also defined, and an invariance principle for hybrid inclusions from Sanfelice, Goebel, and Teel [5] are presented as a tool to determine the set of points to which solutions converge. In addition to the methods for studying invariants and stability, conditions that determine whether a function of the state is true or not is proposed following the theory of temporal logic. Operators and associated semantics are defined for solutions to hybrid inclusions. Conditions on their solutions for the satisfaction of specific formulae are given in Section 1.3.4. Further conditions under which the properties of a cyber-physical system modeled as a hybrid inclusion are robust to small perturbations, which are provided in Section 1.3.5. Robustness to small perturbations is mandatory for cyber-physical systems since several of the perturbations introduced during the implementation of the cyber components, while they cannot always be neglected, can be assumed to be small, for example, quantization and discretization effects. Finally, a brief overview of a tool for simulating hybrid inclusions in general, and the class of cyber-physical systems considered here in particular, is discussed in Section 1.3.6.

1.2 DYNAMICAL MODELS OF CYBER-PHYSICAL SYSTEMS

In this chapter, the temporal evolution of the variables of a cyber-physical system is captured using dynamical models. The state of the physical components is typically determined by variables that change according to physical time (or ordinary time) and take values from a dense set (e.g., real numbers). On the other hand, the state of the cyber components is usually defined by variables that change within the code, which are executed at events, and that take values from discrete sets. Unavoidably, this heterogeneous combination of variables and notions of time requires dynamical models that combine continuous and discrete variables as well as notions of time.

In this chapter, we advocate that hybrid dynamical system models can be employed to capture the behavior of cyber-physical systems. More precisely, the evolution of continuous variables is captured by *differential inclusions*, while the evolution of discrete variables is captured by *difference inclusions*. These inclusions (or equations) are typically nonlinear due to the complexity of the dynamics of these variables. Furthermore, conditions that determine the change of the continuous and discrete variables according to the said equations/inclusions can be conveniently captured by functions of the variables, inputs, and outputs. A particular advantage provided by hybrid models of cyber-physical systems is that a notion of time is automatically imposed on the cyber component, due to the fact that a dynamical model of a cyber-physical system will naturally have a concept of solution attached to it. In fact, a solution to such models will be parameterized by a notion of time so as to determine the change of the continuous variables, according to the physical components, as well as the discrete variables, according to the cyber components. In the next sections, these models are introduced in detail and illustrated in examples.

1.2.1 MODELS OF PHYSICAL COMPONENTS

The physical components of a cyber-physical system include the analog elements, physical systems, and the environment. Among the many possible models available, we will capture the dynamics of the physical components by differential equations or inclusions. This type of semantics is very common in the modeling of the so-called *environment* in embedded systems [6], the system to study in dynamical systems theory [7], and the plant to control in control theory [8]. The proposed model consists of a continuous-time system, in which typically, the time variable t parameterizes the variables of the system, which are called *states*. For example, differential equations arise when studying the physics of systems, such as analog devices, electromechanical systems, and chemical systems. Differential inclusion models allow for the derivative of the state of a system to take values from a set (rather than being assigned by a single-valued function). Such models are useful to model systems that exhibit nondeterminism, perhaps due to the uncertainty on its dynamics or parameters, to capture families of trajectories with a single dynamical model, and to model regularizations of nonsmooth systems.

We denote the state variable of the physical components by z with the state space given by the Euclidean space $\mathbb{R}^{n_P}$. Its dynamics are defined by a differential inclusion with the right-hand side defined by the set-valued map F_P. We let $u \in \mathbb{R}^{m_P}$ denote the input signals affecting the physical components and $y \in \mathbb{R}^{r_P}$ to be the output defined by the output function h, which is a function of the state z and of the input u. With these definitions, the mathematical description of the physical components is given by

$$\dot{z} \in F_P(z, u), \quad y = h(z, u) \tag{1.1}$$

where F_P and h are functions from $\mathbb{R}^{n_P} \times \mathbb{R}^{m_P}$ mapping into $\mathbb{R}^{n_P}$ and $\mathbb{R}^{r_P}$, respectively. The notation $\dot{z}$ represents the derivative of $t \mapsto z(t)$ with respect to $t \in [0, \infty) =: \mathbb{R}_{\geq 0}$, that is, $\dot{z} = \frac{d}{dt}z$. The inclusion in Equation 1.1 indicates that at each (z, u) the time derivative of z is taken from the set $F_P(z, u)$. If F_P is single valued, then Equation 1.1 reduces to a differential equation.

In certain cases, it would be needed to impose restrictions on the state and inputs to the physical component. Such is the case when it is needed to limit the range of their allowed values (e.g., the state belongs to a manifold or the input is constrained) or to impose conditions that they should satisfy so as to evolve continuously according to Equation 1.1. Such conditions can be modeled using sets, namely,

$$(z, u) \in C_P \subset \mathbb{R}^{n_P} \times \mathbb{R}^{m_P} \tag{1.2}$$

The model of the physical components is given by Equations 1.1 and 1.2.

The following two examples illustrate the model proposed in Equations 1.1 and 1.2. While the examples omit *such* features, the model in Equations 1.1 and 1.2 can easily incorporate the dynamics of analog filters, sensors, and actuators.

Example 1.1:

A linear time-invariant model of the physical component is defined by

$$F_P(z, u) = A_P z + B_P u, \quad h(z, u) = M_P z + N_P u$$

where A_P, B_P, M_P, and N_P are matrices of appropriate dimensions. State and input constraints can directly be embedded into the set C_P. For example, the constraint that z has all of its components as nonnegative and that u has its components with *norm less be directly or equal than one* is captured by

$$C_P = \{(z, u) \in \mathbb{R}^{n_P} \times \mathbb{R}^{m_P} : z_i \geq 0 \ \forall i \in \{1, 2, \dots, n_P\}\}$$
$$\cap \{(z, u) \in \mathbb{R}^{n_P} \times \mathbb{R}^{m_P} : |u_i| \leq 1 \ \forall i \in \{1, 2, \dots, m_P\}\}$$

For example, the evolution of the temperature of a room with a heater can be modeled by a linear time-invariant system with state z denoting the temperature of the room and with input $u = (u_1, u_2)$, where u_1 denotes whether the heater is turned on ($u_1 = 1$) or turned off ($u_1 = 0$), while u_2 denotes the temperature outside the room. The evolution of the temperature is given by

$$\dot{z} = -z + \begin{bmatrix} z_\Delta & 1 \end{bmatrix} \begin{bmatrix} u_1 \\ u_2 \end{bmatrix} \qquad \text{when } (z, u) \in C_P = \left\{ (z, u) \in \mathbb{R} \times \mathbb{R}^2 : u_1 \in \{0, 1\} \right\} \tag{1.1.1}$$

where z_Δ is a constant representing the heater capacity.   □

Example 1.2:

A widely used kinematic model of a ground vehicle moving on the plane is the so-called *Dubins vehicle*. The vehicle is assumed to be a particle on the plane that describes smooth paths and has the capability of making turns satisfying a minimum turning radius constraint (similar to a car). Denoting its position by $(z_1, z_2) \in \mathbb{R}^2$ and its orientation by $z_3 \in \mathbb{R}$ (with respect to the vertical axis), the dynamics of the particle defining a Dubins vehicle are given by

$$\dot{z}_1 = v \sin z_3, \quad \dot{z}_2 = v \cos z_3, \quad \dot{z}_3 = u$$

where v is the velocity of the vehicle, $u \in [-\bar{u}, \bar{u}]$ is the angular velocity input and $\bar{u} = \frac{v}{\rho}$, and ρ is the minimum turning radius constraint. Assuming, for simplicity, that v is a constant, this mathematical model can be captured as in Equations 1.1 and 1.2 with

$$F_P(z, u) = \begin{bmatrix} v \sin z_3 \\ v \cos z_3 \\ u \end{bmatrix}, \quad C_P = \left\{ (z, u) \in \mathbb{R}^3 \times \mathbb{R} : u \in [-\bar{u}, \bar{u}] \right\}$$

□

The model in Equations 1.1 and 1.2 captures the continuous dynamics of a hybrid system with inputs, which extends the model Goebel, Sanfelice, and Teel [1] and is required to be able to define an interconnection between the physical components and the other subsystems; see also the models Sanfelice [2].

1.2.2 MODELS OF CYBER COMPONENTS

The cyber components of a cyber-physical system include those which are in charge of performing computations, implementing algorithms, and transmitting digital data over networks. The tasks performed by the code (at the software level) and the logic-based mechanisms (at the circuit level) involve variables that only change at discrete events, not necessarily periodically. Furthermore, unlike most of the quantities involved in the physical components, such variables may take values from discrete sets rather than from a continuum. Due to these unique features, models of cyber components have state variables, typically discrete valued, that are updated at discrete events. In this chapter, we capture the dynamics of such variables using differential equations/inclusions.

We denote the state variable of the cyber components by $\eta \in \Upsilon$, where $\Upsilon \subset \mathbb{R}^{n_c}$ is the state space. The dynamics of η are defined by a difference inclusion with the right-hand side defined by the set-valued map G_C. We let $v \in \mathcal{V} \subset \mathbb{R}^{m_c}$ denote the input signals affecting the cyber components and $\zeta \in \mathbb{R}^{r_c}$ to be the output defined by the output function κ, which is a function of the state η and of the input v. With these definitions, the general mathematical description of the cyber component is

$$\eta^+ \in G_C(\eta, v), \quad \zeta = \kappa(\eta, v) \tag{1.3}$$

In certain cases, *it* would be needed to impose restrictions on the state and inputs to the cyber component. Such conditions can be modeled imposing that η and v belong to a subset of their state and input space, respectively, namely,

$$(\eta, v) \in D_C \subset \Upsilon \times \mathcal{V} \tag{1.4}$$

The model of the cyber components is given by Equations 1.3 and 1.4. Next, we provide specific constructions of models of cyber components.

1.2.2.1 Pure Finite-State Machines

An FSM or deterministic finite automaton is a system with inputs, states, and outputs taking values from discrete sets that are updated at discrete transitions (or jumps) triggered by its inputs. At every jump, the states and the outputs of the FSM are updated. Let v denote the inputs, q denote the states (or mode), and r denote the outputs of the FSM. Following the definition Hopcroft and Ullman [3], an FSM consists of the following objects:

- An input alphabet Σ where v takes values from
- A finite set of states Q where q takes values from
- A set of output symbols Δ where r takes values from
- An output function $\kappa : Q \to \Delta$
- A transition function $\delta : Q \times \Sigma \to Q$

The initial state of the FSM is denoted by q_0, while, at times, a set of final states is imposed and denoted as $Q_\infty \subset Q$. The output function is defined for the current value of the state and input, which, in particular, permits modeling Mealy machines (one way to model Moore machines is to include in the model an additional variable capturing the output and updating it at transitions using the value of the state and input before the transitions). The transition function is defined for each state $q \in Q$ and each input $v \in \Sigma$; moreover, by convention, and assuming that the empty set is an element of the input alphabet, $\delta(q, \emptyset) = q$, which is known as the basis condition, and $\delta(q, ab) = \delta(\delta(q, a), b)$ for each $a, b \in \Sigma$, which implies that δ can be evaluated for an input string, that is, δ satisfies the properties of an *extended transition function*. Note that the model is deterministic since the transition and output functions are uniquely defined for each value of their arguments. Also, while functions δ and κ of the FSM are assumed to be defined for each element in $Q \times \Sigma$, that is, they are *total functions*, an FSM with functions that do not define for all points, that is, *partial functions*, can be modeled similarly.

Then, given an FSM and an initial state $q_0 \in Q$, a transition to a state $q_1 = \delta(q_0, v)$ is performed when an input $v \in \Sigma$ is applied to it. After the transition, the output of the FSM is updated to $\kappa(q_1)$. This mechanism can be captured by the difference equation

$$q^+ = \delta(q, v) \quad \zeta = \kappa(q) \quad (q, v) \in Q \times \Sigma \tag{1.5}$$

This model corresponds to the model of the cyber components in Equations 1.3 and 1.4 with

$$\eta = q, \quad \Upsilon = Q, \quad \mathcal{V} = \Sigma, \quad G_c = \delta, \quad D_c = \Upsilon \times \mathcal{V}$$

Note that there is no notion of time associated with the aforementioned FSM model.

Example 1.3:

The FSM given in Figure 1.1 has two modes and one input. Its output is equal to the current mode. It is given by the difference equation in Equation 1.5 with

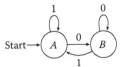

FIGURE 1.1 A finite-state machine with two modes and one input.

$$Q = \{A, B\}, \quad \Sigma = \{0, 1\}, \quad \delta(q, v) = \begin{cases} A & \text{if } v = 1 \\ B & \text{if } v = 0 \end{cases}, \quad \kappa(q) = q$$

Note that this FSM does not terminate. □

The nondeterministic case of an FSM can be treated similarly by using a set-valued transition function $\delta : Q \times \Sigma \rightrightarrows Q \times \Delta$. Note that in this case,

$$\delta(q, ab) = \bigcup_{q' \in \delta(q,a)} \delta(q', b)$$

for each $a, b \in \Sigma$.

1.2.2.2 Finite-State Machines with Conditional Structures as Guards

In many applications, it is desired that the jumps of the FSM are triggered based on conditional structures, for example, perform a transition when $v < 0$. To model conditional structures within a pure FSM, the input alphabet Σ has to be of infinite size. Conditional structures can be added to a pure FSM by allowing for an infinite alphabet and including the conditional structure as a *guard*, that is, a boolean-valued expression that evaluates to *true* when the transition is enabled and to *false* otherwise. As pointed out Girault, Lee, and Lee [4], the dataflow model of computation is appropriate to trigger transitions in an FSM according to conditional structures; for example, the comparison $v < 0$ amounts to externally compute (in a dataflow model) the conditional statement and based on its result, either true or false, triggers a jump of the FSM.

To define an FSM with transitions according to conditional structures, let the function $\ell : Q \times \Sigma \times \Delta \to \mathbb{R}$ be a testing function for the condition on the transition for each mode $q \in Q$. Assume that the value of $\ell(q, v, \zeta)$ is larger than zero if the conditional expression to implement is not satisfied for the current values of (q, v, ζ), and that is less or equal than zero if it is satisfied. Then, an FSM with transitions triggered by the conditional structure modeled by ℓ is given by

$$q^+ = \delta(q, v) \quad \zeta = \kappa(q) \quad \ell(q, v, \zeta) \le 0, \ (q, v) \in Q \times \Sigma \tag{1.6}$$

This model corresponds to the model of the cyber components in Equations 1.3 and 1.4 with

$$\eta = q, \quad \Upsilon = Q, \quad \mathcal{V} = \Sigma, \quad G_c = \delta, \quad D_c = \{(q, v) \in Q \times \mathcal{V} : \ell(q, v, \kappa(q)) \le 0\}$$

The FSM with dataflow is a powerful model as it can be used to describe a control dataflow graph (CDFG). CDFGs are a common intermediate representation in software compilation. Many programming languages are translated into a CDFG during their translation to assembly code. The CDFG is also a common model in hardware compilation; it is frequently used as an entry model into behavioral/high-level synthesis.

Note that as in the pure FSM model in Section 1.2.2.1, there is no notion of time associated with the model in Equation 1.6.

Finally, similar mathematical models can be derived for other machines used in practice, such as Turing machines, as well as Büchi and pushdown automata.

1.2.2.3   Models of Computer Computations and Discrete-Time Algorithms

Computations performed on a computer can be modeled as a purely discrete system that, after one or a series of steps, provides the outcome of the computations. The computation to be performed may require inputs v, and the result of the computation could be used to determine the output ζ of the model. Computations that can be performed in one step of the discrete system can be modeled by the static system

$$\zeta = \widetilde{\kappa}(v) \tag{1.7}$$

where the function $\widetilde{\kappa}$ models the computations being performed. This model corresponds to the model of the cyber components in Equations 1.3 and 1.4 with

$$\eta = \emptyset, \quad \Upsilon = \emptyset, \quad \mathcal{V} = \Sigma, \quad G_C = \emptyset, \quad D_C = \mathcal{V}, \quad \kappa = \widetilde{\kappa}$$

Example 1.4:

The check of the expression $v < 0$ in Section 1.2.2.2 can be modeled by

$$\widetilde{\kappa}(v) := \begin{cases} 1 & v < 0 \\ 0 & \text{otherwise} \end{cases}$$

□

Example 1.5:

The computation of the factorial of an integer $a > 0$ is given by the evaluation at $v = a$ of the function*

$$\widetilde{\kappa}(v) := v(v-1)(v-2)\ldots(v-(v-2))1 \tag{1.5.1}$$

□

Iterative implementations of computations require a number of steps to reach an answer and, at times, additional variables. Denoting the additional variables as $m \in \mathbb{R}^{n_c-1}$ and the counter as $k \in \{0, 1, 2, \ldots, k^*\}$, $k^* \in \{0, 1, 2, \ldots\} =: \mathbb{N}$, we define a discrete system that performs k^* iterations to provide the final outcome of the computations. Denoting $\eta = [m^\top k]^\top$ as the state, v as the input, and $\widetilde{\kappa}$ as the routine performing the computations at each iteration, the model is given by

$$\eta^+ = \begin{bmatrix} \widetilde{\kappa}(m, k, v) \\ k+1 \end{bmatrix} \qquad \zeta = m \qquad k \in \{0, 1, 2, \ldots, k^*-1\}, \quad m \in \mathbb{R}^{n_c-1}, v \in \mathcal{V} \tag{1.8}$$

Example 1.6:

The one-step computation of the factorial of the integer a in Equation 1.5.1 can be performed iteratively by initializing m to a and k to one, setting $k^* = a$, and defining

$$\widetilde{\kappa}(m, k, v) = m(m-k)$$

Note that for $k < k^*$, ζ is equal to the intermediate result $a! - (a-k)!$

□

---

* This operation can be performed recursively by defining the function $\text{fact}(v) = v\,\text{fact}(v-1)$ when $v > 1$, and $\text{fact}(1) = 1$, and using $\widetilde{\kappa}(v) = \text{fact}(v)$.

The model in Equation 1.8 corresponds to the model of the cyber components in Equations 1.3 and 1.4 with

$$\eta = \begin{bmatrix} m \\ k \end{bmatrix}, \quad \Upsilon = \mathbb{R}^{n_c-1} \times \{0, 1, 2, \ldots, k^*\}, \quad \mathcal{V} = \Sigma, \quad G_C = \begin{bmatrix} \widetilde{\kappa}(m, k, v) \\ k+1 \end{bmatrix}$$

and

$$D_C = \mathbb{R}^{n_c-1} \times \{0, 1, 2, \ldots, k^* - 1\}, \quad \kappa(\eta) = m \quad \forall \eta \in \Upsilon$$

Discrete-time algorithms are naturally modeled using difference equations. For instance, feedback controllers in a difference equation form are obtained after a discretization of a controller designed using continuous-time control design tools or from directly designing a controller for a discretized model of the plant. Such discrete-time algorithms can be written as

$$\eta^+ = G_C(\eta, v) \quad \zeta = \kappa(\eta) \tag{1.9}$$

where G_C is obtained from discretizing the control algorithm.

Example 1.7:

If the original controller is given by the system in state-space form

$$\dot{\xi} = A_K \xi + B_K v, \quad \zeta_K = M_K \xi + N_K v$$

with $\xi \in \mathbb{R}^{n_c}$ and $v \in \mathbb{R}^{m_c}$, then its discretization with period T^* is given by the discrete-time algorithm

$$\eta^+ = A_K^d \eta + B_K^d v, \quad \zeta = M_K^d \eta + N_K^d v$$

where

$$A_K^d = \exp(A_K T^*), \quad B_K^d = \left(\int_0^{T^*} \exp(A_K s) \, ds \right) B_K, \quad M_K^d = M_K, \quad N_K^d = N_K$$

which is already in the form of Equation 1.9. Also, this model is already in the form of the model of the cyber components in Equations 1.3 and 1.4, in which $\Upsilon = \mathbb{R}^{n_c}$ and $\mathcal{V} = \mathbb{R}^{m_c}$. Note that while this model does not have a notion of time associated to it, at times, like for the discretized controller, the model should be executed at specific time instants. □

The examples in this section illustrate how these models can be used to describe operations performed by the algorithms implemented in the software of cyber-physical systems. Next, we consider the models of the systems that enable the interconnection between the physical and cyber components.

1.2.3 MODELS OF SYSTEMS AT THE INTERFACE BETWEEN PHYSICAL AND CYBER COMPONENTS

The models describing the behavior of the physical and the cyber components have significantly different dynamics: the former has continuos dynamics while the latter has discrete dynamics. As a result, their interconnection requires interfaces that condition and convert the signals appropriately. Next, we propose mathematical models for some of the most widely used interfaces. In the section to follow, these models will be used to define a complete model of a cyber-physical system.

1.2.3.1 Analog-to-Digital Converters

Analog-to-digital converters, or simply sampling devices, are commonly used to provide measurements of the physical systems to the cyber components. Their main function is to sample their input, which is usually the output of the sensors measuring the output y, at a given periodic rate T_s^* and to make these samples available to the embedded computer. A basic model for a sampling device consists of a timer state and a sample state. When the timer reaches the value of the sampling time T_s^*, the timer is reset to zero, and the sample state is updated with the inputs to the sampling device.

The model for the sampling device we propose has both continuous and discrete dynamics. If the timer state has not reached T_s^*, then the dynamics are such that the timer state increases continuously with a constant, unitary rate. When T_s^* is reached, the timer state is reset to zero, and the sample state is mapped to the inputs of the sampling device. To implement this mechanism, we employ a timer state $\tau_s \in \mathbb{R}_{\geq 0}$ and a sample state $m_s \in \mathbb{R}^{r_p}$. The input to the sampling device is denoted by $v_s \in \mathbb{R}^{r_p}$. The model of the sampling device is

$$\dot{\tau}_s = 1, \quad \dot{m}_s = 0 \quad \text{when} \quad \tau_s \in \left[0, T_s^*\right] \tag{1.10}$$

$$\tau_s^+ = 0, \quad m_s^+ = v_s \quad \text{when} \quad \tau_s \geq T_s^* \tag{1.11}$$

In practice, there exists a time, usually called the *ADC acquisition time*, between the triggering of the ADCs sampling event and the update of its output. Such a delay limits the number of samples per second that the ADC can provide. Additionally, an ADC can only store in the sample state finite-length digital words, which causes quantization. The aforementioned model omits acquisition delays and quantization effects, but these can be incorporated if needed. In particular, quantization effects can be added to the model in Equations 1.10 and 1.11 by replacing the updated law for m_s to $m_s^+ = \text{round}(v_s)$, where the map round is such that $\text{round}(v_s)$ contains the closest numbers to v_s that the machine precision can represent.

1.2.3.2 Digital-to-Analog Converters

The digital signals in the cyber components need to be converted to analog signals for their use in the physical world. DACs perform such a task by converging digital signals into analog equivalents. One of the most common models for a DAC is the zero-order hold model (ZOH). In simple terms, a ZOH converts a digital signal at its input into an analog signal at its output. Its output is updated at discrete-time instants, typically periodically, and held constant in between updates, until new information is available at the next sampling time. We will model DACs as ZOH devices with dynamics similar to Equations 1.10 and 1.11. Let $\tau_h \in \mathbb{R}_{\geq 0}$ be the timer state, $m_h \in \mathbb{R}^{r_c}$ be the sample state (note that the value of h indicates the number of DACs in the interface), and $v_h \in \mathbb{R}^{r_c}$ be the inputs of the DAC. Its operation is as follows. When $\tau_h \geq T_h^*$, the timer state is reset to zero, and the sample state is updated with v_h (usually the output of the embedded computer). A model that captures this mechanism is given by

$$\dot{\tau}_h = 1, \quad \dot{m}_h = 0 \quad \text{when} \quad \tau_h \in \left[0, T_h^*\right] \tag{1.12}$$

$$\tau_h^+ = 0, \quad m_h^+ = v_h \quad \text{when} \quad \tau_h \geq T_h^* \tag{1.13}$$

1.2.3.3 Digital Networks

The information transfer between the physical and cyber components, or between subsystems within the cyber components, might occur over a digital communication network. The communication links bridging each of these components are not capable of continuously transmitting information, but rather, they can only transmit sampled (and quantized) information at discrete-time instants. Combining the ideas in the models of the converters in the previous sections, we propose a model of a digital network link that has a variable that triggers the transfer of information provided at its input, and that stores that information until new information arrives. We assume that the transmission of information occurs at instants $\{t_i\}_{i=1}^{i^*}$, $i^* \in \mathbb{N} \cup \{\infty\}$, satisfying

$$T_N^{*\min} \leq t_{i+1} - t_i \leq T_N^{*\max} \quad \forall i \in \{1, 2, \ldots, i^* - 1\}$$

where $T_N^{*\min}$ and $T_N^{*\max}$ are constants satisfying

$$T_N^{*\min}, T_N^{*\max} \in [0, \infty]$$

and

$$T_N^{*\min} \leq T_N^{*\max}$$

and i^* is the number of transmission events, which might be finite or infinite. The constant $T_N^{*\min}$ determines the minimum possible time in between transmissions, while the constant $T_N^{*\max}$ defines the maximum amount of time elapsed between transmissions. In this way, a communication channel that allows transmission events at a high rate would have $T_N^{*\min}$ small (zero for infinitely fast transmissions), while one with slow data rate would have $T_N^{*\min}$ large. The constant $T_N^{*\max}$ determines how often transmissions may take place. Note that the constants $T_N^{*\min}$ and $T_N^{*\max}$ can be generalized to functions so as to change according to other states or inputs.

At every t_i, the information at the input v_N of the communication link is used to update the internal variable m_N, which is accessible at the output end of the network and remains constant between communication events. This internal variable acts as an information buffer, which can contain not only the latest piece of information transmitted but also previously transmitted information. A mathematical model capturing the said mechanism is given by

$$\dot{\tau}_N = -1, \quad \dot{m}_N = 0 \quad \text{when} \quad \tau_N \in \left[0, T_N^{*\max}\right] \tag{1.14}$$

$$\tau_N^+ \in \left[T_N^{*\min}, T_N^{*\max}\right], \quad m_N^+ = v_N \quad \text{when} \quad \tau_N \leq 0 \tag{1.15}$$

Note that the update law for τ_N at jumps is given in terms of a difference inclusion, which implies that the new value of τ_N is taken from the set $[T_N^{*\min}, T_N^{*\max}]$. The dimension of the states and the input would depend on the type of components that connect to and from it and also the size of data transmitted and buffered. Similar to the models proposed for conversion, the model of the digital link Equations in 1.14 and 1.15 does not include delays nor quantization, but such effects can be incorporated if needed.

Note that the models in this section are given in terms of differential and difference equations/inclusions. The reason for this is that they are at the interface between physical and cyber components. These type of models are referred to as *hybrid inclusions* and will be the class of models we will employ to study cyber-physical systems in the remainder of this chapter. In general, the interface will be modeled as a hybrid inclusion with state λ, input w, output ψ, and dynamics

$$\dot{\lambda} \in F_I(\lambda, w) \quad \text{when } (\lambda, w) \in C_I \tag{1.16}$$

$$\lambda^+ \in G_I(\lambda, w) \quad \text{when } (\lambda, w) \in D_I \tag{1.17}$$

$$\psi = \varphi(\lambda) \tag{1.18}$$

where F_I defines the continuous dynamics on C_I and G_I the discrete dynamics on D_I of the interface.

1.2.4   COMBINING MODELS OF PHYSICAL AND CYBER COMPONENTS

Mathematical models of certain types of cyber-physical systems can be obtained by appropriately interconnecting the models of physical and cyber components given in Sections 1.2.1 and 1.2.2, respectively, and of the interfaces in Section 1.2.3. Figure 1.2 depicts two specific interconnections of such models, a cascade interconnection in Figure 1.2a and a feedback interconnection in Figure 1.2b.

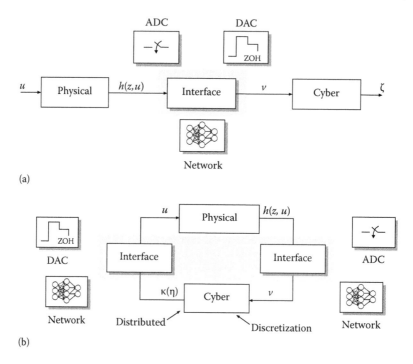

FIGURE 1.2 Cyber-physical systems: Series and parallel interconnections between a plant (part of the physical component), controller (part of the cyber component), and interfaces/converters/signal conditioners (part of both the physical and cyber components). (a) Series topology and (b) feedback topology.

The individual models of the components are used to define a model of the entire cyber-physical system, which, due to the combination of differential and difference equations/inclusions, is in the hybrid inclusion form.

The following examples exercise the models proposed in the previous section for modeling specific cyber-physical systems.

Example 1.8 (Implementation of an FSM):

Transitions of the state of an FSM are triggered by changes of its input v. When the input to the FSM is an external signal, a digital implementation of an FSM would require a conversion of the input, from analog to digital. Due to the conversion occurring at isolated time instants t_i, state transitions of the FSM will only take place at those instants. The resulting system can be modeled as a cascade of two systems, in which an ADC drives the FSM. The ADC is modeled as the hybrid inclusion model Equations 1.10 and 1.11, and the FSM as the difference equation in Equation 1.5. Assuming that the input of the converter is transferred to its output at every conversion event of the converter, and that the output of the converter ψ feeds the input v of the FSM, we do not need to include a memory state m_s in the converter. In this way, when the timer τ_s is within $[0, T_s^*)$, the timer in the converter evolves according to

$$\dot{\tau}_s = 1$$

and the FSM variable q remains constant, which can be written as the trivial differential equation

$$\dot{q} = 0$$

When the timer τ_s reaches T_s^*, then the timer is updated via the difference equation

$$\tau_s^+ = 0$$

and the state of the FSM is updated according to the transition function evaluated at the current input, that is, q is updated according to

$$q^+ = \delta(q, w)$$

The aforementioned model can be summarized as follows:

$$\left. \begin{array}{rl} \dot{\tau}_s &= 1 \\ \dot{q} &= 0 \end{array} \right\} \quad \text{when } (q, v) \in Q \times \Sigma, \tau_s \in [0, T_s^*]$$

$$\left. \begin{array}{rl} \tau_s^+ &= 0 \\ q^+ &= \delta(q, w) \end{array} \right\} \quad \text{when } (q, v) \in Q \times \Sigma, \tau_s \geq T_s^*$$

□

Example 1.9 (Estimation Over a Network):

Consider a physical process given in terms of the state-space model

$$\dot{z} = Az, \quad y = Mz, \quad z \in \mathbb{R}^{n_p}, y \in \mathbb{R}^{r_p} \tag{1.9.1}$$

where z is the state and y is the measured output. The output is digitally transmitted through a network. At the other end of the network, a computer receives the information and runs an algorithm that takes the measurements of y to estimate the state z of the physical process. We consider an estimation algorithm with a state variable $\hat{z} \in \mathbb{R}^{n_p}$, which denotes the estimate of z, that is appropriately reset to a new value involving the information received. More precisely, denoting the transmission times by t_i and L a constant matrix to be designed, the estimation algorithm updates its state as follows:

$$\hat{z}^+ = \hat{z} + L(y - M\hat{z}) \tag{1.9.2}$$

at every instant information is received. In between such events, the algorithm updates its state continuously so as to match the evolution of the state of the physical process, that is, via

$$\dot{\hat{z}} = A\hat{z} \tag{1.9.3}$$

Modeling the network as in Section 1.2.3.3, which, in particular, assumes zero transmission delay; the state variables of the entire system are z, $\tau_N \in \mathbb{R}$, $m_N \in \mathbb{R}^{r_p}$, and $\hat{z}$. Then, transmissions occur when $\tau_N \leq 0$, at which events the state of the network is updated via

$$\tau_N^+ \in [T_N^{*\min}, T_N^{*\max}], \quad m_N^+ = y$$

and the state of the algorithm is updated via Equation 1.9.2. Note that since the state of the physical process does not change at such events, we can use the following trivial difference equation to update it at the network events:

$$z^+ = z$$

In between events, the state of the network is updated as

$$\dot{\tau}_N = -1, \quad \dot{m}_N = 0$$

the state of the algorithm changes continuously according to Equation 1.9.3, and the state of the physical process changes according to Equation 1.9.1. Combining the preceding equations, we arrive at the following model of the system:

$$\left.\begin{aligned} \dot{z} &= Az \\ \dot{\tau}_N &= -1 \\ \dot{m}_N &= 0 \\ \dot{\hat{z}} &= A\hat{z} \end{aligned}\right\} \quad \text{when } z \in \mathbb{R}^{n_P},\ \tau_N \in [0, T_N^{*\max}],\ m_N \in \mathbb{R}^{r_P},\ \hat{z} \in \mathbb{R}^{n_P}$$

$$\left.\begin{aligned} z^+ &= z \\ \tau_N^+ &\in [T_N^{*\min}, T_N^{*\max}] \\ m_N^+ &= y \\ \hat{z}^+ &= \hat{z} + L(y - M\hat{z}) \end{aligned}\right\} \quad \text{when } z \in \mathbb{R}^{n_P},\ \tau_N \le 0,\ m_N \in \mathbb{R}^{r_P},\ \hat{z} \in \mathbb{R}^{n_P}$$

As suggested earlier, a digital implementation of this law would lead to a discrete-time system as in Example 1.7, which when implemented in a computer would require an additional timer and a memory state.

Also, note that the algorithm uses the current measurement of y instead of the latest value held in m_N. This is due to the assumption that there is a zero transmission delay in the network.   □

Example 1.10 (Sample-and-Hold Feedback Control):

Consider a physical process modeled as in Example 1.1 and a control algorithm given in terms of an FSM modeled as in Equation 1.5. The algorithm uses measurements of its output and controls the input of the physical process with the goal of steering its state to zero. Suppose the sampling device is ideal and that the signals produced by the FSM are connected to the plant via a DAC modeled as in Equations 1.12 and 1.13.

The interconnection between the models of the physical process, the sampling device, the FSM, and the DAC has the feedback topology shown in Figure 1.2b. In particular, the output m_h of the DAC is connected to the input u of the physical process, while the input v of the FSM is equal to the output y of the physical process at every sampling instant. The resulting interconnected system leads to the following hybrid inclusion model:

$$\left.\begin{aligned} \dot{z} &= Az + Bm_h \\ \dot{\tau}_s &= 1 \\ \dot{q} &= 0 \\ \dot{\tau}_h &= 1 \\ \dot{m}_h &= 0 \end{aligned}\right\} \quad \text{when } z \in \mathbb{R}^{n_P},\ \tau_s \in [0, T_s^*],\ q \in Q,\ \tau_h \in [0, T_h^*],\ m_N \in \mathbb{R}^{r_P}$$

$$\left.\begin{aligned} z^+ &= z \\ \tau_s^+ &= 0 \\ q^+ &= \delta(q, y) \\ \tau_h^+ &= \tau_h \\ m_h^+ &= m_h \end{aligned}\right\} \quad \text{when } z \in \mathbb{R}^{n_P},\ \tau_s \ge T_s^*,\ q \in Q,\ \tau_h \in [0, T_h^*),\ m_N \in \mathbb{R}^{r_P}$$

$$\left.\begin{aligned} z^+ &= z \\ \tau_s^+ &= \tau_s \\ q^+ &= q \\ \tau_h^+ &= 0 \\ m_h^+ &= \kappa(q) \end{aligned}\right\} \quad \text{when } z \in \mathbb{R}^{n_P},\ \tau_s \in [0, T_s^*),\ q \in Q,\ \tau_h \ge T_h^*,\ m_N \in \mathbb{R}^{r_P}$$

$$\left.\begin{aligned} z^+ &= z \\ \tau_s^+ &= 0 \\ q^+ &= \delta(q, v) \\ \tau_h^+ &= 0 \\ m_h^+ &= \kappa(q) \end{aligned}\right\} \quad \text{when } z \in \mathbb{R}^{n_P},\ \tau_s \ge T_s^*,\ q \in Q,\ \tau_h \ge T_h^*,\ m_N \in \mathbb{R}^{r_P}$$

Note that in this model the events can be triggered by three different conditions, which are parsed independently: expiration of the timer τ_s used for sampling only, expiration of the timer τ_h used in the DAC only, and expiration of both timers (this list of events is in the same order as the difference in equations in the aforementioned model).

There are numerous practical examples of systems that can be modeled within the general model for sample-and-hold feedback control defined earlier. For example, one *classical* example is the control of the temperature of a room by turning on and off a heater so as to keep the temperature within a desired range; see the model in Equation 1.1.1. Another widely known example is the control of the level of water in a tank. $\qquad\qquad\qquad\qquad\qquad\qquad\square$

In general, the models of the cyber-physical systems given earlier can be written as a hybrid inclusion with state $x = (z, \eta, \lambda)$, input $\gamma = (u, v, w)$, and dynamics

$$\dot{x} \in F(x, \gamma) \qquad (x, \gamma) \in C \qquad\qquad\qquad (1.19)$$

$$x^+ \in G(x, \gamma) \qquad (x, \gamma) \in D \qquad\qquad\qquad (1.20)$$

with continuous evolution according to F over C and discrete evolution according to G over D. The map F is the *flow map*, the set C is the *flow set*, the map G is the *jump map*, and the set D is the *jump set*. In fact, the model of the system in Example 1.10 results in a hybrid inclusion as in Equations 1.16 and 1.17 with state

$$x = (z, \tau_s, q, \tau_h, m_h) \in \mathcal{X} := \mathbb{R}^{n_p} \times [0, T_s^*] \times Q \times [0, T_h^*] \times \mathbb{R}^{r_p}$$

with all of the inputs of its subsystems assigned (hence, we can omit γ), and with dynamics

$$\dot{x} = F(x) = \begin{bmatrix} Az + Bm_h \\ 1 \\ 0 \\ 1 \\ 0 \end{bmatrix} \qquad x \in C \qquad\qquad\qquad (1.21)$$

$$x^+ \in G(x) = \begin{cases} G_s(x) & \text{if } x \in D_1 \setminus D_2 \\ G_h(x) & \text{if } x \in D_2 \setminus D_2 \\ G_{sh}(x) & \text{if } x \in D_1 \cap D_2 \end{cases} \qquad x \in D = D_1 \cup D_2 \qquad (1.22)$$

where

$$D_1 = \left\{ x \in \mathcal{X} : \tau_s = T_s^* \right\}, \quad D_2 = \left\{ x \in \mathcal{X} : \tau_h = T_h^* \right\}$$

and, for each $x \in D$,

$$G_s(x) = \begin{bmatrix} z \\ 0 \\ \delta(q, y) \\ \tau_h \\ m_h \end{bmatrix}, \qquad G_h(x) = \begin{bmatrix} z \\ \tau_s \\ q \\ 0 \\ \kappa(q) \end{bmatrix}, \qquad G_{sh}(x) = [G_s(x), G_h(x)]$$

As a difference to the construction in Example 1.10, this definition of the jump map is set valued when both events occur simultaneously. The next section presents control theoretical tools for the analysis and design of cyber-physical systems modeled as in Equations 1.19 and 1.20. The tools permit to study properties of interest for cyber-physical systems, such as asymptotic stability and robustness, which are usually the focus in control theory, as well as set invariants, temporal logic, and simulation.

1.3   CONTROL THEORETICAL TOOLS FOR ANALYSIS AND DESIGN OF CYBER-PHYSICAL SYSTEM

1.3.1   BASIC CONCEPTS

The systems presented in the previous examples suggest that the physical components may have variables that change continuously as ordinary (or physical) time evolves. Some of the examples also indicate that the cyber components may include variables that only change at discrete events. As a consequence, a notion of time parameterizing the evolution of the variables of a cyber-physical system has to keep track of both continuously and discretely changing variables. It is very tempting to define a notion of time that parameterizes the evolution of the variables by a collection of ordinary time intervals in which the upper boundary value of (potentially, the closure of) each interval corresponds to an instant at which an event occurs. However, such a notion arbitrarily prioritizes the continuous evolution of the variables, does not explicitly determine the number of events after a specific amount of continuous evolution, and, to avoid ambiguity at each time instant, forces to use combinations of open to the right (or open to the left) interval and left-continuous (or right-continuous) functions defining solutions. These disadvantages and the implications on robustness of such a notion of time were discussed in detail Sanfelice, Goebel, and Teel [9].

In this chapter, the evolution of the variables of cyber-physical systems are conveniently parameterized by ordinary time

$$t \in [0, \infty) = \mathbb{R}_{\geq 0}$$

which is incremented continuously as continuous evolution of the variables occurs, and a counter

$$j \in \{0, 1, 2, \ldots\} = \mathbb{N}$$

which is incremented at unitary steps when events occur. Then, pairs (t, j) define a hybrid time instant, which takes values on subsets of $\mathbb{R}_{\geq 0} \times \mathbb{N}$ called *hybrid time domains*. A set

$$E \subset \mathbb{R}_{\geq 0} \times \mathbb{N}$$

is a hybrid time domain if, for each $(T, J) \in E$, the set

$$E \cap ([0, T] \times \{0, 1, \ldots, J\})$$

can be written in the form

$$\bigcup_{j=0}^{J} \left([t_j, t_{j+1}] \times \{j\}\right)$$

for some finite sequence of times $0 = t_0 \leq t_1 \leq t_2 \leq \ldots \leq t_{J+1}$. The set

$$E \cap ([0, T] \times \{0, 1, \ldots, J\})$$

defines a compact hybrid time domain since it is bounded and closed. Then, solutions to a hybrid inclusion are given by functions $(t, j) \mapsto (\phi(t, j), \gamma(t, j))$ defined on hybrid time domains that, over intervals of flow, satisfy

$$\frac{d}{dt}\phi(t, j) \in F(\phi(t, j), \gamma(t, j))$$

$$(\phi(t, j), \gamma(t, j)) \in C$$

and, at jump times,

$$\phi(t,j+1) \in G(\phi(t,j), \gamma(t,j))$$
$$(\phi(t,j), \gamma(t,j)) \in D$$

These functions are such that ϕ is a hybrid arc and γ is a hybrid input. A hybrid arc ϕ is a function with domain dom ϕ, where dom ϕ is a hybrid time domain, and that, for each fixed j, the function $t \mapsto \phi(t,j)$ has a derivative, at least for almost every point in $I_j = \{t : (t,j) \in \mathrm{dom}(\phi,\gamma)\}$. More specifically, a hybrid arc ϕ is such that for each $j \in \mathbb{N}$, $t \mapsto \phi(t,j)$ is absolutely continuous on intervals of flow I_j with nonzero Lebesgue measure. Similarly, a hybrid input γ is a function on a hybrid time domain that, for each $j \in \mathbb{N}$, $t \mapsto \gamma(t,j)$, is Lebesgue measurable and locally essentially bounded on I_j.

The conditions determining whether a hybrid arc and a hybrid input define a solution to a hybrid inclusion as well as when it can flow or jump are determined by the data of the system, namely, (C, F, D, G). A solution to a hybrid system $\mathcal{H}$ is given by a pair (ϕ, γ) with dom $\phi =$ dom $\gamma (= \mathrm{dom}(\phi,\gamma))$ and satisfying the dynamics of the hybrid inclusion, where ϕ is a hybrid arc and γ is a hybrid input. More precisely, a hybrid input $\gamma : \mathrm{dom}\,\gamma \to \mathbb{R}^m$ and a hybrid arc $\phi : \mathrm{dom}\,\phi \to \mathbb{R}^n$ define a *solution pair* (ϕ, γ) *to the hybrid inclusion* in Equations 1.19 and 1.20 if the following conditions hold:

(S0) $(\phi(0,0), \gamma(0,0)) \in \overline{C} \cup D$ and dom $\phi =$ dom γ
(S1) For each $j \in \mathbb{N}$ such that $I_j = \{t : (t,j) \in \mathrm{dom}(\phi,\gamma)\}$ has nonempty interior int(I_j), we have

$$(\phi(t,j), \gamma(t,j)) \in C \text{ for all } t \in \mathrm{int}(I_j)$$

and for almost all $t \in I_j$, we have

$$\frac{d}{dt}\phi(t,j) \in F(\phi(t,j), \gamma(t,j))$$

(S2) For each $(t,j) \in \mathrm{dom}(\phi,\gamma)$ such that $(t, j+1) \in \mathrm{dom}(\phi,\gamma)$, we have

$$(\phi(t,j), \gamma(t,j)) \in D$$

and

$$\phi(t,j+1) \in G(\phi(t,j), \gamma(t,j))$$

Solutions can be classified in terms of their hybrid time domains. A solution (ϕ, γ) to the hybrid inclusion Equations 1.19 and 1.20 is said to be

- *Nontrivial* if dom(ϕ, γ) has at least two points
- *Continuous* if dom$(\phi, \gamma) \subset \mathbb{R}_{\geq 0} \times \{0\}$
- *Discrete* if dom$(\phi, \gamma) \subset \{0\} \times \mathbb{N}$
- *Complete* if dom(ϕ, γ) is unbounded
- *ZENO* if it is complete and the projection of dom(ϕ, γ) onto $\mathbb{R}_{\geq 0}$ is bounded
- *Dwell time* if there exists a positive constant c such that for each $j > 0$ such that $(t,j) \in$ dom ϕ for some t, there exist $(t',j), (t'',j) \in$ dom ϕ with $t'' - t' \geq c$
- *Maximal* if there does not exist another pair $(\phi, \gamma)'$ such that (ϕ, γ) is a truncation of $(\phi, \gamma)'$ to some proper subset of dom$(\phi, \gamma)'$

For more details about solutions to hybrid inclusions as in Equations 1.19 and 1.20, see Sanfelice [10].

Example 1.11:

The FSM model in Section 1.2.2.1 is a purely discrete system, which can be modeled as a hybrid inclusion with state $x = q$, input $\gamma = v$, and data

$$C = \emptyset, \qquad F = \star, \qquad D = Q \times \Sigma, \qquad G(x, \gamma) = \delta(x, \gamma)$$

where $\star$ indicates an arbitrary choice. Note that while the model in Section 1.2.2.1 does not involve a notion of time, this hybrid inclusion model confers to it the hybrid time notion. Since δ is a function that maps from D into Q. we have that after every jump, $G(x, \gamma)$ maps to points in Q. Then, since v takes values from Σ, for every input γ, we have that every solution (ϕ, γ) to the hybrid inclusion is discrete. Furthermore, every maximal solution is complete.   □

Example 1.11 motivates the search for conditions under which solutions to a general hybrid inclusion exist and fall under the particular types listed in the preceding text. Such conditions necessarily involve the data of the system. The conditions given in the following, most of which are given in Goebel, Sanfelice, and Teel [1: Propositions 2.10 and 6.10], provide insight on the existence and the type of solutions for such systems. They apply to hybrid inclusions without input, in which case C and D are subsets of $\mathbb{R}^n$ and $F, G : \mathbb{R}^n \rightrightarrows \mathbb{R}^n$.

1. Let $\xi \in \overline{C} \cup D$. If $\xi \in D$ or
   Visibility Condition (VC) here exist $\varepsilon > 0$ and an absolutely continuous function z : $[0, \varepsilon] \to \mathbb{R}^n$ such that $z(0) = \xi, \dot{z}(t) \in F(z(t))$ for almost all $t \in [0, \varepsilon]$ and $z(t) \in C$ for all $t \in (0, \varepsilon]$,
   then there exists a nontrivial solution ϕ to the hybrid inclusion Equations 1.19 and 1.20 with $\phi(0, 0) = \xi$.
2. If (VC) holds for every $\xi \in \overline{C} \setminus D$, then there exists a nontrivial solution to the hybrid inclusion Equations 1.19 and 1.20 from every point of $\overline{C} \cup D$.
3. If C is nonempty and D is empty, then every solution is continuous.
4. If C is empty and D is nonempty, then every solution is discrete. Furthermore, if $G(D) \subset D$, then every maximal solution is complete.
5. If (VC) holds for every $\xi \in \overline{C} \setminus D$, then every maximal solution ϕ satisfies exactly one of the following:
   a. ϕ is complete.
   b. dom ϕ is not complete and *ends with flow*; with $(T, J) = \sup \text{dom } \phi$; the interval I_J has nonempty interior; and either
      i. I_J is closed, in which case $\phi(T, J) \in \overline{C} \setminus (C \cap D)$, or
      ii. I_J is open to the right, in which case $\phi(T, J) \notin \text{dom } \phi$, and there does not exist an absolutely continuous function z : $\overline{I_J} \to \mathbb{R}^n$ satisfying $\dot{z}(t) \in F(z(t))$ for almost all $t \in I_J, z(t) \in C$ for all $t \in \text{int } I_J$, and such that $z(t) = \phi(t, J)$ for all $t \in I_J$.
   c. dom ϕ is not complete and *ends with a jump*: for $(T, J) = \sup \text{dom } \phi$, one has $\phi(T, J) \notin \overline{C} \cup D$.

Furthermore, if $G(D) \subset \overline{C} \cup D$, then item 5(c) does not occur. Moreover, if C is closed and F is a continuous single-valued map, then (VC) given earlier is implied by the following property: for the given $\xi \in \overline{C} \setminus D$, there exists a neighborhood U of ξ such that

$$F(x) \cap T_C(x) \neq \emptyset \qquad \forall x \in U \cap C$$

where $T_C(x)$ is the tangent cone of C at x.

Example 1.12:

A hybrid inclusion model for the sample-and-hold feedback control system in Example 1.10 is given in Equations 1.21 and 1.22. Its state is given by $x = (z, \tau_s, q, \tau_h, m_h)$, and the sets C and D are closed. For each $x = (z, \tau_s, q, \tau_h, m_h) \in D$, we have that either $\tau_s = T_s^*$ or $\tau_h = T_h^*$. If only $\tau_s = T_s^*$, then $G(x) = G_s(x)$, which belongs to $C \setminus D$. If only $\tau_h = T_h^*$, then $G(x) = G_h(x)$, which also belongs to $C \setminus D$. Similarly, when both $\tau_s = T_s^*$ and $\tau_h = T_h^*$, G maps the state to a point in $C \setminus D$. Then, $G(D) \subset C \cup D$. For each $x \in C \setminus D$, we have that $\tau_s < T_s^*$ and $\tau_h < T_h^*$. Then, there exists $\varepsilon > 0$ and a solution to $\dot{x} = F(x)$ such that, over $[0, \varepsilon]$, the solution stays within C—this implies that (VC) holds. Item 2 in the preceding text implies that there exists a nontrivial solution from every point in $C \cup D$. Furthermore, since $G(D) \subset C \cup D$, item 5(c) does not hold and due to the properties F, item 5(b) does not hold either. Then, item 5(a) implies that every maximal solution is complete. □

1.3.2 Invariants

An *invariant* for a system is a set with the property that every solution that starts from it stays in it for all future time. The knowledge of the invariants of a system is key in its design since a wide range of design specifications require its variables to remain within a region. Characterizing the invariants is of paramount importance in the study of reachability, safety, and stability.

A definition of invariant for a hybrid inclusion should account for the fact that solutions may not be unique and that every maximal solution may not be complete. Nonuniqueness of solutions may arise due to F or G being set valued or due to C and D overlapping. In particular, as pointed out in Section 1.2.2.1, nondeterministic FSMs lead to set-valued jump maps G. A set-valued jump map was also used in the model of a digital network in Section 1.2.3.3. The sets C and D may overlap when both flows and jumps should be allowed. Such nondeterminism might be desired to model the behavior of a system or when, due to perturbations, it is not possible to determine whether the perturbed state is in C or D, which is typically the case nearby the boundaries of these sets. Noncompleteness of maximal solutions may be due to, in particular, the flow map F steering the state to points in the boundary of $C \setminus D$ from where flow within C is not possible or due to the jump map G mapping the state to points away from $\overline{C} \cup D$.

Due to the nonuniqueness of solutions, notions of weak and strong invariants can be introduced for hybrid inclusions. In addition, due to the noncompleteness of maximal solutions, notions of invariants that do and do not impose that every maximal solution is complete are plausible. In this chapter, we consider a strong notion that does not insist on every maximal solution being complete. More precisely, a set $K \subset \mathbb{R}^n$ is said to be an invariant for a hybrid inclusion if, for each $\phi_0 \in K$, each solution (ϕ, γ) with $\phi(0, 0) = \phi_0$ satisfies

$$\phi(t, j) \in K \quad \forall (t, j) \in \mathrm{dom}(\phi, \gamma)$$

In other words, over its hybrid time domain, each solution with component ϕ starting in K has to stay in K for any possible input γ. Note that the notion does not insist on solutions to even exist from K.

In general, it is very difficult to directly check that a set is an invariant from the earlier definition, as that would require to explicitly check each solution from the set under study. This motivates the development of solution-independent sufficient conditions for invariants.

The tangent cone to a set will be used in the following conditions. The tangent cone to a set $K \subset \mathbb{R}^n$ at a point $x \in \mathbb{R}^n$, denoted as $T_K(x)$, is the set of all vectors $\omega \in \mathbb{R}^n$ for which there exist sequences $x_i \in K$, $\tau_i > 0$ with $x_i \to x$, $\tau_i \searrow 0$ and

$$\omega = \lim_{i \to \infty} \frac{x_i - x}{\tau_i}$$

See, for example, [1: Definition 5.12]. We will also assume the following. The sets K and C are such that $(K \times \mathbb{R}^m) \cap C$ is closed. For simplicity, F is a single-valued, locally Lipschitz map. Then, the set K is an invariant if

1. $G((K \times \mathbb{R}^m) \cap D) \subset K$
2. For every $(x, y) \in (K \times \mathbb{R}^m) \cap C$,

$$F(x, \gamma) \subset T_{(K \times \mathbb{R}^m) \cap C}(x, \gamma)$$

Example 1.13:

Consider the linear time–invariant model of the temperature of a room in Equation 1.1.1 and the FSM given in Figure 1.3, which can be modeled as in Equation 1.5 with

$$Q = \{ON, OFF\}, \quad \Sigma = \mathbb{R}$$

and for each $(q, v) \in Q \times \Sigma$,

$$\delta(q, v) = \begin{cases} ON & \text{if } v \leq z_{min} \text{ and } q = OFF \\ OFF & \text{if } v \geq z_{max} \text{ and } q = ON \end{cases}, \quad \kappa(q) = \begin{cases} 1 & \text{if } q = ON \\ 0 & \text{if } q = OFF \end{cases}$$

where $z_{min} < z_{max}$.

When $u_2 = z_{out}$ with $z_{out} \in (-\infty, z_{max}]$ and $z_{out} + z_\Delta \in [z_{min}, \infty)$, it can be shown that when an ideal implementation of this FSM is used to control the temperature of the room via the input u_1, every solution to the closed-loop system is such that z reaches the range $[z_{min}, z_{max}]$ in finite time and stays there for all future time. The closed-loop system is modeled as a hybrid inclusion with state $x = (z, q)$ and data

$$C = \left\{ x \in \mathbb{R} \times Q : z \geq z_{min}, q = OFF \right\} \cup \left\{ x \in \mathbb{R} \times Q : z \leq z_{max}, q = ON \right\}$$

$$F(x) = \begin{bmatrix} -z + \begin{bmatrix} z_\Delta & 1 \end{bmatrix} \begin{bmatrix} q \\ z_{out} \end{bmatrix} \\ 0 \end{bmatrix} \quad \forall x \in C$$

$$D = \left\{ x \in \mathbb{R} \times Q : z \leq z_{min}, q = OFF \right\} \cup \left\{ x \in \mathbb{R} \times Q : z \geq z_{max}, q = ON \right\}$$

$$G(x) = \begin{bmatrix} z \\ \delta(q, z) \end{bmatrix} \quad \forall x \in D$$

The latter property can be established by checking that the conditions proposed earlier for $K = [z_{min}, z_{max}] \times Q$ to be invariant hold. In fact, for each point $x \in K \cap D$ with $z \in [z_{min}, z_{max}]$, we have that after the jump, z is mapped back to $[z_{min}, z_{max}]$ and, in turn, x is mapped back to K. Furthermore, for every point in $\xi \in K \cap C$, the flow map F is such that the z component of the solutions from nearby ξ stays in $[z_{min}, z_{max}]$. $\quad\square$

1.3.3 STABILITY AND ATTRACTIVITY

Stability is a notion of particular interest in the study of dynamical systems as it captures the property that solutions starting nearby a point (or region) stay nearby. This property is desired for the executions of a cyber-physical system as, in particular, it would guarantee that the evolution of the physical states and the variables in the code within the cyber components remain nearby steady-state

FIGURE 1.3   A finite-state machine to control the temperature of a room.

values when initialized nearby. For models of such systems given in terms of hybrid inclusions as in Equations 1.19 and 1.20 and without inputs, stability is defined as follows:

A closed set $\mathcal{A} \subset \mathbb{R}^n$ is said to be *Lyapunov stable* if for each $\varepsilon > 0$ there exists $\delta > 0$ such that each maximal solution ϕ to the hybrid inclusion in Equations 1.19 and 1.20 without inputs and with

$$|\phi(0,0)|_{\mathcal{A}} \leq \delta$$

satisfies

$$|\phi(t,j)|_{\mathcal{A}} \leq \varepsilon$$

for all $(t,j) \in \mathrm{dom}\,\phi$.

The notion reduces to stability of a point when the set $\mathcal{A}$ is a singleton, and $|\cdot|_{\mathcal{A}}$ is the distance to the point $\mathcal{A}$. The general closed set $\mathcal{A}$ in the definition allows to study the stability of sets that have more than one point, in which case $|x|_{\mathcal{A}} = \inf_{y \in \mathcal{A}} |x - y|$. For example, the set of interest in Example 1.13 would be

$$\mathcal{A} = [z_{\min}, z_{\max}] \times Q$$

Stability of a set $\mathcal{A}$ can be certified using Lyapunov functions. A Lyapunov function is a function for a hybrid inclusion without inputs that are given by functions $V : \mathrm{dom}\,V \to \mathbb{R}$ that are defined on $\mathrm{dom}\,V$ containing $\overline{C} \cup D \cup G(D)$ and that are continuously differentiable on an open set containing the closure of C. The following sufficient condition for stability of a closed set $\mathcal{A}$ can be established [1: Theorem 3.18]: suppose that V is positive definite with respect to $\mathcal{A}$ and such that

$$\langle \nabla V(x), f \rangle \leq 0 \qquad \forall x \in C,\, f \in F(x) \tag{1.23}$$

$$V(g) - V(x) \leq 0 \qquad \forall x \in D,\, g \in G(x) \tag{1.24}$$

then $\mathcal{A}$ is stable.

A property that is also of interest when studying dynamical systems is attractivity (or convergence). In particular, attractivity of a set $\mathcal{A}$ captures the property of complete solutions converging to $\mathcal{A}$.

More precisely, a set $\mathcal{A} \subset \mathbb{R}^n$ is said to be *attractive* if there exists $\mu > 0$ such that every maximal solution ϕ to the hybrid inclusion in Equations 1.19 and 1.20 without inputs and with

$$|\phi(0,0)|_{\mathcal{A}} \leq \mu$$

is bounded and, if it is complete, satisfies

$$\lim_{(t,j) \in \mathrm{dom}\,\phi,\, t+j \to \infty} |\phi(t,j)|_{\mathcal{A}} = 0$$

Attractivity of a set can also be established using Lyapunov functions. The conditions on V plus a weak invariance property of a set can be used to characterize the set of points to which complete and bounded solutions converge to. In the following, we say that a set $K \subset \mathbb{R}^n$ is *weakly forward invariant* if for each $\phi(0,0) \in K$ there exists at least one complete solution ϕ with $\phi(t,j) \in K$ for all $(t,j) \in \mathrm{dom}\,\phi$; *weakly backward invariant* if for each $x' \in K$, $N > 0$, there exist $\phi(0,0) \in K$ and at least one solution ϕ such that for some $(t^*,j^*) \in \mathrm{dom}\,\phi$, $t^* + j^* \geq N$, we have $\phi(t^*,j^*) = x'$ and $\phi(t,j) \in K$ for all $(t,j) \preceq (t^*,j^*)$, $(t,j) \in \mathrm{dom}\,\phi$; and it is *weakly invariant* if it is both weakly forward invariant and weakly backward invariant.

The invariance principle Sanfelice, Goebel, and Teel [5] states that under the conditions on the earlier given V and further conditions on its data (see Section 1.3.5), every complete and bounded solution to the hybrid inclusion converges to the largest weakly invariant set contained in the set of points

for some $r \geq 0$, where $V^{-1}(r) :=$

$$\left(\left\{ x \in \mathbb{R}^n : \max_{f \in F(x)} \langle \nabla V(x), f \rangle = 0 \right\} \bigcup \left\{ x \in \mathbb{R}^n : \max_{g \in G(x)} V(g) - V(x) = 0 \right\} \right) \bigcap \{ x \in \mathbb{R}^n : V(x) = r \}$$

Alternatively, under further conditions on its data (see Section 1.3.5), attractivity of a set $\mathcal{A}$ can be certified by strengthening the aforementioned Lyapunov conditions. Suppose that V is positive definite with respect to $\mathcal{A}$ and such that Equations 1.23 and 1.24 hold, and

$$\langle \nabla V(x), f \rangle < 0 \qquad \forall x \in C \setminus \mathcal{A}, f \in F(x) \tag{1.25}$$

$$V(g) - V(x) < 0 \qquad \forall x \in D \setminus \mathcal{A}, \ g \in G(x) \tag{1.26}$$

then $\mathcal{A}$ is attractive. Note that these conditions also imply that $\mathcal{A}$ is stable, which in turn implies that $\mathcal{A}$ is asymptotically stable (both stable and attractive).

Example 1.14:

The state estimation problem over a network in Example 1.9 can be solved by asymptotically stabilizing the set

$$\mathcal{A} = \{ (z, \hat{z}) \in \mathbb{R}^{2n} : z = \hat{z} \} \tag{1.14.1}$$

As usual in estimation problems, we consider the estimation error to be defined as

$$\varepsilon := z - \hat{z}$$

In these coordinates, the system composed by the state ε and the timer variable τ_N can be represented by the following hybrid inclusion:

$$\left. \begin{array}{l} \dot{\varepsilon} = A\varepsilon \\ \dot{\tau}_N = -1 \end{array} \right\} (\varepsilon, \tau_N) \in C$$

$$\left. \begin{array}{l} \varepsilon^+ = (I - LM)\varepsilon \\ \tau_N^+ \in [T_N^{*\,min}, T_N^{*\,max}] \end{array} \right\} (\varepsilon, \tau_N) \in D \tag{1.14.2}$$

with the flow set and the jump set defined as

$$C = \{ (\varepsilon, \tau) \in \mathbb{R}^{n+1} : \tau_N \in [0, T_N^{*\,max}] \}$$

$$D = \{ (\varepsilon, \tau) \in \mathbb{R}^{n+1} : \tau_N = 0 \} \tag{1.14.3}$$

It follows that if there exist a symmetric positive definite matrix P and a matrix L such that

$$(I - LM)^\top \exp(A^\top s) P \exp(As)(I - LM) - P < 0 \qquad \forall s \in [T_N^{*\,min}, T_N^{*\,max}] \tag{1.14.4}$$

then the set $\mathcal{A}$ defined in Equation 1.14.1 is asymptotically stable for the hybrid inclusion Equations 1.14.2 and 1.14.3; see Ferrante et al. [11] for more details.   □

1.3.4   TEMPORAL LOGIC

Temporal logic allows for the definition of conditions that the executions to a system need to satisfy over time. The conditions are given in terms of a language that employs logical and temporal connectives (or operators) applied to propositions, which are functions of the state of the system and are used to define formulae. In particular, *linear temporal logic* (LTL) can be efficiently employed to determine *safety*, that is, "something bad never happens," and *liveness*, that is, "something good eventually happens." Verification of these properties can be performed using model-checking tools. For example, safety can be ruled out by finding an execution that, in finite time, violates the assertion defining the safety property.

Following Karaman, Sanfelice, and Frazzoli [12], the standard syntax of LTL language can be defined recursively as follows. An atomic proposition is a statement given in terms of the state x that, for each possible value of x, is either `True` (1 or $\top$) or `False` (0 or $\bot$). The following logical operators are defined:

- $\neg$ is the *negation* operator
- $\vee$ is the *disjunction* operator
- $\wedge$ is the *conjunction* operator
- $\Rightarrow$ is the *implication* operator
- $\Leftrightarrow$ is the *equivalence* operator

In addition, temporal operators are defined as follows:

- $\bigcirc$ is the *next* operator
- $\square$ is the *always* operator
- $\diamondsuit$ is the *eventually* operator
- $\mathcal{U}$ is the *until* operator

Informally speaking, the temporal operator *until* when used as $f_1 \mathcal{U} f_2$ implies that f_2 will eventually become `True` and f_1 will keep being `True` until f_2 becomes `True`. In the case of the operator *next*, $\bigcirc f_1$ implies that f_1 will be `True` at the next time instance. Every atomic proposition is an LTL formula and if f_1 and f_2 are formulae then so are

$$\neg f_1, \qquad f_1 \vee f_2, \qquad \bigcirc f_1, \qquad f_1 \mathcal{U} f_2$$

Note that given the operators *negation* and *disjunction*, the operators *conjunction* ($\wedge$), *implication* ($\Rightarrow$), and *equivalency* ($\Leftrightarrow$) can be defined as

$$f_1 \wedge f_2 = \neg(\neg f_1 \vee \neg f_2)$$
$$f_1 \Rightarrow f_2 = \neg f_1 \vee f_2$$
$$f_1 \Leftrightarrow f_2 = (f_1 \Rightarrow f_2) \wedge (f_2 \Rightarrow f_1)$$

respectively. Note that the operators *eventually* ($\diamondsuit$) and *always* ($\square$) can be defined as

$$\diamondsuit f_1 = \text{True} \, \mathcal{U} f_1$$

and

$$\square f_1 = \neg \diamondsuit \neg f_1$$

respectively.

With the aforementioned grammar, we define a semantics for cyber-physical systems given in terms of hybrid inclusions, to which we refer as $\mathcal{H}$. For simplicity, we consider the case of no inputs and state-dependent atomic propositions $\mathfrak{a}$. The value of a solution $(t,j) \mapsto \phi(t,j)$ assigning an atomic proposition $\mathfrak{a}$ the value $\texttt{True}$ at time (t,j) is denoted

$$\phi(t,j) \Vdash \mathfrak{a}$$

while if the value assigned is $\texttt{False}$, we denote it

$$\phi(t,j) \not\Vdash \mathfrak{a}$$

(equivalently, $\phi(t,j) \Vdash \mathfrak{a}$ can be written as $\mathfrak{a}(\phi(t,j)) = \texttt{True}$). Similarly, a formula $\mathfrak{f}_1$ being satisfied by a solution $(t,j) \mapsto \phi(t,j)$ at some time (t,j) is denoted by

$$(\phi, (t,j)) \vDash \mathfrak{f}_1$$

Now, we define a semantics for solutions to $\mathcal{H}$. Let $\mathfrak{a}$ be an atomic proposition, and $\mathfrak{f}_1$ and $\mathfrak{f}_2$ be two formulae given in terms of the LTL syntax described earlier. Given a solution $(t,j) \mapsto \phi(t,j)$ to $\mathcal{H}$, we define the following operators:

$$(\phi, (t,j)) \vDash \mathfrak{a} \quad \text{iff} \quad \phi(t,j) \Vdash \mathfrak{a} \tag{1.27}$$

$$(\phi, (t,j)) \vDash \neg \mathfrak{f}_1 \quad \text{iff} \quad (\phi, (t,j)) \not\vDash \mathfrak{f}_1 \tag{1.28}$$

$$(\phi, (t,j)) \vDash \mathfrak{f}_1 \vee \mathfrak{f}_2 \quad \text{iff} \quad (\phi, (t,j)) \vDash \mathfrak{f}_1 \text{ or } (\phi, (t,j)) \vDash \mathfrak{f}_2 \tag{1.29}$$

$$(\phi, (t,j)) \vDash \bigcirc \mathfrak{f}_1 \quad \text{iff} \quad (\phi, (t,j+1)) \vDash \mathfrak{f}_1 \tag{1.30}$$

$$(\phi, (t,j)) \vDash \mathfrak{f}_1 \mathcal{U} \mathfrak{f}_2 \quad \text{iff} \quad \exists (t',j') \in \operatorname{dom} \phi, t'+j' \geq t+j \text{ such that } (\phi, (t',j')) \vDash \mathfrak{f}_2, \tag{1.31}$$

$$\text{and for all } (t'',j'') \in \operatorname{dom} \phi \text{ s.t. } t+j \leq t''+j'' < t'+j', (\phi, (t'',j'')) \vDash \mathfrak{f}_1$$

Even though these completely define the semantics, the following operators are used for convenience. Let $\mathfrak{a}_1$ and $\mathfrak{a}_2$ be two atomic propositions. Given a solution $(t,j) \mapsto \phi(t,j)$ to $\mathcal{H}$, we have the following additional operators:

$$(\phi, (t,j)) \vDash \mathfrak{a}_1 \wedge \mathfrak{a}_2 \quad \text{iff} \quad \phi(t,j) \Vdash \mathfrak{a}_1 \text{ and } \phi(t,j) \Vdash \mathfrak{a}_2 \tag{1.32}$$

$$(\phi, (t,j)) \vDash \square \mathfrak{f}_1 \quad \text{iff} \quad (\phi, (t',j')) \vDash \mathfrak{f}_1 \text{ for all } t'+j' \geq t+j, (t',j') \in \operatorname{dom} \phi \tag{1.33}$$

$$(\phi, (t,j)) \vDash \lozenge \mathfrak{a}_1 \quad \text{iff} \quad \exists (t',j') \in \operatorname{dom} \phi, t'+j' \geq t+j \text{ such that } (\phi, (t',j')) \vDash \mathfrak{a}_1 \tag{1.34}$$

We say that a solution $(t,j) \mapsto \phi(t,j)$ satisfies the formula $\mathfrak{f}_1$ if and only if $(\phi, 0) \vDash \mathfrak{f}_1$. With these semantics, we propose conditions to check whether a given formula is true or not at a particular time (t,j) and, more generally, over the whole domain of the definition, namely, whether a solution satisfies a formula.

A solution $(t,j) \mapsto \phi(t,j)$ to $\mathcal{H}$ satisfies the formula

$$\mathfrak{f}_1 = \lozenge \mathfrak{a}_1$$

if there exists N^* such that for some $t+j \geq N^*$, $(t,j) \in \text{dom } \phi$, we have that $\phi(t,j)$ satisfies a_1. Given a_1 and the data (C,F,D,G) of $\mathcal{H}$, define the system $\mathcal{H}_{\cap \neg a_1}$ as the system with data $(C_{\cap \neg a_1}, F, D_{\cap \neg a_1}, G)$, where

$$C_{\cap \neg a_1} = C \cap \{x \in \mathbb{R}^n : a_1(x) = \texttt{False}\}, \qquad D_{\cap \neg a_1} = D \cap \{x \in \mathbb{R}^n : a_1(x) = \texttt{False}\}$$

Let $\phi_0 = \phi(0,0)$. The formula $f_1 = \Diamond a_1$ is true for every maximal solution from ϕ_0 when the following two properties hold:

1. Every maximal solution to $\mathcal{H}$ from ϕ_0 is complete.
2. Every maximal solution to $\mathcal{H}_{\cap \neg a_1}$ from ϕ_0 is not complete.

In fact, since every maximal solution from ϕ_0 is complete, maximal solutions from ϕ_0 to $\mathcal{H}_{\cap \neg a_1}$ would not be complete if there exists (t',j') from which either flow within $C_{\cap \neg a_1}$ or jumps back to $C_{\cap \neg a_1} \cup D_{\cap \neg a_1}$ are not possible, which corresponds to the formula eventually being satisfied. (Note that a complete solution ϕ to the system $\mathcal{H}_{\cap \neg a_1}$ would satisfy $\phi(t,j) \in C_{\cap \neg a_1} \cup D_{\cap \neg a_1}$ for all $(t,j) \in \text{dom } \phi$, which, by construction of $C_{\cap \neg a_1}$ and $D_{\cap \neg a_1}$, implies that $(\phi, (t,j)) \not\models a_1$ for all $(t,j) \in \text{dom } \phi$.) Hence, under the aforementioned conditions 1 and 2, $f_1 = \Diamond a_1$ is satisfied for any solution to $\mathcal{H}$ from ϕ_0.

According to the definition of the $\Box$ operator, a solution $(t,j) \mapsto \phi(t,j)$ to $\mathcal{H}$ satisfies the formula

$$f_1 = \Box a_1$$

if for each $(t,j) \in \text{dom } \phi$, we have that $\phi(t,j)$ satisfies a_1. Along similar lines for the $\Diamond$ operator, given a_1 and the data (C,F,D,G) of $\mathcal{H}$, consider the system $\mathcal{H}_{\cap \neg a_1}$ as the system with data $(C_{\cap \neg a_1}, F, D_{\cap \neg a_1}, G)$ given in the preceding text. The formulae $f_1 = \Box a_1$ is true for every solution to $\mathcal{H}$ when the following properties hold:

- $D \cap D_{\cap \neg a_1} = \emptyset$
- $G(D) \cap D_{\cap \neg a_1} = \emptyset$
- $\overline{C} \cap \overline{C_{\cap \neg a_1}} = \emptyset$

In fact, If $D \cap D_{\cap \neg a_1}$ is nonempty, then there exists a point in D for which a_1 is not true, which would violate the satisfaction of $\Box \cap a_1$. Similarly, if there is a point in D from where, after a jump, the state is mapped to $D_{\cap \neg a_1}$ (which is possible when $G(D) \cap D_{\cap \neg a_1} \neq \emptyset$), then $\Box a_1$ will not be satisfied. Finally, the last condition rules out the possibility of solutions to $\mathcal{H}$ flowing to a boundary point of C that is not in C and at which a_1 is not true.

Regarding the satisfaction of general formulae, the satisfaction of the formula f_1 by a solution ϕ to $\mathcal{H}$ from ϕ_0 is implied by the following property of maximal solutions:

$$\mathcal{S}_{\mathcal{H}_{\cap f_1}}(\phi_0) = \mathcal{S}_{\mathcal{H}}(\phi_0)$$

where the system $\mathcal{H}_{\cap f_1}$ has data $(F, C_{\cap f_1}, G, D_{\cap f_1})$ with

$$C_{\cap f_1} = C \cap \{x \in \mathbb{R}^n : f_1(x) = \texttt{True}\}, \qquad D_{\cap f_1} = D \cap \{x \in \mathbb{R}^n : f_1(x) = \texttt{True}\}$$

and $\mathcal{S}_{\mathcal{H}_{\cap f_1}}(\phi_0)$ and $\mathcal{S}_{\mathcal{H}}(\phi_0)$ denote the collection of maximal solutions to $\mathcal{H}$ and to $\mathcal{H}_{\cap f_1}$ from ϕ_0, respectively. Similar conditions can be derived for other operators and formulae.

The aforementioned conditions can be checked using those in Section 1.3.1 characterizing existence and type of solutions to hybrid inclusions. Note that the time-varying formulae $f_1(x,t,j)$ can be considered by incorporating τ and k as states that keep track of flow time t and jump time j, respectively.

1.3.5 ROBUSTNESS

In this section, we present conditions on the data of a hybrid inclusion that guarantee that small perturbations do not significantly change the behavior of solutions. In Goebel, Sanfelice, and Teel [1], these conditions are shown to be key in assuring that perturbed solutions are close to some unperturbed solution and that asymptotic stability is robust to small perturbations. Robustness to small perturbations is of particular interest for cyber-physical systems due to the fact that several of the perturbations introduced during the implementation of the cyber components can be assumed to be small. Such is the case of quantization and discretization effects, which, while cannot always be neglect, can be assumed to be small.

We consider hybrid inclusions without inputs, perhaps due to the result of assigning their inputs via a function of the state. If the set $\mathcal{A}$ is compact and the data of the hybrid inclusion are such that the resulting flow and jump maps are "*continuous*" and the flow and jump sets are closed, then the asymptotic stability property is robust to small perturbations. To formally state this result, we introduce the following notions for a set-valued map. A set-valued map $S : \mathbb{R}^n \rightrightarrows \mathbb{R}^m$ is *outer semicontinuous* at $x \in \mathbb{R}^n$ if for each sequence $\{x_i\}_{i=1}^{\infty}$ converging to a point $x \in \mathbb{R}^n$ and each sequence $y_i \in S(x_i)$ converging to a point y, it holds that $y \in S(x)$ [13; Definition 5.4]. Given a set $X \subset \mathbb{R}^n$, it is *outer semicontinuous relative* to X if the set-valued mapping from $\mathbb{R}^n$ to $\mathbb{R}^m$ defined by $S(x)$ for $x \in X$ and $\emptyset$ for $x \notin X$ is outer semicontinuous at each $x \in X$. It is *locally bounded* if, for each compact set $K \subset \mathbb{R}^n$, there exists a compact set $K' \subset \mathbb{R}^n$, such that $S(K) := \cup_{x \in K} S(x) \subset K'$.

Then, following [14, Theorem 6.6], if the data of the hybrid inclusion satisfies

(A1) C and D are closed sets
(A2) $F : \mathbb{R}^n \rightrightarrows \mathbb{R}^n$ is outer semicontinuous, locally bounded, and $F(x)$ is nonempty and convex for all $x \in C$
(A3) $G : \mathbb{R}^n \rightrightarrows \mathbb{R}^n$ is outer semicontinuous, locally bounded, and $G(x)$ is a nonempty subset of $\mathbb{R}^n$ for all $x \in D$

and the compact set $\mathcal{A} \subset \mathbb{R}^n$ is asymptotically stable for the hybrid inclusion, then there exists a $\mathcal{KL}$ function* β such that for each $\varepsilon > 0$ and each compact set $K \subset \mathbb{R}^n$, there exists $\delta > 0$ such that every maximal solution ϕ to a perturbation of the hybrid inclusion starting from K satisfies

$$|\phi(t,j)|_{\mathcal{A}} \leq \beta(|\phi(0,0)|_{\mathcal{A}}, t+j) + \varepsilon \quad \forall (t,j) \in \text{dom } \phi \tag{1.35}$$

where the perturbation of the hybrid inclusion is given by

$$\begin{aligned} \dot{x} &\in F_\delta(x) \quad x \in C_\delta \\ x^+ &\in G_\delta(x) \quad x \in D_\delta \end{aligned} \tag{1.36}$$

with

$$F_\delta(x) := \overline{\text{co}} F(x + \delta\mathbb{B}) + \delta\mathbb{B}$$

$$G_\delta(x) := \left\{ \eta : \eta \in x' + \delta\mathbb{B}, x' \in G(x + \delta\mathbb{B}) \right\}$$

$$C_\delta := \{x : (x + \delta\mathbb{B}) \cap C \neq \emptyset\}$$

$$D_\delta := \{x : (x + \delta\mathbb{B}) \cap D \neq \emptyset\}$$

---

* A function $\beta : \mathbb{R}_{\geq 0} \times \mathbb{R}_{\geq 0} \to \mathbb{R}_{\geq 0}$ is said to belong to class $\mathcal{KL}$ ($\beta \in \mathcal{KL}$) if it is continuous, nondecreasing in its first argument, nonincreasing in its second argument, and $\lim_{s \searrow 0} \beta(s,r) = \lim_{r \to \infty} \beta(s,r) = 0$.

The $\mathcal{KL}$ estimate in Equation 1.35 guarantees that when the data of the former hybrid inclusion is perturbed by δ, every solution $(t,j) \mapsto \phi(t,j)$ to it is such that it approaches $\mathcal{A} + \varepsilon\mathbb{B}$ when $t+j$, $(t,j) \in \mathrm{dom}\,\phi$, grows unbounded.

1.3.6 SIMULATION

The coupling between physics and computations in cyber-physical systems makes their simulation difficult. A simulator for cyber-physical systems has to be capable of computing the solution while evolving according to the physics of the system while monitoring the variables for a potential event triggering a discrete transition in the cyber components. A few software packages are available for numerical simulation of systems with such behavior, including Modelica [15], Ptolemy [16], Charon [17], HYSDEL [18], and HyVisual [19], among others. In this chapter, we summarize a recent software package developed for the simulation of hybrid equations within MATLAB®/Simulink®, namely, the Hybrid Equations (HyEQ) Toolbox [20]. Due to its capabilities to model interconnections of hybrid systems given in terms of hybrid equations, which are the single-valued version of hybrid inclusions (their flow maps and jump maps are single valued), within the HyEQ Toolbox, one can model the physical components, the cyber components, and the subsystems used to interconnect them separately within Simulink and then interconnect them to define the entire cyber-physical system.

Over a finite amount of flow and a finite number of jumps, the HyEQ Toolbox computes an approximation of a solution φ by evaluating the flow condition imposed by C and the jump condition imposed by D and, according to the result of this evaluation, by appropriately discretizing the differential equation defining the flows in Equation 1.19 or computing the new value of the state after jumps using Equation 1.20. In this way, the HyEQ Toolbox returns a discrete version of φ and its hybrid time domain dom φ. More precisely, given an input γ, the computed version of the solution (ϕ, γ) is given by

$$\phi_s : \mathrm{dom}\,\phi_s \to \mathbb{R}^n, \quad \gamma_s : \mathrm{dom}\,\gamma_s \to \mathbb{R}^m$$

which we call a *simulated solution* of the hybrid inclusion and satisfies

$$x_s^+ = F_{P,s}(x_s, \gamma_s) \quad (x_s, \gamma_s) \in C_{P,s} \tag{1.37}$$

over the intervals of flow and, at jumps, satisfies the discrete dynamics

$$x_s^+ = G_{P,s}(x_s, \gamma_s) \quad (x_s, \gamma_s) \in D_{P,s} \tag{1.38}$$

The input γ_s is the discretization of γ. The function $F_{P,s}$ is the resulting discretized flow map obtained when employing an integration scheme for the differential equation $\dot{x} = F_P(x, \gamma)$. For instance, when the integration scheme is given by the forward Euler integration scheme, $F_{P,s}(x_s, \gamma_s) = x_s + sF_P(x_s, \gamma_s)$ with $s > 0$ denoting the step size for integration. Similarly, $G_{P,s}$, $C_{P,s}$, and $D_{P,s}$ are the discretized version of G_P, C_P, and D_P, respectively. Formal definitions of simulated solutions and dynamical properties of the discretization given as Equations 1.37 and 1.38 of a hybrid inclusion can be found Sanfelice and Teel [21].

Within this framework for simulation, the HyEQ Toolbox includes two scripts to compute solutions to a hybrid equation:

1. A MATLAB script for simulation of hybrid equations within MATLAB's workspace, called *Lite HyEQ Simulator*
2. A Simulink library and associated MATLAB scripts for simulation of hybrid equations within Simulink, called the *HyEQ Simulator*

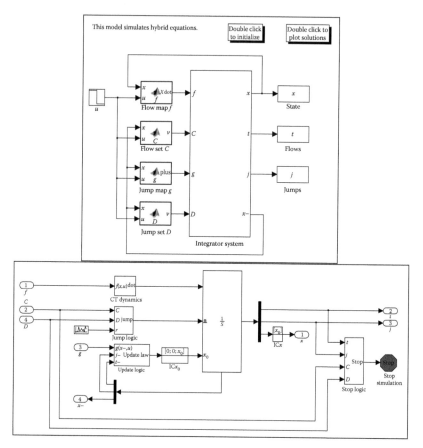

FIGURE 1.4   MATLAB/Simulink implementation of a hybrid inclusion with data (C, f, D, g) with inputs (left). Internals of integrator system (right).

Figure 1.4 shows the Simulink implementation of a hybrid inclusion, in which the user needs to enter the data (C_P, F_P, D_P, G_P). Several examples of systems that can be simulated within this toolbox, including some specific cyber-physical systems can be found Sanfelice [22]. In addition, the toolbox includes predesigned blocks to implement FSMs, ADCs, ZOHs, networks among others.

1.4   CONCLUSION

In this chapter, we showed that a class of cyber-physical systems can be treated as hybrid dynamical systems within the framework of hybrid inclusions. Modeling and analysis tools for such systems were summarized and illustrated in several examples. These tools are amenable for the mathematical analysis and design of cyber-physical systems. Additional tools that were not covered due to space limitations include the design of feedback controllers using control Lyapunov functions [2], passivity and passivity-based control [23], and the input/output stability tools [24,25].

ACKNOWLEDGMENTS

This research has been partially supported by the National Science Foundation under Grant No. ECS-1150306 and by the Air Force Office of Scientific Research under Grant No. FA9550-12-1-0366. The author would like to thank Ryan Kastner for former discussions on models of computations and to the students in his Fall 2014 class on cyber-physical systems who provided useful feedback in the initial versions of this chapter.

REFERENCES

1. R. Goebel, R. G. Sanfelice, and A. R. Teel. *Hybrid Dynamical Systems: Modeling, Stability, and Robustness.* Princeton University Press, Princeton, NJ, 2012.

2. R. G. Sanfelice. On the existence of control Lyapunov functions and state-feedback laws for hybrid systems. *IEEE Transactions on Automatic Control*, 58(12):3242–3248, December 2013.

3. J. Hopcroft and J. Ullman. *Introduction to Automata Theory, Languages, and Computation.* Addison-Wesley, Boston, MA. 1979.

4. A. Girault, B. Lee, and E. A. Lee. Hierarchical finite state machines with multiple concurrency models. *IEEE Transactions on Computer-Aided Design of Integrated Circuits and Systems*, 18:742–760, 1999.

5. R. G. Sanfelice, R. Goebel, and A. R. Teel. Invariance principles for hybrid systems with connections to detectability and asymptotic stability. *IEEE Transactions on Automatic Control*, 52(12):2282–2297, 2007.

6. E. A. Lee. Computing for embedded systems. *Proceedings of IEEE Instrumentation and Measurement*, 3:1830–1837, 2001.

7. P. Hartman. *Ordinary Differential Equations.* Birkhauser, Boston, MA. 1982.

8. H. K. Khalil. *Nonlinear Systems.* Prentice Hall, Boston, MA. 3rd edn., 2002.

9. R. G. Sanfelice, R. Goebel, and A. R. Teel. Generalized solutions to hybrid dynamical systems. *ESAIM: Control, Optimisation and Calculus of Variations*, 14(4):699–724, 2008.

10. R. G. Sanfelice. Results on input-to-output and input-output-to-state stability for hybrid systems and their interconnections. In *Proceedings of 49th IEEE Conference on Decision and Control*, Atlanta, GA, pp. 2396–2401, 2010.

11. R. Ferrante, F. Gouaisbaut, R. G. Sanfelice, and S. Tarbouriech. An observer with measurement-triggered jumps for linear systems with known input. In *Proceedings of the 19th IFAC World Congress*, 140–145, Cape Town, South Africa, August 2014.

12. S. Karaman, R. G. Sanfelice, and E. Frazzoli. Optimal control of mixed logical dynamical systems with linear temporal logic specifications. In *Proceedings of 47th IEEE Conference on Decision and Control*, Cancun, Mexico, pp. 2117–2122, 2008.

13. R. T. Rockafellar and R. J.-B. Wets. *Variational Analysis.* Springer, Berlin Heidelberg, Germany, 1998.

14. R. Goebel and A. R. Teel. Solutions to hybrid inclusions via set and graphical convergence with stability theory applications. *Automatica*, 42(4):573–587, 2006.

15. H. Elmqvist, S. E. Mattsson, and M. Otter. Modelica: The new object-oriented modeling language. In *Proceedings of 12th European Simulation Multiconference*, Manchester, UK. 1998.

16. J. Liu and E. A. Lee. A component-based approach to modeling and simulating mixed-signal and hybrid systems. *ACM Transactions on Modeling and Computer Simulation*, 12(4):343–368, 2002.

17. R. Alur, T. Dang, J. Esposito, Y. Hur, F. Ivancic, V. Kumar, I. Lee, P. Mishra, G. J. Pappas, and O. Sokolsky. Hierarchical modeling and analysis of embedded systems. *Proceedings of the IEEE*, 91:11–28, 2003.

18. F. D. Torrisi and A. Bemporad. HYSDEL — A tool for generating computational hybrid models for analysis and synthesis problems. *IEEE Transactions on Control Systems Technology*, 12:235–249, 2004.

19. E. A. Lee and H. Zheng. Operational semantics for hybrid systems. In *Hybrid Systems: Computation and Control: Eighth International Workshop*, pp. 25–53, Zurich, Switzerland, March 9–11, 2005.

20. R. G. Sanfelice. Hybrid Equations (HyEQ) toolbox. https://hybrid.soe.ucsc.edu/software, 2013.

21. R. G. Sanfelice and A. R. Teel. Dynamical properties of hybrid systems simulators. *Automatica*, 46(2):239–248, 2010.

22. R. G. Sanfelice. Online blog of the Hybrid Equations (HyEQ) toolbox. http://hybridsimulator.wordpress.com, 2013.

23. R. Naldi and R. G. Sanfelice. Passivity-based control for hybrid systems with applications to mechanical systems exhibiting impacts. *Automatica*, 49(5):1104–1116, May 2013.

24. R. G. Sanfelice. Input-output-to-state stability tools for hybrid systems and their interconnections. *In IEEE Transactions on Automatic Control*, 140–145, Cape Town, South Africa, August 2014.

25. R. G. Sanfelice. Interconnections of hybrid systems: Some challenges and recent results. *Journal of Nonlinear Systems and Applications*, 2(1–2):111–121, 2011.

2 Compositional Design of Cyber-Physical Systems Using Port-Hamiltonian Systems

Siyuan Dai, Zsolt Lattmann, and Xenofon Koutsoukos

CONTENTS

2.1 INTRODUCTION

The advancement of technology over the last few decades has led to the ubiquity of embedded systems in everyday lives. Traditional embedded system technology emphasizes the computational elements and how to design these elements to perform a certain task. Cyber-physical systems (CPS) are engineering systems that integrate computational, communication, and control elements with the physical dynamics of the system [38]. As the computational components in modern CPS increase in both number and complexity, great challenges arise for their integration with the physical domain [34]. Consequently, rigorous engineering methods are needed for the integration of computational components with the physical system in order to achieve predictable, correct behavior. Designing these CPS requires the use of well-formulated software tools in order to simulate, analyze, and identify design flaws in the system before prototyping. Additionally, component-based design and modeling is desirable because it allows for scalability, substitutability, and reusability when modeling large-scale systems.

Traditionally, composition is done in an ad hoc way in which components are individually designed and integrated into the system with the objective of making it work. As CPS continue to grow and evolve, more problems emerge from the interaction of heterogeneous domains, which leads to a loss of compositionality, hampering the system integration process [38]. In software systems, processes execute in sequential steps and are commonly represented as difference equations. In physical systems, processes exist in parallel and are captured as differential equations [10]. A software tool needs to be carefully formulated in order to integrate these contrasting domains of CPS. Hybrid dynamics and nonlinearities also present significant challenges. A hybrid system exhibits both continuous-time and discrete-time behavior and is used to model a large class of CPS. Most physical dynamics are nonlinear by nature, and nonlinear systems do not obey the superposition principle.

This chapter describes a software tool/language developed to model and simulate CPS while addressing the challenges of compositionality, nonlinearity, and hybrid dynamics. Section 2.2 presents the related works on modeling and simulation of CPS. We discuss bond graphs, which is a domain-independent modeling tool for physical systems. A basic bond graph is an acausal model in which edges describe energy flow; a simulation environment such as Modelica can simulate an acausal bond graph since the equations denote equality rather than assignment. Causality assignment on a bond graph forces the power variables on the bonds and elements into an input–output format (causal); a simulation environment such as Simulink can simulate a causal bond graph using block diagrams or signal flow graphs.

Section 2.3 presents a software tool used to model CPS, specifically a domain-specific modeling language (DSML) based on port-Hamiltonian systems. The unique structure of port-Hamiltonian systems allows for the modeling of CPS with nonlinearities, hybrid dynamics, and heterogeneous domains. The syntax and semantics of Port-Hamiltonian Systems Modeling Language (PHSML) are discussed. Constraints that are placed among elements in the language are then discussed because a port-Hamiltonian system must obey certain rules. By obeying these constraints, a system built using PHSML is guaranteed to be a port-Hamiltonian system, thereby ensuring certain system properties. An engine-dynamometer model is used in this section to illustrate how the modeling language deals with nonlinear components, hybrid dynamics, and composition.

Section 2.4 presents two component-based simulation methods for port-Hamiltonian systems. Both methods/interpreters use the Dirac structure of the port-Hamiltonian model to derive equations for simulation. Interactions between domains occur through ports that are captured by the Dirac structure and constituent equations. A model transformation method designed for PHSML generates acausal representations of the port-Hamiltonian system (PHS) model and is simulated in a Modelica environment. A second model transformation method generates causal representations of the port-Hamiltonian system (PHS) model and is simulated in Simulink. Both transformation methods are applied to the engine-dynamometer model to generate the mathematical structures and the simulation results.

Section 2.5 presents theory on passivity and how it relates to port-Hamiltonian systems. The basic property of a port-Hamiltonian system is energy balanced; the system only dissipates or stores energy and will never generate energy on its own. This property implies that port-Hamiltonian systems are passive as long as their storage functions are valid. Passivity guarantees that the system is both stable and minimum phase, which are important properties for design. Passivity analysis is applied to the engine-dynamometer example and shows that passive components yield a passive composed system.

2.2   RELATED WORKS

Component-based modeling of CPS is a challenging problem because of the inherent heterogeneity within CPS [37]. Components are implemented as well-defined Models of computation and abstract semantics are used to define interactions between them [10]. The intricacies of both acausal and causal modeling can be captured using a DSML [36]. Software tools typically formalize interactions between the physical and computational components through unidirectional physical signals and computational signals [18]. In recent years, research has focused on using the exchange of energy to merge the abstractions between the computational and physical domains [30].

Port-Hamiltonian systems are formulated through bond graphs, which is a domain-independent modeling tool based on energy exchange. Nodes in a bond graph represent physical phenomena, and edges (bonds) represent idealized energy connections; different physical domains such as electrical, mechanical rotational, hydraulic, and thermodynamics are all described in the same way through a physical exchange of energy. Its major distinction from tools such as signal-flow graphs is that every arc (bond) in a bond graph model represents a bidirectional exchange of energy, rather than a unidirectional flow of information. Bond graphs exhibit strong topological and graphical similarities to schematic models of their corresponding physical systems (e.g., free-body diagrams, electrical circuit diagrams) [5].

Bond graphs allow for the recursive decomposition of a system into subsystems, where each node is represented by a bond graph model [26]. Each bond is represented by a half arrow, where the direction of the arrow denotes the flow of positive energy; power is a ubiquitous concept in all physical domains and is the inner product of two physical variables [3]. The transfer of energy between different subsystems and physical domains occur through real-world engineering interactions such as mechanical actuators, electrical wires, and pulleys. Bond graph elements either store energy, dissipate energy, or conserve energy. Active elements are categorized into sources and sinks; they are not part of the bond graph model but represent the boundary conditions in which the bond graph model interacts with the environment [8]. Sources either impose an effort or a flow value onto the rest of the bond graph model, depending on if it is a source of effort or source of flow. Sources can either be constant or time dependent.

A bond graph in its most basic form is acausal, which allows it to preserve the physical reality of the modeled system. This leads to the formulation of a generalized bond graph, in which the order of causality between power conjugate variables is purposely ignored. Keeping a bond graph in its acausal form allows for the preservation of the physical reality of the modeled system; using a set of declarative equations, physically intuitive relationships are defined among the variables. Acausal models are simulated through a set of differential and algebraic equations (DAEs) [19]; a DAE is a more general form of an Ordinary Differential Equations (ODEs), where variables are implicitly defined and reducing its index involves the use of the Pantelides algorithm [33]. Modelica is an example of an acausal simulation environment that simulates systems using DAE solvers [22].

Modelica can be used to model and simulate acausal bond graph models. Modelica was developed in 1996 with the goal of modeling physical systems by components through a standardized format. Modelica is an acausal object-oriented language; equations denote equality rather than assignments [19]. There are many Modelica software tools including OpenModelica [32], JModelica [25], and Dymola [14]. Modelica has become the standard for acausal modeling of physical systems; the language and standard library continues to expand as it becomes widely adopted in both industry and academia [1].

System modeling in Modelica is done through the Modelica Standard Library (MSL) [2], which is an open-source library developed by the Modelica Association for the modeling of mechanical, electrical, thermal, fluid, and control systems. Models are decomposed into a set of interconnected components and are represented graphically using the graphical editor of a Modelica modeling environment [16]. In order to assist with the integration with other software environments, variable-type definitions in the MSL reflect the naming conventions for physical quantities given by the

International System of Units [15]. Components are defined in the MSL and are referenced through the library hierarchy. Connections specify the interconnection between components.

Each Modelica model begins as a .mo file containing a series of interconnected components. The translator *flattens* the equations of the .mo file, which includes processes such as type checking, inheritance matching, and import statements. The optimizer then applies Pantelides algorithm to the flatten model equations in an effort to reduce the index of the DAE system [33]. The optimized flatten equations are then used by the code generator to generate source code (e.g., C code) needed for simulation, which the compiler transforms into an executable [7].

Equations in a bond graph can also be derived by assigning causality to the graph. Each bond in the model is interpreted as a bidirectional signal flow; through a process called *causality analysis*, the signal directions in each bond can be determined, which allows for the generation of equations [4]. A causal bond graph contains causal strokes, which indicate the direction of the imposition of effort. The *Sequential Causal Assignment Procedure* is an algorithm that assigns causal strokes to every bond in a bond graph model [26]. Once causality analysis is applied to the bond graph model, a series of causal equations can be derived. In the case that there are no causality conflicts in the model, a set of ordinary differential equations are derived [42]. Different bond graph elements have different constraints regarding the imposition of effort or flow, which results in different causality rules for different elements [5]. Simulink is an example of a causal simulation environment used to simulate a bond graph model and obtain simulation results using either a fixed-step or variable-step ODE solver.

2.3   PORT-HAMILTONIAN SYSTEM MODELING LANGUAGE

The use of models as an initial step in the development of complex systems is prevalent in all fields of science and engineering. A model is an abstraction of occurrences in the physical world and represents it through a set of concepts. In engineering, models provide a way of representing systems using a specific set of mathematical constructs. The difference in abstraction between physical dynamics and computational elements poses a significant challenge for the modeling of CPS [40]. The abstractions regarding computational and networking elements will need to be redefined in a way that merge the layers of the CPS [35]. This section presents a DSML called PHSML; it is based on port-Hamiltonian systems and uses principles of model-integrated computing [27].

In order to accurately model the interactions of different components in the systems, interfaces need to be well formulated [31]. The concept of a port is a result of interaction between different elements and submodels within a model. Port-based modeling has garnered attention over the years since it captures the underlying physics of the system. For physical systems, this interaction is depicted as an exchange of energy through the power bond. A power bond represents a bilateral relationship between the power ports of two interacting elements or sub models. Each bond contains two power conjugate variables (called *effort* and *flow*), which relate to power by a product relationship shown in Definition 2.1.

Definition 2.1 [13]   *Let F be an n-dimensional linear space of flow variables f, and F* be its dual space of effort variables e, power is defined as*

$$P = \langle e, f \rangle, (f, e) \in F \times F^*. \tag{2.1}$$

Port-based modeling of lumped-parameter complex physical systems leads to a framework of generalized bond graphs, which are different from traditional bond graphs in two ways: use of symplectic gyrators and absence of causality. Conventional physical domains usually have two types of storage elements associated with two types of energy (e.g., capacitors store electrical energy, while inductors store magnetic energy) [35]. In generalized bond graphs, symplectic gyrators are attached to I-storage elements, which inverts the roles of effort and flow and allow I-storage elements to be

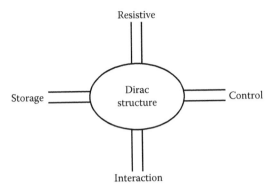

FIGURE 2.1   Port-Hamiltonian systems. (From Duindam, V. et al., *Modeling and Control of Complex Physical Systems: The Port-Hamiltonian Approach*, Springer, 2009 [13].)

treated like C-storage elements [3]. The purpose of symplectic gyrators is to decompose the two types of energy of physical domains to a unified type of energy. The lack of causality within generalized bond graphs allow the preservation of the underlying physics of the model. In a generalized bond graph, storage, resistive, and source ports connect through a generalized junction structure, which includes junctions, transformers, and gyrators [20].

A generalized bond graph can be summarized by its energy-storing elements, resistive elements, and power-conserving elements connected by bonds. Collectively, the power-conserving elements form the Dirac structure and connect all other non-power-conserving elements in the model. This extends to the formulation of port-Hamiltonian systems, represented by Figure 2.1, where a set of ports (control, interaction, resistive, and storage) are interconnected through a Dirac structure [13]. A port-Hamiltonian system with a Hamiltonian function H, energy storage ports S, resistive ports R, control ports C, interconnection ports I, and a Dirac structure D can be written in a formal model in Equation 2.2,

$$\Sigma = (H, R, C, I, D), \tag{2.2}$$

which leads to the implicitly defined port-Hamiltonian system dynamics:

$$\Sigma : \left(-x'(t), \frac{\partial H}{\partial x}(x(t)), f_R(t), e_R(t), f_C(t), e_C(t), f_I(t), e_I(t) \right) \in D. \tag{2.3}$$

Energy storage and resistive elements make up the internal ports of a port-Hamiltonian system; control and interaction elements make up the external ports. The internal ports and the interaction ports make up the physical portion of the CPS where the physical dynamics are captured through Equation 2.3; the control ports make up the cyber portion of the CPS and describe the interaction between a port-Hamiltonian system and a controller; control software is developed to affect one of the control port's power conjugate variables in order to perform an action. The interaction through control ports forms the fundamental concept of compositionality; a port-Hamiltonian controller connects with a port-Hamiltonian system results in another port-Hamiltonian system. The Hamiltonian function of the composed system is the sum of the Hamiltonian functions of its components; the Dirac structure of the composed system is determined by the Dirac structure of its components as well.

2.3.1   SYNTAX AND SEMANTICS

The PHSML is developed using generic modeling environment (GME), based on the model-integrated computing tool suite developed at the Institute for Software and Integrated Systems at Vanderbilt University [17]. GME provides a metamodeling environment to create DSMLs and allows for the construction of domain models through a collection of objects and connectors. GME ensures that the domain-specific instance models comply with the specified metamodel. Within the metamodel, there are different concepts to represent an object. An object can either be an atom, which is the basic indivisible object that cannot contain other objects; a connection, which can connect two objects; or a model, which can contain other objects such as atoms, connections, or models. Hierarchical composition is supported through model elements. Within model elements, objects can be connected to each other based on the rules defined in the metamodel. Each connection has a source object (denoted by *src*) and a destination object (denoted by *dst*).

PHSML encodes *all* modeling and connection rules that are specific for port-Hamiltonian systems. The model is developed in GME as a PHSML model; a model translator, processes the structural information of the PHSML model and creates a port-Hamiltonian system model characterized by Dirac structures and constituent equations of ports. The top-level hierarchy of PHSML is called a *component assembly*, which consists of components that serve as the objects, and bonds that serve as the connections. A component is defined as an object consisting of a Dirac structure, a set of internal ports, and a set of external ports. PHSML defines two types of bonds (i.e., connections): internal bonds connect different sub objects within a component, while external bonds connect external ports of different components. The PHSML model provides a set of structural data for each object, which are enumerated as follows:

1. *ID*, the unique tag generated for each object in the PHSML model
2. *Name*, the name of the object as given in the PHSML model
3. *Type*, the type of the object
4. *Equation*, the constituent equation of the object, defined by the user in the PHSML model
5. *Ratio*, the constant ratio value of the object
6. *Bond*, lists the unique IDs of all bonds that are attached to the component

2.3.1.1 Bonds and Components

The metamodel of PHSML is summarized in Figure 2.3. Bonds form the connection semantics of the PHSML. Each bond in PHSML, shown in Figure 2.2, contains a pair of power conjugate variables, effort, and flow, whose inner product yields power. Bonds are used to connect components and objects together. Each component is instantiated as a model and contains a series of sub-components given by *Dirac Structure*, *Internal Port*, and *External Port*.

Dirac structure: The most important modeling element to each component is the Dirac structure, which describes the power-conserving interconnection structure of each component. The Dirac structure is a collection of four different types of modeling elements: *gyrator, transformer, zero junction,* and *one junction*. Gyrators and transformers contain a ratio attribute, which denotes the transformer or gyrator ratio. Zero junctions and one junctions denote either a parallel or series interconnection between other elements respectively.

Internal ports: The two internal ports (which are also denoted as atoms) both have dynamics that relates their power conjugate variables through constituent equations. The equation attribute allows the user to specify the constituent equations of the port with respect to the power conjugate variables, effort, and flow. A *linear resistive port atom relates* the power conjugate variables through an equation in the form of $e = Rf$, where R is a constant; a linear storage port atom relates the power conjugate variables through an equation in the form of $der(e) = Sf$, where S is a constant. Nonlinear resistive and storage ports do not contain a formulaic expression, but still relate the effort and flow variables through a function.

External ports: The two types of external ports are fundamentally similar. Control ports also contain the equation attribute, because we classify bond graph elements such as sources and sinks in control ports. Through the *equation* attribute, the constituent equations of sources and sinks can be specified. Interaction ports allow for different components to connect to each other and contain no additional attributes.

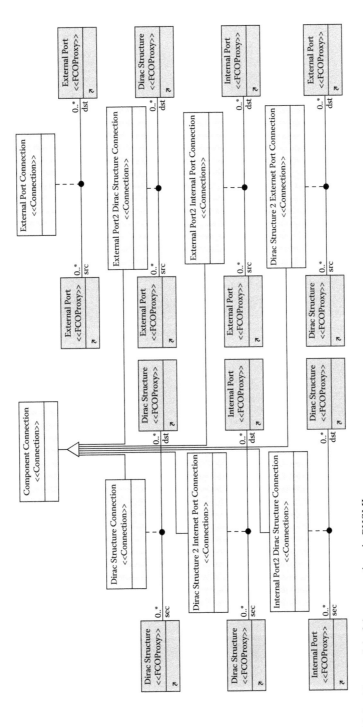

FIGURE 2.2   Metamodel of the connections in PHSML.

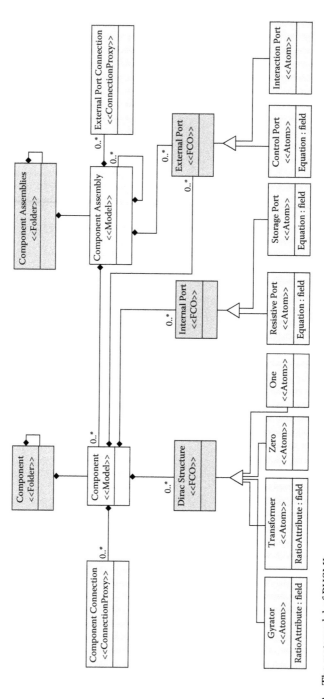

FIGURE 2.3 The metamodel of PHSML.

2.3.2 CONSTRAINTS

The most important goal of PHSML is to ensure that models are valid port-Hamiltonian systems. Constraints are implemented in the PHSML metamodel to ensure that certain interconnections between PHSML modeling elements, which will violate the underlying property of port-Hamiltonian systems, are not allowed. There are two types of constraints: (1) directional connections and connections between modeling types are implemented as UML diagram shown in Figure 2.2 and (2) a model transformation software component defines a set of rules, which require a specific number of connections for each object. A detailed list of allowable connections between elements is detailed as follows:

1. *Junction (includes both one and zero junctions)*: There are no constraints to the number of bonds that connect to these objects. However, at least one bond must have this junction object as destination and at least one bond must have this junction object as source. Because of power conservation property of junctions, power has to come into the junction and also has to leave.
2. *Transformer and gyrator*: There must be exactly two bonds connecting to this object. One bond must have this object as destination and the other bond must have this object as source.
3. *Internal port (includes both resistive and storage ports)*: There must be exactly one bond connection to this object. The bond can have this object as source or destination. Power is generally dissipated or stored, which results in power generally flowing into a resistive or storage port. However, cases can arise in which the storage port or resistive port can deliver power. A discharging capacitor is a good example of a storage port providing power; an effort source that is dependent on flow is an example of a resistive port providing power. An internal port cannot be connected to another internal port (Figure 2.2).
4. *External port (includes both control and interaction ports)*: There must be one bond connection to this object. The bond can have this object as either destination or source. These ports denote boundary conditions either between components or with the environment, such as energy sources and sinks. External ports can be connected to internal ports, but Internal Ports cannot be connected to external ports directly (Figure 2.2).

2.3.3 ENGINE DYNAMOMETER

Throughout this chapter, we will use a running example of an engine-dynamometer system. The engine-dynamometer (dyno) model is inspired by bond graph models of physical systems presented [9]. The model consists of three interconnected components shown in Figure 2.4: engine, starter, and dynamometer. The engine-dynamometer model is a physical system described by three interacting port-Hamiltonian system components. Control ports on each component describe the cyber portions which allow controllers to interface with the model.

The engine component is modeled by a one junction and a transformer connecting a throttle control port, interaction ports to the starter and dyno components, and a storage port that is the inertia of the engine. The engine either delivers a positive amount of calculated torque to the inertia load (when throttle is engaged) or it delivers a negative amount of calculated torque (negative throttle indicates braking). The throttle-controlled torque map is described as a nonlinear transformer that links the throttle control port with the junction. A bond graph model of the engine built using PHSML is

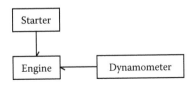

FIGURE 2.4 Diagram of the engine-dynamometer model.

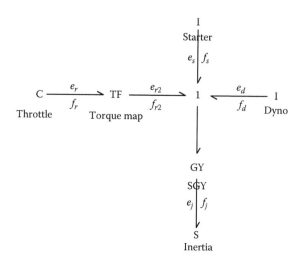

FIGURE 2.5 Generalized bond graph model of the engine.

presented in Figure 2.5; variable names are labeled for bonds in order to describe the constituent equations. e_r and f_r are the power conjugate variables of the throttle control port; e_s and f_s are the power conjugate variables of the interaction port to the starter motor; e_j and f_j are the power conjugate variables of the engine inertia; e_d and f_d are the power conjugate variables of the interaction port to the dynamometer.

The nonlinearity of the engine component is caused by the nonlinear transformer that models the throttle control port. The variables e_r, f_r, e_{r2}, and f_{r2} are related by an interpolation mapping; the interpolation function uses a series of if-then-else statements, based off of real torque map data from a Cummins Inc. engine [24]. The transformer ratio, T, is a function of the engine speed; the mapping of f_{r2} (engine speed) to T is illustrated by the torque map in Figure 2.6:

$$e_{r2} = Te_r,$$
$$f_r = Tf_{r2}. \tag{2.4}$$

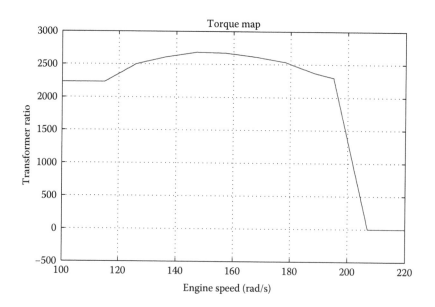

FIGURE 2.6 Torque map of the nonlinear transformer.

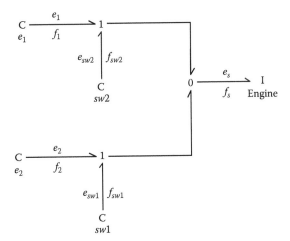

FIGURE 2.7 Generalized bond graph model of the starter motor.

The inertia from the engine has the following constituent equation:

$$\frac{d}{dt}e_j = \frac{f_j}{J},\tag{2.5}$$

where J is the moment of inertia value.

Hybrid dynamics is central to the operation of the starter motor. Hybrid dynamics are implemented as switch networks. Given that one of the constituent variables of a switch is always equal to zero, a switch will never contribute any power to the system. A generalized bond graph model of the starter motor is presented in Figure 2.7. The amount of torque that the starter motor provides is governed by a Boolean signal s. If $s = 1$ is true, the switch $SW2$ becomes active and the switch $SW1$ becomes inactive; if $s = 0$ is true, the switch $SW1$ becomes active and the switch $SW2$ becomes inactive (Figures 2.7 and 2.8). The two torque values are modeled as constant sources of efforts e_1 and e_2. Their flow outputs are f_1 and f_2, respectively.

The hybrid constituent equation of $SW1$ is

$$
\begin{aligned}
e_{sw1} &= 0 \quad \text{if } s = 1, \\
f_{sw1} &= 0 \quad \text{if } s = 0.
\end{aligned}
\tag{2.6}
$$

The hybrid constituent equation of $SW2$ is

$$
\begin{aligned}
e_{sw2} &= 0 \quad \text{if } s = 0, \\
f_{sw2} &= 0 \quad \text{if } s = 1.
\end{aligned}
\tag{2.7}
$$

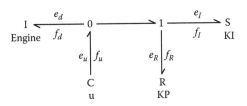

FIGURE 2.8 Generalized bond graph model of the dynamometer.

The dynamometer simulates road load to the engine and is a system without hybrid dynamics or nonlinearities. It is modeled as a proportional-integral (PI) system, which takes the angular velocity of the engine as a measured signal and a step source as the set point signal. The error signal is then calculated by subtracting the set point signal and measured signal. The output of the PI controller is a simulated torque that goes back to the engine as a load. A generalized bond graph model of the dynamometer is presented in Figure 2.8. e_u and f_u are the power conjugate variables of the set point control port; e_R and f_R are the power conjugate variables of the proportional controller; e_I and f_I are the power conjugate variables of the integral controller.

The proportional component has the following constituent equation:

$$e_R = K_R f_R, \tag{2.8}$$

where K_R is the proportional gain value. The integral component has the following constituent equation:

$$e_I = K_I \int_{t_0}^{t} f_I d\tau, \tag{2.9}$$

where K_I is the integrator gain value.

2.4 SIMULATION

The model interpreter generates the Dirac structure equations and the constituent equations of the power ports from a PHSML model. In particular, the Dirac structure is generated as a kernel representation, defined by the E and F matrices. The kernel representation is an acausal model, and one of the advantages of acausal models is independent analysis; the kernel representation of each component can be derived without knowledge of its neighboring components or the global model. Equations defined inside the Modelica language have no inherent causality and can model the kernel representation of Dirac structures. The transition from port-Hamiltonian systems to Modelica is intuitive because both frameworks define interactions in terms of power ports.

2.4.1 ACAUSAL SIMULATION

The kernel representation is an acausal representation of the Dirac structure where there are no explicit input–output relationships between the effort and flow variables.

Definition 2.2 [13] *Every Dirac structure $D \subset F \times F^*$ can be represented in a kernel representation as*

$$D(x) = \{(f,e) \in F \times F^* : F(x)f + E(x)e = 0\}, \tag{2.10}$$

where the matrices $E(x)$ and $F(x)$ satisfy the following two conditions:

1. $\mathrm{rank}[F(x)|E(x)] = \dim F$.
2. $E(x)F^T(x) + F(x)E^T(x) = 0$.

Derivation of the Dirac structure focuses on the power-conserving elements in the PHS, which consists of objects with *type* of zero, one, transformer, or gyrator. The method sequentially combines all power-conserving elements to generate its kernel representation equations by recognizing that there is a pattern to the Dirac structure of every power-conserving element. A transformer or a gyrator only has one possible kernel representation due to its SISO (one bond in, one bond out) nature.

$$\begin{bmatrix} -1 & r \\ 0 & 0 \end{bmatrix} \begin{bmatrix} e_i \\ e_o \end{bmatrix} + \begin{bmatrix} 0 & 0 \\ -1 & \frac{1}{r} \end{bmatrix} \begin{bmatrix} f_i \\ f_o \end{bmatrix} = 0, \tag{2.11}$$

$$\begin{bmatrix} -1 & 0 \\ 0 & \frac{1}{k} \end{bmatrix} \begin{bmatrix} e_i \\ e_o \end{bmatrix} + \begin{bmatrix} 0 & k \\ -1 & 0 \end{bmatrix} \begin{bmatrix} f_i \\ f_o \end{bmatrix} = 0. \tag{2.12}$$

These describe the kernel representations of a transformer and a gyrator, respectively, where r is the transformer turning ratio and k is the gyrator ratio. A zero or one junction, on the other hand, can have an infinite number of kernel representations due to its MIMO nature (multiple bonds in, multiple bonds out). However, there is a pattern to the matrices; for a zero and one junction with n bonds, their kernel representations are shown in Equations 2.13 and 2.14 as follows:

$$\begin{bmatrix} 1 & 1 & 0 & \cdots & \cdots & 0 \\ 1 & 0 & 1 & 0 & \cdots & 0 \\ 1 & 0 & 0 & \ddots & \cdots & 0 \\ \vdots & \vdots & \vdots & \cdots & \ddots & \vdots \\ 1 & 0 & 0 & \cdots & 0 & 1 \\ 0 & 0 & 0 & \cdots & 0 & 0 \end{bmatrix} \begin{bmatrix} e_1 \\ e_2 \\ e_3 \\ \vdots \\ e_{n-1} \\ e_n \end{bmatrix} + \begin{bmatrix} 0 & 0 & 0 & \cdots & 0 & 0 \\ 0 & 0 & 0 & \cdots & 0 & 0 \\ 0 & 0 & 0 & \cdots & 0 & 0 \\ \vdots & \vdots & \vdots & \ddots & \vdots & \vdots \\ 0 & 0 & 0 & \cdots & 0 & 0 \\ 1 & 1 & 1 & \cdots & 1 & 1 \end{bmatrix} \begin{bmatrix} f_1 \\ f_2 \\ f_3 \\ \vdots \\ f_{n-1} \\ f_n \end{bmatrix} = 0, \tag{2.13}$$

$$\begin{bmatrix} 1 & 1 & 1 & \cdots & 1 & 1 \\ 0 & 0 & 0 & \cdots & 0 & 0 \\ 0 & 0 & 0 & \cdots & 0 & 0 \\ \vdots & \vdots & \vdots & \ddots & \vdots & \vdots \\ 0 & 0 & 0 & \cdots & 0 & 0 \\ 0 & 0 & 0 & \cdots & 0 & 0 \end{bmatrix} \begin{bmatrix} e_1 \\ e_2 \\ e_3 \\ \vdots \\ e_{n-1} \\ e_n \end{bmatrix} + \begin{bmatrix} 0 & 0 & 0 & \cdots & \cdots & 0 \\ 1 & 1 & 0 & \cdots & \cdots & 0 \\ 1 & 0 & 1 & 0 & \cdots & 0 \\ \vdots & \vdots & & 0 & \ddots & \vdots \\ \vdots & \vdots & \vdots & \cdots & 1 & 0 \\ 1 & 0 & 0 & \cdots & 0 & 1 \end{bmatrix} \begin{bmatrix} f_1 \\ f_2 \\ f_3 \\ \vdots \\ f_{n-1} \\ f_n \end{bmatrix} = 0. \tag{2.14}$$

2.4.1.1   Implementation of Acausal Simulation

The model interpreter reads structural data from a PHSML model. A model transformation method is then applied to the data, which transforms a port-Hamiltonian system into an acausal computational model (in the form of a MATLAB m-script). We developed an algorithm that constructs the kernel representation of every power-conserving element inside the component using Equations 2.11 through 2.14. The algorithm then combines all individual kernel representations together using equation substitution methods. A summary of the algorithm is listed in Algorithm 2.1.

Algorithm 2.1 Generalized Bond Graph to a Dirac Structure

-**for all** Components
-   **for all** Power-conserving elements
-     Find E and F based on pattern;
-   **while** Interconnections left
-     Substitute Dirac structure equations to form combined Dirac structure;

Algorithm 2.2 Generalized Bond Graphs to Modelica

- Model Hierarchy Section;
- Write Dirac structure matrices;
-**for all** Power ports on the Dirac structure
- Write indexed effort and flow variables;
- Model Equation Section;
-**for all** Power ports on the Dirac structure
- Write Equation attribute;
- end model;

For all nonpower-conserving elements in the PHS, constituent equations are denoted by the *equation* parameter of the data structure. Compiling these equations with the Dirac structure derived by Algorithm 2.1 provides an acausal computational model. The hierarchy section of the Modelica code contains system parameters and power conjugate variables. The equation section of the Modelica code contains the Dirac structure, written as a matrix equation, and the set of constituent equations for each power port. The process of writing the Modelica code is detailed by Algorithm 2.2.

2.4.1.2 Acausal Simulation of the Engine Dynamometer

Applying Algorithm 2.1 to the engine dynamometer example, we obtain the kernel representation of the Dirac structure of the engine (Equation 2.15), starter motor (Equation 2.16), and dynamometer (Equation 2.17):

$$\begin{bmatrix} -1 & -\frac{1}{T} & 0 & -\frac{1}{T} \\ 0 & 0 & 0 & 0 \\ 0 & 0 & T & 0 \\ 0 & 0 & 0 & 0 \end{bmatrix} \begin{bmatrix} e_r \\ e_s \\ e_j \\ e_d \end{bmatrix} + \begin{bmatrix} 0 & 0 & \frac{1}{T} & 0 \\ 1 & -T & 0 & 0 \\ 1 & 0 & 0 & 0 \\ 1 & 0 & 0 & -T \end{bmatrix} \begin{bmatrix} f_r \\ f_s \\ f_j \\ f_d \end{bmatrix} = 0, \tag{2.15}$$

$$\begin{bmatrix} -1 & -1 & -1 & -1 & 0 \\ -1 & -1 & 0 & 0 & 1 \\ 0 & 0 & 0 & 0 & 0 \\ 0 & 0 & 0 & 0 & 0 \\ 0 & 0 & 0 & 0 & 0 \end{bmatrix} \begin{bmatrix} e_1 \\ e_{sw2} \\ e_2 \\ e_{sw1} \\ e_s \end{bmatrix} + \begin{bmatrix} 0 & 0 & 0 & 0 & 0 \\ 0 & 0 & 0 & 0 & 0 \\ 1 & -1 & 0 & 0 & 0 \\ 1 & 0 & -1 & 0 & 1 \\ 0 & 0 & 1 & -1 & 0 \end{bmatrix} \begin{bmatrix} f_1 \\ f_{sw2} \\ f_2 \\ f_{sw1} \\ f_s \end{bmatrix} = 0, \tag{2.16}$$

$$\begin{bmatrix} -1 & 1 & 0 & 0 \\ -1 & 0 & 1 & 1 \\ 0 & 0 & 0 & 0 \\ 0 & 0 & 0 & 0 \end{bmatrix} \begin{bmatrix} e_u \\ e_d \\ e_R \\ e_l \end{bmatrix} + \begin{bmatrix} 0 & 0 & 0 & 0 \\ 0 & 0 & 0 & 0 \\ 1 & 1 & 1 & 0 \\ 1 & 1 & 0 & 1 \end{bmatrix} \begin{bmatrix} f_u \\ f_d \\ f_R \\ f_l \end{bmatrix} = 0. \tag{2.17}$$

The .mo file that is generated from Algorithm 2.2 can be used to simulate the port-Hamiltonian system using a Modelica simulation environment. We simulate the engine-dyno model with parameters of $J = 20$, $K_R = 200$, and $K_I = 100$. Using control inputs of $f_u = 20$, $e_1 = 600$, $e_2 = 0$, and e_r (8 for the first second and 2 for the rest), the simulation results of a system output f_r plotted with 3 s are shown in Figure 2.9.

2.4.2 CAUSAL SIMULATION

Simulation of causal models necessitates the assignment of causality to the acausal models. Causality assignment on the bond graph requires causal strokes to be assigned to every bond in the model. A bond graph with no causality conflicts can be transformed into state-space equations and block

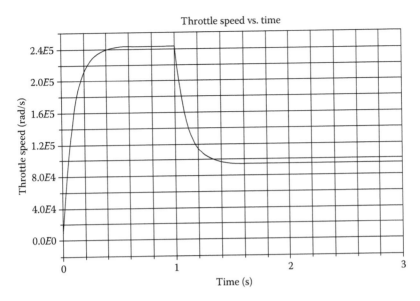

FIGURE 2.9 Modelica simulation results.

diagrams for simulation [5]. We developed a method that exploits the Dirac structure of the port-Hamiltonian system to generate component block diagram subsystems, which can connect together to form the full system. Using the kernel representation of the Dirac structure, a second model transformation method derives its respective hybrid input–output representations.

The hybrid input–output representation of a PHS, which is a causal computational model, is created by causality assignment of the kernel representation. In causality analysis, there can be no causal conflict, and the causality of every bond must be determined. Suppose that the effort and flow variables are split as $\begin{bmatrix} e_1 \\ e_2 \end{bmatrix}$ and $\begin{bmatrix} f_1 \\ f_2 \end{bmatrix}$. Correspondingly, the matrices $E(x)$ and $F(x)$ are also split as $\begin{bmatrix} E_1(x) & E_2(x) \end{bmatrix}$ and $\begin{bmatrix} F_1(x) & F_2(x) \end{bmatrix}$.

Definition 2.3 [13] *Every Dirac structure $D \subset F \times F^*$ can be represented in a hybrid input–output representation as*

$$D(x) = \left\{ (f,e) \in F \times F^* : \begin{bmatrix} e_1 \\ f_2 \end{bmatrix} = J(x) \begin{bmatrix} f_1 \\ e_2 \end{bmatrix} \right\}, \tag{2.18}$$

where $J(x)$ is a skew-symmetric matrix given by

$$J(x) = -J^T(x) = - \begin{bmatrix} E_1(x) & F_2(x) \end{bmatrix}^{-1} \begin{bmatrix} F_1(x) & E_2(x) \end{bmatrix}. \tag{2.19}$$

We developed a method based on elementary column operations of $E(x)$ and $F(x)$ in order obtain the hybrid input–output representation. We illustrate the elementary column operations with a simple example. Consider the following kernel representation of a system with only two power ports:

$$\begin{bmatrix} E_1 & E_2 \end{bmatrix} \begin{bmatrix} e_1 \\ e_2 \end{bmatrix} + \begin{bmatrix} F_1 & F_2 \end{bmatrix} \begin{bmatrix} f_1 \\ f_2 \end{bmatrix} = 0. \tag{2.20}$$

Mathematically, it is equivalent to the following representation:

$$\begin{bmatrix} E_1 & F_2 \end{bmatrix} \begin{bmatrix} e_1 \\ f_2 \end{bmatrix} + \begin{bmatrix} F_1 & E_2 \end{bmatrix} \begin{bmatrix} f_1 \\ e_2 \end{bmatrix} = 0. \tag{2.21}$$

The difference lies in the fact that the positions of E_2 and F_2 are interchanged. This elementary column operation interchanges a column in $E(x)$ with a corresponding column in $F(x)$. From bond graph theory, we know that source elements and nonlinear elements have fixed causality; sources of effort have effort-out-flow-in causality, sources of flow have flow-out-effort-in causality, and nonlinear elements have a causality that is dependent on their constituent equations. As a result, the Dirac structure must solve the flow of a source of effort given its effort (backward for sources of flow). In order to do so, we need to apply the elementary column operation to all source of flow power ports and necessary nonlinear elements.

Although storage elements have preferred causality, we still want to fix their causality to integral causality, due to the issues that arise with implementing differentiators. As a result of this constraint, we do not need to apply the elementary column operation to storage power ports. The difficult part of this process lies in determining column exchanges for linear resistive and interaction power ports. Linear resistive power ports can be either effort-out-flow-in or effort-in-flow-out; there is no *a priori* way of determining how each resistor will behave. Similarly, there are two possible causality configurations for interaction power ports. In order to determine the causality of an interaction power port on a component, we must examine the component along with all neighboring components.

2.4.2.1 Implementation of Causal Simulation

The implementation of causal simulation takes advantage of the indifferent causality nature of linear resistors. Because of its indifferent causality, a linear resistor can never be the source of a causal conflict, which allows for the analysis of each component with its resistive columns intentionally left out. With linear resistive power ports ignored for the time being, the only power ports with questionable causality are interaction power ports. We propose an interaction port propagation assignment algorithm, which iterates through each component sequentially and determines the causality of every interaction power port.

Similar to assigning causal strokes to a bond graph model, we select a component with the least number of interaction ports and most number of constraint ports (sources, storage, and nonlinear) with which to begin propagation of interaction port assignments. We analyze the ranks of the $F(x)$ matrix of the starting component and perform combinations of elementary column operations until it has full rank; additionally, do the same thing for the neighboring components. The process gets repeated until every interaction port is assigned. A summary of the algorithm is listed in Algorithm 2.3:

Now that all interaction ports are assigned, we must determine the causality of every linear resistive power port. In order to do so, we analyze the rank of $F(x)$ of individual components; we then apply the elementary column operation to the appropriate resistive power ports. Note that

---

Algorithm 2.3 Interaction Port Propagation Assignment
-**for all** Components
- Swap all source of flow columns;
- Remove all resistive columns;
- **while** There are still unassigned interactions
-      Find component with greatest interaction to constraint ratio;
-      Propagate interaction assignment;
- Put back all resistive columns;

Algorithm 2.4 Kernel Representation to a Hybrid Input–Output Representation

-for all Components
- Determine which resistive columns to interchange;
- Swap said resistive columns to obtain full ranked F;
- $J \leftarrow -F^{-1}E$;

---

for components with multiple resistors, there can be more than one unique solution, but any of the solutions will work in simulation. A summary is listed in Algorithm 2.4:

In Simulink, every block has a clearly defined input–output relationship. With each component's hybrid input–output representation, we can derive their corresponding block diagrams and simulate them inside Simulink. The interpreter uses Simulink to create the block diagrams, where each component is modeled with the subsystem block. Inside each component subsystem, there is another subsystem (the Dirac structure of the component) along with its attached ports; the Dirac structure subsystem consists of a series of sums and gains, depicting the dynamics of the Dirac structure equations. Power ports are modeled as follows:

1. *Source of effort*: Effort modeled as a step input (i.e., a constant voltage source); flow sunk to a scope.
2. *Source of flow*: Flow modeled as a step input (i.e., a constant current source); effort sunk to a scope.
3. *Linear storage port*: Flow goes into an integrator and a gain (inverse of the storage value) to obtain effort.
4. *Linear resistive port*: Effort and flow have a linear relationship, which is represented by a gain, whose value depends on the orientation of the resistive element.
5. *Nonlinear resistive or storage ports*: Effort and flow have a nonlinear relationship, which is represented by a MATLAB function.
6. *Interaction port*: Effort and flow modeled as inport and outport.

Algorithm 2.5 creates block diagrams using the hybrid input–output representation of the Dirac structure and the constituent equations of the power ports.

2.4.2.2 Causal Simulation of the Engine Dynamometer

Applying Algorithm 2.4 to the engine-dynamometer example, we obtain the hybrid input–output representation of the Dirac structure of the engine (Equation 2.22):

$$\begin{bmatrix} f_r \\ f_s \\ f_j \\ f_d \end{bmatrix} = \begin{bmatrix} 0 & 0 & -T & 0 \\ 0 & 0 & -1 & 0 \\ T & 1 & 0 & 1 \\ 0 & 0 & -1 & 0 \end{bmatrix} \begin{bmatrix} e_r \\ e_s \\ e_j \\ e_d \end{bmatrix}, \tag{2.22}$$

Algorithm 2.5 Dirac Structure to Block Diagrams

-for all Components
- **for all** Power ports
-     **if** Noninteraction port
-         Model according to rules;
-     **if** Interaction port
-         Model as Simulink inports/outports;
-Connect all components;

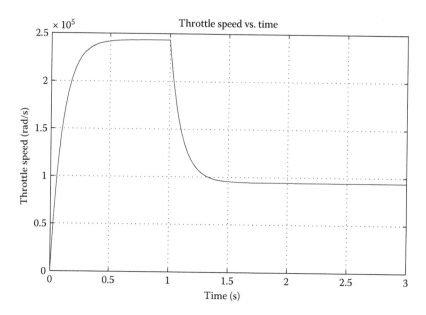

FIGURE 2.10 Simulink simulation results.

Integrating the switch signal s into the Dirac structure results in the following hybrid input–output representation of the starter motor. The $J(x)$ matrix in Equation 2.23 is modulated by the Boolean signal s, which shows the hybrid dynamics of the starter motor system:

$$\begin{bmatrix} e_s \\ f_1 \\ f_2 \end{bmatrix} = \begin{bmatrix} 0 & s & 1-s \\ -s & 0 & 0 \\ s-1 & 0 & 0 \end{bmatrix} \begin{bmatrix} f_s \\ e_1 \\ e_2 \end{bmatrix}. \tag{2.23}$$

Finally, the hybrid input–output representation for the dynamometer is shown in (Equation 2.24):

$$\begin{bmatrix} e_u \\ e_d \\ f_R \\ f_I \end{bmatrix} = \begin{bmatrix} 0 & 0 & 1 & 1 \\ 0 & 0 & 1 & 1 \\ -1 & -1 & 0 & 0 \\ -1 & -1 & 0 & 0 \end{bmatrix} \begin{bmatrix} f_u \\ f_d \\ e_R \\ e_I \end{bmatrix}. \tag{2.24}$$

Simulating the Simulink model of the engine dyno with the same control inputs and parameters, we can view the results of the system output f_r plotted with 3 seconds as shown in Figure 2.10.

2.5 PASSIVITY ANALYSIS

Passivity is commonly defined in the state-space approach, where the energy stored in the system is related to the external energy coming into the system [41]. A passive system is guaranteed to be stable if a positive-definite storage function is used [39]. Therefore, passivity provides a sufficient condition for the stability of a system; a passive system, when unforced, is Lyapunov stable [23]. In the case when the storage function used is positive semidefinite, additional conditions on zero-state detectability is needed to ensure that the passive system remains stable [6]. In addition to stability, passive systems also contain phase properties [29]; passive systems have minimum phase and possess a low relative degree [11,28].

2.5.1   PASSIVITY OF PORT-HAMILTONIAN SYSTEMS

Regardless of the representation, Dirac structures establish the power balancing equation that is fundamental to port-Hamiltonian systems:

$$\frac{d}{dt}H = e_R^T f_R + e_C^T f_C + e_I^T f_I. \tag{2.25}$$

H in Equation 2.25 denotes the Hamiltonian function of the port-Hamiltonian system. The Hamiltonian function represents the energy stored in the system; the flows of the energy storage are given by the rate x' of the energy state variables x; the efforts are given by the coenergy variables $\partial H / \partial x$ [21]. e_R and f_R denote the effort and flow variables associated with internal energy dissipation, where the relationship between the power conjugate variables obey a static resistive relationship of the form

$$R(f_R, e_R) = 0. \tag{2.26}$$

e_C and f_C denote effort and flow variables associated with external control; these control port variables are accessible to controller action; one variable acts as the actuator and the conjugate variable acts as the sensor. The final terms in Equation 2.25, e_I and f_I, denote effort and flow variables associated with interactions of the port-Hamiltonian system with the environment. Equation 2.3 simplifies down to an input–state–output port-Hamiltonian system when there are no algebraic constraints on the state variables [12]:

$$\Sigma : \begin{cases} x' = [J(x) - R(x)]\frac{\partial H}{\partial x}(x) + g(x)u + k(x)d, \\ y = g^T(x)\frac{\partial H}{\partial x}(x), \\ z = k^T(x)\frac{\partial H}{\partial x}(x). \end{cases} \tag{2.27}$$

Its Dirac structure is summarized in the following skew-symmetric matrix equation:

$$D : \begin{bmatrix} -J(x) & -g_R(x) & -g(x) & -k(x) \\ g_R^T(x) & 0 & 0 & 0 \\ g^T(x) & 0 & 0 & 0 \\ k^T(x) & 0 & 0 & 0 \end{bmatrix}, \tag{2.28}$$

where the term $g_R(x)$ denotes the resistive structure of the system.

In PHSML, the storage function of a component is associated with its storage elements. The storage function is described by the following square equation:

$$H(x) = \sum_{i=1}^{n} \frac{1}{2k_i}x_i^2, \tag{2.29}$$

where
  n is the total number of storage elements in the component
  k_i is the parameter constant of the ith storage element
  x_i is the state variable of the ith storage element

Passivity of the component can be stated as

$$\int_{t_0}^{t} e_I^T f_I d\tau + \int_{t_0}^{t} e_C^T f_C d\tau = H(x) - \int_{t_0}^{t} e_R^T f_R d\tau. \tag{2.30}$$

Algorithm 2.6 Storage Function Generation
-for all Components
-    **for all** Power ports
-        **if** Storage ports
-            Write out state variables;
-    **for all** Storage ports
-        Sum up Hamiltonian functions;

Equation 2.30 implies that the component is passive with respect to a supply energy of $\int_{t_0}^{t} e_I^\top f_I d\tau +$ $\int_{t_0}^{t} e_C^\top f_C d\tau$ and dissipated energy of $\int_{t_0}^{t} e_R^\top f_R d\tau$ as long as the storage function is positive definite. Since Equation 2.29 is a sum of squares, we can conclude that any system built using PHSML is passive.

2.5.1.1 Implementation

The basis of passivity analysis revolves around the storage function. The state variable of each storage port is the product of the effort variable and the storage element value. The Hamiltonian function from Equation 2.29 is then calculated from the state variables and can then be plotted inside a Modelica simulation environment. The implementation is summarized in Algorithm 2.6 (Figure 2.11).

2.5.2 PASSIVITY ANALYSIS OF THE ENGINE DYNAMOMETER

To illustrate how passivity analysis works with PHSML, we continue to use the engine-dynamometer example. Given that there is one storage element in the engine component, there exists one state. The state variable, angular momentum of the engine inertia p, is used as the energy variable for the port-Hamiltonian description $p = Je_j$, leading to the following storage function for the engine component:

$$H_p(p) = \frac{1}{2J}p^2. \tag{2.31}$$

where H_p is the storage function of the engine component. The rate of the energy variable and coenergy variable are $\dot{p}$ and p/J, respectively. Using in Equations 2.4, 2.5, 2.15, and 2.31, we can derive a port-Hamiltonian representation for the engine component:

$$\dot{p} = \begin{bmatrix} T(\frac{p}{J}) & 1 & 1 \end{bmatrix} \begin{bmatrix} e_r \\ e_s \\ e_d \end{bmatrix},$$

$$\begin{bmatrix} f_r \\ f_s \\ f_d \end{bmatrix} = \begin{bmatrix} T(\frac{p}{J}) \\ 1 \\ 1 \end{bmatrix} \frac{\partial H_p}{\partial p}. \tag{2.32}$$

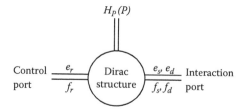

FIGURE 2.11 Port-Hamiltonian system representation of the engine.

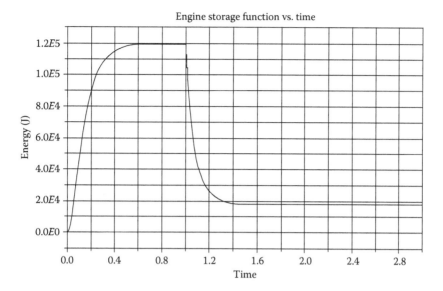

FIGURE 2.12 Storage function of the engine.

The nonlinearity inside the torque map manifests itself in the nonlinear Dirac structure in Equation 2.32. A graphical port-Hamiltonian description of the engine component is shown in Figure 2.11. The control port (e_r, f_r) allows the engine system to interface with control software; the control software takes f_r as input from the engine and provides e_r as output to the engine. Combining the inputs and outputs of the control port and interaction ports, we can assert that the engine system is passive with respect to inputs $\{e_r, e_s, e_d\}$ and outputs $\{f_r, f_s, f_d\}$, with the positive-definite storage function shown in Equation 2.31 if $J \geq 0$. Substituting the inputs and outputs shown in Equation 2.32 into the passivity equation (Figure 2.12):

$$\int_{t_0}^{t} u^{\mathsf{T}}(\tau)y(\tau)d\tau = \int_{t_0}^{t} \begin{bmatrix} e_r & e_s & e_d \end{bmatrix} \begin{bmatrix} f_r \\ f_s \\ f_d \end{bmatrix} d\tau$$

$$= \int_{t_0}^{t} f_j e_j d\tau = \frac{1}{J} \int_{f_j(t_0)}^{f_j(t)} (Te_r + e_s + e_d)d(Te_r + e_s + e_d) = \frac{(Te_r + e_s + e_d)^2}{2J}.$$

The fraction is guaranteed to be positive, and in order for the storage function to be bounded from below, the gain J must be positive.

$$\frac{(Te_r + e_s + e_d)^2}{2J} \geq 0 \to J \geq 0.$$

There are no dissipation in the system, and since all incoming energy becomes stored energy, the system is passive. The storage function of the engine is shown in Figure 2.16.

There are no energy storage or energy dissipation in the starter, which makes the passivity conditions relatively simple. The control ports (e_1, f_1) and (e_2, f_2) allows the starter system to interface with control software; the control software takes f_1 and f_2 as inputs from the starter and provides e_1 and e_2 as outputs to the starter. Combining the inputs and outputs of the control ports and interaction port, we can assert that the starter system is passive with respect to inputs $\{f_s, e_1, e_2\}$ and outputs $\{e_s, f_1, f_2\}$, with the positive-definite storage function of 0. Substituting the inputs and outputs shown in Equation 2.23 into the passivity equation:

$$\int_{t_0}^{t} u^{\mathsf{T}}(\tau) y(\tau) d\tau = \int_{t_0}^{t} \begin{bmatrix} f_s & e_1 & e_2 \end{bmatrix} \begin{bmatrix} e_s \\ f_1 \\ f_1 \end{bmatrix} d\tau = 0.$$

Incoming energy from ports (e_1, f_1) and (e_2, f_2) leaves the system through port (e_s, f_s). The dyno component has one storage element and thus one state. The state variable of the integral controller, q, is used as the energy variable for the port-Hamiltonian description $q = \int_{t_0}^{t}(f_u - f_d)d\tau$, leading to the following storage function for the dyno component:

$$H_q(q) = \frac{K_I}{2} q^2. \tag{2.33}$$

where H_q is the storage function of the dyno component. The rates of the energy variable and coenergy variable are $\dot{q}$ and q/K_I, respectively. Using Equations 2.8, 2.9, 2.17, and 2.33, we can derive a port-Hamiltonian representation for the dyno component:

$$\dot{q} = \begin{bmatrix} 1 & -1 \end{bmatrix} \begin{bmatrix} f_u \\ f_d \end{bmatrix},$$

$$\begin{bmatrix} e_u \\ e_d \end{bmatrix} = \begin{bmatrix} 1 \\ 1 \end{bmatrix} \frac{\partial H_q}{\partial q} + \begin{bmatrix} K_R & -K_R \\ K_R & -K_R \end{bmatrix} \begin{bmatrix} f_u \\ f_d \end{bmatrix}. \tag{2.34}$$

A graphical port-Hamiltonian description of the dyno component is shown in Figure 2.13. The control port (e_u, f_u) allows the dyno system to interface with control software; the control software takes e_u as the input from the dyno and provides f_u as the output to the starter. Combining the inputs and outputs of the control port and interaction port, we can assert that the dynamometer system is passive with respect to inputs $\{f_u, f_d\}$ and outputs $\{e_u, e_d\}$, with the positive-definite storage function shown in Equation 2.33 if $K_R, K_I \geq 0$. Substituting the inputs and outputs shown in Equation 2.34 into the passivity equations (Figure 2.14):

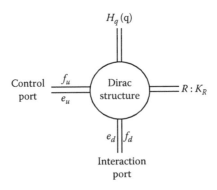

FIGURE 2.13 Port-Hamiltonian system representation of the dyno.

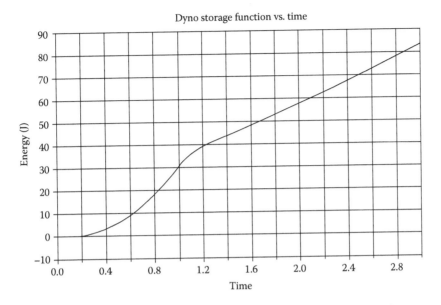

FIGURE 2.14 Storage function of the dynamometer.

$$\int_{t_0}^{t} u^{\mathsf{T}}(\tau)y(\tau)d\tau = \int_{t_0}^{t} \left[\begin{bmatrix} f_u & f_d \end{bmatrix} \begin{bmatrix} K_I \int_{\tau_0}^{\tau}(f_u - f_d)d\omega + K_R(f_u - f_d) \\ K_I \int_{\tau_0}^{\tau}(f_u - f_d)d\omega + K_R(f_u - f_d) \end{bmatrix} \right] d\tau$$

$$= K_I \int_{f_u(t_0)-f_d(t_0)}^{f_u(t)-f_d(t)} (f_u - f_d)d(f_u - f_d) + K_R \int_{t_0}^{t}(f_u - f_d)(f_u - f_d)d\tau.$$

The term K_I denotes the storage function in Equation 2.33. The integral is guaranteed to be positive, and in order for the storage function to be bounded from below, the gain must be positive:

$$K_I \int_{f_u(t_0)-f_d(t_0)}^{f_u(t)-f_d(t)} (f_u - f_d)d(f_u - f_d) \geq 0 \rightarrow K_I \geq 0.$$

The term K_R denotes the dissipation in the system, which must be positive in order for the system to remain passive:

$$K_R \int_{t_0}^{t}(f_u - f_d)(f_u - f_d)d\tau \geq 0 \rightarrow K_R \geq 0.$$

The storage function of the dynamometer is shown in Figure 2.16.

The port-Hamiltonian system representation of the composed system is tersely summarized in Figure 2.15, where the engine, starter, and dyno components are interconnected through the interaction ports of (e_s, f_s) and (e_d, f_d). The control ports (e_r, f_r), (e_1, f_1), (e_2, f_2), and (e_u, f_u) allow the engine dynamometer system to interface with control software; the control software takes f_r, f_1, f_2, and e_u as

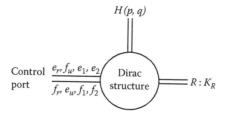

FIGURE 2.15   Port-Hamiltonian system representation of the closed-loop system.

inputs and provides e_r, e_1, e_2, and f_u as outputs. The composition of the three components results in a global storage function, which is the sum of its constituent storage functions:

$$H(p,q) = H_p(p) + H_q(q) = \frac{1}{2J}p^2 + \frac{K_I}{2}q^2. \tag{2.35}$$

Combining the port-Hamiltonian representations of Equations 2.23, 2.32, and 2.34 results in a port-Hamiltonian representation for the composed system:

$$\begin{bmatrix} \dot{p} \\ \dot{q} \end{bmatrix} = \begin{bmatrix} -K_R & 1 \\ -1 & 0 \end{bmatrix} \begin{bmatrix} \frac{\partial H}{\partial p} \\ \frac{\partial H}{\partial q} \end{bmatrix} + \begin{bmatrix} T(\frac{p}{J}) & K_R & s & 1-s \\ 0 & 1 & 0 & 0 \end{bmatrix} \begin{bmatrix} t_e \\ f_u \\ e_1 \\ e_2 \end{bmatrix},$$

$$\begin{bmatrix} t_f \\ e_u \\ f_1 \\ f_2 \end{bmatrix} = \begin{bmatrix} T(\frac{p}{J}) & 0 \\ K_R & 1 \\ s & 0 \\ 1-s & 0 \end{bmatrix} \begin{bmatrix} \frac{\partial H}{\partial p} \\ \frac{\partial H}{\partial q} \end{bmatrix}. \tag{2.36}$$

The composed system is passive with respect to inputs $\{e_r, f_u, e_1, e_2\}$, outputs $\{f_r, e_u, f_1, f_2\}$, and a storage function shown in Equation 2.35 if the engine, starter motor, and dynamometer are all passive. Substituting the inputs and outputs shown in Equation 2.36 into the passivity equation:

$$\int_{t_0}^{t} u^T(\tau)y(\tau)d\tau = \int_{t_0}^{t} \begin{bmatrix} e_r & f_u & e_1 & e_2 \end{bmatrix} \begin{bmatrix} f_r \\ e_u \\ f_1 \\ f_2 \end{bmatrix} d\tau$$

$$= \int_{t_0}^{t} f_j e_j d\tau + \int_{t_0}^{t} f_I e_I d\tau + \int_{t_0}^{t} f_R e_R d\tau$$

$$= \frac{1}{J} \int_{f_j(t_0)}^{f_j(t)} f_j df_j + K_I \int_{f_I(t_0)}^{f_I(t)} f_I df_I + K_R \int_{t_0}^{t} f_R^2 d\tau.$$

The terms J and K_I denote the storage function in Equation 2.35. The integrals are guaranteed to be positive, and in order for the storage function to be bounded from below, the gains must be positive:

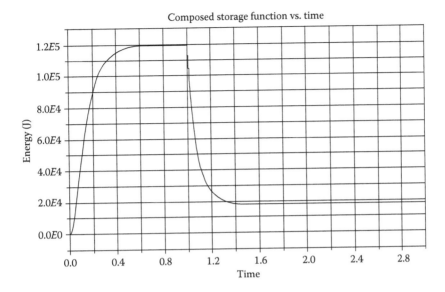

FIGURE 2.16 Storage function of the composed system.

$$\frac{1}{J} \int_{f_j(t_0)}^{f_j(t)} f_j df_j \geq 0 \rightarrow J \geq 0.$$

$$K_I \int_{f_I(t_0)}^{f_I(t)} f_I df_I \geq 0 \rightarrow K_I \geq 0.$$

The term K_R denotes the dissipation in the system, which must be positive in order for the system to remain passive:

$$K_R \int_{t_0}^{t} f_R^2 d\tau \geq 0 \rightarrow K_R \geq 0.$$

The storage function of the composed system is shown in Figure 2.16.

2.5.2.1 Implementation of Controllers

Controllers or cyber components can be connected to the engine-dynamometer model through the four control ports as shown in Equation 2.36. The controller that provides a throttle command to the engine system is implemented on port (e_r, f_r) as

$$e_r = \begin{cases} 8 & \text{for time} \leq 1, \\ 2 & \text{for time} > 1. \end{cases}$$

The controller that provides the positive and zero starter torque for the starter system is implemented on ports (e_1, f_1) and (e_2, f_2) as

$$\begin{bmatrix} e_1 \\ e_2 \end{bmatrix} = \begin{bmatrix} 600 \\ 0 \end{bmatrix}.$$

The controller that provides the set point signal for the dyno system is implemented on port (e_u, f_u) as $f_u = 20$. The controllers implemented in this example are relatively simple. More complicated control software can be developed in similar ways by using the power conjugate variables on the power ports in order to achieve certain system behavior.

2.6 CONCLUSION

The increased growth and advancements in the area of CPS have resulted in many research topics in the modeling and simulation of CPS. Research in CPS will continue to grow and expand as new techniques and methods are introduced. The goal of this chapter is to address the challenges of nonlinearities and hybrid dynamics in the development of CPS. Port-Hamiltonian systems are able to resolve these challenges by employing strict rules in the formulation and composition of systems and components. We have shown that the port-Hamiltonian system is an effective tool for the modeling and simulation of component-based CPS.

ACKNOWLEDGMENTS

This work is supported in part by the National Science Foundation (CNS-1035655, CCF-0820088).

REFERENCES

1. Modelica Association. Modelica—A unified object oriented language for systems modeling, Language specification. http://www.Modelica.org/. September, 2014.
2. Modelica Association. Modelica standard library. https://github.com/modelica/Modelica. September, 2014.
3. P. Breedveld. Multibond graph elements in physical systems theory. *Journal of the Franklin Institute*, 319(1–2):1–36, 1985.
4. P. Breedveld. Modeling and simulation of dynamic systems using bond graphs. In *Control Systems, Robotics and Automation—Modeling and System Identification I*, vol. 18-4-1 of *Encyclopedia of Life Support Systems*, pp. 128–173. EOLSS Publishers, Oxford, UK, 2008.
5. J. Broenink. Introduction to physical systems modeling with bond graphs. In *The SiE Whitebook on Simulation Methodologies*, pp. 18–27, 1999.
6. C. Byrnes, A. Isidori, and J. Willems. Passivity, feedback equivalence, and global stabilization of minimum phase nonlinear systems. *IEEE Transactions on Automatic Control*, 36:1228–1240, 1991.
7. E. Carpanzano. Order reduction of general nonlinear DAE systems by automatic tearing. *Mathematical and Computer Modeling of Dynamical Systems*, 6(2):145–168, 2000.
8. F. Cellier. Hierarchical nonlinear bond graphs: A unified methodology for modeling complex physical systems. *Simulation*, 58:230–248, April 1992.
9. S. Das. *Mechatronic Modeling and Simulation Using Bond Graphs*. CRC Press, Boca Raton, FL, 2009.
10. P. Derler et al. Modeling cyber-physical systems. *Proceedings of IEEE*, 100:13–28, January 2012.
11. C. Desoer and M. Vidyasagar. *Feedback Systems: Input–Output Properties*. Academic Press, New York, 1975.
12. A. Donaire and S. Junco. Derivation of input-state-output port-Hamiltonian systems from bond graphs. *Simulation Modelling Practice and Theory*, 17:137–151, 2009.
13. V. Duindam, A. Macchelli, A. Stramigioli, and H. Bruyninckx. *Modeling and Control of Complex Physical Systems: The Port-Hamiltonian Approach*. Springer, New York, pp. 61–87, 2009.
14. Dymola, http://www.3ds.com/products-services/catia/products/dymola, Dassault Systemes, 2014.
15. H. Elmqvist and S. Mattsson. An introduction to the physical modeling language Modelica. *Proceedings of the Ninth European Simulation Symposium*, Passau, Germany, October 1997.
16. H. Elmqvist, S. Mattsson, and M. Otter. Modelica—The new object-oriented modeling language. *12th European Simulation Multiconference*, pp. 127–132, Manchester, UK. June, 1998.
17. Akos Ledeczi, Miklos Maroti, Arpad Bakay, Gabor Karsai, Jason Garrett, Charles Thomason, Greg Nordstrom, Jonathan Sprinkle, and Peter Volgyesi. *The Generic Modeling Environment*, The 2nd International Workshop on Intelligent Signal Processing, Budapest, Hungary. May 24–25, 2001.

18. S. Feng and L. Zhang. Integrated approach for modeling cyber-physical systems. (Eds.) Yueh-Min Huang, Han-Chieh Chao, Der-Jiunn Deng, and James Park. In *Advanced Technologies, Embedded and Multimedia for Human-Centric Computing*, Springer, Netherlands, pp. 371–376. 2014.

19. P. Fritzson. *Principles of Object-Oriented Modeling and Simulation with Modelica 2.1*. John Wiley & Sons, New York, pp. 145–182, January 2004.

20. G. Golo, P.C. Breedveld, B.M. Maschke, and A.J. Schaft van der. Geometric formulation of generalized bond graph models—Part I: Generalized junction structures, Technical Report from Robotics and Mechatronics Research Group, University of Twente Department of Electrical Engineering, *Mathematics, and Computer Science*. 2000.

21. G. Golo, A. van der Schaft, P. Beedveld, and B. Mascheke. Hamiltonian formulation of bond graphs: Nonlinear and hybrid systems in automotive control, In *Nonlinear and Hybrid Systems in Automotive Control*. Springer, London, UK, pp. 351–372, 2003. ISBN 9781852336523.

22. D. Henriksson and H. Elmqvist. Cyber-physical systems modeling and simulation with modelica. *International Modelica Conference*, Dresden, Germany. March 20–23, 2011.

23. D. Hill and P. Moylan. The stability of nonlinear dissipative systems. *IEEE Transactions on Automatic Control*, 21:708–711, 1976.

24. Cummins Inc. Basic engine model KTA19-M3, engine configuration D193080MX02, September 2009.

25. Johan Åkesson, Karl-Erik Årzén, Magnus Gäfvert, Tove Bergdahl, and Hubertus Tummescheit. Modeling and optimization with Optimica and JModelica.org—Languages and tools for solving large-scale dynamic optimization problems, *Computers and Chemical Engineering*, 34(11):1737–1749, November 2010.

26. D. Karnopp, D. Margolis, and R. Rosenberg. *System Dynamics: Modeling and Simulation of Mechatronic Systems*. John Wiley & Sons, New York, pp. 17–33, 2000.

27. G. Karsai, J. Sztipanovits, A. Ledeczi, and T. Bapty. Model-integrated development of embedded software. *Proceedings of the IEEE*, 91(1):145–164, 2003.

28. H. Khalil. *Nonlinear Systems*. Prentice Hall, Upper Saddle River, NJ, 2002.

29. N. Kottenstette and P. Antsaklis. Relationships between positive real, passive dissipative, and positive systems. *American Control Conference*, pp. 409–416, Baltimore, MD. June 30–July 2, 2010.

30. L. Lee. Cyber-physical systems: Design challenges. *International Symposium on Object, Component, and Service-Oriented Real-Time Distributed Computing (ISORC)*, pp. 363–369. Orlando, FL. May 5–7, 2008.

31. J. Lin, S. Sedigh, and A. Miller. Towards integrated simulation of cyber-physical systems: A case study on intelligent water distribution. *Eighth International Conference on Dependable, Autonomic and Secure Computing*, pp. 690–695. Chengdu, China. December 12–14, 2009.

32. OpenModelica System Documentation, version 1.9.1, September 2014. http://www.openmodelica.org/.

33. C. Pantelides. The consistent initialization of differential-algebraic systems. *Journal of Science and Statistical Computing*, 9(2):213–231, March 1987.

34. R. Rajkumar, I. Lee, L. Sha, and J. Stankovic. Cyber-physical systems: The next computing revolution. *Proceedings of the 47th Design Automation Conference*, pp. 731–736, Anaheim, CA. June 13–18, 2010.

35. S. Sakai and S. Stramigioli. Port-Hamiltonian approaches to motion generation for mechanical systems. *IEEE International Conference on Robotics and Automation*, Rome, Italy. April 10–14, 2007.

36. G. Simko, T. Levendovsky, M. Maroti, and J. Sztipanovits. Towards a theory of cyber-physical systems modeling. *CyPhy*, pp. 56–61, Philadelphia, PA, 2013.

37. J. Sztipanovits. Composition of cyber-physical systems. *Proceedings of the 14th Annual IEEE International Conference and Workshops on the Engineering of Computer-Based Systems*, pp. 3–6, Tucson, AZ, March 26–29, 2007.

38. J. Sztipanovits, X. Koutsoukos, G. Karsai, N. Kottenstette, P. Antsaklis, V. Gupta, B. Goodwine, J. Baras, and S. Wang. Toward a science of cyber-physical system integration. *Proceedings of IEEE*, 100:29–44, January 2012.

39. A. van der Schaft. *L2-Gain and Passivity in Nonlinear Control*. Springer-Verlag, New York/Secaucus, NJ, 1999.

40. A. van der Schaft. Port-Hamiltonian systems: An introductory survey. *Proceedings of the International Congress of Mathematicians*, pp. 1339–1365, Madrid, Spain. August 22–30, 2006.

41. J. Willems. Dissipative dynamical systems, Part II: Linear systems with quadratic supply rates. *Archive for Rational Mechanics and Analysis*, 45:352–393, 1972.

42. L. Yu and X. Qi. Bond-graph modeling in system engineering. *2012 International Conference on Systems and Informatics*, pp. 376–379, Yantai, China. May 19–20, 2012.

3 Controller Redundancy Design for Cyber-Physical Systems

Jianguo Yao, Xue Liu, Guchuan Zhu, and Lui Sha

CONTENTS

3.1　INTRODUCTION

Systems where the physical subsystem is tightly integrated with the cyber subsystem are usually referred as cyber-physical systems (CPS) [1–5]. CPS applications can be found in many areas, such as automated manufacturing, health care, civil infrastructure, avionics, aircraft, and vehicular communications [6]. Networked control systems (NCS) are considered to be a prominent subcategory of CPS, in which networks provide a means for communication among sensors, actuators, and controllers. As networking technology continues to increase in bandwidth capability and decrease in price, NCS will become more pervasive and have the potential to radically change the way the physical systems interact with each other and their environment. Communication networks have been used to exchange information among sensors, controllers, and actuators in digital control systems since the 1970s, originally motivated by the need of reducing costs for cabling, modularizing, and integrating system flexibly in the automobile industry. Today, many other applications of NCS can be easily found in diverse industries. For example, Avionics Full-Duplex Switched Ethernet (AFDX) is used in control systems of the Airbus A380 and Boeing B787 [7], and a new automotive network communications protocol, FlexRay was developed for vehicle control and is used in automobiles, such as the BMW 7 Series (F01), Audi A8, and Rolls-Royce Ghost [8]. Employing communications networks in control systems may considerably reduce integration complexity and make system architecture more scalable.

Different network protocols for real-time control will introduce different forms of communication delays, which can be static, time varying, or even bounded random delays. Roughly speaking, network protocols for control applications can be categorized into the following sets: fieldbuses (e.g., FIP, PROFIBUSI, CAN, MIL-STD-1553B, and TTP), machine buses (e.g., TTEthernet and AFDX), and general purpose networks (e.g., Ethernet and ATM networks) [9]. It is worth noting that fieldbuses are usually used for the connection of low-level, small-scale devices and machine buses are more suitable for the connection of subsystems. Network-induced delays depend greatly on the employed medium access control (MAC) protocols, which can be grouped into random access and scheduling access [10]. For example, carrier sense multiple access (CSMA) is usually used in the random access protocols such as Ethernet; time division multiple access (TDMA) is usually used in scheduling protocols such as FlexRay, TTP, and TTEthernet.

In the design of the traditional CPS, the cyber components (such as control software) and the physical components (such as the plant) are often treated separately, while little effort has been put into the connection and the interaction between these two sets of components. As the interaction between cyber and physical components are not modeled and analyzed and the complexity may introduce more possibilities for faults, reliability is usually difficult to scale in highly complex embedded systems. Reliability represents a more serious issue in CPS-employing networks, because network-induced delays and information loss will impact system performance and make control system design and analysis more complex. In order for CPS respond more quickly and precisely in a highly efficient and reliable mode, many efforts have been devoted to the development of new design paradigms for CPS [1–6]. Nevertheless, highly reliable networked control remains an open and challenging issue in CPS.

A suitable framework for dealing with complex safety-critical CPS is the *Simplex* architecture [11,12], which was developed for reliable embedded control systems. This paradigm is built upon the concept of analytic redundancy, which is widely used in fault-tolerant control. *Simplex* typically consists of two sets of controllers: high-assurance control (HAC) and high-performance control (HPC). HAC is designed to ensure the basic performance requirements with high reliability, while HPC should provide a much better system performance at the expense of possibly lower reliability. The *Simplex* architecture achieves fault tolerance by using the redundant HAC to guard the HPC. However, the stability property of such systems may be very complex and difficult to analyze.

The original *Simplex* architecture was developed for continuous linear time-invariant (LTI) systems, while many cyber-physical systems are networked controlled linear parameter-varying (LPV) systems. To bring *Simplex* into these systems, one of the most important aspects is the computation of a maximum stability region (MSR) with varying parameters and network delays (NetMSR). From the perspective of fault-tolerant control, the NetMSR calculation is the key (1) providing a larger permissible state space for the deployment of HPCs and (2) for the detection of HPC faults and the timely switch to the HAC. To the best of our knowledge, the authors are not aware of publications on how to compute NetMSR in NCS that address the effects of gain scheduling and network delays. In brief, this chapter introduces a complete design process for a *NetSimplex*-based controller redundancy design for CPS [13].

The remainder of this chapter is organized as follows. Section 3.2 describes the *NetSimplex* architecture. Section 3.3 deals with the analysis of MSR for LPV systems. Section 3.4 presents the interpolation state feedback control design of LPV systems with network-induced delay. Section 3.5 is dedicated to stability analysis of the switched system from NetMSR and reach set perspective. An example of satellite altitude control using a controller area network (CAN) bus is presented in Section 3.6 to verify the proposed approach and illustrate the applicability of the main results. Finally, the conclusion and future work are provided in Section 3.7.

3.2   NETSIMPLEX ARCHITECTURE

In this section, we will formulate the *NetSimplex* reliable control problem for networked cyber-physical systems [13].

3.2.1   NETWORKED CONTROL SYSTEM MODEL

Assume that the plant to be controlled is governed by a continuous-time LPV system described by a state-space model of the form

$$\dot{x}(t) = A(\rho)x(t) + B(\rho)u(t), \tag{3.1}$$

where
   $x \in \mathbb{R}^N$ is the system state
   $u \in \mathbb{R}^M$ is the control input
   $A(\rho) \in \mathbb{R}^{N \times N}$, and $B(\rho) \in \mathbb{R}^{N \times M}$ are the corresponding system and control matrices
   ρ is a varying parameter with the constraint $\rho \in [\rho_{min}, \rho_{max}]$

Note that $A(\rho)$ and $B(\rho)$ are linear matrix functions of ρ, that is, $A(\rho) = A_0 + \rho A_1$ and $B(\rho) = B_0 + \rho B_1$. The HAC is typically designed as a state feedback controller, that is, $u(t) = Kx(t)$, where K is the corresponding controller gain. Using the continuous-time state feedback controller, the closed-loop control system is given by

$$\dot{x}(t) = A_c(\rho)x(t), \tag{3.2}$$

where $A_c(\rho) = A_0 + \rho A_\rho + (B_0 + \rho B_\rho)K$ is the closed-loop system matrix.

For the purposes of simplifying the design and analysis while facilitating the implementation, we convert the continuous-time system model to the discrete-time form. Due to the discretization of continuous-time controllers on networks, the sampling and the control of the continuous-time system (Equation 3.1) may be at different discrete-time points. The continuous-time system with time-delayed control is given by

$$\dot{x}(t) = A(\rho)x(t) + B(\rho)u(t^+), \tag{3.3}$$

where

$$u(t^+) \triangleq K(t)x(t - \tau_k), \quad t \in kh + \tau_k, k = 0, 1, 2, \ldots, \tag{3.4}$$

where
  h is the sampling period
  τ_k is the delay from sensors to actuators

The control input vector $u(t^+)$ is piecewise continuous obtained by using zero-order hold (ZOH) and its value changes only at the moment τ_k. The system can then be expressed as [14]

$$x(k + 1) = \Phi(\rho)x(k) + G_0(\rho, \tau_k)u(k) + G_1(\rho, \tau_k)u(k - 1), \tag{3.5}$$

where $x(k) \triangleq x(kh)$ is the system state at the kth sampling period and

$$\Phi(\rho) = e^{A(\rho)h}, \tag{3.6}$$

$$G_0(\rho, \tau_k) = \int_0^{h-\tau_k} e^{A(\rho)s}B(\rho)ds, \tag{3.7}$$

$$G_1(\rho, \tau_k) = \int_{h-\tau_k}^{h} e^{A(\rho)s}B(\rho)ds. \tag{3.8}$$

The aforementioned functions can be computed numerically using multiplication of series truncated at a predefined order:

$$\Phi(\rho) = I + A(\rho)h + \frac{1}{2!}(A(\rho)h)^2 + \frac{1}{3!}(A(\rho)h)^3 + \cdots, \tag{3.9}$$

$$G_0(\rho, \tau_k) = \left[I(h - \tau_k) + \frac{1}{2!}A(\rho)(h - \tau_k)^2 + \frac{1}{3!}(A(\rho))^2(h - \tau_k)^3 + \cdots \right] B(\rho), \tag{3.10}$$

$$G_1(\rho, \tau_k) = \left[I\tau_k + \frac{1}{2!}A(\rho)(\tau_k)^2 + \frac{1}{3!}(A(\rho))^2(\tau_k)^3 + \cdots \right] B(\rho). \tag{3.11}$$

Note that for NCS with multiple inputs, different control channels may have different delays. Therefore, each column of $G_0(\rho, \tau_k)$ and $G_1(\rho, \tau_k)$ should be computed separately using the corresponding column in $B(\rho)$ with respect to the input delay. For example, if the delay of the first input is τ_k^1, then the first column of $G_0(\rho, \tau_k)$ is given by

$$g_1(\rho, \tau_k^1) = \int_0^{h-\tau_k^1} e^{A(\rho)s}b_1(\rho)ds, \tag{3.12}$$

where $b_1(\rho)$ is the first column of $B(\rho)$. Using this method, $G_0(\rho, \tau_k)$ and $G_1(\rho, \tau_k)$ can be derived for any input delays.

3.2.2 NETSIMPLEX ARCHITECTURE AND PROBLEM FORMULATION

NetSimplex architecture for networked CPS is shown in Figure 3.1, in which communication networks are used to exchange information among sensors, controllers, and actuators. In this architecture, the system runs with the following logics [14]:

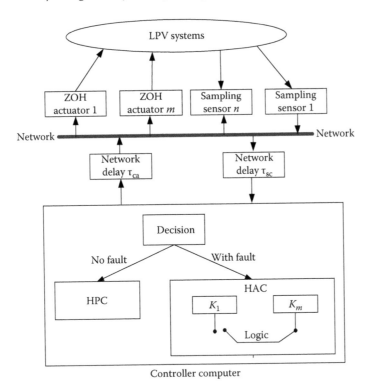

FIGURE 3.1 *NetSimplex* architecture for a networked cyber-physical system.

- Using time-driven sampling in sensors to produce the measurement of system outputs at every sampling instant, where time driven means that each node conducts its task at a prespecified time
- Using event-driven control computation in a controller by an interrupt mechanism each time a new measurement arrives, where event driven means that each node conducts its task upon the occurrence of events
- Using event-driven control inputs update at the actuators by an interrupt mechanism each time a new control signal arrives

Note that ZOH is usually used in practice for signal reconstruction, which can be built in a conventional digital-to-analog converter (DAC).

In this chapter, we extend *Simplex* architecture to enhance the reliability of NCS (*NetSimplex*). In a *NetSimplex*-based design, the NetMSR calculation for HAC is a key step in view of providing a large permissible subspace for HPC and determining criteria for fault detection at the HPC. The following problems should then be solved in the implementation of *NetSimplex* for NCS of LPV systems:

- MSR calculation for continuous-time LPV systems in HAC. The MSR calculation should take into account the range of a varying parameters so that the HAC is designed with the largest MSR.
- MSR shrinkage for discretization and network delay (NetMSR). The discrete implementation of HAC in computer and data transmissions in networks makes the MSR shrink. The NetMSR is the final region in the *NetSimplex* implementation.

For implementing the proposed control scheme, it is assumed that ρ is available. That means that the varying parameter ρ can be measured or estimated online. The worst case of delay τ_k

can be estimated based on the network protocol and system configuration and schedulability analysis.

Gain-scheduling state feedback control is chosen to be the candidate for the HAC because it is a proven technique for LPV system control in many applications. Traditionally, these "self-scheduling" controllers without the state or input constraints can guarantee closed-loop system stability via a common Lyapunov function or more commonly using a parameter-dependent Lyapunov function with less conservatism [15]. However, these techniques may not be applicable to constrained-control and delayed-control systems. This problem is addressed in our work in the computation of the NetMSR.

In networked gain-scheduling HACs, the NetMSR changes at the switching point between two controllers. The switching makes the system behavior more complex and creates a potential instability that must be addressed. For example, the HAC controller is running when *Controller i* switches to *Controller i*+1. Since the system with *Controller i* has different NetMSR from that of *Controller i*+1, the current state that is inside the MSR of *Controller i* may be outside the NetMSR of *Controller i*+1. In this situation, switching may destabilize the system. In this chapter, we propose to use an interpolation control to make the gain-scheduling HAC controllers so that they are always quadratically stable, so that the HAC can offer the capability to ensure stability and reliability of LPV systems.

3.3   MAXIMUM STABILITY REGION ANALYSIS OF LPV SYSTEMS

In this section, we present the MSR analysis of continuous-time LPV systems and the NetMSR computation of networked LPV systems with delays.

3.3.1   MAXIMUM STABILITY REGION ANALYSIS FOR CONTINUOUS TIME

In real-life applications, physical systems usually have limitations on the system states and/or control inputs. These constraints can be described as

$$a_i^T x \leq 1, \ i = 1, \ldots, l, \tag{3.13}$$

$$b_j^T u \leq 1, \ j = 1, \ldots, r, \tag{3.14}$$

where a_i and b_j are constraint vectors. Using the state feedback controller $u(t) = Kx(t)$, input constraints can be transformed to state constraints and the constraints given in Equations 3.13 and 3.14 can be represented as a polytope together in the system state space

$$\alpha_m^T x \leq 1, \ m = 1, \ldots, q, \tag{3.15}$$

where
$\alpha_m^T = a_m^T$, for $m = 1, \ldots, l$; $\alpha_m^T = b_j^T K$, for $m = l+1, \ldots, q$
$q = l + r$

The states within the polytope are called *admissible states*, since they obey the system states and input constraints.

For constraint control, the MSR is a subset of the states within the polytope of state constraints. If the closed-loop system starts from a state inside of the MSR, the future trajectory of the system will always stay in this MSR and finally converge to the control set points. For Lyapunov stability, $x^T P x < 1$ actually represents an MSR within the states-constraint polytope. Such an MSR is an n-dimensional ellipsoid located in the n-dimensional polyhedron of the system state constraints. Note that the states in the ellipsoid avoid saturation. For a closed-loop continuous-time control system, the MSR under the specific Lyapunov matrix P with state constraints is defined by the following ellipsoid:

$$\text{MSR} \triangleq \left\{ x \middle| x^T P x < 1 \right\}, \tag{3.16}$$

where P is a solution of the Lyapunov equation satisfying the state constraints Equation 3.15. The stability region can be computed by linear matrix inequality (LMI) method [16]. Note that a Lyapunov matrix corresponding to a stability region is not unique for the closed-loop system. However, we are only interested in the MSR. The existence of the MSR for continuous-time LPV systems can be determined as follows [16].

Lemma 3.1   *Given system (Equation 3.2) with the state constraints (Equation 3.15), the stability region defined in (Equation 3.16) satisfies the constraints (Equation 3.15) iff $\alpha_m^T Q \alpha_m \leq 1$, $m = 1, \ldots, q$, where $Q = P^{-1}$.*

Proof: Please refer to Seto and Sha [17, Lemma 4.1].
    Note that the stability region is defined as an ellipsoid geometrically in the state space. Maximizing the area of the stability region can be formed as the following LMI problem [18].

Problem 3.1   *The HAC controller $u(t) = Kx(t)$ is solved through maximizing the MSR 3.16:*

$$\min \ \log \det Q^{-1} \qquad\qquad (3.17)$$
$$\text{s.t.} : \ A(\rho)Q < 0, \ B(\rho)Z < 0, \ Q > 0,$$
$$\alpha_m^T Q \alpha_m \leq 1, \ m = 1, \ldots, q,$$
$$\begin{bmatrix} I & b_j^T \\ Z^T b_j & Q \end{bmatrix} \geq 0, j = 1, \ldots, r,$$
$$\rho \in [\rho_{\min}, \rho_{\max}],$$

where $Z = KQ$.

Remark 3.1   *The solution of Problem 3.1 results in a control gain K that will maximize the MSR for all $\rho \in [\rho_{\min}, \rho_{\max}]$.*

    Problem 3.1, as well as Problem 3.2 described next, is derived by solving the single Lyapunov matrix P, which simplifies the MSR computation using LMIs while providing conservative results compared with the parameter-dependent Lyapunov stability theory. The conservatism of the single Lyapunov matrix P can be effectively decreased through gain-scheduling controller design. The above LMI problem can be solved using *YALMIP* with the multiparametric toolbox (MPT) in MATLAB [19,20].
    The MSR can be derived using the parameter-dependent LMI method. It is clear that the MSR varies with different spans of ρ. Consequently, the area of MSR decreases as the span of ρ increases. To obtain a large MSR, we split the span of ρ into small intervals and use the gain-scheduling linear feedback control to design the HAC controllers.
    LPV gain-scheduling control is formed by a family of linear controllers, each of which provides a satisfactory control performance for a specific interval in the span of ρ. One or more observable parameters ρ, called the *scheduling variables*, are used to determine which operating scenario the system is currently in and to activate the corresponding controllers. For example, in an aircraft flight control system, the altitude and Mach number might be chosen as scheduling variables.

3.3.2 MAXIMUM STABILITY REGION ANALYSIS WITH DELAYS (NETMSR)

Essentially, there are three sources of delays in network control systems as shown in Figure 3.2:

1. Communication delay between the sensor and the controller, τ_k^{sc}
2. Computational delay in the controller, τ_k^c
3. Communication delay between the controller and the actuator, τ_k^{ca}

The subscript k indicates a possible time dependence of the delay. Without a loss of generality, a computational delay can be absorbed into τ_k^{ca}. Therefore, we can consider only the delays induced by networks: sensor-to-controller delay, τ_k^{sc}, and controller-to-actuator delay, τ_k^{ca}. The total delay in the loop is then $\tau_k = \tau_k^{sc} + \tau_k^{ca}$, which is the time interval between sensor sampling and actuation. Note that network delays are variables. But the worst-case delays are constants.

Consider the feedback control $u(k) = Kx(k)$ given in Section 3.3.1 and define a new state vector $z(k) = [x^T(k) \ x^T(k-1)]^T$. The closed-loop form of the augment system of $z(k)$ then takes the form

$$z(k+1) = \bar{\Phi}(\rho, \tau_k)z(k), \tag{3.18}$$

with

$$\bar{\Phi}(\rho, \tau_k) = \begin{bmatrix} \Phi(\rho) + G_0(\rho, \tau_k)K & G_1(\rho, \tau_k)K \\ I_{N \times N} & O_{N \times N} \end{bmatrix}, \tag{3.19}$$

where

$I_{N \times N}$ is an N-dimensional identical matrix
$O_{N \times N}$ is an $N \times N$ zero matrix

We then define an NetMSR under a new Lyapunov function for the augment system (Equation 3.18) as

$$\text{NetMSR} \overset{\Delta}{=} \left\{ z(k) | z(k)^T \bar{P} z(k) < 1 \right\}. \tag{3.20}$$

Lemma 3.2 [21] *A discrete-time system (Equation 3.18) is robustly stable if and only if for all ρ, there exists $\bar{P} = \bar{P}^T > 0$ such that*

$$\bar{\Phi}^T(\rho, \tau_k)\bar{P}\bar{\Phi}(\rho, \tau_k) - \bar{P} < 0, \quad \forall \rho \in [\rho_{\min}, \rho_{\max}], \tau_k \in (0, h]. \tag{3.21}$$

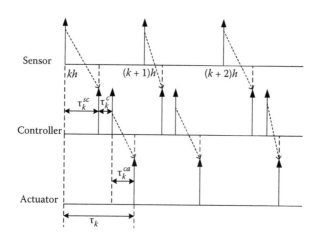

FIGURE 3.2 Illustration of delays in NCS.

Problem 3.2 *Given the HAC $u(k) = Kx(k)$ with K in Problem 3.1, the NetMSR with delays is solved through*

$$\min \ \log \det \bar{P}, \tag{3.22}$$

$$s.t. : \ \bar{P} > 0, \bar{\alpha}_m^T \bar{P} \bar{\alpha}_m \leq 1, \ m = 1, \dots, q,$$

$$\begin{bmatrix} \bar{P} & \bar{\Phi}(\rho, \tau_k)\bar{P} \\ \bar{P}^T \bar{\Phi}^T(\rho, \tau_k) & \bar{P} \end{bmatrix} > 0, \tag{3.23}$$

$$\rho \in [\rho_{\min}, \rho_{\max}], \ \tau_k \in (0, h].$$

where $\bar{\alpha}_m = [\alpha_m^T, O_{1 \times N}]^T$ and $\bar{\alpha}_m = [O_{1 \times N}, \alpha_m^T]^T$ are constraints with respect to $x(k)$ and $x(k-1)$, respectively.

Remark 3.2 *To derive the NetMSR of NCS, the estimation of the worst-case network delay is needed. To this end, network calculus [22] can be employed. This is a theoretical tool based on min-plus and max-plus algebra for performance analysis of computer networks. This topic is beyond the scope of this book, and interested readers are referred to Le Boudec and Thiran [22] for more details.*

Remark 3.3 *Equation 3.19 is the system matrix in the case where the network delay is less than one sampling period. The general form of the system matrix corresponding to a network delay longer than one period, that is, $(l-1)h < \tau_k < lh$ and $l > 1$ for each k, is given in Equation 3.24, for which the corresponding augment state vector is $z(k) = [x^T(k), x^T(k-1), \dots, x^T(k-l)]^T$, and $\tau'_k = \tau_k - (l-1)h$. Then, Lemma 3.2 is still applicable if $0 < \tau'_k < h$.*

$$\bar{\Phi}(\rho, \tau_k) = \begin{bmatrix} \Phi(\rho) & O_{N \times N} & \cdots & O_{N \times N} & G_0(\rho, \tau'_k)K(k-l+1) & G_1(\rho, \tau'_k)K(k-l) \\ I_{N \times N} & O_{N \times N} & \cdots & O_{N \times N} & O_{N \times N} & O_{N \times N} \\ \vdots & \vdots & \ddots & O_{N \times N} & \vdots & \vdots \\ O_{N \times N} & O_{N \times N} & \cdots & O_{N \times N} & I_{N \times N} & O_{N \times N} \end{bmatrix}. \tag{3.24}$$

3.3.3 A CASE STUDY

Different MSRs with different spans of parameter ρ and network delays in gain-scheduling controller design are shown in this case study. Consider a simple LPV system

$$\frac{d}{dt} \begin{bmatrix} x_1(t) \\ x_2(t) \end{bmatrix} = \begin{bmatrix} \rho & 1 \\ 0 & 1 \end{bmatrix} \begin{bmatrix} x_1(t) \\ x_2(t) \end{bmatrix} + \begin{bmatrix} 0 \\ 1 \end{bmatrix} u(t), \tag{3.25}$$

where $\rho \in [-2, 2]$ is a varying parameter. Suppose that the states and the inputs are constrained by

$$-0.2 \leq x_1(t), x_2(t) \leq 0.2, \ -5 \leq u(t) \leq 5. \tag{3.26}$$

In gain-scheduling controller design, there is a controller for every different interval of parameter ρ. We choose 0.2 as the interval length, that is, $\rho \in [-2, -1.8], [-1.8, -1.6], \dots, [1.8, 2]$. To show that different intervals can result in different MSRs, we compute the MSRs when the intervals are chosen as $\rho \in [-0.2, 0.2]$ and $\rho \in [-2, 2]$, respectively—the corresponding MSRs are drawn in

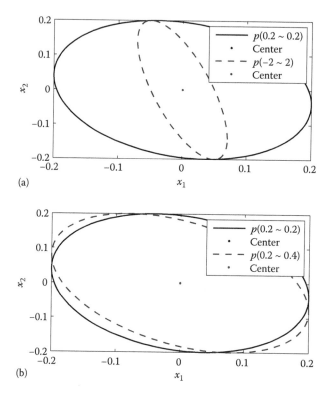

FIGURE 3.3 Illustration of MSRs for continuous systems: (a) Under different intervals of the varying parameter ρ and (b) under gain scheduling with the varying parameter ρ.

Figure 3.3a. It can be seen that the larger the interval, the smaller the MSR. In the HAC design, we prefer to reach large MSR so as to obtain a more reliable control system. That is why we employ gain-scheduling control. Figure 3.3b shows the MSRs when the HAC switches between two controllers that are designed based on $\rho \in [-0.2, 0.2]$ and $\rho \in [0.2, 0.4]$, respectively. We can clearly observe that when the controllers switch, the MSRs change and the states in the MSR of the former controller maybe outside the MSR of the successive one, that is, switching may introduce instability.

To show that different delays can result in different NetMSRs, we compute the NetMSRs when the delays are chosen as $\tau_k \in (0, 0.001]$ and $\tau_k \in (0, 0.121]$ with the same sampling period 0.2 s, and the three NetMSRs are projected on $x_1(k) - x_2(k)$, $x_1(k-1) - x_2(k-1)$, and $x_1(k-1) - x_2(k)$, as illustrated in Figure 3.4a through c, respectively. Note that there should be six projections on the states, however, only three are shown due to limited space. It can be seen that the NetMSRs projected on $x_1(k) - x_2(k)$ and $x_1(k-1) - x_2(k)$ vary with different network delays, while those projected on $x_1(k-1) - x_2(k-1)$ are almost the same with different network delays. These results can be explained using the structure of $\bar{\Phi}$ in Equation 3.19, in which the third and fourth rows are independent of τ_k. When we design *NetSimplex*-based highly reliable NCS, the NetMSR should be derived with network delays.

Note that the area of the MSR/NetMSR is proportional to the determinant of matrix P^{-1} (denoting the stability index). Figure 3.5 shows the stability index versus delay. Actually, the NetMSR is smaller for the system with the larger delay. Hence, as the delay decreases, the system becomes more stable. Furthermore, it has been shown in Liu et al. [16] that the NetMSR is smaller for the system with the larger sampling period. In order to have a larger NetMSR for HAC, the system should run faster and have a smaller delay.

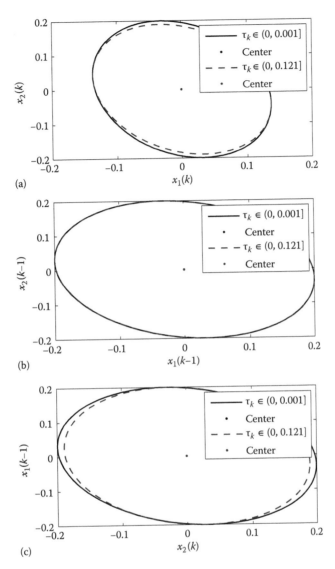

FIGURE 3.4 Illustration of MSRs under different network delays (a) projected on $x_1(k) - x_2(k)$; (b) projected on $x_1(k-1) - x_2(k-1)$; and (c) projected on $x_1(k-1) - x_2(k)$.

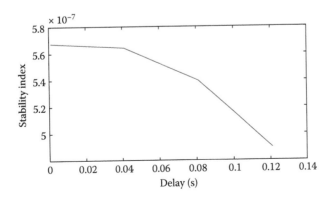

FIGURE 3.5 Stability index versus delay.

3.4 INTERPOLATION GAIN-SCHEDULING CONTROLLER DESIGN OF LPV SYSTEMS WITH NETWORK DELAYS

In this section, we present an interpolation gain-scheduling controller design for LPV systems with delays.

3.4.1 INTERPOLATION GAIN-SCHEDULING CONTROLLER DESIGN OF LPV CONTINUOUS SYSTEMS

Figure 3.6 shows the gain-scheduling controllers indexed by the varying parameter ρ. Since parameter ρ varies continuously, the switching happens only between the neighboring controllers. Suppose that *Controller i* is designed as a linear state feedback control

$$u(t) = K_i x(t), \ \rho \in [\rho_i, \bar{\rho}_i]. \tag{3.27}$$

The MSR for *Controller i* is defined as

$$\text{MSR}(i) \triangleq \left\{ x | x^T P_i x < 1 \right\}. \tag{3.28}$$

Controller i + 1 is also designed as a linear state feedback control

$$u(t) = K_{i+1} x(t), \ \rho \in [\bar{\rho}_{i+1}, \rho_{i+1}]. \tag{3.29}$$

The MSR for *Controller i* + 1 is defined as

$$\text{MSR}(i+1) \triangleq \left\{ x | x^T P_{i+1} x < 1 \right\}. \tag{3.30}$$

Controller$(i, i+1)$ is an interpolation linear state feedback control [23,24] given by

$$u(t) = K_{i,i+1}(\rho)x(t), \ \rho \in (\bar{\rho}_i, \bar{\rho}_{i+1}), \tag{3.31}$$

where

$$K_{i,i+1}(\rho) = \left[\frac{\bar{\rho}_{i+1} - \rho}{\bar{\rho}_{i+1} - \bar{\rho}_i} K_i P_i + \frac{\rho - \bar{\rho}_i}{\bar{\rho}_{i+1} - \bar{\rho}_i} K_{i+1} P_{i+1} \right]$$
$$\times \left[\frac{\bar{\rho}_{i+1} - \rho}{\bar{\rho}_{i+1} - \bar{\rho}_i} P_i + \frac{\rho - \bar{\rho}_i}{\bar{\rho}_{i+1} - \bar{\rho}_i} P_{i+1} \right]^{-1}.$$

The MSR for *Controller*$(i, i+1)$ is defined as

$$\text{MSR}(i, i+1) = \left\{ x | x^T P_{i,i+1} x < 1 \right\}, \ \rho \in (\bar{\rho}_i, \bar{\rho}_{i+1}), \tag{3.32}$$

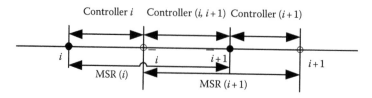

FIGURE 3.6 Gain-scheduling controllers indexed by a varying parameter.

where

$$P_{i,i+1} = \frac{\bar{\rho}_{i+1} - \rho}{\bar{\rho}_{i+1} - \bar{\rho}_i} P_i + \frac{\rho - \bar{\rho}_i}{\bar{\rho}_{i+1} - \bar{\rho}_i} P_{i+1}.$$

To analyze the stability property of switched linear systems, we introduce the following lemma.

Lemma 3.3 *For a varying parameter $\rho \in (\bar{\rho}_i, \bar{\rho}_{i+1})$, we suppose that K_i and K_{i+1}, given in Equations 3.27 and 3.29, respectively, both stabilize $(A(\rho), B(\rho))$ with the MSRs given in Equations 3.28 and 3.30. If there exist $\gamma > 1$ and $P_j A_{c,j}^T + A_{c,j} P_j < -\gamma I$, P_j are symmetric positive-definite matrices and $j = i, i+1$, where $A_{c,j} = A(\rho) + B(\rho) K_j$, then for every $\mu = (\rho - \bar{\rho})/(\bar{\rho}_{i+1} - \bar{\rho}_i)$, the interpolation control*

$$K_{i,i+1}(\rho) = [(1 - \mu) K_i P_i + \mu K_{i+1} P_{i+1}] P^{-1} \tag{3.33}$$

stabilizes $(A(\rho), B(\rho))$, if ρ satisfies the following condition

$$|\dot{\rho}(t)| < \min \frac{|\bar{\rho}_{i+1} - \bar{\rho}_i|}{\|P_{i+1} - P_i\|}, \quad t \geq 0, \tag{3.34}$$

where

$$P = (1 - \mu) P_i + \mu P_{i+1}. \tag{3.35}$$

Furthermore, the corresponding MSR is given by

$$MSR(i, i+1) \triangleq \{x | x^T P x < 1\}. \tag{3.36}$$

Note that a similar proof on the stability has been given in Stilwell and Rugh [24], and the result can be easily extended to systems with multiple varying parameters. The complete proof of Lemma 3.3 is given in Appendix 3.A.

3.4.2 MSR of Interpolation Gain-Scheduling Controller with Delays

Suppose that *Controller i*, $u(k) = \bar{K}_i z(k)$, and *Controller $i + 1$*, $u(k) = \bar{K}_{i+1} z(k)$, stabilize the augment system with the network delay (Equation 3.18), where $\bar{K}_i = [K_i \ O]$ and $\bar{K}_{i+1} = [K_{i+1} \ O]$. Denoted by

$$z(k + 1) = \bar{\Phi}_i(\rho, \tau_k) z(k), \quad \rho \in [\rho_i, \bar{\rho}_i], \tag{3.37}$$

$$z(k + 1) = \bar{\Phi}_{i+1}(\rho, \tau_k) z(k), \quad \rho \in [\bar{\rho}_{i+1}, \rho_{i+1}], \tag{3.38}$$

the closed-loop augment systems with different spans of varying parameters.

The NetMSRs for *Controller i* and *Controller $i + 1$* are defined as

$$\text{NetMSR}(i) \triangleq \{z | z^T \bar{P}_i z < 1\}, \tag{3.39}$$

$$\text{NetMSR}(i + 1) \triangleq \{z | z^T \bar{P}_{i+1} z < 1\}, \tag{3.40}$$

where

$$\bar{\Phi}_i(\rho, \tau_k) = \begin{bmatrix} \Phi(\rho) + G_0(\rho, \tau_k) K_i & G_1(\rho, \tau_k) K_i \\ I & O \end{bmatrix}, \tag{3.41}$$

$$\bar{\Phi}_{i+1}(\rho, \tau_k) = \begin{bmatrix} \Phi(\rho) + G_0(\rho, \tau_k) K_{i+1} & G_1(\rho, \tau_k) K_{i+1} \\ I & O \end{bmatrix}. \tag{3.42}$$

Lemma 3.4 *For a varying parameter $\rho \in (\bar{\rho}_i, \bar{\rho}_{i+1})$, suppose Controller i and Controller $i + 1$, the closed-loop system (Equations 3.37 and 3.38) with the NetMSRs given in (Equations 3.39 and 3.40). Then for every $\mu = (\rho - \bar{\rho})/(\bar{\rho}_{i+1} - \bar{\rho}_i)$, the interpolation control*

$$\bar{K}_{i,i+1}(\rho) = [(1 - \mu)\bar{K}_i \bar{P}_i + \mu \bar{K}_{i+1} \bar{P}_{i+1}]\bar{P}^{-1} \tag{3.43}$$

stabilizes the augment system (Equations 3.18) for $\rho \in (\bar{\rho}_i, \bar{\rho}_{i+1})$, where

$$\bar{P} = (1 - \mu)\bar{P}_i + \mu \bar{P}_{i+1}. \tag{3.44}$$

The proof of Lemma 3.4 is given in Appendix 3.B.

Remark 3.4 *Based on Equations 3.43, the interpolation Controller$(i, i+1)$ is of the form $\bar{K}_{i,i+1}(\rho) = [K^1_{i,i+1}(\rho) \quad K^2_{i,i+1}(\rho)]$ and the state feedback control signal in the original coordinate will be $u(k) = K^1_{i,i+1}(\rho)x(k) + K^2_{i,i+1}(\rho)x(k-1)$. Note that the interpolation controller switching needs two samplings of successive state measurements in the augment system.*

3.5 STABILITY PROPERTY OF SWITCHED LINEAR SYSTEMS FROM NetMSR AND REACH SET PERSPECTIVE

This section presents the reach set of the plant with network delays and switch logic. Stability analysis of switched linear systems will then be carried out from an NetMSR and reach set (RS) perspective.

3.5.1 REACH SET OF STATES WITH DELAY

Due to network-induced delays, a *NetSimplex*-based scheme may introduce up to one period of recovery delay when a fault occurs. The basic idea on how to use reach set to handle extra network delays is illustrated in Figure 3.7. Suppose that at time t, a fault is detected in the HPC. In the framework of *NetSimplex*, since the HAC and the HPC are running in parallel, upon the detection of a fault, the HAC control command is expected to be immediately available and to take action right away. However, in the presence of a network-induced delay, the HAC needs to carry out control computations and transmissions. Consequently, the control command becomes available only after up to one period, that is, at time $t + \tau_k$. As a result, compared to the recovery procedure in the original *Simplex*, the one in the NCS will incur an extra delay up to one period. The delay due to recovering can be compensated using a reach set computation [25–27]. More specifically, at every decision time, the decision module will compute the reach set of the states within τ_k. If this reach set is still inside the NetMSR

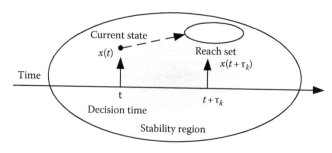

FIGURE 3.7 Illustration of using reach set to handle an extra network delay.

associated with the HAC, the HPC can still run; otherwise, the HAC will be activated to take over the control.

3.5.2 SWITCH LOGIC

Consider a gain scheduling HAC with a switch logic as shown in Figure 3.8, which works as follows:

- When the varying parameter $\rho \in [\rho_i, \bar{\rho}_i]$ and the predicted reach set of states x for the next τ_k in HPC control is outside NetMSR(i), the HAC *Controller i* will be activated for HAC control.
- When the varying parameter $\rho \in [\bar{\rho}_{i+1}, \rho_{i+1}]$ and the predicted reach set of states in the next τ_k for HPC control is outside NetMSR($i+1$), *Controller i + 1* is activated for HAC control.
- When the varying parameter $\rho \in (\bar{\rho}_i, \bar{\rho}_{i+1})$ and the predicted reach set of states in the next τ_k for HPC control is outside NetMSR($i, i+1$), *Controller(i, i + 1)* will be activated for HAC control.

Suppose that the current state is $x(k)$, the current controller is $u(k) = K_i x(k)$, and the corresponding NetMSR is described as NetMSR(i) $\triangleq \{z|z^T \bar{P}_i z < 1\}$. NetMSR($i, i+1$) is varying with parameter ρ since parameter μ is a parametrical function of ρ. In fact, NetMSR($i, i+1$) is smoothly varying from NetMSR(i) to NetMSR($i+1$). We then obtain the following.

Theorem 3.1 NetSimplex-*based highly reliable control using interpolation feedback given in Equation 3.33 can ensure the stability when all switches conforming to Equation 3.34 occur for HAC with a switch logic as described in Figure 3.8.*

Proof: In the *NetSimplex* architecture, the HAC controller is activated when the predicted reach set of states in the next τ_k for HPC control is outside the corresponding NetMSR. The switch logic guarantees that the initial states for HAC are always inside the NetMSR. Based on Lemmas 3.3 and 3.4, HAC is stable. We then conclude Theorem 3.1.

3.5.3 STABILITY ANALYSIS

In stability analysis for switched systems, an important question is whether the system is stable when the switching signals are activated. Usually, we assume that all the subsystems are asymptotically stable. However, this is not a sufficient condition for guaranteeing the overall stability of the system for

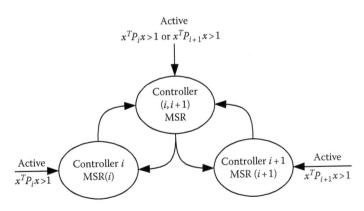

FIGURE 3.8 *NetSimplex*-based highly reliable control switch logic.

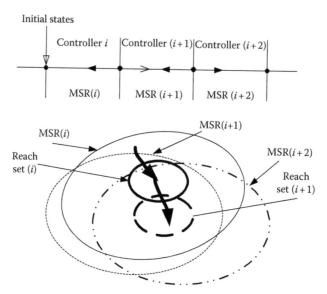

FIGURE 3.9   Illustration of switching MSR and reach set.

all switching signals. The common Lyapunov function method is one possible way for stability analysis of the system under switching [15,28]. It is already proven that, if there is a common Lyapunov function for all the subsystems, the stability of the system is guaranteed. However, this method may not be applicable when there are constraints on states or inputs.

To overcome this difficulty, we assess the stability based on NetMSR and RS analyses. Without loss generality, we consider the situation where the system switches among three subsystems. Figure 3.9 shows the stability analysis using the NetMSR perspective, which can be described as follows:

- The system is initialed in a state inside NetMSR(i) and the reach set RS(i) for subsystem i is computed using the time at which *Controller i* runs.
- RS(i) is inside NetMSR($i, i+1$) and it indicates that the subsystem ($i, i+1$) is stable when the subsystem i switches to the subsystem ($i, i+1$). RS($i, i+1$) for subsystem ($i, i+1$) is computed using the time at which *Controller ($i, i+1$)* runs.
- RS($i, i+1$) is inside either NetMSR($i+1$) or NetMSR(i) and it indicates that subsystem $i+1$ or subsystem i is stable when the subsystem ($i, i+1$) switches to it.

Theorem 3.2   *The stability of a switched system consisting of a set of subsystems can be ensured if*

- *Each closed-loop subsystem is stabilized.*
- *Initial states of a subsystem belong to its corresponding NetMSR.*
- *The reach set for a subsystem within the running time is inside the NetMSR of the neighbor subsystems, where the running time denotes the time interval for a subsystem before switching to another subsystem.*

Proof: Each subsystem is stale if the initial states are inside its corresponding NetMSR. The last item ensures that the initial states are always inside the NetMSR of a subsystem when it is activated. Then we conclude Theorem 3.2.

Remark 3.5   *The running time is important for stability under switching. The longer the running time, the smaller the reach set for the stable closed-loop subsystem. A smaller reach set can be of a larger likelihood inside NetMSR for the switched subsystem. This property is consistent with the existing analysis using average dwell time [29], as well as the stability with slow-switching signals [15]. For interpolation controller design, NetMSR is the same at the switching point. Therefore, the reach set for a controller will always be inside the neighbor NetMSR. This clearly indicates that the stability is ensured using interpolation controller design.*

3.6   EVALUATIONS

This section presents simulation studies of a satellite attitude control using the proposed *NetSimplex*-based highly reliable networked control.

3.6.1   SATELLITE ALTITUDE CONTROL MODEL

The state-space model of the satellite system consisting of two rigid bodies joined by a flexible link is given by

$$\dot{x}(t) = A(\rho_1, \rho_2)x(t) + Bu(t), \tag{3.45}$$

where
$\dot{x}(t) = [\ \theta_1 \quad \theta_2 \quad \dot{\theta}_1 \quad \dot{\theta}_2 \ ]^T$ is the state vector
θ_1 and θ_2 are the yaw angles for the main body and the sensor module
u is the control torque
the dependent parameters ρ_1, ρ_2 are torque constant and viscous damping

The system matrix and input matrix are described as follows:

$$A(\rho_1, \rho_2) = \begin{bmatrix} 0 & 0 & 1 & 0 \\ 0 & 0 & 0 & 1 \\ -\rho_1 & \rho_1 & -\rho_2 & \rho_2 \\ \rho_1 & -\rho_1 & \rho_2 & -\rho_2 \end{bmatrix}, \ B = \begin{bmatrix} 0 \\ 0 \\ 0 \\ 1 \end{bmatrix}.$$

The parameters in the model vary in the intervals

$$\rho_1 \in [0.09, 0.4], \quad \rho_2 \in [0.0038, 0.04].$$

The states and the input are constrained by

$$|\theta_1| \leq 20°, \ |\theta_2| \leq 20°, \ \left|\dot{\theta}_1\right| \leq 300°/s, \ \left|\dot{\theta}_2\right| \leq 300°/s, \ |u| \leq 5.$$

3.6.2   SIMULATION CONFIGURATION

The simulation is carried out on a platform of real-time NCS simulation software named TrueTime [30], which is a MATLAB Simulink-based simulator. This simulator includes two main interface modules, TrueTime kernel and TrueTime network. In a cosimulation, the controller task is usually executed in TrueTime kernel and data transmission is simulated by TrueTime network. TrueTime kernel can simulate the execution of several independent tasks (periodic or nonperiodic). TrueTime network can simulate various network protocols, including CSMA/CD (Ethernet), CSMA/AMP (CAN), switched Ethernet, TDMA, FDMA, as well as the round robin scheduling algorithm.

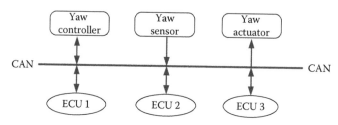

FIGURE 3.10   Configuration of a CAN-based satellite system.

It can also simulate diverse wireless networks (802.11b/g WLAN and 802.15.4 ZigBee) and battery-powered devices. In our simulation for NCS, TrueTime kernel is used as a processor to compute control the law and the TrueTime network module is used as the network employed by this system.

We consider an example of controlling the yaw angles of a satellite shown in Gahinet et al. [20]. In this system, CAN bus is employed to exchange information among sensors, actuators, and controllers. CAN is a multimaster broadcast vehicle bus standard for communicating among electronic control units (ECUs) and is widely used by automobiles, small-scale unmanned aerial vehicles (UAVs), small satellite systems [31], and so on.

In the simulation, the yaw attitude (angle and angle rate) is measured by the inertial measurement unit (IMU). The sampling period for measurement and control is 10 ms (100 HZ). The actuator is a reaction/momentum wheel that is triggered by actuation signals. The sensor, controller, and actuator are located at different places in the satellite. All of them, as well as other ECUs, are connected by a CAN2.0A bus with a bit rate of 1 Mb/s. The configuration of this simulation is shown in Figure 3.10. The assigned priority order for elements in the control loop is the sensor, controller, and actuator. There are three ECUs with higher priority. It is assumed that the three ECMs transmit 10 frames, 20 frames, and 20 frames with a maximum payload at a frequency of 50 HZ. To estimate the worst-case delay, the network calculus approach is employed. The obtained worst-case delays from sensor to controller and from controller to actuator are 5.772 ms with nonpreemptive transmission. Assuming that the computation delay of the control law is 2 ms in the controller, the total delay in the control loop is 13.544 ms.

3.6.3   NetMSR Evaluations

Under different varying parameter regions, the projections of MSRs on state space are illustrated in Figure 3.11. The regions are defined as follows:

$$\text{Region } 1 \rightarrow \rho_1 \in [0.09, 0.25], \ \rho_2 \in [0.0038, 0.025]$$

$$\text{Region } 2 \rightarrow \rho_1 \in [0.18, 0.4], \ \rho_2 \in [0.018, 0.04]$$

$$\text{Region } 3 \rightarrow \rho_1 \in [0.09, 0.25], \ \rho_2 \in [0.018, 0.04]$$

$$\text{Region } 4 \rightarrow \rho_1 \in [0.18, 0.4], \ \rho_2 \in [0.0038, 0.025]$$

The corresponding control gains with MSR are obtained as

$$K_1 = [-0.9326 \ -14.7841 \ -21.8367 \ -2.8976]$$

$$K_2 = [-1.8907 \ -13.8959 \ -11.9937 \ -2.7541]$$

$$K_3 = [-1.5179 \ -14.2543 \ -20.0288 \ -2.8620]$$

$$K_4 = [-1.5336 \ -14.2187 \ -12.8992 \ -2.7760]$$

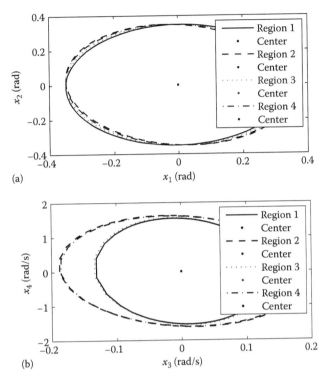

FIGURE 3.11 MSR projections for varying parameter regions: (a) on θ_1 and θ_2; and (b) on $\dot{\theta}_1$ and $\dot{\theta}_2$.

Employing Equation 3.33, the corresponding interpolation controllers can be obtained. The shrunken MSRs (NetMSR) with a variable delay bounded by the worst network delay of 13.544 ms are shown in Figure 3.12. Note that there are 28 (derived from the combination C_{2N}^2, where N is the number of the states) projections on a 2D plane. Due to space limitations, we only show three in this chapter.

We consider the design of a 2D interpolation state feedback controller using two varying parameters, ρ_1 and ρ_2, which conducts three steps:

1. Using Region 1 and Region 4, the 1D interpolation state feedback controller is derived with respect to ρ_1 and the new varying parameters, Region (1,4), will be $\rho_1 \in [0.18, 0.25]$ and $\rho_2 \in [0.0038, 0.025]$.
2. Using Region 2 and Region 3, another 1D interpolation state feedback controller is derived with respect to ρ_1 and the new varying parameters, Region (2,3), will be $\rho_1 \in [0.18, 0.25]$ and $\rho_2 \in [0.018, 0.04]$.
3. Using Region (1,4) and Region (2,3), the final 2D interpolation state feedback controller is derived with respect to ρ_1 and ρ_2, and the varying parameter regions will be $\rho_1 \in [0.18, 0.25]$ and $\rho_2 \in [0.018, 0.025]$.

To obtain high performance, the HPC is designed based on the H_2-norm [32], which amounts to finding a linear parameter varying state feedback control $u(t) = K(\rho)x(t)$ with the minimum H_2-norm control performance cost. The objective of HPC control is to find a solution to minimize the control tracking error while not explicitly considering the aspect of robustness.

It can be seen from Figure 3.11a and b that the MSR will vary as the span of the varying parameter changes. Note that the system starts from an initial state belonging to the MSR and the state trajectory always stays within the MSR.

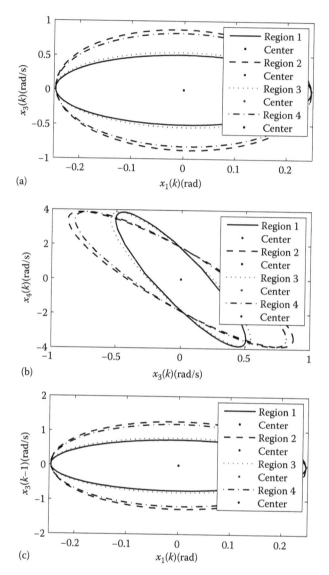

FIGURE 3.12 NetMSR projections for different delays: (a) on $\theta_1(k)$ and $\dot{\theta}_1(k)$; (b) on $\dot{\theta}_1(k)$ and $\dot{\theta}_2(k)$; and (c) on $\theta_1(k)$ and $\dot{\theta}(k-1)$.

3.6.4 RELIABILITY EVALUATIONS

To compare the performance of different types of controllers, simulations with one fault in the satellite system are conducted. In simulations, the initial state is chosen as $x_0 = \begin{bmatrix} 8° & 8° & 0 & 0 \end{bmatrix}^T$, which is inside the corresponding NetMSR, and the simulation time is 60 s. From the time of 20–20.3 s, there is one fault, such as divided by the zero (DIVzero) bug, which makes the HPC control input jump to a very large value, driving the system trajectory outside of the corresponding NetMSR. The simulation results using HAC control are shown in Figure 3.13. Note that the control performance is conservative while the control system runs safely without violating the constraints. For the HPC control, the results are summarized in Figure 3.14. We can see that, although the control performance is very good, the state constraints are not respected when the fault occurs. This will make the system run a dangerous situation and lead to loss of the satellite. To evaluate the performance of the proposed

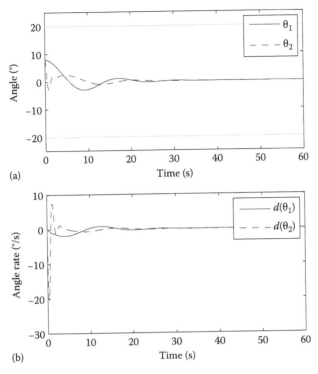

FIGURE 3.13 HAC control for a satellite system: (a) angle responses; and (b) angular rate responses.

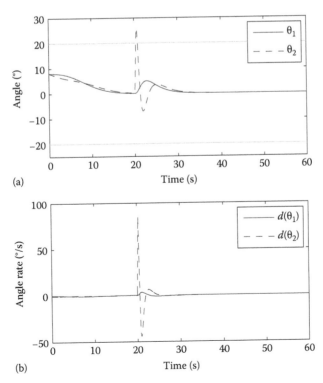

FIGURE 3.14 HPC control for a satellite system: (a) angle responses; and (b) angular rate responses.

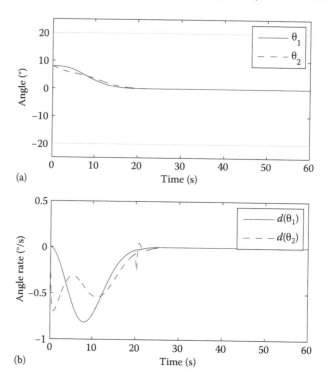

(a)

(b)

FIGURE 3.15 *NetSimplex*-based control for satellite system: (a) angle responses and (b) angular rate responses.

control, a simulation is conducted and the results are given in Figure 3.15. It is clearly shown that the control performance is as good as the one given by HPC control and the safety is ensured as the constraints are respected. Hence, the proposed *NetSimplex*-based control does take advantage of both the merits of HAC and HPC.

3.7 CONCLUSION AND FUTURE WORK

NetSimplex extends the original *Simplex* architecture from a single-node environment to a networked control environment. This architecture supports the controller redundancy design for cyber-physical systems. In order to enhance reliability by maximizing the MSR, interpolation gain-scheduling state feedback control is employed in the HAC, which is easy to apply and allows for the incorporation of a network-induced delay in the MSR computation. The stability analysis based on the NetMSR shows that with appropriately designed decision logic, the overall system stability can be guaranteed when switching among the gain-scheduling HAC controllers. Simulation results confirmed the viability and the effectiveness of the *NetSimplex*. Finally, as discrete-time models may be more convenient for the development of NCS, one of the envisaged future works will be to formulate the entire *NetSimplex* in the framework of digital control. This would further enhance the reliability and the robustness of networked cyber-physical systems.

ACKNOWLEDGMENTS

This work was supported in part by the Program for NSFC (No. 61303013) and the Aero-Science Fund (No. 2013ZC18003).

3.A APPENDIX

3.A.1 PROOF OF LEMMA 3.3

Since both K_i and K_{i+1} stabilize $(A(\rho), B(\rho))$, the symmetric positive-definite matrices P_i and P_{i+1} verify the following stability property:

$$P_i A_{c,i}^T + A_{c,i} P_i < -\gamma I, \tag{3.46}$$

$$P_{i+1} A_{c,i+1}^T + A_{c,i+1} P_{i+1} < -\gamma I, \tag{3.47}$$

where γ is a positive constant. We obtain get from Equations 3.46 and 3.47

$$(1 - \mu)[P_i A_{c,i}^T + A_{c,i} P_i] + \mu[P_{i+1} A_{c,i+1}^T + A_{c,i+1} P_{i+1}] < 0, \tag{3.48}$$

where $\mu = (\rho - \bar{\rho})/(\bar{\rho}_{i+1} - \bar{\rho}_i)$ and satisfies $0 \le \mu \le 1$.

Define

$$K_{i,i+1}(\rho) = [(1 - \mu)K_i P_i + \mu K_{i+1} P_{i+1}]P^{-1}, \tag{3.49}$$

with

$$P = (1 - \mu)P_i + \mu P_{i+1}. \tag{3.50}$$

We can rewrite Equation 3.48 as

$$P A_{c,i-i+1}^T + A_{c,i-i+1} P < -\gamma I. \tag{3.51}$$

where $A_{c,i-i+1} = A(\rho) + B(\rho)K_{i,i+1}$.

Since $(\bar{\rho}_i, \bar{\rho}_{i+1})$ is compact, there exists $\delta_w > 0$ such that $P \ge \delta_w I$. Define $\hat{P}$ as the approximation P defined in Equation 3.35. Then we have $\|P - \hat{P}\| < \varepsilon$, where

$$\varepsilon < \min\left(\frac{\gamma - 1}{2 \max \|A_{c,i-i+1}\|}, \delta_w\right). \tag{3.52}$$

Hence, the approximation error $\tilde{P} = P - \hat{P}$ satisfies

$$\max \|\tilde{P}\| < \varepsilon < \frac{\gamma - 1}{2 \max \|A_{c,i-i+1}\|}, \tag{3.53}$$

which implies that

$$\|\tilde{P} A_{c,i-i+1}^T + A_{c,i-i+1} \tilde{P}\| < \gamma - 1. \tag{3.54}$$

Substituting the above inequation in Equation 3.51, we have

$$\hat{P} A_{c,i-i+1}^T + A_{c,i-i+1} \hat{P} < -I. \tag{3.55}$$

According to Lemma I.3 in Stilwell and Rugh [24], we have $\hat{P} > 0$. To prove the stability, consider the Lyapunov function candidate $V(x) = x^T \hat{P}^{-1} x$. Then, the time derivative of V is given by

$$\frac{d}{dt} V(x) = x^T \left(A_c \hat{P}^{-1} + \hat{P}^{-1} A_c + \frac{d}{dt} \hat{P}^{-1} \right)^T x. \tag{3.56}$$

Substituting Equation 3.55 to Equation 3.56 yields

$$\frac{d}{dt}V(x) < x^T \hat{P}^{-1} \left(-I - \frac{d}{dt}\hat{P} \right) \hat{P}^{-1}x. \tag{3.57}$$

There exists $\delta_1 > 0$ such that $\frac{d}{dt}V(x) \leq -\delta_1\|x\|$ if $\max\left\|\frac{d}{dt}\hat{P}\right\| < 1$. According to Lemma I.3 in Stilwell and Rugh [24], we have

$$\left\|\frac{d}{dt}\hat{P}\right\| \leq \max\left\|\frac{d}{dt}P\right\| = \frac{\dot{\rho}}{|\bar{\rho}_{i+1} - \bar{\rho}_i|} \max\|P_{i+1} - P_i\|. \tag{3.58}$$

Hence, the condition given in Equation 3.34 is obtained. $K_{i,i+1}(\rho)$ also stabilizes $(A(\rho), B(\rho))$ if the condition Equation 3.34 is satisfied.

We now verify the states and inputs constraints conforming to $MSR(i, i + 1)$. Note that these constraints conform to $MSR(i)$ and $MSR(i + 1)$:

$$\alpha_m^T P_i \alpha_m \leq 1, \quad m = 1, \ldots, q, \tag{3.59}$$

$$\alpha_m^T P_{i+1} \alpha_m \leq 1, \quad m = 1, \ldots, q. \tag{3.60}$$

From Equations 3.59 and 3.60, we can get

$$(1 - \mu)\alpha_m^T P_i \alpha_m + \mu\alpha_m^T P_{i+1}\alpha_m \leq 1, m = 1, \ldots, q, \tag{3.61}$$

that is,

$$\alpha_m^T P\alpha_m \leq 1, m = 1, \ldots, q. \tag{3.62}$$

The stability region for controller (Equation 3.33) is then given by (Equation 3.36) with P given in (Equation 3.35).

3.A.2 PROOF OF LEMMA 3.4

Since both *Controller i* and *Controller i + 1* result in the closed-loop form augment systems Equations 3.37 and 3.38, by employing robust Lyapunov stability theory [21] (Lemma 3.2), we have the following inequalities:

$$\bar{\Phi}_i^T(\rho)\bar{P}_i\bar{\Phi}_i(\rho) - \bar{P}_i < 0, \quad \rho \in [\rho_i, \bar{\rho}_i], \tag{3.63}$$

$$\bar{\Phi}_{i+1}^T(\rho)\bar{P}_{i+1}\bar{\Phi}_{i+1}(\rho) - \bar{P}_{i+1} < 0, \quad \rho \in [\bar{\rho}_{i+1}, \rho_{i+1}], \tag{3.64}$$

which imply that the following LMIs obtained from Schur complement on condition $\bar{P}_i$ and $\bar{P}_{i+1}$ are all positive definite:

$$\begin{pmatrix} \bar{P}_i & \bar{\Phi}_i(\rho)\bar{P}_i \\ \bar{P}_i^T \bar{\Phi}_i^T(\rho) & \bar{P}_i \end{pmatrix} > 0, \tag{3.65}$$

$$\begin{pmatrix} \bar{P}_{i+1} & \bar{\Phi}_{i+1}(\rho)\bar{P}_{i+1} \\ \bar{P}_{i+1}^T \bar{\Phi}_{i+1}^T(\rho) & \bar{P}_{i+1} \end{pmatrix} > 0. \tag{3.66}$$

Multiplying $(1 - \mu)$ and μ to Equations 3.65 and 3.66, respectively, and then summing up these two terms, we get

$$\begin{pmatrix} \bar{P} & \bar{\Phi}_{i,i+1}(\rho)\bar{P} \\ \bar{P}^T \bar{\Phi}_{i,i+1}^T(\rho) & \bar{P} \end{pmatrix} > 0, \tag{3.67}$$

where
$\bar{\Phi}_{i,i+1}(\rho)$ is the closed-loop system matrix with the new controller $\bar{K}_{i,i+1}(\rho) = [(1-\mu)\bar{K}_i\bar{P}_i + \mu\bar{K}_{i+1}\bar{P}_{i+1}]\bar{P}^{-1}$
$\bar{P} = (1-\mu)\bar{P}_i + \mu\bar{P}_{i+1}$

The LMI Equation 3.67 implies that

$$\bar{\Phi}_{i,i+1}^T(\rho)\bar{P}\bar{\Phi}_{i,i+1}(\rho) - \bar{P} < 0, \quad \rho \in [\bar{\rho}_i, \bar{\rho}_{i+1}]. \tag{3.68}$$

Considering the Lyapunov function candidate for the system with the new controller (Equation 3.43) of the form

$$V_{i,i+1}(k) = z^T(k)\bar{P}z(k) \tag{3.69}$$

and computing the difference of V yield

$$\Delta V_{i,i+1}(k) = z^T(k+1)\bar{P}z(k+1) - z^T(k)\bar{P}z(k)$$

$$= z^T(k)(\bar{\Phi}_{i,i+1}^T\bar{P}\bar{\Phi}_{i,i+1} - \bar{P})z(k). \tag{3.70}$$

Employing the parameter-dependent Lyapunov stability theory [21] (Lemma 3.1), we can conclude from (Equation 3.68) that $\Delta V_{i,i+1}(k) < 0, \forall z(k) \neq 0$, which implies that the new controller (Equation 3.43) should stabilize the augment system (Equation 3.18) for $\rho \in (\bar{\rho}_i, \bar{\rho}_{i+1})$.

REFERENCES

1. E. Lee, Cyber physical systems: Design challenges, in *Proceedings of 11th IEEE International Symposium on Object Oriented Real-Time Distributed Computing (ISORC)*, Orlando, FL, May 2008, pp. 363–369.
2. L. Sha, S. Gopalakrishnan, X. Liu, and Q. Wang, Cyber-physical systems: A new frontier, in *Proceedings of IEEE International Conference on Sensor Networks, Ubiquitous and Trustworthy Computing (SUTC)*, Taichung, Taiwan, June 2008, pp. 1–9.
3. A. Cheng, Cyber-physical medical and medication systems, in *Proceedings of 28th International Conference on Distributed Computing Systems (ICDCS) Workshops*, Beijing, China, June 2008, pp. 529–532.
4. P. Antsaklis, From hybrid to networked cyber-physical systems, in *Proceedings of American Control Conference*, St. Louis, MO, June 2009, pp. 6–7.
5. P. Tabuada, Cyber-physical systems, in *Proceedings of National Science Foundation Workshop on Cyber-Physical Systems*, Cyber-Physical Systems: Position Paper. Austin, Texas, October, June 2006, pp. 1–3.
6. Y. Tan, S. Goddard, and L. C. Prez, A prototype architecture for cyber-physical systems, *ACM SIGBED Review*, 5(1), 1–2, 2008.
7. I. Moir and A. Seabridge, *Aircraft Systems: Mechanical, Electrical and Avionics Subsystems Integration*. AIAA Education Series. New York: Wiley, 2001.
8. A. Hagiescu, U. Bordoloi, S. Chakraborty, P. Sampath, P. Ganesan, and S. Ramesh, Performance analysis of flexray-based ECU networks, in *Proceedings of 44th ACM/IEEE Design Automation Conference (DAC)*, San Diego, CA, June 2007, pp. 284–289.
9. J. Nilsson, Real-time control systems with delays. PhD dissertation, Department of Automatic Control, Lund Institute of Technology, Lund, Sweden, 1998.
10. J. Spragins, J. Hammond, and K. Pawlikowski, *Telecommunications: Protocols and Designs*. Reading, MA: Addison-Wesley Longman Publishing Co., 1991.
11. L. Sha, Dependable system upgrade, in *Proceedings of 19th IEEE Real-Time Systems Symposium (RTSS)*, Madrid, Spain, December 1998, pp. 440–448.
12. D. Seto, B. Krogh, L. Sha, and A. Chutinan, Dynamic control system upgrade using the simplex architecture, *IEEE Control Syst. Mag.*, 18(4), 72–80, August 1998.
13. J. Yao, X. Liu, G. Zhu, and L. Sha, NetSimplex: Controller fault tolerance architecture in networked control systems, *IEEE Trans. Ind. Inform.*, 9(1), 346–356, February 2013.

14. W. Zhang, M. Branicky, and S. Phillips, Stability of networked control systems, *IEEE Control Syst. Mag.*, 21(1), 84–99, February 2001.
15. H. Lin and P. Antsaklis, Stability and stabilizability of switched linear systems: A survey of recent results, *IEEE Trans. Automat. Control*, 54(2), 308–322, February 2009.
16. X. Liu, Q. Wang, S. Gopalakrishnan, W. He, L. Sha, H. Ding, and K. Lee, ORTEGA: An efficient and flexible online fault tolerance architecture for real-time control systems, *IEEE Trans. Ind. Informat.*, 4(4), 213–224, November 2008.
17. D. Seto and L. Sha, An engineering method for safety region development, CMU SEI, Technical Report, Pittsburgh, PA, 1999.
18. S. Boyd, L. E. Ghaoui, E. Feron, and V. Balakrishnan, *Linear Matrix Inequalities in System and Control Theory*. Philadelphia, PA: SIAM, 1994.
19. J. Lofberg, YALMIP: A toolbox for modeling and optimization in MATLAB, in *Proceedings of IEEE International Symposium on Computer-Aided Control Systems Design*, Taipei, Taiwan, pp. 284–289, September 2004.
20. P. Gahinet, A. Nemirovski, A. J. Laub, and M. Chilali, *MATLAB LMI Control Toolbox*. Natick, MA: The MathWorks, 1995.
21. M. de Oliveira, J. Bernussou, and J. Geromel, A new discrete-time robust stability condition, *Systems & Control Letters*, 37(4), pp. 261–265, 1999.
22. J.-Y. Le Boudec and P. Thiran, *Network Calculus: A Theory of Deterministic Queuing Systems for the Internet*. Berlin, Germany: Springer-Verlag, 2001.
23. D. Stilwell and W. Rugh, Interpolation of observer state feedback controllers for gain scheduling, in *Proceedings of American Control Conference (ACC)*, Philadelphia, PA, vol. 2, June 1998, pp. 1215–1219.
24. D. Stilwell and W. Rugh, Interpolation of observer state feedback controllers for gain scheduling, *IEEE Trans. Autom. Control*, 44(6), 1225–1229, June 1999.
25. A. Kurzhanskiy and P. Varaiya, Ellipsoidal toolbox, EECS Department, University of California, Berkeley, CA, Technical Report UCB/EECS-2006-46, May 2006.
26. A. Kurzhanskiy and P. Varaiya, *Ellipsoidal Calculus for Estimation and Control*, 1st edn. Boston, MA: SCFA, 1997.
27. A. Kurzhanskiy and P. Varaiya, Ellipsoidal toolbox (ET), in *Proceedings of 45th IEEE Conference on Decision and Control (CDC)*, San Diego, CA, December 2006, pp. 1498–1503.
28. D. Cheng, L. Guo, Y. Lin, and Y. Wang, Stabilization of switched linear systems, *IEEE Trans. Autom. Control*, 50(5), 661–666, May 2005.
29. J. Hespanha and A. Morse, Stability of switched systems with average dwell-time, in *Proceedings of 38th IEEE Conference on Decision and Control (CDC)*, vol. 3, 1999, pp. 2655–2660.
30. A. C. M. Ohlin and D. Henriksson, *TrueTime 1.5—Reference Manual*, 1st edn. Lund, Sweden: Department of Automatic Control, Lund University, 2007.
31. S. Montenegro, K. Briess, and H. Kayal, Cyber-physical systems: A new frontier, in *Proceedings of Conference Data Systems in Aerospace (DASIA)*, Berlin, Germany, May 2006, pp. 51–55.
32. W. Xie, H_2 gain scheduled state feedback for LPV system with new LMI formulation, *IEE Proc. Control Theor. Appl.*, 152(6), 693–697, November 2005.

Section II

Modeling and Design

4 Foundations of Compositional Model-Based System Designs

Stavros Tripakis

CONTENTS

4.1 INTRODUCTION

4.1.1 COMPLEXITY OF DESIGNING CYBER-PHYSICAL SYSTEMS

Computers these days are no longer just stand-alone machines such as our desktops or laptops. They are parts of *systems* like airplanes, trains, and nuclear power plants, but also cars, pacemakers, "smart buildings," and the electric power grid. Such systems, called *cyber-physical systems* (CPS) integrate *cyber* parts (the computer or computers in the system), *physical* parts (e.g., the car chassis, engine, wheels, etc., or the human heart), and various other interfacing and connecting components (sensors, actuators, networks). The computer (both hardware and software) is truly the *brain* of these systems, as it performs the critical and often complex control logic functions.

Designing CPS is not a trivial task, for many reasons. One of the difficulties is that most of these systems are *safety critical*. The failure of a cell phone may not be considered catastrophic, but the failure of a car-braking system or a pacemaker can be a matter of life or death.

Another difficulty is that CPS are often very large. A single modern car may have dozens of computers on it, running millions of lines of code. The sheer size of such a piece of software makes it hard to understand, maintain, and have confidence in. This is partly due to the fundamental limitations of software analysis such as the undecidability of the halting problem, which make automation a challenge. But even small pieces of software can be difficult to reason about, even for humans. For example, consider the following simple program:

```
int n;
begin
  n := read_input();
  while (n > 1)
    if (n modulo 2 = 0) then
      n := n/2;
    else
      n := 3*n+1;
end
```

The program reads an integer number from the console and then performs a series of arithmetic transformations on the number, depending on whether it is even or odd. Does this program terminate for any input? This simple question represents the famous *Collatz conjecture*, an unsolved problem in mathematics. If simple programs like the one above can be difficult to understand, we can imagine how daunting the task can be for real programs of millions of lines of code, spread across many components and libraries, developed by many programmers over many years. Software is perhaps the most complex artifact that humans create today.

In addition to software, a CPS involves hardware (including sensors and actuators) and parts of the physical world. The foundations for software and computing hardware are to be found in logic and the theory of computation. On the other hand, the foundations for capturing the physical world can be found in analysis and continuous system theory, together with mechanics, physics, biology, and other disciplines depending on the application domain. Although these fields are all well established and can be used separately for the cyber and physical parts of a CPS, they often fail to capture the subtle interactions between the two. For this, a unified CPS theory is needed. Attempts toward such a unification include the theory of *hybrid systems* [17,34].

4.1.2 MODEL-BASED DESIGN

Today, we do not really know how to design CPS in a satisfactory way. On the one hand, we can build safety-critical systems such as airplanes and nuclear power plants that are relatively reliable, but very expensive. On the other hand, we can build cheap and relatively complex systems such as consumer electronics, but such systems are not very reliable (would we trust a cell phone to run a pacemaker?). What we do not know how to do is build systems that are both reliable and cost-effective.

Today, systems are often designed by *trial and error*. This approach is prevalent, especially in software design (think of the daily bug fixes applied to the operating system and other applications running on your laptop). Designing a CPS by trial and error is not an option, due to cost and safety concerns. An alternative approach that has been constantly gaining ground in the last several years is *model-based design* (MBD).

The main idea behind the MBD approach is to design a system primarily with the help of *models*, rather than prototypes. Building a model (sometimes called a *virtual system*) is much easier, cheaper, and safer, than building a (physical) prototype. Testing a model (say, by simulation) is also much easier than testing a prototype. As a result, the design process can be made much more efficient by building and testing models of the desired system, finding problems in the design (e.g., bugs or perfor-mance defects), correcting them (by fixing the model), and repeating the process until a satisfactory design is achieved. The success of the MBD approach relies critically on the existence of power-ful automated implementation techniques (e.g., *automated code generation*), which can synthesize system implementations automatically from the system models.

It is important that these automated implementation techniques are *semantics preserving*, mean-ing that they guarantee that the synthesized implementation preserves the properties of the original model. This is essential in order to avoid, as much as possible, having to retest the synthesized implementation: recall that avoiding to test prototypes is the goal of MBD in the first place. With semantics-preserving implementation techniques at hand, MBD offers designers a powerful method-ology, which guarantees that the synthesized systems are *correct by construction* (assuming the original model is "correct" in a certain sense).

In summary, the effort behind implementing an MBD methodology and related tools can be seen as an effort of building a *system compiler*. A system compiler can be seen as a system-level version of a standard compiler, as illustrated in Figure 4.1. A standard compiler, for, say, C or Java, provides three things: (1) an input language (C or Java), for programmers to express their programs; (2) a set of analysis tools (e.g., type checker, debugger) to help programmers get their programs correct; and (3) a back end for machine code generation and optimization. A system compiler should provide the same types of tools, but at the more general (and demanding) system level. System compilers do not exist as such today, but parts of them do exist, for example, within frameworks that focus on software and middleware generation [10], or others that focus on mechanical and control elements [41]. The MBD vision can be seen as meeting the challenge to turn these parts into a consistent whole.

4.1.3 COMPOSITIONALITY AND COMPOSITIONAL MODEL-BASED DESIGN

The need for compositionality comes from the fact that real-life systems are large and complex and therefore cannot be handled *monolithically*. In fact, most real-life systems are built *compositionally* in the sense that they are made of subsystems or components, connected in some way. Design theory and tools must exploit this natural decomposition, to make system design easier and scalable.

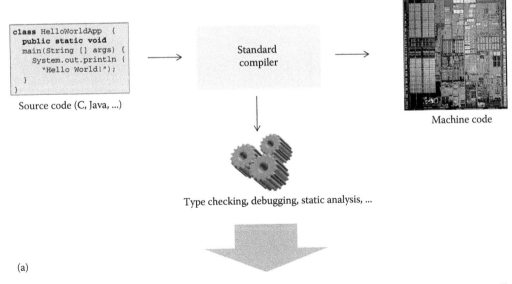

(a)

FIGURE 4.1 The vision of a model-based design: from a standard compiler (a) to a *system compiler*.
(Continued)

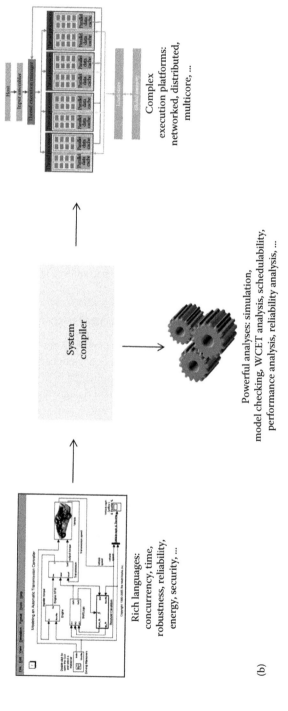

Rich languages: concurrency, time, robustness, reliability, energy, security, ...

System compiler

Powerful analyses: simulation, model checking, WCET analysis, schedulability, performance analysis, reliability analysis, ...

Complex execution platforms: networked, distributed, multicore, ...

(b)

FIGURE 4.1 (*Continued*)   The vision of a model-based design: from a standard compiler (b). (Simulink image copyright of MathWorks).

We will not attempt a precise definition of compositionality, as the term is used in many different situations, with various meanings. Generally speaking, however, a compositional framework should have some notion of a *component* or *module*, as well as the mechanisms for *composing* components. Beyond these basic requirements and depending on its specific goals, a framework may provide methods for compositional analysis (e.g., analyzing each component independently and then somehow combining the results to reason about the whole system), compositional implementation (e.g., compiling components independently from one another), and so on. A general principle is the principle of *divide and conquer*, as a means of mastering complexity.

In software engineering, compositionality exists primarily in the fact that large pieces of software are organized in components, modules, packages, and so on. Software engineering tools support this decomposition. For instance, compilers allow for the compiling of different source files separately and then link them to produce an executable. This allows for not to having to recompile the entire program when only a small part of it changes. It also enables software libraries that can be "black boxes" (object code is available but not source code) to protect the intellectual property of the authors. Such a state of the art has not yet been reached in systems engineering. This is the goal of *compositional MBD*.

Compositionality touches all aspects of MBD, from modeling, to analysis, to implementation and synthesis. This chapter focuses on work that the author has been involved in and by no means attempts to give an exhaustive account of the field of compositional MBD. In particular, this chapter does not cover compositional frameworks such as the real-time calculus [36], compositional real-time scheduling [33], Giotto [19], and Behavior, Interaction, Priority (BIP) [5]. Nevertheless, we will examine some fundamental problems on the topic of compositionality from all areas of MBD. As we shall see, a key concept that helps address all these problems is that of an *interface*. We will examine different types of interfaces, having different properties and used for different purposes. Despite the differences, an underlying principle that is common to all these types of interfaces is that of *abstraction*: interfaces keep the essential attributes of a component, especially those that concern how the component interacts with its environment, while hiding irrelevant internal details.

4.2 COMPOSITIONAL MODELING AND MODULAR SIMULATION

4.2.1 GRAPHICAL MODELING LANGUAGES WITH ACTORS AND HIERARCHY

The fact that larger systems are often assembled from smaller components is reflected also in state-of-the-art modeling languages and tools, such as Simulink from The MathWorks and Ptolemy from the University of California, Berkeley. Both Simulink and Ptolemy provide a graphical user interface (GUI), in which the designer can conveniently use to build models. Both tools provide a library of predefined model components and mechanisms for the user to assemble and connect these components to build larger models, in a hierarchical manner. In Simulink, these components are called *blocks*. In Ptolemy, they are called *actors*. Components formed hierarchically by composing simpler components are called *subsystems* in Simulink and *composite actors* in Ptolemy.

An illustration of the capability to model systems graphically and hierarchically is shown in Figure 4.2. The figure shows a hierarchical Ptolemy model. At the top level, the system is modeled as a state machine, each state of which corresponds to a discrete "operation mode." Each state of the machine includes (at the next lower level of hierarchy) a submodel, which in this example is a continuous-time model (differential equations) modeling the dynamics governing each mode. This example illustrates another capability of modern modeling and simulation tools. The fact that a single model can combine different modeling notations (e.g., state machines and differential equations) is often termed *heterogeneity*.

Hierarchy is a useful modeling feature, which can be seen as the primary modularization mechanism in graphical languages, playing a role similar in principle to the role played by *functions* and *procedures* in programming languages. Hierarchy allows for the structuring of a model into submodels and hides the details of these submodels from the higher level of abstraction. It allows for the building of complex models (with, say, 100s or 1000s or blocks) but still manages this complexity,

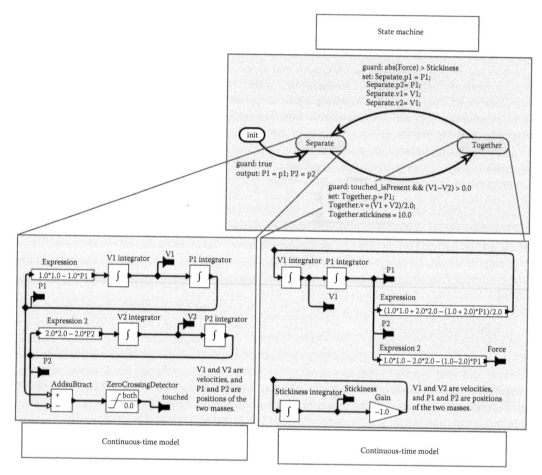

FIGURE 4.2 A hierarchical Ptolemy model. (By Edward A. Lee. University of California, Berkeley.)

thanks to the fact that each node in the hierarchy only contains a few blocks (e.g., 5–10). A hierarchically structured model is much more understandable than a "flat" model that contains all the atomic blocks at a single level.

4.2.2 MODULAR SIMULATION WITH STANDARD APIS

Modularity in tools like Simulink and Ptolemy goes beyond the GUI and the hierarchical modeling it provides. Modularity also extends to the way these tools perform simulation. In a nutshell, actors in these tools* are implemented as software components that implement a certain standard application programming interface (API). For instance, in Ptolemy, every actor is implemented as a Java class that `implements` (in the Java sense) a Java interface called `executable`. All actors in Ptolemy implement this interface, which includes methods such as `fire` to generate the outputs of an actor and `postfire` to update the state of an actor (see [14,40] and the documentation available from the Ptolemy website: http://ptolemy.eecs.berkeley.edu/). These two methods are similar to the output and transition functions, respectively, of a state machine such as a Mealy machine [21]. Ptolemy also includes other methods, used in particular to model the passage of time, which are not typically available in "untimed" models such as Mealy machines.

The Simulink architecture is also based on the same idea. The API of Simulink blocks contains functions to provide the `outputs` of a block and to `update` its state. In addition, Simulink is a

---

* We use the term *actors* generically, to cover Ptolemy actors, Simulink blocks, and other similar model components.

tool that targets simulating continuous-time systems, some blocks implement a method that returns the `derivatives` of continuous-state variables.

Although the APIs of Simulink and Ptolemy are slightly different, the principle is the same, namely, that of *modular simulation*. Having all actors implement a standard API allows for the decoupling of the two elements of a simulation tool: models on the one hand and simulation engines on the other. In essence, a model component (block or actor) does not need to "know" anything about the simulation engine that may be used to execute the component. In turn, a simulation engine does not need to know what is "inside" an actor. The actor appears to the simulation engine as a *black box*. All that is required is for this black box to implement a certain API.

This allows for a library of blocks/actors, which can be maintained independently from the simulation engines. New actors can be added to the library without having to modify the simulation code. Existing actors can also be modified (e.g., optimized) as long as the API does not change. On the other hand, the tool typically also has a library of more than one simulation engines, which can be applied interchangeably to the same model. Such an engine library might include, for instance, different numerical solvers (e.g., fixed step or variable step). But the engine library might also consist of simulation engines that implement quite different semantics for the same model, such as synchronous vs. asynchronous execution, as is the case with the Ptolemy tool (in Ptolemy, a simulation engine is called a *director*).

Modular simulation of this type raises a number of interesting questions, such as the following:

- What is the "right" API, and what exactly does "right" mean?
- What kind of modeling formalisms can be captured by a certain API, and which ones cannot?
- Given an API, how can we develop simulation engines with "good" properties, in terms of performance, simulation accuracy, etc.?

For instance, we might want to ask whether the Ptolemy API is "better" than the Simulink API and if so in what sense. We may also want to know to what extent a certain modeling formalism can be supported by a given API. For example, can the API support models such as *timed automata* [1] or *hybrid automata* [17,18]? That is, given a timed or hybrid automaton, can we create an actor that implements the given API and behaves as defined by this automaton? If a certain API cannot support a modeling formalism in its full generality, we may want to know whether it can support subsets of that formalism. For instance, even if an API cannot support hybrid automata, it may still be able to support the restricted model of timed automata.

For the most part, these are open research questions. We next describe some recent work toward answering some of these questions. The first deals with the functional mock-up interface (FMI), a recently developed standard API aimed at interconnecting heterogeneous modeling languages and tools. The second deals with the actor API of Ptolemy.

4.2.3 Cosimulation with the FMI Standard

The functional mock-up interface (FMI) is an international standard for model exchange and cosimulation (see https://fmi-standard.org/ and documentation available from that page). Its goal is to allow different modeling and simulation tools to interoperate, in particular, by exchanging models between tools as well as by being able to execute (or coexecute) models coming from different tools. For instance, one might wish to export a Simulink (sub)model, then import this in Ptolemy within a Ptolemy model, and then execute (simulate) the combined model.

FMI is based on the same principle of using standard APIs for modular simulation, discussed previously. FMI defines the API, or set of C functions, that a model component must implement in order to conform to the standard. Components implementing the FMI API are called *functional mock-up units* (FMUs). In fact, FMI defines two different APIs, one for model exchange and another for cosimulation. The main functions of the FMI API for cosimulation have been formalized in Broman et al. [6] and are the following:

$$\begin{aligned} \texttt{init} &: \mathbb{R}_{\geq 0} \to S \\ \texttt{set} &: S \times U \times \mathbb{V} \to S \\ \texttt{get} &: S \times Y \to \mathbb{V} \\ \texttt{doStep} &: S \times \mathbb{R}_{\geq 0} \to S \times \mathbb{R}_{\geq 0} \end{aligned}$$

where

$\mathbb{R}_{\geq 0}$ denotes the set of nonnegative reals
S denotes the set of states of a given FMU
U (respectively, Y) denotes the set of input (respectively, output) ports of that FMU
$\mathbb{V}$ denotes the set of all possible values that these input and output ports may take

The meaning of these functions is the following:

- `init` initializes the state of the FMU. It returns the state $s(t)$ for given time $t \in \mathbb{R}_{\geq 0}$.
- `set`(s, u, v) sets input port u to the given value v. That is, this corresponds to an assignment of the form $u := v$.
- `get`(s, y) returns the value of output port y in state s.
- `doStep`(s, δ) performs a simulation step, updating the state of the FMU from a current state $s(t)$ at time t to some state $s(t + \delta')$ at time $t + \delta'$, and returns the new state $s(t + \delta')$ and the achieved time step δ', where $0 \leq \delta' \leq \delta$.

It is beyond the scope of this chapter to cover all the subtleties of this formalization. But it is worth pointing out one interesting feature of FMI, namely, that FMUs may *accept* or *reject* a proposed time step. In particular, when `doStep`(s, δ) is called, the FMU may return $\delta' = \delta$, that is interpreted as the FMU having accepted the proposed time step δ, or it may return some $\delta' < \delta$. In the latter case, the FMU has managed to make only partial progress and managed to advance time only by δ' instead of δ. This possibility leads to a number of interesting problems, having to do with the possibility of FMUs becoming *nonsynchronized* during a simulation run, due to the fact that some may accept and others may reject a certain time step. Because of this, developing a simulation algorithm for FMI that behaves in a *determinate* manner (meaning that the simulation results depend only on the model and not on arbitrary factors such as the order in which independent FMUs in the model are executed) is a nontrivial problem. Some solutions to this problem are proposed in Broman et al. [6]. Follow-up work tackles the question of what kinds of modeling formalisms can be captured in FMI and how [7,37].

4.2.4 Modular Formal Semantics for Ptolemy

Ptolemy is also based on the modular simulation architecture. This architecture has recently been formalized in Tripakis et al. [40]. Specifically, Tripakis et al. [40] propose a formal model for actors (i.e., model components) as well as directors (i.e., simulation engines). According to this model, an actor is a tuple*

$$(I, O, S, s_0, F, P, D, T)$$

where

I is a set of inputs
O is a set of outputs
S is a set of states, s_0 being the initial state

---

* For the sake of readability, we use a simpler formulation than the one in Tripakis et al. [40].

F is the *fire* function, with signature $F : S \times I \to O$
P is the *postfire* function, with signature $P : S \times I \to S$
D is the *deadline* function, with signature $D : S \times I \to \mathbb{R}_{\geq 0} \cup \{\infty\}$
T is the *time-update* function, with signature $T : S \times I \times \mathbb{R}_{\geq 0} \to S$

F and P are essentially the same as the output and transition functions of a Mealy machine. These functions allow for the capture of untimed actors, but cannot capture actors whose behavior depends on time. The functions D and T fill this gap. D enables an actor to state what is the maximum time step it can accept (∞ if the actor can accept any time step). T is similar to the doStep function of FMI. However, there is a crucial difference: doStep includes the possibility of an FMU rejecting the given time step (by returning $\delta' < \delta$), whereas T has no such option (T only returns the updated state). The reason we do not need to worry about actors rejecting a time step in this formalization is thanks to the deadline function. The idea is that a director calls D on all actors in the model *before* it calls T. The deadline function of each actor i returns a time step δ_i. The director computes the smallest time step $\delta_{\min} = \min_i \delta_i$ and then calls $T(\_, \_, \delta_{\min})$ on each actor. Since $\delta_{\min} \leq \delta_i$ for all i and since actor i declared that it can accept δ_i, it is expected that it will also accept $\delta_{\min}$.

But what are directors, formally? In Tripakis et al. [40], directors are formalized as *composition operators*. A director takes as input a model, that is, a network of interconnected actors, and returns a new actor. The latter can be seen as an atomic actor and can therefore be further composed with any other actor. In fact, in this view, there is no difference between atomic and composite actors: They are all black boxes and the only thing we know about them is that they implement the F, P, D, T interface.

This formalization is elegant and can capture hierarchical models in a uniform way. However, the F, P, D, T interface is different from what is actually implemented in the Ptolemy software. The latter does not use the D and T functions, but instead uses a function called fireAt to capture timed models. This function is implemented by a director and called by an actor to indicate at which future time it wants to be fired. The motivation here is simulation efficiency. For actors that fire infrequently, it is inefficient for the director to have to consider these actors at every time step (e.g., by calling D). Instead, it is better for such actors to be able to directly indicate to the director their future firing times. Formalizing the fireAt interface, comparing it to the F, P, D, T interface, and proving a semantic equivalence between the two is an interesting, open problem.

4.3 MODULAR CODE GENERATION AND COMPOSITIONALITY OF HIERARCHICAL NOTATIONS

4.3.1 MODULAR CODE GENERATION

In Section 4.2, we saw how interfaces play a key role in modular simulation, by allowing for the characterization of a model component as a black box implementing a standard API. These black box components can be used for simulation, but they can also be used in other situations as well. For instance, they can be used as code embedded in a CPS (e.g., an automotive controller). This brings up the problem of *code generation*: how to generate executable code from a model (e.g., C or Java code that implements a state machine). Code generation is also the fundamental problem behind problems such as how to export an FMU that implements a given Simulink or Ptolemy model. Ideally, we would like this exporting mechanism to be fully automatic, so that users do not write FMUs "by hand." This is both for reasons of efficiency (presumably a "push-button" exporting mechanism is much faster than a manual design process) as well as of correctness. Indeed, once automatic code generation methods are available and assuming they are bug free, we can have a true MBD methodology as described earlier, where the model represents the source code ("golden reference"), while the FMU is simply a compilation artifact, similar to *object code*.

For many of the modeling formalisms used in CPS applications, the code generation problem is well understood. However, the current code generation techniques available in tools like Simulink

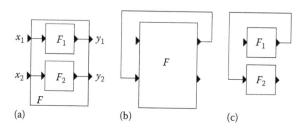

FIGURE 4.3 A hierarchical block diagram (a), a possible way to connect macro block F (b), and the same connection after flattening F (c).

are limited. Some of their main limitations come from the way they deal with hierarchy. Sometimes hierarchical models need to be *flattened*, meaning that all internal blocks of a subsystem, subsubsystem, and so on, all the way down to the basic blocks, need to be exposed at the upper level. In other cases, the model is not flattened, but information is lost that prevents being able to handle some models, which could be handled otherwise (for instance, if flattening was used).

To illustrate the problems that arise, consider, for example, the hierarchical block diagram shown on the left in Figure 4.3. Suppose that blocks F_1 and F_2 represent two stateless functions, say, with signatures

$$F_1 : A_1 \rightarrow B_1 \quad \text{and} \quad F_2 : A_2 \rightarrow B_2$$

where A_1, A_2, B_1, B_2 are some input and output domains. Block F is a *composite* block formed by encapsulating F_1 and F_2: essentially F represents the *parallel composition* of F_1 and F_2. It is tempting to view F as a new stateless function whose input and output domains are the Cartesian products of the input and output domains of F_1 and F_2, respectively:

$$F : (A_1 \times A_2) \rightarrow (B_1 \times B_2) \tag{4.1}$$

This is problematic, however, as we then lose the information that the output of F_1 does not depend on the input of F_2, and vice versa. Such information turns out to be critical when using F in certain contexts. For instance, if we connect F in a feedback configuration, as shown in the middle of Figure 4.3, we obtain a diagram with a cyclic dependency: the input of F depends on its output.

State-of-the-art tools like Simulink handle such situations in two possible ways. Either they flatten the model, as shown on the right in Figure 4.3. Flattening exposes the fact that there is no true cyclic dependency in this case. In fact, this model is equivalent to the cascade (serial) composition of F_1 followed by F_2. Once flattening reveals this, the model can be executed without problem. But flattening is not satisfactory from a *modularity* point of view. Indeed, imagine the case where we want to treat F as a black box (e.g., as a new component in a user-defined block library). Then we do not know the internals of F, and we cannot flatten it.

Another option is to reject the model in the middle, warning the user that the model has a cyclic dependency. Essentially, this corresponds to taking a conservative approach: since we know nothing about the internals of P, let us assume that *every output depends on every input*. Although this assumption is wrong in this example, the user is still forced to change the model, because the tool cannot handle it. This is essentially how Simulink handles so-called `CodeReuse Subsystems` (whereas it flattens normal `Subsystems`). Other hierarchical block-diagram-based tools, such as SCADE, behave similarly. SCADE forces the user to add so-called *unit-delay blocks* in every feedback loop at every level of the hierarchy, whether these unit delays are necessary or not. Unit delays change the timing behavior of a model and generally have unpredictable effects on the behavior of the system (e.g., the stability of a control algorithm). The designer should not have to add such blocks simply because the tool is not able to deal with the complexity of the model. An analogous situation in programming would amount to the programmer having to change his or her algorithm because the compiler cannot compile his or her (otherwise valid) program.

4.3.2   MONOLITHIC INTERFACES AND NONCOMPOSITIONALITY

The main problem with current code generation approaches is that they treat blocks in a *monolithic* way, meaning that they assume that all blocks must conform to the same interface. In the case of the example in Figure 4.3, the specific problem is that all the blocks are assumed to have a *single* output function that accepts all inputs and returns all outputs. A single output function is enough when every output depends on every input, but it is not rich enough to express more complex input–output dependencies. In particular, it is not able to capture the input-output (I–O) dependencies of composite block F.

But the problem is more fundamental and has to do with the fact that many hierarchical modeling formalisms are fundamentally *noncompositional*, in the sense that a composite model in the formalism does not admit an equivalent representation as an atomic model in the formalism. This applies to formalisms as basic as (stateless) functions, or finite-state machines (FSMs). For reasons explained earlier, in the case of functions, the parallel composition of two functions F_1 and F_2 is not a new function F as defined as (4.1). Similarly, the parallel composition of FSMs is generally not an FSM. For instance, consider the standard and widespread model types of FSMs, Moore or Mealy [21]. Such a machine can be represented as a tuple:

$$M = (I, O, S, s_0, \delta, \lambda)$$

where
   I is the set of input symbols
   O is the set of output symbols
   S is the set of states
   $s_0 \in S$ is the initial state
   $\delta : S \times I \rightarrow S$ is the transition function

λ is the output function, which has a signature of $\lambda : S \rightarrow O$ in the case of a Moore machine and $\lambda : S \times I \rightarrow O$ in the case of a Mealy machine. Now, consider two Mealy machines:

$$M_1 = (I_1, O_1, S_1, s_0^1, \delta_1, \lambda_1) \quad \text{and} \quad M_2 = (I_2, O_2, S_2, s_0^2, \delta_2, \lambda_2).$$

What is the parallel composition of M_1 and M_2? The standard reply is that it is another Mealy machine, say, M, often called the *parallel product* of M_1 and M_2, and defined as

$$M = (I_1 \times I_2, O_1 \times O_2, S_1 \times S_2, (s_0^1, s_0^2), \delta, \lambda)$$

where $\delta : (S_1 \times S_2) \times (I_1 \times I_2) \rightarrow (S_1 \times S_2)$ and $\lambda : (S_1 \times S_2) \times (I_1 \times I_2) \rightarrow (O_1 \times O_2)$ are defined as

$$\delta((s_1, s_2), (x_1, x_2)) = (\delta_1(s_1, x_1), \delta_2(s_2, x_2)) \quad \text{and} \quad \lambda((s_1, s_2), (x_1, x_2)) = (\lambda_1(s_1, x_1), \lambda_2(s_2, x_2)).$$

However, this definition suffers from the same problem as the one raised in Figure 4.3, namely, that it treats inputs and outputs monolithically and does not reveal the fact that some outputs do not depend on all the inputs. As a result, M cannot be used in some contexts where M_1 and M_2 could be used. This shows that Mealy machines are not closed under parallel composition.*

4.3.3   NONMONOLITHIC INTERFACES AND COMPOSITIONALITY

The solution to this problem is relatively straightforward: allow blocks to have more general, *nonmonolithic* interfaces. In our example, the interface of block F requires *two* functions instead of one. Similarly, in order for Mealy machines to be compositional, they should be allowed to have generally

---

* Moore machines, on the other hand, *are* closed under parallel composition.

more than one output functions. Then, the parallel composition of two Mealy machines M_1 and M_2 as defined earlier could be defined as the *generalized Mealy machine*:

$$M' = (\{S_1 \times S_2, (s_0^1, s_0^2), \delta : (I_1 \times I_2 \times S_1 \times S_2) \to (S_1 \times S_2), \{\lambda_1 : I_1 \to O_1, \lambda_2 : I_2 \to O_2\}).$$

where

λ_1, λ_2 are exactly the output functions of M_1 and M_2
δ is defined as in the case of machine M earlier

This new definition reveals that M' has two input ports and two output ports, where the first output depends only on the first input and the second output only on the second input. Therefore, λ_1 and λ_2 together represent accurately the I–O dependencies of the composite machine, and no information is lost during composition.

This generalized definition raises a number of interesting questions:

- Is this type of nonmonolithic interface powerful enough to ensure compositionality of arbitrary hierarchical block diagrams? If not, what is the right interface?
- Can nonmonolithic interfaces be generated automatically from arbitrary hierarchical block diagrams, and at what cost?
- How many output functions does a generalized machine require in the worst case?
- Is there a canonical representation with a minimal number of functions?

It is beyond the scope of this chapter to delve into these questions. The reader is referred to [24,26] where these and other related questions are studied in detail.

4.3.4 COMPOSITIONALITY IN OTHER HIERARCHICAL MODELS

In our discussion so far in this section, we have implicitly assumed a model of *synchronous* hierarchical block diagrams, where blocks are state machines composed in a synchronous manner. However, the problems of modular code generation and compositionality are not limited to the model of synchronous hierarchical block diagrams. The same problems arise in several other cases of hierarchical modeling languages. Unfortunately, compositionality is often not a primary concern in the design of such languages, and as a result, many of them are noncompositional.

One example is a simple extension of synchronous block diagrams, called *multirate* synchronous diagrams. Multirate diagrams extend synchronous diagrams by allowing different blocks to fire at different rates (periods). As shown in Lublinerman and Tripakis [25], representing the period information of a block with a single number is not enough. In particular, it does not allow for the composing of blocks with nonharmonic periods and obtain a period that accurately describes the behavior of the composite block.

For instance, consider again a hierarchical diagram like the one shown in Figure 4.3 and assume that block F_1 has period 3, while block F_2 has period 2. The question is, *What should be the period of the parent block F?"* Typically, the period of F is taken to be the greatest common divisor (gcd) of the periods of all children blocks of F, or in this case, $\gcd(3, 2) = 1$. Notice, however, that this is conservative, as it implies that F needs to fire at every synchronous step. In fact, F needs to fire only when at least one of its children needs to fire, that is, only at steps $0, 2, 3, 4, 6, 8, \cdots$, but not at steps $1, 5, \ldots$.

What the aforementioned example illustrates is that the composition of two periods (i.e., two numbers) is not always a period (i.e., it cannot always be represented as a number). An alternative, automata-based representation is proposed in Lublinerman and Tripakis [25] to overcome this problem. The idea is to represent a period by a *deterministic unary automaton*, that is, a deterministic automaton over a 1-letter alphabet. In our example, the two automata representing periods 3 and 2 are shown in Figure 4.4 (top). The automaton for 3 has three states, and the automaton for 2 has two

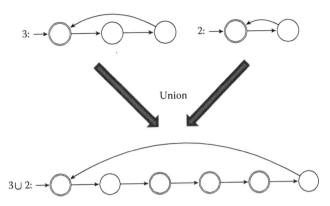

FIGURE 4.4   A compositional representation of periods using deterministic unary automata.

states. Since the automata are deterministic and unary, there is exactly one transition from every state to a unique next state.

Each automaton also has one or more accepting states (drawn with double circles in the figure), which model the points in time at which the corresponding block needs to fire. That is, the automaton for 3 represents the set of times $\{0, 3, 6, \ldots\}$, while the automaton for 2 represents the set $\{0, 2, 4, \ldots\}$. Then, the goal is to compute a new automaton that represents the *union* of these sets: this is exactly the set of times when the parent F needs to fire (i.e., when either of its children need to fire). The automaton corresponding to the union of 3 and 2 is shown at the bottom of Figure 4.4.

Another example of a noncompositional hierarchical modeling language is the hierarchical version of *synchronous dataflow* (SDF). SDF is a classic model for signal processing applications [22], and hierarchical SDF models can be constructed in tools such as Ptolemy. As shown in Tripakis et al. [38], SDF is noncompositional: an SDF graph cannot always be represented by an equivalent atomic SDF actor. For example, consider the hierarchical SDF graph shown in Figure 4.5. This graph contains two atomic SDF actors, A and B, contained within a composite actor P. Actor A consumes 1 token and produces 2 tokens every time it fires. Actor B consumes 3 tokens and produces 1 token every time it fires.

Now suppose we want to represent the composite actor P as an atomic SDF actor. The standard way to do this is to view P as an actor that consumes 3 tokens and produces 2 tokens every time it fires. This corresponds to firing A three times and B twice, which means the total number of tokens produced by A equals the total number of tokens consumed by B, namely 6.

Unfortunately, this solution does not always work, as it can result in deadlocks. This is illustrated in Figure 4.6. Suppose we connect P with another SDF actor C, as shown to the left of Figure 4.6. The black dots on the links represent initial tokens, that is, tokens that exist already at the beginning of execution of the system, before any actor fires. Thus, the queue from C to P has two initial tokens. The queue from P to C has a single initial token.

The model to the left of Figure 4.6 is deadlocked. Actor C cannot fire, since it requires at least two input tokens, but only one is available. P (viewed as an atomic actor as earlier) cannot fire either,

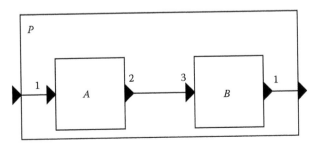

FIGURE 4.5   Example of a hierarchical SDF graph.

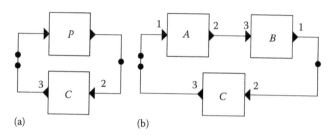

(a)   (b)

FIGURE 4.6   (a) Using the composite actor P of Figure 4.5 in an SDF graph with feedback and initial tokens. (b) The same graph after flattening P.

since it requires three or more input tokens, but only two are available. But this deadlock is the result of treating P as an atomic SDF actor. If, instead, P was flattened as shown to the right of Figure 4.6, we would obtain a deadlock-free model. Indeed, in that model, the firing sequence A, A, B, C, A, B can be repeated indefinitely.

This example shows that SDF is not closed under composition. The problem here is similar to the monolithic composition of Mealy machines. There, as we saw, a single output function is not rich enough to express complex input–output dependencies. In the case of SDF, a single firing rule is not rich enough to express an SDF graph. A compositional extension of SDF is proposed in Tripakis et al. [38], which can also serve as nonmonolithic interface for modular code generation purposes. We refer the reader to [38] for the details.

4.4   COMPONENT THEORIES WITH COMPATIBILITY AND REFINEMENT

In Section 4.2, we saw how interfaces are used for modular simulation and in Section 4.3 how they are used for modular code generation. In this section, we will discuss how interface formalisms can be used for compositional design and analysis. In particular, we will discuss *interface theories*, which are formalisms allowing to reason about the interactions of components among them and with their environment. To motivate these theories, let us consider an *incremental design* scenario.

4.4.1   INCREMENTAL DESIGN WITH INTERFACE THEORIES

Consider a control system, for instance, a *drive-by-wire* system such as the one depicted on the top of Figure 4.7. The system receives inputs from sensors attached to the steering wheel of a car and outputs control commands to actuators attached to the wheels of the car. In between, a set of components A, B, C do the processing and decision making. Suppose this system has been designed and verified, so that we know that it satisfies a certain set of requirements. For instance, we might know that it satisfies a certain *safety property*, for example, that the voltage v applied to an actuator is always within a safety range $[v_{min}, v_{max}]$, or a certain *performance property*, for example, that the end-to-end latency between input and output is at most 10 ms.

Now, suppose that, for some reason, the design team decides to replace one of the components in the system, namely, component B, with a new component, Z, as depicted in Figure 4.8. Since the

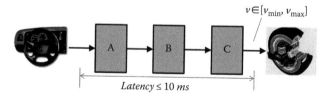

FIGURE 4.7   A steer-by-wire system satisfying safety and performance properties.

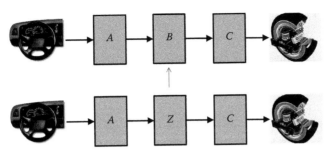

FIGURE 4.8   An incremental design scenario, where component *B* is replaced by another component *Z*.

system has changed, there is no guarantee that the new system preserves the properties of the original system, namely, safety and end-to-end latency. The question that the design team faces is this: Can we avoid having to reverify the new system from scratch? Verification is a difficult and time-consuming process, and the larger the system, the more expensive this process gets. It is therefore key to be able to avoid it whenever possible. In this case, the original system is known to be correct (in the sense of satisfying the safety and latency properties), and the new system is *almost* the same as the old one, with the exception of *Z* having substituted *B*. Would not it be great if we could have a way to ensure that correctness is preserved in the new system without having to verify the entire system from scratch? This is the problem of incremental design and analysis.

Interface theories provide a methodology to support this type of incremental design. In a nutshell, an interface theory provides the following:

1. A formal model to specify the interactions of a component with its environment (i.e., the *interface* of that component).
2. A notion of interface *compatibility* that captures when two or more components are compatible, that is, can be composed.
3. A set of composition operators, each taking two or more interfaces as input and producing a new interface as output, provided the given interfaces are compatible.
4. A notion of interface *refinement* that captures when a component can replace another one while preserving its properties.
5. A set of compositionality guarantees (theorems). There are typically two types of compositionality guarantees:
   a. *Preservation of refinement by composition*, which states that if *A* refines *A'* and *B* refines *B'*, then the composition of *A* and *B* refines the composition of *A'* and *B'*.
   b. *Preservation of properties by refinement*, which states that if *A* refines *A'*, then any property (in a certain class of properties) satisfied by *A'* is also satisfied by *A*.

To see how these can be used, consider again the steer-by-wire example. In order to prove that the new system (with *A, Z, C*) preserves the properties of the old system (with *A, B, C*), it suffices to prove that *Z* refines *B*. Then, we can reason as follows: First, employing preservation of refinement by composition, it follows that the new system (i.e., the composition of *A, Z, C*) refines the old system (i.e., the composition of *A, B, C*). Second, employing preservation of properties by refinement, it follows that the properties satisfied by the old system will also be satisfied by the new system (which refines the old). As we can see, using these compositionality theorems, we have managed to reduce a complex verification problem, namely, proving that the *A, Z, C* system satisfies the safety and latency properties, to a simpler verification problem, namely, proving that *Z* refines *B*. The latter is a simpler problem because it involves only two components, whereas the former problem involves the entire system (which in general might consist of a large number of components).

4.4.2   Interface Theories as Behavioral-Type Theories

The term *interface theories* originates from the work of de Alfaro and Henzinger [12,13] and has been followed up in numerous other works. There are different types of theories, for asynchronous or synchronous systems, untimed or timed, targeting correctness or performance properties, and so on. In addition, interface theories are related to the problem of *compositional verification* that seeks to simplify large verification problems by decomposing them into smaller problems. It is beyond the scope of this chapter to give an account of this vast body of research. Instead, we will focus on a specific characteristic of interface theories that distinguishes them from other compositional frameworks (e.g., [2,9,27]), namely, the notion of *compatibility* that interface theories typically provide.

Roughly speaking, two or more components are compatible if they can be composed. Otherwise, they are incompatible, meaning that their composition does not "make sense," or is *illegal*. These notions are similar to notions of *type correctness* found in programming languages and type theory. Consider, for example, a strictly typed language where we declare a function with an integer input and an integer output:

```
function f( x : int ) returns (y : int);
```

Then, if we attempt to call this function by passing a non integer value, say, as in

```
tmp := f(3.5);
```

the compiler will reject our program, notifying us of a type error. Type checking is a very useful tool that allows for the catching of some bugs at compile time. It can be seen as a form of *lightweight verification*. It is lightweight in the sense that the programmer does not need to specify any special property that his or her program must satisfy. The program must simply type check.

Interface theories can be seen as *behavioral-type theories*, or type theories for dynamical systems, where compatibility plays the role of type checking. Let us illustrate this using an example borrowed from [39].

Consider a component performing division, denoted *Div*, shown in Figure 4.9. *Div* has two input variables, x and y and one output variable, z. The interface specification of *Div* can be written as the logical formula:

$$y \neq 0 \wedge z = \frac{x}{y} \tag{4.2}$$

which states that the denominator, y, must be nonzero and that the output z is equal to x divided by y.

The use of the logical conjunction operator, $\wedge$, in Formula 4.2, is not obvious. Why not use implication instead?

$$y \neq 0 \rightarrow z = \frac{x}{y} \tag{4.3}$$

It is explicitly stated in (4.2) that the inputs where $y = 0$ are *illegal*. On the other hand, Formula 4.3 allows inputs where $y = 0$, since an implication $A \rightarrow B$ is trivially satisfied when A is false. In that case, the output y may take any value (and *must* take *some* value).

Using the conjunction form, Formula 4.2, allows for the incompatibility of components to be expressed during composition. For example, suppose we connect *Div* to a component that outputs

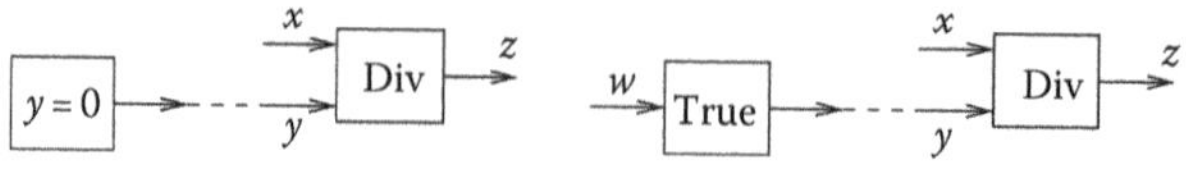

FIGURE 4.9   A division component *Div* connected to other components.

$y = 0$, as shown to the left of Figure 4.9. Clearly, this composition is illegal, since $y = 0$ contradicts the assumption $y \neq 0$ that *Div* makes. This contradiction is obvious when we take the conjunction of $y = 0$ with Formula 4.2. No immediate contradiction arises, however, when we take the conjunction of $y = 0$ with the implication form, Formula 4.3.

This example might suggest that composition is equivalent to conjunction. However, this is not always the case. To see this, suppose we connect *Div* to a component with interface *true*, as shown to the right of Figure 4.9. How are we to interpret the component with interface *true*? This interface is *nondeterministic*. It makes no guarantees on the output y of that component, which may take any value. Such an interface may appear meaningless; however, it may arise from the fact that we have no knowledge about the output function of a certain component.

Should the composition to the right of Figure 4.9 be considered valid? In other words, is *true* compatible with *Div*, where *Div* is specified by Formula 4.2? Taking the conjunction of *true* and Formula 4.2 yields Formula 4.2, which indicates no incompatibility. But this is misleading. The reason is that, since *true* provides no guarantees, $y = 0$ is *one possible* output for this component. But $y = 0$ is not a legal input for *Div*. As a result and because of the fact that we have no way of ensuring that the *true* component will not output $y = 0$, we must declare these two components incompatible.

Additional evidence to this incompatibility is provided when we consider *refinement*. Refinement is a binary relation between interface specifications, aimed at capturing substitutability, as explained in Section 4.4.1. Specifically, consider two components with interface specifications ϕ_1 and ϕ_2. If ϕ_1 refines ϕ_2, denoted $\phi_1 \sqsubseteq \phi_2$, then we should be able to replace ϕ_2 by ϕ_1 in any context. This is similar to the notion of *subtyping* in programming languages. For example, `Integer` is a subtype of `Real`, which implies that a value of type `Integer` can be used wherever a value of type `Real` can be used.

Returning to our example, it is reasonable to expect $y = 0 \sqsubseteq true$ to hold, that is, $y = 0$ to be a valid refinement of *true*. This is because $y = 0$ provides more information (is *more deterministic*) than *true*. Indeed, *true* makes no guarantees about its output, so any output value is *a priori* to be expected. $y = 0$ simply chooses one of the possible outputs.

Then, assuming that we accept that $y = 0 \sqsubseteq true$ holds, we are forced to declare *true* and *Div* incompatible. Indeed, in accordance with the *preservation of properties by refinement* theorem, we expect that if ϕ_2 is compatible with ϕ and $\phi_1 \sqsubseteq \phi_2$, then ϕ_1 is also compatible with ϕ. Therefore, if we declare *true* compatible with *Div*, by the fact that $y = 0 \sqsubseteq true$, we must also declare $y = 0$ compatible with *Div*. Since the latter is false, we must conclude that *true* and *Div* are also incompatible.

Now that we know that *true* and *Div* are incompatible, how can we catch this incompatibility formally? The answer is to define the composition of two interface specifications ϕ_1 and ϕ_2 not just as the conjunction $\phi_1 \wedge \phi_2$, but as the formula

$$\phi_1 \wedge \phi_2 \wedge \forall y : (\phi_1 \rightarrow \exists z : \phi_2) \tag{4.4}$$

Here, we assumed that y is the output of ϕ_1 connected to the input of ϕ_2 with the same name and that z is the output of ϕ_2. For simplicity, assume that these are the only inputs and outputs of ϕ_2. The last conjunct in Formula 4.4, namely, $\forall y : (\phi_1 \rightarrow \exists z : \phi_2)$, states that *every* possible output value of ϕ_1 must be a legal input value of ϕ_2. Applying this conjunct to our example gives $\forall y : (true \rightarrow \exists z : y \neq 0 \wedge z = \frac{x}{y})$. The latter formula is equivalent to $\forall y : \exists z : y \neq 0 \wedge z = \frac{x}{y}$, which in turn is equivalent to $\forall y : (y \neq 0 \wedge \exists z : z = \frac{x}{y})$, which is *false*, because $\forall y : y \neq 0$ is *false*. As we can see, we reached a contradiction similar to the contradiction we reach when conjoining $y = 0$ with *Div*. This new contradiction can be interpreted as incompatibility of *true* and *Div*.

In addition to detecting incompatibility, the aforementioned definition allows for deriving the *weakest input conditions*, similar to the weakest preconditions in program semantics [3,14]. For example, suppose that, instead of *true*, we connect *Div* to a component with input w, output y, and interface specification $y \geq w$. Then, $\forall y : (\phi_1 \rightarrow \exists z : \phi_2)$ reduces to $\forall y : y \geq w \rightarrow y \neq 0$, which in turn reduces to $w > 0$. The condition $w > 0$ is the weakest condition on input w that ensures compatibility between $y \geq w$ and *Div*. This condition can be seen as a new input condition derived for the composite interface, characterizing the composition of the two components.

In practice, how can we check compatibility or refinement? Or how can we derive the weakest input conditions like the aforementioned? Ideally, we want to have "push-button," fully automated, efficient tools to solve these problems, in accordance with our system compiler vision. To what extent we can achieve such automation depends on the specific mathematical formalism we use for interface specifications. For instance, if we use simple propositional logic (Boolean expressions where all variables are of type Boolean), then we need in principle only an SAT (satisfiability) solver or a binary decision diagram package. Note that quantifier elimination in case of purely Boolean formulas can be reduced to disjunction. If we use a richer logic, for instance, one that contains integer or real variables and arithmetic operators, more powerful tools are needed, such as satisfiability modulo theory solvers that can also handle quantified formulas. Finally, if we use *temporal logic* or automata-based formalisms, which are typically used for formal specification and verification [4,28], *model checkers* and similar types of tools may be needed. Providing satisfactory solutions to algorithmic problems that arise in interface theories that deal with such advanced models is a topic of current research (e.g., see Preoteasa and Tripakis [31]).

Another research topic is extending the types of properties that can be handled by interface theories, in particular, with respect to the *preservation of properties by refinement* theorem. Many existing theories, for instance, *interface automata* [12], can only handle *safety* properties, which, informally speaking, state that something bad never happens. Safety is a very important class of properties, but is not sufficient to specify systems completely: to trivialize, a system that does nothing is safe, because it never does anything wrong! In order to specify that the system must actually *do* something, we need another type of properties, called *liveness* [4,28]. Informally speaking, liveness properties state that something good eventually happens.

Liveness is notoriously more difficult to handle than fairness, especially compositionally. In the case of interface theories, some recent work that appears promising is the one presented in Preoteasa and Tripakis [31]. This work is inspired by the classic compositional framework of *refinement calculus* [3], which applies to sequential programs. Preoteasa et al. [31] extends this framework to the setting of *reactive systems*, that is, discrete dynamical systems described by formalisms such as automata, state machines, or temporal logic. Using this framework, we can do similar tasks like the ones we described in the examples earlier, but using more powerful formalisms. For instance, we can use *linear temporal logic* (LTL) [30], which can express liveness, as well as safety, properties.

Let us illustrate this with an example, borrowed from [31]. Consider the composition of two components A and B shown in Figure 4.10. The interfaces of both components are specified in LTL. The LTL formula $\square (x \geq 0)$ states that x is *always* (i.e., at every execution step) nonnegative. This is a guarantee that A makes about its output. The LTL formula $\square \lozenge (x = 1)$ states that x must be equal to 1 *infinitely often* (i.e., at infinitely many execution steps). This is a requirement that B imposes on its input. The two components are incompatible, because the guarantee of A is not strong enough to meet the input requirement of B. For example, if A always outputs $x = 0$, then the formula $\square (x \geq 0)$ is satisfied, but the formula $\square \lozenge (x = 1)$ is violated.

Another example, also borrowed from [31], shows how this framework can be used to derive the weakest input conditions. This is illustrated in Figure 4.11. The LTL specification of component C is a so-called *request–response* property, which states that whenever the input y is equal to 1 (the request), the output x will also eventually be equal to 1 (the response). In other words, every request will eventually be followed by a response. C is compatible with B, which requires x to be infinitely often 1, provided the external environment can guarantee that y will be 1 infinitely often. This newly

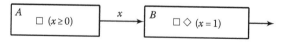

FIGURE 4.10   Two incompatible systems specified in temporal logic.

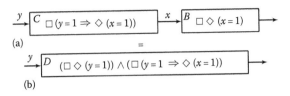

FIGURE 4.11 Two compatible systems (a) and their composition (b).

derived input requirement on y is explicitly part of the interface specification of composite block D, which represents the composition of C and B.

4.5 MULTIVIEW MODELING

Thus far, we have focused on rather traditional notions of compositionality, which deal with compositions of interfaces, components, modules, subsystems, and so on. These entities have in common the fact that they are clearly *separated*. They *interact*, but they do not *overlap*. In this section, we discuss a different form of compositionality, among *overlapping* entities called *views*.

The motivation for this study comes from modern systems engineering practice. As systems become larger and more complex, they involve a large number of designers and stakeholders, often organized in several teams. These teams often come from different disciplines, have different goals and concerns, and use different modeling languages and tools to capture and analyze their design ideas [8,29]. For instance, the design of an automotive control system may involve a control engineering team, concerned with controller performance and stability, and a software engineering team, concerned with software execution times and schedulability. Some concerns, for example, digital arithmetic precision, may be shared by both teams.

In such a scenario, each team has its own viewpoint about the system to be designed and expresses this viewpoint by constructing a system model, using its language and tool of choice. We call this model a *view*. Since there are many teams, there are many models, that is, many views. Moreover, all these views refer to the same system, but from different perspectives. This raises the problem of ensuring *view consistency*: if different views of the system are captured by different models and these models have some degree of overlap, how can we guarantee that the models are consistent, that is, that they do not contradict each other? View consistency is the key problem in *multi view modeling*.

To provide some intuition of what could go wrong when multiple views are involved, consider the example shown in Figure 4.12. Figure 4.12a depicts a collection of boxes forming a 3D structure. Each of the three 2D projections of this structure is a view. If we know the 3D structure, we can generate the 2D views. But what if we start from three arbitrary 2D views and we want to build a 3D structure that "conforms" to them, that is, that could generate those views? In that case, there is the possibility that the 2D views are inconsistent, namely, that there is no 3D structure that could generate them. In the static world of simple geometric objects, it is easy to imagine a tool that performs simple calculations to check whether three given 2D views are consistent. We would like to build similar tools for models of dynamical systems.

As mentioned, views are models, and in a dynamical system context, they are models of dynamical systems. Given a collection of such models, we would like to check whether they are consistent with each other. But how can we formulate consistency as a mathematical problem? This question is studied in Reineke and Tripakis [32]. Let us next summarize the elements of the approach taken in that work.

First, consider a set $\mathcal{U}$, called the *system domain*, representing the set of all possible system models. Next, consider a collection of *view domains* $\mathcal{D}_1, \mathcal{D}_2, ..., \mathcal{D}_n$, each representing a set of view models. Finally, for each view domain $\mathcal{D}_i$, let $a_i : \mathcal{U} \rightarrow \mathcal{D}_i$ be an *abstraction function*, which maps a given system $S \in \mathcal{U}$, to a view $a_i(S) \in \mathcal{D}_i$. In the case of the geometric example previously, $\mathcal{U}$ could be the set of all 3D structures, and we could have three identical view domains, $\mathcal{D}_1 = \mathcal{D}_2 = \mathcal{D}_3$, each being

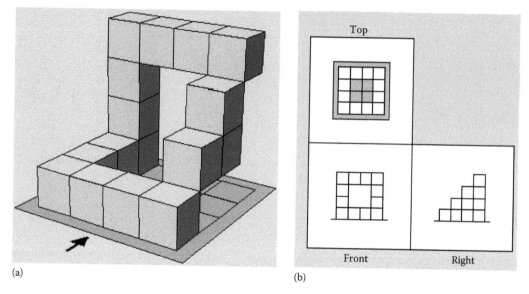

(a)   (b)

FIGURE 4.12 A 3D structure (a) and three views of it (b)—image produced using this tool: http://www.fi.uu.nl/toepassingen/02015/toepassing_wisweb.en.html.

the set of all 2D structures. Then, a_1, a_2, a_3 could be the three 3D-to-2D projections representing the views from the top, front, and side, respectively.

Given this formal framework, we can define view consistency in the following way: Consider a set of n views, $V_1 \in \mathcal{D}_1, V_2 \in \mathcal{D}_2, \ldots, V_n \in \mathcal{D}_n$. We will say that $V_1, V_2, \ldots, V_n$ are *consistent* iff there exists a system $S \in \mathcal{U}$ such that for all $i = 1, \ldots, n$, we have $a_i(S) = V_i$. What this definition states is that a set of models are consistent with each other iff there exists a system whose views in each domain are exactly these models. In other words, a set of views are inconsistent, iff there exists no system that could "generate" these views.

4.5.1 VIEW CONSISTENCY BETWEEN SCENARIOS AND STATE MACHINES

Let us illustrate the definition of view consistency with an example. Suppose there are two teams designing a client–server application. The application relies on a simple message-passing protocol between a *client* process and a *server* process. One team designs a set of *scenarios* of protocol behaviors, in the form of *message sequence charts* (MSCs)* like the one shown in Figure 4.13a. MSCs are an intuitive notation often used in industrial documents to express and communicate design ideas [35]. In the case of Figure 4.13, the MSC describes a scenario where the client process sends a request message to the server process, which responds by sending a response message back to the client. However, the response message is lost and never reaches the client. An external time-out message is then received by the client, which resends the request.

Suppose that the other team designs a pair of state machines, one for the client and one for the server, as shown in Figure 4.13b. This type of state machine model can be used for verification purposes, for example, to check that the system does not deadlock [4]. In the case of Figure 4.13, ! denotes sending and ? denotes receiving a message, so the transition labeled request! in the client machine corresponds to the client sending a request message to the server. The transition labeled request? in the server machine corresponds to the server receiving a request message from the client. For simplicity, we assume zero-delay communication here, so transitions with complementary labels a! and a? *synchronize* and result in a single transition labeled a when the machines are composed.

---

* Message sequence charts (MSCs) are a family of notations, including the ITU MSC standard [20], related notations such as *sequence diagrams* in UML, and several formal models. A survey can be found in Harel and Thiagarajan [16].

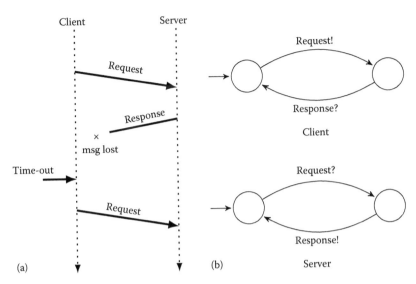

FIGURE 4.13   (a) Scenario for a simple client/server application described as a message sequence chart. (b) State machines for the client (top) and server (bottom).

The two notations shown in Figure 4.13, the MSC, on the one hand, and the two state machines, on the other hand, constitute two different views of the client–server system. We would like to ensure that these two views are consistent with each other. Let us see how we can use the formal framework mentioned earlier to check whether this is indeed the case.

We start by defining the system domain $\mathcal{U}$. This will be the set of all real client–server protocols we are interested in. Each of these protocols might include real code (e.g., in Java or C) for the two processes, together with library functions or other mechanisms for interprocess communication via message passing.

The state machine view abstracts this complex piece of software into simple FSMs like the ones in Figure 4.13. Only the events related to message transmissions and receptions, and other interesting events such as time-outs, are kept. Denoting this abstraction function by a, for a given system $S \in \mathcal{U}$, and a given state machine view V_m, we want $V_m = a(S)$. That is, we require that the abstracted set of all possible behaviors of the system, $a(S)$, exactly matches the set of behaviors of the combined state machine model, V_m.

For the MSC view, the abstraction function a can be the same, since the MSC also refers only to the same abstract events as the state machines. However, if S is a system and V_c is an MSC, we require $V_c \subseteq a(S)$, instead of $V_c = a(S)$. The reason for this is that V_c contains only *some* of the possible scenarios and not *all* possible scenarios (typically there is an infinite number of them, since the operation of the client–server application is not supposed to ever stop). Therefore, V_c generally captures only a *subset* of the abstracted set of all possible behaviors of the system.

Let V_c and V_m be the MSC view and the state machine view of Figure 4.13, respectively. We claim that these two views are inconsistent, that is, there is no system S such as both $V_m = a(S)$ and $V_c \subseteq a(S)$ hold. To see this, observe that such a system S must contain at least one behavior that, when abstracted, results in the following sequence of events:

$$\sigma = \text{request, response, time-out, request}$$

This is because σ is generated by the MSC of Figure 4.13, that is, σ is contained in V_c and therefore must also be contained in $a(S)$. On the other hand, $a(S)$ is required to be equal to V_m, and thus, we must have $\sigma \in V_m$. But V_m contains only the prefixes of the behavior

$$\text{request, response, request, response, } \cdots$$

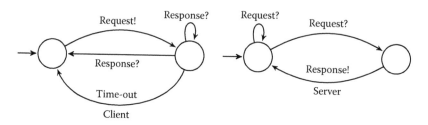

FIGURE 4.14   Modified state machines for the client–server example to account for message loss.

which is the only behavior arising from the interaction of the client and server automata of Figure 4.13. Therefore, $\sigma \notin V_m$. As we reached a contradiction, there can be no witness system S, and therefore, we conclude that the two views are inconsistent.

The problem here can be seen to lie within the state machine view, which does not account for the possibility of message loss. On the other hand, the MSC scenario explicitly specifies that a retransmission must occur after a message loss and the resulting time-out. A way to fix the inconsistency is to modify the state machine view to account for message loss, as shown in Figure 4.14. As can be seen, the new state machines can generate many more behaviors, including the previous behavior σ previously, which now ensures consistency of the two views.

4.5.2   VIEW CONSISTENCY IN AN EMBEDDED CONTROL SOFTWARE APPLICATION

The client–server example illustrates consistency checking in multiview modeling in a simple way. However, this example is relatively limited as it involves only discrete systems in both cases, with the same semantic domain (sequences of events). As a result, this example does not illustrate the full power of the multiview modeling framework presented earlier. We now present a more elaborate example of *heterogeneous* views, coming from different semantic domains.

Consider an embedded control software application involving three design teams and corresponding views:

- *Algorithm view*: This view concerns the high-level control algorithm.
- *Software view*: This view concerns the software implementation of the control algorithm.
- *Worst-case execution time (WCET) analysis view*: This view concerns estimation of the WCET of each software task of the software view.

The algorithm view could be captured in a block diagram model such as the one shown in Figure 4.15. The model shows two blocks, A and B, annotated by their corresponding periods, of 10 and 20 ms, respectively. The intuitive meaning is that A and B execute, periodically, reading inputs and producing outputs (and potentially also updating their internal state). It is not important for this discussion what exactly A and B do. What matters is the semantics of their interaction. In particular, according to the standard synchronous semantics of tools such as Simulink from MathWorks, the behavior of this diagram is as follows:

- A produces a sequence of values, $a_0, a_1, a_2, \cdots$, at corresponding times $0, 10, 20, \ldots$.
- B consumes the value a_0 and produces a value b_0 at time 0. Then B consumes the value a_2 and produces a value b_2 at time 20, and so on.

FIGURE 4.15   Block diagram model of the control algorithm view.

It is important to realize that B executes only once for every two executions of A and therefore "misses" the outputs $a_1, a_3, \cdots$ of A. Also, each output of B depends on the corresponding output of A, that is, b_0 depends on a_0 and b_2 depends on a_2. Note that B may have internal state that is also modified at every execution of B. The output will generally depend on the internal state, which in turn generally depends on the history of inputs. Therefore, b_2 will generally depend not only on a_2 but also on a_0, b_4 will depend on a_4, a_2, a_0, and so on.

Let us now turn to the software view. We assume that this application runs on a single-processor execution platform, equipped with some multitasking operating system. Typically, multiperiodic applications such as this one are implemented using multiple tasks, running at different periods. In the case of our example, there will be one software task implementing A and running with a period of 10 ms and another task implementing B and running at 20 ms. Note that an alternative implementation would use a single task running at 10 ms, with an internal modulo 2 counter and if-then statement, in order to execute A every time, but B only every two times. The reason for using two tasks instead of a single task is that the WCET of a single task might be larger than its period, that is, larger than 10 ms. This would violate the *deadline* of a task that is typically equal to its period. Using a multitasking implementation and a real-time scheduling algorithm with preemption, the problem can be avoided. We refer the reader to Caspi et al. [11] for details.

Once the design team decides on a multitasking implementation with two tasks, task T_A for A and task T_B for B, it proceeds with designing the scheduling and intertask communication mechanism. Typically, a *rate monotonic scheduling* scheme is used, where the task with the highest frequency is assigned the highest priority [23]. In our example, T_A will have a higher priority than T_B. Regarding intertask communication, we will assume a simple mechanism here, that consists of a buffer that can hold a single value, written by T_A and read by T_B. The buffer is protected using some mechanism that ensures atomicity of reads and writes. This ensures, for instance, that T_A cannot preempt T_B while T_B is in the process of reading from the buffer, but must wait until T_B is done reading.

Note that, beyond the atomicity property mentioned above, this intertask communication implementation does not ensure that the order of values read by T_B is the same as the order prescribed by the algorithmic view. In fact, this is precisely a type of inconsistency problems we may expect a multiview modeling framework to be able to reveal. We return to this point later.

In order to ensure that the deadlines of each task are met, that is, that every task instance finishes execution before the next task instance becomes ready to execute, that is, within its period, we can apply techniques from real-time scheduling theory (e.g., [23]). In order to apply such tools, the WCET of each task needs to be known. Obviously, if the WCET is larger than the period of the task, we cannot guarantee that the task will meet its deadline. But even if it is smaller, interference with higher-priority tasks might result in a task missing its deadline. For our purpose, we will assume that the WCET analysis view consists simply of two numbers, W_A and W_B, representing the WCETs of T_A and T_B, respectively.

We are now ready to capture the aforementioned example formally in the previous formal multiview modeling framework. We need to define the system domain, view domains, and abstraction functions. The system domain will consist of sets of complete behaviors of the embedded control system. These behaviors capture everything that happens during a system execution, including the beginning and end of execution of a task, its preemption by another task, the times at which these events occur, and which values are written and read, and when.

The abstraction functions map these complete behaviors into different kinds of *abstract* behaviors, depending on the view domain. Specifically, for the algorithm view domain, the abstract behaviors are simply sequences of writes and reads, such as

$$A.w(a_0), B.r(a_0), A.w(a_1), A.w(a_2), B.r(a_2), \ldots \qquad (4.5)$$

which can be read as "A writes value a_0, B reads a_0, B writes b_0, and so on.

In the software view domain, abstract behaviors consist of tasks becoming ready, starting, ending, and being preempted events and their corresponding times, such as the following:

$$T_A(r, t_0), T_A(s, t_1), T_B(r, t_2), T_A(e, t_3), T_B(s, t_4), \ldots$$

which can be read as "T_A became ready at time t_0, started execution at time t_1, and ended execution at time t_3. In the meantime, T_B became ready at t_2 but since it has lower priority than T_A, only started execution at time t_4, and so on.

The elements of the WCET analysis view domain are simply numbers, denoting WCETs. These numbers are computed from complete system executions by taking the *maximum* (i.e., worst-case) execution time encountered for a given task, over the entire set of system executions.

Given the aforementioned, suppose the design teams come up with the following three views:

1. Algorithm view: The block diagram model of Figure 4.15
2. Software view: As described earlier
3. WCET analysis view: $W_A = 5$ and $W_B = 8$

Then, we claim that these three views are inconsistent. That is, according to our definition of view consistency, there is no system that, when abstracted using each of the three abstraction functions previously, yields these three views. To prove this claim, let us suppose that such a *witness* system exists. We will derive a contradiction.

Assuming that a witness system exists, we first observe that, in order for the witness system to be consistent with the WCET analysis view, it must be able to generate behaviors where the execution time of task T_A is 5 and that of T_B is 8. One such behavior is shown in Figure 4.16. In the figure, we use A and B instead of T_A and T_B, for simplicity. Tasks A and B become ready to execute at times 0, 20, 40, and so on. In addition, A becomes ready also at times 10, 30, and so on. This corresponds to A and B being triggered periodically with periods 10 and 20 ms, respectively. According to the software view, A has higher priority than B; therefore, A must execute first. A executes and finishes at time 5. Assuming there are no other tasks in the system, B is then ready to execute, and it does so for 5 time units, until it is preempted by A at time 10, when A is triggered again. Assuming A takes again 5 time units to complete its execution, B resumes at time 15, executes for the remaining $3 = 8 - 5$ time units, and completes.

We claim that the behavior shown in Figure 4.16 should be included in the set of behaviors of the witness system. Note that there are other behaviors that result in no preemption, yet still give the same WCETs for A and B, for instance, a behavior where A first executes for 5 time units, then B for 3 time units, then A for 1 time unit, then B for 8 time units, and so on. However, we may require that the sets of system behaviors are *closed* under arbitrary permutations of execution times, so that we cannot avoid the behavior of Figure 4.16. This is realistic, as in a real system, one cannot generally control the execution time that a certain task exhibits.

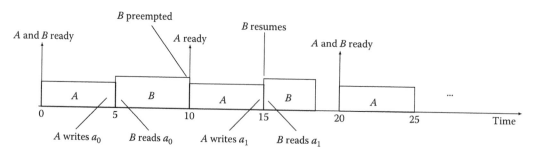

FIGURE 4.16   Beginning of system execution where the WCET of task T_A is 5 and the WCET of task T_B is 8.

Given the aforementioned, consider the abstraction of the behavior of Figure 4.16 onto the algorithm view. This abstraction results in the following sequence of reads and writes:

$$A.w(a_0), B.r(a_0), A.w(a_1), B.r(a_1), \dots \tag{4.6}$$

where B first reads the correct value, a_0, but, after it is preempted, reads the wrong value a_1. This is because, in general, we cannot forbid B to read from its input buffer at any point during its execution. In particular, B may not know that the value a_0 produced by A during its first execution has been overwritten by the value a_1 produced by A during the second execution. In general, B may not save the value a_0 internally (for large data, this may even be prohibitive) and attempt to read it again from its input buffer, using the latter as the only storage.

Sequence 4.6 is inconsistent with the algorithm view, which can only produce a sequence of type as in Sequence 4.5. This indicates that the three views are inconsistent.

ACKNOWLEDGMENTS

The author thanks Martin Törngren, Jan Reineke, and Christos Stergiou for the useful discussions and feedback.

This work was partially supported by the Academy of Finland and by the National Science Foundation (NSF) via projects COSMOI: Compositional System Modeling with Interfaces and ExCAPE: Expeditions in Computer Augmented Program Engineering. This work was also partially supported by IBM and the United Technologies Corporation (UTC) via the iCyPhy Consortium at UC, Berkeley.

REFERENCES

1. R. Alur and D. Dill, A theory of timed automata, *Theoretical Computer Science*, 126, 183–235, 1994.
2. R. Alur and T. Henzinger, Reactive modules, *Formal Methods in System Design*, 15, 7–48, 1999.
3. R.-J. Back and J. Wright, *Refinement Calculus*, Springer, 1998.
4. C. Baier and J.-P. Katoen, *Principles of Model Checking*, MIT Press, 2008.
5. A. Basu, M. Bozga, and J. Sifakis, Modeling heterogeneous real-time components in BIP, in *Software Engineering and Formal Methods*, September 2006, pp. 3–12.
6. D. Broman, C. Brooks, L. Greenberg, E. A. Lee, S. Tripakis, M. Wetter, and M. Masin, Determinate composition of FMUs for co-simulation, in *Proceedings of the 13th ACM & IEEE International Conference on Embedded Software (EMSOFT'13)*, 2013.
7. D. Broman, L. Greenberg, E. A. Lee, M. Masin, S. Tripakis, and M. Wetter, Requirements for hybrid cosimulation, EECS Department, University of California, Berkeley, Technical Report UCB/EECS-2014-157, August 2014.
8. D. Broman, E. Lee, S. Tripakis, and M. Törngren, Viewpoints, formalisms, languages, and tools for cyberphysical systems, in *Sixth International Workshop on Multi-Paradigm Modeling (MPM'12)*, 2012.
9. M. Broy and K. Stølen, *Specification and Development of Interactive Systems: Focus on Streams, Interfaces, and Refinement*, Springer, 2001.
10. P. Caspi, A. Curic, A. Maignan, C. Sofronis, S. Tripakis, and P. Niebert, From Simulink to SCADE/Lustre to TTA: A layered approach for distributed embedded applications, in *Proceedings of the 2003 ACM SIGPLAN Conference on Languages, Compilers, and Tools for Embedded Systems (LCTES'03)*. ACM, pp. 153—162, June 2003.
11. P. Caspi, N. Scaife, C. Sofronis, and S. Tripakis, Semantics-preserving multitask implementation of synchronous programs, *ACM Transactions on Embedded Computing Systems (TECS)*, 7(2), pp. 1–40, February 2008.
12. L. de Alfaro and T. Henzinger, Interface automata, in *Foundations of Software Engineering (FSE)*. ACM Press, 2001.
13. L. de Alfaro and T. Henzinger, Interface theories for component-based design, in *EMSOFT'01*, Springer, LNCS 2211, 2001.
14. E. W. Dijkstra, *A Discipline of Programming*. Prentice Hall, Upper Saddle River, NJ, 1976.
15. J. Eker, J. Janneck, E. Lee, J. Liu, X. Liu, J. Ludvig, S. Neuendorffer, S. Sachs, and Y. Xiong, Taming heterogeneity—The Ptolemy approach, *Proceedings of the IEEE*, 91(1), 127–144, January 2003.

16. D. Harel and P. Thiagarajan, Message sequence charts, in *UML for Real*, L. Lavagno, G. Martin, and B. Selic, Eds., Springer, pp. 77–105, 2003.

17. T. A. Henzinger, The theory of hybrid automata, in *11th IEEE Symposium on Logic in Computer Science, ser. LICS '96*, 1996.

18. T. A. Henzinger, P. W. Kopke, A. Puri, and P. Varaiya, What's decidable about hybrid automata? *Journal of Computer and System Sciences*, 373–382, 1995.

19. T. A. Henzinger, C. M. Kirsch, and S. Matic, Composable code generation for distributed Giotto, pp. 21–30, 2005.

20. *ITU-R Recommendation Z.120, Message Sequence Charts (MSC '96)*, ITU Telecommunication Standardization Sector, May 1996.

21. Z. Kohavi, *Switching and Finite Automata Theory*, 2nd edn. McGraw-Hill, 1978.

22. E. Lee and D. Messerschmitt, Synchronous data flow, *Proceedings of the IEEE*, 75(9), 1235–1245, 1987.

23. C. Liu and J. Layland, Scheduling algorithms for multiprogramming in a hard real-time environment, *Journal of the ACM*, 20(1), 46–61, January 1973.

24. R. Lublinerman, C. Szegedy, and S. Tripakis, Modular code generation from synchronous block diagrams—Modularity vs. code size, in *36th ACM SIGPLAN-SIGACT Symposium on Principles of Programming Languages (POPL'09)*. pp. 78–89, January 2009.

25. R. Lublinerman and S. Tripakis, Modular code generation from triggered and timed block diagrams, in *14th IEEE Real-Time and Embedded Technology and Applications Symposium (RTAS'08)*. IEEE CS Press, pp. 147–158, April 2008.

26. R. Lublinerman and S. Tripakis, Modularity vs. reusability: Code generation from synchronous block diagrams, in *Design, Automation, and Test in Europe (DATE'08)*. ACM, pp. 1504–1509, March 2008.

27. N. Lynch and M. Tuttle, An introduction to input/output automata, *CWI Quarterly*, 2, 219–246, 1989.

28. Z. Manna and A. Pnueli, *The Temporal Logic of Reactive and Concurrent Systems: Specification*, Springer-Verlag, New York, 1991.

29. M. Persson, M. Törngren, A. Qamar, J. Westman, M. Biehl, S. Tripakis, H. Vangheluwe, and J. Denil, A characterization of integrated multi-view modeling for embedded systems, in *Proceedings of the 13th ACM & IEEE International Conference on Embedded Software (EMSOFT'13)*, 2013.

30. A. Pnueli, A temporal logic of concurrent programs, *Theoretical Computer Science*, 13, 45–60, 1981.

31. V. Preoteasa and S. Tripakis, Refinement calculus of reactive systems, in *Proceedings of the 14th ACM & IEEE International Conference on Embedded Software (EMSOFT'14)*, 2014.

32. J. Reineke and S. Tripakis, Basic problems in multi-view modeling, in *Tools and Algorithms for the Construction and Analysis of Systems—TACAS 2014*, 2014.

33. I. Shin and I. Lee, Compositional real-time scheduling framework, *ACM Transactions on Embedded Computing Systems (TECS)*, 2008.

34. P. Tabuada, *Verification and Control of Hybrid Systems: A Symbolic Approach*, Springer, 2009.

35. M. Talupur and M. R. Tuttle, Going with the flow: Parameterized verification using message flows, in *Formal Methods in Computer-Aided Design, 2008. FMCAD '08*, November 2008, pp. 1–8.

36. L. Thiele, S. Chakraborty, and M. Naedele, Real-time calculus for scheduling hard real-time systems, in *Circuits and Systems, ISCAS*, 2000.

37. S. Tripakis and D. Broman, Bridging the semantic gap between heterogeneous modeling formalisms and FMI, EECS Department, University of California, Berkeley, CA, Technical Report UCB/EECS-2014-30, April 2014.

38. S. Tripakis, D. Bui, M. Geilen, B. Rodiers, and E. A. Lee, Compositionality in synchronous data flow: Modular code generation from hierarchical SDF graphs, *ACM Transactions on Embedded Computing System*, 12(3), 83:1–83:26, March 2013.

39. S. Tripakis, B. Lickly, T. A. Henzinger, and E. A. Lee, Theory of synchronous relational interfaces, *ACM Transactions on Programming Languages and Systems (TOPLAS)*, 33(4), July 2011.

40. S. Tripakis, C. Stergiou, C. Shaver, and E. A. Lee, A modular formal semantics for Ptolemy, *Mathematical Structures in Computer Science*, 23, 834–881, August 2013.

41. H. Van Brussel, P. Sas, I. Nemeth, P. De Fonseca, and P. den Braembussche, Towards a mechatronic compiler, *IEEE/ASME Transactions on Mechatronics*, 6(1), 90–105, March 2001.

5 Hybrid Stochastic Synchronization and Flow Models for Decision Making in Cyber-Physical Systems

Mengran Xue and Sandip Roy

CONTENTS

5.1 INTRODUCTION

Most traditional physical devices and systems have a single authority that is responsible for system-level decision making. This authority—whether a physical device (e.g., a control circuit), a cybertechnology (an algorithm or piece of software), or a human—holds primary responsibility for overall strategic control and management of system dynamics. The embedding of cybertechnologies into the physical world, and the consequent networking of *things*, is changing the traditional

paradigm: modern cyber-physical systems (CPS) and networks often contain multiple heterogeneous intelligences, which must coordinate for decision making [1–3]. These intelligent agents
involved in decision making are often quite diverse, involving heterogeneous physical-world systems,
cybertechnologies, and humans. Decision making for the Internet of Things thus crucially requires
new frameworks, and associated algorithms and models, for coordination of diverse physical-world,
cyber, and human agents.

The development of distributed algorithms and process models for cyber–physical–human systems entails several new challenges. Centrally, hybrid algorithms and models (ones with mixed
continuous-valued and discrete-valued actions or states) are needed, to capture physical-world
processes (which are often continuous valued) along with computational and human constructs
(which are often discrete). This requires frameworks that can tractably represent discrete-valued and
continuous-valued state evolutions, as well as mixed interfaces and actions, at a network's components [1]. Further, cyber–physical–human networks usually encompass a mixture of highly structured
deterministic dynamics (e.g., power-flow swing dynamics in the electric power grid) and stochastic
network processes (e.g., information-dissemination processes in the associated Supervisory Control and Data Acquisition (SCADA) system). Thus, frameworks that capture hybrids of stochastic
and deterministic processes are also essential. In addition to hybridization, distributed algorithms for
CPS must address (1) the multiple timescales inherent to diverse agents and (2) varied communication
capabilities and media available to the agents, among other challenges.

This book chapter explores the development of stochastic hybrid algorithms and models that can
assist with decision making in-networked cyber–physical–human platforms. Specifically, the main
purpose of the chapter is to introduce a stochastic quasi-linear modeling framework, which encompasses a number of discrete-valued, continuous-valued, and hybridized algorithms and models. The
chapter focuses particularly on (1) synchronization or consensus algorithms, which are often needed
for coordination, and (2) flow and spread models/algorithms, which are relevant to information dissemination for decision making. While flow and synchronization dynamics only represent a subset
of the algorithms that may be needed for decision making, they constitute evocative canonical examples that have direct application to CPS decision making. These evocative examples indicate that
the proposed framework may be broadly relevant for developing hybrid algorithms/models for CPS
decision making.

The remainder of this chapter is organized as follows. First, some background is given on network flow and synchronization models and their use in representing algorithms in the cyber world
and dynamical processes in the physical world (Section 5.2). Next, encompassing models for
flow and synchronization processes, which can capture stochastic hybrid dynamics, are introduced
(Sections 5.3 and 5.4). We show that common discrete-valued and continuous-valued algorithms or
processes can be represented using the framework and introduce some new hybrid algorithms/pro-
cesses that also can be captured. Section 5.5 pursues basic statistical characterizations of the described
models, with a focus on graph-theoretic analysis of asymptotics and model comparisons. Finally, in
Section 5.6, the synchronization dynamics of one model subclass is explored further.

5.2   FLOW AND SYNCHRONIZATION PROCESSES: BACKGROUND
         AND CONTRIBUTION

A wide variety of network processes have been described using dynamical models defined on graphs
(see, e.g., [4–9]). While these network models vary in their specifics, two common themes thread
together several of the models:

1. *Synchronization* phenomena, wherein states or opinions of networked autonomous agents
   come to a common value through interaction and communication, are widely captured.
2. Alternately, many models describe *network flow* or *spread dynamics*, that is, the movement
   of material or items among a network's components.

Many network models for synchronization and flow have been motivated by applications from different research communities. Relevant to our development here, both classes of models are descriptive of important distributed algorithms for cyber systems. In particular, network synchronization models have been widely used as distributed consensus-building algorithms (e.g., [10–13]), wherein distributed agents reach a common opinion via local communications; these algorithms are finding wide applications in important decision-making problems in networks, for instance, for cooperative spectrum sensing in cognitive radios [14]. Likewise, flow and spread models are descriptive of some distributed and network algorithms that underlie decision making: these include classical Markov chain–derived algorithms for searching [15]. Additionally, both synchronization and flow models are widely used to represent physical-world processes in networks, such as dissemination of goods/products through a transportation system (flow), coupling of electrical oscillators (synchronization), disease transmission (flow/spread), and swing dynamics in electric power networks (synchronization) (e.g., [16,17]). These models are often also used in real-time decision making and strategic design, for instance, disease-spreading models can be used to develop and adapt intervention policies for emerging disease threats [18].

Flow and synchronization models from different application domains have significant differences in their dynamics. For instance, they may have continuous-valued or discrete-valued states, stochastic or deterministic updates that may be clocked or asynchronous, and may exhibit complex state dependencies in their evolution. Yet, the models have in common that the graph topology plays a central role in the model's dynamics. For some of these models, connections between the graph topology and both asymptotic and transient characteristics of the dynamics have been obtained, and classes of graphs that yield desirable or undesirable dynamical properties have been found. More recently, tools for estimation and design of dynamics that exploit the graph topology also have been developed (see, e.g., [6,13,19,20], among many others). However, in many of these domains, graph-based analysis and estimation/design of network dynamics remain areas of active research, and indeed different models have been characterized to different extents.

In our ongoing research on automated decision management in cyber-physical infrastructure networks, we are using several different synchronization and flow network models to represent both network algorithmic processes and physical-world dynamics (e.g., [8,9,21–23]). These include both discrete-valued and continuous-valued synchronization models for consensus dynamics and clustering processes, queueing network and other stochastic flow models for traffic, and Markov chain models, among others (e.g., [21]). In using these algorithms and models, however, we are encountering two challenges. First, we often require algorithms and models that can capture mixtures of cyber, physical, and human platforms. For instance, fast fault-detection algorithms for the electric power grid may require modeling of (1) the swing dynamics of the physical grid, (2) information dissemination of synchrophasor data through SCADA systems, and (3) cooperative decision-making (consensus) algorithms. These mixed cyber-physical algorithms/models often require hybrid states and interactions/interfaces (e.g., mixtures of analog averaging dynamics and discrete selection/communication actions), as well as mixtures of deterministic and stochastic dynamics. Second, in studying various cyber-world and physical-world algorithms/models, we are encountering common problems in network parameterization, simulation/analysis of dynamics, and state estimation for each model, which seem to admit similar but not identical solutions. For some of the models, aspects of the desired analysis are well established, but these results often do not directly transfer to the other network models. Meanwhile, some analyses have not been studied thoroughly for any of the models.

This chapter takes a step toward addressing these challenges, by (1) introducing encompassing models for flow and synchronization processes that can capture hybrid states and interactions and (2) pursuing preliminary analyses of these general models. We believe that the framework is promising in that it can capture a wide range of hybrid and mixed stochastic/deterministic processes, yet is sufficiently structured to permit efficient simulation and statistical characterization of the dynamics, so as to overcome the *curse of dimensionality* that is present in complex stochastic network models.

Broadly, the flow and synchronization model classes studied here are discrete-time, Markov models defined on graphs. The novelty in our approach is to impose certain implicit linearity conditions on the expectation of the model's next state conditioned on its current one, which are generic enough to capture many hybrid and stochastic dynamics but structured enough to permit formal analysis. In introducing the stochastic modeling frameworks for flow and synchronization, we pursue two key outcomes:

1. We illustrate that a broad family of network flow and synchronization processes, including stochastic and deterministic processes, discrete-valued and continuous-valued dynamics, and even certain apparently nonlinear processes, admit a common representation and common graphical analyses. This common representation, which only enforces Markovianity and a certain conditional linearity in the state, also permits us to compare the dynamics of different flow and synchronization models.

2. We give an introductory graph-theoretic analysis of the model classes. We focus primarily on statistically characterizing the asymptotics of the network models and on comparing/relating various flow and synchronization processes. For an example hybrid model, a characterization of whether synchronization is achieved in a mean square and probability 1 sense is also presented. Our methodology demonstrates that graphical analysis methods that have been developed for a particular model can provide new insights into more general algorithms and processes. For instance, our work illustrates that the graphical asymptotic analysis of Markov chains translates to the broad flow and synchronization networks introduced here and gives interesting new insights into, for example, probabilistic routing and voter models.

We point out that the models introduced here are similar in favor to certain quasi-linear models for discrete-valued network processes known as influence and voter models [24–27] and to a class of network models with moment-linearity structures [28]. However, in contrast to these works, we specifically focus here on model structures for synchronization and flow processes, which are prominent in CPS decision-making contexts, and obtain a keener statistical analysis for these processes.

A notation for graphs: At several points, our model formulation requires definition of graphs from state matrices describing linear dynamics on networks. The following notation will be used for such graph definitions: for an $n \times n$ state matrix A, the notation $\Gamma(A)$ will be used to describe a weighted and directed graph with n vertices, labeled $1, \ldots, n$. An edge will be drawn from vertex i to vertex j if and only if $a_{ji} > 0$, with the weight of the edge equal to a_{ji}. We note that edges from a vertex to itself will also be drawn according to this definition, based on the diagonal entries of A.

5.3 LINEAR STOCHASTIC FLOW NETWORK MODEL

Flow network models are concerned with tracking the movement of items or material or information among network components or nodes. These models have found wide application in fields ranging from sensor networking to traffic engineering and cell biology (e.g., [8,22]). The focus of this section is to introduce a tractable framework for modeling linear flow network dynamics defined on a graph (Section 5.3.1). We then confirm that the model encompasses several classical stochastic and deterministic models for network flows (Section 5.3.2) and also describe hybrid-valued and mixed stochastic/deterministic dynamics that can be captured in the framework.

5.3.1 MATHEMATICAL FORMULATION

A network with n *components* (or nodes or agents), labeled $1, \ldots, n$, is considered. We are concerned with tracking in discrete time a scalar state associated with each network component, which represents an amount of material or number of items (of a single type) at the component. Specifically, we use the notation $r_i[k]$ for the scalar state of component i ($i = 1, \ldots, n$) at discrete time k ($k = 0, 1, 2, \ldots$) and refer to this state variable as the *local quantity* of component i at time k.

The local quantities of the network components are modeled as evolving in discrete time, due to conservative flows between the components. We first introduce a general stochastic model for the flows and discuss a linear-algebraic and graph-theoretic representations of the model dynamics. Subsequently, we impose a weak conditional linearity condition on the model dynamics that facilitates analysis.

First, let us describe the general (possibly nonlinear) *stochastic flow network* model. In this model, we view the local quantities of the components as being updated by a two-stage process at each time step, that is, a *flow-determination* stage followed by a *flow-combination* stage. First, in the flow-determination stage, each local quantity $r_i[k]$ is represented as forming *flows* $f_{ij}[k]$, $j = 1, \ldots, n$, to the components in the network (including the component i). Each flow is assumed nonnegative ($f_{ij}[k] \geq 0$), and the total of the flows equals the local quantity ($\sum_{j=1}^{n} f_{ij}[k] = r_i[k]$). Flow determination from local quantities is assumed to be general: it may be either deterministic or stochastic, and the flows $f_{ij}[k]$ may depend on $r_i[k]$ in an arbitrary way. We permit correlation between stochastic flow determinations originating from different components but do assume that flow determinations at time k are independent of the system's past history given the time-k local quantities. Second, in the flow-combination stage, the incoming flows to each component are summed to determine the local quantity at the component at the next time step. That is, we compute $r_i[k+1]$ as follows: $r_i[k+1] = \sum_{j=1}^{n} f_{ji}[k]$. We have thus specified the stochastic flow network update.

We develop a matrix-theoretic formulation of the stochastic flow network, to facilitate graphical analysis of the model. To do so, let us define a *quantity vector* as $\mathbf{r}[k] = \begin{bmatrix} r_1[k] & \cdots & r_n[k] \end{bmatrix}^T$ and a *flow vector* for each component i as $\mathbf{f}_i[k] = \begin{bmatrix} f_{i1}[k] & \cdots & f_{in}[k] \end{bmatrix}^T$. To continue, we note that the flow-determination stage of the flow network's update enforces that a fraction of each local quantity is directed to each component as a flow. Thus, it is automatic that the flow vector $\mathbf{f}_i[k]$ can be written as $\mathbf{f}_i[k] = \mathbf{p}_i[k] r_i[k]$, where the $n \times 1$ *flow fraction vector* $\mathbf{p}_i[k]$ has entries that are nonnegative and sum to 1 (formally, $\mathbf{p}_i[k] \geq 0$ and $\mathbf{1}^T \mathbf{p}_i[k] = 1$). Here, the vectors $\mathbf{p}_i[k]$ may be stochastically determined, and further the $\mathbf{p}_i[k]$ may depend on $r_i[k]$. Next, using this expression for flows together with the flow-combination update, the quantity vector at time $k + 1$ can be expressed in terms of the vector at time k as $\mathbf{r}[k+1] = P[k]\mathbf{r}[k]$, where the *flow state matrix* $P[k] \overset{\triangle}{=} \begin{bmatrix} \mathbf{p}_1[k] & \cdots & \mathbf{p}_n[k] \end{bmatrix}$ is a column-stochastic matrix that (in general) is stochastically determined and dependent on $\mathbf{r}[k]$. Thus, the matrix representation has been achieved.

The aforementioned matrix notation suggests one graphical representation of the flow network dynamics. The (possibly randomly generated) matrix $P[k]$ indicates flows of material/items at time k and hence naturally admits a graphical interpretation. Thus, viewing $P[k]$ as an instantiation of the flow network dynamics, we draw a corresponding *flow instantiation graph* $\Phi[k]$ as $\Phi[k] = \Gamma(P[k])$ for time k. The instantiation graph captures the particular splitting of local quantities that occur to form flows at time k, that is, an edge is drawn from vertex i to j if material/items flow from component i to j at that time, and the weight captures the fraction of the quantity at component i that flows in this direction. We stress that a flow network may have many possible instantiations, and so instantiation graphs, at each time k.

Next, let us introduce a notion of linearity in the flow of network dynamics that facilitates many graph-theoretic characterizations yet allows representation of several interesting dynamics. To introduce this notion of linearity, we first note that the flow network dynamics described earlier are Markovian, in the sense that the quantity vector at time $k+1$ can be determined only from the quantity vector at time k, given the whole past history of the network. Based on this Markovian structure, we can specify the model in terms of the conditional distribution for the quantity vector at time $k+1$ given the quantity vector at time k, for each k. In turn, conditional statistics for the time-$(k + 1)$ quantity vector given the time-k quantity vector can be envisioned. Here, let us define linearity of the model implicitly in terms of the first-moment conditional statistics, that is, the conditional mean for next quantity vector given the current one, or $E(\mathbf{r}[k+1] \mid \mathbf{r}[k])$. Specifically, we will view the stochastic flow network as *linear*, if $E(\mathbf{r}[k+1] \mid \mathbf{r}[k])$ is a purely linear function of $\mathbf{r}[k]$ for all k or equivalently if $E(\mathbf{r}[k+1] \mid \mathbf{r}[k]) = Q[k]\mathbf{r}[k]$ for a fixed matrix $Q[k]$ (which does not depend on $\mathbf{r}[k]$) for each k. Some remarks on linear stochastic flow networks are needed:

1. The condition for linearity does not require that the $P[k]$ be independent of $\mathbf{r}[k]$, only that the conditional mean of the next quantity vector is a linear function of the current one. One example in Section 5.3.2 shows that even some dynamics with such state dependences may be linear.
2. For linear stochastic flow networks, the matrix $Q[k]$ permits us to specify another graphical representation. First, since the matrix $Q[k]$ maps the quantity vector at time k to expected flows and hence the expected quantity vector at time $k+1$, we refer to $Q[k]$ as the *flow expectation matrix*. Second, we define a weighted and directed *flow expectation graph* $\overline{\Phi}[k] = \Gamma(Q[k])$. The flow expectation graph captures whether or not, on average, there is a flow between each pair of vertices at each time.
3. Some of the results that we obtain depend on time invariance in addition to linearity of the quantity vector's conditional expectation. If $E(\mathbf{r}[k+1] \mid \mathbf{r}[k]) = Q\mathbf{r}[k]$ for some fixed Q for all k, we will refer to the stochastic flow network model as a linear time-invariant (LTI) one. A single flow expectation graph $\overline{\Phi} = \Gamma(Q)$ can be defined for an LTI stochastic flow network.

5.3.2   EXAMPLES

One of our primary goals in this chapter is to illustrate that the stochastic flow network model (and analogous synchronization model) encompasses stochastic and deterministic processes/algorithms, as well as discrete-valued and continuous-valued dynamics. To this end, we briefly demonstrate that three classical flow network models can be posed as linear stochastic flow networks. Also, we introduce a new hybrid flow model that falls within the framework.

5.3.2.1   Conservative Linear Fluid Flow Model

Classical fluid flow models track continuous-valued quantities (i.e., amounts of material) at network components. In a conservative linear model, deterministic or stochastic fractions of the quantity at each component are viewed as flowing to multiple components over each time interval (e.g., [29]). These models capture a wide range of physical-world dynamics, including water distribution processes, voltage dynamics in circuits, flow of materials via transportation systems, and large-scale population dynamics.

The update equation of such a fluid flow model can be expressed in the stochastic flow network formalism as follows: $\mathbf{r}[k+1] = P[k]\mathbf{r}[k]$, where the flow state matrix $P[k]$ may be deterministic or randomly selected from a finite sample space, but has no dependence on $\mathbf{r}[k]$ or any other previous state. We note here that the jth entry in $\mathbf{p}_i[k]$, that is, the flow $f_{ij}[k]$, is a precise fraction of the local $r_i[k]$ (specifically, $f_{ij}[k] = P_{ji}[k]r_i[k]$, where $P_{ji}[k]$ here represents the entry in row j and column i of P). We stress that only the selection of $P[k]$ may be stochastic (independently of the past history of the process), whereupon exact fractional flows of the local quantities occur in the network.

It is straightforward to ascertain that the model is a linear stochastic flow network. To this end, let us define $\bar{P}[k]$ as the expectation of $P[k]$ (or $E(P[k]) = \bar{P}[k]$). In this notation, we have $E(\mathbf{r}[k+1] \mid \mathbf{r}[k]) = E(P[k]\mathbf{r}[k] \mid \mathbf{r}[k]) = E(P[k] \mid \mathbf{r}[k])\mathbf{r}[k] = E(P[k])\mathbf{r}[k] = \bar{P}[k]\mathbf{r}[k]$. Thus, the model is linear, with $Q[k] = \bar{P}[k]$.

5.3.2.2   Markov Chain

Finite-state Markov chains are widely and classically used to represent discrete state-transition processes in both the cyber-world and physical-world Markov chains, and their generalizations also can be used effectively as algorithms for search, partitioning, and so forth. Let us verify that a finite-state Markov chain can be captured in the stochastic flow network framework. Specifically, let us consider an n state Markov chain, with states or *statuses* labeled $1, \ldots, n$. We denote the Markov chain's transition matrix by $A[k]$ (where, specifically, the entry $a_{ji}[k]$ is the transition probability status i to status j at time k). The initial value of the Markov chain's state is assumed to be specified according to a probability distribution.

Let us now argue that the Markov chain can be formulated as a linear stochastic flow network with n components. In the flow network formulation, we will say that the quantity $r_i[k]$ of component i is 1 if the value of the Markov chain is i (i.e., the single object is at location i) and is 0 otherwise. From this definition, we see that the quantity vector $\mathbf{r}[k]$ is the $0 - 1$ *indicator vector* of i (typically denoted $\mathbf{e}_i$) when the Markov chain is in status i. In this case, the instantiation of $\mathbf{p}_i[k]$ should be an indicator vector for the next state of the Markov chain. The evolution of the Markov chain from time k to time $k + 1$ can be captured as follows using the flow network formalism: each column i of the matrix $P[k]$ (equivalently, the flow vector $\mathbf{p}_i[k]$) is chosen independently to be an indicator vector, specifically to equal the vector $\mathbf{e}_j$ with probability $a_{ji}[k]$. Then we notice that $\mathbf{r}[k + 1] = P[k]\mathbf{r}[k]$ equals $\mathbf{e}_j$ with probability $a_{ji}[k]$ if $\mathbf{r}[k] = \mathbf{e}_i$ for all i and j, which matches exactly the Markov chain's update. In other words, this update captures that a unit quantity at component i will flow as a whole to a component j, with appropriate probability. Thus, we have phrased the state update as a stochastic flow network. To check linearity, we first note that $E(\mathbf{p}_i[k]) = \begin{bmatrix} a_{1i}[k] & \cdots & a_{ni}[k] \end{bmatrix}^T$ and hence that $E(P[k]) = A[k]$. Further noting the independence of $P[k]$ from the past quantity vector $\mathbf{r}[k]$, we have $E(\mathbf{r}[k + 1] \mid \mathbf{r}[k]) = E(P[k]\mathbf{r}[k] \mid \mathbf{r}[k]) = E(P[k]|\mathbf{r}[k])\mathbf{r}[k] = E(P[k])\mathbf{r}[k] = A[k]\mathbf{r}[k]$. This verifies that the stochastic flow network is linear, with expectation flow matrix $Q[k]$ equal to $A[k]$.

5.3.2.3 Probabilistic Routing Model

A common probabilistic routing or queueing model can also be viewed as a stochastic flow network. In this model, component quantities comprise integral numbers of discrete units, each of which independently flow through the network in a Markovian fashion. Dynamics of this form are captured in infinite-server queueing network models and have found application in modeling information dissemination in communication networks and traffic densities in air transportation networks, among many other domains (e.g., [8,22]). In general, these models are representative of flow processes in which small numbers of discrete items transition among network nodes. The canonical example that we describe here, specifically, can be viewed as a discrete-time Markov probabilistic routing model or $M/M/\infty$ queueing network model.

Let us describe the dynamics of the probabilistic routing model directly in the stochastic flow network modeling framework. The quantity variables $r_i[k]$ that we track represent the number of discrete units at each component i at time k and are constrained to be integral. Let us begin by describing the flow-determination stage at each component i. At time k, each of the $r_i[k]$ units at component i is independently routed to other components (possibly including itself) with certain probabilities, which are specified in a probability vector $\mathbf{d}_i[k]$ (i.e., the jth entry in $\mathbf{d}_i[k]$ is the probability with which each unit at component i will flow to component j). Therefore, the flow vector $\mathbf{f}_i[k]$ follows the *multinomial distribution* with parameters $r_i[k]$ and $\mathbf{d}_i[k]$ ([30]) and has $\begin{pmatrix} n + r_i[k] - 1 \\ n - 1 \end{pmatrix}$ possible values (with probabilities specified by the multinomial distribution). The flow fraction vector $\mathbf{p}_i[k]$ is given by $\mathbf{p}_i[k] = \mathbf{f}_i[k]/r_i[k]$ and so is simply a scaled version of the multinomially distributed vector $\mathbf{f}_i[k]$. For example, let us assume that $n = 2$, $r_i[k] = 3$ and $\mathbf{d}_1[k] = \begin{bmatrix} 0.2 & 0.8 \end{bmatrix}^T$. Then, $\mathbf{p}_1[k]$ can be one of four possible values, which are $\begin{bmatrix} 0 & 1 \end{bmatrix}^T$, $\begin{bmatrix} \frac{1}{3} & \frac{2}{3} \end{bmatrix}^T$, $\begin{bmatrix} \frac{2}{3} & \frac{1}{3} \end{bmatrix}^T$, and $\begin{bmatrix} 1 & 0 \end{bmatrix}^T$, with probabilities $(0.8)^3$, $3(0.2)(0.8)^2$, $3(0.2)^2(0.8)$, and $(0.2)^3$, respectively. Since we have described the generation rule for the flow fraction vector $\mathbf{p}_i[k]$, we have thus specified the instantiation matrix $P[k]$ of the probabilistic routing model.

Now, let us discuss the linearity of the model. We first note that the instantiation matrix $P[k]$ (which is $\begin{bmatrix} \mathbf{p}_1[k] & \cdots & \mathbf{p}_n[k] \end{bmatrix}$) actually depends on $\mathbf{r}[k]$, since the distribution of $\mathbf{p}_i[k]$ depends on $r_i[k]$. However, the model itself is still linear even with such a state dependence. Specifically, since the conditional first moment of the scaled multinomially distributed vector $\mathbf{p}_i[k]$ is $E(\mathbf{p}_i[k] \mid \mathbf{r}[k]) = \mathbf{d}_i[k]$ (which is in fact independent on $\mathbf{r}[k]$), we have $E(\mathbf{r}[k+1] \mid \mathbf{r}[k]) = E(P[k]\mathbf{r}[k] \mid \mathbf{r}[k]) = E(P[k] \mid \mathbf{r}[k])\mathbf{r}[k] = \begin{bmatrix} \mathbf{d}_1[k] & \cdots & \mathbf{d}_n[k] \end{bmatrix} \mathbf{r}[k]$. Thus, the stochastic flow network is linear with $Q[k] = \begin{bmatrix} \mathbf{d}_1[k] & \cdots & \mathbf{d}_n[k] \end{bmatrix}$.

5.3.2.4 Hybridized Splitter–Router Model

Finally, a hybridized stochastic flow model is considered, which captures discretized routing by a subset of network components and fluid flow dynamics at the remaining components. This model is evocative of some dissemination processes in cyber-physical networks, where physical-world processes and some network nodes lead to proportional distribution of resources, while control or communication elements at other nodes route entire flows in specific directions.

Formally, the network model is assumed to consist of two types of nodes or components, namely, *splitters* and *routers*. Without loss of generality, the first n_1 network components (labeled $1, \ldots, n_1$) are assumed to be splitters, while the remaining $n - n_1$ network components (labeled $n_1 + 1, \ldots, n$) are assumed to be routers. The splitters serve to transmit proportions of the local quantity to neighboring components at each time step. Specifically, for each splitter i, the flow $f_{ij}[k]$ at time step k is determined as $f_{ij}[k] = z_{ij}r_i[k]$, where $z_{ij} \geq 0$ and $\sum_j z_{ij} = 1$; thus, the splitters operate according to a conservative fluid flow model. Meanwhile, the routers serve to move the entire local quantity to a graphical neighbor at each time step. Specifically, each router i chooses a neighboring component j with probability z_{ij} at each time step, independently of all past selections. The selected component, say $\hat{j}$, receives the entirety of the local quantity. That is, $f_{i\hat{j}}[k] = r_i[k]$ and $f_{ij}[k] = 0$ for all $j \neq \hat{j}$. We refer to the model as the splitter–router model.

Combining the arguments described earlier for the conservative fluid flow model and Markov chain, we can easily verify that the splitter–router model is a linear stochastic flow network with flow expectation matrix given by $P = [z_{ij}]^T$.

5.4 STOCHASTIC SYNCHRONIZATION NETWORK MODEL

Models for *synchronization* (or *consensus*) among networked autonomous agents have also found application in diverse fields, including distributed computing, analog circuit design, cell biology, and particle physics, among many others (e.g., [9,10,23,24]). Fundamentally, synchronization models are often concerned with tracking either physical or information states of essentially autonomous agents, which through decentralized interaction or communication achieve identical state values. Many distributed decision-making paradigms make use of synchronization/consensus algorithms, since coming to a consensus is often a prerequisite to decision and action. Although synchronization and flow dynamics are typically of interest in different application domains, the topological structures and analysis of flow and synchronization networks are similar, and a common theory can be developed for both classes of models. Our purpose here is to introduce a general discrete-time stochastic linear modeling framework for synchronization (Section 5.4.1) and to show that the framework encompasses several synchronization models in the literature as well as new hybrid models (Section 5.4.2).

5.4.1 Mathematical Formulation

In analogy with the flow networks, let us first define stochastic synchronization models in some generality and then focus on the linear case. Formally, we consider a network of n agents, labeled $1, \ldots, n$, and associate with each agent i a scalar variable $s_i[k]$, which we call the *opinion* of agent i. In contrast with the flow model, state variables may be positive or negative for the stochastic synchronization network. We find it convenient to assemble the opinions into a single *opinion vector* $\mathbf{s}[k] = \begin{bmatrix} s_1[k] & \cdots & s_n[k] \end{bmatrix}^T$.

Each agent is viewed as updating its opinion in discrete time, in such a way that its neighbors' current opinions are incorporated. Specifically, we model each agent's next opinion as a weighted nonnegative *unitary* linear combination of multiple agents' current opinions. That is, agent i's opinion is updated as $s_i[k + 1] = \mathbf{g}_i^T[k]\mathbf{s}[k]$, where the *influence vector* $\mathbf{g}_i[k]$ is entrywise nonnegative and sums to 1 ($\mathbf{g}_i^T \mathbf{1} = 1$). Let us stress here that each influence vector $\mathbf{g}_i[k]$ may be a deterministic or stochastic quantity, which may be correlated with $\mathbf{g}_j[k]$ ($j \neq i$) and may even depend on $\mathbf{s}[k]$ (but otherwise must be independent of the past history of the network). Assembling the opinion update equations of the n agents as a single vector equation, we obtain $\mathbf{s}[k + 1] = G^T[k]\mathbf{s}[k]$, where $G[k] = \begin{bmatrix} \mathbf{g}_1[k] & \cdots & \mathbf{g}_n[k] \end{bmatrix}$ is a column-stochastic *influence matrix* that in general may be stochastic and concurrent-state dependent.

Let us define graphs that describe the interactions among agents in the stochastic synchronization model's opinion update. In particular, we define an n-vertex *influence instantiation graph* $\Lambda[k]$ at time k as $\Gamma(G^T[k])$, that is, as the weighted and directed graph associated with the state matrix $G^T[k]$ at time k. The graph can be interpreted as follows: an arrow is drawn from vertex i to vertex j if component i's current opinion influences the next opinion of component j, with the weight equal to the strength of the influence. We notice that an influence instantiation graph may be stochastic, time varying, and state dependent but always has the property that the sum of the weights on edges entering each vertex is 1.

Finally, let us identify a subset of the stochastic synchronization network models defined earlier, which are specially structured in that their expected dynamics are linear. Specifically, we call the aforementioned synchronization model a *linear stochastic synchronization network*, if the state dynamics satisfy $E(\mathbf{s}[k + 1] \mid \mathbf{s}[k]) = H^T[k]\mathbf{s}[k]$, where the *influence expectation matrix* $H^T[k]$ is a state-independent matrix, for each k. In other words, we say that the synchronization network is linear, if the expectation of the next opinion vector given the current one is a linear function of the current opinion vector. If, further, the expected influence matrix $H^T[k]$ is time independent (say equal to H^T), we shall call the model a "LTI stochastic synchronization network." For linear and LTI stochastic synchronization networks, we find it convenient also to define graphs based on the influence expectation matrix. In particular, we define the *influence expectation graph* at time k for the linear model (or simply the influence expectation graph for the LTI model) as $\overline{\Lambda}[k] = \Gamma(H^T[k])$ (respectively, $\overline{\Lambda} = \Gamma(H^T)$).

Remark 5.1 *Trivially, the transposing of an influence matrix is a flow matrix, and vice versa. Deeper consideration exposes a tighter duality, that is, a certain time reversal of a synchronization process is a flow network dynamics. Details are omitted in the interest of space.*

5.4.2 EXAMPLES

To illustrate the scope of our framework, two classical synchronization algorithms/models are phrased as linear stochastic synchronization networks. Also, two hybrid models are defined that fall within the framework. The examples show that interesting deterministic and stochastic dynamics, as well as discrete-valued and continuous-valued processes, can be captured within the framework.

5.4.2.1 Distributed Averaging Algorithm

Distributed averaging algorithms, in which network agents have continuous-valued opinions that evolve according to purely linear updates, have been extensively studied. Such models have been widely used to represent distributed decision-making processes and are classically the dynamics considered in synchronization models ([10,11,31]). They are representative of a number of physical-world coordination processes, such as synchronization of mechanical and electrical oscillators. In these algorithms, each agent is assumed to originally have a continuous-valued opinion, which is updated through averaging with the opinions of its graphical neighbors. Under broad conditions on the interaction topology, the agents can be shown to achieve a common opinion.

Let us give a precise formulation of a distributed averaging algorithm as a stochastic synchronization network. To this end, a network model with n agents is considered, where agent i has associated with a scalar opinion $s_i[k]$ at time k. In a distributed averaging algorithm, agent i's next opinion, $s_i[k+1]$, is generated as a linear combination of all agents' current opinions. That is, the next opinion is computed as $s_i[k+1] = \mathbf{g}_i^T[k]\mathbf{s}[k]$, where the jth entry of $\mathbf{g}_i[k]$ is the weight of influence from agent j at time k. We note here that the influence vector $\mathbf{g}_i[k]$ can be either deterministic or independently selected from a sample space at each time.

Using a similar approach to the linearity analysis of the fluid flow model, we can show that the distributed averaging algorithm is linear, with $H^T[k] = E(G^T[k])$.

5.4.2.2 Voter Model

A voter model is a network model in which each agent independently stochastically chooses one other agent (maybe itself) according to a probability vector and then copies the chosen agent's current opinion as its next opinion ([9,23–25]). The voter model has been used to represent decision making/voting and communication patterns among humans [24,32] and environmental phenomena, such as convective weather propagation [9,23]. It has also been advanced as a distributed algorithm for consensus building and self-partitioning among networked agents [11,33]. In contrast with the distributed averaging algorithm, the voter model maintains discreteness of opinions/states and captures updates based on discrete selection rather than averaging.

Let us formulate the voter model as a stochastic synchronization network. We assume that each agent's opinion is initially an arbitrary real scalar. (Often, binary voter models, in which the opinions are constrained to be 0 or 1, are studied; however, our formulation permits propagation of more than two opinions.) At each time step, agent i's next opinion is determined as follows: the agent i selects a neighbor j with probability $c_{ji}[k]$ (where $c_{ji}[k] > 0$ and $\sum_j c_{ji}[k] = 1$), whereupon it copies the current status of the neighbor (i.e., $s_i[k+1] = s_j[k]$). Equivalently, we can write the updating dynamics as $s_i[k+1] = \mathbf{g}_i^T[k]\mathbf{s}[k]$, where $\mathbf{g}_i[k] = \mathbf{e}_j$ with probability $c_{ji}[k]$. Thus, we have posed the dynamics as that of a stochastic synchronization network.

Next, let us verify the conditional linearity of the voter model. According to the aforementioned description, we obtain that $E(\mathbf{g}_i[k]) = \mathbf{c}_i[k]$, where $\mathbf{c}_i[k] \triangleq \begin{bmatrix} c_{1i}[k] & \cdots & c_{ni}[k] \end{bmatrix}^T$ (and $\mathbf{c}_i^T[k]\mathbf{1} = 1$). Since $\mathbf{g}_i[k]$ is independent of the previous and current opinions, we then have $E(\mathbf{s}[k+1] \mid \mathbf{s}[k]) = E(G^T[k]\mathbf{s}[k] \mid \mathbf{s}[k]) = E(G^T[k] \mid \mathbf{s}[k])\mathbf{s}[k] = E(G^T[k])\mathbf{s}[k] = \begin{bmatrix} \mathbf{c}_1[k] & \cdots & \mathbf{c}_n[k] \end{bmatrix}^T\mathbf{s}[k]$, which indicates that the voter model is a linear synchronization network model, with $H^T[k] = \begin{bmatrix} \mathbf{c}_1[k] & \cdots & \mathbf{c}_n[k] \end{bmatrix}^T$.

One further note is worthwhile, in the case that the voter model is a binary one. For this case, we note that the expected opinion vector contains the *probabilities* that each agent has status 1, and hence local occupancy probabilities are tracked by the expected opinion vector.

5.4.2.3 Mixed Model

Next, let us introduce an apparently nonlinear stochastic synchronization network model that is a state-dependent mixture of a distributed averaging algorithm and a voter model. This type of model may be useful for complex algorithmic tasks in CPS, where discretized choices must be made in one operating regime, while continuous-valued opinions are possible in another operating regime (e.g., if the agents' opinions are positive, then the mean must be found, but a negative opinion indicates a failure whereupon the failure mode must be agreed on). Here, a specific model of this type is considered.

In this specific example, each agent's opinion at each time is assumed to be a real-valued scalar. We also associate with each agent i a threshold value, say b_i. At each time k, agent i updates its state as follows: the agent compares its current opinion $s_i[k]$ with its threshold b_i. If $s_i[k] \geq b_i$, agent i randomly picks one other agent j with probability g_{ij} and then copies agent j's current opinion as its next opinion. For this case, the influence vector is $\mathbf{g}_i[k] = \mathbf{e}_j$ with probability g_{ij}. If $s_i[k] < b_i$, agent i computes a weighted average of the agents' current opinions based on weights $g_{i1}, \cdots, g_{in}$ and chooses this average to be its next opinion (i.e., $s_i[k+1] = \sum_{j=1}^n g_{ij}s_j[k]$). For this case, the influence vector $\mathbf{g}_i[k]$ is simply $\begin{bmatrix} g_{i1} & \cdots & g_{in} \end{bmatrix}^T$. We note that this stochastic synchronization network model is actually a mixture of a voting model and a distributed averaging model. We note that the instantiation matrix $G[k]$ (i.e., $G[k] = \begin{bmatrix} \mathbf{g}_1[k] & \cdots & \mathbf{g}_n[k] \end{bmatrix}$) depends on the current opinions of the agents.

Although the two synchronization mechanisms in the two operating regimes (i.e., above threshold and below threshold) are completely different, the model is stochastically linear. Let us use notation $\bar{G}[k]$ where the i,jth entry of $\bar{G}[k]$ has value g_{ji}. For the case that $s_i[k] \geq b_i$ (voting model phase), we can easily obtain that $E(\mathbf{s}[k+1] \mid \mathbf{s}[k]) = \bar{G}^T[k]\mathbf{s}[k]$. For the case that $s_i[k] < b_i$ for any i (distributed averaging model phase for agents i), we also obtain that $E(\mathbf{s}[k+1] \mid \mathbf{s}[k]) = \bar{G}^T[k]\mathbf{s}[k]$. Therefore, the model dynamics satisfy the linearity condition.

5.4.2.4 Averager–Voter Algorithm

Finally, we introduce a second hybrid model for network synchronization dynamics, which captures coordination among *voter-type* agents and *averaging* agents. This model is motivated by coordination tasks in cyber-physical networks, which involve (1) physical-world devices with states that are updated through (continuous) physics-based dynamics and (2) cyber/human agents whose states are updated via communication. Coordination in these cyber-physical networks can be achieved naturally using consensus algorithms, wherein some agents engage in averaging, while other agents use a stochastic voting process. Here, we consider algorithms of this sort that can be modeled as linear stochastic synchronization processes.

Formally, a network is considered with n_1 *averaging* agents (labeled $i = 1, \ldots, n_1$) and $n - n_1$ *voting* agents (labeled $i = n_1 + 1, \ldots, n$). The agents seek to synchronize their scalar opinions to a common value, by iteratively updating their opinions according to a clocked model; the opinion of agent i at time k is denoted by $s_i[k]$. To this end, each averaging agent computes a simple weighted average of its neighbors' opinions at each time step (whether the neighbors are averagers or voters). Specifically, each agent $i \in 1, \ldots, n_1$ computes its next opinion as $s_i[k+1] = \sum_{j=1}^{n} g_{ij} s_j[k]$, where the weights g_{ij} satisfy $g_{ij} \geq 0$ and $\sum_{j=1}^{n} g_{ij} = 1$. Meanwhile, the voting agents select a neighboring agent (whether it is an averager or voter) probabilistically and simply copy the opinion of this neighbor. Specifically, at each time step, each agent $i \in n_1 + 1, \ldots, n$ chooses another agent $j \in 1, \ldots, n$ with probability g_{ij}, independently of the whole past history of the network. The agent then simply copies the opinion of the chosen neighbor $\hat{j}$: $s_i[k+1] = s_{\hat{j}}[k]$.

It is easy to check that the described stochastic synchronization process is linear. Specifically, for both averaging and voting agents, it follows immediately that $E(s_i[k+1] \mid \mathbf{s}[k]) = \sum_{j=1}^{n} g_{ij} s_j[k]$. We thus recover that $E(\mathbf{s}[k+1] \mid \mathbf{s}[k]) = \bar{G}^T[k]\mathbf{s}[k]$, where the i,jth entry of $\bar{G}[k]$ has value g_{ji}.

Remark 5.2 *It is worth highlighting the strong analogy between the first two examples of flow models (the fluid flow and Markov chain models) and the corresponding examples of synchronization models (the distributed averaging and voter models, respectively). We also note the possibility for state-dependent update processes that are nevertheless linear in both settings, as indicated by the third and fourth examples in each case.*

5.5   GRAPH-THEORETIC ANALYSIS OF THE MODELING FRAMEWORKS

The cyber-physical network applications that motivate this study require a comprehensive suite of analysis, inference, and design tools for flow and synchronization models. In this first work, we provide graph-theoretic analyses of the modeling frameworks, with a particular focus on characterizing the models' asymptotic behaviors in a statistical sense and comparing models. Although these analyses are basic, they enhance known characterizations of several particular algorithms/models (e.g., for the probabilistic routing model described earlier) and yield interesting comparisons among and between flow and synchronization processes. These statistical characterizations and comparisons provide a baseline for using the models in CPS applications, by giving basic insights into hybrid network dynamics in terms of the network's graph. They are also important as a foundation for more intricate graph-theoretic analysis and estimation of complex network dynamics (e.g., [20]).

The section is organized as follows: we begin by noting basic invariants of the flow and synchronization models (Section 5.5.1). The main focus of the section is on graph-theoretic characterizations for the models' statistics in the asymptote (Section 5.5.2). In developing these characterizations, we develop and draw on a couple of interesting connections between flow and synchronization processes. Finally, we briefly pursue a comparison among flow models that fall within the broad class defined here (Section 5.5.3).

5.5.1 PRELIMINARY OBSERVATIONS: INVARIANCES

As a preliminary step, let us formalize complementary invariants of the (general, possibly nonlinear) flow and synchronization models' dynamics. These invariants at their essence are simply formalizations of the principles underlying conservative flow and synchronization, respectively. Here are the results:

1. For the stochastic flow model, the sum of entries in the quantity vector remains unchanged with time. That is, for any trajectory of the stochastic flow model, $\mathbf{1}^T \mathbf{r}[k]$ is identical for $k = 0, 1, 2, \ldots$. For convenience, let us use r_s to represent the total quantities in the flow network ($r_s = \mathbf{1}^T \mathbf{r}[k]$).

2. For the stochastic synchronization network, an opinion vector whose entries are identical is an invariant of the dynamics. That is, if $\mathbf{s}[k_0] = c\mathbf{1}$ for some scalar c and some time step k_0, then $\mathbf{s}[k] = \mathbf{s}[k_0] = c\mathbf{1}$ for all $k > k_0$.

5.5.2 ASYMPTOTIC STATISTICS

The development of graph-theoretic characterizations of the asymptotics of LTI stochastic flow and synchronization networks will now be discussed. Specifically, we first give preliminary characterizations of the expected state dynamics in terms of the expectation graphs. We then build on the characterization of the expected state dynamics, to relate statistical properties of the network model's persistent stochastic dynamics (e.g., ergodicity) and selected settling properties to the graph.

Let us first provide basic characterizations of the asymptotics of the expected state. A graphical characterization for both models can be achieved, using classifications of the graph topology that are analogous to the ones used in analyzing Markov chains [34,35]. This analysis depends on the fact that an LTI flow network's expectation graph is identical to a finite-state Markov chain's transition graph, and in fact the precise classification used for finite-state Markov chains will suffice for flow networks. Meanwhile, the state matrix for a synchronization network's expected state dynamics is the transpose of that for a flow network. Thus, the graph edges are reversed, and a slightly different classification is needed for this case. These analyses are important in that they specify the persistent behavior of linear flow and synchronization processes, including the hybrid and mixed stochastic/deterministic processes that are of interest in CPS applications.

Let us begin by defining some terminologies for the LTI stochastic flow network's expectation graph. The expectation flow graph is said to have a *path* from vertex i to vertex j if it has a sequence of directed edges from vertex i to vertex j. If there is a path from i to j, and also from j to i, the two vertices i and j are said to *cotransport*. A set of vertices that cotransport with each other, and do not cotransport with any vertex outside the set, is called a *flow class*. A flow class is called *absorbing*, if there is no path from a vertex in the flow class to the one outside. In other words, a flow class is absorbing if material/items cannot flow out of the corresponding network components. The vertices within an absorbing flow class are called *absorbing vertices*. A flow class that is not absorbing is called *transient*. The vertices within a transient flow class are called *transient vertices*. For an absorbing vertex i, a notion of periodicity needs to be defined. Specifically, the greatest common factor among the path lengths from vertex i back to itself is termed the *period* of the vertex; the vertex is called *aperiodic* if the period is 1 and periodic otherwise. The vertices in an absorbing class can be shown to have the same period, so the period measure (and periodicity) can be associated with the whole flow class.

We also define complementary terminologies for the LTI stochastic synchronization network, in terms of its expectation graph. Unlike the notions for flow processes (which are classical in Markov chain analysis), the complementary terminologies are not widely used in studying synchronization processes; however, they are similar to the notions developed for a stochastic automation known as the influence model [25,26]. The expectation synchronization graph is said to have an *influence path* from vertex i to vertex j if the graph has a sequence of directed edges from vertex i to vertex j. If there is a path from vertex i to vertex j, and also from vertex j to vertex i, the two vertices i and j are said to *coinfluence*. A set of vertices that coinfluence with each other, and do not coinfluence with any vertex outside the set, is called an *influence class*. An influence class is called *autonomous*, if there is no path from a vertex outside the influence class to the one inside. In other words, an influence class is autonomous if the expected opinions of the corresponding network components are not dependent on previous opinions of other network components. The vertices within an autonomous influence class are called *autonomous vertices*. An influence class that is not autonomous is called *dependent*. The vertices within a dependent influence class are called *dependent vertices*. For an autonomous vertex i, a notion of periodicity needs to be defined. Specifically, the greatest common factor among the path lengths from vertex i back to itself is termed the *period* of the vertex; the vertex is called *aperiodic* if the period is 1 and periodic otherwise. All the vertices in an autonomous influence class have the same period, so the period measure (and the periodicity concept) can be associated with the whole influence class.

We are now ready to present results regarding the asymptotics of the expected state dynamics. For both networks, we will characterize the asymptotics according to a classification of the graph topology, showing that qualitatively different asymptotics result depending on the topology. We begin with the flow network result, which admits the well-known characterization of Markov chains as a special case. The proof of the result follows rather simply from the formulation of the stochastic flow network, together with classical analysis of stochastic matrices for Markov chains; however, we include it here to highlight the relevance of these methods to the much broader class of flow processes considered here.

Theorem 5.1 *Consider an LTI stochastic flow network with expectation flow graph Γ. Then the asymptotics of the expected quantity vector $E(\mathbf{r}[k])$ are as follows:*

(1) *The expected quantities for the network components associated with the transient vertices of Γ asymptotically approach 0, that is, $\lim_{k\to\infty} E(r_i[k]) = 0$ if i is a transient vertex.*

(2) *Consider vertices in Γ that are in absorbing aperiodic flow classes. The expected quantity for each corresponding network component reaches a limit asymptotically, that is, $\lim_{k\to\infty} E(r_i[k])$ exists for all such i. If in fact Γ only has a single absorbing class that is aperiodic, then the asymptotic expectations at network components associated with this class are fixed positive fractions of the expected total quantity at the initial time. That is, for each vertex i in the absorbing aperiodic flow class, we have $\lim_{k\to\infty} E(r_i[k]) = v_i r_s$, where $v_i > 0$ and the sum of v_i over the vertices i in the absorbing aperiodic flow class is 1.*

(3) *In general, the expected quantities at network components corresponding to absorbing periodic vertices are not guaranteed to converge asymptotically. However, these expected quantities sampled at intervals equal to the period are convergent. That is, for an absorbing q-periodic vertex i in Γ, we have $\lim_{k\to\infty} E(r_i[qk + z])$ exists, for $z = 0, 1, \ldots, q - 1$.*

Proof: From the law of iterated expectations, we immediately find that $E(\mathbf{r}[k+1]) = QE(\mathbf{r}[k])$ for the LTI stochastic flow network and hence $E(\mathbf{r}[k]) = Q^k E(\mathbf{r}[0])$. Let us use the eigen analysis of the matrix Q in an analogous way to the corresponding analysis for Markov chains, to prove the result.

To begin, let us assume without loss of generality that the sites are ordered such that Q is a block lower-triangular matrix (with each diagonal block describing flows within a flow class). We notice that the columns of Q corresponding to absorbing flow classes are nil except for the diagonal blocks, since the flow out of these classes is not possible. From the block upper-triangular structure, it is apparent that eigen analysis of each diagonal block can permit characterization of the asymptotics.

First, let us characterize the expected quantities in the transient classes. It is clear that these expected quantities, say $E(\tilde{\mathbf{r}}[k])$, can be found as $E(\tilde{\mathbf{r}}[k]) = \tilde{Q}^k E(\tilde{\mathbf{r}}[0])$, where $\tilde{Q}$ is a block lower-triangular submatrix of Q whose diagonal blocks capture flows within the transient classes. Noticing that each of these diagonal blocks is a columnwise substochastic matrix corresponding to a connected graph, it immediately follows from the Frobenius–Perron theorem that the eigenvalues of these diagonal blocks and hence of $\tilde{Q}$ lie strictly within the unit circle. Hence, $\tilde{Q}^k$ approaches a nil matrix asymptotically, and the expected quantities at transient network components are seen to approach 0 asymptotically.

Next, let us characterize expected quantities for the absorbing aperiodic classes. To do so, let us consider each diagonal block of Q associated with an absorbing aperiodic class. For any such block, say $\widehat{Q}$, the Frobenius–Perron theorem immediately yields that $\widehat{Q}$ has an eigenvalue at 1 that strictly majorizes in magnitude all the other eigenvalues; further, the left and right eigenvectors of $\widehat{Q}$ associated with the unity eigenvalue are $\mathbf{1}^T$ and $\mathbf{v}$, respectively, where $\mathbf{v}$ is strictly positive and its entries sum to unity. Next, noticing that the expected quantities at the transient network components approach 0 asymptotically, we see that the expected quantities $E(\widehat{\mathbf{r}}[k])$ at network components associated with this absorbing class satisfy $E(\widehat{\mathbf{r}}[k+1]) = \widehat{Q}E(\widehat{\mathbf{r}}[k])$ asymptotically. From the eigen decomposition of $\widehat{Q}$, it immediately follows that the each expected quantity in an aperiodic absorbing component approaches a limit. In the case that this absorbing flow class is the only one, we note further that $\mathbf{1}^T\widehat{\mathbf{r}}[k]$ must approach $\mathbf{1}^T E(\mathbf{r}[0])$ asymptotically (since the expected quantities at all other components approach zero and the total quantity and expected quantity remain constant). It thus follows that asymptotic expected quantity at a vertex i in the absorbing class approaches $(v_i)\mathbf{1}^T E(\mathbf{r}[0])$, that is, a positive fraction of the total quantity is in component i of the absorbing class on average asymptotically.

Finally, let us consider periodic absorbing classes. We again use the fact that, asymptotically, $E(\widehat{\mathbf{r}}[k+1]) = \widehat{Q}E(\widehat{\mathbf{r}}[k])$ for an absorbing class, where $\widehat{Q}$ describes the expected flows within the class. The eigenanalysis of $\widehat{Q}$ in the periodic case, via the Frobenius–Perron theorem ([36]), immediately yields the result: we omit the details.

Let us present complementary results for the LTI stochastic synchronization model.

Theorem 5.2 *Consider an LTI stochastic synchronization network with expectation synchronization graph Λ. Then the following are the asymptotics of the expected opinion vector $E(\mathbf{s}[k])$:*

(1) The expected opinions of network components associated with an autonomous aperiodic class each converge to a limit asymptotically, and further they synchronize (become equal) asymptotically. That is, for a vertex i in an autonomous class, $\lim_{k\to\infty} E(s_i[k])$ exists; also, for two vertices i and j in an autonomous class, we have $\lim_{k\to\infty} E(s_i[k]) - E(s_j[k]) = 0$. Further, the asymptotic value of the expected opinions is a positive unitary linear combination of the initial expected opinions of the agents. That is, $\lim_{k\to\infty} E(s_i[k]) = \sum_{j\in V_s} w_j E(s_j[0])$, where $w_j > 0$, $\sum_{j\in V_s} w_j = 1$, and V_s contains all the vertices in the autonomous class.

(2) If all the autonomous classes in the graph are aperiodic, then the expected opinions of all network components converge to a limit asymptotically (i.e., $\lim_{k\to\infty} E(s_i[k])$ exists for all

vertices i). The expected opinion of each network component associated with a dependent vertex converges to a unitary nonnegative linear combination of the limiting expected opinions of the autonomous classes. If in fact the graph has a single autonomous class that is aperiodic, then all network components synchronize asymptotically, to the same limiting expected opinion as in the autonomous class.

(3) *The expected opinions at network components corresponding to autonomous periodic vertices are not guaranteed to converge asymptotically. However, these expected opinions sampled at intervals equal to the period are convergent. That is, for a autonomous q-periodic vertex i in* Γ*, we have that* $\lim_{k\to\infty} E(s_i[qk+z])$ *exists, for* $z = 0, 1, \ldots, q-1$. *When one or more autonomous classes are periodic, then the expected opinions at the network components associated with dependent vertices also vary periodically, with period equal to the least-common multiple of the periods of some or all of the autonomous classes' periods.*

Proof: Again, we recover from the law of iterated expectations that $E(\mathbf{s}[k+1]) = H^T E(\mathbf{s}[k])$ and hence $E(\mathbf{s}[k+1]) = (H^T)^k E(\mathbf{s}[k])$.

Let us begin by characterizing the expected opinions for agents associated with the autonomous classes. We note that the expected opinions $E(\hat{\mathbf{s}}[k])$ for an autonomous class satisfy $E(\hat{\mathbf{s}}[k+1]) = \widehat{H}^T E(\hat{\mathbf{s}}[k])$, where $\widehat{H}^T$ is a submatrix of H^T and is itself a row-stochastic matrix. Now, for an autonomous aperiodic class, the Frobenius–Perron theorem immediately yields that $\widehat{H}^T$ has a single unity eigenvalue that strictly dominates the others, with corresponding right eigenvector of $\mathbf{1}$ and left eigenvector $\mathbf{w}^T$ whose entries are strictly positive and sum to 1. It thus immediately follows that $\lim_{k\to\infty} E(\hat{\mathbf{s}}[k]) = \mathbf{1}\mathbf{w}^T E(\hat{\mathbf{s}}[0])$. Thus, we see that the expected opinions not only converge but reach consensus asymptotically, with the consensus value equal to a positive unitary combination of the agents' expected initial opinions. For an autonomous q-periodic class, the Frobenius–Perron analysis also immediately yields that the expected opinions converge to a periodic signal of period q.

Let us now characterize the expected state asymptotics for agents associated with the dependent classes. To do so, we first note that the expected state dynamics, without loss of generality, can be written as

$$\begin{bmatrix} E(\bar{\mathbf{s}}[k+1]) \\ E(\tilde{\mathbf{s}}[k+1]) \end{bmatrix} = \begin{bmatrix} H_a^T & 0 \\ H_{ad}^T & H_d^T \end{bmatrix} \begin{bmatrix} E(\bar{\mathbf{s}}[k]) \\ E(\tilde{\mathbf{s}}[k]) \end{bmatrix},$$

where $E(\bar{\mathbf{s}}[k])$ contains the expected statuses associated with the autonomous classes, and $E(\tilde{\mathbf{s}}[k])$ contains those associated with the dependent classes.

From the Frobenius–Perron theory, we immediately recover that H_d^T has eigenvalues strictly inside the unit circle and hence that the expected opinions of all agents asymptotically do not depend on the initial expected opinions of the agents in the dependent classes. However, the expected opinion of each agent at any time is necessarily a nonnegative unitary linear combination of some agents' initial states. Hence, we recover that the expected opinions of dependent classes' agents are unitary linear combinations of the expected initial opinions of the autonomous classes' agents. Hence, if all autonomous classes are aperiodic, we obtain that the dependent agents' states also converge asymptotically, with their opinions given by a unitary combination of the consensus opinions of the autonomous classes. In the case where some autonomous classes are periodic, the opinions of the dependent classes' agents are linear combinations of periodic signals, and hence we recover that these opinions also approach periodic signals whose periods are least-common multiples of those of the autonomous classes.

Several remarks about the preceding results are needed.

Remark 5.3 *The aforementioned basic characterizations of flow and synchronization processes recover classical results for Markov chains [36]. At their essence, they follow simply from well-known properties of reducible nonnegative matrices with unity eigenvalues. The results are interesting from that standpoint that they inform the statistical analysis of several other models for stochastic network dynamics, including infinite-server, queueing network, and voter models, in terms of a network graph. The results also highlight the duality between flow and synchronization process asymptotics.*

Remark 5.4 *For synchronization networks, it is the autonomous classes' initial states that impact the asymptotic expected dynamics. The number of these autonomous classes and their periodicity structures modulate the asymptotic expected dynamics just as for flow networks, albeit with some subtle differences. One particularly interesting difference is the fact that periodic orbits in the expected opinion vector that are multiples of the periods of individual autonomous classes may result; to the best of our knowledge, synchronization algorithms that asymptotically achieve such periodic orbits have not been developed or analyzed. It is useful to know that coordination algorithms with more sophisticated asymptotics of this sort can be constructed.*

Remark 5.5 *The thesis [10] (and numerous papers following on it) provides a comprehensive study of distributed averaging in the time-varying case. These approaches can permit some generalization of the aforementioned results to the time-varying case.*

Building on the basic analysis of the expected state, we also characterize asymptotics of the networks' stochastic state dynamics, focusing primarily on second-moment characterizations. Specifically, we present graph-theoretic results on mean-square convergence of the state, as well as on ergodicity of state dynamics. We again give complementary results for synchronization and flow networks. Here are the two results for LTI stochastic flow networks:

Theorem 5.3 *Consider an LTI stochastic flow network. The quantities $r_i[k]$ at network components associated with transient vertices i converge to 0 in a mean-square sense.*

Proof: In order to prove the theorem, we need to show that $\lim_{k \to \infty} E((r_i[k])^2) = 0$ for a transient vertex i. For a stochastic flow network, we have $\mathbf{1}^T \mathbf{r}[k] = r_s$ for all k. Also, because each local flow quantity $r_i[k]$ should not be smaller than the total amount of flow r_s, we have $0 \le r_i[k] \le r_s$.

Next, given that $\lim_{k \to \infty} E(r_i[k]) = 0$ for a transient vertex i, we have $E((r_i[k])^2) \le r_s E(r_i[k])$, which immediately yields that $\lim_{k \to \infty} E((r_i[k])^2) = 0$.

Theorem 5.4 *Consider an LTI stochastic flow network that has a single absorbing class, which is aperiodic. Given the total quantity $\mathbf{1}^T \mathbf{r}[0]$, the quantity vector process $\mathbf{r}[k]$ is mean-square ergodic.*

Proof: Preliminarily, let us recall that $\lim_{k \to \infty} E(r_i[k]) = v_i r_s$ for each vertex i in the aperiodic absorbing class, where $v_i > 0$ and $\sum_{i=1}^{n} v_i = 1$. Defining a column vector $\mathbf{v} = [v_1, v_2, \cdots, v_n]^T$, we also have $\lim_{k \to \infty} E(\mathbf{r}[k]) = r_s \mathbf{v}$.

To verify ergodicity of $\mathbf{r}[k]$, we need to show that the time average $\frac{1}{N} \sum_{k=1}^{N} \mathbf{r}[k]$ converges to the ensemble average $\lim_{k \to \infty} E(\mathbf{r}[k])$ as N goes to infinity, in a mean-square sense. Mathematically, it is equivalent to show that

$$\lim_{N \to \infty} E\left(\left(\frac{1}{N} \sum_{k=1}^{N} \mathbf{r}[k] - r_s \mathbf{v} \right)^T \left(\frac{1}{N} \sum_{k=1}^{N} \mathbf{r}[k] - r_s \mathbf{v} \right) \right) = 0.$$

Hybrid Stochastic Synchronization and Flow Models

Let us now prove the aforementioned statement through a convergence analysis. Throughout the analysis, we assume that the initial total quantity $\mathbf{1}^T\mathbf{r}[0]$ is given (but suppress the conditioning notation to simplify the presentation).

To begin, we rewrite $E((\frac{1}{N}\sum_{k=1}^{N}\mathbf{r}[k] - r_s\mathbf{v})^T(\frac{1}{N}\sum_{k=1}^{N}\mathbf{r}[k] - r_s\mathbf{v}))$ as follows:

$$E\left(\left(\frac{1}{N}\sum_{k=1}^{N}\mathbf{r}[k] - r_s\mathbf{v}\right)^T\left(\frac{1}{N}\sum_{k=1}^{N}\mathbf{r}[k] - r_s\mathbf{v}\right)\right)$$

$$= \frac{1}{N^2}E\left(\sum_{k=1}^{N}\sum_{l=1}^{N}\mathbf{r}^T[k]\mathbf{r}[l]\right) - \frac{2r_s}{N}\mathbf{v}^T\sum_{k=1}^{N}E(\mathbf{r}[k]) + r_s^2\mathbf{v}^T\mathbf{v}. \tag{5.1}$$

Since $\lim_{k\to\infty}E(\mathbf{r}[k]) = r_s\mathbf{v}$, let $E(\mathbf{r}[k]) = r_s\mathbf{v} + \delta[k]$, where $\delta[k]$ is a vector whose norm is getting smaller exponentially with time (i.e., $\lim_{N\to\infty}\frac{1}{N}\sum_{k=1}^{N}\delta[k] = 0$). Then, we have

$$\lim_{N\to\infty}\frac{2r_s}{N}\mathbf{v}^T\sum_{k=1}^{N}E(\mathbf{r}[k])$$

$$= \lim_{N\to\infty}\frac{2r_s}{N}\mathbf{v}^T\left(Nr_s\mathbf{v} + \sum_{k=1}^{N}\delta[k]\right)$$

$$= 2r_s^2\mathbf{v}^T\mathbf{v}. \tag{5.2}$$

Therefore, as N goes to infinity, Equation 5.1 becomes

$$\lim_{N\to\infty}E\left(\left(\frac{1}{N}\sum_{k=1}^{N}\mathbf{r}[k] - r_s\mathbf{v}\right)^T\left(\frac{1}{N}\sum_{k=1}^{N}\mathbf{r}[k] - r_s\mathbf{v}\right)\right)$$

$$= \lim_{N\to\infty}\frac{1}{N^2}E\left(\sum_{k=1}^{N}\sum_{l=1}^{N}\mathbf{r}^T[k]\mathbf{r}[l]\right) - r_s^2\mathbf{v}^T\mathbf{v}. \tag{5.3}$$

Without loss of generality (WLOG), let us consider the term $E(\mathbf{r}^T[k]\mathbf{r}[l])$, for $k \leq l$:

$$E\left(\mathbf{r}^T[k]\mathbf{r}[l]\right) = E\left(E(\mathbf{r}^T[k]\mathbf{r}[l]|\mathbf{r}[k])\right)$$

$$= E\left(\mathbf{r}^T[k]E(\mathbf{r}[l]|\mathbf{r}[k])\right)$$

$$= E\left(\mathbf{r}^T[k]Q^{l-k}\mathbf{r}[k]\right). \tag{5.4}$$

Notice that for an LTI stochastic flow network, column-stochastic matrix Q has a dominant eigenvalue at 1, that is, $\lim_{k\to\infty}Q^k = \mathbf{v}\mathbf{1}^T$. Hence, Q^k can be written as $Q^k = \mathbf{v}\mathbf{1}^T + \Delta[k]$, where the norm of matrix $\Delta[k]$ becomes small exponentially with k. Then, Equation 5.4 becomes

$$E\left(\mathbf{r}^T[k]\mathbf{r}[l]\right) = E\left(\mathbf{r}^T[k](\mathbf{v}\mathbf{1}^T + \Delta[l-k])\mathbf{r}[k]\right)$$

$$= E\left(\mathbf{r}^T[k]\mathbf{v}\mathbf{1}^T\mathbf{r}[k]\right) + E(\tilde{\Delta}[l-k])$$

$$= r_s\mathbf{v}^T E\left(\mathbf{r}[k]\right) + E(\tilde{\Delta}[l-k]), \tag{5.5}$$

where $\tilde{\Delta}[l-k] = \mathbf{r}^T[k]\Delta[l-k]\mathbf{r}[k]$, and $E(\tilde{\Delta}[l-k])$ decays exponentially with $l-k$.

Thus, we have $\lim_{N\to\infty} \frac{1}{N^2} E\left(\sum_{k=1}^{N}\sum_{l=1}^{N}\tilde{\Delta}[l-k]\right) = 0$. Finally, we rewrite Equation 5.1 as follows:

$$\lim_{N\to\infty} E\left(\left(\frac{1}{N}\sum_{k=1}^{N}\mathbf{r}[k] - r_s\mathbf{v}\right)^{T}\left(\frac{1}{N}\sum_{k=1}^{N}\mathbf{r}[k] - r_s\mathbf{v}\right)\right)$$

$$= \lim_{N\to\infty}\frac{1}{N^2}\left(Nr_s\mathbf{v}^{T}E\left(\sum_{k=1}^{N}\mathbf{r}[k]\right) + E\left(\sum_{k=1}^{N}\sum_{l=1}^{N}\tilde{\Delta}[l-k]\right)\right) - r_s^2\mathbf{v}^{T}\mathbf{v}$$

$$= \lim_{N\to\infty}\frac{1}{N}r_s\mathbf{v}^{T}E\left(\sum_{k=1}^{N}\mathbf{r}[k]\right) - r_s^2\mathbf{v}^{T}\mathbf{v}$$

$$= r_s\mathbf{v}^{T}(r_s\mathbf{v}) - r_s^2\mathbf{v}^{T}\mathbf{v}$$

$$= 0, \tag{5.6}$$

which completes the proof.

The LTI stochastic synchronization network admits dual characterizations of its asymptotic state dynamics, in terms of the underlying expectation graph structure. Specifically, the asymptotic characterizations of the synchronization network are concerned with the dependence of an opinion at some time on initial or previous opinions of all the agents. In particular, we note that the opinion of each agent i at time k, or $s_i[k]$, can always be written as a unitary linear combination of the opinion vector at any previous time $\bar{k} < k$: $s_i[k] = \mathbf{w}_i^{T}[k,\bar{k}]\mathbf{s}[\bar{k}]$, where the *opinion-influence vector* for site i $\mathbf{w}_i[k,\bar{k}]$ is nonnegative and satisfies $\mathbf{w}_i^{T}[k,\bar{k}]\mathbf{1} = 1$. Here, let us present two results on the time evolution of the opinion-influence vector:

Theorem 5.5   *Consider an LTI stochastic synchronization network. For fixed initial time $\bar{k}$, any element of an opinion-influence vector corresponding to a dependent vertex in the expectation graph converges to 0 with respect to the time index k, in a mean-square sense.*

Theorem 5.6   *Consider an LTI stochastic synchronization network that has a single autonomous, aperiodic class. Then the opinion-influence vectors' dynamics for the network are ergodic in the following sense: for fixed k_0 and any i, $\frac{1}{k-k_0+1}\sum_{z=k_0}^{k}\mathbf{w}_i[k,z]$ approaches $E(\mathbf{w}_i[k,k_0])$ in a mean-square sense, as k is made large.*

The ergodicity results are particularly informative, in that they show that unknown stochastic flow/synchronization processes, including hybrid ones, can be partially characterized from time histories. These results thus also provide a starting point toward developing tools for process characterization from recorded data, which is often of importance in heterogeneous cyber-physical systems. Several further remarks about the convergence and ergodicity results are worthwhile:

Remark 5.6   *It is interesting that ergodicity results can be proved, even though only a first-moment conditional linearity condition is assumed. In general, the mean-square calculus for stochastic flow and synchronization networks is difficult because second moments may not satisfy closure properties that allow explicit statistical computation; this is the case, for example, for the nonlinear/hybrid models that we have defined (e.g., the mixed flow model with thresholding behavior). However, even in these cases, first-moment linearity together with certain nonnegativity constructs permits bounding of the second moments in a way that allows verification of ergodicity. Thus, the hybrid* *algorithms/processes in our framework that are relevant to CPS applications are guaranteed to have pleasant asymptotic characteristics (ergodicity), which facilitate analysis and design.*

Remark 5.7 *The mean-square convergence results also immediately imply convergence in probability; however, almost-sure convergence is not guaranteed.*

Remark 5.8 *The proofs of convergence and ergodicity indicate that the rate of approach to the asymptote is governed by the subdominant eigenvalue of the flow and synchronization expectation matrices, which are stochastic matrices (or transposes of stochastic matrices). Numerous graph-theoretic results on stochastic matrices' subdominant eigenvalues have been developed (e.g., [37,38]), which thus characterize this approach rate. It is worth stressing that the convergence and ergodicity results, as well as the approach-rate characterization, encompass the models captured in our framework. For instance, based on the preceding results, the asymptotics of the voter model and probabilistic routing model can be characterized entirely from their expectation graphs. This sharply contrasts with a full statistical analysis of the models, which requires a* master Markov chain *whose dimension is exponential in the number of network components.*

5.5.3 COMPARISONS AMONG MODELS

The broad framework introduced here also permits us to compare stochastic flow or synchronization models with different update rules and to develop general bounds for their dynamical properties. Here, let us present a first result of this sort, for linear stochastic flow networks. In particular, we bound the variability of a linear stochastic flow network's state with given expectation matrices $Q[k]$ and argue that a Markov chain-type model achieves the maximum variability among all flow networks with these expectation matrices. This result is useful in that it gives a worst-case analysis for state variability and indicates comparisons among different hybrid models within the class of stochastic flow networks. Here is the result:

Theorem 5.7 *Consider a linear stochastic flow network, with expected initial state $E(\mathbf{r}[0])$ and expectation flow matrices $Q[0], Q[1], \ldots$. Consider the variability of the quantity vector $\mathbf{r}[k]$. In particular, consider the measure $L = \sum_{i=1}^{n} var(r_i[k])$, where $var()$ represents the variance of a random variable. Then $L \leq (\mathbf{1}^T E(\mathbf{r}[0]))^2 - \mathbf{z}^T \mathbf{z}$, where $\mathbf{z} = Q[k] \times \ldots \times Q[0]E(\mathbf{r}[0])$. The maximum variability is achieved by a Markov chain-type dynamics. Specifically, a model in which the total quantity is placed at a single component at the initial time (with the component chosen probabilistically so that the expected initial state is as specified) and in which the flow state matrix at each time k is that of a Markov chain with transition matrix $Q[k]$ achieves the bound.*

Proof: Let us first verify the bound and then show that the Markov chain–type dynamics achieves the bound. To verify the bound, let us first develop a bound on the variability of the quantity at each vertex i, that is, let us upper bound $var(r_i[k])$. To do so, we note that (1) $0 \leq r_i[k] \leq \mathbf{1}^T\mathbf{r}[0]$ and (2) $E(r_i[k]) = z_i$. Thus, we see that $E(r_i^2[k]) \leq E((r_i[k])(\mathbf{1}^T\mathbf{r}[0])) = \mathbf{1}^T\mathbf{r}[0]E(r_i[k]) = (\mathbf{1}^T\mathbf{r}[0])z_i$. Thus, we recover that $var(r_i[k]) = E(r_i^2[k]) - (E(r_i[k]))^2 \leq z_i(\mathbf{1}^T\mathbf{r}[0]) - z_i^2$. It thus immediately follows that $L = \sum_{i=1}^{n} var(r_i[k]) \leq \mathbf{1}^T\mathbf{r}[0] \sum_{i=1}^{n} z_i - \mathbf{z}^T\mathbf{z} = (\mathbf{1}^T\mathbf{r}[0])^2 - \mathbf{z}^T\mathbf{z}$, and we have proved the bound.

Next, let us consider the Markov chain–type dynamics described in the theorem. From the description, we recover immediately that $r_i[k] = \mathbf{1}^T\mathbf{r}[0]$ with probability $\frac{z_i}{\mathbf{1}^T\mathbf{r}[0]}$ and is 0 otherwise. Thus, we immediately recover that $var(r_i[k]) = z_i(\mathbf{1}^T\mathbf{r}[0]) - z_i^2$, whereupon it follows that this dynamics achieves the bound (and hence achieves maximum variability).

We have thus shown that the unitized variability of the linear stochastic flow network is upper bounded by a Markov chain model. In a similar fashion, the influence model can be shown to be the most uncertain among the stochastic synchronization network models (in a normalized sense); details of the argument are omitted.

5.6 MORE DETAILED ANALYSES OF ALGORITHM OUTCOMES: CHALLENGES AND INITIAL RESULTS

Thus far, the formal analyses of the stochastic flow and synchronization models have largely focused on statistical characterizations. In many cases, more detailed analyses of the computations/outcomes achieved by the algorithms are needed. For a stochastic synchronization process, we naturally would like to characterize whether state synchronization is achieved and how the asymptotic state depends on the initial opinions of the agents. Likewise, we may wish to analyze the asymptotic distribution of the flowing quantity in the stochastic flow network. Such characterizations are often especially relevant for algorithm design, to ensure that developed algorithms achieve performance specifications with certainty. For instance, the implementation of the hybrid algorithm for cyber-physical consensus introduced in Section 5.4 requires a performance analysis ascertaining the synchronization is achieved and characterizing the synchronized state.

Generic analysis of model outcomes for the defined stochastic flow and synchronization processes is challenging, because of the breadth of these model classes. For instance, the stochastic synchronization models encompass linear consensus processes that are traditionally analyzed via spectral and monotonicity-based arguments [10]. The model class also encompasses voter models, which can be shown to achieve synchronization via characterization of a master Markov chain [11,25]. Some hybrid algorithms that fall within the class of stochastic synchronization processes can also be analyzed using similar techniques as for Markov jump linear systems [39]. However, these analyses depend on certain second-moment-closure properties, which do not necessarily hold even for linear stochastic synchronization processes (for which only closure of the first moment is guaranteed). The variety of dynamics captured in the stochastic flow and synchronization models thus complicates their performance analysis. Perhaps the most promising approach toward a formal performance analysis is to impose a second-moment-closure restriction and then to use mean-square calculus to extract performance characterizations; this approach has the benefit of encompassing continuous-valued, discrete-valued, and hybrid cases. We are still pursuing a formalization of this approach, and so a general treatment is not included here.

As a step toward a general analysis of algorithm/process outcomes, we here illustrate the analysis for the hybrid averager–voter algorithm described in Section 5.4. This algorithm may be a useful tool for decision making in cyber-physical networks, provided that (1) the hybrid agents in the model indeed do reach synchronization and (2) the synchronization state (the asymptotic value taken by the agents) can be characterized. Here, we formalize that the algorithm achieves synchronization subject to connectivity conditions on the influence expectation graph and illustrate the approach to synchronization via simulation of a small-scale example.

Let us begin with the formal result:

Theorem 5.8 *The averager–voter algorithm achieves synchronization in both a mean-square sense and a probability 1 sense, provided that the synchronization expectation graph is strongly connected (i.e., there is a path between any two vertices). That is, the sequences $d[k] = s_j[k] - s_1[k]$ converge to 0 in a mean-square sense and a probability 1 sense, for $j = 2, \ldots, n$, given strong connectivity.*

Sketch of the proof: The proof exploits a linear recursion for the second moment of the state for the averager–voter model; see Roy [28] for the development of similar recursions for hybrid models using Kronecker-product formalisms. This recursion together with the recursion on the expected state

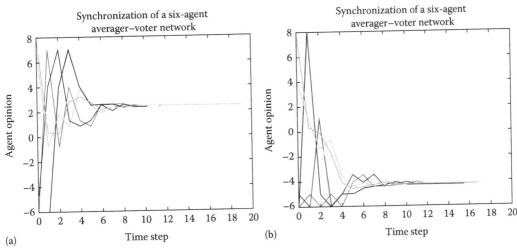

FIGURE 5.1 Two simulations of an example averager–voter network are shown, for a common initial condition. Synchronization is achieved, but the synchronization value is different in the two cases.

permits characterization of $E((s_j[k] - s_1[k])^2)$. In particular, the state matrices of the first- and second-moment recursions can be shown to have a unique dominant eigenvalue at 1 given strong connectivity, and their corresponding left and right eigenvectors can be characterized, using similar arguments to those given in Asavathiratham [25]. Based on the spectral characterization, $E((s_j[k] - s_1[k])^2)$ can be shown to be upper bounded by a decaying geometric function, with the decay factor specified by the subdominant eigenvalue of the second-moment recursion. The mean-square convergence result follows. Monotonicity of the sequence $(s_j[k] - s_1[k])^2$ together with mean-square convergence then guarantees convergence with probability 1. The details of the proof require some amount linear-algebraic development using Kronecker-product formalisms, which we exclude here in the interest of conciseness.

5.6.1 ILLUSTRATIVE EXAMPLE

The synchronization of the averager–voter model is illustrated. A network with six agents was considered. The network's graph was chosen to be a complete graph with different fixed edge weights. Specifically, each edge weight was first chosen to be uniform on [0, 1], and then the weights for each agent (i.e., averaging weights or copying probabilities) were scaled to sum to 1.

The model satisfies the strong connectivity condition of Theorem 5.8, and hence synchronization should be achieved with probability 1. Two simulations of the stochastic model are shown in Figure 5.1. As expected, the model achieves synchronization in each case. However, the synchronization value is different for the two model instances, reflecting the stochasticity of the update rule.

ACKNOWLEDGMENTS

This work was completed at the School of Electrical Engineering and Computer Science, Washington State University, Pullman, WA (E-mails: morashu@gmail.com,sroy@eecs.wsu.edu). This project was partially supported by NSF grants ECS-0901137, CNS-1035369, and CNS-1058124.

REFERENCES

1. R. Baheti and H. Gill, Cyber-physical systems, in *The Impact of Control Technology* (T. Samad and A. N. Annaswamy eds.), IEEE Control Systems Society, pp. 161–166, 2011.

2. E. A. Lee, Cyber-physical systems-are computing foundations adequate, Position Paper for *NSF Workshop On Cyber-Physical Systems: Research Motivation, Techniques and Roadmap*, Austin, TX, vol. 2, 2006.
3. I. Stojmenovic, Large-scale cyber-physical systems: Distributed actuation, in-network processing and machine-to-machine communications, in *Second Mediterranean Conference on Embedded Computing (MECO 2013)*, IEEE, Budva, Montenegro, pp. 21–24, 2013.
4. S. Strogatz, Exploring complex networks, *Nature*, 410, 268–276, 2001.
5. X. Li and G. Chen, Synchronization and desynchronization of complex dynamical networks: An engineering viewpoint, *IEEE Transactions on Circuits System I*, 50(11), 1381–1390, 2003.
6. Y. Wan, S. Roy, and A. Saberi, A new focus in the science of networks: Towards methods for design, *Proceedings of the Royal Society A*, 464(2091), 513–535, 2008.
7. M. Krajci and P. Mrafko, Computer simulations of the structure of amorphous systems by the Markov chain method, *Journal of Physics F: Metal Physics*, 14(6):1325–1332, June 1984.
8. S. Roy, B. Sridhar, and G. C. Verghese, An aggregate dynamic stochastic model for an air traffic system, in *Proceedings of the Fifth Eurocontrol/Federal Aviation Agency Air Traffic Management Research and Development Seminar*, Budapest, Hungary, June 2003.
9. M. Xue, S. Roy, A. Saberi, and B. Lesieutre, On generating sets of binary random variables with specified first- and second-moments, in *Proceedings of the 2010 American Control Conference*, Baltimore, MD, pp. 1133–1138, June 30 to July 2, 2010.
10. J. N. Tsitsiklis, Problems in decentralized decision making and computation, PhD thesis, Department of Electrical Engineering and Computer Science, Massachusetts Institute of Technology, Cambridge, MA, 1984.
11. S. Roy, A. Saberi, and K. Herlugson, A control-theoretic perspective on the design of distributed agreement protocols, *International Journal of Robust and Nonlinear Control*, 17, 1034–1066, 2007.
12. W. Ren and R. W. Beard, Consensus seeking in multiagent systems under dynamically changing interaction topologies, *IEEE Transactions on Automatic Control*, 50(5), 655–661, 2005.
13. L. Xiao and S. Boyd, Fast linear iterations for distributed averaging, *Systems and Control Letters*, 53(1), 65–78, 2004.
14. Z. Li, F.R. Yu, and M. Huang, A distributed consensus-based cooperative spectrum-sensing scheme in cognitive radios, *IEEE Transactions on Vehicular Technology*, 59(1), 383–393, 2010.
15. L. Page, S. Brin, R. Motwani, and T. Winograd, The PageRank citation ranking: Bringing order to the web, Stanford University Technical Report, Stanford, CA, 1998.
16. P. M. Anderson and A. A. Fouad, *Power System Control and Stability*, John Wiley & Sons, Hoboken, NJ, 2008.
17. S. Riley et al., Transmission dynamics of the etiological agent of SARS in Hong Kong: Impact of public health interventions, *Science*, 300(5627), 1961–1966, 2003.
18. S. Roy, T. F. McElwain, and Y. Wan, A network control theory approach to modeling and optimal control of zoonoses: Case study of brucellosis transmission in sub-Saharan Africa, *PLoS Neglected Tropical Diseases*, 5(10), e1259, 2011.
19. M. Xue, and S. Roy, Spectral and graph-theoretic bounds on steady-state-probability estimation performance for an ergodic Markov chain, *Journal of the Franklin Institute*, 348(9), 2448–2467, November 2011.
20. M. Xue, E. Yeung, A. Rai, S. Roy, Y. Wan, and S. Warnick, Initial-condition estimation in network synchronization processes: Graphical characterizations of estimator structure and performance, *Complex Systems*, 21(4):297–333, 2013.
21. M. Xue, S. M. Zobell, S. Roy, C. P. Taylor, Y. Wan, and C. Wanke, Using stochastic, dynamic weather-impact models in strategic traffic flow management, in *Proceedings of 91st American Meteorological Society Annual Meeting*, Seattle, WA, January 2011.
22. A. Ribeiro, Z. Q. Luo, N. D. Sidiropoulos, and G. B. Giannakis, Modelling and optimization of stochastic routing for wireless multihop networks, in *Proceedings of the 26th Annual Joint Conference of the IEEE Computer and Communications Societies (INFOCOM)*, Anchorage, AK, pp. 1748–1756, May 2007.
23. M. Xue, S. Roy, Y. Wan, and A. Saberi, Designing asymptotics and transients of linear stochastic automaton networks, in *Proceedings of the AIAA Guidance, Navigation and Control Conference*, Toronto, Ontario, Canada, August 2010.
24. T. M. Liggett, *Stochastic Interacting Systems: Contact, Voter and Exclusion Processes*, Springer, Berlin, Germany, 1999.

25. C. Asavathiratham, The influence model: A tractable representation for the dynamics of networked Markov chains, PhD thesis, Massachusetts Institute of Technology, Cambridge, MA, October 2000.
26. C. Asavathiratham, S. Roy, B. Lesieutre, and G. Verghese, The influence model, *IEEE Control Systems Magazine*, 21(6), 52–64, 2001.
27. R. A. Holley and T. M. Liggett, Ergodic theorems for weakly interacting infinite systems and the voter model, *The Annals of Probability*, 3(4):643–663, 1975.
28. S. Roy, Moment-linear stochastic systems and their applications, PhD dissertation, Massachusetts Institute of Technology, Cambridge, MA, 2003.
29. A. Berman, M. Neumann, and R. J. Stern, *Nonnegative Matrices in Dynamical Systems*, Wiley Interscience, New York, 1989.
30. J. E. Mosimann, On the compound multinomial distribution, the multivariate β-distribution, and correlations among proportions, *Biometrika*, 49(1–2), 65–82, 1962.
31. V. D. Blondel, J. M. Hendrickx, A. Olshevsky, and J. N. Tsitsiklis, Convergence in multiagent coordination, consensus, and flocking, in *Proceedings of the Joint 44th IEEE Conference on Decision and Control and European Control Conference (CDC-ECC'05)*, Seville, Spain, pp. 2996–3000, December 2005.
32. S. Basu, T. Choudhury, B. Clarkson, and A. Pentland, Learning human interactions with the influence model, in *Proceedings of the Neural Information Processing Systems (NIPS) Conference*, Vancouver, Canada, 2001.
33. Y. Wan, S. Roy, A. Saberi, and B. Lesieutre, A stochastic automation-based algorithm for flexible and distributed network partitioning, in *Proceedings of the 2005 IEEE Swarm Intelligence Symposium*, Pasadena, CA, pp. 273–280, 2005.
34. J. G. Kemeny and J. L. Snell, *Finite Markov Chains*, Springer-Verlag, New York, 1976.
35. P. Brémaud, *Markov Chains: Gibbs Fields, Monte Carlo Simulation, and Queues*, Springer, New York, 1999.
36. R. G. Gallager, *Discrete Stochastic Processes*, Kluwer Academic, Boston, MA, 1996.
37. D. J. Hartfiel and C. D. Meyer, On the structure of stochastic matrices with a subdominant eigenvalue near 1, 272(1), 193–203, 1998.
38. U. G. Rothblum and C. P. Tan, Upper bounds on the maximum modulus of subdominant eigenvalues of nonnegative matrices, *Linear Algebra and Its Applications*, 66, 45–86, 1985.
39. O. L. V. Costa, M. D. Fragoso, and R. P. Marques, *Discrete-Time Markov Jump Linear Systems*, Springer, New York, 2006.

6 Multi-View, Multi-Domain, and Multi-Paradigm Approaches for Specification and Modeling of Big Data Driven Cyber-Physical Systems

Lichen Zhang

CONTENTS

6.1 INTRODUCTION

Big data as a phase have been among the biggest trends of the four years, giving rise to an upsurge of research and development, as well as industry and government applications. Data are deemed a powerful and useful raw material that can impact multidisciplinary research endeavors as well as government, industry, and business performance. Everyone is talking about big data, and it is believed that science, business, industry, government, society, and so forth will undergo a thorough change with the influence of big data [1]. In fact, we now live in the era of big data as huge and complex data sets, which are being produced and collected for diverse purposes through all kinds of technologies or approaches including mobile devices, remote sensing technologies, software logs, wireless sensor networks, and social media. The big data sets tend to be more unstructured, distributed, and complex than ever before [2]. Thus, big data are defined as huge amount of data that require new technologies and architectures so that it becomes possible to extract value from them by capturing and analysis process. Due to such large size of data, it becomes very difficult to perform effective analysis using the existing traditional techniques. Big data due to their various properties like volume, velocity, variety, variability, value, and complexity put forward many challenges [3]. Under the great increase in global data, the term big data is constantly used to describe huge data sets. Compared with traditional data sets, big data typically compose of masses of unstructured data, sensor data, and stream data that need more real-time processing. Big data are characterized by three aspects: (1) data are numerous, (2) data cannot be categorized into regular relational databases, and (3) data are generated, captured, and processed rapidly. Moreover, big data are transforming health care, transportation, aerospace, science, engineering, finance, business, and eventually, the society [4]. In addition, big data also give rise to new opportunities for discovering new values, make us gain an in-depth understanding of the hidden values, and also bring new challenges, for example, how to effectively specify, model, capture, transfer, organize, and manage such data sets [5]. Big data aim to handle high-volume, high-velocity, high-variety data to extract intended data value and ensure high veracity of original data and obtained information that demand cost-effective, innovative forms of data and information processing for enhanced insight, decision making, and process control; all of these demand new data models and new infrastructure services and tools that allow also obtaining from a variety of sources and delivering data in a variety of forms to different data and information consumers and devices [6].

Cyber-physical systems (CPS) feature a tight combination of, and coordination between, the system's computational and physical elements and integration of computer and information-centric physical and engineered systems [7]. The CPS (cyber-physical system) is composed of sensing, monitoring, computing, control, and communication systems. The physical world modeling, hardware, software, and networking technology are efficiently integrated together rather than computation alone. In a typical CPS system, embedded computers and communication networks are sort of computational resources, while physical units (cars, engines, aircraft, and trains), actuators, and sensors are viewed as physical resources. Today, computational resources and physical resources are strongly coupled and interacted. That is, computational resources schedule and control physical resources; on the other hand, the proper functioning of computational resources requires steady operation of physical resources [8]. A vehicular ad hoc network (VANET) is a typical cyber-physical system, to enhance passenger comfort, traffic efficiency, safety of passengers and so on, by forwarding upcoming traffic information in a timely manner [9]. Most of the safety-related applications (such as collision warning, road merging, and emergency braking) require quick computation and exchange of messages [10]. Thus, cyber-physical systems such as transportation cyber-physical systems are systems that have both a physical world and cyber world, which are integrated tightly, and the cyber world must interact with the physical world timely. Cyber-physical systems must directly capture and record physical data using sensors; monitor, affect, and control physical processes using actuators; evaluate, processes, and save recorded data; and actively or reactively interact with both the physical and cyber worlds. Cyber-physical systems are connected with one another and in global networks via different communication facilities, and use globally available data and services [11]. A very important issue for cyber-physical systems is the collection and acquisition of data such as parallel data collection via sensors, data transfer and communication, data store, data fusion, data enquiry, and processing of physical data from the environment, locally, globally, and in real time. Data from sensors combined with data fusion, data mining, and interpretation enable physical control and awareness of systems efficiently. The real-time data management in cyber-physical systems can be seen as all the necessary resources to meet requirements (collection, storage, and exploitation) of real-time applications working on data from sensors. These real-time applications differ from other applications by taking into account time constraints, compliance with which is as important as the accuracy of the result. In other words, these applications should not just deliver accurate results, but within the deadlines [12]. The interpretation aims at regarding achievements of goals and targets of cyber-physical systems. In the acquisition and interpretation models, methods and techniques of deduction are useful unified with the prediction of physical environment dynamic change, faults, obstacles, and risks. A very important feature of cyber-physical systems is the interaction between CPS and their context consisting of users, the physical environment, and systems and services from the cloud. This demands interoperation, integration, rules for and control of CPS components and functions in globally distributed, networked real-time control, and regulation [13]. The availability of efficient embedded systems and real-time systems in combination with new communication technologies such as mobile communication technologies has accelerated and increased the application of CPS. Such systems contain software modules, network, communication, and physical components such as airplanes, cars, trains, and medical devices that can be considered as the basis and platform for addressing many upcoming application domains and challenges: sustainable transportation concepts, production systems, medical processes, logistic processes, coordination processes and management processes, intelligent and adaptive power grids, self-optimizing platforms for managing buildings, and flexible manufacturing processes for the industry. An important requirement for tackling these scenarios is the availability of standardized networked components as well as the intelligent application of CPS data. This requires techniques for the data specification, modeling, capture, transfer, management, and algorithms for their (real-time) collection, transfer, analysis, storage, and processing [14].

Big data [15] driven cyber-physical systems make use of large amounts of data to determine their operation. Data are clearly crucial for the correct operation of these big data driven cyber-physical systems. The design of big data driven cyber-physical systems requires the introduction of new concepts to model classical data structures, multiple V's features, time constraints and spatial constraints, and the dynamic continuous behavior of the physical world. In this chapter, we propose a multi-view, multi-domain, and multi-paradigm approach for specification and modeling of big data driven cyber-physical systems. A view-oriented approach allows designers from different views to develop and evaluate design alternatives within the context of formalisms relevant to selected aspects of the system. Each view highlights certain features and occludes others to make a tractable analysis and to focus on particular attributes of big data driven CPS. It is clear that an appropriate specification and modeling approach cannot be fitted into the limitations of one formalism at a time, particularly in the early stages of the design process. Hence, it is necessary to take a combination of different methodologies in a multi-formalism approach from different views. In this chapter, we propose a model-based approach to model big data driven CPS based on the integration of AADL [16], ModelicaML [17], clock theory [18], Modelica [19], RCC [20], UML [21], SysML [22], and Hybrid Relation Calculus [23]; we illustrate our approach by specifying and modeling VANET [24].

6.2   ISSUES AND CHALLENGES FOR BIG DATA DRIVEN CYBER-PHYSICAL SYSTEM

First, big data driven CPS have big data characteristics. The term big data is usually described by the following aspects: volume, velocity, variety [25], veracity, validity, value, and volatility (also known as the *seven Vs*).

Volume—The amount of big data is very large, in the order of terabytes or larger.

Velocity—The capture, the transfer, the computation, the store, and the access for big data are needed in real time. Big data require fast processing. Time factor plays a very crucial role in CPS.

Variety—Variety refers to the many sources and types of data both structured and unstructured. The data include unstructured and multiple types, possibly including texts, stream data such as videos and log files.

Veracity—The veracity aspect deals with uncertain or imprecise data. The veracity aspect includes data inconsistency, incompleteness, ambiguities, latency, deception, and approximations.

Validity—The validity aspect includes not only ensuring that the measurements are correct, but also transparency of the assumptions and connections behind the process. The validity can be further divided into content validity, structural fidelity, criterion validity, and consequential validity.

Value—The business value of big data is in how the data and technology are applied, the same as with Business Intelligence (BI). Big data's value derives from capabilities that enable new and broader uses of information or that removes limitations in the current environment. All that available data will create a lot of value for organizations, societies, and consumers. Big data means big business, and every industry will reap the benefits from big data.

Volatility—Big data volatility refers to how long is data valid and how long should it be stored. In this world of real-time data, you need to determine at what point is data no longer relevant to the current analysis.

Second, big data driven CPS have special characteristics, constraints, and requirements that must be met during system development. In CPS, a huge amount of networking sensors are embedded into various devices and machines in the physical world. Such sensors deployed in different fields may capture various kinds of data, such as speed, temperature, environmental data, geographical data, astronomical data, people health data, and logistic data. Mobile equipment, transportation facilities, public facilities, and home appliances could all be data acquisition equipment [5].

CPS data processing challenges in real time are very complex. The multiple V characteristics of big data are defined as volume, velocity, veracity, and variety; and a Hadoop-like system can process the volume and variety parts of it. In addition to the volume and variety, the CPS must process the velocity of the data as well. And meeting the requirements of the velocity of big data in CPS is not an easy task. First, the system should be able to understand the physical data, acquire physical data, and collect the data generated by the physical world and real-time event stream coming in at a rate of millions of events per seconds. Second, it needs to handle the parallel or distributed processing of these data in real time as and when it is being collected. Third, it should perform real-time data mining from moving data stream and spatial–temporal historic data.

Spatiotemporal requirements—CPS not only must meet real-time requirements, but also must meet spatial requirements. Spatiotemporal data have many different forms and include data such as georeferenced time series, remote sensing images, or moving object trajectories. Trajectory data are data containing the movement history of mobile objects. The specification, modeling, processing, and analysis of trajectory data are in an interdisciplinary research field and involve communities from geographic information science, software engineering, moving database technology, sensor networks, distributed systems, transportation science, as well as privacy [26]. Spatiotemporal data are very large. Especially, the data of moving physical entities are increased in geometry growth over time. Spatiotemporal data captured through remote sensors are always big data [27]. In CPS, every data acquisition device (sensor) is located at a specific geographic position, and every piece of data has a time stamp. The time and space correlations are very important properties of data for CPS. During data analysis and processing phases, time and space information are also very important dimensions for data evaluation, store, access, and analysis. Many data sets in CPS have certain levels of heterogeneity in type, structure, semantics, organization, granularity, and accessibility. Data representation and model aim to make data more meaningful for data processing, store, access, and analysis. Efficient data representation shall reflect data structure, class, and type, spatial–temporal data, as well as integrated technologies, so as to enable efficient operations on different data sets [5].

The big data driven CPS should be not only a low-latency system so that the computation can happen very fast with near real-time response capabilities, but also the big data driven CPS should meet spatial requirements when the physical objects move. A moving object represents continuous evolution of a spatial object over time. Regarding big data modeling, an important question is how to represent a moving object. In contrast to static objects, moving objects [28] are difficult to represent and model. The efficiency of modeling methods for moving objects is highly affected by the chosen method to represent and analyze the continuous nature of the moving object. The data model defined for modeling continuously changing locations over time should be simple, extensive, though expressive, and be easy to use and implement. In order to realize full benefits of big spatiotemporal data, one has to overcome both computational and I/O challenges. Not only do we need new models that explicitly model spatial and temporal constraints efficiently, but also further research is required in the area of physical world modeling, moving object behavior analysis, moving object database, and model integration and transformation.

Traditional moving object database technologies cannot fully discover some features and patterns of moving objects that hide under complicated spatial–temporal data, which provide more valuable support for scientific decision making. Using traffic systems, for example, which are situated in some big cities, would be expected to handle everyday phenomena of detecting, predicting, and avoiding traffic jams and finding relations between these jams. It is a great challenge to discover spatial–temporal characteristics, relationships, unknown patterns, and decision-making knowledge from big data of moving object databases. Moving objects data mining is currently meeting two key scientific issues. Because of multiple sources, heterogeneous, dynamic, complicated moving objects data, it is required to analyze moving objects characteristics, studying data model, and semantics representation, including location, topology, and semantic description. On the other hand, it is necessary to study spatial–temporal granularity, clustering, association and evolution, discovering, and interpreting behavior patterns of moving objects over time [29].

Real-time communication in big data environment—CPS must provide guarantees that huge data in the network are delivered according to their real-time deadlines. Some mechanism is needed to assign priorities to the data that are to be sent on the sensor network or physical objects. The environment in CPS is often harsher and noisier and thus has more stringent requirements on reliable and real-time communication. Missing or delaying the real-time data may severely degrade. Because big data set sizes are growing exponentially, it is important to use available data transfer protocols that are the most efficient.

MapReduce has been widely used as a big data processing platform. As it gets popular, its scheduling becomes increasingly important. In particular, since many cyber-physical applications require real-time data processing, scheduling realtime applications in MapReduce environments has become a significant problem [30]. Current scheduling techniques in the Hadoop MapReduce framework are not adequate, as they either adopt fair scheduling or capacity scheduling. These strive to balance the load across the resources rather than meeting real-time demands of the tasks. Supporting real-time tasks on MapReduce platforms is particularly challenging due to the heterogeneity of the environment, the load imbalance caused by skewed data blocks, as well as real-time response demands imposed by the big data driven cyber-physical applications [31].

Cloud computing services provide a flexible platform for realizing the goals of the big data driven CPS. The National Institute of Standards and Technology defines cloud computing as "a model for enabling convenient, on–demand network access to a shared pool of configurable computing resources (e.g., networks, servers, storage, applications, and services) that can be rapidly provisioned and released with minimal management effort or service provider interaction." Applying the characteristics of cloud computing to change big data driven CPS, any device (sensor, actuator, data archive, and computation resource) is available as a service and can rapidly be provisioned for use in a virtual big data driven CPS to meet the user demands. Cloud-based services allow big data driven CPS devices to be used in multiple virtual systems at the same time providing more efficient use of those resources, resiliency during unexpected situations, and flexibility to change and expand as users need change [32]. Traditional virtualization technology typically implements machine or process models that are deterministic only in terms of functional behavior. Repeated execution of sequential programs in these models is guaranteed to compute the same output for the same input. However, other nonfunctional properties such as execution time and energy consumption may vary and may even be unbounded (mostly due to complex resource sharing). Virtualization technology for big data driven CPS based on cloud computing services must therefore include solutions that provide deterministic behavior beyond computing mathematical functions. Determinism with respect to nonfunctional properties such as time, space, and energy is key to compositionality and thus scalability of the involved engineering methods and tools [33].

With the rise of industrial big data, and smart infrastructure (cities, transportation, homes, etc.), this is now a challenge for CPS modeling across multiple domains. Tackling this challenge requires an interdisciplinary approach that combines performance modeling and system building with designing novel algorithms [34]. In order to meet the challenge of big data driven CPS design, we need to realign abstraction layers in design flows and develop semantic foundations for composing heterogeneous models and modeling languages describing different physics and logics. We need to develop new understanding of compositionality in heterogeneous systems that allow us to take into account both physical and computational properties. One of the fundamental challenges in research related to big data driven CPS is accurate modeling and representation of these systems. The main difficulty lies in developing an integrated model that represents both cyber and physical aspects with high fidelity.

6.3   VIEW-ORIENTED APPROACH TO SPECIFY AND MODEL CYBER-PHYSICAL SYSTEMS

Big data driven CPS are the integration of cyber systems and physical systems. Modeling the heterogeneous composition of physical world part, computational part, and communication part of aerospace CPS is an enormous challenge in engineering aerospace CPS because of the major sources of heterogeneity such as safety, fault tolerance, causality, time semantics, spatial reasoning, and different physical domains. The physical world of aerospace CPS follows classical physical laws that express continuous-time dynamics; thus, the behavior of physical systems is inherently depicted by continuous-time differential equations. Computational part and communication part are expressed by the notion of causality and discrete-time semantics. Integrating the physical world and cyber part is challenging and call for proper dimensioning of mechanical, electronic and embedded control subsystems from different views. Usually, the physical world part and the cyber part are modeled independently of each others. This modeling approach does not exploit many potential advantages of an integrated specification, analysis and design process, which are lost in separate points of view of different engineering domains. In fact, the physical properties and the dynamical behavior of physical world parts play a central role in aerospace CPS development process. Thus, significant improvements to overall system performances can be obtained by early integrating the physical world and the cyber part development. It is clear that an appropriate specification and modeling approach cannot be fitted into the limitations of one formalism at a time, particularly in the early stages of the design process. Hence, it is necessary to take a combination of different methodologies in a multi-formalism approach from different views.

View-oriented specification, modeling, and design methods enable designers from different views to develop and evaluate design alternatives within the context of formalisms relevant to selected aspects of the aerospace CPS. Each view highlights certain features and occludes others to make analysis tractable and to focus on particular attributes of aerospace CPS. A given view typically expresses either the cyber or the physical aspect well, but not both. For example, differential equation models represent physical processes well, but do not represent naturally the details of computation or data communication.

A view is the representation of a system from the perspective of a related set of concerns. A concern will be related to one or more aspects of the systems. A viewpoint is a specification of the conventions for using a view, by establishing the purposes and audience for a view, and the techniques for its creation and analysis [35]. Multi-view modeling refers to a system designer constructing distinct and separate models of the same system to model different (semantic) aspects of a system. Multi-view modeling can be applied at different levels and in very different ways [36].

Using multi-view modeling, it is possible to describe different aspects of an aerospace CPS being designed and thus support system development process phases such as requirements analysis, design, implementation, verification, validation, and integration. Especially, multi-view modeling approach supports mathematical modeling with equations since equations specify the behavior of physical world of big data driven CPS.

In the specification, analysis, design and implement of big data driven CPS, multi-views on the system to be developed are often used. These views are typically composed of models in different formalisms. Different views usually are suitable to various partial aspects of the complex big data driven CPS in a multi-view approach; individual views are simpler than a single model representing all aspects of the system. As such, multi-view modeling, like modular, hierarchical modeling, simplifies system development. Most importantly, it becomes possible for individual experts on different aspects of a system develop and work in isolation on individual, possibly domain-specific

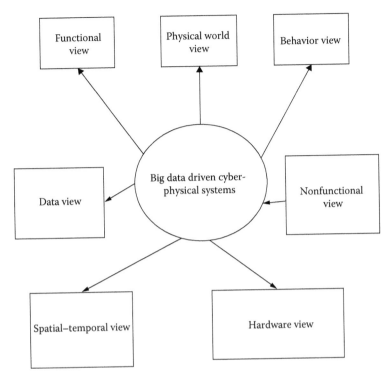

FIGURE 6.1 Seven complementary views approach for big data driven cyber-physical systems.

views without being encumbered with other aspects. These individual experts can work mostly independently, thereby considerably speeding up of the development process [37].

The most obvious difference between current methodologies and our approach is that we propose a methodology that explicitly avoids the use of a simple framework to design complex systems. Instead, this new methodology is proposed according to seven complementary views as shown in Figure 6.1: physical world view, function view, data view, behavior view, spatial–temporal view, nonfunctional view, and hardware view.

The goals of a multi-view approach are the following:

- Propose an integrated framework in which system-level models can be composed of subsystem models in different disciplines from multi-views.
- Formalize a multi-domain modeling paradigm for physical world of big data driven CPS: a representation of the models that allows the system model to evolve with the system devolvement process.
- Realize model transformation to integrate different models with the design environment so that consistency between the artifacts and their corresponding models is automatically kept at all times.

Big data driven CPS must be dependable, secure, safe, and efficient and operate in real time. They must also be scalable, cost-effective, and adaptive. They must seamlessly integrate operations in the cyber-computing and communications domain with actions and events in the physical world. The design of the physical and computational aspects is an integrated activity. Design decisions made in one aspect interacts with the physical component and have profound consequences on the dynamic properties of the entire system. The design of CPS must be accomplished by taking an integrated view and codesigning the physical parts with the computational parts. A number of methodologies and tools have been proposed for the design of big data driven CPS. However, they are still evolving because systems and software engineers do not yet completely understand the aerospace CPS requirements analysis process, and the complex dynamic behavior of aerospace CPS is difficult to capture in a simple understandable representation. Consequently, development time for many of these systems extends beyond expectations. Furthermore, neither many methods are tested properly, nor do they meet customer requirements or expectations. There are no good formal representations and tools capable of expressing and integrating multiple viewpoints and multiple aspects of big data driven CPS. This includes lack of robust specification and modeling methods of multiple abstraction layers from physical processes through various layers of the information processing hierarchy and their cross-layer analyses.

Big data driven CPS use computations and communication deeply embedded in and interacting with physical processes to add new capabilities to physical systems. Embedded computer systems that are an integral part of a physical system have to react quickly and correctly to very complex sequences of external events. The entities that produce these events are called the "physical world." The behavior of a system must change as the behavior of environment changes. Consequently, an accurate description and analysis of the behavior of the physical world is a very important step in the specification of embedded systems. Insufficient analysis of the external environment will lead to the development of a system that does not meet the basic expectations of users and developers. It is important to note that the computer is not an intelligent device until now; it can only do what you tell it to do. When it encounters an unknown environment, it fails. This situation may cause a catastrophe. Unfortunately, the methodologies and tools developed for the analysis of real-time systems do not adequately address these specification issues because they do not allow for the description of complex dynamic external environments. The current methodologies confuse the analysis of behavior of the physical world with the internal behavior of the system. The earlier facts explain why we propose a view of a *physical world* of a system and a stage of the system environment analysis. We use Modelica [38] and ModelicaML [39] to make an environment analysis for the physical world of aerospace CPS. Modelica is an object-oriented modeling language that facilitates the physical modeling paradigm. It supports a declarative (noncausal) model description, which permits better reuse of the models.

Notice that the objective to construct a system lies in the realization of functionality that a user demands. So we propose a view of function of the system, asking any questions like "What exactly should the system do for a user?" Generally, functional requirements describe input–output behavior of the software. Functional analysis in our methodology involves a multilevel hierarchical specification. It usually yields a good hierarchical design. Based on the software requirements given, it is a common practice to generate functional diagrams that depict pictorially the functionality to be incorporated into the candidate system. Integrating the descriptive power of UML, SysML, and ModelicaML models with the analytic and computational power of Modelica models provides a capability that is significantly greater than that provided by UML or Modelica individually. UML and Modelica are two complementary languages supported by two active communities. By integrating UML and Modelica, we combine the very expressive formal language for differential algebraic equations and discrete events of Modelica with the very expressive UML constructs for requirements, structural decomposition, logical behavior, and corresponding crosscutting constructs.

A data view captures the static relation between the object and data. Data view emphasizes the data structure of the system using objects, attributes, operations, and relationships. Data processing can be generally divided into four phases: data generation, data acquisition, data storage, and data analysis. Data generation is the first step of big data. As the second phase of the big data system, big data acquisition includes data collection, data transmission, and data preprocessing. Data collection is to utilize special data collection techniques to acquire raw data from a specific data generation environment. Upon completion of raw data collection, data will be transferred to a data storage infrastructure for processing and analysis. Big data storage refers to the storage and management of large-scale data sets while achieving reliability and availability of data accessing.

The storage infrastructure needs to provide information storage service with reliable storage space; it must provide a powerful access interface for query and analysis of a large amount of data. The analysis of big data mainly involves analytical methods for traditional data and big data, analytical architecture for big data, and software used for mining and analysis of big data. Data analysis is the final and the most important phase in the value chain of big data, with the purpose of extracting useful values, providing suggestions or decisions [5]. We can use AADL, SysML, UML, and ModelicaML to model the data structure of the system. AADL is an architecture description language developed to describe embedded systems. AADL, which is a modeling language that supports text and graphics, was approved as the industrial standard AS5506 in November 2004. Component is the most important concept in AADL. The main components in AADL are divided into three parts: software components, hardware components, and composite components. Software components include data, threads, thread groups, processes, and subprograms. Hardware components include processors, memory, buses, and devices.

The changes of an internal behavior of system are caused by the changes of the behavior of the external environment of the physical world. So we propose a *behavioral view* in order to help analyze the behavior of real-time systems. The behavioral view captures the dynamics of the system, that is, the conditions under which the functionality is performed. Safety-critical systems in the aforementioned industries typically have a joint discrete and continuous state space. There, system behavior depends on both, the state in a discrete controller and continuous physical world. Hybrid systems represent a mathematical model for those dynamic systems of interacting discrete and continuous behavior. Frequently, in hybrid systems in the past, the event-driven dynamics were studied separately from the time-driven dynamics, the former via automata or Petri net models and the latter via differential or difference equations. To understand fully the system's behavior and meet high-performance specifications, one needs to model all dynamics together with their interactions. For example, flight calculations are made quite precisely for space missions, taking into account such factors as the earth's oblateness and nonuniform mass distribution; gravitational forces of all nearby bodies, including the moon, the sun, and other planets; and 3D flight path. We can use an approach to integrate Modelica, AADL behavior Annex, UML, and Hybrid Relation Calculus for specifying and verifying dynamic behavior of big data driven CPS. The Hybrid Relation Calculus is such a formal formwork. On the one hand, it can deal with continuous-time systems based on sets of ordinary differential equations. On the other hand, it can deal with discrete-event systems, without continuous variables or differential equations. Hybrid Relation Calculus models physical world as well as its interaction with the control program.

A spatiotemporal view captures the spatiotemporal requirements [41] of CPS. CPS are spatiotemporal, in the sense that correct behavior will be defined in terms of both space and time. As CPS, computations must be completed within established response times, for which we will need new notions of timeliness, but they may also have varying temporal requirements based on the current frame of time. As spatial systems [42], the computations performed and their timeliness will be dependent on

1. The location of the platform in its environment
2. The velocity with which the platform is moving
3. The number of objects in the environment
4. The velocity vectors of the objects in the environment

The motion is a key notion in our understanding of spatial–temporal relations; the question remains of how to describe motion, and more specifically the interaction between moving objects, adequately within a qualitative calculus. The existing theory for specifying objects in both space and time is not applicable to big data driven CPS for at least three reasons: (1) the theory ignores physical attributes and constraints on the objects; (2) the interaction of objects in space and time is not addressed; and (3) the range and precision of time resolution is not adequate for real-time spatiotemporal systems [23]. In this chapter, we propose a spatiotemporal sequence diagram that integrates Region Connection Calculus (RCC), clock theory, a and UML sequence diagram to describe spatial configurations that change over time. We aim to specify, model, and analyze some properties of space and time in a formalism, which allows for the representation of relations between moving entities over time.

A nonfunctional view addresses important issues of quality and restrictions for CPS, although some of their particular characteristics make their specification and analysis difficult: first, nonfunctional requirements can be subjective, since they can be interpreted and evaluated differently by different people; second, nonfunctional requirements can be relative, since their importance and description may vary depending on the particular domain being considered; third, nonfunctional requirements can be interacting, since the satisfaction of a particular nonfunctional requirement can hurt or help the achievement of other nonfunctional requirements. Dependability is that property of a system that justifies placing one's reliance on it. The dependability of a system is the collective term that describes the availability performance of a system and its influencing factors: reliability, safety, maintainability, and maintenance support performance. These nonfunctional properties are highly important for aerospace CPS as they are designed to operate in environments where failure to provide functionality or service can have enormous cost both from financial, influential, or physical aspects.

A hardware view aims at supporting, fitting the specification on a particular target hardware environment. A developer must decide what hardware should be used and how it is to be used to implement the specification, so we propose a *hardware view* that helps improve the implementation. Current methodologies have gaps between specification, software design, and implementation level, and they do not detail the implementation stage. In fact, this stage is very important for multi-core trustworthy aerospace CPS because it maps the application onto a parallel configuration. A parallel configuration has become crucial in many critical-time applications. This configuration is generally based on an allocation of tasks on multiprocessors in order to improve the response time.

The development of a big data driven CPS such as intelligent transportation systems involves the integration of many different disciplines. We advocate a multidimension, multi-domain, and multi-paradigm approach to the development of CPS, based on the identification of complementary dimensions and views on the system as starting point for modeling and design as shown in Figure 6.2.

Specification and design method integration actually involves many types of integrations at different levels [43]:

- *Model integration* defines how the entities in the dataflow model should relate to the parts of the method.
- *Principle integration* ensures that the model's principles are consistent with the existing method's principles.
- *Process integration* integrates the process for making the diagrams fit into the overall process of the existing method.
- *Representation integration* integrates the modeling representation into the existing method's representation.

Model transformation is a central concept in model-driven development approaches and integration development approaches in CPS, as it provides a mechanism for automating the manipulation of models. Assuming that the model transformation is correctly defined, it can be used to ensure consistency between different models. Model transformations can be used for various tasks in model-driven development, for example, for modifying, creating, adapting, merging, weaving, or filtering models. The reuse of information captured in models is common to all of these tasks. Instead of creating

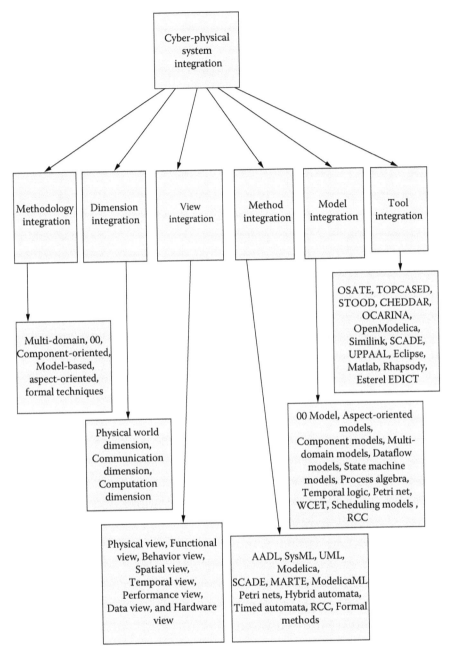

FIGURE 6.2 The multi-domain, multidimension, multi-view, multi-paradigm integrated approach to developing CPS.

artifacts from scratch, model transformations enable the use of information that was once captured as a model and build on it [44].

Model transformations are touted to play a key role in model-driven development of CPS. Their intended applications include [45] the following:

- Generating lower-level models, and eventually code, from higher-level models
- Mapping and synchronizing among models at the same level or different levels of abstraction

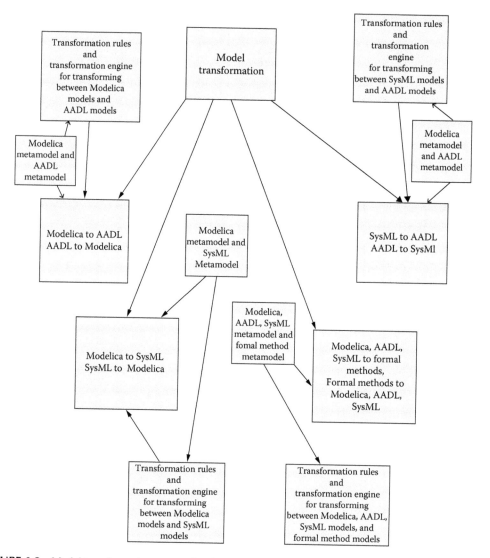

FIGURE 6.3 Model transformation approaches for cyber-physical systems.

- Creating query-based views on a system
- Model evolution tasks such as model refactoring
- Reverse engineering of higher-level models from lower-level ones

We propose a model transformation approach for CPS as shown in Figure 6.3.

6.4 BIG DATA DRIVEN CYBER-PHYSICAL SYSTEM DESIGN METHOD

Big data driven cyber-physical applications that require real-time processing of high-volume data streams are pushing the limits of traditional design methods. Big data driven CPS not only meet big data 7V feature requirements, but also have to meet time constraints and spatial constraints of CPS. In this chapter, we propose a model-based approach to model big data driven CPS based on integration of AADL, ModelicaML, and clock theory as shown in Figure 6.4

AADL provides the data component to model data types and data abstraction [46]. We can use the data component to model the big data of CPS. The data component category supports representing

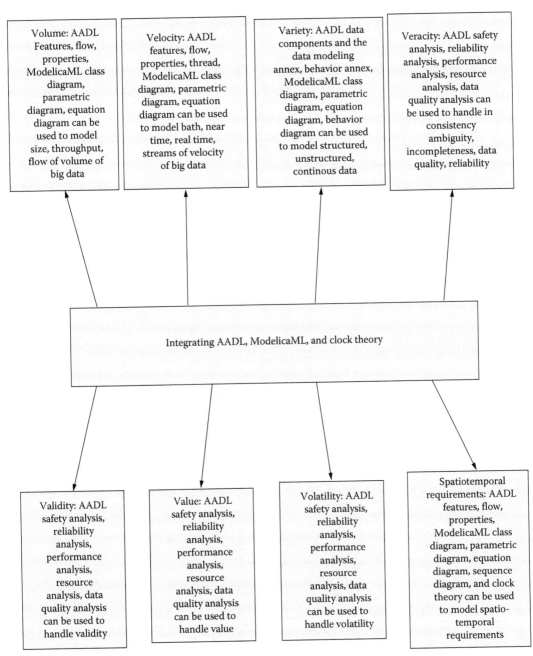

FIGURE 6.4 The proposed approach to model big data driven cyber-physical systems based on AADL.

data types and data abstractions in the source text at the appropriate level of abstraction for the modeling effort. The data *type* is used to type ports to specify subprogram parameter types. Data type inheritance can be modeled using the AADL component-type extension (e.g., extends) mechanism. Class abstractions can be represented by using subprogram as a feature of a data type. Provided data features of a data component are shareable using a specified concurrency control property. A shareable data component instance is specified in a required subclause of the component type. The AADL supports modeling of three kinds of interactions between components: directional flow of data and/or control through data, event, and event data port connections; call/return interaction on subprogram entry points; and through access to a shared data component (see data component mentioned earlier). Threads, processors, and devices, and their enclosing components (process and system) have in-ports and out-ports declared. Data ports communicate unqueued state data; event ports communicate events that are raised in their implementation, their associated source text, or actual hardware; and event data ports represent queued data whose arrival can have event semantics. Arrival of an event at a thread results in the dispatch of that thread—with semantics defined via property values and hybrid automata for event arrival while the thread is active. For data port connections, the data are communicated upon the completion of execution (an immediate connection with the effect of mid-frame communication for periodic threads) or upon thread deadlines (a delayed connection with the effect of a phase delay for the periodic threads).

AADL proposes two interesting extension mechanisms: property sets and extends [47]. Property sets allow one to define custom properties to extend standard ones. This is the path taken by the "Data modeling annex document" [48] that allows one to model precisely data types to be manipulated. AADL extends mechanism that allows one to reuse and extend one component type. If a component type extends another component type, then features, flows, and property associations can be added to those already inherited. A component type extending another component type can also refine the declaration of inherited feature and flow declarations by more completely specifying partially declared component classifiers and by associating new values with properties. The AADL language can be extended through annex languages. They offer the possibility to attach additional considerations to an AADL component like behavioral specification [49].

The main aim of ModelicaML is to enable an efficient and effective way to use both Modelica and UML/SysML models reusing notations that are also used for software modeling. ModelicaML is defined as a graphical notation that facilitates different views (composition, inheritance, behavior) on system models. ModelicaML is based on a subset of the OMG UML and reuses concepts from the OMG SysML. ModelicaML is designed toward the generation of Modelica code from graphical models. Since the ModelicaML profile is an extension of the UML metamodel, it can be used for both: Modeling with standard UML and with SysML Modelica [50], SysML and UML are object-oriented modeling languages. They provide methods to model a system as objects and to specify its internal structure and behavior. SysML is a specific UML profile for systems modeling and design. This modeling language facilitates efficient capturing of relevant system requirements, design, or test data by means of graphical formalisms, crosscutting constructs and views (diagrams) on the model data. Modelica is defined as a textual language with standardized graphical annotations for model icons and is designed for efficient modeling and analysis of physical system dynamic behavior [51]. By using Modelica and UML/SysML together, UML/SysML's advantages in graphical and descriptive modeling are enforced with Modelica's formal executable modeling for analyses and simulation. Vice versa, Modelica will benefit from the capacity of software analysis and design and from using the selected subset of the UML/SysML graphical notation (visual formalisms) for editing and maintaining Modelica models [52]. The ModelicaML extends the models in UML to represent continuous dynamical physical elements and their coupling. Using ModelicaML, we can develop CPS and get a unified representation of both physical and cyber entities and their interactions in the same modeling environment. We can use architecture as the common system representation to relate the structure and semantics of heterogeneous models.

6.5 MODELING SPATIOTEMPORAL REQUIREMENTS OF BIG DATA DRIVEN CYBER-PHYSICAL SYSTEMS

The existing theory for specifying objects in both space and time is not applicable to Intelligent Transportation Systems for at least three reasons: (1) the theory ignores physical attributes and constraints on the objects; (2) the interaction of objects in space and time is not addressed; and (3) the range and precision of time resolution is not adequate for real-time spatiotemporal systems. The spatiotemporal representation and reasoning methods should be powerful enough to express categories of motion that can be useful in a qualitative context and be kept as simple as possible so that characterizing its

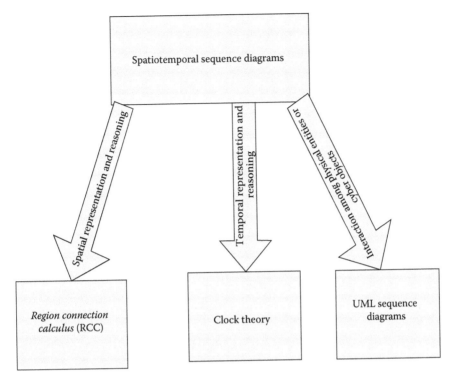

FIGURE 6.5 The proposed spatiotemporal sequence diagram.

properties is still possible in a precise way. Only then can we have a principled representation for motion that could unify the various needs for spatiotemporal representation and reasoning of transportation CPS. The expressive power of the combined spatiotemporal formalisms of the proposed spatiotemporal sequence diagram lays in the expressivity of the spatial requirements, the expressivity of the temporal requirement, and the interaction between the two components allowed in the combined logic. In this chapter, we propose a spatiotemporal sequence diagram that integrates RCC, clock theory, a and UML sequence diagram as shown in Figure 6.5; such a combined spatiotemporal formalism permits us to describe spatial configurations that change over time. We aim to specify, model, and analyze some properties of space and time in a formalism that allows for the representation of relations between moving entities over time as shown in Figure 6.6.

RCC is a logic theory for qualitative spatial representation and reasoning. The fundamental approach of RCC is that extended spatial entities, that is, *regions* of space, are taken as primary rather than dimensionless points of traditional geometry; and the primitive relation between regions— giving the language and the ability to represent the structure of spatial entities—is called a *connection*. The RCC is an axiomatization of certain spatial concepts and relations in first-order logic. The basic theory assumes just one primitive dyadic relation: C(x, y) read as "x connects with y." Individuals (x, y) can be interpreted as denoting spatial regions. The relation C(x, y) is reflexive and symmetric. Of the defined relations, disconnected (DC), externally connected (EC), partially overlaps (PO), equal (EQ), tangential proper part (TPP), nontangential proper part (NTPP), tangential proper part inverse (TPPi), and nontangential proper part inverse (NTPPi) have been proven to form a jointly exhaustive and pairwise disjoint set, which is known as RCC-8. Similar sets of one, two, three, and five relations are known as RCC-1, RCC-2, RCC-3, and RCC-5 [20]. The syntax of RCC-8 contains eight binary predicates:

- DC(X, Y)—Regions X and Y are disconnected
- EC(X, Y)—X and Y are externally connected

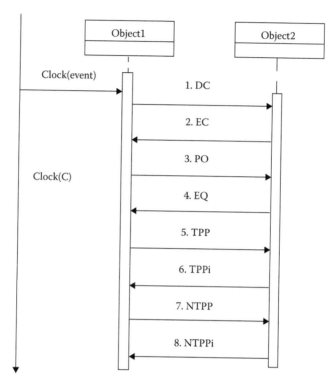

FIGURE 6.6   The representation of spatiotemporal diagram in RCC-8 with the clock theory.

- EQ(X, Y)—X and Y are equal
- PO(X, Y)—X and Y partially overlap
- TPP(X, Y)—X is a tangential proper part of Y
- NTPP(X, Y)—X is a nontangential proper part of Y
- The inverses of the last two—TPPi(X, Y) and NTPPi(X, Y)

Clock theory [18] describes temporal relations in the event in physical world by using a clock, and can analyze, records the event by clock. To use clock theory to specify CPS, the time description is clearer to every event and can link continuous world with discrete world better.

6.6   CASE STUDY: BIG DATA DRIVEN VEHICULAR CYBER-PHYSICAL SYSTEMS DESIGN

VANETs [53] capture, collect, possess, and distribute traffic information to improve traffic congestion and to massively reduce the number of accidents by warning drivers about the danger before they actually face it. VANETs contain sensors and on-board units installed in the vehicle as well as roadside units (RSUs). The data captured and collected from the sensors on the vehicles can be delivered and displayed to the driver, sent to the RSU and traffic control centers, or even broadcasted to other vehicles depending on its nature and importance. The traffic control center also can send traffic information to vehicles and RSUs and control the states of vehicles and RSUs, vehicles can communicate with other vehicles, and RSUs can distribute these data, along with data from road sensors, weather centers, traffic control centers, and so on, to the vehicles and can also provide commercial services such as parking space booking, Internet access, and gas payment [54].

VANETs [55] are supported by a large amount of data that are captured and collected from various resources, are systems that would enable users to efficiently utilize data resources that pertain

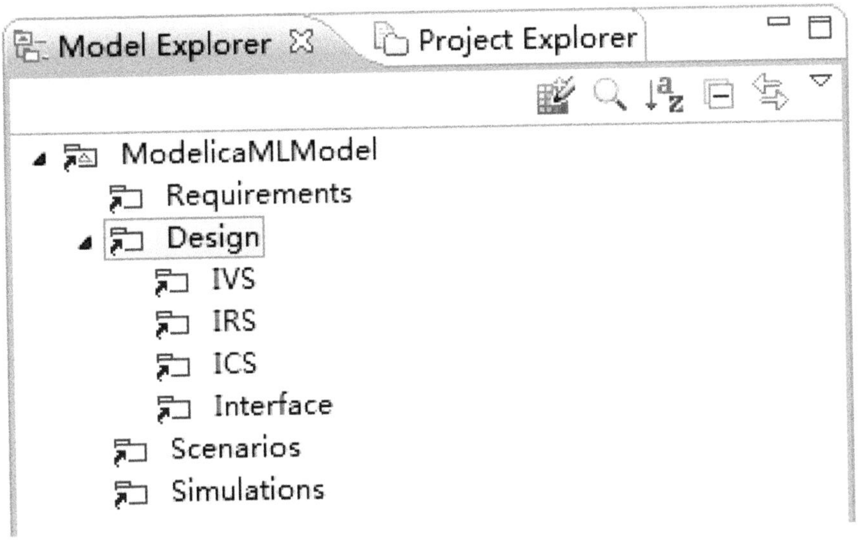

FIGURE 6.7 The top Modelica ML model of VANET.

to Intelligent Transportation Systems, and can access and employ data through more convenient and reliable services to improve the performance of transportation systems. The data in VANETs include information, such as geographic space, location, speed, count, and behavior patterns, which are combined to obtain estimates of traffic conditions. As traffic data continue to grow, the average monthly data for traffic information have now reached 100 TB. The traffic volume, speed and occupancy data, and position of vehicles over time have been regarded as important features in traffic control and information management systems such as VANETs. For example, in an urban area with hundreds of thousands of vehicles, drivers and passengers in these vehicles want to get information relevant to their trip. They require that the location information is displayed on screen and at any time, the available parking spaces around the current location of the vehicle are informed. The driver may be interested in the traffic conditions 1 mile ahead. Such information is important for drivers to optimize their travel, to alleviate traffic congestion, or to avoid wasteful driving [56]. There are enormous amount of data in VANETs [57]. Spatiotemporal data for moving vehicles in VANETs are very large. Especial, the data of moving vehicles are increased in geometry growth over time. Spatiotemporal data captured through remote sensors in VANETs are always big data.

The VANET includes ITS vehicle station systems, IVS, ITS roadside station system, and ITS control station system. The top ModelicaML model of VANET is shown in Figure 6.7.

Figure 6.8 represents the requirements of VANET.

Figure 6.9 represents the system connection model of the IVS subsystem in ModelicaML.

AADL can model static data and map model data on hardware architecture. Figure 6.10 represents the system composition architecture model of a vehicle station IVS.

The component in_control can be modeled in AADL text form as follows:

```
system in_control
features
    cc_system_on_off: in data port;
    brake_status: in data port bool_type;
    resume: in data port;
    decrease_speed: in data port;
```

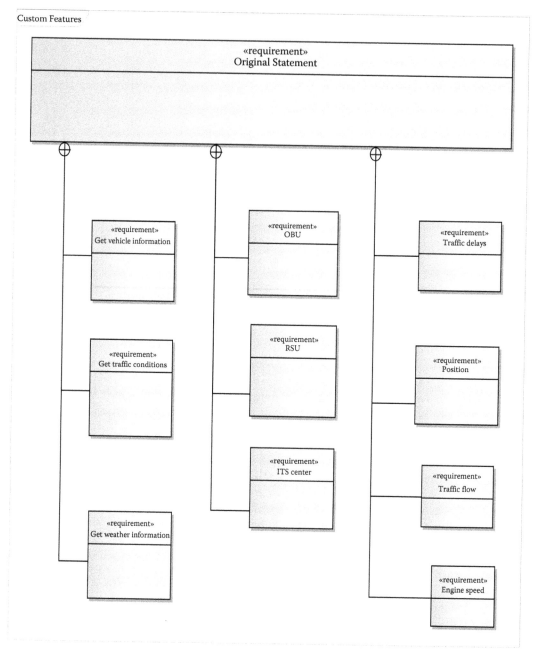

FIGURE 6.8   The requirements of VANET.

```
    increase_speed: in data port;
    set_speed: in data port;
    engine_status: in data port;
    ok_to_run: out data port;
flows
    Flows1: flow path brake_status -> ok_to_run; {Latency
        => 20 ms;};
end in_control
```

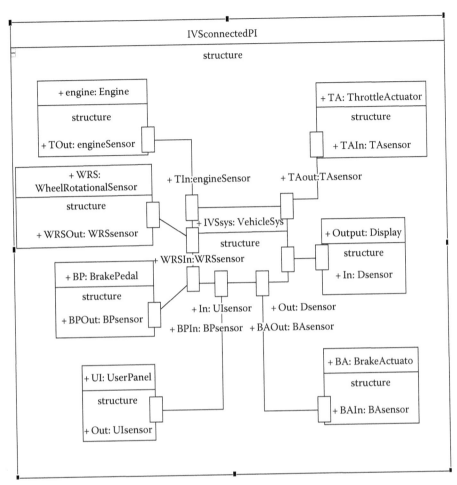

FIGURE 6.9 System connection model of the IVS subsystem in ModelicaML.

The brake pedal is one data source, which can be modeled using AADL as follows:

```
device brakepedal
Features
BrakeInfo:out data port bool_type;
Flows
Flow_one:flow source BrakeStatus{Latency=>20ms;};
end brake brakepedal;
```

We can also use the ModelicaML class diagram to model static data of big data driven CPS. Figure 6.11 represents the static data model of traffic control in the class diagram of ModelicaML.

The Modelica model of continuous data of the optimizer in traffic light control is as follows:

```
within ModelicaMLModel.Scenarios.ICS;
model Optimizer
    Real S;
    output Real Ty;
    Real broom;
    Boolean safe;
```

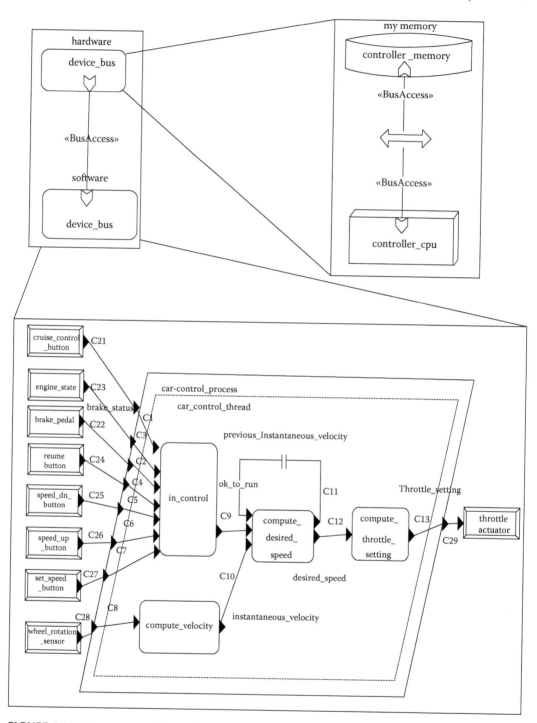

FIGURE 6.10 System composition architecture model of a vehicle station IVS.

```
ModelicaMLModel.Scenarios.Interface.speedSensor VIn;
ModelicaMLModel.Scenarios.Interface.BrakeSensor BIn;
ModelicaMLModel.Scenarios.Interface.LaneSensor XIn;
equation
broom=VIn.V/2*BIn.B;
```

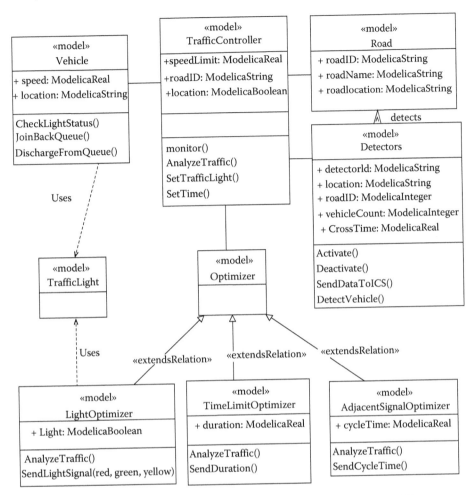

FIGURE 6.11 The static data model of traffic control in the class diagram of ModelicaML.

```
der(XIn.X)=VIn.V;
Ty = if (XIn.X/VIn.V>broom) then 0 else XIn.X/VIn.V;
S=VIn.V*VIn.V/2*BIn.B;
safe= if (XIn.X>S) then true else false;
end Optimizer;
```

Flow analysis and delay analysis of dataflow from the brake pedal to the throttle actuator are shown in Figure 6.12.

```
flows
    ETE_F1: end to end flow BRAKE.Flow1
    -> C22 -> CC.brake_flow_1
    -> C29 -> TA.Flow1
    {
    Latency => 200 Ms;
    };
flows
    brake_flow_1: flow path brake_status
    -> C2 -> I_C.FS1
    -> C9 -> C_D_S.FS1
```

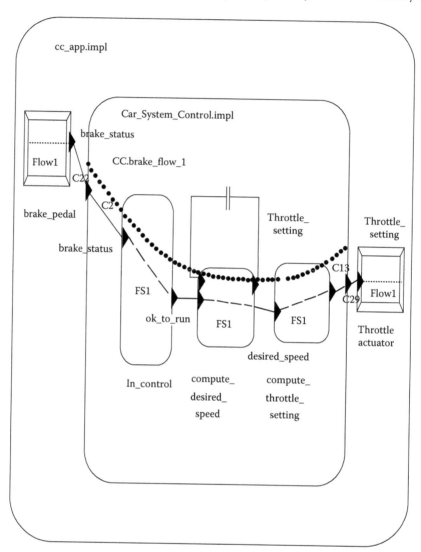

FIGURE 6.12 Flow analysis from brake pedal to throttle actuator.

```
    -> C12 -> C_T_S.FS1
    -> C13 -> throttle_setting{
    Latency => 130 Ms;
};
```

The end-to-end flow delay is predefined as 200 ms; we make a flow analysis by the tool OSATE [58], and we obtained the delay 150 ms, which is less than the delay of 200 ms. The test result is shown in Figure 6.13.

Vehicle-to-Vehicle (V2V) Communications [59] for Safety is the dynamic wireless exchange of data between nearby vehicles that offers the opportunity for significant safety improvements. By exchanging anonymous vehicle-based data regarding position, speed, and location (at a minimum), V2V communications enable a vehicle to sense threats and hazards with a 360° awareness of the position of other vehicles and the threat or hazard they present; calculate risk; issue driver advisories or warnings; and take preemptive actions to avoid and mitigate crashes.

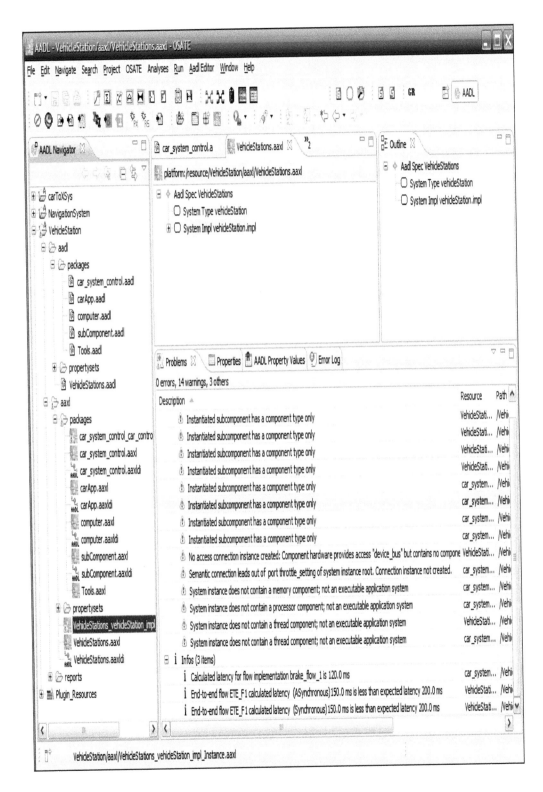

FIGURE 6.13   Simulation results of flow analysis by the OSATE tool.

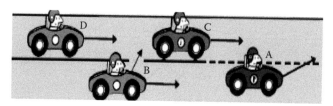

FIGURE 6.14 Vehicles in parallel lanes.

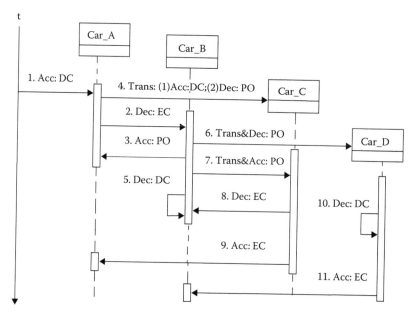

FIGURE 6.15 The spatiotemporal sequence diagram of vehicle relations in parallel lanes.

The two basic objectives of any driver on road is to avoid collision with nearby vehicles, pedestrians, and traffic infrastructure [60], and share the road with other drivers. Various driving tasks include lane-driving, lane-changing, braking, passing other vehicles, and making turns. It can be observed that some of these, such as lane-driving, braking, and lane-changing, are basic tasks. Lane-driving means moving along the mid-lane curve (t) of a lane without collision with other neighboring vehicles. This is one of the most basic driving tasks on road.

When we specify spatiotemporal relations of vehicle to vehicle using spatiotemporal sequence diagram, we need to walk a fine line between realism and a suitable level of abstraction. Since automobile driving is a physically complex process, realistic models are often also complex. We define four basic operations: Acc: accelerate, Dec: decelerate, Trans: transform lanes, and Turn: turn lanes.

In parallel lanes, we not only need to consider the spatiotemporal relations between vehicles, but also need to consider that vehicles change lanes from one lane to another. In Figure 6.14, Car A and Car B are in one lane, Car C and Car D are in another lane. Car A and Car B turn to the lane of Car C and Car D.

We can represent and reason spatiotemporal relations between vehicles in parallel lanes using a spatiotemporal sequence diagram as shown in Figure 6.15.

```
climb(Car_B, Car_A ) Ö clock(Dec) Ö clock(Acc) Ö
clock(Trans)Ö climb(Car_B, Car_C) Ö clock(Acc)
Ödrop(Car_B, Car_D)
```
   Suppose that V is the speed and d is the distance.

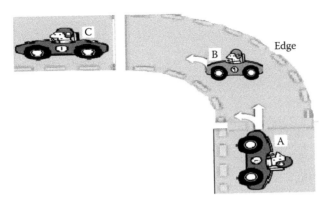

FIGURE 6.16 Same-lane vehicles turning a corner.

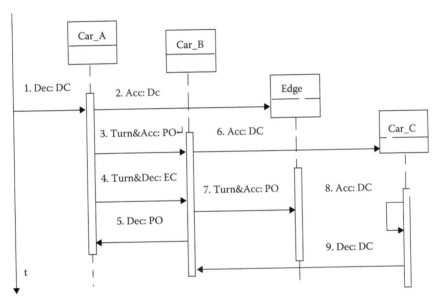

FIGURE 6.17 The spatiotemporal sequence diagram of same-lane vehicle relations in turning a corner.

$\rho(\text{climb}(\text{Car\_B}, \text{Car\_A}), \text{clock}(\text{Dec})) \leq d_{BA}/V_B$
$\rho(\text{clock}(\text{Dec}), \text{clock}(\text{Acc})) \leq d_{BA}/V_B$
$\rho(\text{clock}(\text{Acc}), \text{climb}(\text{Car\_C}, \text{Car\_A})) \leq d_{CA}/V_C$
$\rho(\text{clock}(\text{Acc}), \text{clock}(\text{Trans})) \leq d_{CA}/V_C$
$\rho(\text{clock}(\text{Trans}), \text{climb}(\text{Car\_B}, \text{Car\_C})) \leq d_{BC}/V_B$
$\rho(\text{climb}(\text{Car\_B}, \text{Car\_C}), \text{clock}(\text{Acc})) \leq d_{BC}/V_B$
$\rho(\text{clock}(\text{Acc}), \text{drop}(\text{Car\_B}, \text{Car\_D})) \leq d_{BD}/V_D$

Although the vehicles are in the same lane, the distance between the vehicles may not be known when the vehicles turn a corner; there are some places that the driver cannot see as shown in Figure 6.16.

We can represent and reason spatiotemporal relations between same-lane vehicles in turning a corner using a spatiotemporal sequence diagram as shown in Figure 6.17.

```
climb(Car_A, Car_B ) Ö clock(Dec) Ö clock(Acc)) Ö
clock(Turn)Ö climb(Car_B, Car_C) Ö clock(Dec) Ö clock(Acc)
Ödrop(Car_C, Car_B) Ö clock(Turn)
```

Suppose that V is the speed and d is the distance.

$\rho(\text{climb}(\text{Car\_A, Car\_B}), \text{clock}(\text{Dec})) \leq d_{AB}/V_A$

$\rho(\text{clock}(\text{Dec}), \text{clock}(\text{Acc})) \leq d_{AB}/V_A$

$\rho(\text{clock}(\text{Acc}), \text{clock}(\text{Turn})) \leq d_{AB}/V_A$

$\rho(\text{clock}(\text{Turn}), \text{climb}(\text{Car\_B, Car\_C})) \leq d_{BC}/V_B$

$\rho(\text{climb}(\text{Car\_B, Car\_C}), \text{clock}(\text{Dec})) \leq d_{BC}/V_B$

$\rho(\text{clock}(\text{Dec}), \text{clock}(\text{Acc})) \leq d_{BC}/V_B$

$\rho(\text{clock}(\text{Acc}), \text{drop}(\text{Car\_C, Car\_B})) \leq d_{BC}/V_C$

$\rho(\text{drop}(\text{Car\_C, Car\_B}), \text{clock}(\text{Turn})) \leq d_{BC}/V_C$

6.7 CONCLUSION

The design methods for big data driven CPS not only meet multiple V's characteristics, but also must meet special characteristics of CPS such as spatiotemporal requirements and real-time communication requirements. In this chapter, we propose a multi-view, multi-domain, and multi-paradigm approach for specification and modeling of big data driven CPS. View-oriented approach allow designers from different views to develop and evaluate design alternatives within the context of formalisms relevant to selected aspects of the system. Each view highlights certain features and occludes others to make analysis tractable and to focus on particular attributes of big data driven CPS. It is clear that an appropriate specification and modeling approach cannot be fitted into the limitations of one formalism at a time, particularly in the early stages of the design process. Hence, it is necessary to take a combination of different methodologies in a multi-formalism approach from different views. In this chapter, we propose a model-based approach to model big data driven CPS based on the integration of Modelica, ModelicaML, AADL, RCC, clock theory, UML, SysML, and Hybrid Relation Calculus; we illustrate our approach by specifying and modeling VANETs. Future work will focus on the following aspects.

First, a moving object represents the continuous evolution of a spatial object over time. Regarding big data modeling, an important question is how to represent a moving object. In the future work, we shall focus our work on specification and modeling of moving object spatiotemporal behavior.

Second, real-time communication in big data environment–CPS must provide guarantee that huge data in the network are delivered according to their real-time deadlines. Because big data set sizes are growing exponentially, it is important to develop data transfer protocols that are the most efficient available.

Third, it is important to integrate big data, CPS, and cloud computing into a single framework that can provide the collection data from sensors, transfer big data on real-time communication channels, and process and analyze big data on a cloud platform.

Fourth, we shall develop an automatic transformation tool, which can make the transformation among AADL, UML, SysML, ModelicaML, and Modelica.

Fifth, MapReduce is developed for batch and high-throughput job execution, and it is suitable for jobs that process large volumes of data in a long time. However, big data driven CPS have new demands like real-time store, queries, sensor data, and stream data that cannot be handled efficiently by batch-based frameworks like Hadoop. We shall develop real-time scheduling algorithms for MapReduce.

Sixth, we need to develop an integration tool to support the specification, modeling, analysis, design, simulation, test, and verification of big data driven CPS.

ACKNOWLEDGMENTS

This work is supported by the National Natural Science Foundation of China under Grant No. 61370082 AND No. 61173046) and the Natural Science Foundation of Guangdong Province under Grant No. S2011010004905). This work is also supported by the Shanghai Knowledge Service Platform Project (No. ZF1213).

REFERENCES

1. Z.-H. Zhou et al. Big data opportunities and challenges: Discussions from data analytics perspectives, *IEEE Computational Intelligence Magazine* 9(4), 63–74 (November 2014).

2. Z. Wu, O.B. Chin. From big data to data science: A multi-disciplinary perspective, *Big Data Research* 1, 1 (2014).

3. A. Katal, M. Wazid, R.H. Goudar. Big data: Issues, challenges, tools and good practices, *2013 Sixth International Conference of Contemporary Computing (IC3)*, Noida, India, pp. 404–409 (2013).

4. I.A.T. Hashem et al. The rise of "big data" on cloud computing: Review and open research issues, *Information Systems* 47, 98–115 (2015).

5. M. Chen, S. Mao, Y. Liu. Big data: A survey, *Mobile Networks and Application* 19(2), 171–209 (2014).

6. C.L. Philip Chen, C.Y. Zhang. Data-intensive applications, challenges, techniques and technologies: A survey on big data. *Information Sciences* 275, 314–347 (August 10, 2014).

7. I. Stojmenovic. Machine-to-machine communications with in-network data aggregation, processing and actuation for large-scale cyber-physical systems, *IEEE Internet of Things Journal* 1(2), 122–128 (April 2014).

8. Z. Huang, C. Wang, A. Nayak, I. Stojmenovic. Small cluster in cyber physical systems: Network topology, interdependence and cascading failures, *IEEE Transactions on Parallel and Distributed Systems* 1(1), (July 17, 2014).

9. D. B. Rawat, G. Yan, B. B. Bista and V. Chandra. Wireless Network Security. In *Building Next-Generation Converged Networks: Theory and Practice* (Eds. Dr. Pathan et al.), CRC Press/Taylor & Francis Group, Boca Raton, FL, pp. 199–220, ISBN: 1466507616. doi: 10.1109/SC.2008.5215771.

10. G. Yan, D. B. Rawat and B. B. Bista. Provisioning Vehicular ad hoc networks with quality of service. *The International Journal of Space-Based and Situated Computing (IJSSC)*, 2(2), pp. 104–111, 2012. doi: 10.1504/IJSSC.2012.047467

11. Acatech, National Academy of Science and Engineering (Ed.). *Cyber-Physical Systems* (2011) Springer-Verlag, Berlin/Heidelberg, Germany.

12. O. Diallo, J.J.P.C. Rodrigues, Mbaye Sene. Real-time data management on wireless sensor networks: A survey, *Journal of Network and Computer Applications* 35(3), 1013–1021 (2012).

13. M. Broy. Engineering cyber-physical systems: Challenges and foundations, in *Complex Systems Design & Management*, (Eds. Aiguier, M., Caseau, Y., Krob, D., Rauzy, A.). Springer, Berlin/Heidelberg, Germany, pp. 1–13, (2013).

14. Call For Papers—CPSData. Big data technologies for the analysis and control of complex Cyber-Physical Systems. www.systematic-paris-region.org/.../call-papers-cps-data-big-da. October 4, 2014.

15. F. Tekiner, J.A. Keane. Big data framework, *Proceedings—2013 IEEE International Conference on Systems, Man, and Cybernetics, SMC*, pp. 1494–1499 (2013). Manchester, UK.

16. J. Hugues. AADLib, a library of reusable AADL models, SAE Technical Papers, v 7, 2013, SAE 2013 AeroTech Congress and Exhibition, Aerotech (2013). Montreal, Canada.

17. W. Schamai. Modelica Modeling Language (ModelicaML): A UML profile for Modelica, Technical report 2009:5, EADS IW, Linkoping University, Linkoping, Germany (2009).

18. H. Jifeng. A clock-based framework for construction of hybrid systems, *Lecture Notes in Computer Science*, *ICTAC 2013*, 8049, 22–41 (2013).

19. S.E. Mattsson, H. Elmqvist, M. Otter. Physical system modeling with Modelica, *Control Engineering Practice* 6, 501–510 (1998).

20. D. Randell, Z. Cui, A. Cohn. A spatial logic based on regions and connection, *Proceedings of the Third International Conference on Knowledge Representation and Reasoning* pp. 165–176 (1992). Cambridge, MA.

21. OMG. OMG Unified Modeling Language TM (OMG UML) Version 2.2 (2009). http://www.omg.org/spec/UML/. January 2, 2009.

22. OMG. OMG Systems Modeling Language (OMG SysML™) Version 1.1 (2008). http://www.omg.org/spec/SysML/, January 11, 2008.

23. J. He. Hybrid relation calculus, *2013 18th International Conference on Engineering of Complex Computer Systems (ICECCS)*, IEEE, Singapore, p. 2 (2013).

24. D.B. Rawat, G.J. Heijenk, M.C. Weigle, B.B. Bista, Y.-S. Chen. Special issue of ad hoc networks on recent advances in vehicular communications and networking, *Ad Hoc Networks* 11(7), 1989–1991 (2013).

25. M. Obitko, V. Jirkovský, J. Bezdíěk. Big data challenges in industrial automation, HoloMAS 2013, LNAI 8062, pp. 305–316(2013).

26. C. Körner, M. May, S. Wrobe. Spatiotemporal modeling and analysis—Introduction and overview, *Künstliche Intelligenz* 26, 215–221 (2012).

27. M. Vazirgiannis, O. Wolfson. A spatiotemporal model and language for moving objects on road networks, 7th International Symposium, SSTD 2001, Redondo Beach, CA, July 12–15, 2001, Jensen, C.S., Schneider, M., Seeger, B., Tsotras, V.J. (Eds.), *Lecture Notes in Computer Science*, Vol. 2121, pp. 20–35, Springer.

28. P. Sistla, O. Wolfson, S. Chamberlain, S. Dao. Modeling and Querying Moving Objects, *Proceedings of the 13th International Conference on Data Engineering (ICDE13)*, Birmingham, UK, April 1997, pp. 422–432.

29. R. Xie, X. Luo, Key issues and theoretical framework on moving objects data mining, *ADMA 2010, Part I, LNCS 6440*, pp. 577–584 (2010).

30. C He, Y Lu, D Swanson. Real-time scheduling in MapReduce clusters, *2013 IEEE International Conference on High Performance Computing and Communications & 2013 IEEE International Conference on Embedded and Ubiquitous Computing*, Zhangjiajie, China, pp. 1536–1544 (2013).

31. N. Zacheilas, V. Kalogeraki. Real-time scheduling of skewed MapReduce jobs in heterogeneous environments, *11th International Conference on Autonomic Computing (ICAC'14)*, Philadelphia, PA, pp. 189–200 (June 18–20, 2014).

32. E.D. Simmon et al. A vision of cyber-physical cloud computing for smart networked systems, NIST Interagency/Internal Report (NISTIR)—7951 http://www2.nict.go.jp/univ-com/isp/doc/NIST. IR.7951.pdf, (2013), January 12, 2013.

33. S.S. Craciunas et al. Information-acquisition-as-a-service for cyber-physical cloud computing, *HotCloud'10 Proceedings of the 2nd USENIX Conference on Hot Topics in Cloud Computing*, Boston, MA, pp. 14–14, 2010.

34. A.B. Sharma et al. Modeling and analytics for cyber-physical systems in the age of big data, *ACM SIGMETRICS Performance Evaluation Review* 41(4), pp. 74–77 (March 2014).

35. M. Persson et al. A characterization of integrated multi-view modeling in the context of embedded and cyber-physical systems, *International Conference on Embedded Software (EMSOFT)*, Montreal, Quebec, Canada (2013).

36. R. von Hanxleden, E.A. Lee, C. Motika, H. Fuhrmann. Multi-view modeling and pragmatics in 2020, 17 Monterey Workshop 2012, Oxford, UK, March 19–21, 2012, *LNCS 7539*, pp. 209–223 (2012).

37. J de Lara, T Levendovszky, PJ. Mosterman, H Vangheluwe. Current issues in multi-paradigm modeling, in *Models in Software Engineering*, Holger and Giese (Eds.), pp. 237–246, Lecture notes in computer science, vol. 5002, Springer, Berlin, Germany (2008).

38. H. Elmqvist, S.E. Mattsson. An introduction to the physical modeling language Modelica, *ESS'97 European Simulation Symposium*, Passau, Germany (October 19–22, 1997).

39. A. Pop, D. Akhvlediani, P. Fritzson, P. Towards unified systems modeling with the ModelicaML UML profile, *International Workshop on Equation-Based Object-Oriented Languages and Tools*, Linköping University Electronic Press, Berlin, Germany (2007).

40. P.H. Feiler, D.P. Gluch, J.J. Hudak. *The Architecture Analysis & Design Language (AADL): An Introduction*. CMU/SEI-2006-TN-001, Carnegie Mellon University and Software Engineering Institute, Pittsburgh, PA (2006).

41. S. Goddard, J.S. Deogun. Future mobile cyber-physical systems: Spatio-temporal computational environments. varma.ece.cmu.edu/cps/Position-Papers/Goddard-2.pdf. August 9, 2006.

42. J.S. Deogun, S. Goddard. Reasoning about time and space: A cyber physical systems perspective, *14th IEEE Real-Time and Embedded Technology and Application Symposium (RTAS'08), Work in Progress (WIP) Proceedings*, St. Louis, MO, pp. 1–4 (April 2008).

43. X. Song. Systematic integration of design methods, *IEEE Software* 14(2), 107–117 (March 1997).

44. M. Biehl. Literature study on model transformations, Technical Report ISRN/KTH/MMK/R-10/07-SE, Royal Institute of Technology, Sweden, (July 2010).

45. K. Czarnecki, S. Helsen. Feature-based survey of model transformation approaches, *IBM Systems Journal, special issue on Model-Driven Software Development* 45(3), 621–645 (2006).

46. J. Hudak, P. Feiler. Developing AADL models for control systems: A practitioner's guide, CMU SEI, Tech. Rep. CMU/SEI-2007-TR-014, 2007, www.aadl.info/aadl/documents/06tr019.ControlSystemEngin, January 12, 2007.

47. P.H. Feiler, B. Lewis. The SAE Avionics Architecture Description Language (AADL) standard: A basis for model-based architecture-driven embedded systems engineering, Software Engineering Institute (2003). http://resources.sei.cmu.edu/library/asset-view.cfm?assetID=29547, January 12, 2013.
48. SAE/AS2-C. Data Modeling Annex document for the Architecture Analysis & Design Language v2.0 (AS5506A), October 2009.
49. O. Gilles, J. Hugues. Expressing and enforcing user-defined constraints of AADL models, *Proceedings of the 5th UML and AADL Workshop (UML and AADL 2010)*. 15th IEEE International Conference on Engineering of Complex Computer Systems, pp. 337–342, 22–26, March 2010, Oxford, UK.
50. Modelica Association. Modelica: A unified object-oriented language for physical systems modeling: Language specification version 3.0 (September 2007). www.modelica.org.
51. W. Schamai, P. Fritzson, C. Paredis, A. Pop. Towards unified system modeling and simulation with ModelicaML.: Modeling of executable behavior using graphical notations, *Proceedings Seventh Modelica Conference*, Como, Italy, pp. 612–621 (September 20–22, 2009).
52. A. Pop, D. Akhvlediani, P. Fritzson. Towards unified systems modeling with the ModelicaML UML profile, *International Workshop on Equation-Based Object-Oriented Languages and Tools*, Linköping University Electronic Press, Berlin, Germany (2007).
53. H.S. Adam, O. Gmbh, O., B.M. Simtd. A car-to-X system architecture for field operational tests, *IEEE Communications Magazine* 48(5), 148–154 (2010).
54. D.B. Rawat, B.B. Bista, G. Yan, S. Olariu. Vehicle-to-vehicle connectivity and communication framework for vehicular Ad-Hoc networks, *CISIS*, Birmingham, UK, pp. 44–49 (2014).
55. S. Yousefi, M.S. Mousavi, M. Fathy. Vehicular Ad Hoc Networks (VANETs): Challenges and perspectives, *Proceedings of Sixth International Conference on ITS Telecommunications*, Chengdu, China, pp. 761–766 (2006).
56. G. Yan, D.B. Rawat, B.B. Bista, Provisioning vehicular ad hoc networks with quality of service, *Proceedings of the Fifth International Conference on Broadband and Wireless Computing, Communication and Applications (BWCCA-2010)*, Fukuoka, Japan, pp. 102–107 (November 2010).
57. G. Yan, D.B. Rawat, B.B. Bista. Towards Secure Vehicular Clouds, *CISIS*, Sanpaolo Palace Hotel, Palermo, Italy, pp. 370–375 (2012).
58. B. Lewis, P.H Feiler, J. Hugues. AADL—OSATE status and future. http://gforge.enseeiht.fr/docman/view.php/52/4296/A3-ISAE.pdf. June 10, 2012.
59. S.M. Loos, A. Platzer. Adaptive cruise control: Hybrid, distributed, and now formally verified, *Lecturer Notes in Computer Science* 6664, 42–56 (2011).
60. S. Mitsch, S.M. Loos, A. Platzer. Towards formal verification of freeway traffic control, in C. Lu, ed., *ACM/IEEE Third International Conference on Cyber-Physical Systems* Beijing, China, pp. 171–181 (April 17–19).

Section III

Communications and Signal Processing

7 Algorithms for Control with Limited Processing and Communication Resources

Daniel E. Quevedo, Vijay Gupta, and Wann-Jiun Ma

CONTENTS

7.1   INTRODUCTION

In networked and embedded systems, it is often the case that the communication and processing resources are shared among multiple processes or control loops. Thus, the availability of these resources for any particular loop is usually time-varying. Even though the availability *on average* may be sufficiently large, at particular time steps, only limited communication bandwidth or processor attention may be available. This is the reason for the recent interest in the area of control with limited communication and processing resources.

Control design when the sensor or controller data are transmitted across communication channels that impose significant delays or distortion in the transmitted data is studied under the rubric of networked control systems. Of particular interest to the present work is the literature on control across analog erasure channels. Such a channel model is used to model stochastic data loss due to effects such as fading, as induced by wireless channels [1,2]. The same model is also popular for modeling the effect of congestion in a communication network. Indeed, if the network protocols detect congestion in the network, the most common prescribed action is to erase the data packets in the buffers at various routers. Much work has analyzed and designed controllers for various configurations when such a channel model is used (see, e.g., [3–5]).

On the other hand, the problem of control in the presence of time-varying processor (i.e., computation) availability has also gained much attention [6–8]. These works can be in the form of bounding the processing time that is required to solve the control input to a specified accuracy and calculating control inputs aperiodically to reduce the load on the processor through event-triggered and

171

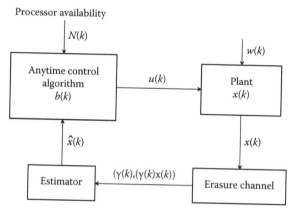

FIGURE 7.1 Anytime control with an unreliable communication channel.

self-triggered algorithms [9,10] and anytime algorithms for control [11–14]. In keeping with the general philosophy of anytime algorithms that are popular in real-time systems community, such algorithms calculate a coarse control input as soon as the processor becomes available and then progressively refine it if the processor is still available. In our previous work [14], we had considered a sequence-based anytime control in which the controller calculates a sequence of control inputs whenever the processor is available. The number of inputs in the sequence is time-varying since it depends on the time for which the processor was available at that instant, but no control input is calculated very rarely. Thus, the system evolves without a calculated control input very infrequently.

However, relatively few works have considered control design under both limited communication and processing resources. We consider such a setting in this chapter. Specifically, we consider a plant in which the sensor data are transmitted across an analog erasure channel to the controller (see Figure 7.1). The channel erases the packets at any time step in a stochastic manner. The controller uses a state estimator and implements the sequence-based anytime control algorithm from [14]. In our previous work [15], we had analyzed a related control loop and shown that the anytime algorithm significantly improves the closed-loop performance. However, one drawback in the solution proposed in that paper was that if the erasure probability of the sensor to controller channel is high, then the controller will often run out of data to feed to the plant actuator. In the present work, we propose one possible way to counteract this problem. Specifically, we propose that the controller be coupled to an estimator that estimates the state based on the last received information from the sensor and the control inputs that the controller calculated at previous time steps (and that are being applied to the plant).* We show that the presence of such an estimator improves the control performance.

The remainder of this chapter is organized as follows. Section 7.2 presents the control architecture. The anytime control algorithm is described in Section 7.3. Section 7.4 derives an analytical model of the resulting closed loop. Numerical results are included in Section 7.5. Section 7.6 draws conclusions.

Notation: We write $\mathbb{N}$ for $\{1, 2, \ldots\}$ and $\mathbb{N}_0$ for $\mathbb{N} \cup \{0\}$, whereas $\mathbb{R}$ represents the real numbers. The $p \times p$ identity matrix is denoted via I_p, $0_p \triangleq 0_{p \times p}$, and $\mathbf{0}_p \triangleq 0_{p \times 1}$. The superscript T refers to transpose. The Euclidean norm of a vector x is denoted via $|x| = \sqrt{x^T x}$.

---

* In the present setup (see Figure 7.1), the information about the control inputs is available since the controller can buffer the inputs transmitted to the plant.

7.2 CONTROL WITH LIMITED COMPUTATION AND COMMUNICATION RESOURCES

We consider nonlinear plant models sampled periodically with the sampling interval $T_s > 0$ that are described in discrete time via

$$x(k + 1) = f(x(k), u(k), w(k)), \quad k \in \mathbb{N}_0. \tag{7.1}$$

where in Equation (7.1),

$x(k) \in \mathbb{R}^n$ is the plant state
$w(k) \in \mathbb{R}^m$ is an unmeasured disturbance
$u(k) \in \mathbb{U} \subseteq \mathbb{R}^p$ is the (possibly constrained) plant input

As depicted in Figure 7.1, the plant is equipped with a sensor, which has direct access to the plant state at the sampling instants $k \in \mathbb{N}_0$.

Throughout this work, we will assume that a suitable controller $\kappa : \mathbb{R}^n \to \mathbb{U}$ has been designed such that the nominal (and ideal) closed loop

$$x(k + 1) = f(x(k), \kappa(x(k)), w(k)), \quad k \in \mathbb{N}_0, \tag{7.2}$$

exhibits the desired performance specifications.

The main purpose of this chapter is to illustrate how the underlying control law κ can be implemented in a situation where computational and communication resources are limited.

7.2.1 COMPUTATIONAL ISSUES

When designing discrete-time control systems such as Equation 7.2, it is generally assumed that the processing resources available to the controller are always such that the control law can be evaluated within a fixed (and small) time delay, say $\delta \in (0, T_s)$.* However, in practical networked and embedded systems, the computational resources (e.g., processor execution times) available for control calculations may vary and, at times, be insufficient to generate a control input within the prescribed time delay δ. One possible remedy for this issue would be to redesign the control system for a worst-case scenario by choosing larger values of δ and, possibly, T_s. However, such an approach will, in general, lead to unnecessary conservativeness and associated poor performance (see also [8]). In the present work, we adopt an anytime control paradigm to seek favorable trade-offs between processor availability and control performance.

7.2.2 COMMUNICATION ISSUES

We focus on a loop where transmission of sensor data is across an erasure channel that introduces random packet dropouts. To keep communication costs low, the controller does not send acknowledgments back to the sensor and no retransmissions are allowed. To describe the situation, we introduce the binary transmission success process $\{\gamma(k)\}_{k \in \mathbb{N}_0}$, which can be related to channel fading and interference (see, e.g., [1,2]) and serves to describe packet loss: a successful transmission at time k is denoted by $\gamma(k) = 1$ and a packet erasure by $\gamma(k) = 0$.

In order to implement the control law κ of Equation 7.2 in the presence of unreliable communications, the algorithms to be studied will use a simple state estimator of the form (cf. [17])

$$\hat{x}(k) = \begin{cases} x(k) & \text{if } \gamma(k) = 1, \\ f(\hat{x}(k - 1), u(k - 1), \mathbf{0}_m) & \text{if } \gamma(k) = 0. \end{cases} \tag{7.3}$$

---

* Recall that fixed delays can be easily incorporated into the model Equation 7.1 by aggregating the previous plant input to the plant state (see also [16]). For ease of exposition, throughout this work, we will adopt the standard discrete-time notation as in Equation 7.1.

Such an estimator can be readily implemented at the controller node, since we assume that the controller has direct access to the actuator node (see Figure 7.1).

7.2.3 BASELINE ALGORITHM

A direct implementation of the control law κ, when processing and communication resources are time-varying and the estimator Equation 7.3 is used, results in the following *baseline* algorithm:

$$u(k) = \begin{cases} \kappa(\hat{x}(k)), & \text{if sufficient computational resources to evaluate} \\ & \qquad \kappa(\hat{x}(k)) \text{ are available between } kT_s \text{ and } kT_s + \delta, \\ \mathbf{0}_p, & \text{otherwise,} \end{cases} \tag{7.4}$$

where the symbol $u(k)$ with $k \in \mathbb{N}_0$ denotes the plant input that is applied during the interval* $(kT_s + \delta, (k+1)T_s + \delta)$. While the baseline algorithm (7.4) is intuitive and simple, it is by no means clear that it cannot be outperformed by more sophisticated control formulations. In the following section, we will combine the anytime control algorithm of [14] with the estimator Equation 7.3. The aim is to make more efficient use of the processing resources available for control, when compared to the baseline algorithm (7.4).

7.3 ANYTIME CONTROL THROUGH CALCULATION OF CONTROL SEQUENCES

Throughout this work, we will assume that the controller requires processor time to carry out mathematical computations, such as evaluating the control function κ. However, simple operations at a bit level, such as writing data into buffers, shifting buffer contents, and setting values to zero, do not require processor time. Furthermore, for the ease of exposition, we will assume that evaluating (7.3) does not require processor time either. Similarly, input–output operations, that is, A/D and D/A conversion, are triggered by external asynchronous loops with a real-time clock and do not require that the processor be available for control. As in regular discrete-time control, these external loops ensure that, in the absence of communication loss, state measurements are available at the instants $\{kT_s\}_{k\in\mathbb{N}_0}$ and that the controller outputs are passed on to the plant actuators at times $\{kT_s + \delta\}_{k\in\mathbb{N}_0}$, where δ is fixed; see, e.g., [18].

A standing assumption is that if the processor were fully available for control, then calculating the desired plant input $u(k) = \kappa(\hat{x}(k))$ would be possible within the preallocated timeframe: $t \in (kT_s, kT_s + \delta)$. Issues arise when, at times, processor availability does not permit the desired plant input to be calculated. To take care of the associated performance loss, in the present work we will adopt an anytime control algorithm (named Algorithm A_2 in Quevedo and Gupta [14]).

The control algorithm implements the following basic idea: At time intervals when the controller is provided with more processing resources than are needed to evaluate the current control input, the controller uses its current state estimate $\hat{x}(k)$ to calculate a sequence of tentative future plant inputs, say $\vec{u}(k)$. The sequence is stored in a local buffer and may be used when, at some future time steps, the processor availability precludes any control calculations.

Figure 7.2 outlines the algorithm. In this figure, we denote the buffer states via $\{b(k)\}_{k\in\mathbb{N}_0}$, where

$$b(k) = \begin{bmatrix} b_1(k) \\ b_2(k) \\ \vdots \\ b_\Lambda(k) \end{bmatrix} \in \mathbb{R}^{\Lambda p}, \quad k \in \mathbb{N}_0$$

---

* If sufficient computational resources are not available, then one could alternatively hold the previous control value and set $u(k) = u(k-1)$. Extensions are left to the interested reader.

Step 1: At time $t = 0$,
$\quad$ SET $b(-1) \leftarrow \mathbf{0}_{\Lambda p}, k \leftarrow 0$

Step 2: IF $t \geq kT_s$, THEN
$\quad$ INPUT $\hat{x}(k)$ from the estimator, see (7.3);
$\quad$ SET $\chi \leftarrow \hat{x}(k), j \leftarrow 1, b(k) \leftarrow Sb(k-1)$;

$\quad$ END

Step 3: WHILE "sufficient processor time is available" and $j \leq \Lambda$ and time $t < (k+1)T_s$,
$\quad$ EVALUATE $u_j(k) = \kappa(\chi)$;
$\quad$ IF $j = 1$, THEN
$\quad\quad$ OUTPUT $u_1(k)$;

$\quad$ END

$\quad$ SET $b_j(k) \leftarrow u_j(k)$;
$\quad$ IF "sufficient processor time is not available" or $t \geq (k+1)T_s$, THEN
$\quad\quad$ GOTO Step 5;

$\quad$ END

$\quad$ SET $\chi \leftarrow f(\chi, u_j(k), \mathbf{0}_m), j \leftarrow j+1$;

$\quad$ END

Step 4: IF $j = 1$, THEN
$\quad$ OUTPUT $b_1(k)$;

$\quad$ END

Step 5: SET $k \leftarrow k+1$ and GOTO Step 2;

FIGURE 7.2 Anytime control algorithm.

for a given value $\Lambda \in \mathbb{N}$ and where each $b_j(k) \in \mathbb{R}^p, j \in \{1, \ldots, \Lambda\}$. The symbol S refers to the shift matrix

$$S \triangleq \begin{bmatrix} 0_p & I_p & 0_p & \cdots & 0_p \\ \vdots & \ddots & \ddots & \ddots & \vdots \\ 0_p & \cdots & 0_p & I_p & 0_p \\ 0_p & \cdots\cdots\cdots & 0_p & I_p \\ 0_p & \cdots\cdots\cdots\cdots & 0_p \end{bmatrix} \in \mathbb{R}^{\Lambda p \times \Lambda p}.$$

It is worth noting that the algorithm does not require prior knowledge of future processor availability for control. This opens the possibility to employ the algorithm in shared systems, where the controller task can be preempted by other computational tasks carried out by the processor (see also [19,20]). As in other anytime algorithms, there exists a compromise between resultant closed-loop performance and processor availability. Understanding this trade-off forms the bulk of this work.

Remark 7.1 (Relationship to Model Predictive Control) *The basic idea of the algorithm, that is, calculation of future control inputs, is reminiscent of the philosophy behind receding horizon control [21–23]. However, there are some differences between the two methods. Most importantly, the algorithm in Figure 7.2 does not calculate a control input to optimize a given cost function. Further, the number of time steps for which the control input is calculated is time-varying and externally imposed by the available processor time.*

7.4 CLOSED-LOOP MODEL

With the anytime control algorithm presented in Section 7.3, any extra processing time available is used to calculate additional elements of the tentative plant input sequence $\vec{u}(k)$. When compared to the baseline algorithm (7.4), this provides a higher-quality result, that is, a control sequence that better safeguards against performance loss at *future* time instances where processor availability may be insufficient. The algorithm in Figure 7.2 amounts to a dynamical system that is driven by the dropout process and the processor availabilities. We will next further elucidate the situation.

During each iteration of the while-loop in Step 3, the current state estimate $\hat{x}(k)$ (see Equation 7.3), is used to calculate a tentative control, namely, $u_j(k)$. We shall denote by $N(k)$ the total number of iterations of the while-loop that are carried out during the interval $t \in (kT_s, (k+1)T_s)$ and note that $N(k) \in \{0, 1, \ldots, \Lambda\}$. Thus, if $N(k) \geq 1$, then the entire sequence of tentative controls calculated is given by

$$\vec{u}(k) = \begin{bmatrix} u_0(k) \\ u_1(k) \\ \vdots \\ u_{N(k)-1}(k) \end{bmatrix}.$$

This control sequence is provided by evaluating the nominal controller κ on either the current plant state or, if unavailable, using the plant state estimate:

$$\vec{u}(k) = \begin{cases} \bar{F}_{N(k)}(x(k)), & \text{if } \gamma(k) = 1 \\ \bar{F}_{N(k)}(\hat{x}(k)), & \text{if } \gamma(k) = 0, \end{cases} \tag{7.5}$$

where

$$\bar{F}_{N(k)}(\chi) \triangleq \begin{bmatrix} \kappa(\chi) \\ \kappa(\bar{f}(\chi)) \\ \vdots \\ \kappa(\bar{f}^{(N(k)-1)}(\chi)) \end{bmatrix},$$

and $\bar{f}^j(\chi)$ denotes the j-fold composition of the nominal controlled plant model

$$\bar{f}(\chi) \triangleq f(\chi, \kappa(\chi), \mathbf{0}_m),$$

evaluated at $\chi \in \mathbb{R}^n$. The plant actuator value $u(k)$ is then given by either $\kappa(x(k))$ or $\kappa(\hat{x}(k))$, depending on whether a dropout occurred at time k or not.

If $N(k) = 0$, then the processor was not available for control, and the actuator value is taken as the first element of the (shifted) buffer state. This gives

$$u(k) = \begin{cases} \kappa(x(k)), & \text{if } \gamma(k) = 1 \text{ and } N(k) \geq 1 \\ \kappa(\hat{x}(k)), & \text{if } \gamma(k) = 0 \text{ and } N(k) \geq 1 \\ e_1^T Sb(k-1), & \text{if } N(k) = 0, \end{cases} \tag{7.6}$$

where

$$e_1 \triangleq \begin{bmatrix} I_p \\ 0_p \\ \vdots \\ 0_p \end{bmatrix} \in \mathbb{R}^{\Lambda p \times p}.$$

In order to describe the buffer dynamics, it is convenient to introduce the matrices

$$M_i \triangleq (I_{\Lambda p} - D_i)S, \quad i \in \{1, 2, \dots, \Lambda\},$$

where

$$D_i \triangleq \begin{cases} \mathrm{diag}(I_{ip}, 0_{(\Lambda-i)p}), & \text{if } i \in \{1, 2, \dots, \Lambda - 1\} \\ I_{\Lambda p}, & \text{if } i = \Lambda. \end{cases}$$

In terms of the notation introduced earlier, we have

$$b(k) = \begin{cases} Sb(k-1), & \text{if } N(k) = 0, \\ \begin{bmatrix} \vec{u}(k) \\ 0_{(\Lambda - N(k))p} \end{bmatrix} + M_{N(k)}b(k-1), & \text{if } N(k) \geq 1, \end{cases}$$

expression that, together with Equations 7.1, 7.3, 7.5, and 7.6, characterizes the closed loop.

Remark 7.2 (Closed-Loop Stability)   *Under suitable conditions on plant model, nominal control law, and communication and computation resources, the random-time drift analysis methods described in our recent papers [15,24] can be extended to derive stochastic stability guarantees for the situation of interest in the current work. A key point here is to recall that, in the absence of disturbances w(k), the state estimates in Equation (7.3) are error free and therefore the system dynamics only depends on whether the buffer is empty or not. The associated effective buffer length process $\{\lambda(k)\}_{k\in\mathbb{N}}$, which quantifies how many values that stem from the tentative control sequences $\{\vec{u}(k-\ell)\}$, $\ell \in \mathbb{N}_0$ are contained in the current buffer state b(k), follows the recursion*

$$\lambda(k) = \max\{N(k), \lambda(k-1) - 1\}, \quad k \in \mathbb{N}_0,$$

and therefore corresponds to a finite Markov chain [25].   □

7.5 SIMULATION STUDY

We consider an open-loop unstable nonlinear system:

$$\begin{bmatrix} x_1(k+1) \\ x_2(k+1) \end{bmatrix} = \begin{bmatrix} x_2(k) + u_1(k) \\ -\mathrm{sat}(x_1(k) + x_2(k)) + u_2(k) \end{bmatrix} + \begin{bmatrix} w_1(k) \\ w_2(k) \end{bmatrix}$$

where

$$\mathrm{sat}(\mu) = \begin{cases} -10, & \text{if } \mu < -10, \\ \mu & \text{if } \mu \in [-10, 10], \\ 10, & \text{if } \mu > 10, \end{cases}$$

(see Quevedo and Gupta [14, Example 2]). The initial condition $x(0)$ and the disturbance $w(k)$ are zero-mean i.i.d. Gaussian with covariance matrices $10I_2$. Suppose that the buffer length is given by $\Lambda = 4$ and that the process $\{N(k)\}_{k\in\mathbb{N}_0}$ is i.i.d. uniformly distributed on $\{0, \dots, 4\}$. Similarly,

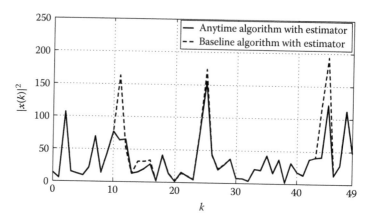

FIGURE 7.3 Realization of the state trajectories (measured by the squared norm of the state) generated by the anytime algorithm and the baseline algorithm with estimator.

packet dropouts between sensor and estimator/controllers are i.i.d. The control law κ is taken as $\kappa(x) = [-x_2 \quad 0.505\mathrm{sat}(x_1 + x_2)]^T$, $x \in \mathbb{R}^2$. The reader is referred to our recent works[15,24] for a more comprehensive discussion of the controller design.

In Figure 7.3, typical realizations of the state trajectory (measured by the squared norm of the state) are shown. The successful transmission probability is set to 0.8. We can see in Figure 7.4 that at every time step the state trajectory generated by the anytime algorithm of Section 7.3 has a smaller squared norm than the one generated by the baseline algorithm (7.4).

To give a more thorough performance evaluation, we will focus on the empirical cost

$$J \triangleq \frac{1}{50} \sum_{k=0}^{49} |x(k)|^2.$$

By averaging over 10^4 realizations, we obtain Figure 7.4, which illustrates achieved performance as a function of the transmission success probability. In this figure we also include the empirical cost achieved by the anytime and baseline algorithms when no estimator is used. To be more specific, in

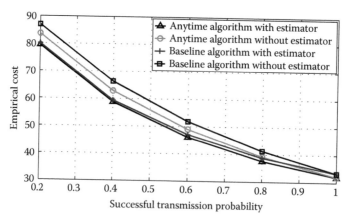

FIGURE 7.4 Empirical cost versus successful transmission probability for the algorithms with and without an estimator.

the anytime algorithm without the estimator, when a packet erasure happens, the actuator value is taken as the first element of the shifted buffer state, that is,

$$u(k) = \begin{cases} \kappa(x(k)), & \text{if } \gamma(k) = 1 \text{ and } N(k) \geq 1 \\ e_1^T Sb(k-1), & \text{if } \gamma(k) = 0 \text{ and } N(k) \geq 1 \\ e_1^T Sb(k-1), & \text{if } N(k) = 0, \end{cases}$$

(cf. Equation 7.6). In the baseline algorithm, when a dropout occurs, the plant input is set to zero.

The numerical results documented in Figure 7.4 clearly show that, for the cases examined, the proposed anytime control algorithm gives better empirical cost than the baseline algorithm. If no estimator is used, then the anytime algorithm still gives better empirical cost than the baseline algorithm; however, both algorithms without estimator provide worse empirical cost than the algorithms with estimator. This verifies the effectiveness of the estimator for both the baseline and the anytime algorithm.

7.6 CONCLUSIONS

We considered a closed-loop system in which a sensor transmits to the controller across an analog erasure channel. Further, the processor availability for executing the control algorithm is time-varying in an *a priori* unknown fashion. To improve the closed-loop performance over baseline algorithms, we adapt a sequence-based anytime control algorithm of [14]. In the present work, we show that the performance can be further improved by using an estimator in conjunction with the controller. The estimator provides an estimate of the current state of the plant using the information last received across the analog erasure channel and the control inputs calculated at the previous time steps. By making use of knowledge at the controller side and respecting processor availability, the performance of the system can be safeguarded against instances when computational resources are unavailable. Numerical simulations illustrate potential performance gains of the method presented.

REFERENCES

1. D. E. Quevedo, J. Østergaard, and A. Ahlén, Power control and coding formulation for state estimation with wireless sensors, *IEEE Trans. Control Syst. Technol.*, 22(2), 413–427, March 2014.
2. D. E. Quevedo, A. Ahlén, and K. H. Johansson, State estimation over sensor networks with correlated wireless fading channels, *IEEE Trans. Automat. Control*, 58(3), 581–593, March 2013.
3. V. Gupta, A. F. Dana, J. P. Hespanha, R. M. Murray, and B. Hassibi, Data transmission over networks for estimation and control, *IEEE Trans. Automat. Control*, 54(8), 1807–1819, August 2009.
4. O. C. Imer, S. Yüksel, and T. Başar, Optimal control of LTI systems over unreliable communication links, *Automatica*, 42(9), 1429–1439, September 2006.
5. L. Schenato, B. Sinopoli, M. Franceschetti, K. Poolla, and S. S. Sastry, Foundations of control and estimation over lossy networks, *Proc. IEEE*, 95(1), 163–187, January 2007.
6. L. K. McGovern and E. Feron, Closed-loop stability of systems driven by real-time dynamic optimization algorithms, in *Proceedings of the IEEE Conference Decision and Control*, vol. 4, Phoenix, AZ, pp. 3690–3696, December 1999.
7. D. Henriksson and J. Åkesson, Flexible implementation of model predictive control using sub-optimal solutions, D of Automatic Control, Lund University, Lund, Sweden, Technical Report Internal Report No. TFRT-7610-SE, 2004.
8. P. Andrianiaina, A. Seuret, and D. Simon, Robust system control method with short execution deadlines, European Patent Application EP 2 568 346 A1, Airbus Operations Toulouse, France, March 2013.
9. A. Cervin, M. Velasco, P. Martí, and A. Camacho, Optimal online sampling period assignment: Theory and experiments, *IEEE Trans. Control Syst. Technol.*, 19(4), 902–910, July 2011.
10. P. Tabuada, Event-triggered real-time scheduling of stabilizing control tasks, *IEEE Trans. Automat. Control*, 52(9), 1680–1685, September 2007.

11. R. Bhattacharya and G. J. Balas, Anytime control algorithms: Model reduction approach, *AIAA J. Guid. Control Dyn.*, 27(5), 767–776, September–October 2004.
12. L. Greco, D. Fontanelli, and A. Bicchi, Almost sure stability of anytime controllers via stochastic scheduling, in *Proceedings of the IEEE Conference Decision and Control*, New Orleans, LA, pp. 5640–5645, December 2007.
13. V. Gupta and F. Luo, On a control algorithm for time-varying processor availability, *IEEE Trans. Automat. Control*, 58(3), March 2013.
14. D. E. Quevedo and V. Gupta, Sequence-based anytime control, *IEEE Trans. Automat. Control*, 58(2), 377–390, Febuary 2013.
15. D. E. Quevedo, V. Gupta, W. Ma, and S. Yüksel, Stochastic stability of event-triggered anytime control, *IEEE Trans. Automat. Control*, 59(12), 3373–3379, December 2014.
16. J. Nilsson, B. Bernhardsson, and B. Wittenmark, Stochastic analysis and control of real-time systems with random time delays, *Automatica*, 34(1), 57–64, 1998.
17. B. Sinopoli, L. Schenato, M. Franceschetti, K. Poolla, M. I. Jordan, and S. S. Sastry, Kalman filtering with intermittent observations, *IEEE Trans. Automat. Control*, 49(9), 1453–1464, September 2004.
18. K. J. Åström and B. Wittenmark, *Computer Controlled Systems: Theory and Design*, 2nd edn. Englewood Cliffs, NJ: Prentice Hall, 1990.
19. M. Caccamo, T. Baker, A. Burns, and G. Buttazzo, Real-time scheduling for embedded systems, in *Handbook of Networked and Embedded Systems*, D. Hristu-Varsakelis and W. S. Levine, (eds.). Birkhäuser, Boston, MA, 2005.
20. D. Liu, X. Hu, and M. D. Lemmon, Scheduling tasks with Markov-chain based constraints, in *Proceedings of the 17th Euromicro Conference on Real-Time Systems (ECRTS'05)*, pp. 157–166, 2005.
21. J. M. Maciejowski, *Predictive Control with Constraints*. Englewood Cliffs, N: Prentice-Hall, 2002.
22. J. B. Rawlings and D. Q. Mayne, *Model Predictive Control: Theory and Design*. Madison, WI: Nob Hill Publishing, 2009.
23. L. Grüne and J. Panne, *Nonlinear Model Predictive Control: Theory and Algorithms*. Springer-Verlag, London, 2011.
24. D. E. Quevedo, W.-J. Ma, and V. Gupta, Anytime control using input sequences with Markovian processor availability, *IEEE Trans. Automat. Control*, 60(2), 515–521, February 2015.
25. J. G. Kemeny and J. L. Snell, *Finite Markov Chains*. D. Van Nostrand Company, Inc., Princeton, NJ, 1960.

8 Signal and Data Processing Techniques for Industrial Cyber-Physical Systems

*George Tzagkarakis, Grigorios Tsagkatakis,
Daniel Alonso, Cesar Asensio, Eugenio Celada,
Athanasia Panousopoulou, Panagiotis Tsakalides,
and Baltasar Beferull-Lozano*

CONTENTS

8.1 INTRODUCTION

Cyber-physical systems (CPS) are large-scale interconnected systems of heterogeneous, yet collaborating, components that are envisioned to provide integration of computation with physical processes. Nowadays, a first generation of CPS can be found in areas as diverse as aerospace, civil infrastructures, energy, health care, manufacturing, transportation, entertainment, and consumer appliances.

Unlike traditional *embedded systems*, a full-fledged CPS is typically designed as a network of interacting elements with physical input and output instead of being simply a combination of standalone devices. The inherent heterogeneity and integration of different components pose new challenges to traditional data analysis, communication, control, and software theories. This often makes system design inefficient with current technologies.

Advances in the cyber world, such as communications, networking, sensing, computing, storage, and control, as well as in the physical world, such as materials, hardware, and renewable energy sources, are all converging rapidly to dramatically increase the adaptability, autonomy, efficiency, functionality, reliability, safety, and usability of CPS. This aspires to broaden the economic and societal potential of such highly collaborative computational systems in various dimensions, such as intervention (e.g., collision avoidance), precision (e.g., robotic manufacturing), operation in dangerous or inaccessible environments (e.g., search and rescue), efficiency (e.g., energy and cost reduction in water treatment plants), and augmentation of human capabilities (e.g., health care monitoring and delivery).

A major difference between CPS and a typical control system or an embedded system is the use of communications, which adds reconfigurability and scalability, as well as complexity and potential instability. Furthermore, CPS have significantly more intelligence in sensors and actuators, as well as substantially stricter performance and energy constraints, which are critical for its efficiency and lifetime. On the other hand, cyber capabilities are embedded in every physical process and component, networking is employed at multiple scales, complexity lies at multiple temporal and spatial scales, and high heterogeneity is seen across devices and protocols.

The design and realization of the complex interface between cyber and physical worlds for seamless interactions is by no means a nontrivial task. Implacable, concurrent laws of physics govern the physical world, as opposed to the discrete and asynchronous nature of the cyber world. Timing and spatial precision, uninterrupted connectivity, predictability, and repeatability are extremely critical for the cyber-physical interface. It is hence vital to build new theoretical foundations, scientific models, abstractions, and explicit dissociation between cyber and physical worlds for the interface, and rethink or reinvent interface functions, such as coordination, integration, monitoring, and control.

Focusing on a demanding paradigm of interrelation between the cyber and physical worlds, *industrial control systems* are widely used to provide autonomous control through appropriate control loops dedicated to performing specific tasks. Such systems monitor an industrial environment through sensors deployed around the product line and interact with the various processes through proper actuators. Moreover, the complexity of modern industrial settings is usually simplified by dividing the overall infrastructure into individual subsystems containing separate processing and control modules. When interactions between distinct subsystems are required, usually skilled system operators or simple communication methods are exploited.

To this end, CPS can provide broad control over complex and large industrial environments through heterogeneous network architectures of sensors, actuators, and processors. However, coverage and connectivity should be redefined in the framework of *industrial cyber-physical systems* (iCPS). Such systems will usually consist of both wired and wireless sensor and actuation networks with different capacities and reliability. Furthermore, emphasis is put on real-time operations, whereas sensing, processing, communication, and actuation will be handled by different components in the iCPS infrastructure. To model such heterogeneous issues, an innovative technological turn is necessary.

To compromise all those critical aspects of iCPS, sophisticated signal and data processing techniques, coupled with novel network and communication protocols, should be designed to provide unprecedented performance and efficiency levels for iCPS.

In this chapter, we present an integrated framework of signal and data processing for treating different layers of information abstraction, ranging from raw samples at the front end of the cyber-physical space to data semantics for extracting high-level patterns. By taking into account the limitations and the imperfections of the sensor network infrastructure employed for iCPS, we focus on three complementary, yet equally important aspects: (1) signal processing-driven performance optimization for industrial sensor networks; (2) in-network signal processing techniques for estimation, detection, and tracking for iCPS; and (3) knowledge management for detecting behavior variations in the recorded iCPS data.

Concerning the first aspect, a novel technique is introduced in the framework of matrix completion for recovering missing information due to network or sensor failures, as well as in the case of adapting to higher temporal resolution than the operating resolution of the available sensors. With respect to the second aspect, an efficient iterative consensus method is introduced for distributed estimation and tracking, by employing a cross-layer link scheduling protocol. Finally, regarding the third objective, an uncertainty-aware framework for extreme events detection, in conjunction with a fast and robust correlations monitoring are analyzed. Doing so, we improve upon existing data management systems by proposing an integrated framework spanning all the stages of the data acquisition and processing in iCPS infrastructures.

While the corresponding methods reflect on different angles of the iCPS architecture, the common factor is the exploitation of the inherent information content of raw sensor streams, toward the formulation of an integrated mathematical and algorithmic framework for signal and data processing in iCPS. With a strong emphasis on providing the essential theoretical background, the efficacy of the resulting framework will be ultimately evaluated in different aspects of real-life iCPS, designed for the autonomous monitoring and decentralized control of water treatment plants [1]. It is anticipated that the methods herein presented and the accompanying discussions on real-life results will yield novel directions for iCPS standardization.

The rest of the chapter is organized as follows: Section 8.2 introduces the main concepts of data-driven architectures for industrial CPS, while Section 8.3 describes novel signal processing methods based on the theories of compressed sensing and matrix completion for recovering missing information in sensor streams. Section 8.4 presents techniques for performing in-network signal processing tasks, such as parameter estimation and tracking, in a distributed fashion, and discusses their main performance issues. In Section 8.5, a set of novel algorithmic tools is introduced for producing intelligent reasoning over the data by supporting advanced operations, such as, querying, uncertainty-aware high-level data analysis, and alerting. Finally, in Section 8.6, the performance of the introduced signal and data processing techniques is evaluated in the real-world scenario of the HYDROBIONETS project, whose key objective is the autonomous monitoring and control of industrial water treatment and desalination plants, while Section 8.7 summarizes the main achievements and gives directions for further enhancements.

8.2 DATA-DRIVEN ARCHITECTURES FOR INDUSTRIAL CPS

The transition from cyber to cyber-physical architectures is dictated by the tight coordination between cyber and physical resources; while traditional cyber systems observe a constantly changing physical world, in CPS the information processing components are inseparable from the physical procedures. As such, despite the inherited heterogeneity and complexity, CPS architectures should respond effectively to unexpected conditions, while exhibiting an increased level of resiliency and adaptability to system failures.

This necessity intensifies for the case of industrial processes as the harshness of the operational environment is combined with strict performance requirements in terms of data acquisition, estimation, and control. In particular, the practical realization of wireless sensor-actuator networks (WSAN), which are considered the enablers of CPS architectures [2], is characterized by periods of severe impairments due to propagation phenomena, noise, and interference, dictated by the characteristics of the operational space [3]. In RF-harsh environments, like those met in industrial plants, the surrounding environment becomes even less mindful of the wireless communications performance. The sources of noise and interference increase significantly due to the presence of heavy machinery, obstacles with large volume and highly reflective characteristics [4]. In such operational spaces, the phenomenon of link disruption due to inter-symbol interference has, along with multipath fading and signal absorption, a direct impact on the network performance that cannot be treated with conditional methods, such as increased transmission power [5]. In addition, while electromagnetic interference affects the hardware characteristics of the network components [6], the presence of impulsive noise can lead to short periods of excessively weak channel conditions [7].

These limitations have a direct impact on the performance and reliability of the network, which are typically expressed in terms of packet losses and excessive delays. Modern, networked approaches for industrial processes that rely on WSAN consider these imperfections of the network performance as an inseparable aspect of the control procedures. As such, the respective, system-wide architectures emphasize on guarding the control requirements against the uncertainties imposed by the network. Subsequently, data-driven schemes for iCPS often coincide with networked control systems (NCS) [8,9], which feature the communication of spatially distributed sensors, actuators, and controllers over a shared data network (Figure 8.1a).

During the past decade, NCS have faced several challenges related to the quality of service (QoS) provided by the underlying network backbone. Representative examples include the preservation of the required bandwidth, the network synchronization, and the compensation from information losses. Depending on the network parameter that is modeled as part of the industrial process, the control loop can be viewed as an *asynchronous dynamic system* for which the stability margins are guaranteed by retaining the information loss below the threshold dictated by the characteristics of the open-loop system [10].

The recent advances on wireless communication standards for industrial processes, such as WiHart [11], ISA100 [12], and IEEE 802.15.4 [13], have given a different spin in the area of data-driven information processing and control. The scale and the complexity of the problem have been dramatically increased, thereby seeking cross-layer communication and control solutions, employing multihop topologies, and, ultimately, distributing the intelligence of the system across different parts of the iCPS architecture. Along this direction, in Cao et al. [14] an integrated control-communication framework is presented. It comprises of a communication protocol that enables self-triggered control actions and an optimization algorithm based on simulated annealing, while considering both packet losses, as well as the physically constrained actions of the actuators. In parallel, Alur et al. [15] propose a mathematical framework that can support the operation of an iCPS over a multihop WSAN. The resulting architecture is abstracted as a switched system, while link failures introduce random switching signals. Transiting this approach into a real-world industrial process provides a scalable architecture according to which each control loop is analyzed separately and associated to the maximum delay between sensing and actuation. All control loops are then integrated and a scheduling policy is applied, allowing data transmission from the sensors to the controller and from the controller to the actuators for each control loop, within the specified time bounds. This approach is characterized as compositional as it enables applying the set of all schedules that satisfy the temporal boundaries for a single control loop to additional control loops, added at a later stage in the iCPS architecture.

While the aforementioned approaches emphasis is on the extreme ends of the iCPS architecture, and specifically how the controller can treat the network imperfections as part of the process under control, recent theoretical results have indicated that increasing the intelligence on the sensors and intermediate networked components can improve the stability and performance of the entire system [16,17]. Building upon this statement, while considering a multihop, arbitrary topology between the sensor and the remote estimator/controller, an iCPS architecture that enables in-network processing is proposed in Gupta et al. [18], originated by the point of sensor and recursively adopted by each node located on the network path between the sensor and the estimator/controller (Figure 8.1b). In specific, under the assumption of perfect synchronization between all involved nodes, when the sensor collects a new measurement its encoder applies a Kalman filter for updating the estimation of the physical process, updates the time instant, and transmits the updated estimation to all outgoing edges. Upon reception of this information, each intermediate node checks the timestamp on the incoming data and keeps the one corresponding to the latest measurement, which is in turn redirected to its outgoing edges. The destination node located at the side of estimator/controller will collect all incoming

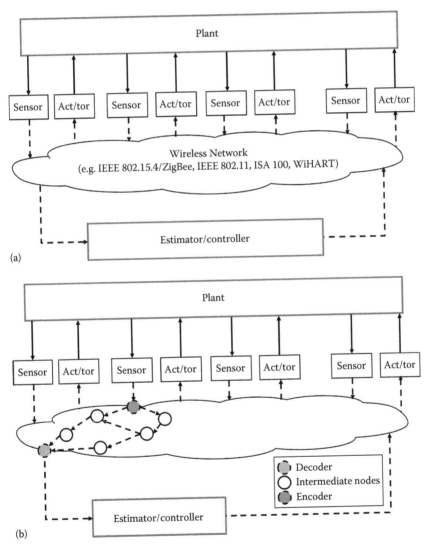

FIGURE 8.1 Data-driven iCPS architectures: (a) traditional NCS approach—sensors and actuators communicate with the controller/estimator of the system over a wireless network; (b) a more sophisticated approach—encoders and decoders collaborate with the intermediate wireless nodes for improving the high-level control/estimation procedure [18] over multihop network paths. *(Continued)*

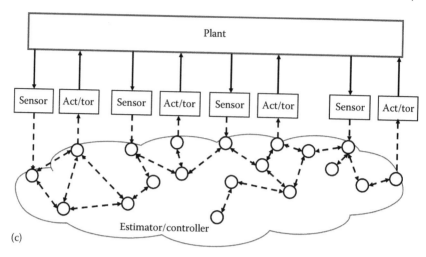

(c)

FIGURE 8.1 (Continued) Data-driven iCPS architectures: (c) architecture of the WCN ([19,20])—control/estimation process is distributed inside the fully connected network.

information and retain the one corresponding to the most recent measurement for either calculating the best possible estimate in terms of minimum mean square error or minimizing a quadratic cost.

The mechanisms described so far rely on the existence of both a single sensor–actuator pair, as well as a designated estimator/controller, which is located somewhere in the industrial plant. While this convention corresponds to many of existing industrial processes, its practical realization is often subject to the computational and energy constraints of the embedded platforms, which are often employed for implementing iCPS architectures due to their cost-effectiveness and miniatured size. In addition, modern industrial processes, such as Smart Grid and Smart Water Networks, require multiple sensor and actuator points, distributed in a wide geographical area and characterized by heterogeneous signals travelling from and to different points of interest. Driven by this motivation, Pajic et al. introduced [19,20] a novel paradigm of iCPS architectures that enables the entire network acting as a controller, instead of assigning this role to designated nodes of the network. The resulting architecture, entitled wireless control network (WCN), is considered an extension of traditional approaches that builds upon the collaboration of the operational nodes with their one-hop neighbors. As shown in Figure 8.1c, under the assumption of *a priori* known topology, some nodes have access to the sensor measurements, while others are located in the communication range of the actuators. Based on a linear iterative strategy, each node periodically updates its state, which results from the linear combination of the states of its one-hop neighbors. Similarly, the actuators apply linear combinations of the states of the nodes in their neighborhood. The resulting overall scheme has been proven capable of controlling continuous-time physical processes, while preserving the stability of the system under the existence of independent link failures.

The discussion thus far highlights the emphasis that has been given on improving feedback control laws against the presence of network imperfections in industrial processes. While this is an essential requirement for iCPS, challenges associated to the system-wide *information management*, capable of covering the entire span of the industrial procedure, remain open.

More specifically, the utilization of inexpensive sensing nodes is crucial for the proliferation of CPS in everyday environments. In industrial settings, the challenges associated with this specific type of environment, such as excessive temperature or humidity, can hinder the performance of even high-end sensing nodes. As a consequence, node failures are frequent and the systems must be able to handle the dynamic insertion and removal of sensors. Although catastrophic failures do happen, node failure is typically attributed to energy depletion. The design of iCPS platforms must therefore maximize the amount of information that can be extracted from the data, while minimizing the usage of valuable resources, such as power, bandwidth, and storage.

In addition to the fragility of the infrastructure, real-life architectures are characterized by periods of communication failures due to changes in the environmental conditions, such as the occurrence of moving objects, interference from other devices, and conjunction due to network traffic overload. As a consequence, packets that encode the captured data will be lost, while disseminating information through the network to other nodes or local sinks. System designers must therefore account for such communications impairments and provide robustness without consuming unnecessarily the networks' resources.

These properties and characteristics suggest that modern data-driven architectures iCPS should consider the following:

- Data-driven techniques for improving low-level network and sensing imperfections in the industrial environment. According to the current trend, such shortcomings on the joint sensing-communicating performance are assimilated into the controlled process.
- Mechanisms capable of distributing the intelligence of the system between sensors, and thereby extracting correlations associated to the quality of sensing in the industrial plant. Related approaches that consider a distributed approach elaborate on collaboration between sensing and pure network components for achieving the higher-level control and estimation objectives.
- Algorithms for high-level decision making capable of keeping the end user in the control loop, based on heterogeneous sources of information. According to [21], such mechanisms are essential for the system-wide reliability and robustness of CPS.

These aspects are considered a necessity for modern iCPS architectures, capable of adopting a knowledge engineering approach [22]; they would enable the layered contextualization of real-time streamed data according to the level of abstraction and the different perspectives of the same infrastructure. As such, in parallel to the control perspective, three additional views and respective levels of data abstractions are recognized:

1. Front-end data representations, provided by sensors deployed at physical frontier of the industrial process and capable of improving data acquisition and sampling
2. In-network correlations, enabled by the decentralized collaboration between sensing components, for extracting useful observations on the quality of sensing and the industrial-driven relationships between physically distant components of the same industrial process
3. High-level data abstractions, resulting from distilling information from raw data streams into laconic notifications on the status and quality of the underlying sensing and actuation procedures

With these considerations in mind, in the remaining of this section, we will introduce our framework for signal and data processing for iCPS, capable of treating different layers of information abstraction, ranging from raw samples at the front end of the cyber-physical space to data semantics for extracting high-level patterns.

8.2.1   AN INTEGRATED FRAMEWORK FOR SIGNAL AND DATA PROCESSING FOR iCPS

The herein proposed framework yields essentially a virtual multilayered architecture, corresponding in a straightforward manner to the data abstraction layers described earlier. In an attempt to magnify the benefits of sensing at the industrial processes, our emphasis is explicitly on the signal and data processing techniques that have impact on improving the *intelligence* of the information that flows from the sensors toward the controlling processes. Therefore, our approach can act as a complementary tool for sophisticated networked controlled mechanisms, while decoupling the limited and imperfect access to the sensor data streams from the specific characteristics of the controlled plant and/or the design of the controller.

In a nutshell, our framework considers the following levels of information processing:

- *Level 1: Signal modeling of the front-end industrial data representations*, focusing on two recently developed yet extremely influential, signal processing paradigms, namely compressed sensing and matrix completion
- *Level 2: In-network processing* for estimation, detection, and tracking for iCPS, while taking into account the impact of network imperfections
- *Level 3: High-level data analysis and early warning*, focusing on uncertainty management, notification mechanisms for extreme events, and extraction of high-level pairwise correlations for improving decision support/making systems in industrial processes.

The proposed architecture is presented in Figure 8.2, highlighting the positioning of each information processing tier with respect to the generalized industrial control process. It is considered important to highlight the particular interest on providing to the system administrator the necessary means for qualitative supervision of the industrial procedure. At the level of raw sensing from the plant, the proposed framework considers the front-end data handlers, which are responsible for improving the data acquisition and sampling according to the statistical and mathematical attributes of the iCPS data. These handlers are in turn combined to sensing agents (Level 2), which travel within the wireless network for the characterization of the quality of sensing, the recovery of an accurate sample from the target field, and tracking of time varying signals. This approach operates in a distributed manner, thereby transiting the tasks of noise detection, signal estimation and tracking to the level of low-level sensing, and increasing the intelligence of the iCPS architecture, in order to account for failures at the level of sensing and/or networking. Finally, the sensor streams are fed into Level 3, which is responsible for further data analysis toward the qualitative and quantitative characterization of the industrial sensing process. The result can be directed toward both the controller components for adjusting the control actions in the presence of unexpected conditions, as well as the user-oriented decision support systems for facilitating the supervision of the overall system.

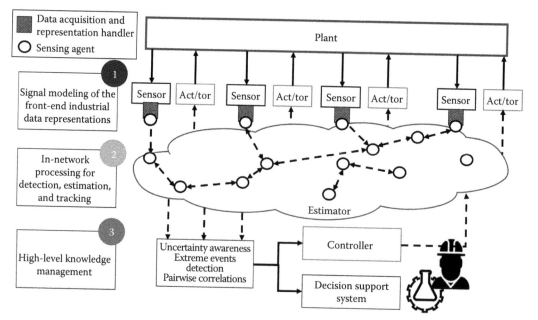

FIGURE 8.2 The proposed framework for signal and data processing in industrial CPS, featuring the three information processing tiers.

While the corresponding methods reflect on different angles of the iCPS architecture, their common factor is the exploitation of mathematical and statistical characteristics of the information, and not the information itself, thereby remaining agnostic with respect to the data-driven details for each industrial process. As such, while addressing aspects complementary to the control of industrial processes, it is anticipated that the proposed framework will provide a universal methodology for signal and data processing techniques that can be employed at different industrial scenarios.

In the remaining of this chapter, we will elaborate on the theoretical background of each information processing tier, followed by a discussion on the evaluation of the proposed framework in different aspects of real-life iCPS, designed for the autonomous monitoring and decentralized control of water treatment plants.

8.3 INDUSTRIAL DATA REPRESENTATIONS: FROM DENSE TO SPARSE SAMPLING

8.3.1 PROPERTIES AND ISSUES OF iCPS DATA

Data acquisition and processing in iCPS relies on the presence of a WSN composed of distributed nodes that communicate their measurements to other nodes with similar or higher capabilities. Efficient acquisition and communication of these measurements is a critical aspect of WSN systems that determine directly the lifetime and usability of the infrastructure. To address the issues associated with data sampling and processing, one must consider both the physical constraints of such networks, ranging from node failures to communications breakdowns, as well as the properties and recovery capabilities of cutting-edge signal sampling and processing algorithms.

To achieve the strict requirements and overcome the limitations of such environments, one can exploit data redundancies and employ signal processing algorithms to guarantee the performance for numerous types of signals. In the following, we will focus on two particular characteristics of iCPS data, namely sparsity and low rankness. These characteristics are linked directly with properties of WSN architectures including spatiotemporal correlations, predictable behavior, and physical constraints. Sparsity can refer to either the presence of a very small number of large-valued measurements or to the ability in expressing a complex signal using a small number of representative examples. While the former case reflects a specific type of signals, such as biological, seismic, and astronomic data, the latter corresponds to a rather large number of signals that can be accurately represented using a sparse collection of fundamental signals encoded in a so-called *dictionary*. Intelligent exploitation of sparsity can offer numerous advantages from a WSN point of view.

Furthermore, we will consider the low rankness of various matrices that can be found in WSNs, such as measurement matrices, where each row of the matrix corresponds to a specific sensor, and each column represents a sampling instance in time. The rank of such matrices is indicative of the amount of correlation that exists within the data since highly correlated measurements lead to low-rank measurement matrices. The rank of a matrix is manifested by the number of nonzero singular values, in which case low-rank matrices can be described by sparse singular values. By exploiting the low-rank property, significant benefits can be achieved for WSN architectures, such as efficient sampling and robust storage.

The sparsity and low rankness of iCPS data is demonstrated in Figure 8.3, which presents iCPS data collected by a WSN (part of the Intel-Berkeley data set*) along with certain properties of the matrix. More specifically, Figure 8.3b shows the magnitude of the signal values. One can observe that a very small number of singular values capture most of the signals' energy, while the rest correspond to noise and outliers. Furthermore, Figure 8.3c shows the magnitude of the representations coefficients for three sampling instances, based on a mapping in a dictionary generated using data (measurements vectors) collected from the previous day. One may also observe that only a small number of coefficients are nonzero leading to a sparse representation of the new vectors in terms

---

* P. Bodik, C. Guestrin, W. Hong, S. Madden, M. Paskin, and R. Thibaux, http://select.cs.cmu.edu/data/labapp3/index.html.

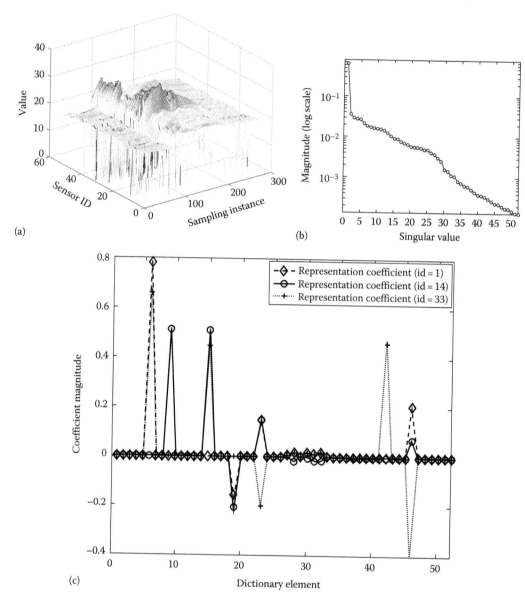

FIGURE 8.3 Graphical illustration of (a) a WSN measurements matrix; (b) the corresponding singular values; and (c) the representation coefficients. We observe that (i) a small number of singular values captures most of the signals' energy and (ii) only a small number of non zero coefficients suffice for representing the measurement vectors.

of dictionary atoms. Two innovative signal processing algorithms are presented, namely, *compressed sensing* (CS) and *matrix completion* (MC), which can exploit efficiently the sparsity and low rankness of iCPS data captured via WSNs.

8.3.2 COMPRESSED SENSING

CS is a radically novel approach in signal acquisition and processing [23,24]. The main underlying concept of CS is that a complex signal can be recovered from a small number of random measurements, far below the traditional Nyquist–Shannon limit. The key assumption in CS is that either the signal itself is sparse or that it can be *sparsely* represented in an appropriate dictionary, and that enough *random* measurements are collected. Formally, a signal $s \in \mathbb{R}^N$ is called k-sparse if $\|s\|_0 < k$, where $\|s\|_0 =$ # nonzero elements of s. This signal can be reliably recovered from a low-dimensional representation $y = \Psi s \in \mathbb{R}^M$, where $M \ll N$ by solving an ℓ_0-constrained minimization problem given by

$$\min \|s\|_0 \quad \text{subject to} \quad y = \Psi s. \tag{8.1}$$

To guarantee the stable recovery of the original signal, the $M \times N$ sensing matrix Ψ must satisfy the so-called *restricted isometry property* (RIP). A sensing matrix $\Psi \in \mathbb{R}^{M \times N}$ satisfies the RIP with isometry constant $0 \leq \delta < 1$ if for all k-sparse signals, s, it holds that

$$(1 - \delta)\|s\|_2^2 \leq \|\Psi x\|_2^2 \leq (1 + \delta)\|s\|_2^2 . \tag{8.2}$$

Designing such a sensing matrix is proven to be a challenging task. However, it has been proven that matrices whose elements are drawn randomly from appropriate distributions satisfy the RIP with high probability. Examples of such distributions include normalized mean bounded variance Gaussian [23] and Rademacher [25] distributions.

The formulation of CS expressed by Equation 8.1 assumes that the signals in question are naturally sparse, that is, they consist of a small number of nonzero elements. However, a large class of signals do not belong to this category. To tackle this issue, the CS theory has been extended through the use of a dictionary of elementary examples as a sparsifying transform. During the early stages of CS theory formulation, well-known orthogonal transforms, including the discrete Fourier transform (DFT), the discrete cosine transform (DCT), and wavelets were employed as sparsifying dictionaries. In Candes et al. [24], it was shown that the theory of CS is also applicable in cases where the signal is sparse over coherent and redundant dictionaries, including overcomplete DFT, wavelet frames, and concatenations of multiple orthogonal bases. By incorporating the dictionary in the reconstruction problem, the dictionary-based ℓ_0 minimization is formulated according to

$$\min \|s\|_0 \quad \text{subject to} \quad y = \Psi Ds. \tag{8.3}$$

Even though solving the ℓ_0 minimization in Equations 8.1 and 8.3 will produce the correct solution, this is an NP-hard problem and therefore impractical for moderate-sized scenarios. To address this issue, greedy methods, such as the orthogonal matching pursuit (OMP) [26], have been proposed among other approaches for solving Equation 8.3. OMP tries in a greedy way to identify the elements of the dictionary that contain most of the signal energy by selecting iteratively the element of the dictionary exhibiting the highest correlation with the residual, and updating the current residual estimate. One of the main breakthroughs of the CS theory is that under the sparsity constraint and the incoherence of the sensing matrix, the reconstruction of the original signal, x, and the coefficient vector, s, from y, can be found by solving a tractable ℓ_1 optimization problem, the so-called *Basis Pursuit*, given by

$$\min \|s\|_1 \quad \text{subject to} \quad y = \Psi Ds. \tag{8.4}$$

For compressible signals, the goal is not the exact reconstruction of the signal, but the reconstruction of a close approximation of the original signal. In this case, the problem is reduced to a *Basis Pursuit Denoising* and Equation 8.4 takes the following form:

$$\min \|s\|_1 \quad \text{subject to} \quad \|y - \Psi Ds\|_2 < \epsilon, \tag{8.5}$$

where ϵ is a bound on the residual error of the approximation, which is related to the amount of noise in the data. The optimization in Equation 8.5 can be solved efficiently via the LASSO [27] algorithm for sparsity regularized least squares.

The number of measurements required for the signal reconstruction is dictated by the *mutual coherence* between the sensing matrix $\mathbf{\Psi}$ and the dictionary $\mathbf{D}$, which is defined as the maximum of the inner product between columns of the dictionary and the sampling matrix:

$$\mu(\mathbf{\Psi}, \mathbf{D}) \doteq \max_{\substack{1 \le i \le M \\ 1 \le j \le N}} \sqrt{N} \left| \langle \mathbf{\psi}_{.,i}, \mathbf{d}_{j,.} \rangle \right|, \tag{8.6}$$

where $\mathbf{\psi}_{.,i}$ and $\mathbf{d}_{j,.}$ denote the ith column of $\mathbf{\Psi}$ and the jth row of $\mathbf{D}$, respectively. For a specific mutual coherence, recovery is possible from $M \ge C \cdot \mu^2(\mathbf{\Psi}, \mathbf{D}) \cdot K \cdot \log(N)$ random measurements. As a consequence, having low coherence between the dictionary and the sampling matrix is beneficial in terms of performance.

8.3.3 Matrix Completion

Matrix completion (MC) [28–30] is a recently proposed framework, which builds on the concepts of CS and extends the sparsity framework to the case of subsampled matrix-valued data. More specifically, MC considers a measurement matrix $\mathbf{M} \in \mathbb{R}^{N \times S}$, which can encode various CPS data, such as sensors measurements over time or spatial data in a single time instance, where a large number of its entries are missing. In general, one cannot recover the $N \cdot S$ entries of the matrix $\mathbf{M}$ from a smaller number of K entries, where $K \ll N \cdot S$, unless some characteristics about the measurement matrix are known. MC theory suggests that such a recovery is possible if the matrix is characterized by a rank smaller than its dimensions and a sufficient number of randomly selected entries of the matrix is available. The rank of the matrix indicates the number of linearly independent columns (or rows) and thus serves as a proxy to the correlations that exists within the data.

More specifically, one can recover an accurate approximation $\mathbf{X}$ of the matrix $\mathbf{M}$ from $K \ge C \cdot N^{6/5} \cdot r \cdot \log(N)$ random measurements, where $rank(\mathbf{M}) = r$, by solving the following minimization problem:

$$\min_{\mathbf{X}} rank(\mathbf{X}) \quad \text{subject to} \quad \mathcal{A}(\mathbf{X}) = \mathcal{A}(\mathbf{M}), \tag{8.7}$$

where $\mathcal{A}$ is, in general, a linear map from $\mathbb{R}^{S \times N} \mapsto \mathbb{R}^K$. The theory of matrix completion suggests that recovery is possible when the linear map $\mathcal{A}$ is defined as a random sampling operator that records a small number of entries from the matrix $\mathbf{M}$, that is, $\mathcal{A}_{ij} = \{1, \text{if } (ij) \in \mathcal{S} \mid 0, \text{otherwise}\}$, where $\mathcal{S}$ is the sampling set. In the context of WSN sampling, the set $\mathcal{S}$ specifies the collection of sensors that are active at each specific sampling instance. In general, the solution of the MC problem requires the linear map $\mathcal{A}$ to satisfy a modified restricted isometry property, which is the case when uniform random sparse sampling is employed in both rows and columns of matrix $\mathbf{M}$ [31].

Unfortunately, the rank minimization in Equation 8.7 is an NP-hard problem and therefore cannot be directly employed for data recovery. According to MC, one can resort to a relaxation capable of producing arbitrary accurate results by replacing the rank constraint by the tractable nuclear norm, which represents the convex envelope of the rank. The minimization in Equation 8.7 can then be reformulated as follows:

$$\min_{\mathbf{X}} \|\mathbf{X}\|_* \quad \text{subject to} \quad \mathcal{A}(\mathbf{X}) = \mathcal{A}(\mathbf{M}), \tag{8.8}$$

where the nuclear norm is defined as $\|\mathbf{X}\|_* = \sum \|\lambda_i\|_1$, that is, the sum of absolute values of the singular values. For the noisy case, an approximate version is given by

$$\min_{\mathbf{X}} \|\mathbf{X}\|_* \quad \text{subject to} \quad \|\mathcal{A}(\mathbf{X}) - \mathcal{A}(\mathbf{M})\|_F^2 \le \epsilon, \tag{8.9}$$

where

$\|\mathbf{X}\|_F^2 = \sum \lambda_i^2$ denotes the Frobenius norm

ϵ is the approximation error

To solve the nuclear norm minimization problem (Equations 8.8 and 8.9), various distinct approaches have been proposed, including the singular value thresholding (SVT) [32], the augmented Lagrange multipliers (ALM) [33], and the so-called *OptSpace* [34]. In the following, the technique based on the ALM is reviewed briefly since it has been shown to offer exceptional performance both in terms of processing complexity and reconstruction accuracy, and because it serves as a basis for our extended scheme.

8.3.4 APPLICATIONS IN iCPS

CS and MC have been successfully employed in various tasks related to iCPS data acquisition, processing, and management. This success can be attributed to various characteristics of these algorithms. Both CS and MC employ lightweight encoders, while shifting the computational complexity and the associated resources to the decoder side. Furthermore, both CS and MC offer scalable signal recovery capabilities, where more measurements contribute positively to the reconstruction performance. The benefits of CS have been explored for efficient compression and transmission of many complex cyber-physical data, such as video and audio in wireless multimedia sensor networks [35,36], vehicle information in vehicular networks [37], and ECG in wireless body sensor networks [38] to name a few. Moreover, CS offers the ability to perform independent encoding and joint decoding of the data, while MC does not require a specific sampling architecture but instead relies on the random subsampling of the measurements themselves.

- *iCPS Data Sampling and Compression*: CPS data acquisition is a prominent case, where intelligent signal processing can greatly support network operation and increase the usability of the sensing infrastructure. The concept of MC has been successfully employed in the efficient sampling of the spatiotemporal iCPS data acquired by WSNs [39,40], as well as data from Internet of Things platforms [41]. For instance, [42] investigates the co-design of the sampling pattern with the network channel access for recovering spatiotemporal fields monitored by WSNs. MC has been also considered for the coupled reconstruction of missing measurements and data classification of WSN data [43], where it was shown that both objectives can be achieved through the introduction of a dictionary in the low-rank matrix estimation process. Recently, a robust compression scheme was introduced and evaluated on real WSN data [44] based on the introduction of both CS and MC for data compression and recovery of lost packets. On the other hand, certain properties of CS, such as the lightweightness and the universality of the encoding stage, make it a good candidate for efficient distributed data compression in WSNs [45,46]. In this setting, spatial transformations including DFT, graph wavelets, and diffusion wavelets can be utilized for storage and retrieval of network data. CS was also investigated as a rateless distributed coding scheme, offering reduced communications cost independently of the routing algorithm and the network topology [47]. A CS-guided architecture for decentralized recovery of sparse signals in WSNs was proposed in Ling and Tian [48], where the authors considered a random node sleeping pattern, in conjunction with a consensus algorithm for achieving global signal reconstruction from the local estimates. Introducing CS into data sampling and compression has also been supported by novel hardware architectures [49].
- *Aggregation and Routing*: CS has been also applied recently for data aggregation in multihop WSNs, where typically the objective is to collect the full set of measurements to a centralized location, such as a sink node or a gateway. CS-driven data aggregation techniques utilize a random encoding process as nodes forward measurements to a central processing unit, reducing the amount of packets that have to be communicated [50,51]. While CS requires a

specific random encoding process, MC, which relies on randomized subsampling, has been recently explored as an alternative data gathering scheme [52], where MC was supported from a temporal prediction process to recover completely empty measurement vectors. The authors also provided evidence that a large class of CPS signals, such as temperature and humidity, are indeed characterized by the low-rank property, while other signals, such as illumination conditions, cannot be reliably approximated by low-rank measurement matrices.

The efficient interaction between CS encoding/decoding and routing in WSNs was investigated in Quer et al. [53], where the authors showed that the high coherence between the data sparsifying transform and the routing can limit significantly the straightforward applicability of CS in networked data. A remedy to this issues was proposed in Caione, Brunelli, and Benini [54], where an optimal adaptive forwarding scheme was considered for network lifetime maximization. The CS framework has also been combined with another communications paradigm, the network coding. NetCompress [55], for instance, proposes the simultaneous transmission of measurement packets and the encoding via the random projection step of CS.

- *Sensor localization*: Localization information is an important task in a large class of iCPS monitoring scenarios in both outdoors and indoors scenarios. For instance, one can consider the situation where sensing nodes are mobile, or the case where the data collection unit possesses mobility capabilities. Cutting-edge signal processing paradigms, such as CS and MC, can offer substantial benefits with respect to training time, positioning accuracy, algorithmic complexity, and adaptability. One class of approaches is founded upon the low-rank property of the Euclidean distance matrix, and employ MC for the recovery of the complete set of distances from a small number of noisy measurements [56–58] and allow tracking of mobile devices [59]. Another class utilizes CS and MC in order to improve the efficiency of fingerprinting approaches in challenging environments by reducing the training requirements and offering dynamic adaptation mechanisms [60,61].

- *iCPS Data Classification and Event Detection*: Recently, a combination of low rank and sparse signal recovery was introduced for traffic anomaly detection in large-scale networks [62], distributed temporal pattern detection in WSNs [63], and target localization and counting [64] among others. The concept of decentralized estimation of missing data has also been investigated in Ling et al. [65] and is applicable to WSNs.

8.3.5 POTENTIAL OF INTELLIGENT SIGNAL PROCESSING IN iCPS

Despite their youth, CS and MC have shown great potential in iCPS data acquisition and processing via WSNs. Performance gains have been observed in various aspects including sampling, compression, routing, and detection to name a few. The connection between WSN and high-performance distributed computing platforms such as *cloud computing* can serve as an excellent paradigm for the next generation of CPS architectures. Issues related to the coupled design of such infrastructure include the robust distributed storage and the practical implementation of linear random sampling measurements acquisition.

When considering iCPS, one must not only consider the sensing platform, but the actuation components as well. In wireless sensing and actuation networks, the physical environment is responsible for closing the loop between sensing and actuation. Application of CS and MC for control of CPS processes remains an open scientific and technical challenge that calls for immediate attention.

8.4 IN-NETWORK PROCESSING: DISTRIBUTED ESTIMATION AND TRACKING FOR iCPS

Throughout this section, we present how various in-network signal processing tasks, such as parameter estimation and tracking, can be accomplished in an industrial environment. Since these approaches are based on *iterative average consensus*, we focus in the distributed implementation of this specific approach, while also considering the generic case of *nonuniform* deployment of the sensing devices. A distributed framework eliminates the need of performing all the computations at one or more sink nodes, thus reducing congestion around them and increasing the robustness of the WSN. Moreover, we study how the randomness and asymmetry of instantaneous communications, occurred in real iCPS, affect the performance of both estimation and tracking tasks. To alleviate this impact, we present a cross-layer approach based on a link scheduling protocol that deals with the particularities of the industrial environment, providing a suitable framework for the in-network processes to be executed with a reduced error.

8.4.1 STATISTICAL SIGNAL PROCESSING BACKGROUND

In general, the sensors that compose an iCPS are covering the monitored area by being placed on a nonuniform grid, while observing a distorted version of the target data, which is usually corrupted by random noise. Let $\mathbf{y} = f(\mathbf{x}, \mathbf{w})$ be the vector whose elements are the distorted observations of the S nodes, $\mathbf{x} \in \mathbb{R}^M$ is the parameter vector of interest, and $\mathbf{w} \in \mathbb{R}^S$ is a random vector modeling the noise component. Then, the data observed by the sensors can be expressed as follows:

$$\mathbf{y} = \mathbf{Hx} + \mathbf{w}, \tag{8.10}$$

where
   the observation matrix $\mathbf{H}$ models the spatial distortion of the data
   $\mathbf{w}$ is a spatially uncorrelated additive noise, which is modeled as a zero-mean Gaussian with covariance matrix $\mathbf{Q}_w = \sigma^2\mathbf{I}$, where $\mathbf{I}$ denotes the identity matrix

Each component of the vector $\mathbf{s} \triangleq \mathbf{Hx}$ can be viewed as the signal or field present at the location of each node. Based on these noisy and random observations, the iCPS needs to draw conclusions and actuate accordingly within the control loop. Thus, the more accurate the estimation of the target data is, the better this actuation can be performed. In general, the estimation process can be viewed as a problem of data selection from a continuous space that minimizes a certain cost function.

   If the data to be estimated is a time-invariant quantity, the process is reduced to a *parameter estimation* problem. On the other hand, if the data evolves in time according to a stochastic equation, the process corresponds to a *state estimation* problem. Although the term *tracking* may be applied only to the specific estimation of the state of a moving object, throughout this section we will use the terms state estimation and tracking indistinctly.

   For the state estimation problem, the objective is to estimate the value of a deterministic data vector as accurately as possible, without relying on previous information. The general approach is to maximize the probability density function (PDF) of the observations conditioned on the data, which is simply the maximum likelihood estimate given by

$$\mathbf{x}_{ML} \triangleq \underset{\mathbf{x}}{\mathrm{argmax}}\, p(\mathbf{y}|\mathbf{x}).$$

If we assume the linear Gaussian model of Equation 8.10 for the observations, it can be shown that the maximum likelihood estimator is obtained as follows [66]:

$$\mathbf{x}_{ML} = (\mathbf{H}^T\mathbf{H})^{-1}\mathbf{H}^T\mathbf{y}. \tag{8.11}$$

Furthermore, it can be shown that this estimator is unbiased, that is, $\mathbb{E}[\mathbf{x}_{ML}] = \mathbf{x}$, with covariance matrix

$$\mathbf{C}_{ML} = \sigma^2(\mathbf{H}^T\mathbf{H})^{-1}. \tag{8.12}$$

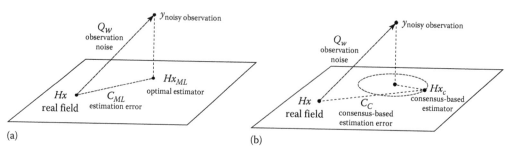

(a)   (b)

FIGURE 8.4 Geometrical interpretation of the parameter estimation when the observation matrix is orthonormal, for $M = 2$. (a) the optimal estimation is the orthogonal projection of the observations onto the subspace; (b) the projection is computed in a distributed way by means of iterative average consensus, yielding a suboptimal estimator.

This estimation process can be explained in a geometrical sense; specifically, from Equation 8.11, we can deduce that the optimal estimation of s, given the observations y, is the projection of y onto the subspace spanned by H, with the projection matrix defined as follows:

$$P = H(H^T H)^{-1} H^T$$

Moreover, if $H^T H = I$, that is, if H is an orthonormal matrix, then, the estimation process is the orthogonal projection of the observation vector onto the earlier mentioned subspace with $P = HH^T$. Figure 8.4a illustrates this geometrical representation for $M = 2$. Besides, it can be seen that the estimator is always improving the initial observations, which is stated formally as $\text{trace}(C_{ML}) = M\sigma^2 \leq \text{trace}(Q_w) = S\sigma^2$, as long as the dimension of the target vector M is smaller than the number of observations S.

On the other hand, if the parameters to be estimated evolve discretely and stochastically across time, the parameter estimation problem reduces to a state estimation process. To this end, successive observations are acquired, $y[k] = H[k]x[k] + w[k]$, and the prior information $(p(x))$ about the evolution of the process is used to refine the estimations in a Bayesian framework. Specifically, the system is assumed to evolve according to a Markov–Gaussian model as follows:

$$x[k + 1] = A[k]x[k] + v[k], \tag{8.13}$$

where

$x[k]$ is the state vector at time k
$A[k]$ is a $M \times M$ time-varying matrix that rules the evolution of the state
$v[k]$ is the noise of the system, which is considered to be white, Gaussian and uncorrelated with $w[k]$, with covariance matrix $Q_v[k]$

Under the assumption of a Markov–Gaussian model, the optimal estimator can be computed recursively by means of recursive least squares. This is obtained via the dynamics of a Kalman filter [67], which is given by

$$\hat{x}[k + 1] = A[k]\hat{x}[k] + K[k]\left(y[k] - H[k]A[k]\hat{x}[k]\right). \tag{8.14}$$

The gain, $K[k]$, of the filter is given by

$$K[k] = M[k]H^T[k]Q_w^{-1}[k], \tag{8.15}$$

where $M[k]$ is defined as

$$M^{-1}[k] = P^{-1}[k] + H^T[k]Q_w^{-1}[k]H[k]. \tag{8.16}$$

where $Q_w[k] = \sigma^2 I$.

In the aforementioned expression, $\mathbf{P}[k]$ denotes the covariance error of the estimator, whose dynamic evolution is described by

$$\mathbf{P}[k+1] = \mathbf{A}[k]\mathbf{M}[k]\mathbf{A}^T[k] + \mathbf{Q}_v[k]. \tag{8.17}$$

The basis of this filter is to achieve a trade-off between the optimal state estimator (MLE) computed with each new observation and the previous estimation, or between the previous estimation and the innovation computed from the new observations. The weight assigned to each term of the trade-off is given by the gain of the filter, which is computed in order to minimize the covariance error of the estimator at each time step. As stated before, this methodology can be tackled in the general case of nonuniformly deployed sensing devices, working in a distributed fashion by employing an iterative consensus algorithm. However, an industrial environment imposes certain communication constraints that must be taken into account in a real implementation.

8.4.2 Distributed Average Consensus under Realistic Constraints

The harsh environmental conditions of an industrial scenario provoke random packet losses in the communications between the nonuniformly located iCPS devices, thus affecting the performance of consensus-based applications, such as estimation and tracking. This performance loss is mainly due to the resulting randomness and asymmetry of the links, which affect the convergence time and error of the process.

Let $\mathcal{S}$ be a set of S autonomous nodes with initial measurements $x_i[0]$, $i = 1, \ldots, S$ following a normal distribution with mean x_{avg} and variance σ^2. Then, the *distributed consensus* (or agreement) problem consists of successive iterations, where each node i refines its own value by exchanging information only with those nodes belonging to the set of its neighbors $\mathcal{S}_i$. This procedure continues until the nodes agree asymptotically on a global common value α, where the asymptotics is expressed in terms on infinite time, $\lim_{k \to \infty} \mathbf{x}[k] = \alpha \mathbf{1}$, where $\mathbf{1} \in \mathbb{R}^N$ is the vector of all ones. Let $\mathbf{W}$ be the weight matrix that rules the mixing of information at each iteration. Then, the state evolution is expressed by the following process:

$$\mathbf{x}[k] = \mathbf{W}\mathbf{x}[k-1] = \mathbf{W}^k\mathbf{x}[0] = \mathbf{M}_k\mathbf{x}[0],$$

with $[\mathbf{W}]_{ij} \neq 0$ if and only if $j \in \{\mathcal{S}_i \cup i\}$. Moreover, the Perron–Frobenius theorem states that if $\mathbf{W}$ is row-stochastic, that is, $\mathbf{W}\mathbf{1} = \mathbf{1}$, and irreducible, then $\lim_{k \to \infty} \mathbf{W}^k = \mathbf{1}\mathbf{q}^T/\mathbf{q}^T\mathbf{1}$, where $\mathbf{q}$ is the left eigenvector of $\mathbf{W}$ corresponding to the eigenvalue 1. Consequently, all rows of the matrix $\mathbf{M}$ are asymptotically equal to a vector $\mathbf{m}$, where $m_i = q_i / \sum_{i=1}^{S} q_i$. Therefore, the nodes achieve a consensus, which corresponds to the value

$$\alpha = \sum_{i=1}^{S} m_i x_i[0]. \tag{8.18}$$

If, in addition, $\mathbf{W}$ is column-stochastic, that is, $\mathbf{1}^T\mathbf{W} = \mathbf{1}^T$, then $\mathbf{m} = \frac{1}{S}\mathbf{1}$, and $\alpha = \frac{1}{S}\sum_{i=1}^{S} x_i[0]$, which is exactly the average of the initial values. Nevertheless, since the weight matrix $\mathbf{W}$ should be compatible with the underlying topology of the network, we have that in a realistic industrial scenario, where interference, fading, and packet losses may occur, each instantaneous topology is totally random, and, in general, different. Based on that, we also consider instantaneous matrices, $\mathbf{W}[k]$, in the time-evolving state equation:

$$\mathbf{x}[k] = \mathbf{W}[k] \ldots \mathbf{W}[0]\mathbf{x}[0] = \mathbf{M}[k]\mathbf{x}[0]. \tag{8.19}$$

By construction, we can force every $\mathbf{W}[k]$ to be row-stochastic and guarantee the nodes to reach a consensus, $\lim_{k \to \infty} \mathbf{M}[k] = \mathbf{1}\mathbf{m}^T$. However, in this case $\mathbf{m}$ becomes a random vector which is,

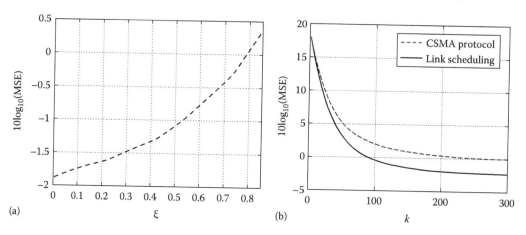

FIGURE 8.5 (a) MSE of the average consensus as a function of the asymmetry of the connection probability matrix. (b) Evolution of the MSE across the iterations of the consensus process. The proposed link scheduling protocol outperforms the general CSMA when computing distributed average consensus. The parameter ξ represents the asymmetry of the underlying graph.

in general, different from $\frac{1}{s}\mathbf{1}$, and whose first two moments are computed in Tahbaz-Salehi and Jadbabaie [68]. The moments of $\mathbf{x}[k]$ can be asymptotically computed as follows:

$$\mathbb{E}[\mathbf{x}] = x_{\text{avg}}\mathbf{1}$$
$$\mathbf{C}_x = \sigma^2 \mathbb{E}[\mathbf{m}^T\mathbf{m}]\mathbf{1}\mathbf{1}^T. \tag{8.20}$$

Thus, $\lim_{k\to\infty} x_i[k]$ can be viewed as the unbiased estimator of x_{avg} computed by node i, with variance $\sigma_x^2 = \sigma^2 \mathbb{E}[\mathbf{m}^T\mathbf{m}]$.

If $\mathbf{\Sigma}$ is the matrix whose entry ij is the activation probability of the link between nodes i and j, then the variance σ_x^2 can be reduced by enforcing $\mathbf{\Sigma}$ to be as symmetric as possible. This is motivated by the fact that its symmetry entails the symmetry of matrix $\mathbb{E}[\mathbf{W}[k]]$, yielding $\sum_{i=1}^{S} \mathbb{E}[m_i]^2 = \frac{1}{s}$, which corresponds to the minimum possible value. Figure 8.5a shows how the mean squared error (MSE), which is equal to the variance since the estimator is unbiased, increases with the asymmetry of $\mathbf{\Sigma}$. This means that the randomness and asymmetry of communications, occurring in an industrial environment due to packet losses, affect the error of the consensus.

The approach described in Asensio-Marco and Beferull-Lozano [69] deals with this imperfection of communications, ensuring, on average, the symmetry of the links. This implies having less consensus errors than by applying traditional approaches, such as the protocol implemented by default in most of the motes, which follows a CSMA strategy and, as a consequence, only focus on reducing collisions. Moreover, the protocol introduced in Asensio-Marco and Beferull-Lozano [69] employs connectivity patterns that are as dense as possible, so as to favor the convergence time, which is also crucial towards enabling the iCPS to actuate as fast as possible. The implementation of this new protocol is based on a cross-layer scheme in which the decisions taken by the MAC layer about whether to transmit or not, besides providing collision avoidance, favors the performance of the consensus process. To this end, at each scheduling step, the protocol activates randomly a link and creates an associated inhibition area that contains the links that are inhibited when the current link is activated. In order to ensure that this inhibition radius guarantees a collision-free communication pattern, the worst-case scenario is assumed as follows: (1) every transmitter is at maximum distance from its intended receiver and (2) every interferer is assumed to be as close as possible to the receivers. By locating these inhibition areas at the center of each link, every pair of nodes includes the same number of potential inhibitors inside those areas, leading to the same probability of inhibition. Moreover, since every link in the network presents the same probability of being considered for activation, symmetric probabilities of connection for each pair of nodes are also ensured.

Figure 8.5b shows the efficiency of this protocol, which outperforms the general CSMA protocol when computing the average consensus. This cross-layer approach is applied in the following sections to reduce the error of the proposed in-network processing algorithms for iCPS, which are all based on iterative average consensus.

8.4.3 CONSENSUS-BASED IN-NETWORK PROCESSING

The cross-layer protocol proposed in the previous section supports the implementation of consensus-based estimation techniques in iCPS with reasonably low error. In this section, we describe in detail how the distributed parameter estimation and the distributed state estimation techniques are both favored by the application of this cross-layer technique, demonstrating its performance in the case of reconstructing a generic two-dimensional field.

8.4.3.1 Distributed Parameter Estimation

The maximum likelihood estimator given by Equation 8.11 can be also expressed as follows:

$$\mathbf{x}_{\mathrm{ML}} = \left(\sum_{i=1}^{S} \mathbf{h}_i \mathbf{h}_i^T \right)^{-1} \sum_{i=1}^{S} \mathbf{h}_i y_i, \tag{8.21}$$

where $\mathbf{h}_i$ is the ith row of $\mathbf{H}$. It is straightforward to show that each node is able to compute both the matrix $\frac{1}{S} \sum_{i=1}^{S} \mathbf{h}_i \mathbf{h}_i^T$ and the vector $\frac{1}{S} \sum_{i=1}^{S} \mathbf{h}_i y_i$ in a distributed fashion by means of two iterative average consensus processes, and consequently to compute the ML estimate asymptotically. We emphasize again that our proposed approach does not make any assumption for the network topology, thus it can be used successfully in both uniform and nonuniform sensor deployments. Nevertheless, due to the randomness of the iterative processes, there exists a deviation from the average, with the actual estimator being given by

$$\mathbf{x}_c \triangleq \left(\sum_{i=1}^{S} m_i \mathbf{h}_i \mathbf{h}_i^T \right)^{-1} \sum_{i=1}^{S} m_i \mathbf{h}_i y_i = (\mathbf{H}^T \Delta_m \mathbf{H})^{-1} \mathbf{H}^T \Delta_m \mathbf{y}, \tag{8.22}$$

where
  $\mathbf{m} = \{m_1, \ldots, m_S\}$ is a random vector
  Δ_m is the diagonal random matrix with the elements of $\mathbf{m}$ on its main diagonal

The covariance matrix of this estimator takes the following form:

$$\mathbf{C}_c = \sigma^2 \mathbb{E} \left[(\mathbf{H}^T \Delta_m \mathbf{H})^{-2} \mathbf{H}^T \Delta_m^2 \mathbf{H} \right] . \tag{8.23}$$

The expression of $\mathbf{x}_c$ in Equation 8.22 may be seen as a noisy (random) version of the expression of $\mathbf{x}_{\mathrm{ML}}$ in Equation 8.11, since the deviation from the average in the iterative consensus involves an additional random error. This is shown in Figure 8.4b, where following the geometrical interpretation of the estimation, the consensus-based estimation error is the sum of the optimal estimation error and the consensus error. It can be shown that trace($\mathbf{C}_c$) $\geq$ trace($\mathbf{C}_{\mathrm{ML}}$), where equality holds if and only if the average in both consensus processes is always achieved. Therefore, the consensus-based estimator is suboptimal in probabilistic terms. In fact, it may be the case that the consensus process is so inaccurate that the consensus-based estimator is worse than the initial observations, that is, trace($\mathbf{C}_c$) $\geq$ trace($\mathbf{Q}_w$) $= \sigma^2 N$.

However, interestingly, from Figure 8.4b it can be also seen that there exists a possibility that some realizations of the consensus-based estimator improve the MLE. Specifically, for $M = 1$ the probability is 0.50, but it decreases quickly as M increases. The performance of the consensus-based estimator can be improved if the connection probability matrix approximates a symmetric matrix. To accomplish that, the link scheduling protocol defined in Asensio-Marco and Beferull-Lozano [69] can be applied in a cross-layer scheme at the link layer, instead of the standard CSMA protocol, when a distributed parameter estimation is performed using iterative consensus. As an illustration, Figure 8.6 shows the distributed reconstruction of a 2D field by a network of $S = 49$ sensors deployed uniformly

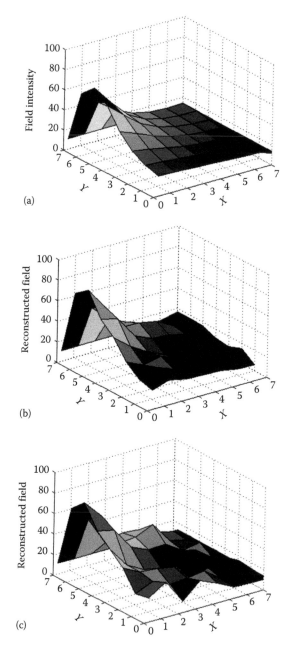

FIGURE 8.6 Two-dimensional field reconstruction using distributed parameter estimation based on iterative average consensus. The nodes form a 7×7 grid. (a) Real field and (b) reconstructed field using the link scheduling protocol proposed in Asensio-Marco and Beferull-Lozano [69]. (c) Reconstructed field using a generic CSMA protocol.

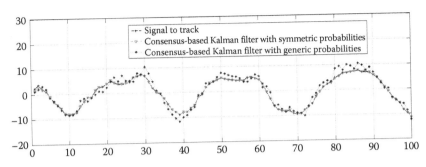

FIGURE 8.7 Consensus-based distributed Kalman filter. The performance of the filter obtained via the algorithm in Asensio-Marco and Beferull-Lozano [69] is compared against the performance of the filter obtained when generic CSMA is used.

on a grid structure. Notably, it can be seen that the symmetry of Σ improves the performance of the estimator.

8.4.3.2 Distributed State Estimation

For the distributed computation of the Kalman filter, notice that Equation 8.14 can be expressed as [70]

$$\hat{\mathbf{x}}[k+1] = \mathbf{A}[k]\hat{\mathbf{x}}[k] + \frac{1}{\sigma^2}\mathbf{M}[k]\left(\mathbf{H}^T[k]\mathbf{y}[k] - \mathbf{H}^T[k]\mathbf{H}[k]\mathbf{A}[k]\hat{\mathbf{x}}[k]\right).$$

The necessary computations for the distributed implementation of this filter are very similar to the ones performed in the parameter estimation problem. In fact, at each time k, every node has to compute $\mathbf{H}^T[k]\mathbf{y}[k] = \sum_{i=1}^{S} \mathbf{h}_i y_i$ and $\mathbf{H}^T[k]\mathbf{H}[k] = \sum_{i=1}^{S} \mathbf{h}_i \mathbf{h}_i^T$, which are the two terms that are present in Equation 8.21. Furthermore, each node has also to compute $\mathbf{M}[k]$ in order to obtain the gain of the filter, and perform the corresponding weighting between the optimal estimation at the current time instant and the previous estimation. From Equations 8.16 and 8.17, we deduce that for the computation of $\mathbf{H}^T[k]\mathbf{H}[k]$ it suffices for each node to compute $\mathbf{M}[k]$ in a distributed way at time k. Thus, the estimator at time k is expressed as a function of the individual observations as follows:

$$\hat{\mathbf{x}}[k+1] = \mathbf{A}[k]\hat{\mathbf{x}}[k] + (\sigma^2 S \mathbf{P}^{-1}[k] + \sum_{i=1}^{S} \mathbf{h}_i \mathbf{h}_i^T)^{-1}\left(\sum_{i=1}^{S} \mathbf{h}_i y_i - \sum_{i=1}^{S} \mathbf{h}_i \mathbf{h}_i^T \mathbf{A}[k]\hat{\mathbf{x}}[k]\right), \qquad (8.24)$$

which can be computed in a completely decentralized way. Again, in practice we have to consider a deviation from the average in the computation of the iterative consensus, thus obtaining a suboptimal filter:

$$\hat{\mathbf{x}}[k+1] = \mathbf{A}[k]\hat{\mathbf{x}}[k] + (\sigma^2 \mathbf{P}^{-1}[k] + \mathbf{H}^T[k]\Delta_m \mathbf{H}[k])^{-1}\left(\mathbf{H}^T[k]\Delta_m \mathbf{y}[k] - \mathbf{H}^T[k]\Delta_m \mathbf{H}[k]\mathbf{A}[k]\hat{\mathbf{x}}[k]\right).$$

Concerning the performance of this filter, all the statements presented in the previous section with respect to the parameter estimation are also valid here. Consequently, the performance of this suboptimal filter can be improved significantly by employing the same cross-layer scheme described in Asensio-Marco and Beferull-Lozano [69], which enforces Σ to approximate a symmetric matrix, as shown in Figure 8.7.

8.5 HIGH-LEVEL DATA ANALYSIS AND EARLY WARNING

In an industrial CPS setting, the associated distributed autonomous sensing introduced before is further exploited to produce intelligent reasoning over the data by supporting advanced operations, such as, querying, high-level analysis, and alerting. In particular, a *high-level data management and analysis* (HDMA) module is an integral part toward an efficient decision making. Typically, an HDMA component comprises of collaborating computational nodes, which observe and control distinct physical entities and dynamic phenomena. Rather than relying on single sensor stream statistics, such as average and standard deviation, which is the customary approach in most data analysis systems, an efficient HDMA module focuses on finding and extracting inherent information for detecting behavioral variations in the acquired data. This is crucial, especially in an industrial CPS framework, since the accurate and timely detection of abnormal changes in sensor measurements will enable early actuation aiming at minimizing operational and maintenance costs.

Usually, WSN nodes do not handle any quality aspect of physical device data, but rather interface with a high-level representation of the sensed physical world. In practice, the recorded sensor measurements are often incomplete, imprecise, or even misleading, thus impeding the task of an accurate and reliable decision making. Motivated by this, a powerful HDMA system should also cope with what we call *uncertain* data. *Uncertainty-aware data management* [71] presents numerous challenges in terms of collecting, modeling, representing, querying, indexing, and mining the sensor data. Since many of these issues are interrelated, we address them jointly wherever possible. In contrast to most of the existing industrial CPS, a versatile HDMA module considers uncertainty as an additional source of information that could be valuable during data analysis and thus it should be preserved.

Another major functionality assigned to a modern data analysis system is to perform *high-level operations*, such as the notification of extreme events from raw sensor data. Since the detection of abnormal behavior is affected by the underlying uncertainty, incorporation of its estimated value is expected to yield more meaningful results. More specifically, widely used methods for extreme events detection can be enhanced by incorporating the inherent data uncertainty, yielding an integrated uncertainty-aware HDMA (U-HDMA) system capable of identifying, quantifying, and combining the individual uncertainties corresponding to the most significant sources of uncertainty for providing *early warning notifications* of extreme events.

On the other hand, extracting *highly correlated* pairs of data streams acquired by distinct sensors is another important issue. Doing so, we aim at revealing interrelations between seemingly independent physical quantities, or guaranteeing the validity of a detected extreme event. Whereas traditional statistical machine learning provides well-established mathematical tools for monitoring and analyzing multiple data streams by exploiting potential pairwise correlations [72,73], their performance is limited when processing heterogeneous and uncertain data streams. More specifically, [74] studies the problem of maintaining data stream statistics over sliding windows, with the focus being only on single stream statistics, while [75] introduced an extension for monitoring the statistics of multiple data streams, but the computation of correlated aggregates is limited to a small number of monitored sensor streams. On the other hand, [76] introduced a successful data stream monitoring system, which enables the computation of single- and multiple-stream statistics. However, its performance diminishes in an industrial environment, since the sensor streams we manage describe dynamic phenomena, whose distribution is not known *a priori*. Such limitations of previous approaches can be overcome by designing an appropriate stream correlation engine based on a computationally efficient similarity function, which enables fast and accurate monitoring of pairwise correlations between time-synchronized (high-dimensional) sensor data streams.

Figure 8.8 summarizes our utmost goal in this section, which is to provide an insight into the design and implementation principles of efficient and robust U-HDMA systems, integrating the above functionalities for industrial monitoring and surveillance applications, while emphasizing the importance of accounting for the underlying data uncertainty as an additional source of information, which should be preserved across all stages of the data processing chain. In particular, a generic U-HDMA module consists of the three building blocks shown in Figure 8.8, namely, (1) uncertainty estimation, (2) correlations extraction, and (3) detection of extreme events. Appropriate data services are provided to manipulate the sensor measurements, as well as to characterize the generated data quality. Computationally efficient extraction of correlations from uncertain data streams is then coupled with modified

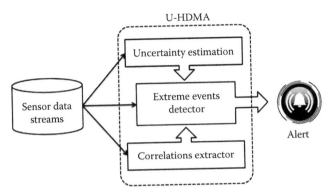

FIGURE 8.8   Building blocks of an uncertainty-aware high-level data analysis system.

uncertainty-aware extreme event detectors to enable higher-level analysis, which form the basis for the development of an integrated U-HDMA system for monitoring dynamic sensor networks and providing early warning notifications in case of abnormal events.

8.5.1   Managing Uncertainty in Sensor Measurements

In practice, the raw sensor data acquired by distinct sensors distributed across an industrial infrastructure are often unreliable, imprecise, or even misleading. This yields results of unknown quality, which may impede the task of an accurate and reliable decision making. To this end, the notion of *measurement uncertainty* arises as an indicator of measurement quality. Speaking formally, the uncertainty is "a parameter associated with the result of a measurement that characterizes the dispersion of the values that could reasonably be attributed to the measurand," where a measurand refers to a quantity to be measured.

The underlying uncertainty may arise due to several distinct sources, such as an incomplete definition of the observed quantities, sampling effects and interferences, varying environmental conditions, or hardware defections of the equipment. The effects of all these factors can be observed and quantified from the recorded sensor data only. For this purpose, a set of ordered steps need to be performed in order to obtain an estimate of the uncertainty associated with a measurement result. Figure 8.9 presents the processing flow, which starts by identifying the measurands to be monitored and returns the overall estimated uncertainty.

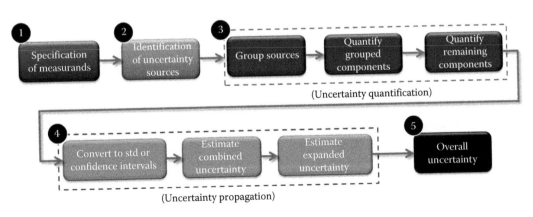

FIGURE 8.9   Flow diagram for uncertainty estimation in sensor data streams.

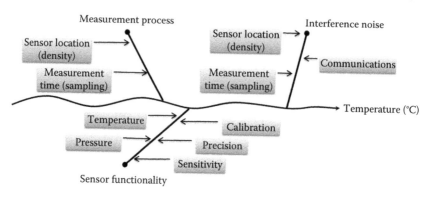

FIGURE 8.10   Cause and effect diagram for a temperature sensor.

Having specified appropriate measurands associated with our industrial application, such as, temperature (°C), pressure (bar), capacitance (F), and current (A), the next step is to identify the potential sources of uncertainty. A very convenient way to determine the most dominant uncertainty sources, along with their potential interdependencies, is to exploit the so-called *cause and effect* (or Ishikawa) diagram [77]. This diagram also ensures comprehensive coverage, while helping to group similar sources and avoid double counting. Figure 8.10 shows a typical cause and effect diagram for a temperature sensor. Its performance may be affected by several distinct factors, such as its sensitivity and precision, calibration, and operating temperature. Furthermore, the accuracy of the acquired measurements depends also on the deployment density and location of the sensors, as well as on the sampling process. Possible misplacement or a very sparse time-sampling is expected to increase the uncertainty, especially when the monitored variables vary rapidly across time.

Despite the pervasive nature of computational analysis in nowadays engineering practice, an objective establishment of the confidence levels in the measurement procedure, as well as in the subsequent numerical processing, still remains a difficult task. This is due to the differences between a real device and the corresponding numerical models, and the lack of knowledge associated with the underlying physical processes. *Uncertainty quantification*, which is the third step in the estimation chain, plays a fundamental role aiming at developing a rigorous framework to characterize the impact of variability and lack of knowledge on the final quantity of interest and provides the basis for certification in high-consequence decisions. However, it is important to notice that not all of the sources will make a significant contribution to the combined uncertainty. In practice, it is often likely that only a small number of them will contribute the major portion of the overall uncertainty. If possible, an initial estimate of the contribution of each separate source, or groups of sources, to the uncertainty could be made, so as to eliminate the less significant ones.

Toward assessing the underlying uncertainty component in a given raw sensor data stream, we recall its distinction into two separate categories, namely, *type A* (aleatoric, statistical, or irreducible) and *type B* (epistemic, systematic, or reducible) uncertainty [78]. For instance, the physicochemical properties of substances concentration, the operating conditions of the sensors, and their manufacturing tolerances are typical examples associated with type A uncertainties, which cannot be reduced. On the other hand, the mathematical models, the calibration methods, and the inference techniques from experimental observations are typical sources of type B uncertainties, which can be reduced by improving the accuracy of our physical models or calibration methods.

Without going into too much detail, in the following, we introduce the main approaches for carrying out steps 3 and 4 in Figure 8.9. Specifically, uncertainties of type A are characterized by the estimated variances σ_i^2 (or the standard deviations σ_i), which are obtained by statistical analysis of the observations in the raw sensor data streams. This is equivalent to obtaining a *standard uncertainty* from a probability density function (pdf) derived from an observed frequency (empirical) distribution. Let $\mathbf{y} = \{y_1, \ldots, y_N\}$ be a set of N sensor measurements, which correspond to a specific observed variable. Then, the standard uncertainty of $\mathbf{y}$, which is denoted by $u(\mathbf{y})$, is expressed in terms of the corresponding standard deviation σ_y, estimated directly from the observations y_i, as follows:

$$u(\mathbf{y}) = \frac{\sigma_y}{\sqrt{N}}. \tag{8.25}$$

For uncertainties of type B, the estimated "variance" s_j^2 is obtained from an assumed probability density function based on our prior knowledge for the corresponding source of uncertainty, which may include (1) data from previous measurements; (2) experience or knowledge of the properties of instrumentation and materials used; (3) manufacturers specifications; and (4) calibration data. In general, concerning type B uncertainties, the quantification is performed either by means of an external information source, or from an assumed distribution.

Typical assumptions for the prior distributions include the Gaussian (e.g., when an estimate is made from repeated observations of a randomly varying process, or when the uncertainty is given as a standard deviation or a confidence interval), the uniform (e.g., when a manufacturers specification, or some other certificate, gives limits without specifying a confidence level and without any further knowledge of the distributions shape), and the triangular distribution (e.g., when the measured values are more likely to be close to a value α than near the bounds of an interval with mean equal to α) [79]. For instance, if a manufacturer's specification, or some other certificate, gives limits in the form of a maximum range, $y \pm \alpha$, without any further knowledge of the distributions shape, then the estimated standard uncertainty is equal to $u(\mathbf{y}) = \alpha/\sqrt{3}$, while if the maximum range is described by a symmetric distribution then $u(\mathbf{y}) = \alpha/\sqrt{6}$.

Having expressed the individual uncertainties as standard uncertainties, the next step (see Figure 8.9 Step 4) is to combine them in the form of a *combined standard uncertainty*. Although in practice there may exist correlations between the individual uncertainty sources, however, it is usually impossible to compute those correlations accurately. For this purpose, it is more convenient to rely on an assumption of independence between the individual uncertainty sources. In the following, let $y = f(x_1, \ldots, x_L)$ be an observed variable, which depends on L input variables x_l through a functional relation $f(\cdot)$. Then, the combined standard uncertainty of y, for independent input variables x_l, $l = 1, \ldots, L$, is given by

$$u_c(y) = \sqrt{\sum_{l=1}^{L} \left(\frac{\partial f}{\partial x_l}\right)^2 u^2(x_l)}, \tag{8.26}$$

where $u(x_l)$ denotes the standard uncertainty of the input variable x_l (either of type A, or of type B), while the partial derivatives $\partial f/\partial x_l$, the so-called *sensitivity coefficients*, quantify how much the output y varies with changes in the values of the input variables x_l. It is also important to note that, before the evaluation of $u_c(y)$, we have to ensure that all the distinct standard uncertainties are expressed in the same units.

However, in practice, even for the modern sensing devices, it is usually extremely difficult to calculate the sensitivity coefficients accurately. To this end, the easiest way is either to consider a weighted scheme, that is, $\partial f/\partial x_l = w_l$ with $\sum_{l=1}^{L} w_l = c$, where c is a predefined constant, while the degree of contribution (or, equivalently, the weight value) of the individual input variables (uncertainty sources) is set in a rather empirical fashion. More details about the complex case of correlated input variables can be found in Farrance and Frenkel [78], we emphasize though that this assumption is usually avoided in the industrial practice due to the difficulties in computing accurately the interdependencies among the identified uncertainty sources. On the other hand, if such an assumption is not valid explicitly, the correlations themselves can be avoided if the common influences are introduced as additional independent input variables.

TABLE 8.1

Coverage Factor as a Function of a Confidence Level for the Gaussian Distribution

Coverage Factor (k)	Confidence Level (%)
$k = 1$	67
$k = 1.96$	95
$k = 2.576$	99
$k = 2.3$	99.7

Finally, the combined standard uncertainty, which may be thought of as equivalent to one standard deviation, is transformed into an overall *expanded uncertainty*, U, which is the final output, via multiplication with a coverage factor k, that is,

$$U(y) = k \cdot u_c(y), \tag{8.27}$$

where the value of k is determined in terms of the desired confidence level of a Gaussian distribution as shown in Table 8.1.

The computation of U completes the first building block of the U-HDMA system shown in Figure 8.8. In the following, we focus on the second building block, namely, the detection of extreme events by employing appropriate extreme event detectors in order to account for the inherent data uncertainty.

8.5.2 Uncertainty-Aware Alerting Notifications

When working in an industrial environment, various distinct rarely occurring "events" can be devastating to the proper operation of the whole infrastructure. For instance, in manufacturing industries we are interested in preventing potential defections of the production engines by monitoring critical structural parameters, while in a large-scale water treatment industrial plant a key task is to support early detection of high concentrations for several chemicals which may be harmful for the public health. Thus improving our understanding of such *extreme events* will further help to mitigate their effects.

Extreme events can occur at any phase and time instant of the infrastructure's lifecycle, which necessitates its continuous and efficient monitoring to achieve early detection of abnormal behavior. Although a typical industrial setting is generally intended to operate autonomously, however, in extreme events it is of high significance to anticipate the impact of the detected events by triggering appropriate actuators in time. To this end, designing fast and accurate extreme events detectors for providing *early warning notifications* is a strong demand in order to guarantee the smooth operation of critical industrial infrastructures.

Among the several approaches, which have been introduced in the literature, *extreme value theory* (EVT) provides efficient algorithmic tools to assess, from a given ordered sample of a given random variable, the probability of events that are more extreme than any previously observed. Two approaches are the most widely used in practice for extreme value analysis, namely, the method of *block maxima* (BMax) [80] and the method of *peaks-over-threshold* (POT) [81,82]. Depending on the application, each method has its own advantages and limitations. For instance, for the method of block maxima, theoretical assumptions are less critical in practice, while it is also easier to apply. However, estimation errors can be large for relatively small block sizes. On the other hand, the method of peaks-over-threshold yields more independent exceedances than block maxima, along with tighter confidence intervals. Its main drawback is that an independence assumption is critical, which may not hold in practice, and also the choice of an appropriate threshold is somewhat ambiguous in practice resulting in a less easier implementation.

Furthermore, a common characteristic is that, in both cases, the detection of extreme events is based on the raw data without accounting for their underlying uncertainty. In addition, given our major requirement for providing timely notifications of abnormal behavior, the selected extreme events detection method must have a small computational complexity, without sacrificing the detection accuracy. The simplest approach to satisfy both requirements, that is, to exploit the inherent data uncertainty while being computationally efficient, is obtained by modifying an alternative widely used method, the so-called *compliance with operating limits* (COL).

Without loss of generality, in the following, we restrict ourselves in the case of an upper operating limit; however, the same remarks are straightforward when compliance with a lower operating limit is required. More specifically, let l_u denote an upper operating limit dictated by a manufacturer or a specification standard. In addition, let $\tilde{y} = y \pm U$ be a measurement augmented by its associated expanded uncertainty interval. In contrast to the typical COL method, for which only two cases exist when checking for compliance between the raw measurement y and the upper limit l_u, as shown in Figure 8.11, there are two additional cases for its uncertainty-aware counterpart, hereafter denoted as U-COL. Specifically, the four possible cases of U-COL are as follows: (i) both the measurement and the expanded uncertainty interval are above the upper limit l_u; (ii) the measurement is larger than l_u and the expanded uncertainty interval contains l_u; (iii) the measurement is lower than l_u and the expanded uncertainty interval contains l_u; and (iv) both the measurement and the expanded uncertainty interval are below l_u.

Among them, only case (i) triggers clearly an alerting notification for the occurrence of an extreme event, while (iv) is the only one that is in compliance with the specifications. On the other hand, in cases (ii) and (iii) we could not infer with absolute certainty whether an alert should appear or not. Nevertheless, in applications with profound social impacts, such as, for instance, water treatment, a system operator should classify cases (ii) and (iii) as possible divergences from normal operation, and thus draw more attention on the associated monitored variables.

The importance of accounting for the inherent data uncertainty in order to increase the efficiency of an extreme events detector is demonstrated in Figure 8.12. More specifically, this figure shows the extreme events detection performance of (a) the original COL (upper plot) and the uncertainty-aware U-COL (bottom plot) method for a temperature sensor by setting $l_u = 17\,°C$. The black dots correspond to detected extreme events, while the gray dots correspond to potential extreme events. Although both these methods achieve to identify the extreme peaks in the recorded temperature measurements, however, their key difference is that COL notifies for an extreme event only when we

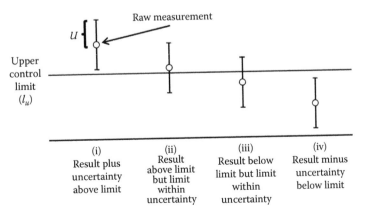

FIGURE 8.11 Compliance of uncertainty-augmented measurements with a predetermined upper operating limit.

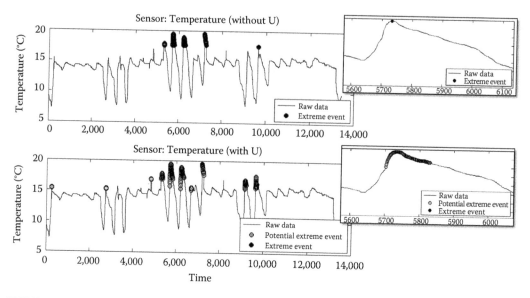

FIGURE 8.12 Extreme events detection without and with uncertainty for a temperature sensor ($l_u = 17\,^{\circ}\text{C}$).

reach the peak of the curve. On the contrary, U-COL starts notifying for a potential deviation from "normal" behavior when the curve of the uncertainty-augmented measurements exceeds the predefined threshold. Indeed, this can be seen clearly in the two zoomed regions shown in Figure 8.12. This observation reveals the increased tolerance of U-COL in detecting extreme events, when compared to its simple COL counterpart. The benefit of a system operator from using U-COL is that the U-HDMA system will start sending notifications prior to the occurrence of an event.

8.5.3 FAST AND EFFICIENT MONITORING OF PAIRWISE DATA STREAM CORRELATIONS

Efficient discrimination between occasional and extreme events is a major issue in the design of robust data management systems. It is of great importance to ensure that a true extreme event occurs and not some coincidence, or system or network failure. On the other hand, the degree of correlation between two or more sensor data streams characterizes their interrelations and dependencies. To this end, timely identification of highly correlated streams can be further exploited as a guarantee to verify the existence of a detected extreme event. Though, the degree of "high correlation" is related to the specific application and the end user, who has the flexibility to define how much strict this degree will be.

Extraction of pairwise correlations yields a partition of the set of available sensors into subsets of highly correlated sensors. This clustering facilitates the monitoring of the overall infrastructure by a system operator, who focuses only on a subset of sensors, where an abnormal behavior has been detected for at least one of its members. In the following, let $\mathbf{x} \in \mathbb{R}^N$, $\mathbf{y} \in \mathbb{R}^N$ be two sensor streams of length N, and $\mathbf{x}_w = (x_{t_1}, \ldots, x_{t_w})$, $\mathbf{y}_w = (y_{t_1}, \ldots, y_{t_w})$ be two time-synchronized windows of size w). The typical approach for extracting pairwise sensor stream correlations is by means of the Pearson's correlation coefficient, which is given by

$$\text{corr}(\mathbf{x}_w, \mathbf{y}_w) = \frac{\sum_{i=1}^{w} x_{t_i} y_{t_i} - w \bar{x}_w \bar{y}_w}{(w-1)\sigma_{x_w} \sigma_{y_w}}, \tag{8.28}$$

where

$\bar{x}_w, \bar{y}_w$ are the means of $\mathbf{x}_w$ and $\mathbf{y}_w$, respectively
$\sigma_{x_w}, \sigma_{y_w}$ denote their corresponding standard deviations

From a computational perspective, the main limitation is that the correlation coefficient has to be recalculated for each newly acquired measurement, which increases the computational burden, especially for high-dimensional data streams or for a large number of sensors. To this end, a computationally efficient solution was proposed based on the use of the discrete Fourier transform (DFT). Working in a DFT framework, each sample x_{t_k} (similarly y_{t_k}) can be expressed as a linear combination of exponential functions

$$x_{t_k} \approx \frac{1}{\sqrt{w}} \sum_{f=0}^{K-1} X_f e^{i2\pi f k/w}, \quad k = 1, \ldots, w, \tag{8.29}$$

where X_f ($f = 0, \ldots, K-1$) is the set of K DFT coefficients, with $K < w$. Doing so, the computation of the correlation coefficient in Equation 8.28 is performed in terms of DFT coefficients.

Most importantly, the fast and efficient computation of DFTs enables the fast monitoring of synchronized sensor streams, whose correlation exceeds a predefined threshold. This is dictated by the following lemma, which gives a correspondence between the correlation coefficient and the Euclidean distance between DFTs.

Lemma 8.1 [83] *Let $\hat{\mathbf{x}}_w, \hat{\mathbf{y}}_w$ be the normalization to mean zero and variance one of $\mathbf{x}_w$ and $\mathbf{y}_w$, respectively. In addition, let $\hat{\mathbf{X}}_w = \mathcal{F}\{\hat{\mathbf{x}}_w\}, \hat{\mathbf{Y}}_w = \mathcal{F}\{\hat{\mathbf{y}}_w\}$ be their corresponding DFTs. Then,*

$$\text{corr}(\mathbf{x}_w, \mathbf{y}_w) \geq \epsilon \Rightarrow d_M\left(\hat{\mathbf{X}}_w, \hat{\mathbf{Y}}_w\right) \leq \sqrt{2w(1-\epsilon)}. \tag{8.30}$$

In Equation 8.30, ϵ is a predefined threshold and $d_M(\hat{\mathbf{X}}_w, \hat{\mathbf{Y}}_w)$ is the Euclidean distance between the corresponding *truncated* DFTs, which are obtained by keeping the first $M \leq w/2$ DFT coefficients with the largest amplitudes. The validity of this approach is based on the *compactness* of DFT representations, that is, the concentration of the main portion of the energy for a given sensor stream in the first few high-amplitude DFT coefficients.

The lemma implies that by focusing on those sensor pairs, whose associated truncated DFTs are "close" enough, we get a set of *likely correlated* sensor pairs. Notice that this constitutes a superset of the correlated sensors, without false negatives. Furthermore, in our U-HDMA system, we are interested in identifying and tracking highly correlated sensor pairs in an online fashion by also incorporating the estimated data uncertainty. Aiming at improving the computational performance of the DFT-based approach, while maintaining its accuracy, in our U-HDMA system the problem of extracting highly correlated pairs of sensors is translated into a problem of identifying highly *similar* sensors, where the similarity is measured by an appropriately designed function.

Let $\mathbf{x}$ be the reference sensor stream and $(\mathbf{y}_1, \ldots, \mathbf{y}_C)$ the set of candidate streams. At the core of our fast and robust "correlation extractor" is an efficient *peak similarity* function. Given two windowed, yet time-synchronized, data streams $\mathbf{x}_w, \mathbf{y}_{i,w}$ ($i = 1, \ldots, C$) the corresponding expanded uncertainties $U_{x_w}, U_{y_{i,w}}$ are estimated first. Then, the uncertainty-augmented windows are formed: $\mathbf{x}_w^U = \mathbf{x}_w + U_{x_w}$ (or $\mathbf{x}_w^U = \mathbf{x}_w - U_{x_w}$), $\mathbf{y}_{i,w}^U = \mathbf{y}_{i,w} + U_{y_{i,w}}$ (or $\mathbf{y}_{i,w}^U = \mathbf{y}_{i,w} - U_{y_{i,w}}$). After their normalization to mean zero and variance one, $\hat{\mathbf{x}}_w^U$ and $\hat{\mathbf{y}}_{i,w}^U$, respectively, the M-sized ($M \ll w/2$) truncated DFTs are computed, $\hat{\mathbf{X}}_w^U = \mathcal{F}\{\hat{\mathbf{x}}_w^U\}, \hat{\mathbf{Y}}_{i,w}^U = \mathcal{F}\{\hat{\mathbf{y}}_{i,w}^U\}$. Finally, our *uncertainty-aware peak similarity* function is defined as

$$p_{sim,U}(\mathbf{x}_w, \mathbf{y}_{i,w}) = \frac{1}{M} \sum_{j=1}^{M} \left[1 - \frac{\left|\hat{\mathbf{X}}_{w;j}^U - \hat{\mathbf{Y}}_{i,w;j}^U\right|}{2 \cdot \max\left(|\hat{\mathbf{X}}_{w;j}^U|, |\hat{\mathbf{Y}}_{i,w;j}^U|\right)} \right], \tag{8.31}$$

where $\hat{\mathbf{X}}_{w;j}^U$ denotes the jth element of $\hat{\mathbf{X}}_w^U$ (similarly for $\hat{\mathbf{Y}}_{i,w}^U$).

Our U-HDMA system reports as "highly-similar" those sensor pairs for which $p_{sim,U}(\mathbf{x}_w, \mathbf{y}_{c,w}) > \epsilon_U$. In order to account for the potential loss of information caused by the truncation of the set of DFT coefficients, as well as for the incorporation of the underlying uncertainties, special attention should be given on the selection of the threshold ϵ_U. An extensive experimental evaluation on real measurements acquired by various distinct sensors, we observed that by selecting an "elastic" enough threshold ϵ_U, the subset of sensor streams $\mathbf{y}_c$, $c \in \{1, \dots, C\}$, with the highest peak similarity values with $\mathbf{x}$ will also contain the highly correlated streams with $\mathbf{x}$ (i.e., those with correlation coefficient above ϵ as stated in Lemma 8.1). Our evaluation showed that $p_{sim,U}$ achieves at least the performance of corr by setting $\epsilon_U = \epsilon + \epsilon_{offset}$, where ϵ_{offset} is a small positive number. Although our experimentation showed that it suffices to set $\epsilon_{offset} < 0.05$, however, an automatic and adaptive rule to select an optimal threshold ϵ_U needs a more thorough investigation.

To illustrate the computational efficiency of $p_{sim,U}$, its performance is compared against the typical correlation coefficient and two state-of-the-art methods, namely, BRAID [84] and StatStream [76]. BRAID can handle data streams of semi-finite length, incrementally, quickly, and can estimate lag correlations with little error. On the other hand, StatStream resembles more the design principles of $p_{sim,U}$ by finding high correlations among sensor pairs based on DFTs and a three-level time interval hierarchy. Figure 8.13 compares the execution times of $p_{sim,U}$ with the aforementioned three alternatives, as a function of the window size. The results reveal a significant improvement in execution time achieved by $p_{sim,U}$, which is more prominent for higher window lengths. Most importantly, we observe that the execution time of $p_{sim,U}$ remains almost constant over the whole range of selected lengths, in contrast to the naive (corr) and BRAID methods, whose execution times increase rapidly for increasing window length.

The BRAID algorithm, for which we set the correlation lag to zero, is characterized by gradual increase for increasing window size, since it employs all the values in the observed time interval. On the other hand, StatStream is based on a simple hash function of the mean of each sensor window. Keeping the integer part of the means, the data windows are mapped to appropriate cells in a grid structure. Doing so, only the correlations between neighboring cells are computed. The increased

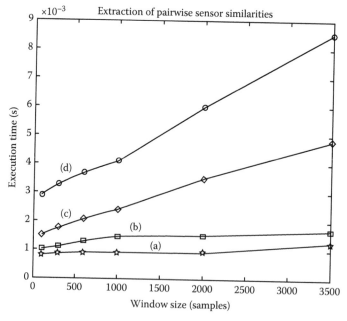

FIGURE 8.13 Comparison of execution times, as a function of the window size, between (a) ⋆, uncertainty-aware peak similarity ($p_{sim,U}$); (b) □, StatStream; (c) ◇, BRAID; and (d) ∘, correlation coefficient (corr).

execution time of StatStream, compared to $p_{sim,U}$, is due to the hash function, which involves more computations for the mapping. It is expected though that the performance of StatStream could be enhanced by designing a more efficient hash function.

8.6 A USE CASE SCENARIO: THE HYDROBIONETS PLATFORM FOR SEAWATER DESALINATION

The proposed framework, comprised of the techniques described in the previous sections, has been applied in an iCPS designed for the microbiological monitoring of water quality in industrial plants. The specific use case considers desalination plants and focuses on the procedure of reverse osmosis, which is a widely adopted technique across Europe and worldwide. Desalination by means of reverse osmosis relies on *osmotic membranes* that allow water to pass through at much higher rates than the dissolved salt.

The use of such membranes during seawater desalination suffers from the phenomenon of biofouling, which is related to the accumulation of unwanted bacterial matter on their surface. Biofouling is considered a very complex phenomenon that can be affected negatively by several variables, such as organic matter, pH, and temperature of feed water. As such, the combination of existing quantitative indices (e.g., temperature, conductivity, and pH) with novel sensing technology, capable of monitoring the presence and growth of bacteria in different locations of a desalination plant, is considered critical for the early detection of biofouling.

This has been the primary motivation of the HYDROBIONETS project [1] for the design and development of an iCPS platform responsible for the autonomous monitoring of the biofouling phenomenon in industrial desalination plants. The resulting platform, henceforth called the *HYDROBIONETS platform*, combines a multitier network architecture and novel wireless biofouling sensors [85], with the existing, wired sensing infrastructure for optimizing the cleaning and maintenance of the osmotic membranes, thereby increasing their lifetime.

The main components of the HYDROBIONETS platform, illustrated in Figure 8.14, include the following [86].

- The Wireless BioMEM Network (WBN) is comprised of miniaturized, computationally limited, and energy-autonomous sensor platforms that are responsible for monitoring the growth of biofouling bacteria at designated locations in the desalination plant. Each WBN

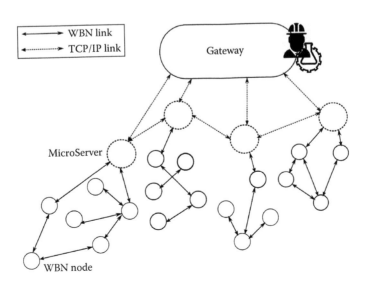

FIGURE 8.14 The architecture of the HYDROBIONETS platform.

node implements a sophisticated protocol stack that builds upon the scheduling mechanism described in Section 8.4.2, in order to effectively address the network and communication challenges often met in industrial environments.

- μServer devices are portable platforms with increased computational capabilities, which are responsible for the management of a cluster of WBN nodes in the industrial plant. Each μServer undertakes the network configuration, management, and adaptation mechanisms for its appointed cluster of WBN nodes, while implementing functionalities that cannot be computationally supported by the WBN nodes.
- Gateway is considered the endpoint between the WBN network and the existing infrastructure. It is the component of the platform from where the system administrator can interact with the biofouling sensor nodes, while allowing interconnectivity with the existing infrastructure of the industrial plant.

From the perspective of signal and data processing, the WBN nodes combine the characteristics of Tier 1 and Tier 2 of the proposed data-driven framework for iCPS architectures. As such, the WBN nodes located at the front-end sensing, and in coordination with their assigned μServer, are also capable of communicating with their peer components. Subsequently, the WBN nodes can directly handle the front-end signal modeling of the biofouling sensor data (see Section 8.3). The extraction in-network correlation (see Section 8.4) for the growth of bacteria at different locations can exploit both the direct links between WBN nodes, as well as the exchange of information between different μServers. Finally, analysis based on the U-HDMA frame (see Section 8.5) for both biofouling data, as well as relevant sensing indices, such as pH, temperature, and conductivity, is undertaken by the Gateway, which collects all information conveyed over the HYDROBIONETS platform.

8.6.1 Experimental Studies

The HYDROBIONETS platform has been deployed at a desalination pilot plant, located at La Tordera, Spain, and owned by Acciona Agua, which is a worldwide industrial leader in water treatment. Snapshots of the industrial plant are presented in Figure 8.15.

Driven by the small dimensions of the industrial plant, the HYDROBIONETS platform is comprised of 10 WBN nodes, assigned to one μServer and the Gateway. The WBN nodes are placed accordingly to the specifications of the end user, at three different locations in the plant, namely (a) the phase of reverse osmosis, (b) the seawater intake, and (c) the phase of pretreatment.

(a)   (b)   (c)

FIGURE 8.15 Different perspectives of the La Tordera desalination pilot plant: (a) reverse osmosis filters, (b) seawater intake, and (c) ultrafiltration tanks.

(a)   (b)

FIGURE 8.16 The technology employed for the realization of the WBN network at the pilot plant: (a) the CS5000-SMA mote [89] employed for the realization of the protocol stack on the WBN node; (b) the casing and the biofouling sensor cell of the WBN node.

The WBN protocol stack has been implemented in Contiki OS [87,88], featuring at the Routing Layer the IETF standard for low power and lossy networks. The resulting stack has been deployed on CM5000-SMA or XM1000 motes [89] (Figure 8.16a), integrated with the biofouling sensor over serial interface. Each biofouling sensor exploits an array of capacitive microelectrodes, which merge into the treated water. The concentration bacteria in the treated water changes permittivity of the microelectrodes, therefore modifying the magnitude of the capacitance. Subsequently, measuring the impedance of the sensor is considered sufficient for characterizing the phenomenon of bio-fouling [85]. A snapshot of the employed sensors mounted at the desalination plant is shown in Figure 8.16b.

While the qualitative characterization of the biofouling phenomenon based on impedance values remains in progress, data collected both by the wireless biofouling sensors as well as existing wired sensing infrastructure are employed for evaluating the efficacy of our proposed data-driven frame-work. In the remaining part of this section, the respective experimental results are presented and accompanying discussions are made.

8.6.2 IN-NETWORK DISTRIBUTED PROCESSING

An illustration of the distributed parameter estimation and field reconstruction based on iterative consensus, as described in Section 8.4.3, is presented in this section. In particular, the spatial field $s(x, y, t)$ to be estimated at time t and coordinates (x, y) is the superposed temperature originated by M distinct heat sources, each one emitting with a different time-varying power. The spatial diffusion of the heat coming from each individual source is modeled as a Cauchy bell. Doing so, the temperature at spatial location (x, y) and time t is given by

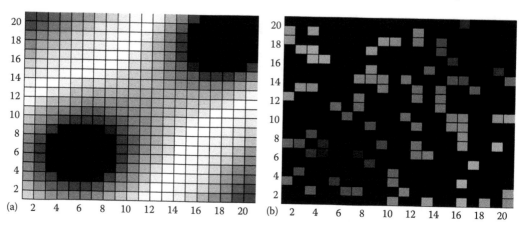

FIGURE 8.17 Temperature field at a given time instant formed by two heat sources located in a 20 m × 20 m square area, each emitting a heat power whose propagation is modeled as a Cauchy bell. (a) Temperature field to be estimated at a single time instant and (b) noisy observations of the nodes at a single time instant.

$$s(x, y, t) = \sum_{i=1}^{M} \frac{p_i(t)}{1 + ((x - x_i)^2 + (y - y_i)^2)/\beta_i},$$

where each source i is located at coordinates (x_i, y_i) and emits a time-varying heat power $p_i(t)$ with spread β_i. Note that this expression shows the value of the field $s(t)$ at each location (x, y) as a spatial distortion of the state vector $\mathbf{p}(t) = [p_1(t) \ldots p_M(t)]$ at (x, y), following a linear observation model. Figure 8.17a shows a snapshot of this field at a single time instant over a 20 m × 20 m grid with $M = 2$ heat sources located at coordinates $(20, 20)$ and $(60, 60)$ and emitting heat powers of $p_1 = 50$ and $p_2 = 20$, with spreads $\beta_1 = 20$ and $\beta_2 = 30$, respectively. In this example, no assumption is made regarding the evolution of the power $\mathbf{p}$ across time.

To monitor the temperature field, a set of $S = 100$ nodes are deployed randomly throughout the area. This is also in agreement with the nonuniform deployment adopted by the HYDROBIONETS platform, while the importance for monitoring temperature fields stems from the fact that temperature affects the operating conditions of the sensor network, as well as the industrial process.

Following our observation model given as Equation 8.10, data obtained by the nodes are corrupted by additive zero-mean Gaussian noise, that is, the measurement of node i at time t is given by $y_i(t) = s_i(t) + w_i(t)$, where $s_i(t) = s(x = x_i, y = y_i, t)$. A snapshot of the measurements of all nodes at the same time instant is shown in Figure 8.17b for noise variance $\sigma^2 = 30$. The communication range area of each node has a radius of 8 m. Since we have no knowledge about the evolution of the field, we cannot make any inference from the previous estimations. For this purpose, the optimal parameter estimator is computed at each time instant separately.

Based on the acquired noisy observations, the nodes start an iterative process to exchange information with the neighbors inside their range, as explained in Section 8.4.3. The final aim is to compute the maximum likelihood estimator of the field. The evolution of the MSE of the distributed estimator (solid line) along the iterations is shown in Figure 8.18. More specifically, the MSE between the original and estimated temperature values is computed for each sensor in each iteration, which is then averaged over all the $S = 100$ sensors. In this plot, the dashed line corresponds to the average MSE between the initial sensor measurements and the real values of the temperature field at the location of each sensor, whereas the dash-dotted line corresponds to the average MSE between the centralized (MLE) estimator and the real values of the temperature field. A comparison among the three curves reveals that as the nodes reach a consensus, the distributed estimation error decreases and approximates closely the error of the centralized estimator. Figure 8.19 visualizes the

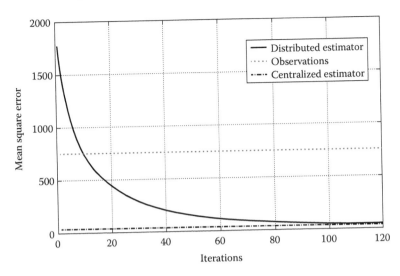

FIGURE 8.18   Evolution of the MSE of the distributed estimation of the field across the iterations of the consensus process.

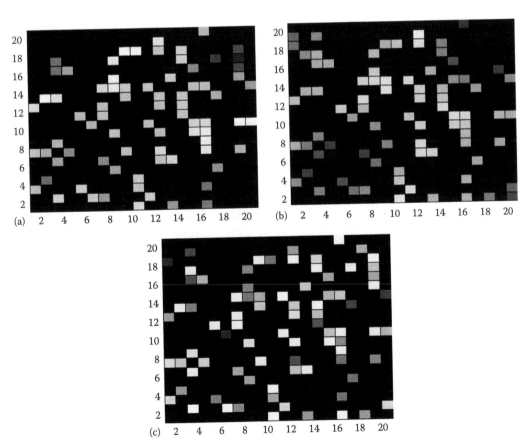

FIGURE 8.19   Successive screenshots of the iterative temperature field reconstruction until a consensus is reached ($S = 100$, $\sigma^2 = 30$). (a) Iteration 40; (b) Iteration 80; and (c) Iteration 120.

evolution of the distributed estimation process over the 2D temperature field, and across the iterations, by employing the noisy observations and the distributed exchange of information among the nodes.

8.6.3 iCPS Data Recovery via MC

To validate the merits of introducing intelligence into the signal acquisition and processing, we present a few indicative results from the testbed developed within the HYDROBIONETS platform. We consider a collection of five sensor nodes, each one capable of measuring impedance in 1 out of 10 frequency channels, leading to a total of 50 sensing units. Each measurement is communicated via the μServer and the Gateway and stored locally in a database. To generate the corresponding measurements matrix, one must select the temporal resolution of the sampling process, which dictates the number of rows of this matrix. For instance, assuming one measurement every hour, a measurements matrix corresponding to a single day will have 24 rows.

Figure 8.20 presents an indicative collection of impedance measurements acquired by a set of biofilm sensors over a period of 3 days with a temporal sampling rate equal to one measurements every 3 h (180 min resolution), along with the recovery performance of the measurements matrix from a subset of its entries. In Figure 8.20a, the existence of a spatiotemporal correlation can be observed in the data, while Figure 8.20b demonstrates the relationship between the recovery error and the sampling rate. Specifically, we observe that 25% of the measurements are sufficient in order to reduce the recovery error to less than 25% of the original signal's magnitude. Furthermore, it can be seen that by increasing the sampling rate beyond 40% has no effect on the reconstruction quality. This phenomenon is attributed to the noise that corrupts the data acquisition process and increases artificially the rank of the matrix, setting a lower bound on the recovery error.

Whereas missing entries are typically attributed to lost packets and node failures, this situation can also arise by increasing the temporal resolution of the sampling process. This can be easily observed in Figure 8.21a,c, where the same number of stored values as in Figure 8.20 is employed, however, at different temporal resolutions. More specifically, while in Figure 8.20 each entry corresponds to a period of 3 h, Figures 8.21a,c present the measurements matrix generated by considering one entry every 2 and 1 h, respectively. One can see that an increase of the temporal resolution will yield the introduction of zero-valued entries due to the lack of measurements for the predefined time intervals. Figure 8.21b,d present the estimated measurements matrices for the two cases, where zeros have been replaced with approximated values estimated via MC. While it is not possible to provide

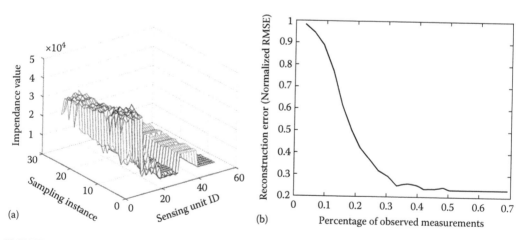

FIGURE 8.20 Impedance measurements of biofilm sensors: (a) complete set of measurements over a period of 3 days at a sampling rate of one measurement every 3 h; (b) reconstruction error for a given sampling rate. MC is able to achieve good performance even at sampling rates as low as 30% of the total measurements.

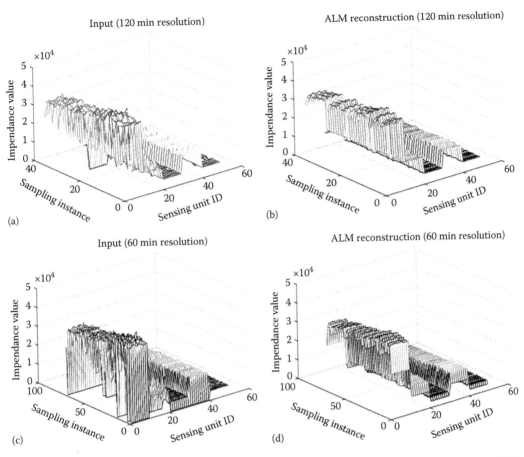

Input (120 min resolution)

ALM reconstruction (120 min resolution)

(a)

(b)

Input (60 min resolution)

ALM reconstruction (60 min resolution)

(c)

(d)

FIGURE 8.21 Illustration of recovery performance in the case of zero-valued measurements due to artificial increase of sampling rate. (a) and (c): Correspond to the same measurement period as in Figure 8.20a, albeit at higher sampling rates, leading to an increase in the number of missing entries. (b) and (d): Present the performance of MC in removing artificially introduced zero entries, and the output data which is both consistent with the measurements, as well as less abrupt due to the zeros.

a quantitative performance evaluation, since these values were never recorded in practice, from a visualization perspective it is much easier for a system operator to monitor the iCPS infrastructure conditions and perform the necessary actions.

8.6.4 HIGH-LEVEL SENSOR DATA ANALYSIS AND ALERTING

In the following, the performance of the U-HDMA system introduced in Section 8.5 is evaluated on a real data set in terms of managing the underlying data uncertainty and providing early warnings. In particular, the high-level analysis and early warning is performed for a data set provided by Acciona Agua, which consists of 22 sensors of several types (pressure, temperature, conductivity, turbidity, pH, flow, and redox) deployed in La Tordera's desalination plant. The corresponding measurements cover a period of 1 month (April 2–29, 2013) at a sampling rate of one measurement every three minutes. Full sensor specifications, such as sensor precision, sensitivity, and resolution, along with the corresponding measurements are provided for each individual sensor.

The inherent uncertainty of the acquired sensor data is estimated over sliding windows. If not stated explicitly otherwise, in the subsequent experimental evaluation the window size is set equal to 80 samples, which corresponds to a time interval of approximately 4 h, while the step size is fixed at 1 sample corresponding to a time-step of about 3 min. The expanded uncertainty is computed by fixing the coverage factor in Equation 8.27 at $k = 1.96$, which is equivalent to a 95% confidence level.

The performance of the spreadsheet-based approach (Reference 90) for estimating the underlying uncertainty in several distinct sensor streams is illustrated first. Figure 8.22 shows the estimated expanded uncertainty for four randomly chosen sensors in our data set. An additional potential,

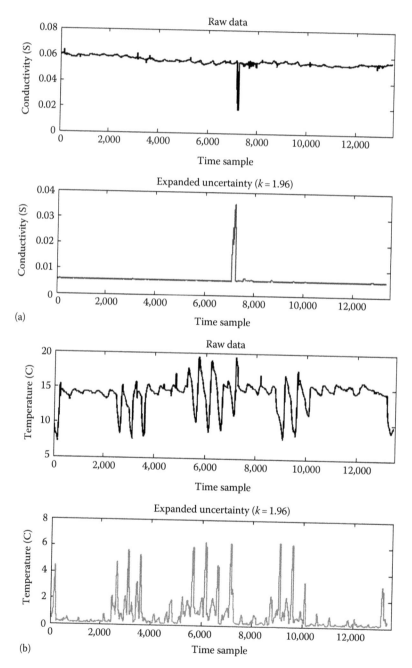

FIGURE 8.22 Raw data and estimated expanded uncertainties for four distinct electrochemical sensors: (a) conductivity and (b) temperature.

(*Continued*)

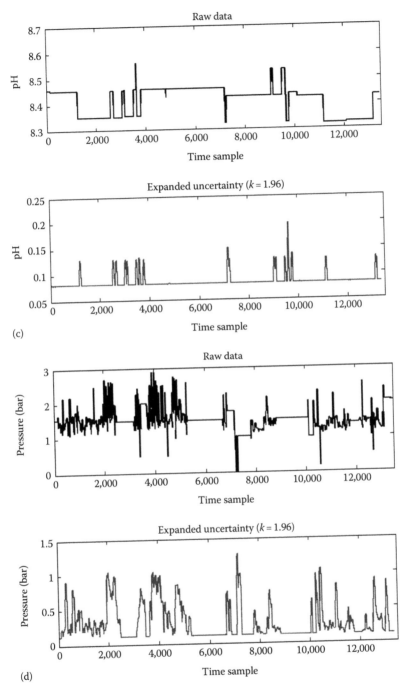

FIGURE 8.22 (Continued) Raw data and estimated expanded uncertainties for four distinct electrochemical sensors: (c) pH and (d) pressure (window size = 80, step size = 1, k = 1.96).

which is revealed by this figure, is the use of the estimated uncertainty as an indicator of abnormal behavior. Indeed, the time instants where the uncertainty presents a peak coincides with the time windows where the corresponding sensor measurements vary significantly compared to the previously recorded values. However, a more thorough study is required towards the design of an efficient extreme event detector based on expanded uncertainties.

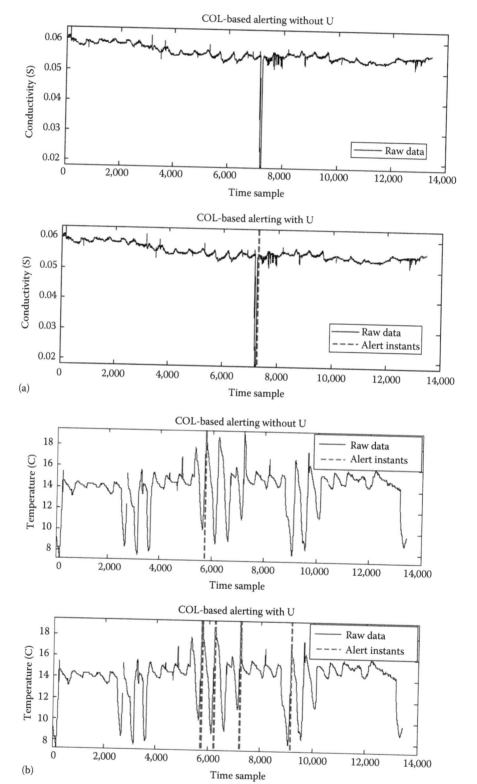

FIGURE 8.23 Raw data and identified alerting instants using the typical COL and the uncertainty-aware U-COL methods for: (a) conductivity sensor and (b) temperature sensor (window size = 80, step size = 1, $k = 1.96$, $N_c = 20$).

As a final illustration, we evaluate the performance of the uncertainty-aware extreme event detector, U-COL, introduced in Section 8.5.2. To this end, Figure 8.23 shows the identified instants (dashed vertical lines) for which an alerting notification is sent by the U-HDMA system for the conductivity and temperature sensors shown in Figure 8.22. More specifically, the typical COL and the uncertainty-aware U-COL methods are employed for early warning about an abnormal behavior in the acquired sensor data using the following rule: "an alerting notification is sent by the system if N_c consecutive measurements exceed a predefined operational upper limit". Notice that in the case of U-COL, the measurements are augmented with their corresponding estimated expanded uncertainty. In this example, we set $N_c = 20$, that is, the system operator is notified for a potential alert when the 20 most recent sensor measurements satisfy the above alerting rule. Besides, for demonstration purposes, for both sensors the upper limit is set to max{sensor data} $- 0.01 \cdot$ max{sensor data}, that is, 10% below the maximum recorded value. Clearly, accounting for the inherent data uncertainty improves the early warning performance, as it can be seen in Figure 8.23 for both sensors. Indeed, in both cases, U-COL is able to detect the occurrence of abnormal behavior in the sensor data, even if the recorded raw measurements do not strictly exceed the corresponding operational upper limit.

8.7 CONCLUSION AND FUTURE RESEARCH

In this chapter, the main architectural characteristics towards designing efficient data-driven industrial cyber-physical systems were analyzed. Furthermore, an integrated framework of signal and data processing techniques was presented, for treating different layers of information abstraction. By also accounting for the potential limitations and imperfections of the associated sensor network infrastructure employed to observe various physical parameters of the industrial environment, we focused on three major aspects, namely, (1) signal processing–driven performance optimization for industrial sensor networks, (2) in-network signal processing for distributed estimation and tracking of spatio-temporal fields, and (3) high-level analysis and early warning by employing the recorded iCPS data.

Along with a strong emphasis on providing the essential theoretical background, the effectiveness of the resulting framework was evaluated on a real iCPS. In particular, the signal and data processing techniques described herein were applied on real sensor data recorded by several distinct sensors deployed for monitoring a water desalination plants. Comparison with well-established and state-of-the-art methods for sensor data processing and distributed in-network inference revealed a superior performance of the integrated multitier iCPS architecture as described in this chapter.

Further extensions and enhancements, spanning the whole extent of the data processing chain in the iCPS setting presented herein, could yield additional improvements in the overall performance. Concerning the representation of industrial data based on the concepts of compressed sensing and matrix completion, as presented in Section 8.3, innovative signal processing and learning algorithms can provide elegant solutions to issues that hinder the efficacy of iCPS. However, the majority of work in this area still relies on deliberately introducing these algorithms in various stages of data acquisition, processing, and understanding. We expect that significantly more profound benefits can arise from the intelligent design and integration of these algorithms into the hardware platforms. Such designs will consider the end-to-end architectures, optimizing the overall performance of iCPS, instead of individual stages. Furthermore, by introducing domain-expert knowledge into the recovery process, we expect a surge in the number of iCPS applications due to the clear and measurable benefits that are associated with these methods.

Concerning the parameter and state estimation problems described in Section 8.4, linear observation and process models are assumed, along with zero-mean Gaussian noise. However, in practice, such assumptions are satisfied rarely. To this end, we have to focus on the design of distributed implementations for optimal parameter and state estimation in the general case of nonlinear observation models and non-Gaussian noise. Furthermore, a thorough investigation of the performance of random and asymmetric network topologies also has to be carried out.

Finally, the efficiency of high-level data analysis and early warning methods presented in Section 8.5 can be further improved in several directions. First, statistical dependencies among distinct sources of uncertainty usually exist in practice. Although it is often a very difficult task to quantify such dependencies in practice, however, an increased accuracy in estimating the underlying sensor data uncertainty is expected by employing sensitivity coefficients (see Equation 8.26), which better approximate the input–output interrelation for a given sensor. A second extension is related to the performance of the similarity function employed for measuring pairwise sensor correlations. Specifically, an incremental implementation of the DFT-based peak similarity function, given by Equation 8.31, as new sensor measurements are acquired, could further reduce its computational complexity, and subsequently the execution time, without sacrificing its effectiveness in identifying high correlated pairs of sensors. Moreover, the design of more sophisticated uncertainty-aware extreme event detectors, capable of simultaneously exploiting information even from heterogeneous, distinct sensors, could achieve a superior performance in terms of accurate detection of extreme events in different, yet correlated, sensor streams.

It is highly anticipated that the presented methods and the accompanying illustrations of real-life results will act as a fertile ground for further enhancements and adaptations to distinct use cases, as well as for yielding novel directions for iCPS standardization.

ACKNOWLEDGMENTS

This work is supported by the HYDROBIONETS Project (ICT-2011-7) funded by the European Commission in FP7 (GA-2011-287613) and the PEFYKA Project within the KRIPIS Action of the GSRT, Greece. We are also grateful to Acciona Agua* for providing the premises of La Tordera's desalination plant, as well as to Ateknea Solutions† and CNM (Centre Nacional de Microelectrònica)‡ for assisting in the collection of real biofouling data.

REFERENCES

1. The hydrobionets project: Autonomous control of large-scale water treatment plants based on self-organized wireless biomem sensor and actuator networks, http://www.hydrobionets.eu/.
2. F.-J. Wu, Y.-F. Kao, and Y.-C. Tseng, From wireless sensor networks towards cyber physical systems, *Pervasive and Mobile Computing*, 7 (4), 397–413, 2011.
3. T. S. Rappaport et al., *Wireless Communications: Principles and Practice*. Prentice Hall PTR, Upper Saddle River, NJ, 1996, vol. 2.
4. K. Remley, G. Koepke, C. Holloway, D. Camell, and C. Grosvenor, Measurements in harsh RF propagation environments to support performance evaluation of wireless sensor networks, *Sensor Review*, 29 (3), 211–222, 2009.
5. J. Ferrer Coll, RF Channel Characterization in Industrial, Hospital and Home Environments, Licentiate thesis, Stockholm: KTH Royal Institute of Technology, 2012., xiii, 65 p. (DiVA: diva2:478976).
6. I.-U.-H. Minhas, Wireless sensor network performance in high voltage and harsh industrial environments, Master's thesis, Blekinge Institute of Technology, School of Engineering, Blekinge, Sweden, 2012.
7. S. Ghadimi, J. Hussian, T. S. Sidhu, and S. Primak, Effect of impulse noise on wireless relay channel, *Wireless Sensor Network*, 4 (6), 167–172, 2012.
8. J. Hespanha, P. Naghshtabrizi, and Y. Xu, A survey of recent results in networked control systems, *Proceedings of the IEEE*, 95 (1), 138–162, January 2007.
9. K. J. Astrom and P. Kumar, Control: A perspective, *Automatica*, 50 (1), 3–43, 2014.
10. W. Zhang, M. S. Branicky, and S. M. Phillips, Stability of networked control systems, *IEEE Control Systems Magazine*, 21 (1), 84–99, February 2001.
11. *Wireless Hart*, The HART Communication Foundation Std. Available at: http://en.hartcomm.org, last accessed: June 2015.

---

* http://www.acciona-agua.com/.
† http://ateknea.com/.
‡ http://www.imb-cnm.csic.es/.

12. ISA100, *Wireless Systems for Automation*, International Society for Automation Std. https://www.isa.org/isa100/, last accessed: June 2015.
13. IEEE standard for local and metropolitan area networks—Part 15.4: Low-rate wireless personal area networks (lr-wpans), *IEEE* Std 802.15.4-2011 (Revision of IEEE Std 802.15.4-2006), pp. 1–314, September 2011.
14. X. Cao, P. Cheng, J. Chen, and Y. Sun, An online optimization approach for control and communication codesign in networked cyber-physical systems, *IEEE Transactions on Industrial Informatics*, 9 (1), 439–450, February 2013.
15. R. Alur, A. D'Innocenzo, K. Johansson, G. Pappas, and G. Weiss, Compositional modeling and analysis of multi-hop control networks, *IEEE Transactions on Automatic Control*, 56 (10), 2345–2357, October 2011.
16. V. Gupta, B. Hassibi, and R. M. Murray, Optimal {LQG} control across packet-dropping links, *Systems & Control Letters*, 56 (6), 439–446, 2007.
17. S. Deshmukh, B. Natarajan, and A. Pahwa, State estimation in spatially distributed cyber-physical systems: Bounds on critical measurement drop rates, in *2013 IEEE International Conference on Distributed Computing in Sensor Systems (DCOSS)*, Cambridge, MA, May 2013, pp. 157–164.
18. V. Gupta, A. Dana, J. Hespanha, R. Murray, and B. Hassibi, Data transmission over networks for estimation and control, *IEEE Transactions on, Automatic Control*, 54 (8), 1807–1819, August 2009.
19. M. Pajic, S. Sundaram, G. Pappas, and R. Mangharam, The wireless control network: A new approach for control over networks, *IEEE Transactions on Automatic Control*, 56 (10), 2305–2318, October 2011.
20. M. Pajic, S. Sundaram, J. Le Ny, G. J. Pappas, and R. Mangharam, Closing the loop: A simple distributed method for control over wireless networks, in *Proceedings of the 11th International Conference on Information Processing in Sensor Networks (IPSN'12)*. New York: ACM, 2012, pp. 25–36.
21. K.-D. Kim and P. Kumar, Cyber physical systems: A perspective at the centennial, *Proceedings of the IEEE*, 100, (Special Centennial Issue), 1287–1308, May 2012.
22. L. Sha, S. Gopalakrishnan, X. Liu, and Q. Wang, Cyber-physical systems: A new frontier, in *IEEE International Conference on Sensor Networks, Ubiquitous and Trustworthy Computing, 2008. SUTC '08.* Taichung, Taiwan, June 2008, pp. 1–9.
23. D. Donoho, Compressed sensing, *IEEE Transactions on Information Theory*, 52 (4), 1289–1306, 2006.
24. E. Candes, Y. Eldar, D. Needell, and P. Randall, Compressed sensing with coherent and redundant dictionaries, *Applied and Computational Harmonic Analysis*, 31 (1), 59–73, 2011.
25. R. Baraniuk, M. Davenport, R. DeVore, and M. Wakin, A simple proof of the restricted isometry property for random matrices, *Constructive Approximation*, 28 (3), 253–263, 2008.
26. J. Tropp and A. Gilbert, Signal recovery from random measurements via orthogonal matching pursuit, *IEEE Transactions on Information Theory*, 53 (12), 4655–4666, 2007.
27. R. Tibshirani, Regression shrinkage and selection via the LASSO, *Journal of the Royal Statistical Society. Series B (Methodological)*, 58, 267–288, 1996.
28. C. Johnson, Matrix completion problems: A survey, in *Proceedings of Symposia in Applied Mathematics*, vol. 40, 171–198, 1990.
29. E. Candès and B. Recht, Exact matrix completion via convex optimization, *Foundations of Computational Mathematics*, 9 (6), 717–772, 2009.
30. E. Candès and Y. Plan, Matrix completion with noise, *Proceedings of the IEEE*, 98 (6), 925–936, 2010.
31. B. Recht, M. Fazel, and P. Parrilo, Guaranteed minimum-rank solutions of linear matrix equations via nuclear norm minimization, *SIAM Review*, 52 (3), 471–501, 2010.
32. J. Cai, E. Candès, and Z. Shen, A singular value thresholding algorithm for matrix completion, *SIAM Journal on Optimization*, 20 (4), 1956–1982, 2010.
33. Z. Lin, R. Liu, and Z. Su, Linearized Alternating Direction Method with Adaptive Penalty for Low-Rank Representation, Advances in Neural Information Processing Systems (NIPS 2011), Granada, Spain, December 12–17, 2011, pp. 612–620.
34. R. Keshavan, A. Montanari, and S. Oh, Matrix completion from a few entries, *IEEE Transactions on Information Theory*, 56 (6), 2980–2998, 2010.
35. S. Pudlewski, A. Prasanna, and T. Melodia, Compressed-sensing-enabled video streaming for wireless multimedia sensor networks, *IEEE Transactions on Mobile Computing*, 11 (6), 1060–1072, June 2012.
36. A. Griffin and P. Tsakalides, Compressed sensing of audio signals using multiple sensors, *Reconstruction*, 3 (4), 5, 2007.

37. X. Yu, H. Zhao, L. Zhang, S. Wu, B. Krishnamachari, and V. O. Li, Cooperative sensing and compression in vehicular sensor networks for urban monitoring, in *2010 IEEE International Conference on Communications (ICC)*, Cape Town, South Africa, May 2010, pp. 1–5.
38. H. Mamaghanian, N. Khaled, D. Atienza, and P. Vandergheynst, Compressed sensing for real-time energy-efficient ECG compression on wireless body sensor nodes, *IEEE Transactions on Biomedical Engineering*, 58 (9), 2456–2466, September 2011.
39. J. Cheng, H. Jiang, X. Ma, L. Liu, L. Qian, C. Tian, and W. Liu, Efficient data collection with sampling in WSNs: Making use of matrix completion techniques, in *Global Telecommunications Conference (GLOBECOM 2010)*, Miami, FL, December 2010, pp. 1–5.
40. A. Majumdar and R. K. Ward, Increasing energy efficiency in sensor networks: blue noise sampling and non-convex matrix completion, *International Journal of Sensor Networks*, 9 (3), 158–169, 2011.
41. S. Li, L. D. Xu, and X. Wang, Compressed sensing signal and data acquisition in wireless sensor networks and Internet of Things, *IEEE Transactions on Industrial Informatics*, 9, no. (4), 2177–2186, November 2013.
42. F. Fazel, M. Fazel, and M. Stojanovic, Random access sensor networks: Field reconstruction from incomplete data, in *Information Theory and Applications Workshop (ITA)*, San Diego, CA, February 2012, pp. 300–305.
43. G. Tsagkatakis and P. Tsakalides, Dictionary based reconstruction and classification of randomly sampled sensor network data, in *2012 IEEE Seventh Sensor Array and Multichannel Signal Processing Workshop (SAM)*, Hoboken, NJ, June 2012, pp. 117–120.
44. A. Fragkiadakis, I. Askoxylakis, and E. Tragos, Joint compressed-sensing and matrix-completion for efficient data collection in WSNs, in *2013 IEEE 18th International Workshop on Computer Aided Modeling and Design of Communication Links and Networks (CAMAD)*, Berlin, Germany, September 2013, pp. 84–88.
45. J. Haupt, W. U. Bajwa, M. Rabbat, and R. Nowak, Compressed sensing for networked data, *Signal Processing Magazine, IEEE*, 25 (2), 92–101, 2008.
46. H. Hu and Z. Yang, Spatial correlation-based distributed compressed sensing in wireless sensor networks, in *2010 Sixth International Conference on Wireless Communications Networking and Mobile Computing (WiCOM)*, Chengdu, China, September 2010, pp. 1–4.
47. M. Sartipi and R. Fletcher, Energy-efficient data acquisition in wireless sensor networks using compressed sensing, in *Data Compression Conference (DCC)*, Snowbird, UT, March 2011, pp. 223–232.
48. Q. Ling and Z. Tian, Decentralized sparse signal recovery for compressive sleeping wireless sensor networks, *IEEE Transactions on Signal Processing*, 58 (7), 3816–3827, 2010.
49. F. Chen, A. P. Chandrakasan, and V. M. Stojanovic, Design and analysis of a hardware-efficient compressed sensing architecture for data compression in wireless sensors, *IEEE Journal of Solid-State Circuits*, 47 (3), 744–756, 2012.
50. L. Xiang, J. Luo, and A. Vasilakos, Compressed data aggregation for energy efficient wireless sensor networks, in *2011 Eighth Annual IEEE Communications Society Conference on Sensor, Mesh and Ad Hoc Communications and Networks (SECON)*, Salt Lake City, UT, June 2011, pp. 46–54.
51. C. Luo, F. Wu, J. Sun, and C. W. Chen, Compressive data gathering for large-scale wireless sensor networks, in *Proceedings of the 15th Annual International Conference on Mobile Computing and Networking*. ACM, Beijing, China, September 2009, pp. 145–156.
52. J. Cheng, Q. Ye, H. Jiang, D. Wang, and C. Wang, Stcdg: An efficient data gathering algorithm based on matrix completion for wireless sensor networks, *IEEE Transactions on Wireless Communications*, 12 (2), 850–861, 2013.
53. G. Quer, R. Masiero, D. Munaretto, M. Rossi, J. Widmer, and M. Zorzi, On the interplay between routing and signal representation for compressive sensing in wireless sensor networks, in *Information Theory and Applications Workshop, 2009*. San Diego, CA, February 2009, pp. 206–215.
54. C. Caione, D. Brunelli, and L. Benini, Distributed compressive sampling for lifetime optimization in dense wireless sensor networks, *IEEE Transactions on Industrial Informatics*, 8 (1), 30–40, 2012.
55. N. Nguyen, D. L. Jones, and S. Krishnamurthy, Netcompress: Coupling network coding and compressed sensing for efficient data communication in wireless sensor networks, in *2010 IEEE Workshop on Signal Processing Systems (SIPS)*, San Francisco, CA, October 2010, pp. 356–361.
56. A. Y. Alfakih, A. Khandani, and H. Wolkowicz, Solving Euclidean distance matrix completion problems via semidefinite programming, *Computational Optimization and Applications*, 12 (1-3), 13–30, 1999.
57. A. Javanmard and A. Montanari, Localization from incomplete noisy distance measurements, *Foundations of Computational Mathematics*, 13 (3), 297–345, 2013.

58. V. N. Ekambaram and K. Ramchandran, Non-line-of-sight localization using low-rank+ sparse matrix decomposition, in *2012 IEEE Statistical Signal Processing Workshop (SSP)*, Ann Arbor, MI, August 2012, pp. 317–320.

59. R. Rangarajan, R. Raich, and A. O. Hero, Euclidean matrix completion problems in tracking and geo-localization, in *IEEE International Conference on Acoustics, Speech and Signal Processing, (ICASSP)* Las Vegas, Nevada, March 30–April 4 2008, pp. 5324–5327.

60. D. Milioris, G. Tzagkarakis, A. Papakonstantinou, M. Papadopouli, and P. Tsakalides, Low-dimensional signal-strength fingerprint-based positioning in wireless LANs, *Ad Hoc Networks*, 12 (100–114), 2014.

61. S. Nikitaki, G. Tsagkatakis, and P. Tsakalides, Efficient recalibration via dynamic matrix completion, in *2013 IEEE International Workshop on Machine Learning for Signal Processing (MLSP)*, Southampton, UK, September 2013, pp. 1–6.

62. M. Mardani, G. Mateos, and G. B. Giannakis, Unveiling anomalies in large-scale networks via sparsity and low rank, in *2011 Conference Record of the 45th Asilomar Conference on Signals, Systems and Computers (ASILOMAR)*, Asilomar, Pacific Grove, CA, November 2011, pp. 403–407.

63. R. Paffenroth, P. Du Toit, R. Nong, L. Scharf, A. P. Jayasumana, and V. Bandara, Space-time signal processing for distributed pattern detection in sensor networks, *IEEE Journal of Selected Topics in Signal Processing*, 7 (1), 38–49, 2013.

64. B. Zhang, X. Cheng, N. Zhang, Y. Cui, Y. Li, and Q. Liang, Sparse target counting and localization in sensor networks based on compressive sensing, in *2011 Proceedings of the IEEE INFOCOM*. Shanghai, China, April 2011, pp. 2255–2263.

65. Q. Ling, Y. Xu, W. Yin, and Z. Wen, Decentralized low-rank matrix completion, in *2012 IEEE International Conference on Acoustics, Speech and Signal Processing (ICASSP)*, Kyoto, Japan, March 2012, pp. 2925–2928.

66. S. M. Kay, *Fundamentals of Statistical Signal Processing: Estimation Theory*. Prentice Hall, New York, 1993.

67. D. Simon, *Optimal State Estimation*. Wiley-Interscience, New York, 2006.

68. A. Tahbaz-Salehi and A. Jadbabaie, Consensus over ergodic stationary graph processes, *IEEE Transactions on Automatic Control*, 55 (1), 225–230, January 2010.

69. C. Asensio-Marco and B. Beferull-Lozano, Link scheduling in sensor networks for asymmetric average consensus, in *2012 IEEE 13th International Workshop on Signal Processing Advances in Wireless Communications (SPAWC)*, Cesme, Turkey, June 2012, pp. 319–323.

70. R. Olfati-Saber, Distributed kalman filter with embedded consensus filters, in *44th IEEE Conference on Decision and Control, 2005 and 2005 European Control Conference. CDC-ECC '05*, Seville, Spain, December 2005, pp. 8179–8184.

71. C. Aggarwal, Managing and Mining Uncertain Data. Advances in Database Systems, 35, Springer, New York, 2009.

72. T. Thanh, P. Liping, D. Yanlei, M. Andrew, and L. Anna, CLARO: Modeling and processing uncertain data streams, *The VLDB Journal*, 21 (5), 651–676, 2012.

73. M.-Y. Yeh, K.-L. Wu, P. Yu, and M.-S. Chen, PROUD: A probabilistic approach to processing similarity queries over uncertain data streams, in *Proceedings of the 12th International Conference on Extending Database Technology: Advances in Database Technology*. Saint Petersburg, Russia: ACM New York, 2009, pp. 684–695.

74. M. Datar, A. Gionis, P. Indyk, and R. Motwani, Maintaining stream statistics over sliding windows, in *Proceedings of the 13th Annual ACM-SIAM Symposium on Discrete Algorithms*. San Francisco, CA: SIAM, 2002, pp. 635–644.

75. J. Gehrke, F. Korn, and D. Srivastava, On computing correlated aggregates over continual data streams, in *Proceedings of the 2001 ACM SIGMOD International Conference on Management of Data*. Santa Barbara, CA: ACM New York, 2001, pp. 13–24.

76. Y. Zhu and D. Shasha, StatStream: Statistical monitoring of thousands of data streams in real time, in *Proceedings of the 28th International Conference on Very Large Data Bases*. Hong Kong, China: VLDB Endowment, 2002, pp. 358–369.

77. K. Ishikawa and J. Loftus, *Introduction to Quality Control*. Tokyo, Japan: 3A Corporation, 1990.

78. I. Farrance and R. Frenkel, Uncertainty of measurement: A review of the rules for calculating uncertainty components through functional relationships, *Clinical Biochemist Reviews*, 33(2), 49–75, 2012.

79. B. Taylor and C. Kuyatt, Guidelines for evaluating and expressing the uncertainty of NIST measurement results, NIST Technical Note 1297, 1994 (http://www.nist.gov/pml/pubs/tn1297/).

80. E. J. Gumbel, Statistics of Extremes. Courier Dover Publications, New York, 2004.

81. J. Pickands, Statistical inference using extreme order statistics, *The Annals of Statistics*, 3 (1), 119–131, 1975.

82. L. de Haan and A. Ferreira, Extreme Value Theory: An Introduction. Springer-Verlag, New York, 2006.

83. D. Rafiei and A. Mendelzon, Similarity-based queries for time series data, in *Proceedings of ACM SIGMOD International Conference on Management of Data*. Tucson, AZ: ACM New York, 1997, pp. 13–25.

84. Y. Sakurai, S. Papadimitriou, and C. Faloutsos, BRAID: Stream mining through group lag correlations, in *Proceedings of ACM SIGMOD International Conference on Management of Data*. Baltimore, MD: ACM New York, pp. 599–610.

85. D. 8.2, Hydrobionets: Demonstration activities of low-scale WBN test-bed, http://www.hydrobionets.eu/index.php/deliverables, Ateknea Solutions, Technical Report, 2014. December 10, 2014.

86. D. 4.2, Hydrobionets: Network protocol design, http://www.hydrobionets.eu/index.php/deliverables, Kungliga Tekniska Hoegskolan (KTH), Technical Report, 2013. December 10, 2014.

87. A. Dunkels, B. Gronvall, and T. Voigt, Contiki—A lightweight and flexible operating system for tiny networked sensors, in *29th Annual IEEE International Conference on Local Computer Networks,* (LCN 2004), Florida. November 2004, pp. 455–462.

88. The Contiki Operating System: Version 2.6, 2012. [Online]. Available: http://www.contiki-os.org/start.html. November 27, 2014.

89. Advanticsys wireless sensor modules, http://www.advanticsys.com/wiki/, 2013. November 27, 2014.

90. A. Seliniotaki, G. Tzagkarakis, V. Christofides, and P. Tsakalides, Stream correlation monitoring for uncertainty-aware data processing systems, in *Proceedings of the Fifth International Conference on Information, Intelligence, Systems and Applications (IISA'14)*. Chania, Greece, July 7–9, 2014.

9 Quality-Guaranteed Data Streaming in Resource-Constrained Cyber-Physical Systems

Kui Wu, Emad Soroush, and Jian Pei

CONTENTS

In many cyber-physical systems, the physical system is monitored continuously, and data streams are transmitted in a network environment. Due to limited communication bandwidths and other resource constraints, a critical and practical demand is to compress data streams online continuously with quality guarantee. Although many data compression and digital signal processing methods have been developed to reduce data volume, their superlinear time and more-than-constant space complexity prevent them from being applied directly on data streams, particularly over resource-constrained cyber-physical systems. This chapter presents methods to tackle the problem of online quality-guaranteed compression of data streams using fast linear approximation (i.e., using line segments to approximate a time series). Technically, two versions of the problem are investigated to explore quality guarantees in different forms. Online algorithms with linear time complexity and constant cost in space are introduced. The presented algorithms are optimal in the sense that they generate the minimum number of segments that approximate a time series with the required quality guarantee. To meet the resource constraints, a fast algorithm is developed that creates connected segments with very simple computation. The low-cost nature of the methods in this chapter results in a unique practical advantage in the applications of massive streaming environment, low-bandwidth networks, and heavily constrained nodes in computational power (e.g., tiny sensor nodes). The algorithms are implemented and evaluated in a real-world test bed and with real-world trace data.

9.1   INTRODUCTION

A CPS [24] is an engineered system that monitors, coordinates, and controls the physical system with sensing, computing, and communication technologies. A CPS normally consists of three core tightly integrated functional modules: communication, computation, and control. Embedded sensors monitor the physical system and communicate the sensing data to the computation module, often via wireless networks; the computation module makes control decisions and sends back signals to the control module to control the physical system. Feedback loops are formed, as changes in physical processes in turn affect the computation. Clearly, information, that is, sensing data and control signals, is the key to smoothly orchestrate the three functional modules. It is critical to guarantee the quality of information for the success of any CPS.

There are diverse types of CPS, ranging from the nanoworld to large-scale wide-area systems [18]. Examples of CPS include medical devices and systems, aerospace systems, manufacture automation, building and environmental control, smart grid, intelligent farming, and robotic systems. While in many cases the CPS has abundant computation and communication resources, a large class of CPS is severely constrained with very limited computing power and/or communication bandwidth. For instance, to save cost and energy, many CPS use low-cost, low-power microprocessors only. In some cases, the resource constraints are posed by the special applications. One such example is virtual fence [3] used for intelligent farming. Typical physical fences are used to keep the animals within a boundary. In the virtual fencing application, however, animal movements are tracked and are controlled by stimuli, such as auditory and mild electric shocks, applied by tiny devices worn by the animals. Due to ethical considerations, the devices worn by the animals must pose negligible inconvenience to animals and thus must be extremely small and lightweight. This special requirement excludes the use of any large devices. The system should be built with resource-constrained tiny devices only. Due to the limited bandwidth and storage size, quality guarantee in information delivery is particularly challenging in this application context. As such, this chapter only focuses on cyber-physical systems that are severely subject to resource constraints.

Substantial research efforts have been devoted to tackling the problem of quantity-guaranteed data transmission in resource-constrained CPS. One solution is to utilize the temporal and spatial correlation in sensing data to reduce the amount of data transmission. The works in References [6, 19,30,34] all belong to this category. Another solution is to utilize the special features in the sensing data for better compressing the data. One broadly used feature is the sparsity of signals, that is, signals after transform such as DCTs, wavelets (Haar, Daubechies, Symlets), and Fourier transforms consist of only a small number of coefficients in the transformed space that have large magnitude [4]. The technique of compressive sensing takes advantage of the sparsity of signals and can greatly reduce the communication overhead and save energy in resource-constrained CPS [25]. Unlike existing work that depends on temporal/spatial correlation and the special features in sensing signals, this chapter tries to address the problem from a different angle by focusing only on the basic building block, data streaming from a single node to the information sink. It presents solutions that are general enough for virtually all CPS applications.

The demand for continuous data streaming with resource-constrained devices can be seen in many applications. The following two examples motivate the research presented in this chapter.

Example 9.1:

In an acoustic monitoring system using wireless sensor networks, sensor nodes are deployed in a target area, while each node contains an acoustic sensor that samples sound signals continuously. The sensor nodes are connected by a wireless network. The acoustic monitoring system has many applications. An appealing scenario is toward "smart conference hall." By analyzing the data collected from an acoustic monitoring system deployed in a large conference setting, people can identify and locate speakers as well as some of their activities. The information can be used to adjust the equipment such as the lighting system, the microphone system, the video monitoring system, and the air conditioning system. Another potential application is bird surveillance in the wilderness. By analyzing bird sounds collected using such a sensor network, ornithologists can study the distribution of birds and their behavior patterns.

To save costs, wireless sensor nodes that integrate sensors, processors, memory, and wireless transceivers often are small and have only very limited computational power and communication bandwidth [32]. For instance, the Chipcon radio chip in the broadly used MICA2 motes [7] has the maximum transmission power of 27 mA and the maximum bandwidth of 38 kbps.

When MICA2 motes are used in such acoustic monitoring system, many challenges arise. One technical challenge is that, although a sensor can sample the acoustic signals frequently, the acoustic data stream cannot be sent out in time due to the low-bandwidth radio channel. Specifically, in order to make the data analysis useful, it is needed to sample human voice with the normal sampling rate of 8 kHZ and 16 bits per sample. This sampling mode requires the bandwidth of 128 kbps for 1 channel (mono) voice, which greatly exceeds the maximum bandwidth of 38 kbps that an MICA2 mote can support. In addition, a large number of samples cannot be stored since the memory size of MICA2 motes is only 512 kb. The only technical solution to the bottleneck is to online compress data streams continuously and send out the compressed streams instead of the original streams through the network. Sending compressed streams can also reduce the power consumption of sensors on communication and thus extend the lifetime of sensors. In large environmental surveillance sensor networks, recharging or replacing batteries of sensor nodes is often very difficult or even impossible after the sensors are deployed. □

Example 9.2:

Animal tracking with lightweight devices has seen great popularity in many scenarios. ZebraNet [35] uses GPS technology to record fine-grained position data in order to track long-term animal migrations. ZebraNet consists of sensor nodes built into collars on zebras, which track zebras' positions. The hardware used for this purpose must be lightweight and self-sustainable in power supply [35]. A similar idea has been applied in virtual fencing applications [3], where animal movements are tracked with GPS and controlled with stimuli. Stimuli are presented to animals that cross a predefined virtual fence boundary. Virtual fences can be very useful in intelligent farming and can save substantial costs when cows or goats roam in a large rural area. In both ZebraNet and virtual fencing applications, the location of animals are required to be continuously transmitted, via one-hop or multihop radios, to an information center. □

Many data compression and digital signal processing methods have been developed to reduce data volume, such as Fourier transforms [1,12,15,33], discrete cosine transforms [27], wavelets [5,9,22,29], and linear predictive coding [2]. However, these methods may not be suitable to data stream compression in the aforementioned two examples due to the high cost of these methods in time and space. Moreover, sensor nodes like MICA2 motes only have very limited computational power. For example, only simple arithmetic operations are supported by TinyOS [8], the operating system (OS) for MICA2 motes. Although it is possible to implement a mathematical module to calculate essential functions like sinusoid and exponential functions or use dedicated DSP chips for audio processing and compression, such complex modules are highly undesirable due to the limited memory size and computational capacity of MICA2 motes as well as the extra energy cost of dedicated DSP chips.

In this chapter, the problem of online compression of data streams in resource-constrained CPS is tackled. Particularly, the fast linear approximation methods (i.e., using line segments to approximate a time series) with quality guarantee are designed. The technical content of the chapter can be summarized as follows.

First, the piecewise linear approximation (PLA) problem is modeled properly for data streams. Different from the conventional situations where the whole time series to be compressed and the required compression rate can be specified, a data stream is potentially unlimited, and the distribution is often unpredictable. The error-bounded PLA problem is proposed to tackle those challenges. *Second*, fast online solutions with linear time complexity and constant cost in space are presented. The algorithms are optimal in the number of segments used to approximate a (potentially unlimited) time series. In other words, the algorithms create the minimal number of line segments *even without knowing the future incoming data*. Moreover, the algorithms have an approximation factor of 2 to the maximum compression ratio using PLA. *Third*, to address the computational challenges in sensor nodes, another online approximation algorithm is developed that is particularly tailored for tiny sensor devices by requiring only very simple computation. The low-cost nature of the methods leads to a unique advantage in the applications of massive and fast streaming environment, low-bandwidth networks, and heavily constrained nodes in computational power (e.g., tiny sensor nodes). *Fourth*, the algorithms are further extended to approximate multidimensional time series. Fifth, all the algorithms are implemented and evaluated in the application of an acoustic wireless sensor network. The empirical evaluation clearly shows that the methods presented in this chapter are highly feasible for resource-constrained CPS applications.

The remainder of the chapter is organized as follows. In Section 9.2, the problem is formulated and analyzed, and the related work is reviewed. Two online algorithms are developed in Section 9.3, and their optimality is studied in Section 9.4. In Section 9.5, an online approximation algorithm is designed and is more economic in computation for tiny sensors. In Section 9.6, the algorithms are extended to approximate multidimensional time series. Section 9.7 includes the implementation and evaluation of the proposed methods in an acoustic wireless sensor network. The chapter is concluded in Section 9.8.

9.2 PROBLEM DEFINITION AND RELATED WORK

9.2.1 PROBLEM FORMULATION

PLA is an effective method to compress a time series. A numeric data stream can be treated as a potentially unlimited time series. Thus, it is natural to explore the possibility of compressing a numeric data stream with the PLA method.

Let $X = x_1 \cdots x_n$ be a time series of n points, and x_i $(1 \leq i \leq n)$ be the value of the ith point of X. A *(line) segment* is a tuple $s = ((i, y_i), (j, y_j))$ where $i < j$ and (i, y_i) and (j, y_j) are two endpoints. $[i, j]$ is called the *range* of s.

Given a time series X, PLA uses a set of line segments as the approximation of the time series. Figure 9.1 elaborates on the general idea where three line segments AA', BB', and CC' are used to approximate a time series. A line segment $s = ((i, y_i), (j, y_j))$ approximates the kth point $(i \leq k \leq j)$ of the time series by value

$$\tilde{x}_k = y_i + \frac{k - i}{j - i}(y_j - y_i).$$

The compression comes from the number of line segments used for approximation that can be much smaller than the number of points in the time series. In the figure, the time series has 18 points.

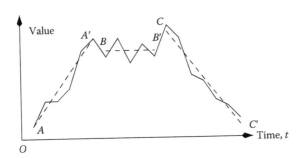

FIGURE 9.1   Piecewise linear approximation.

Three segments are used to approximate the time series, and each segment has two endpoints. Thus, the three line segments only need 6 points to represent. A compression ratio of 3 is achieved. Generally, the endpoints in the segments are not necessarily positioned at some points in the time series (e.g., B, B', and C' in the figure).

Formally, a set of segments $\tilde{X} = \{s_1, \ldots, s_m\}$ is a PLA of X if (1) $s_1, \ldots, s_m$ are segments; and (2) for each index i ($1 \le i \le n$), either i is in the range of exactly one segment in $\tilde{X}$ or there exist two segments $s, s' \in \tilde{X}$ such that s and s' share the same endpoint at index i. Clearly, using the segments, for every index i, $\tilde{X}$ can give a value $\tilde{x}_i$ to approximate x_i.

PLA for static time series has been well studied (e.g., [11,13,17,23]). Most of the previous studies address an optimization problem as follows:

Problem 9.1 (Conventional PLA problem): *Given a time series X of n points and a number $m < n$, find a set of m segments as a PLA of X such that the approximation error is minimized.* ∎

Unfortunately, solutions to the conventional PLA problem are not applicable to data streams. A data stream is potentially unlimited. It is impossible to know in advance the number of points in the stream or to specify the number of segments to be used for approximation. To tackle the stream compression problem, this chapter turns to the *error-bounded PLA problem*.

Problem 9.2 (Error-bounded PLA problem): *Given an error measurement function err() such that err$(X, \tilde{X})$ gives the error that a PLA $\tilde{X}$ approximates X, let ϵ be a user-specified error bound. $\tilde{X}$ is called an ϵ-PLA of X if err$(X, \tilde{X}) \le \epsilon$. An ϵ-PLA $\tilde{X}$ of X is optimal if $|\tilde{X}|$ (i.e., the number of segments in $\tilde{X}$) is minimized.* ∎

Two error measurement functions meaningful for data streams are proposed as follows:
First, the *maxerr* function captures the maximal error between X and $\tilde{X}$ at any index. That is,

$$maxerr(X, \tilde{X}) = \max_{i=1}^{n}\{|x_i - \tilde{x}_i|\}$$

With potentially unlimited streams, the *maxerr* function can guarantee that the approximation quality is consistently bounded at every point.

Second, the *segerr* function checks the error introduced by each segment and captures the maximal error. That is,

$$segerr(X, \tilde{X}) = \max_{s \in \tilde{X}}\left\{\sum_{i \in range(s)} (x_i - \tilde{x}_i)^2\right\}$$

Using the *segerr* function, the error introduced by every segment is bounded.

Using the two error measurement functions, two versions of the error-bounded PLA problem can be formulated.

Problem 9.3 (PLA-PointBound problem): *Given an error-bound ϵ, the PLA-PointBound problem is to find an ϵ-PLA $\tilde{X}$ such that $maxerr(X,\tilde{X}) \le \epsilon$ and $|\tilde{X}|$ is minimized.* ∎

Problem 9.4 (PLA-SegmentBound problem): *Given an error-bound ϵ, the PLA-SegmentBound problem is to find an ϵ-PLA $\tilde{X}$ such that $segerr(X,\tilde{X}) \le \epsilon$ and $|\tilde{X}|$ is minimized.* ∎

It is worth noting that different versions of the ϵ-PLA problem require different solutions. Generally, the PLA-PointBound problem composes a stronger quality requirement than the PLA-SegmentBound problem, and thus the PLA-PointBound problem is more challenging.

A good solution to one problem does not necessarily result in a good solution to the other. For instance, suppose that we have an effective algorithm to solve the PLA-PointBound problem. One may wonder whether we can translate the PLA-SegmentBound problem to the PLA-PointBound problem by setting the error bound (of each point) to $\sqrt{\frac{\epsilon}{n}}$, where n is the number of points approximated by a line segment.

Unfortunately, it is hard to obtain the number of points in a segment. The segments may have different lengths. Moreover, a segment in the PLA-SegmentBound problem might approximate most points well, but permits several exceptional points to have large errors. A naïve translation of the PLA-SegmentBound problem to the PLA-PointBound problem using a uniform error bound on all points excludes such segments. Similarly, the PLA-PointBound problem cannot be easily reduced to the PLA-SegmentBound problem.

9.2.2 RELATED WORK

PLA has been well investigated in References [10,16,17,23,26,31]. The idea behind PLA comes from the fact that a sequence of line segments can be used to represent the time series while preserving a low approximation error. A standard linear regression technique is widely used in most existing PLA algorithms to calculate a line segment approximating the original data with the minimum mean squared error. Many of them target at solving the conventional PLA problem and may not be applicable to streaming data [11,13,17,23].

Despite the substantial research efforts in PLA techniques [11,13,16,17,21,23], existing solutions are not tailored for data streams over resource-constrained sensor networks. They either require complex computation or have high cost in space.

Liu, Wu, and Pei [19] use PLA to estimate a time series. But they do not study the problem for streaming data. Also, the authors put unnecessary constraints on the algorithm, which requires the endpoints to come from the original data set. On the whole, their algorithm can run in $O(n^2 \log n)$ time complexity and takes $O(n)$ space complexity.

In [16], Keogh et al. give a comprehensive review on the existing techniques for segmenting time series. They categorize the solutions into three different groups, namely, sliding window methods, top-down methods, and bottom-up methods. They then take advantage of both sliding window and bottom-up (SWAB) methods and design a SWAB algorithm. The SWAB algorithm uses a moving window to constrain a time period in consideration.

In Palpanas et al. [21], an amnesic function is introduced to give weights to different points in the time series. The PLA-SegmentBound problem is discussed in the context of the Unrestricted Window with Absolute Amnesic problem, but complete solutions to this problem are not provided in Palpanas et al. [21].

A solution to the PLA-PointBound problem is addressed in Manis, Papakonstantinou, and Tsanakas [20] with a different definition of point error bound. The algorithm is claimed to be optimal, but the time complexity is $O(n^3)$, where n is the number of points in the time series. Moreover, no performance evaluation of the solution is presented in the chapter.

In summary, although the error-bounded PLA problem has been investigated before, the problem has not been studied systematically. No solutions applicable to data streams have been developed, let alone solutions for resource-constrained sensor networks.

9.3 ONLINE ALGORITHMS

In this section, two online algorithms are developed to solve the PLA-PointBound and the PLA-SegmentBound problems, respectively. The two algorithms share the same framework.

9.3.1 THE FRAMEWORK

The framework of the algorithms works in a greedy manner. When x_1, the first point in the stream, arrives, it stored in memory. When x_2 arrives, it is also stored since x_1 and x_2 can be compressed by a segment exactly. When x_3 arrives, it needs to check whether x_3 can be compressed together with x_1 and x_2 by a line segment satisfying the error-bound requirement. If so, x_3 is stored. Otherwise, it outputs a line segment compressing x_1 and x_2, removes x_1 and x_2 from the main memory, and stores x_3.

Generally, imagine we have a buffer in main memory storing points $x_i, x_{i+1}, \ldots, x_j$ such that the points in the buffer can be compressed by a line segment satisfying the error-bound requirement. When a new point x_{j+1} arrives, we check whether x_{j+1} can be compressed together with $x_i, \ldots, x_j$ by a line segment satisfying the error-bound requirement. If so, we add x_{j+1} to the buffer and move on to the next point. Otherwise, we output a segment compressing $x_1, \ldots, x_j$ satisfying the error-bound requirement and remove them from the buffer. x_{j+1} is then stored in the buffer.

Although the framework is simple, there are two critical issues that need to be solved carefully in order to make sure that the runtime of the algorithms is linear with respect to the number of points in the streams, and the space size needed by the algorithms is bounded by a constant.

First, how can we store the information about the points we have seen but have not compressed? In the worst case, there can be an unlimited number of such points (e.g., a time series where all points take the same value). How can we summarize them using only constant size memory?

Second, how can we determine whether a newly arrived point can be compressed together with the points already in the buffer that have been seen but have not been compressed? Revisiting those points one by one leads to the runtime quadratic with respect to the number of such points. As explained earlier, there can be an unlimited number of such points. The overall time complexity is quadratic if those points are revisited one by one.

The central idea to tackling the earlier two challenges is the following. Instead of storing the points explicitly, we monitor the range of all possible line segments that can be used to compress the points that have been seen but have not been compressed in a concise way. When a new point arrives, we can check whether the point can be compressed using some line segments in the range. If so, it means that the new point can be compressed together with the points accumulated. We only need to adjust the range of the possible line segments to make sure the new point is also compressed. If not, it means that the new point cannot be compressed together with the points accumulated. A segment should be output.

The remainder of this section introduces the techniques to implement the previous idea for the PLA-PointBound problem and the PLA-SegmentBound problem, respectively.

9.3.2 SOLVING THE PLA-POINTBOUND PROBLEM

A segment $s = ((i, y_i), (j, y_j))$ can also be represented by the left endpoint (i, y_i), the slope $m = \frac{y_j - y_i}{j - i}$, and the index of the right endpoint j.

For two points x_i and x_j in a data stream, if a line segment $s = ((i, y_i), (j, y_j))$ with slope $m = \frac{y_j - y_i}{j-i}$ can approximate x_i and x_j, that is, $|x_i - \tilde{x}_i| \leq \epsilon$ and $|x_j - \tilde{x}_j| \leq \epsilon$ where ϵ is the error bound, s must satisfy the following four conditions:

$$(x_i - \epsilon) \leq y_i \leq (x_i + \epsilon) \tag{9.1}$$

$$m_1 = \frac{(x_j + \epsilon) - y_i}{j - i} \tag{9.2}$$

$$m_2 = \frac{(x_j - \epsilon) - y_i}{j - i} \tag{9.3}$$

$$m_2 \leq m \leq m_1 \tag{9.4}$$

Figure 9.2 illustrates the conditions and their relations. Particularly, m_1 and m_2 are the slopes of the two lines shown in the figure.

Since the line segments are determined by the value of the left endpoint y_i and slope m, we examine the distribution of points (y_i, m) that satisfy Equations 9.1 through 9.4. As illustrated in Figure 9.3, the possible line segments form a polygon $poly(i,j)$.

Lemma 9.1 (PLA-PointBound) *A line segment of left endpoint y_i and slope m can approximate points $x_i, \ldots, x_j$ with maxerr at most ϵ if and only if (y_i, m) is in polygon $poly(i, i + 1) \cap poly(i, i + 2) \cap \cdots \cap poly(i, j)$.*

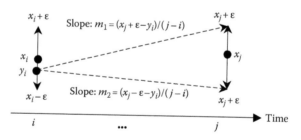

FIGURE 9.2   Ranges of possible line segments.

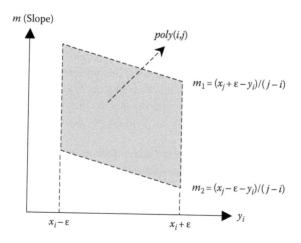

FIGURE 9.3   Polygon $poly(i,j)$.

Proof: The necessity follows with the definition of $poly(i,j)$. For any line segment $s \notin poly$ $(i, i+1) \cap poly(i, i+2) \cap \cdots \cap poly(i,j)$, there exists an index k $(i \le k \le j)$ such that $s \notin poly(i,k)$, that is, s cannot approximate either x_i or x_k.

The sufficiency can be proved by contradiction. Suppose a segment $s \in poly(i, i+1) \cap poly$ $(i, i+2) \cap \cdots \cap poly(i,j)$ but s cannot approximate x_k $(i \le k \le j)$. Two situations may arise. First, $k = i$. Then, $s \notin poly(i, i+1)$ since $|x_i - y_i| > \epsilon$ where y_i is the value of s on index i. Second, $k \ne i$. Then, $s \notin poly(i,k)$. In both cases, we have contradictions. ∎

Based on Lemma 9.1, algorithm PointBound, an online algorithm, is presented in Figure 9.4. The algorithm maintains the intersection of polygons $poly(i, i+1), \ldots, poly(i,j)$, where x_i is the first point that has not been compressed yet in the data stream, and x_j is the last point arrived such that $poly(i, i+1) \cap \cdots \cap poly(i,j) \ne \emptyset$.

When a new point x_{j+1} arrives, we compute $poly(i, j+1)$ and $poly(i, i+1) \cap \cdots \cap poly(i,j) \cap poly(i, j+1)$. If it is $\emptyset$, then a line segment s is randomly chosen to approximate $x_i, \ldots, x_j$ such that (y_i, m) is in $poly(i, i+1) \cap \cdots \cap poly(i,j)$, where y_i is the value of s on index i, and m is the slope of s. s is output, and the intersection of polygon is removed. x_{j+1} and x_{j+2} are used to generate a new polygon $poly(j+1, j+2)$.

If $poly(i, i+1) \cap \cdots \cap poly(i,j) \cap poly(i, j+1) \ne \emptyset$, then the intersection is kept, and the algorithm moves on to the next point in the stream.

For any i and j, $poly(i,j)$ is a parallelogram where there are two edges parallel to the slope axis. It is easy to show that for any i and j, $\cap_{k=i}^{j} poly(i,k)$ is a convex polygon. In the worst case, the edges of the intersection of parallelograms could be up to $2(j - i + 1)$, that is, twice the number of parallelograms intersected. A straightforward method for keeping all edges of the intersection area still has the quadratic time complexity and linear space complexity, which are not applicable to data streams.

Fortunately, we do not need to record all edges of the intersection polygon. Instead, we need to record only up to four edges to determine whether a new point can be compressed together with the points seen but not compressed.

Using Equations 9.1 through 9.4, it is easy to see that each parallelogram has two properties.

First, each parallelogram has two vertical edges and two sloping edges with a negative slope value, as shown in Figure 9.3. The range of y_i is the same for all parallelograms (i.e., $x_i - \epsilon \le y_i \le x_i + \epsilon$).

Second, for $j_2 > j_1 > i$, the absolute slope value of the two sloping edges in $poly(i, j_2)$ is strictly smaller than the absolute slope value of the two sloping edges in $poly(i, j_1)$.

Input: A data stream $X = x_1, x_2, \ldots$ and error-bound ϵ;
Output: A list of line segments $\tilde{X}$ approximating X
      such that $maxerr(X, \tilde{X})) \le \epsilon$;
Method:
```
1:   P = poly(1, 2); i = 1; j = 3;
2:   WHILE (1) DO {
3:       P' = P ∩ poly(i, j);
4:       IF P' ≠ ∅ THEN P = P', j = j + 1;
5:       ELSE {
6:           randomly choose a point (y, m) in P;
7:           output a line segment
                 ((i, y), (j − 1, y + (j − 1 − i) * m));
8:           P = poly(j, j + 1); i = j; j = j + 2;
         }
     }
```

FIGURE 9.4 PointBound, an online algorithm for the PLA-PointBound problem.

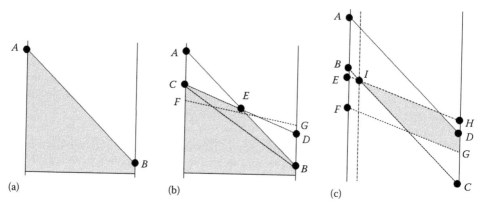

FIGURE 9.5 Using up to four edges to represent the intersection polygon. (a) The upper sloping edge *AB*. (b) The upper sloping edge *CD* cutting *AB* into two parts. (c) Parallelogram *ABCD* intersects with parallelogram *EFGH*.

To keep the discussion simple, let us focus on the intersection points of the upper sloping edge of parallelograms. The case for the lower sloping edges can be analyzed similarly.

The situations are illustrated in Figure 9.5. Suppose that the first parallelogram gives the upper sloping edge *AB* with slope value m_{AB} as in Figure 9.5a. When a new data point arrives, a new parallelogram is formed. In the worst case, the upper sloping edge of the parallelogram *CD* cuts *AB* into two parts. Let *E* be the intersection point between *AB* and *CD*, as shown in Figure 9.5b.

By the second property, we have $|m_{CD}| < |m_{AB}|$. Moreover, the upper sloping edge *FG* of any future parallelogram cannot cut both *CE* and *EB* due to the smaller absolute slope value of *FG* than m_{CD}. In other words, if a future parallelogram intersects with the current intersection polygon, the upper sloping edge of the parallelogram can only cut either *CE*, *EB*, or the right vertical edge. Instead of keeping *CE* and *EB*, we can keep line segment *CB*. Then, a future parallelogram intersects with the current intersection polygon if and only if it cuts *CB*.

Generally, we only need to keep the line segment connecting the leftmost upper corner and the rightmost upper corner for the upper sloping edges. Similarly, we only need to keep the line segment connecting the leftmost lower corner and the rightmost lower corner for the lower sloping edges.

In addition to these two line segments, we need to keep the two vertical edges in the intersection polygon. The reason is that the intersection of two parallelograms may shrink the range of the intersection, as illustrated in Figure 9.5c, where parallelogram *ABCD* intersects with parallelogram *EFGH*. The left vertical edge is shrunk into point *I* right to the original edge.

In summary, only up to four edges are needed to determine whether a new point can be compressed together with the points seen but not compressed. This immediately leads to the following result:

Theorem 9.1 (Complexity—PointBound) *The algorithm PointBound for the PLA-PointBound problem has the time complexity $O(n)$ and the space complexity $O(1)$, where n is the number of points in a time series to be compressed.* ∎

Since algorithm PointBound only looks ahead for one point in the data stream to output a line segment whenever necessary in the PLA, it is an online algorithm and can be applied on data streams.

9.3.3 SOLVING THE PLA-SEGMENTBOUND PROBLEM

A similar result as follows has been reported in Qu, Wang, and Wang [23] without proof.

Lemma 9.2 *Suppose that a line segment s approximates a fragment X of n points $x_1, \ldots, x_n$ in a time series. Then, s minimizes segerr(s, X) if the slope of s is*

$$m = \frac{\left(\sum_{i=1}^{n} i x_i\right) - \frac{1}{n}\sum_{i=1}^{n} i \sum_{i=1}^{n} x_i}{\left(\sum_{i=1}^{n} i^2\right) - \frac{1}{n}\left(\sum_{i=1}^{n} i\right)^2} \tag{9.5}$$

and the left endpoint of s has value

$$m + \frac{\sum_{i=1}^{n}(x_i - i \cdot m)}{n}$$

Proof: Consider a line segment s approximating fragment X. Let the left endpoint of s be $(1, y_1)$ and the slope be m. For each point x_i $(1 \le i \le n)$, the error is $|x_i - \tilde{x}_i| = |x_i - y_1 - m(i-1)|$. Thus,

$$segerr = \sum_{i=1}^{n}(x_i - y_1 - m(i-1))^2 \tag{9.6}$$

Clearly, when $y_1 = m + \frac{\sum_{i=1}^{n}(x_i - i\cdot m)}{n}$, *segerr* reaches the minimum value

$$segerr = \sum_{i=1}^{n} x_i^2 + m^2 \sum_{i=1}^{n} i^2 - 2m \sum_{i=1}^{n} x_i i - \frac{\left(\sum_{i=1}^{n}(x_i - i * m)\right)^2}{n} \tag{9.7}$$

From Equation 9.7, when

$$m = \frac{\left(\sum_{i=1}^{n} i x_i\right) - \frac{1}{n}\sum_{i=1}^{n} i \sum_{i=1}^{n} x_i}{\left(\sum_{i=1}^{n} i^2\right) - \frac{1}{n}\left(\sum_{i=1}^{n} i\right)^2}$$

segerr is minimized. ∎

Lemma 9.2 leads to algorithm SegmentBound, an online algorithm for the PLA-SegmentBound problem as shown in Figure 9.6. Suppose $x_1, \ldots, x_n$ are the points that have not been compressed yet. When a new point x_{n+1} arrives, we check whether the line segment identified by Lemma 9.2 can achieve the segment error bound. If so, then x_{n+1} is added into the buffer, and the algorithm moves on to the next point in the stream. Otherwise, the line segment suggested by Lemma 9.2 for points $x_1, \ldots, x_n$ is output, and $x_1, \ldots, x_n$ are considered compressed. x_{i+n} is added into the buffer.

The time cost is constant to apply Lemma 9.2 to check whether a newly arrived point can be compressed together with the points that have been seen but have not been compressed. When a new data point x_{n+1} arrives, the left endpoint and the slope of the line segment suggested by Lemma 9.2 can be calculated quickly. Technically, Equation 9.5 indicates that we need to calculate $\sum_{i=1}^{n+1} i$, $\sum_{i=1}^{n+1} x_i$, $\sum_{i=1}^{n+1} x_i i$, and $\sum_{i=1}^{n+1} i^2$. Since we already have $\sum_{i=1}^{n} i$, $\sum_{i=1}^{n} x_i$, $\sum_{i=1}^{n} x_i i$, and $\sum_{i=1}^{n} i^2$, the addition of the new point only incurs a constant cost to update the values of m and the left endpoint.

Since each point in the data stream is processed in constant time, and it only needs to maintain the values of $\sum_{i=1}^{n+1} i$, $\sum_{i=1}^{n+1} x_i$, $\sum_{i=1}^{n+1} x_i i$, and $\sum_{i=1}^{n+1} i^2$, the following theorem holds.

Theorem 9.2 (Complexity—SegmentBound) *The algorithm SegmentBound for the PLA-SegmentBound problem has the time complexity $O(n)$ and space complexity $O(1)$, where n is the number of points in a time series to be compressed.* ∎

Input: A data stream $X = x_1, x_2, \ldots$ and error-bound ϵ;
Output: A list of line segments $\tilde{X}$ approximating X
     such that $segerr(X, \tilde{X}) \leq \epsilon$;
Method:
1:   $i = 1; j = 3$
2:   s = the line segment $((1, x_1), (2, x_2))$;
3:   WHILE (1) DO {
4:      s' = the line segment identified in Lemma 9.2 to
        compress $x_i, \ldots, x_j$;
5:      IF $segerr(s', x_i \cdots x_j) \leq \epsilon$ THEN
6:        $s = s'; j = j + 1$;
7:      ELSE {
8:        output s;
9:        $i = j; j = j + 2$;
10:      s = the line segment $((i, x_i), (i + 1, x_{i+1}))$;
      }
  }

FIGURE 9.6   SegmentBound, an online algorithm for the PLA-SegmentBound problem.

Similar to algorithm PointBound, algorithm SegmentBound only looks ahead for one point in the data stream to output a line segment whenever necessary in the PLA. Thus, it is an online algorithm and can be applied on data streams.

9.4 OPTIMALITY

This section addresses the quality of algorithms PointBound and SegmentBound. Recall that, as defined in Section 9.2, the error-bounded PLA problem needs to minimize the number of segments used to compress a time series.

Theorem 9.3 (PLA-PointBound quality) *The PointBound algorithm in Section 9.3.2 produces a minimum number of segments to compress a time series.*

Proof: For a time series $X = x_1, \ldots, x_n$, let $l = \min\{|\tilde{X}|\}$, where $\tilde{X}$ is an ϵ-PLA approximating X (i.e., $maxerr(X, \tilde{X}) \leq \epsilon$). Conducting an induction on l, it can be shown that algorithm PointBound outputs an ϵ-PLA of l line segments.

(Base case) Consider $l = 1$, that is, there exists a line segment that approximates the whole time series. According to Lemma 9.1, $poly(1, 2) \cap \cdots \cap poly(1, n) \neq \emptyset$. Thus, algorithm PointBound finds a line segment s approximating $x_1, \ldots, x_n$ and $maxerr(s, X) \leq \epsilon$.

(Induction) Assume that, when $l \leq k$, algorithm PointBound finds an ϵ-PLA $\tilde{X}$ of l line segments to approximate X. Now, let us consider the case of $l = (k + 1)$, that is, there exists an optimal ϵ-PLA $\tilde{Y} = \{s_1, \ldots, s_{k+1}\}$ that approximates X.

Suppose that s_1 approximates $x_1, \ldots, x_m$. Assume that s_1' output by algorithm PointBound approximates points $x_1, \ldots, x_{m'}$. Due to Lemma 9.1, $poly(1, 2) \cap \cdots \cap poly(1, m) \neq \emptyset$. Thus, s_1' must approximate $x_1, \ldots, x_m$ with the quality guarantee, that is, $maxerr(s_1', x_1 \cdots x_m) \leq \epsilon$. In other words, $m' \geq m$.

If $m = m'$, then points $x_{m+1}, \ldots, x_n$ in X can be approximated by an ϵ-PLA of $(l - 1) = k$ line segments. According to the assumption, algorithm PointBound finds an ϵ-PLA of $(l - 1)$ line segments approximating $x_{m+1}, \ldots, x_n$.

Suppose that $m' > m$. Since $x_{m+1}, \ldots, x_n$ can be approximated by an ϵ-PLA of $(l-1)$ line segments, a proper subset $x_{m'+1}, \ldots, x_n$ must also be approximated by an ϵ-PLA of at most $(l-1) = k$ line segments. It only needs to drop the segments approximating $x_{m+1}, \ldots, x_{m'}$. According to the assumption, algorithm PointBound finds an ϵ-PLA of the minimum number of line segments to approximate points $x_{m'+1}, \ldots, x_n$.

In summary, algorithm PointBound finds an ϵ-PLA of $l = (k+1)$ line segments approximating X. ■

Similarly, it can also show the optimality of the SegmentBound algorithm.

Theorem 9.4 (PLA-SegmentBound quality)   *The SegmentBound algorithm in Section 9.3.3 produces a minimum number of segments to compress a time series.*

Although the number of line segments used to approximate a time series is a good measure on the compression quality, it is not directly translated to compression ratio. For example, in the methods introduced in this chapter, the endpoints of segments are not constrained. Thus, two points are needed to represent a segment. On the other hand, a PLA using connected segments (i.e., two consecutive segments share the same endpoint) may use more segments but achieve a better compression ratio since only one point is needed to represent a segment except for the first segment.

The following result discloses how good compression algorithms PointBound and SegmentBound can achieve.

Theorem 9.5 (approximation factor)   *Algorithms PointBound and SegmentBound have an approximation factor of 2 to the optimum compression factor that an ϵ-PLA can achieve.*

Proof: It only needs to show the case for the PointBound algorithm. The same argument applies to the SegmentBound algorithm.

For any time series X of m points, suppose that the PointBound algorithm approximates X using n line segments. Then, according to Theorem 9.3, any PLA cannot have less than n line segments. To represent n line segments, at least $(n+1)$ points are needed. Thus, the optimum compression ratio using PLA is at most $\alpha_{opt} = \frac{m}{n+1}$.

The line segments generated by the PointBound algorithm may not be connecting. Thus, at most $2n$ points are needed to represent the n line segments. The worst case compression ratio of the PointBound algorithm is $\alpha_{PointBound} = \frac{m}{2n}$.

Clearly, $\frac{\alpha_{opt}}{\alpha_{PointBound}} = \frac{2n}{n+1} < 2$. ■

9.5   PLAZA FOR TINY DEVICES

Although algorithm PointBound is optimal for the PLA-PointBound problem, it still may be too computation intensive for tiny, resource-constrained sensors due to two reasons.

First, algorithm PointBound may generate nonconnected segments such that each segment requires the transmission of two endpoints. As analyzed before, connected line segments reduce the data transmission volume since each segment (except the first one) requires the transmission of only one endpoint. Second, algorithm PointBound has to calculate the intersection of parallelograms. The computation may be too heavy for tiny, resource-constrained sensor nodes.

This section introduces the simple, fast online algorithm piecewise linear approximation with zoning angle (PLAZA) for the PLA-PointBound problem. PLAZA generates connected line segments. Although PLAZA is not optimal in the number of line segments used for approximation, it is light in computation and very effective in compression ratio, as verified with real-world experiments.

9.5.1 PLAZA

PLAZA builds on the concept of zoning angle. Given an error-bound ϵ and two points (i, x_i) and (k, x_k) $(i < k)$, the *zoning angle* from (i, x_i) to (k, x_k), denoted by $\theta^\epsilon_{(i,k)}$, is defined as the angle that has (i, x_i) as the endpoint and $((i, x_i), (k, x_k))$ as the bisector, and has a degree of $2 \arctan \frac{\epsilon}{|x_i x_k|}$, where $|x_i x_k| = \sqrt{(k - i)^2 + (x_k - x_i)^2}$.

Figure 9.7a shows an example of zoning angle $\theta^\epsilon_{(i,k)}$. The zoning angle defines a zone to include any potential line segments that can be used to compress x_i and x_k. The following important results can be obtained.

Lemma 9.3 *For three points x_i, x_k, x_j $(i < k < j)$ in a time series, the line segment $((i, x_i), (j, x_j))$ approximates x_k with error up to ϵ if and only if the line segment $((i, x_i), (j, x_j))$ falls in the zoning angle $\theta^\epsilon_{(i,k)}$.*

Proof: (Necessity) Without loss of generality, assume that (j, x_j) is above the line $((i, x_i), (k, x_k))$, as shown in Figure 9.7a. If $((i, x_i), (j, x_j))$ approximates x_k with error up to ϵ, the vertical distance between point (k, x_k) and line $((i, x_i), (j, x_j))$ must be smaller than ϵ. That is, the line segment $((i, x_i), (j, x_j))$ must fall in the zoning angle $\theta^\epsilon_{(i,k)}$.

(Sufficiency) If the line segment $((i, x_i), (j, x_j))$ falls in the zoning angle $\theta^\epsilon_{(i,k)}$, the degree of angle $\angle x_j x_i x_k$ is smaller than $\arctan \frac{\epsilon}{|x_i x_k|}$. Therefore, the vertical distance between point (k, x_k) and line $((i, x_i), (j, x_j))$ is smaller than ϵ, which means line segment $((i, x_i), (j, x_j))$ approximates x_k with error up to ϵ. ∎

Lemma 9.4 *For three points x_i, x_k, x_j $(i < k < j)$ in a time series, if zoning angle $\theta^\epsilon_{(i,j)}$ has no overlap with zoning angle $\theta^\epsilon_{(i,k)}$, there does not exist a line segment s with (i, x_i) as the left endpoint such that $maxerr(s, x_i \cdots x_k \cdots x_j) \le \epsilon$.*

Proof: The lemma can be proved by contradiction. Assume that there is a line segment s with (i, x_i) as the left endpoint such that s approximates both x_k and x_j with an error up to ϵ. Based on Lemma 9.3, this line segment must fall in both $\theta^\epsilon_{(i,k)}$ and $\theta^\epsilon_{(i,j)}$, which means $\theta^\epsilon_{(i,k)}$ and $\theta^\epsilon_{(i,j)}$ overlap. Contradiction. ∎

Algorithm PLAZA works as follows. Starting from a point x_i, Lemma 9.3 is used to check if there is a line segment approximating points between indexes i and $j (i < j)$. Moreover, Lemma 9.4 is

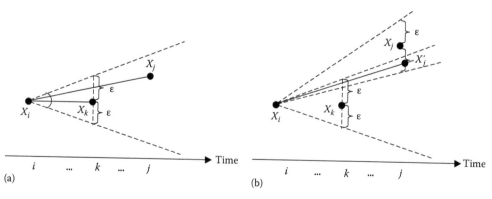

FIGURE 9.7   An example of a zoning angle. (a) Zoning angle $\theta^\epsilon_{(i,k)}$. (b) Overlapping of two zoning angles $\theta^\epsilon_{(i,k)}$ and $\theta^\epsilon_{(i,j)}$.

Input: A data stream $X = x_1, x_2, \ldots$ and error-bound ϵ;
Output: An ϵ-PLA $\tilde{X}$ of a list of *connected* line
      segments, i.e., $maxerr(X, \tilde{X})) \leq \epsilon$;
Method:
1:   $i = 1$; $angle = \theta^\epsilon_{(1,2)}$;
2:   s = line segment $((1, x_1), (2, x_2))$; $j = 3$;
3:   WHILE (1) DO {
4:      $angle = angle \cap \theta^\epsilon_{(i,j)}$;
5:      IF $angle \neq 0$ THEN {
6:         IF segment $((i, x_i), (j, x_j))$ falls in $angle$
7:         THEN s = line segment $((i, x_i), (j, x_j))$;
8:         ELSE {
9:            x'_j = the value of the bisector line of
               $angle$ at index j as shown in Figure 9.7b;
10:        s = the line segment $((i, x_i), (j, x'_j))$;
11:        $x_j = x'_j$;
12:        }
13:      $j = j + 1$;
14:      }
15:      ELSE {
16:         output s;
17:         $i = j - 1$; $x_i = x_{j-1}$; $j = j + 1$;
18:         $angle = \theta^\epsilon_{(i,i+1)}$;
19:         s = line segment $((i, x_i), (i + 1, x_{i+1}))$;
20:      }
21: }

FIGURE 9.8   Algorithm PLAZA.

used to check if searching further in the time series is futile. The pseudocode of PLAZA is shown in Figure 9.8.

Algorithm PLAZA scans each point in a data stream only once and stores only the zoning angle and the current approximating segment in the main memory; the algorithm clearly has linear time complexity and constant space complexity.

Compared to algorithm PointBound, PLAZA requires much simpler computation. The major operation is to construct the zoning angle and to calculate the intersection of zoning angles. The exact degree of each angle is not required. Instead, it only needs to record the endpoint and the two edges of the angle. Similar to algorithm PointBound, PLAZA returns endpoints of segments that may not necessarily belong to the original time series (i.e., the value of x_j may be changed as shown in line 11 in the pseudocode).

9.5.2 BENCHMARKING PLAZA

PLAZA creates connected line segments. Only transmission of one point is needed for each line segment except for the first line segment. This feature distinguishes PLAZA from algorithms Point-Bound and SegmentBound. What is the optimal compression that can be achieved by an ϵ-PLA consisting of only connected line segments? This subsection gives such an optimal algorithm, optimal PLAZA benchmark.

The idea behind the optimal PLAZA benchmark algorithm is similar to that of algorithm Point-Bound. The main difference is that, unlike the PointBound algorithm, the optimal PLAZA benchmark algorithm does not start the new segment with the initial condition $x_i - \epsilon \le y_i \le x_i + \epsilon$, where y_i is the value of the left endpoint of the new segment. Instead, a smaller range on y_i is set to guarantee the connectivity of two consecutive segments. Specifically, to decide the range of y_i, the last nonempty polygon intersection in the previous point is used.

Suppose points $x_1, \ldots, x_{j-1}$ are checked but have not been compressed yet, that is, $poly(1,2) \cap \cdots \cap poly(1, j-1) \ne \emptyset$. The exact intersection of the parallelograms is kept. The intersection is used to confine the range of y_j. Apparently, as the intersection is a convex polygon, only the corners of the polygon need to be checked when looking for the minimum and maximum values. This fact helps in the search of y_j.

The optimal solution can be found by a thorough search. Starting from x_1, it tries all values of j such that $x_1, \ldots, x_j$ can be approximated by a line segment with maximal error ϵ. For each such a subset $x_1, \ldots, x_j$, it computes the intersection of parallelograms $poly(1,2) \cap \cdots \cap poly(1,j)$ and finds a line segment with left endpoint (j, y_j) that can approximate some points $x_{j+1}, \ldots, x_i$, where $j+1 < i$ and y_j is in the range confined by $poly(1,2) \cap \cdots \cap poly(1,j)$. By doing so, the first and the second line segments are connected. A depth-first search is conducted to find an ϵ-PLA consisting of the minimum number of connected line segments.

The optimal PLAZA benchmark is an offline algorithm: it assumes the time series is given and can be scanned multiple times. Its complexity is far above linear due to the thorough search. This algorithm is obviously not suitable for online compression of data streams. It is for comparison purpose only.

9.6 MULTIDIMENSIONAL PLA PROBLEMS

9.6.1 PROBLEM FORMULATION

This section takes a step forward to investigate the PLA problems in multidimensional space, where each point of the time series consists of an array of values. One example application of multidimensional PLA problem is to use line segments to approximate the trace of a moving object in the 3D space. In this case, the endpoints of line segments are points in the 3D space.

Let X be a time series of n points and x_i ($1 \le i \le n$) be the value of the ith point of X in a multidimensional space. Assume that the dimensionality of the space is M. Denote each point x_i as $(x_i^1, x_i^2, \ldots, x_i^M)$. Let $\tilde{X}$ be a PLA of X. The error measurements *maxerr* and *segerr* in Section 9.2 are revised to the M-dimensional case and are denoted as *Mmaxerr* and *Msegerr*, respectively:

$$Mmaxerr(X, \tilde{X}) = \max_{i=1}^{n} \{|x_i - \tilde{x}_i|\} \tag{9.8}$$

where $|x_i - \tilde{x}_i| = \sqrt{\sum_{j=1}^{M} (x_i^j - \tilde{x}_i^j)^2}$.

$$Msegerr(X, \tilde{X}) = \max_{s \in \tilde{X}} \left\{ \sum_{i \in range(s)} (x_i - \tilde{x}_i)^2 \right\} \tag{9.9}$$

where $(x_i - \tilde{x}_i)^2 = \sum_{j=1}^{M} (x_i^j - \tilde{x}_i^j)^2$.

Similar to the PLA-PointBound problem and the PLA-SegmentBound problem, the multidimensional PLA problems could be defined as follows:

Problem 9.5 (Multidimensional PLA-PointBound problem): *Given an error-bound ϵ, the multidimensional PLA-PointBound problem is to find an ϵ-PLA $\tilde{X}$ such that $Mmaxerr(X, \tilde{X}) \le \epsilon$ and $|\tilde{X}|$ is minimized.* ∎

Problem 9.6 (Multidimensional PLA-SegmentBound problem): *Given an error-bound ϵ, the multidimensional PLA-SegmentBound problem is to find an ϵ-PLA $\tilde{X}$ such that $Msegerr(X, \tilde{X}) \leq \epsilon$ and $|\tilde{X}|$ is minimized.* ∎

9.6.2 MULTIDIMENSIONAL PLA-SEGMENTBOUND PROBLEM

It has been shown in Lemma 9.2 that in the 1D case, a line segment s can be easily found that minimizes $segerr(s, X)$ for a time series X of n points. With this Lemma, the M-dimensional PLA-SegmentBound problem can be solved by breaking it to M 1D problems.

Problem 9.7 (kth PLA-SegmentBound problem): *Given an M-dimensional PLA-SegmentBound problem with Msegerr of ϵ, the kth PLA-SegmentBound problem ($1 \leq k \leq M$) is defined as the 1D PLA-SegmentBound problem in the kth dimension with segerr of ϵ_k, where $\epsilon = \sum_{k=1}^{M} \epsilon_k^2$.*

The algorithm M-SegmentBound to solve the multidimensional PLA-SegmentBound problem is shown in Figure 9.9. The basic idea of this algorithm is to calculate the minimum $segerr$ in each individual dimension based on Lemma 9.2 and then check if $Msegerr$ calculated with the $segerr$ values is within the given bound. If it is, we can move ahead to check the next point without starting a new line segment. Otherwise, we output the current segment and start a new line segment. According to Lemma 9.2, $segerr$ is minimized at each dimension. Hence, the $Msegerr$ obtained with the $segerr$ values is also minimized.

It is necessary to explain the operation in Line (7) of the algorithm. A line segment in an M-dimensional space is determined by the two endpoints in the M-dimensional space. For the first (start) point, the value of its kth dimension is the value of the start point of (1D) line segment s'_k; and

Input: A data stream $X = x_1, x_2, \ldots$ in an M-dimensional
space and error-bound ϵ;
Output: A list of line segments $\tilde{X}$ approximating X in the
M-dimensional space such that $Msegerr(X, \tilde{X})) \leq \epsilon$;
Method:
```
1:   i = 1; j = 3
2:   s = the line segment ((1, x₁), (2, x₂));
3:   WHILE (1) DO {
4:      FOR (d = 1; d + +; d ≤ M) DO
5:         s'_d = the line segment in the dth dimension
               identified in Lemma 9.2 to compress x_i^d, ..., x_j^d;
6:      IF ∑_{d=1}^{M} segerr(s'_d, x_i^d ⋯ x_j^d) ≤ ε THEN
7:         Construct line segment s' in the M-dimensional
               space with s'_1, ..., s'_M;
8:         s = s'; j = j + 1;
9:      ELSE {
10:        output s;
11:        i = j; j = j + 2;
12:        s = the line segment ((i, x_i), (i + 1, x_{i+1}));
        }
    }
```

FIGURE 9.9 M-SegmentBound, an online algorithm for the multidimensional PLA-SegmentBound problem.

similarly the value of the second (end) point's kth dimension is the value of the endpoint of (1D) line segment s'_k, where $k = 1, \ldots, M$. The following theorems are thus straightforward.

Theorem 9.6 (Complexity—M-SegmentBound)   *The M-SegmentBound algorithm for the M-dimensional PLA-SegmentBound problem has the time complexity $O(Mn)$ and space complexity $O(1)$, where n is the number of points in a time series to be compressed.* ∎

Theorem 9.7 (Optimality—M-SegmentBound)   *The M-SegmentBound algorithm for the M-dimensional PLA-SegmentBound problem produces a minimum number of segments to compress a time series in the M-dimensional space.* ∎

The optimality comes from the fact that the M-SegmentBound algorithm works in the same way of the SegmentBound algorithm. More specifically, we can calculate the exact value of the smallest *MSegerr* with Equation 9.9 and check if it is within the given bound (Line (6) of the algorithm).

9.6.3 MULTIDIMENSIONAL PLA-POINTBOUND PROBLEM

Unfortunately, the idea in Section 9.6.2 cannot be used to extend the PointBound algorithm (Section 9.3.2) to solve the multidimensional PLA-PointBound problem. The main difference is that the PointBound algorithm checks the existence of approximating line segment in 1D space, and as long as such line segments exist, the PointBound algorithm will be able to find one but does not guarantee the approximating error is minimum. Due to this reason, Equation 9.8 cannot be used to obtain the minimum *Mmaxerr*, and thus people cannot simply check whether it should output current line segments or move ahead to another new point in the multidimensional case. Designing an optimal algorithm for the multidimensional PLA-PointBound problem is left as an open challenge.

This section extends algorithm PLAZA to obtain approximate solution for the multidimensional PLA-PointBound problem. The main idea is to adjust the error bound on each individual dimension so that it can move ahead for the next point, as long as the error *Mmaxerr* calculated with Equation 9.8 is within the given bound. Initially, it equally assigns the allowed error bound to each dimension.

With the initial error bound of each dimension, algorithm PLAZA is performed on each dimension. Assume that at some point, algorithm PLAZA needs to stop and output the current segment on the kth dimension ($1 \leq k \leq M$). The error bounds of each dimension within the constraint of *Mmaxerr* $\leq \epsilon$ are adjusted to check if search ahead is possible. Specifically, the error bound of the kth dimension is increased, and the error bounds on other dimensions are reduced so that algorithm PLAZA can move forward on all dimensions. If the adjustment of error bounds among different dimensions is impossible, the algorithm outputs current segment and starts a new segment with initial error bound on each dimension equally assigned. It is worth noting that the error bound adjustment at a point only impacts the current point because only the overlapping area of zoning angles is considered. That is, all previous points still fall within *Mmaxerror* bounds if approximated using a line segment chosen based on the overlapping angle in each dimension.

There are certainly many ways to assign initial error bound to each dimension and to adjust error bounds among different dimensions. The adjustment method presented in Figure 9.10 is very easy to implement and effectively prevents early output of a line segment, whenever error bound adjustment is still feasible.

9.7 EXPERIMENTAL EVALUATION

This section evaluates the performance of the online algorithms by simulation in MATLAB and by real implementation with MICA2 motes [7].

Input: Error bound on each dimension $\epsilon_1, \ldots, \epsilon_M$;
　　　Current point;
Output: TRUE (adjustment successful) or FALSE (otherwise)
Method:
1:　Sort the bounds $\epsilon_1, \ldots, \epsilon_M$ in increasing order;
2:　FOR $(i = 1; i++; i \le M - 1)$ DO {
3:　　IF (the ith dimension of current point cannot be
　　　approximated with error bound ϵ_i) {
4:　　　Increase ϵ_i by ϵ_δ, where ϵ_δ is the minimum value
　　　so that the current point can be compressed
　　　with previously uncompressed points within the
　　　increased error bound on ith dimension.
5:　　　Decrease ϵ_{i+1} by ϵ_δ so that *Mmaxerr* remain
　　　unchanged;
　　}
　}
6:　IF $(\epsilon_M > 0)$ return TRUE; else return FALSE;

FIGURE 9.10 ErrorBoundAdjustment, the algorithm to adjust the error bound among different dimensions for the PLA-PointBound problem.

9.7.1 Existing Work

The works cited in Keogh et al. and Palpanas et al. [16,21] are related to the online algorithms introduced in this chapter. Nevertheless, it is impossible to get a fair comparison between the online algorithms and those methods due to the following reasons.

First, the SWAB algorithm in Keogh et al. [16] uses a moving window to constrain a time period in consideration. Its performance largely depends on the size of the moving window. As such, SWAB is largely different from the methods in this chapter that do not maintain any window. Furthermore, it is hard to allocate enough space for the moving window in space-limited tiny sensors.

Second, the algorithms in Palpanas et al. [21] use an amnesic function to give weights to different points in the time series. The focus and the application context are different from that in this chapter.

Due to this consideration, the remainder of this section only focuses on the implementation and the evaluation of the online algorithms presented in this chapter.

9.7.2 Experimental Setting

Two audio files are generated for testing. The first file includes a human voice with the sampling rate of 8 khz in a mono channel. The second file includes piano music with the sampling rate of 44 khz in a mono channel. Each file includes 1,000,000 samples, and the size of each sample is 16 bits. Figures 9.11 and 9.12 show the waveform of the human voice data and the waveform of the piano music, respectively. It can be seen that the music data look "smoother" in that at the same time window, it changes values in a smaller range.

The files are used to test the performance of the online algorithms in bandwidth savings, in terms of the following two metrics:

1. *Sample reduction ratio (inverted compression ratio)*. It is defined as the total number of points to represent the ϵ-PLA divided by the total number of points in the original time series.
2. *Distortion*. It is defined as $\frac{\sum_{i=1}^{n}(x_i - \tilde{x}_i)^2}{n}$, where n is the total number of points in the time series, x_i is the original value, and $\tilde{x}_i$ is the approximated value of x_i.

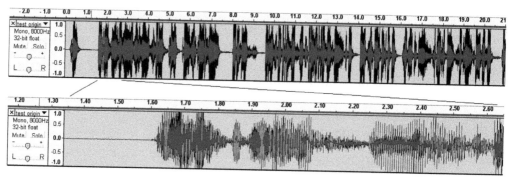

FIGURE 9.11　The waveform of human voice data (the lower part is in a smaller timescale).

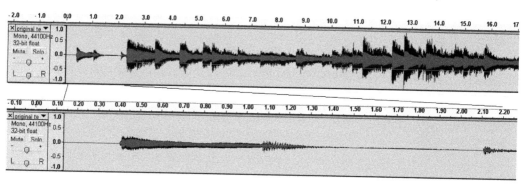

FIGURE 9.12　The waveform of piano music data (the lower part is in a smaller timescale).

In simulation, the online algorithms are applied on the audio files and measure the sample reduction ratio. Simulation results are reported in Sections 9.7.3 and 9.7.4. In the test using MICA2 motes, the original audio files are played on a desktop computer and are monitored and transmitted with a MICA2 mote over wireless channel to a laptop computer. More details are provided in Section 9.7.5.

For verification, the audio files and the audio files recovered from the samples approximated by the online algorithms are available upon request.

9.7.3　Results on Quality

This section tests the sample reduction ratio and distortion using the two audio data sets:

9.7.3.1　Results on Sample Reduction Ratio

Figures 9.13 and 9.14 show the results of algorithms PointBound and SegmentBound, respectively, with respect to various error-bound values. As shown in figures, a higher bandwidth saving can be obtained on piano music than on human voice. As shown in Figures 9.11 and 9.12, the waveform of the piano music is "smoother" than that of the human voice. By replaying the audio files recovered from the samples by the algorithms, the human voice recovered from the samples by the algorithms is fully recognizable with the segment error bound up to 0.4, or with the point error bound up to 0.2. The quality of recovered piano music is acceptable with the segment error bound up to 0.2, or with the point error bound up to 0.1.

Figures 9.13 and 9.14 clearly demonstrate a significant bandwidth saving. With the online algorithms, we only need to transmit around 5% of the original sample size for piano music and around 20% of the original sample size for human voice. As such, both sound files can be transmitted with the current sensor nodes.

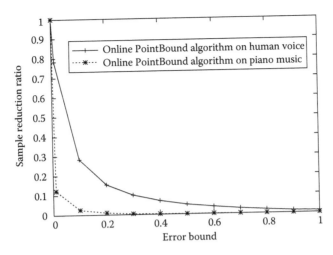

FIGURE 9.13   The sample reduction ratio of PointBound.

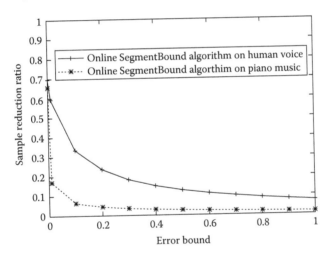

FIGURE 9.14   The sample reduction ratio of SegmentBound.

Figure 9.15 shows the sample reduction ratio of algorithm PLAZA with respect to various point error bounds. A similar phenomenon can be observed as in Figures 9.13 and 9.14. With PLAZA, the recovered human voice is fully recognizable with the (point) error bound up to 0.2, and the quality of recovered piano music is acceptable with the (point) error bound up to 0.1. From Figure 9.15, these above qualities correspond to the bandwidth reduction of nearly 3% of the original data size for piano music and about 15% of the original data size for human voice.

One interesting phenomenon is that the SegmentBound algorithm can reduce a sample transmission volume even if the error bound is set to zero, as shown in Figure 9.14. This is because in the audio files, there are some silent periods where the sample values are all zeros. The SegmentBound algorithm finds a line segment to approximate those situations. This nice feature, however, does not exist in the algorithms for the PLA-PointBound problem. If the error bound is zero, the initial polygon is empty in the PointBound algorithm, and the degree of the initial feasible angle is zero in PLAZA, resulting in no sample reduction.

Figure 9.16 compares algorithms PLAZA, PointBound, and SegmentBound on the human voice data set. The gap between algorithms PLAZA and PointBound is very small when the error bound is less than 0.5. Algorithm PointBound leads to more samples than algorithm PLAZA when the

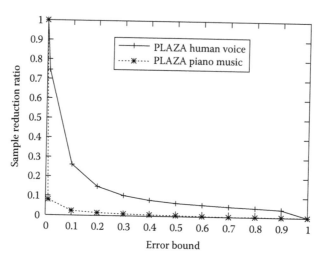

FIGURE 9.15 The sample reduction ratio of PLAZA.

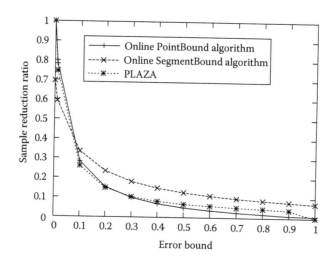

FIGURE 9.16 Comparison of the three algorithms on the human voice data set.

error bound is less than 0.3. The gap between algorithm SegmentBound and the two algorithms for the PLA-PointBound problem comes from the fact that, using the same error-bound value, the PLA-SegmentBound problem puts a tighter error constraint than the PLA-PointBound problem. The similar performance comparison of the three algorithms can be observed on the piano data set.

9.7.3.2 Results on Distortion

Figures 9.17 and 9.18 quantitatively show the distortion of the algorithms on the human voice data set and the piano music data set, respectively. The overall distortion on human voice is larger than that on piano music. With the same error bound, algorithm PLAZA has the largest distortion. Algorithm PointBound is the next. Algorithm SegmentBound has the smallest distortion because the same error bound on the PLA-SegmentBound problem and the PLA-PointBound problem poses a tighter error constraint on the PLA-SegmentBound problem. The smaller distortion, however, comes with the cost of lower bandwidth saving as analyzed before. This is the trade-off between bandwidth saving and quality of recovery.

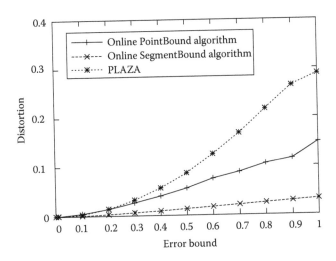

FIGURE 9.17   The distortion on the human voice data set.

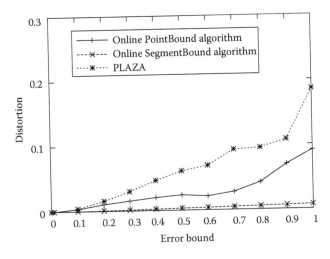

FIGURE 9.18   The distortion on the piano music voice data set.

9.7.4   BENCHMARKING PLAZA

The performance of PLAZA is compared to the optimal solution of its kind (i.e., using connected line segments to tackle the PLA-PointBound problem). Due to the high complexity of the PLAZA Benchmark method, the audio files are too big to obtain the optimal results within a reasonable time. It is only feasible to use a small portion of the audio files for this test.

Interestingly, the PLAZA method and the optimal PLAZA benchmark algorithm generate very similar PLA line segments. The audio files have many silent gaps where sample values are close to 0. Thus, algorithm PLAZA can obtain line segments very similar to those computed by the benchmark algorithm.

To further test algorithm PLAZA with more difficult scenarios, data sets are generated to contain Gaussian noise with the mean equal to 0 and the variance equal to 1. Five Gaussian noise data sets are generated. Each data set has 1000 samples. The simulation is run on the five data sets, and the average sample reduction ratio is calculated as the final result.

Figure 9.19 compares algorithms PLAZA, PointBound, and the PLAZA benchmark. Since algorithm PointBound may generate disconnected line segments, it has to send two endpoints for

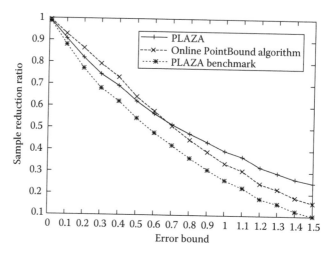

FIGURE 9.19 The comparison between the algorithm PLAZA and the PLAZA benchmark method.

each line segments. In contrast, algorithm PLAZA and the PLAZA benchmark method always generate connected segments, where every segment, except for the first one, requires the transmission of only one endpoint. The figure shows that algorithm PointBound actually needs more endpoints than algorithm PLAZA when the error bound is small, even though algorithm PointBound is optimal in terms of the number of segments. Compared to the PLAZA benchmark algorithm, algorithm PLAZA always generates more samples. The gap between algorithm PLAZA and the PLAZA benchmark method, however, is not significant when the error bound is small.

9.7.5 RESULTS ON REAL SENSORS

The online algorithms are implemented using MICA2 motes [7] from Crossbow Technology Inc. The test bed is illustrated in Figure 9.20.

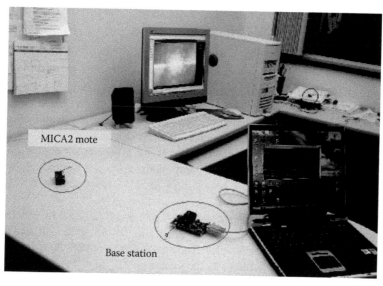

FIGURE 9.20 The test bed using real sensors.

A MICA2 mote includes a radio/processor board and a sensor board. The radio/processor board uses 900 Mhz radio. The sensor board includes a microphone that can be used for sampling sound. The interface of the base station is based on RS232. It is used to program the mote and in the mean time acts as a gateway to connect the laptop and the radio wireless sensor network. The laptop does not have an RS232 port, and thus a USB/RS232 adaptor is used to connect to the base station interface. The original audio files are played on a desktop computer, monitored by a MICA2 mote, and transmitted over wireless channel from the MICA2 mote to the base station.

The results about the sample reduction ratio on the real sensor test bed are close to the simulation results using MATLAB. But the audio quality obtained using the real test bed is worse than that obtained in the MATLAB simulation. The deterioration in audio quality is caused by the major restriction of TinyOS [8], the current OS in MICA2 motes. The OS does not support multiple threads, and thus it cannot perform radio transmission and sound sampling concurrently. Due to this limit, when the data are transmitted to the base station, the sensor board stops sampling and the sound during this period is missed, resulting in small silent gaps in the recovered audio.

It may be easy to carry out the same task with the most recent, more advanced sensor device like MICAz from Crossbow Technology Inc. With a higher price and energy consumption, MICAz sensors support up to 250 Kbps wireless transmission. This task, however, has never been fulfilled with low-end devices like MICA2. To this end, the methods introduced in this chapter break the limit of scarce radio bandwidth and carry out a task that is hard to achieve without the fast online compression methods.

9.7.6   EVALUATION IN OTHER APPLICATIONS

Although the online algorithms are only implemented in an acoustic sensor monitoring system, the algorithms are actually applicable to many other application domains such as electrocardiogram (ECG) monitoring for patients. The algorithms are tested on an ECG data set obtained from the UCR (University of California, Riverside) time series collection [14]. The maximum value on the data set is 2490, and the minimum value is $-8,190$. The online algorithms are tested with an error bound varying from 1 to 100.

Figure 9.21 compares the sample reduction ratio of algorithms PLAZA, PointBound, and SegmentBound on the ECG data set. The performance of algorithms PLAZA and PointBound is very similar. When the error bound is set to over 35, both algorithms can compress the data up to 10% of the original size. The gap between algorithms SegmentBound and PointBound comes from the fact

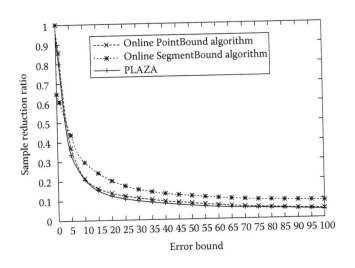

FIGURE 9.21   Results on an ECG data set.

that, using the same value on *segerr* and *maxerr*, the PLA-SegmentBound problem has a tighter error bound than the PLA-PointBound problem.

9.7.7 Experimental Results of Multidimensional PLA Problems

The performance of the multidimensional PLA problems is tested on a 3D accelerometer data from a Sony ERS-210 AIBO Robot, as shown in Figure 9.22. The Sony Aibo is a small, quadruped robot that comes equipped with a triaxial accelerometer. This accelerometer returns data at a rate of 125 Hz. The algorithm is tested on Aibo data set obtained from the UCR time series collection. This data set consists of the track of a robot playing soccer, including chasing after a moving ball, capturing the moving ball, kicking the moving ball, spinning in place searching for the ball, and standing stationary for a brief time period. The attributes of the multidimensional data set are shown in Table 9.1, which shows that the second dimension (Dim2) is harder to approximate due to the high variation in data.

The experimental results are shown in Figure 9.23. Note that the error bound on the X-axis is the whole error bound (i.e., *Mmaxerr* or *Msegerr*) that has to be satisfied. The error bound allocated to each dimension is smaller than this value. For instance, if the specified bound in multidimensional PLA-PointBound problem, *Mmaxerr*, is ϵ, then the initial error bound on each dimension should be $\frac{\epsilon}{\sqrt{M}}$, where M is the size of dimension. As discussed in Section 9.6, the M-SegmentBound algorithm

FIGURE 9.22 Sony ERS-210 AIBO Robot.

TABLE 9.1

Three-Dimensional Accelerometer Data from a Sony ERS-210 AIBO Robot

	Max	Min	Average	Variance
Dim1	0.983	−0.916	0.280	0.020
Dim2	0.643	−0.661	0.016	0.0177
Dim3	−0.168	−2	−0.942	0.0165

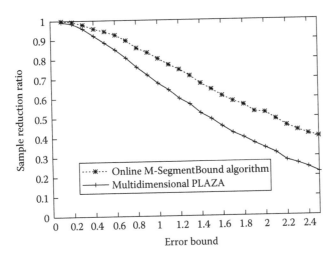

FIGURE 9.23   Multidimensional PLA problem.

for the M-dimensional PLA-SegmentBound problem produces a minimum number of segments to compress a time series in the M-dimensional space. So the result in Figure 9.23 shows the minimum segmentation that is possible for the multidimensional PLA-SegmentBound problem. The multidimensional PLA-PointBound problem is, however, much harder to solve. When $Mmaxerr$ is equal to 1, only 60% bandwidth reduction ratio can be obtained in this experiment.

9.8   CONCLUSION

This chapter studies the problem of online compression of data streams in the resource-constrained network environment, where the traditional data compression techniques cannot apply. Particularly, fast PLA methods are explored to guarantee data quality. Two versions of the problem are formulated to explore quality guarantees in different forms. For the error-bounded PLA problem, fast online algorithms are developed, which run in linear time complexity and require a constant space cost. The online algorithms are also optimal in terms of the number of generated segments. To meet the needs from tiny, resource-constrained sensors, another online algorithm is developed that involves only very simple computation and generates connected line segments. Furthermore, an optimal algorithm is designed to tackle the multidimensional PLA-SegmentBound problem, and an approximate algorithm is presented for the multidimensional PLA-PointBound problem. Simulation and real-world experimental results demonstrate that the fast online linear approximation methods are very effective for data stream compression and transmission over low-bandwidth networks with nodes heavily constrained in computational power.

Equipped with the insights gained in this study, people can see a lot of application opportunities for the introduced methods. Meanwhile, there are also some interesting open challenges for future work. For example, an interesting question is to design an online algorithm that can compute an ε-PLA consisting of connected line segments that have an approximation factor to the optimum.

REFERENCES

1. R. Agrawal, C. Faloutsos, and A. Swami, Efficient similarity search in sequence databases, in *Proceedings of the Fourth International Conference of Foundations of Data Organization and Algorithms (FODO)*, D. Lomet, Ed. Chicago, IL: Springer Verlag, 1993, pp. 69–84. [Online]. Available: citeseer.ist.psu.edu/agrawal93efficient.html.
2. B. S. Atal and L. S. Hanauer, Speech analysis and synthesis by linear prediction of the speech wave, *Journal of the Acoustical Society of America*, 50, 637–655, 1971.

3. G. Bishop-Hurley, D. L. Swain, D. Anderson, P. Sikka, C. Crossman, and P. Corke, Virtual fencing applications: Implementing and testing an automated cattle control system, *Computers and Electronics in Agriculture*, 56(1), 14–22, 2007.

4. E. J. Candès and M. B. Wakin, An introduction to compressive sampling, *Signal Processing Magazine, IEEE*, 25(2), 21–30, 2008.

5. K. Chan and A. W. Fu, Efficient time series matching by wavelets, in *Proceedings of the 15th International Conference on Data Engineering*. Washington, DC, 1999, pp. 126–133. [Online]. Available: citeseer.ist.psu.edu/chan99efficient.html.

6. R. Cristescu and M. Vetterli, On the optimal density for real-time data gathering of spatio-temporal processes in sensor networks, in *Proceedings of the Fourth International Symposium on Information Processing in Sensor Networks*. IEEE Press, Los Angeles, CA, 2005, p. 21.

7. Crossbow, Crossbow technology, MICA2 Mote datasheet, 2007, http://www.datasheetarchive.com/dl/Datasheets-SW21/DSASW00417437.pdf. April 2014.

8. D. Cull, J. Hill, P. Bounadonna, R. Szewczyk, and A. Woo, A network-centric approach to embedded software for tiny devices, in *Proceedings of First International Workshop on Embedded Software (EMSOFT 2001)*. Tahoe City, CA, October 2001.

9. W. Dobson, J. Yang, K. Smart, and F. Guo, High quality low complexity scalable wavelet audio coding, in *1997 IEEE International Conference on Acoustics, Speech, and Signal Processing*, vol. 1. IEEE Computer Society, Munich, Germany, 1997, pp. 327–330.

10. D. H. Douglas and T. K. Peucker, Algorithms for the reduction of the number of points required to represent a digitized line or its caricature, *Canadian Cartographer*, 10(2), 112–122, December 1973.

11. J. G. Dunham, Optimum uniform piecewise linear approximation of planar curves, *IEEE Transactions on Pattern Analysis and Machine Intelligence*, 8(1), 67–75, 1986.

12. C. Faloutsos, M. Ranganathan, and Y. Manolopoulos, Fast subsequence matching in time-series databases, in *SIGMOD '94: Proceedings of the 1994 ACM SIGMOD International Conference on Management of Data*. New York: ACM Press, 1994, pp. 419–429.

13. M. T. Goodrich, Efficient piecewise-linear function approximation using the uniform metric, in *Proceedings of the 10th Annual Symposium on Computational Geometry*. New York, 1994, pp. 322–331.

14. E. Keogh, Ucr time series collection, http://www.cs.ucr.edu/~eamonn/time_series_data. August 2008.

15. E. Keogh, K. Chakrabarti, M. Pazzani, and S. Mehrotra, Dimensionality reduction for fast similarity search in large time series databases, *Knowledge and Information Systems*, 3(3) 263–286, 2001. [Online]. Available: citeseer.ist.psu.edu/keogh00dimensionality.html.

16. E. Keogh, S. Chu, D. Hart, and M. Pazzani, An online algorithm for segmenting time series, in *Proceedings of International Conference on Data Mining*, San Jose, CA, 2001, pp. 289–296. [Online]. Available: citeseer.ist.psu.edu/keogh01online.html.

17. E. Keogh and M. Pazzani, An enhanced representation of time series which allows fast and accurate classification, clustering and relevance feedback, in *Fourth International Conference on Knowledge Discovery and Data Mining (KDD'98)*. New York, 1998, pp. 239–241.

18. K.-D. Kim and P. R. Kumar, Cyber-physical systems: A perspective at the centennial, *Proceedings of the IEEE*, 100 (Special Centennial Issue), 1287–1308, 2012.

19. C. Liu, K. Wu, and J. Pei, An energy efficient data collection framework for wireless sensor networks by exploiting spatiotemporal correlation, *IEEE Transactions on Parallel and Distributed Systems*, 18(7), 1010–1023, July 2007.

20. G. Manis, G. Papakonstantinou, and P. Tsanakas, Optimal piecewise linear approximation of digitized curves, in *Proceedings of 13th International Conference on Digital Signal Processing*, Santorini, Hellas, Greece, 1997, pp. 1079–1081.

21. T. Palpanas, M. V. E. Keogh, D. Gunopulos, and W. Truppel, Online amnesic approximation of streaming time series, in *Proceedings of the 20th IEEE International Conference on Data Engineering*, 2004, pp. 339–349.

22. I. Popivanov and R. Miller, Similarity search over time-series data using wavelets, in *ICDE '02: Proceedings of the 18th International Conference on Data Engineering*. Washington, DC: IEEE Computer Society, 2002, pp. 212–221.

23. Y. Qu, C. Wang, and X. Wang, Supporting fast search in time series for movement patterns in multiples scales, in *Proceedings of the Seventh International Conference on Information and Knowledge Management*, Washington, DC, 1998, pp. 251–258.

24. R. R. Rajkumar, I. Lee, L. Sha, and J. Stankovic, Cyber-physical systems: The next computing revolution, in *Proceedings of the 47th Design Automation Conference*. ACM, Anaheim, CA, 2010, pp. 731–736.
25. R. K. Rana, Addressing three wireless sensor network challenges using spare approximation methods. PhD thesis, University of New South Wales, New South Wales, Australia, 2011.
26. H. Shatkay and S. Zdonik, Approximate queries and representations for large data sequences, in *Proceedings of the 12th IEEE International Conference on Data Engineering*, New Orleans, LA, February 1996.
27. D. Sinha and J. Johnston, Audio compression at low bit rates using a signal adaptive switched filterbank, in *Proceedings of IEEE ICASSP*, Atlanta, GA, 1996, pp. 1053–1056.
28. Sony, Sony ERS-210 AIBO Robot user guide, 2008, http://support.sony-europe.com/AIBO. August 2008.
29. P. Srinivasan and L. Jamieson, High-quality audio compression using an adaptive wavelet packet decomposition and psychoacoustic modelling, in *IEEE Transactions on Signal Processing*, 46, 1085–1093, April 1998.
30. M. C. Vuran, Ö. B. Akan, and I. F. Akyildiz, Spatio-temporal correlation: theory and applications for wireless sensor networks, *Computer Networks*, 45(3), 245–259, 2004.
31. C. Wang and S. Wang, Supporting content-based searches on time series via approximation, in *Proceedings of the 12th International Conference on Scientific and Statistical Database Management*, Berlin, Germany, July 2000.
32. B. Warneke, M. Last, B. Leibowitz, and K. Pister, Smart dust: Communicating with a cubic-millimeter computer, *Ad Hoc Networks Journal*, 34(1), 44–51, January 2001.
33. Y. Wu, D. Agrawal, and A. Abbadi, A comparison of DFT and DWT based similarity search in time-series databases, in *Proceedings of the Ninth International Conference on Information and Knowledge Management*. New York, 2000, pp. 488–495.
34. S. Yoon and C. Shahabi, The clustered aggregation (CAG) technique leveraging spatial and temporal correlations in wireless sensor networks, *ACM Transactions on Sensor Networks (TOSN)*, 3(1), 3, 2007.
35. P. Zhang, C. M. Sadler, S. A. Lyon, and M. Martonosi, Hardware design experiences in ZebraNet, in *Proceedings of the Second International Conference on Embedded Networked Sensor Systems*. ACM, Baltimore, MD, 2004, pp. 227–238.

Section IV

Mobility Issues

10 Robotic Operations Network

Cyber-Physics Framework and Cloud-Centric Software Architecture

*Vladimir Zaborovsky, Alexey Lukashin,
and Vladimir Muliukha*

CONTENTS

10.1 INTRODUCTION

We live in a world that even in everyday life is unthinkable without quantum mechanics, relativity theory, and cybernetics. Cell phones, computers, global positioning system (GPS devices) modern electronics, not to mention light-emitting diodes (LEDs), and semiconductor lasers—are all devices that are based on quantum physics and effects following the theory of relativity. People who use these appliances greatly increase their own physical and communicative abilities, which suggests that the result of the development of science and technology is a radical change in the entire human environment.

This is why the old aphorism holds true: *Fundamental physics today is the technology of tomorrow.* The content of the fundamental physics laws is descriptive in nature and can be viewed as a result of analysis and following mathematical formalization of properties that characterize features of natural objects and processes of matter in motion. Technology is created by humans in accordance with the laws of physics, as opposed to the objects of nature; which is the result of specific engineering works, including the assembly process and the formation of the rules of its applications.

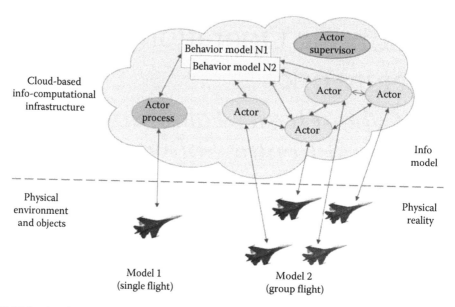

FIGURE 10.1   Aerobatic kiberno.

The process of creative engineering is the process of synthesis of new objects of physical reality. This process is inseparable from different aspects of management, communications, and control that are studied within the framework of cybernetics. It is useful to note that the term *cybernetics* was used by Plato to denote the science of management of the objects containing people. Such objects may be, for example, a city populated by people or Persian soldiers in a chariot or a trireme—an ancient battleship.

According to Plato, a ship that is built and equipped is just a thing, but the ship with the crew and passengers—it is considered to be a *kiberno* (κoβερνω) or *cyber object*. This "cyber object" under the control of a skilled specialist or helmsman obtains fundamentally new properties, which are not possessed by its parts if taken separately. So a sailboat cannot itself go against the wind, but a skilled crew of sailors can cope with this task easily. In Plato's definition, the important word is "skilled," that is, with the experience of successful execution of certain operations. The visual metaphor for the kiberno now is a group of aircraft controlled by experienced pilots who perform synchronous aerobatic maneuvers at the limit of human physical abilities (see 0).

To implement the new properties, first, a *part* of the cyber object must be given special resources—namely, memory, which captures the *correct* sequence of operation leading to the success of the chosen scenario of the action. Unlike simple mechanical manipulations, such as use of the lever of Archimedes, whose actions are defined by clear mathematical formula, this scenario describes complex communication between connected states or parts of the cyber object (Figure 10.1).

The attributes of the physical reality are complex and difficult to describe formally, because this concept expresses not only the causal relations, but also the events, that is, it represents stochastic or information relation. In the nineteenth century, Andre Ampere placed cybernetics on the third position in his own systematization of sciences. Then, Rudolf Clausius formalized the notion of complexity of a reversible thermodynamic process through a new physical quantity—entropy:

$$\Delta S = \frac{\Delta Q}{T},$$

where
  ΔQ is the change in amount of heat
  T is the value of absolute temperature

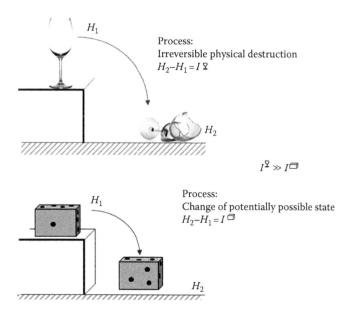

H_1

Process:
Irreversible physical destruction
$H_2-H_1 = I\,\Sigma$

H_2

$I^\Sigma \gg I^\square$

Process:
Change of potentially possible state
$H_2-H_1 = I^\square$

H_1

H_2

FIGURE 10.2 Entropy and information as attributes of physical objects.

Subsequently, based on the analysis of the *behavior* of Maxwell's demon, it was shown that such entropy-variable can be interpreted also as the amount of information that *falls* on one elementary event or statistically independent message. Characterizing the complexity of the physical object by the number of possible states, namely n, the values $H_i = -\log_2 p(i)$, where $i = 1, 2 \ldots n$, were called elementary entropy, or the amount of information characterizing the state of the object, $p(i)$ being the probability of the object achieving ith state (Figure 10.2). In the twentieth century, scientists began to treat cybernetics as a field of research of the general laws, principles, and methods of data processing, communication, and control of complex systems.

With the advent of computers—physical systems, in which all processes take place in accordance with the instructions stored in computer memory—begins a new stage in the development of cybernetics. This new development phase of cybernetics was coupled with the design and creation of automatic and automated control systems. In these systems, the role of Plato's "helmsman" plays a symbolic program matching coded targets and logical operations with controlled states of computer components. As a result, the problem of complexity transformed into a problem of correct operation of computer programs.

At the first *technical* stage of development of cybernetics as a science of control, the interaction between pure physics and cybernetics has not received proper support due to substantial differences in the control objectives. However, to explain effects such as self-organization of dissipative structure (*Bernard cells*) or the occurrence of self-oscillations of chemical reactions (Prigogine's *brusselator*), many physicists began using terms borrowed from pure cybernetics.

In his works, I. Prigogine emphasized the point that dissipative processes can be viewed as a natural carrier of synchronization of chaotic oscillations, and proposed to simply consider these processes as a physical level of natural communication environment that transferred *targeted* signals in accordance with the laws of physics. These signals form the structure of the object and can change its entropy states. In this context, the laws of physics play the role of the *programs* that have been written by the nature but implemented or *calculated* using physical processes. With the development of many core and multithreaded computer technologies and concepts of quantum computing, the problem of control complexity has been recognized as one of the fundamental problems of modern physics. Admittedly, management principles based on feedback, first formulated in cybernetics in order to stabilize the trajectories of technical objects, acquired all scientific significance.

In recent years, the application of scientific results was increasingly focused on applied problems associated with the use of network-centric technology, remote control, and information exchange. Consequently, the interpretation of the physical laws that reflect the characteristics of nonlinear processes, chaotic or quantum phenomena from the standpoint of cybernetic principles, is actively paving the way for the new paradigm of studying the phenomena of nature, which is called cyber-physics. Within this paradigm, there have been successfully solved complex problems of control of chaotic systems and purposeful change of flow bifurcation in the systems with continuous state space.

Although while these considerations touched upon the problems of complexity generated by the sensitivity of the trajectories of dynamical systems to initial conditions or the probabilistic description of the motion of systems with non reparable states, they paid no attention to the original problem of kiberno, which was indicated in the works of Plato, namely, the control of objects whose behavior occurs in the context of the state of their memory. Notable theoretical results associated with the use of Markov chains, although they may be considered a special case of kiberno control, nevertheless do not fulfill the requirements of cyber-physical applications including group robotic actions, multiservice operational infrastructures for the *Internet of Things*, or the organization of Mesh networks between dynamic objects.

Our work is dedicated to the resolution of various applied problems stated above of cyber physics as a new interdisciplinary science. In subsequent sections of this chapter, we focus on the organization of engineering infrastructure and software architecture, which provide a means of communication between physical, mechanical, and information-computing processes arising during network-centric interaction of intelligent robots and their virtual "avatars." Operating space for virtual objects is formed by specific cloud infrastructure based on robust communication channels and Hadoop data repository. The issues of implementation of the engineering infrastructure, and of the concept of cloud computing as applied to embedded software, regarded as the carrier of the control algorithms, constitute the main part of Section 10.3 and 10.4 in this chapter. The effectiveness of the proposed solutions is illustrated by the example of a network-centric management system of planetary robots whose operations at the level of calculating trajectories or maintaining information integrity and functional robustness are controlled from the board of ISS.

10.2   CLASSICAL PHYSICS AND CYBER PHYSICS

In the near future, a new generation of artificial physical devices will be created. They will be characterized by the flexibility, elasticity, and sensitivity that are common to living organisms, but will have greater strength and durability because of the materials used. Such devices will be able to receive, store, and transmit information about their surroundings, which will be used during their operation. Information is transmitted between physical objects and also between objects and the human operator.

Complex engineering tasks concerning control for groups of mobile robots are not yet sufficiently developed. In our work, we use the CPh approach, which extends the range of engineering and physical methods for the design of complex technical objects by researching the informational aspects of communication and interaction between objects and an external environment.

It is appropriate to consider control processes with the cyber-physical perspective because of the necessity for spatiotemporal adaptation to changing goals and characteristics of the operational environment. Thus, the priority task is to organize the reliable and high-performance system of information exchange between all entities involved in the realization of all requirements. Hereinafter, by CPh object, we mean an open system for the information exchange processes. Data in such systems are transmitted through the computer networks, and its content characterizes the target requirements achieved through execution of physical and mechanical operations, energy being supplied by the internal resources of the object (Figure 10.3).

An example of a cyber-physical object is a mobile robot that makes complex spatial movements, controlled by the content of the received information messages that have been generated by a human operator or other robots that form a multipurpose operation network. An ontological model of informationally open cyber-physical object may be represented by different formalisms, such as a set

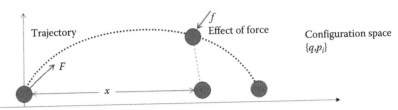

Physics as a science of causality:
movement trajectory of the object is changed by external «force»

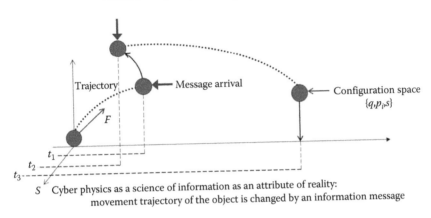

S Cyber physics as a science of information as an attribute of reality:
movement trajectory of the object is changed by an information message

FIGURE 10.3 Physical and cyber-physical motion.

of epistemic logic model operations parameterized by the data of local measurements or messages received from other robots via computer connection.

Although there are different approaches in modern science to the application of information aspects of the physical objects, only within cybernetics, such approaches have structural engineering applications. The conceptual distinction between closed and open systems in terms of information and computational aspects requires the use of new models, which take into account the characteristics of information processes that are generated during the operation of the physical objects and are available for monitoring, processing, and transmission via a computer network.

According to Figure 10.4, the cyber-physical model of a control system can be represented as a set of components, including the following units:

- Information about the characteristics of the environment (observation)
- Analysis of the parameters of the current state for the controlled object (CO; orientation)
- Decision making according to the formal purpose of functioning (decision)
- Implementation of the actions that are required to achieve the goal (action)

The interaction of these blocks using information exchange channels allows us to consider this network structure as a universal platform, which allows us to use various approaches, including the use of algorithms and feedback mechanisms or reconfiguration of the object's structure for the goal's restrictions on entropy reduction or the reduction of the internal processes' dissipation.

Centralized solutions allow for the use of universal means for the organization of information exchange to integrate different technologies for both observed and observable components of the controlled system. The parameters and the structure of such control system can quickly be adjusted according to the current information about the internal state of the object and the characteristics of the environment, which are in a form of digital data.

These features open up new prospects for the development of intelligent CPh systems that will become in the near future an integral part of the human environment in the information space

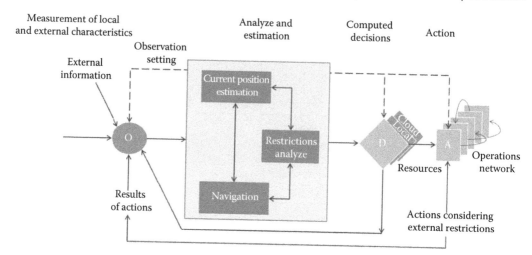

FIGURE 10.4 Cyber-physical interpretation of the John Boyd's OODA loop.

of so-called Internet of Things. According to the estimates [11], network-centric cyber objects in the global information space of the Internet will fundamentally change the social and productive components of people's lives. That will accelerate the knowledge accumulation and the intellectualization of all aspects of human activity.

However, this process requires not only innovative engineering ideas, but also the development of scientific concepts united into a universal scientific paradigm. Within this paradigm, the information should be considered a fundamental concept of objective reality, in which physical reality has *digital* basis and therefore is computable. The idea of integrating physical concepts with computation theory has led to the new conceptual scheme of nature descriptions, known as "it from bit". In this scheme, all physical objects, processes, and phenomena of nature, which are available to be perceived and understood by a person, are inherently informational, and therefore they are isomorphic to some digital computing devices. Within this paradigm, information acts as an objective attribute of matter that characterizes the fundamental distinctiveness of the potential states of the real object. The distinctiveness, according to Landauer's principle, is an energy factor of the object's states, and that is why it gives an explanation of what are the states and how they are perceived by other objects. This distinctiveness appears while creating the systems that are capable of ensuring the autonomy of their existence during the interaction with the external environment by the self-reproduction of their characteristics. It should be noted that on the way to the widespread use of *digital reality* for the control problems, there are some limitations that reflect the requirements for the existence of the special state of physical objects reflecting their changes as a result of information exchange processes.

The selection of cyber-physical systems as a special class of designed objects is due to the necessity of integrating various components responsible for computing, communications, and control processes (≪3C≫—computation, communication, control). Therefore, the description of the processes in such systems is local, and the change of its state can be described by the laws of physics, which are, in its most general form, a deterministic form of the laws of conservation of, for example, energy, mass, and momentum. The mathematical formalization of these laws allows us to determine computationally the motion parameters of the physical systems, using position data on the initial condition, the forces in the system, and the properties of the external environment. Although the classical methodology of modern physics, based on an abstraction of a *closed system*, is significantly modified by studying the mechanisms of dissipation in the so-called open systems, such aspect of reality as the information is still not used to build the control models and to describe the properties of complex physical objects. In the modern world, where the influence of the Internet, supercomputers, and global information systems on all aspects of the human activity becomes dominant, accounting an impact of information on physical objects cannot be ignored, for example, while realizing sustainability due to

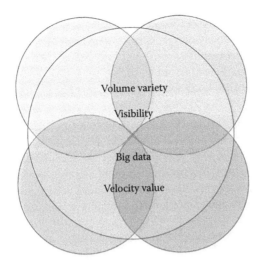

Volume variety

Visibility

Big data

Velocity value

FIGURE 10.5 Big data structure.

the information exchange processes. The use of cyber-physical methods becomes especially important while studying the properties of systems, known as the "Internet of Things," in which robots network cyber objects, and people interact with each other by sharing data in the single information space for the characterization of which are used such concepts as "integrity," "structure," "purposeful behavior," "feedback," "balance," and "adaptability."

The scientific bases for the control of such systems have become data science. The term "big data" describes the process of integration technologies for digital data processing from the external physical or virtual environment, which are used to extract useful information for control purposes. However, the realization of the data science potential in robotics requires creation of new methods for use of the information in control processes based on sending data in real time at the localization points of moving objects (the concept of "*data in motion*").

In general, "*big data*" is characterized by a combination of four main components (four "V"): volume, variety, velocity, and value (see Figure 10.5). The general "V" is the visibility of data, and it is also a key defining characteristic of big data.

As a result, "big data" (Figure 10.5) in modern science has become synonymous to the complexity of the system control tasks, combining such factors of the physical processes that characterize the volume, velocity, variety, and value of data generated by them.

So CPh approach is now often used to describe the properties of the so-called non-Hamiltonian systems in which the processes of self-organization are described by dissipative evolution of the density states matrix. However, the cyber-physical methodology may be successfully used to create complex robotic systems, the components of which are capable of reconfiguration as a result of transmitting and processing digital data or metadata. The control tasks that are considered in this book cover the actual scope of the CPh approach, which is the basis of cloud computing technology and development of the methodology of cybernetics toward the metadata control.

10.3 A SECURE AND HIGH-PERFORMANCE IaaS ENVIRONMENT IMPLEMENTING CYBER-PHYSICAL PROCESSES

Cloud providers, such as Amazon, Rackspace, Heroku, and Google, may provide different services based on the models of infrastructure as a service (IaaS), platform as a service (PaaS), or software as a service (SaaS), whose integration into a specific environment of industrial development is carried out by highly qualified engineers and IT specialists. Thus far, the actual challenge has been to develop cloud services for hybrid cyber-physical environments where robots and humans can operate together.

These services have to provide human resources as well as computing environment. Existing engineering centers are being built today on specially designed software and hardware platforms, which limits their performance and flexibility, or on the IaaS model, which also does not allow you to efficiently solve a variety of engineering problems.

The use of such a heterogeneous computing environment has the following advantages:

- The computing environment allows one to expand the range of information services, which in turn allows one to quickly and cost-effectively implement multidisciplinary projects.
- Virtualization and heterogeneity provide scaling resources to ensure high performance of computation at all stages of the implementation of engineering projects.
- Cloud architecture implements automatic configuration of hardware and software components, versioning of applications, and monitoring the integrity of the computing environment.
- Network services provide the benefits of a network-centric approach in the implementation of complex engineering projects by geographically and logically distributed development teams and specialists.
- A stealth security system implements a common policy in the field of information security.

The heterogeneous cloud platform is the basis of computing infrastructure for cyber-physical centers; the principal difference from the classic data centers is to provide remote access not only for computing resources or applications, but also for intelligent services. Using the resources of modern cloud-based engineering centers, it is possible to create equivalent social networks that bring together multiple agents, which can be potentially distributed over our planet or galaxy to perform coordinated actions, computation, verification of test results based on the use of different materials, virtual prototyping, and data visualization. These problems, from the point of view of the computational algorithms, can be combined into chains, which form a network of operations. Their implementation is provided within a heterogeneous cloud. The components of the platform (Figure 10.6), based on the OpenStack, include IaaS cloud class segment, computing infrastructure within the cluster, and the specialized high-performance hybrid system based on reconfigurable computing nodes.

Virtualization has changed the approach of deploying, managing, and using enterprise resources by providing new opportunities for consolidation and scalability of computational resources available to applications; however, this leads to the emergence of new threats posed by the complexity and dynamic nature of the process of resources provisioning. These threats can lead to the formation of cascade security violations, which traditional data protection systems are unable to deal with. The existing approaches such as "scan and patch" do not work in a cloud environment—network scanners cannot track changes of resource configurations in real time. These approaches do not accurately identify the change in the level of risk and take steps to block dynamically emerging threats.

To solve the problem of controlling access in the cloud, it is necessary to continuously monitor resources, and it cannot be achieved without automatic generation of rules for filtering and firewall log files analysis. Information security management products in a dynamic cloud environment should include mechanisms that provide total control over processes for deploying virtual machines; proactive scanning of the virtual machines for the presence of vulnerabilities and configuration errors; and tracking the migration of virtual machines and system configuration to control access to resources. Therefore, within the center of the "polytechnic," field a series of measures are set out to improve information security resources, namely, the following:

- *Enhanced control of virtual machines.* Virtual machines as active components of the service are activated in the cloud application random moments, and administrator cannot activate or deactivate a virtual machine until the security scanner checks the configuration and evaluates the security risks.

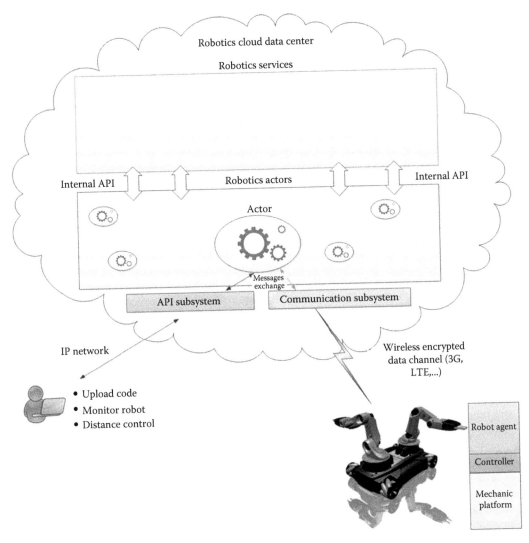

FIGURE 10.6 Robotic cloud platform with heterogeneous computing resources.

- *Automatic detection and scanning.* Information security services are based on the discovery of vulnerabilities in the computing environment. This discovery in turn is based on the current virtual machine configurations and on the reports of potential threats that come from trusted sources, such as antivirus update servers.
- *Migration of virtual machines.* Proactive application migration is an effective method to control security.

In addition, the computing center has access control service for the cloud services protection. Its main features are support of the dynamic infrastructure, scalability, and the ability to support security policies without reference to the composition of resources. This service is built on the technology of stealth traffic filtering and software-defined networks (SDN) and provides the reconfiguration of access isolation system according to the current state of the environment (the question of access control in cloud environments is discussed in detail in Lukashin, Zaborovsky, and Kupreenko and Zaborovskiy et al. [3,4]). Static platform segments adhere to the principle of "rent" under which the filter rules of segment access are formed only by users and services running at any given time.

On the platforms of this type, the situations when user needs a single virtual machine are rare. There-fore, cloud services support the dynamic creation of secure networks with a set of preconfigured virtual machines. Secured networks are connected to the firewalls, which are integrated with the distributed SDN switch, Open vSwitch, or OpenStack Neutron networking service.

Firewalls protecting the dynamically generated cloud networks are created during computing seg-ment initialization. An important feature of the firewall is its ability to function in the address-less mode. It allows implementing invisible protection of a cloud and security system integration not requiring the reconfiguration of a cloud network subsystem. The firewall acts as a virtual machine. The firewall of a network segment filters network traffic according to the rules created by the access policy service. Access policy in a cloud computing environment is based on the role-based access control model of access control. This policy can be represented as a set of the following attributes:

- User IDs, which are involved in the management of virtual machines and information services.
- Privileges that are described in the form of permitted information services (privileges set rules for user access to services; it is possible to change the privileges for the user in the specified virtual machine filtering rules for your firewall, which allow access to a network service).
- Set of roles that can be assigned to users.
- User sessions in a computing environments based on the network connections between subjects and objects.

Access policy is translated to firewall filtering rules according to computing environment state. This state can be represented by a set of IP addresses of computing resources, with assigned user labels. Label represents a user holding the computing resource. When the state of a computing environment changes, then it is necessary to generate a new set of filtering rules and reconfigure firewalls. For this purpose, a method of the dynamic configuration rules has been developed, which consists of substitution of the network address lookup in user-owners' privileges for each virtual machine. This approach formed the rules of access to the services of the computational resource and of computing resource to services of other users.

At least one virtual firewall in each virtualization server and one general bare-metal firewall for protecting cloud services from external threats are required. In a cloud-based system, there is a dedicated management network separated from virtual machines, so this network is used for the information exchange between the components of the access control system and cloud services (Figure 10.7).

OpenStack cloud platform is implemented by using service bus for communication between its components. Service bus is based on Advanced Message Queuing Protocol technology and RabbitMQ service. Access control security service was integrated with OpenStack bus by subscrib-ing its software components to events of OpenStack Compute service, which is managing the life cycle of virtual machines, and OpenStack Neutron service, which is managing the life cycle of cloud networks. When security service receives an event that a new virtual machine is starting, it gener-ates and distributes filtering rules for the firewalls and generates rules for the virtual switch using OpenFlow technology, which redirects traffic from a virtual machine to the firewall. Firewall-based approach allows controlling traffic between instances that are connected to one virtual switch but belong to different users or security groups.

The security service requires additional resources in the cloud. Traffic filtering costs make up about 10% of the virtualization server's resources [4].

For tasks that require heterogeneous computing resources, it is necessary to automate creation of the protected segments. We describe heterogeneous computing system as a set of logical computing resources. Such a segment must be applied to the specified security policy to permit the possibility

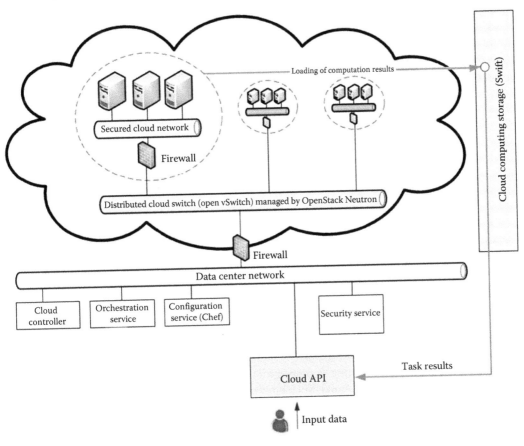

FIGURE 10.7 Components of a cloud platform and security in an SDN infrastructure.

of access to computing resources for the owner, but forbid access to these resources for other users. When the task is complete, the results must be loaded into the data warehouse, and the computing resources are freed. At the same time, it is essential to guarantee access to computing resources in simultaneous execution of multiple tasks.

We used OpenStack for creating groups of virtual machines in a cloud environment service. This service supports description of configurations in an Amazon Cloud Formation format that ensures compatibility with public services such as Amazon AWS. This service allows creating groups of virtual machines according to pattern, virtual networks, cloud-based routers, and other components. The images of virtual machines contain a basic set of services. Any other application-specific packages are installed using the automation services provided by the Opscode Chef framework that provides automated deployment of software configurations in virtual machines and bare-metal servers. When a new computation segment is being created, the security system spawns and configures a virtual firewall, which is filtering the access to newly created network serving the computation. Dynamic network creation is supported by OpenStack Neutron services and by a distributed virtual switch Open vSwitch. After computation is complete and the results are received, the segment is removed, the cloud resources are released, and the results are uploaded to cloud storage to become available to the other consumers of the service. Every operation is automated: there are no steps requiring human intervention.

Reconfigurable segments of the cloud allow solve a wide range of scientific and technical tasks, among them: tasks that operate on large data sets based on the MapReduce technology. Tasks that

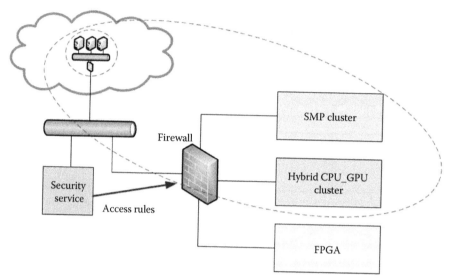

FIGURE 10.8   Reconfiguration of a hybrid supercomputer center using firewalls.

cannot be solved in the cloud virtual machines (e.g., requiring quick access to globally addressable memory and massively parallel or streaming computations) are transferred to dedicated hybrid clusters for high-performance computing, equipped with an internal high-speed communication bus, and nodes accelerators based on FPGA and GPU. Firewalls provide protection from unauthorized access to computing resources in a time of challenge and consolidation of heterogeneous segments (cloud and high-performance) computing resources into a single computation network, which components can communicate with each other, using the allowed protocols.

Built this way, the infrastructure allows to dynamically create secure computing segments and thus provides an opportunity to organize a simultaneous execution of various tasks on a single set of hardware resources (Figure 10.8). This solution implements reconfigurable a federated cloud with one interface and multiple computation segments. A similar approach was used for organizing a mobile cloud for intelligent transport systems [8].

The proposed approach for organizing a cyber-physical data center, which is based on cloud services enables the ability to reconfigure computing resources for different computation tasks. Integrated security services allow sharing computing resources between different users and clients. Reconfiguration of computing resources by using cloud firewalls is not a standard approach. It requires additional resources and makes platform more complex. From the other side, it provides opportunity of reconfiguration of resources on network level. Stealth technology allows leaving applied software without modification. Dynamic computation segments creation service allows to effectively use the IaaS resources on demand.

The proposed IaaS cloud platform, named *Pilgrim*, serves hybrid computing resources in the data center of St. Petersburg Polytechnic University. Figure 10.9 shows the structure of the cloud-centric computing environment for avatars that control robots as a group of cyber-physical objects.

10.4   PaaS ENVIRONMENT FOR CYBER-PHYSICAL OBJECTS

It is necessary to extend IaaS computing environment to organize the operations of cyber-physical systems. It is impossible to create a scalable software environment just by using virtual machines and virtual networks. To organize the cyber-physical operations, the following formalism has been proposed:

$$CPO = <Avatar, Agent, Platform>,$$

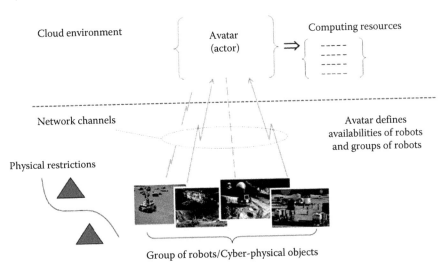

FIGURE 10.9 Computing environment for avatars that manage robots.

where a cyber-physical object (CPO) is decomposed into following three instances:

- The avatar is the brain of an object. It is represented by a set of processes inside a high-performance computing environment.
- The agent is client software that runs on a robot itself. It can be placed on a controller or a small computer. It does not perform complex computation operations. The agent collects data from the robot's sensors and sends control signals to a platform. All data are sent to the avatar, and the avatar sends control messages. However, some agents might have autopilot functions engaged when connectivity with the avatar is temporarily lost.
- The platform is a physical representation of a cyber-physical object (CPO). It could be vehicle, manipulator, unmanned aerial vehicle (UAV), etc. A platform is managed by the agent's controller.

Let's describe the tasks to be performed by an avatar of a cyber-physical object:

- The avatar decides which operation has to be performed by its agent and platform.
- The avatar converses with another avatar using an internal cloud communication bus.
- The avatar converses with the available services: databases, storage services, Internet services (it can even use Wikipedia to get required information).
- The avatar launches computation and data management processes inside a cloud. It could be, for example, big data tasks driven by Apache Hadoop and high-performance computing tasks.

According to the above, an actor-based architecture was proposed. An actor as primitive as a computing unit was proposed in a model developed in 1972 [1]. In this case, the avatar was presented as a set of actors:

$$\text{Avatar} = \{\text{actor1, actor2,..., actorN}\}$$

Each avatar's actor had its own program code, which was presented in high-level language. In our platform implementation, Scala and Akka platforms were used, but, in general, this can be done in any language that supports the actor model Java or Erlang, for instance. Actor models allow one to scale out the cloud environment and organize effective communication between avatars and other services. Message-centric architecture allows for transferring internal messages from avatars to their agents using access gateways [2,5].

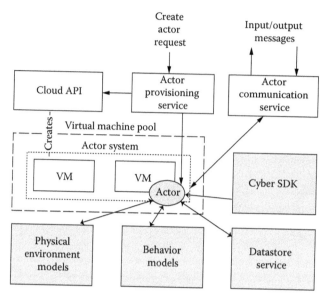

FIGURE 10.10 Cloud-centric software architecture.

Avatars in a cloud have the ability to subscribe to custom events generated by other actors. For instance, one CPO collects images from its camera. The avatar of such cyber-physical objects performs image analysis and, if the results are of some interest to the system, sends an event to a hub, and all the avatars interested in this topic extract the required data from message and generate control data for their agents.

Cloud-centric software architecture is presented in Figure 10.10. An actor provisioning service handles requests from users or agents. Based on the request data, it allocates a new actor in the actor system, which is deployed on multiple virtual machines. If all virtual machines are fulfilled by another actor, it invokes a cloud API and creates a new virtual machine for an actor system. An actor can communicate with an agent by sending and receiving messages through an actor communication service. It translates an actor's messages to ProtoBuf data and sends it over the network. Cyber SDK is a library, which consists of generic robot algorithms, utility methods, and so on. A behavior model is a storage of serialized data, which represents a robot's behavior. It could be neural networks, input data for algorithms, and other parameters. Physical environment models service provides descriptions of physical environments such as maps and routes. An actor could communicate with these services by using their API based on message-driven architecture.

Platforms of such avatars can be distributed all over the world: in this case, all communication is performed inside a data center without sensible delays. Figure 10.11 presents an architecture diagram of a PaaS platform.

A platform is split into eight levels. Levels 1–3 represent an infrastructure layer and are provided by the Pilgrim IaaS platform. In general and with some restrictions, the first three levels might be implemented in another IaaS cloud, for example, in Amazon VPC cloud. Levels 4–6 are the heart of a PaaS platform. It is the avatar's actors placed into their containers and the service's actors that provide an interface to computing and information services. Level 7 is a data transmission layer. Each agent has its own interface for connecting with the data center. It could be a GSM or a WiFi connection, or a radio channel. The last, eighth level, is an agent world. All robots, operators, and other cyber-physical objects are represented by this level.

Such architecture and CPO decomposition have the following advantages:

- A PaaS paradigm helps platform users to write code on high-level language (Scala, Java) and simply load it into a computing environment.

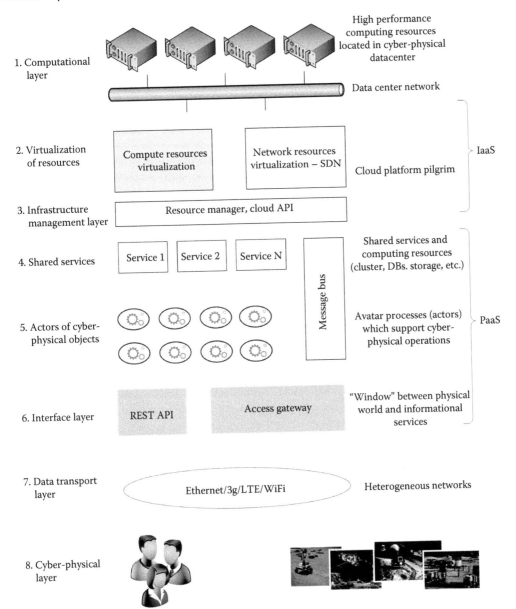

FIGURE 10.11 Architecture diagram of a cloud platform for cyber-physical operations.

- The avatar as a brain of a CPO can be easily replaced during runtime. One robot can play a lot of different roles: everything depends on the avatar.
- There are no cognitive functions inside the agent. So, complex logic cannot be reverse-engineered: the agent is useless without its brain.
- A high-performance computing environment allows one to perform operations that are impossible to implement on relatively low-power mobile hardware [6].
- A platform allows to one perform fast communications between CPOs distributed over long distances.
- Also, such architecture allows one to organize network-centric systems with shared knowledge base and information actualization in real time.

10.5 CYBER-PHYSICAL APPROACH FOR SPACE ROBOTICS

10.5.1 SPACE ROBOTICS BACKGROUND

Currently, in the development of space activities, a major focus is on the use of space robotics. Both during transit and at their destination, astronauts need advanced autonomous and robotic systems to free them from routine tasks and to maximize their valuable scientific exploration time. This is why it is so important to increase autonomous operation time for space robots, which we consider the cyber-physical objects.

It is clear that the current robotic systems, which are based on the cyber-physical approach have serious limitations. Planetary surface rovers have never been designed for real-time interaction with each other or with astronauts. Moreover, controlling the robotic systems currently requires extensive human resources (e.g., a room full of controllers for one Mars rover), far beyond the capabilities of a crew on Mars tasked with real-time operations. We propose modular cyber-physical robotic systems, which can be fielded more quickly, are more reliable, and are easier to control than current systems. Modular designs will allow multiple robots to interact as peers among themselves and with humans, and improve the flexibility of robots. Because space robotic systems have to operate in highly unstructured environments with limited sensory information, limited actuator capability, limited power sources, limited communication bandwidth for control and coordination, and stochastic time delays, space-based research in autonomy and robotics is an incubator of innovative technologies for terrestrial robotics that can improve the quality of life for the earth's growing and aging population, which will have utility in manufacturing and service industries, in medicine, and in disaster management.

10.5.2 ADAPTATION OF AUTONOMOUS CYBER-PHYSICAL OBJECTS FOR LUNAR AND PLANETARY EXPLORATION

This research involves the development of control methods for tasking cyber-physical systems that make it as intuitive and natural to interact with autonomous agents as it is to interact with human teammates. There are four major components in this work:

1. *Collaborative human–robot plan diagnosis*—Reduction of cognitive load on human supervisors will require human and robot teammates to adapt to complex, changing, and sometimes conflicting mission goals. We propose to address this challenge by developing a capability for human and robot teams to collaboratively negotiate in order to diagnose problems with mission requirements, so that a feasible initial plan can be produced using high-performance cloud-centric software. This process will be repeated if circumstances or goals change as the plan is being executed. The communication between humans, actors in cloud environment, and their autonomous cyber-physical teammates must be efficient and natural, as if the humans were communicating with other humans.

2. *Automatic learning of new skills and models*—Using cloud-centric software, we propose to develop learning and execution monitoring capabilities that allow task execution to be more easily adapted to new robotic hardware, mission tasks, and human teammate behavior. Robots are adapted to specific missions by supplying them with a set of models of hardware, allowed robot actions, and user behavior. We propose to reduce the time of developing new models by having robots learn new behavior directly in cloud environment using hardware and task requirements. The ability to recognize actions and task sequences is crucial. In order for a robot to provide assistance, it must understand what the human is trying to accomplish, particularly when operating in an unstructured environment, where new skills may have to be developed on the fly in order to adapt to unforeseen circumstances.

3. *Reactive planning*—In an unstructured environment, execution of a plan will often be disturbed by unforeseen changes in the environment, in the condition of the robots, or in the mission goals themselves. Dealing with unforeseen disturbances requires three capabilities. First, the autonomous system that is monitoring plan execution must recognize whether a

disturbance is severe enough that it will make successful execution infeasible. Second, the autonomous system must be able to quickly change a plan that its estimates will fail due to a disturbance. Third, the autonomous system must understand risk and must be able to predict the likelihood of failure of the current plan. We propose to integrate these capabilities into our cloud environment for motion planning algorithms, as well as our collaborative diagnosis capabilities, so that risk is considered at every decision point.

4. *Perspective space robotic network*—To integrate information, intelligent, and robotics technology within robust human–robot infrastructure, we need to have solid cyber-physics platform that merge together formal models of human knowledge, operation ontology, high-performance computing resource, and control goals into space robotic network. For this purpose, we will a implement perspective cloud computing approach and extend IaaS/PaaS concepts to a robot as a service model.

10.5.3 CLOUD-BASED FUNCTIONAL MODELS OF AGGREGATE-MODULAR ROBOTS FOR SPACE EXPLORATION

The creation of an autonomous robot base on the Moon's surface is the first step to verify the technology of human exploration of the Moon and other celestial bodies. At the moment, research that is focused on robotic lunar exploration are included in the space programs of many countries, including the United States, China, and Russia. For example, the well-known Chinese Lunar Exploration Program incorporates different types of robots: lunar orbiters, landers, rovers, and sample return spacecraft.

According to the regulatory documents of Russian Space Agency, the main component of the robotic lunar base would be a group of robots, which would be capable of solving all kinds of problems arising during Moon exploration, namely, manipulation with objects, transportation of cargo, installation and maintenance of different facilities, construction and installation works, and mining.

There are two main approaches to form such a group of robots:

1. Few multifunctional complex-to-manufacture-and-maintain robots, each of which performs a wide range of tasks.
2. Lots of different robots that are easy to manufacture and operate, each of which performs a limited range of tasks.

The main advantage of the second approach is the simplicity of manufacturing robots' and functional modules' and, as a consequence, a higher reliability of the overall system, which in the context of space exploration is a critical parameter. It should also be noted that the specialized robots, which have been formed from different functional modules, have less redundancy for the task. This improves the efficiency of the robotic system and help to achieve the optimum of the quality criteria, namely, accuracy rate of the operation, cost of operation in terms of energy and fuel, and runtime of the task.

One of the most common methods to implement this approach is the aggregate-modular design of robots. This method is based on the universal interfaces for the information and the mechanical interactions between components of a robot. It gives us the opportunity to construct a specialized robot on the base of limited number of standardized components. Such construction meets the requirements of the current task in the best possible way, so this specific solution has minimal redundancy.

The main purpose of our work is to research and develop an algorithm for automatic construction of the optimal configuration for the robot from the set of standardized modules to implement a task. It is assumed that such standardized modules have uniform mechanical and informational interfaces. The main objectives of our work include the following:

- Classification of robots and their modules
- Development of a model of an aggregate-modular robot

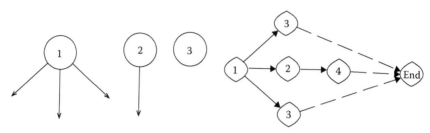

FIGURE 10.12   Various modules and examples of a representation robot as a network: (1)—transport platform with three unified interfaces; (2)—manipulator with one interface; (3,4)—sensors.

- Researching the limits of the robot model based on the requirements for the optimal realization of the task
- Development of a method and algorithm for automatic construction of the optimally configured robot

At the first phase of the research, standardized robotics modules have been classified. The following groups have been identified:

- Handling systems and control devices
- Manipulators
- Transport platforms and movement systems for mobile robots
- Sensors and sensor systems

On the basis of the proposed classification, we have developed a mathematical model of a combined aggregate-modular robot that meets the following requirements:

- Each module is presented as a graph vertex with emanating from him *hanging* outgoing edges that represent unified interfaces that may be attached to other nodes. Examples of various unified modules are shown in Figure 10.12.
- All robotic objects are described as a network graph G (Figure 10.11). A source of such a network is always a transport platform or another movement system for a mobile robot. A drain of this network is a dummy node.
- A mathematical model assumes the existence of a vector estimation function of the current solution—network graph G. This function is named $F(G)$ and should estimate the performance of a combined robot using the characteristics of individual modules.
- A mathematical model assumes the existence of a system of functional limitations $f_i(F(G))$, which is formed according to functional requirements of the robot's task.
- A mathematical model requires an optimization criterion of functioning quality for a particular purpose $J(F(G))$.

The requirements for the problem of finding the optimal configuration of aggregate-modular robots can be formulated as follows. It is necessary to generate a graph G^*, which satisfies the functional limitations $f_i(F(G))$ and gives minimum to the optimization criterion $J(F(G))$.

The proposed mathematical model describes the NP-complete optimization problem. During our research, the following types of methods and algorithms that can be used for solving NP-complete problems were considered:

- Exhaustive search algorithms or brute-force search
- Approximate and heuristic methods that involve the use of *a priori* rates and heuristics to select the elements of the solution
- Branch and bound method, involving discarding obviously suboptimal decisions according to some estimates.

10.5.4 ROBOT CONTROL FROM ISS

We consider the application of the earlier proposed principles to control physical objects: an on-surface robot, the motion of which is set and controlled from an orbital space station (Figure 10.13).

The feature of this Space Experiment (SE) "Kontur-2," which allows us to consider it a class of cyber-physical experiments, is that it uses telepresence technology for the operator to simulate the robotic movement while the parameters of the environment (delays in communication channels and obstacles on the planet's surface) may vary in a random way. The designed control system allows one to achieve considerable results of robotic operations in real time by analyzing the information about the values of the current axes of movement or moments in the joints of the manipulator, which are transmitted via computer communication network with a frequency of 500 packets per second.

The use of the circuit-torque delays sensitization effect allows the operator to adjust the speed and movement direction of the robot, and by using the force feedback effect on the joystick, feel the impact of the network environment and generate an assessment of the environment state in which operates an on-surface robot. In the described control system, the process of the information exchange between the joystick and robot can be decomposed into two processes of local command realizations in hard real time and the command delivery process via a network infrastructure using the TCP/IP stack.

The physical structure of the data streams in such a control system is shown in Figure 10.13 and includes the following:

1. The local loop, in which the software module, the "Joystick Controller" (JC) provides cyclic polling of the current joystick coordinates, calculating and sending in joystick the force vector depending on the current position and velocity of movement of the joystick's handle, as well as feedback information (T'), obtained from the cyber object (CO).

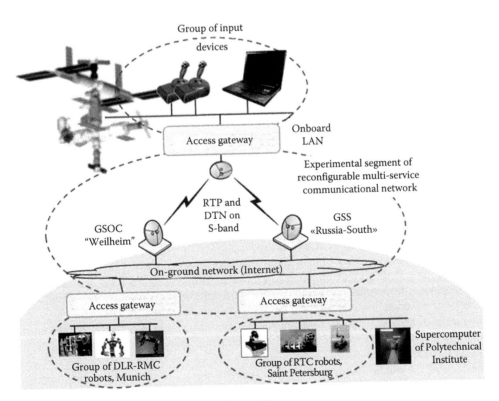

FIGURE 10.13 Scheme of Space Experiment "Kontur-2."

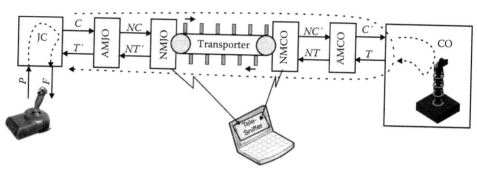

FIGURE 10.14 Data streams in Space Experiment "Kontur-2."

2. The network loop, in which the software components are used to organize the transfer of the control vector (C) and the telemetry (T) between the JC and the CO.

The basis of the network control loop is a software module "Transporter", which consists of the network modules of the joystick and CO (NMJo, NMCO, see Figure 10.14), which are connected by the virtual transport channel, based on UDP. With the end systems (JC and CO), the network modules are connected through the adaptation modules (AMs) to the properties of the communication media (AMJo and AMCO).

The software module "Transporter" delivers data using UDP, providing isochronous communication for local controllers: vector control samples that are uniformly received from the JC should also be uniformly (but, of course, with some latency) delivered to the CO. Similarly, in the opposite direction, the vector data obtained from the telemetry system of the robot are transferred. Thus, the *digital reality* of the physical processes is updated with a sampling frequency equal to the frequency of sending packets. Taking into account that the data delivery delay in the digital communication channels is not constant, some packets may be lost, and the delivery order may be disturbed. The adaptation module provides recovery of the missing packets using an automata model of data transfer processes, the adequacy of which is verified using the probabilistic model of packet delivery.

The "Transporter" module also ensures the delivery of asynchronous messages about events that are relevant to the remote control. The example of such events may be pressing the button on the joystick's handle. These buttons may be used to control the operating mode of the CO or for some other action. Asynchronous event occurs sporadically at any given time. However, it should be guaranteed to reach the operator, saving the time reference to the transmitted isochronous packets' stream.

The computing resources of the adaptation module allow one to implement the methods of predictive modeling and, in the case of insufficient data, to predict the behavior of the control object or operator without any delay in the transmission of the control signals, thereby ensuring smooth control for the robot. The control circuit scheme is presented in Figure 10.15.

In Figure 10.14, there is the separation of the control circuit into three main parts.

The loop of the Joystick (Jo) (the left part of the scheme), whose output is the control vector c, which shows the current position of the Jo handle p_m. The movement of the handle is determined by the force f, which is a sum of the operator's impact forces f_h, force f_c, which is formed by the Joystick, and force f'_e, displaying the impact of the environment f_e on the robot. P input of JC is the mismatch between the current Jo position (p_m) and the display of the current position of the robot (p_s). The controller parameters vector {A} can be changed dynamically.

The means of communication (Transporter), describe the delays in delivery, and the adaptation module counteracts the effect of these delays. In general, the delays in the direct (T_f) and the reverse channels (T_b) can be different.

The loop of the controlled device (the right part of the scheme), generates an impact on the robot using a mismatch between its current position p_s and the displayed position of the Jo p'_m. A robot may also be affected by the environment with force f_e. For the feedback, the vector of current

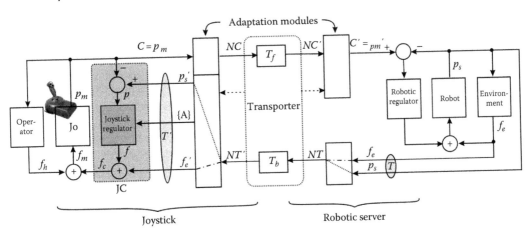

FIGURE 10.15 "Kontur-2" control circuit scheme.

robot position p_s can be used, and if there are appropriate sensors, the force vector of environmental impact f_e can also be used.

For such a network control problem, queuing theory is a powerful tool. The Transporter is implemented using a preemptive queuing system [7–10]. The remote robot control with force-torque sensitization requires that during the control process, that operator has to be able to *feel* the current state of the robot; it also requires a communications network. Therefore, to describe the configuration space of the controlled system is proposed to use a model of the virtual spring that is fixed at one end to the base of the joystick, and the initial position of which coincides with the current position of the controlled robot. In the initial position, while not moving, the robot does not affect the joystick and the operator. However, if the operator starts the control process and moves the joystick's handle from the initial position, the virtual string will try to return it to the initial position. Farther the joystick will be deflected from the initial position, the greater force will the operation of stretching virtual springs require from the operator. Virtual elastic force of the spring is calculated based on the analysis of the current situation, taking into account the current positions of the joystick and information from the robot. It is calculated using the following equation:

$$p = p_s' - p_m,$$ (10.1)

where
   p_s' is a vector of the robot's coordinates, which is currently available for the adaptation module of the joystick
   p_m is the current position of the joystick

The impact force felt by the operator is calculated according to the following equation:

$$f(t) = A_0 \int_0^t p(\tau)d\tau + A_1 p + A_2 \dot{p} + A_3 \ddot{p},$$ (10.2)

where
   A_0 is an integral gain of the system
   A_1 is the virtual spring stiffness
   A_2 is the viscosity of the control medium
   A_3 is a virtual mass of the handle

During the control process, the initial position of the virtual spring will change to coincide with the current position of the robot. The proposed model allows taking into account the information about the state of the communication channel by making adjustments during the process of sensitization. In other words, the virtual stiffness of the spring must be increased not only according to the divergence between positions of the robot and the joystick, but also using the value of the data delay in the communication channel, thus increasing the inertia of the network control loop.

To implement such an interaction mode, we will need to handle the additional information about the state of the communication channel: the sequence of RTT samples that is ordered by time and the percentage of the lost network packets (PLP). Thus, the stiffness of the virtual spring and the moving speed of the robot will vary proportionally average values of RTT and PLP, which are calculated for the time period according to Equation 10.3. This period is commensurate with the time constant for the closed-loop control system.

$$f(t) = \left(A_0 \int_0^t p(\tau)d\tau + A_1 p + A_2 \dot{p} + A_3 \ddot{p} \right) + A_4 \cdot RTT \cdot PLP \cdot \frac{p}{\|p\|}, \qquad (10.3)$$

where A_4 is a coefficient of influence of the environment on the force-torque feedback. As a result, while increasing the latency, the virtual spring will not allow the operator to change robot's position quickly. This circumstance reduces the speed of the robot while moving, but allows the operator to adjust the results of operations, analyzing data of control actions, despite the fact that these data that will be available to the operator are delayed.

The proposed organization of the remote control system was designed for ground testing of algorithms and software debugging of the scientific equipment in the Space Experiment "Kontur-2." On the experimental stand, the parameters of the communication system and software module "Transporter" are modified according to the constraints imposed by S-band communication channel's bandwidth. These constraints are estimated according to the necessity of the telepresence mode, which requires every 2 ms in order to transmit IP packets containing 22 bytes of application data. These requirements helped to modify the format of the messages and to identify the properties that reflect the specifics of network processes. To reduce the bit length of the packet's number field and time stamps, we propose to divide numbers of transmitted packets into even and odd numbers. The even packets are used for the transmission of isochronous traffic (5 parameters in 32-bit floating-point format), and the odd ones are for the transmission of asynchronous messages and proprietary information, which provides the definition of the channel state with the implementation of network protocols and the use of an *lightweight* LwIP stack.

According to the requirements of the bilateral control, the software for a two-loop remote control system is symmetrical, that is, the structure of software modules on the robot side and on the joystick side is the same. The local controllers in joystick and robot are implemented as PID controllers and differ from each other only by settings. Figure 10.16 shows the structure of the developed software for joystick implemented for the 32-bit ARM microcontroller.

JC software module reads the current joystick position and state of its buttons, as well as forms and transmits the force feedback control commands. The module implements the PID controller function, for which the input value and the coefficients in the control law are formed by the software AMJo.

Transporter module provides regular transmission for network packets, requesting the AM to convert control vector and telemetry, as well as to process the current information about communication delays in the channel.

Adaptation module (AMJo) performs the conversion and ensures stability of the closed system and the quality of control. The system software on the ARM microcontroller generates runtime and includes the following components: USB-host, LwIP, FreeRTOS, and a set of drivers for configuring and using network interfaces.

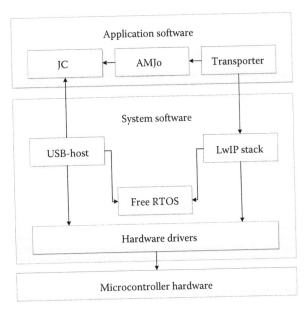

FIGURE 10.16 The structure of developed joystick software.

The FreeRTOS module is used for the sharing of different modules' tasks listed as follows:

- The Robot control server receives commands from a personal computer and forms counter-flow data to control a 3D-model of the robot and to register control process parameters.
- A USB-host provides regular requests for the current joystick's coordinates and transmitting force control commands.
- The LwIP stack performs primary processing of IP packets (checksums, statistics, etc.) and setting them into the input queue of the program module Transporter.
- The Transporter provides a two-way exchange with isochronous packets and asynchronous messages.

The Transporter module is activated at the double frequency of data exchange that allows to call the adapting module and to transmit it alternatively the telemetry vector, received from the robot, and the control vector, obtained from JC. The result from AMJo is transmitted to the JC or sent via the network to robot. If there is no input message at the proper time, in this case, the AMJo generates a control action, predicted using Equation 10.3.

10.6 CONCLUSION

The proposed approach to cyber physics as a new interdisciplinary science focuses on the organization of the engineering infrastructure and software architecture, providing an opportunity for communication between physical, mechanical, and information-computing processes arising during network-centric interaction of intelligent robots and their virtual *avatars*. Using the metaphor of *information ashes* allows us to emphasize the importance of presenting physical connections as one of the forms of information relations underlining the idea that "everything is from bit." Operating space for the virtual image of physical objects can be formed by specific cloud-centric infrastructure based on robust communication channels, big data repository. The designed software architecture of cloud computing infrastructure is the reliable carrier of the control algorithms that form robotic operation network as the constructive model for interactions of cyber-physical objects. The effectiveness of the proposed solutions is illustrated by the example of a network-centric management system of planetary robots whose operations at the level of calculating trajectories or maintenance of information integrity and functional robustness are controlled from the board of ISS in the framework of SE "Kontur-2."

ACKNOWLEDGMENTS

This work uses the results of the International Space Experiment "Kontur-2" carried out by the Russian State Scientific Center for Robotics and Technical Cybernetics and DLR. The chapter is funded by RFBR Grant 13-07-12106 and is done within the framework of the joint research project between Ford Motor Company and St. Petersburg Polytechnic University.

REFERENCES

1. C. Hewitt, P. Bishop, and R. Steiger (1973). A universal modular actor formalism for artificial intelligence. *IJCAI*.
2. V. Zaborovsky, A. Lukashin, S. Kupreenko, and V. Mulukha. Dynamic Access control in cloud services, *The 2011 IEEE International Conference on Systems, Man, and Cybernetics (IEEE SMC 2011)*, Anchorage, Alaska, October 9–12, 2011, pp. 1400–1404.
3. A. Lukashin, V. Zaborovsky, and S. Kupreenko. Access isolation mechanism based on virtual connection management in cloud systems, *13th International Conference on Enterprise Information Systems (ICEIS 2011)*, pp. 371–375.
4. V.S. Zaborovskiy, A. Lukashin, S.G. Popov, and A.V. Vostrov. Adage mobile services for ITS infrastructure, *13th International Conference on Proceedings of ITS Telecommunications (ITST)*, November 5–7, 2013, pp. 127–132.
5. A. Lukashin, L. Laboshin, V. Zaborovsky, and V. Mulukha. Distributed packet trace processing method for information security analysis, *The 14th International Conference on Next Generation Wired/Wireless Advanced Networks and Systems NEW2AN/ruSMART*, 2014, LNCS 8638, pp. 535–543.
6. A. Lukashin and A. Lukashin. Resource scheduler based on multi-agent model and intelligent control system for OpenStack, *The 14th International Conference on Next Generation Wired/Wireless Advanced Networks and Systems NEW2AN*, 2014, St. Petersburg, Russia.
7. V. Zaborovsky, O. Zayats, and V. Mulukha. Priority queueing with finite buffer size and randomized push-out mechanism, *Proceedings of the Ninth International Conference on Networks (ICN 2010), Menuires, the Three Valleys*, French Alps, April 11–16, 2010, Published by IEEE Computer Society, pp. 316–320.
8. V. Zaborovsky, V. Mulyukha, A. Ilyashenko, and O. Zayats. Preemptive priority queueing system with finite buffer size and randomized push-out mechanism, *Modern Traffic and Transportation Engineering Research*, 2012 1(2), 46–53.
9. A. Ilyashenko, O. Zayats, V. Muliukha, and L. Laboshin, L. Further investigations of the priority queuing system with preemptive priority and randomized push-out mechanism, *The 14th International Conference on Next Generation Wired/Wireless Advanced Networks and Systems NEW2AN/ruSMART*, 2014, LNCS 8638, pp. 433–443.
10. V. Muliukha, A. Ilyashenko, O. Zayats, and V. Zaborovsky. Preemptive queueing system with randomized push-out mechanism, *Communications in Nonlinear Science and Numerical Simulation*, Available online September 16, 2014, http://dx.doi.org/10.1016/j.cnsns.2014.08.020. (http://www.sciencedirect.com/science/article/pii/S1007570414004031).
11. V. Zaborovsky, M. Guk, V. Muliukha, and A. Ilyashenko. Cyber-physical approach to the network-centric robot control problems, *The 14th International Conference on Next Generation Wired/Wireless Advanced Networks and Systems NEW2AN/ruSMART*, 2014, LNCS 8638, pp. 619–629.

11 A Cyber–Physical–Social Platform for Intelligent Transportation Systems

Mohamed Elshenawy, Mohamed El-Darieby, and Baher Abdulhai

CONTENTS

11.1 INTRODUCTION AND MOTIVATION

The field of ITS [1] capitalizes on emerging information and communication technologies (ICT) to better manage today's increasingly congested and dynamic transportation networks. ITS involve the application of ICT to improve the performance of transportation systems and to increase the contribution of these systems to economic and social well-being. Typical ITS applications [2] involve a wide variety of interconnected research problems, the nature of which can be (1) technical (sensing, communication, automated vehicle technology, etc.); (2) methodological (control systems theory, operations research, traffic flow theory, artificial intelligence, simulation methods, image processing, etc.); and (3) behavioral (driver–vehicle interactions, user decision making, travel demand management, etc.). These three aspects can also be referred to as physical, cyber, and social, respectively. Thus, transportation-related research is inherently multi- and interdisciplinary in nature. A contemporary challenge to the ITS community is to integrate human knowledge and expertise with technical resources. ITS stakeholders are known to be working in isolated silos.

This chapter describes the high-level design and operational aspects of the ONE-ITS platform. ONE-ITS is a cyber–physical–social computing platform [3] that aims at the collaborative understanding of work processes, knowledge patterns (e.g., best practices, trends, how-to's, insights), and human needs to support the *flow of knowledge* between ITS stakeholders. This is embodied through trying to tackle the following exemplary questions: How can knowledge be extracted from existing software systems and integrated with daily ITS business and technical processes? Can a stakeholder expose, compare, and match the expertise of other stakeholders? Can they synchronize their work processes to allow for a meaningful/purposeful and relevant flow of data and decisions between activities?

11.1.1 Motivational ITS Example

In general, to handle the ever-increasing complexity of ITS applications, ITS business operations involve the execution of many distributed business processes. Real-life ITS operations may cross the organizational boundaries where business processes enacted by different organizations may interact. An operational workflow controls the sequence of execution of these processes.

Figure 11.1 shows an example of an operational workflow that involves various engineers from different organizations. The workflow facilitates (*social*) collaboration among different stakeholders to test new control strategies. As the figure indicates, traffic operators initiate the process by manually configuring the testing environment. Initial configuration defines the control area, number, and location of intersections, number of phases, minimum acceptable green intervals, maximum green intervals, yellow intervals, existence of all-red phases, and so forth. Defined control area is used to initiate a modeling process in which a traffic modeler creates a valid simulation model that runs on the cloud to test various control strategies. Modeling process continues by requesting the available modeling (*physical*) data from traffic surveillance component. Based on the simulated area, the platform automatically incorporates software services from different providers including services of the transportation systems' physical dynamics (*data*). These services include data services (origin–destination

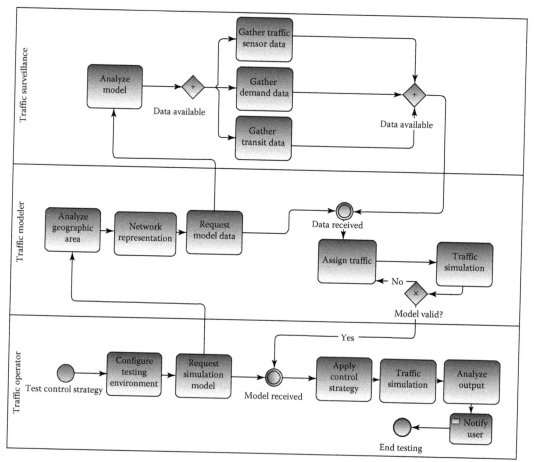

FIGURE 11.1 An example of collaborative operational ITS workflow.

matrices), traffic sensor data services, and transit data services (transit routes, schedules, etc.), which process the following dynamics: location, duration and intensity of congestions, accidents, sports and social events, transit schedules and delays, availability of high-occupancy lanes, availability of parking spots, wind speed, visibility, and icing conditions. Dynamics also includes vehicular information such as vehicle type (transit vehicle, commercial vehicle, emergency vehicle, etc.), vehicle size, vehicle speed, acceleration, and position. Data are then fed back to the modeling process to be used for model calibration. Model calibration minimizes the distance between modeled and observed data and assures the validity of the model. Once a valid model is obtained, the process continues by applying the control strategy, analyzing the output, and notifying the user of the test results.

11.2   ONE-ITS DESIGN

ONE-ITS is a double-sided computing platform with (1) a cyber, physical, and service-oriented architectural (SOA) side and (2) a *social* virtual organization (VO) side. Figure 11.2 shows these dimensions of ONE-ITS in a layered approach, where ONE-ITS software (cyber) services are the glue that links physical ITS infrastructure resources to the ITS social and organizational aspects at the bottom and the top of the figure, respectively. This section describes these features of ONE-ITS.

11.2.1   ONE-ITS: A SERVICE-ORIENTED PLATFORM

On the cyber and physical SOA side, ONE-ITS depends on SOA to allow ITS stakeholders to establish and capture business logic in business models. These business models are realized through ONE-ITS business workflows, similar to the one described in Figure 11.1, which orchestrate the execution of ITS software services. These services, in turn, abstract and manage ITS physical resources (e.g., sensors and CPUs), cyber software modules, and ITS business process models. (This is augmented by ONE-ITS VO that allows sharing of human expertise and knowledge). These are cloud computing technologies respectively known as Software as a Service and Infrastructure as a Service. In this sense, the ONE-ITS SOA can be considered a cloud Platform as a Service [4].

Within ONE-ITS, ITS software and ITS deployed infrastructure are viewed as services with open and standard interfaces. Services are modular software components that are self-contained with the ability to advertise their capabilities and with well-defined interfaces that enable collaboration with other services. SOA is a well-established paradigm for designing and implementing business collaborations within and across organizational boundaries. SOA defines, builds, composes, and orchestrates software services as required by business operations. SOA adopts loose coupling between services in a workflow in order to be able to respond efficiently to changes in business and technology.

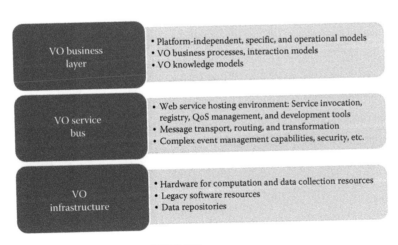

VO business layer
- Platform-independent, specific, and operational models
- VO business processes, interaction models
- VO knowledge models

VO service bus
- Web service hosting environment: Service invocation, registry, QoS management, and development tools
- Message transport, routing, and transformation
- Complex event management capabilities, security, etc.

VO infrastructure
- Hardware for computation and data collection resources
- Legacy software resources
- Data repositories

FIGURE 11.2   A high-level architecture of ONE-ITS.

The real power behind using SOA and web services in ONE-ITS lies in providing novel applications in response to changes in application requirements in a flexible and scalable manner. SOA deploys services that are usable by others to build applications that manipulate data and other services. ITS business applications are composed of executing a set of services. This changes the focus of ITS application development from algorithm implementation to service discovery and composition. An ITS stakeholder can replace one service with another that was recently developed and discovered. Such flexibility allows ITS stakeholders to handle substantial changes in their IT infrastructure with relative ease and at low cost. SOA is flexible enough that it allows ITS stakeholders to take advantage of existing applications through *wrapping* them as services.

Moreover, SOA allows partners, users, and *customers* of an organization, with appropriate authorization and technological means, to access the organization's services. SOA technologies control independently developed services in order to provide the novel application. This enables stakeholders to focus much more on business logic and intelligence than focusing on implementation details. As a side but very important benefit, this approach enables the industry (public departments of transportation, private sector companies) to test and evaluate innovative products of academic research prior to any deployment investment.

11.2.2   ONE-ITS: A VO-BASED PLATFORM

On the VO side, ONE-ITS provides tools, shown in Figure 11.3, and services to enable building VO among ITS stakeholders. A VO typically consists of a set of ITS stakeholders who interact, collaborate, and share resources to achieve a business objective through harnessing their *collective intelligence* [5]. ONE-ITS uses social web technologies to enable the exchange of not only information but also knowledge among ITS stakeholders during the planning, design, and management of ITS systems. This can be realized through a dynamic socio-technical ONE-ITS that allows various ITS stakeholders linking work processes, product, data, and decision makers in various organizations in order to synchronize and integrate elements of ITS projects.

ITS stakeholders may dynamically create temporary VO for business partnerships between multiple stakeholders and maintain the VO only for the required business duration before dissolving it. This involves managing multiple subprocesses within or across organizations in spite of heterogeneity in software platforms and different data models. This architecture has the advantage of enabling individual stakeholders to focus on core businesses, while delegating application integration to IT specialists. In the example described in Figure 11.1, various ITS stakeholders collaborate to build process models and workflows that capture their operational expertise. ONE-ITS VO side enables them to collaboratively design, implement, and test business services such as the best ITS control strategy for a defined control area of their choice.

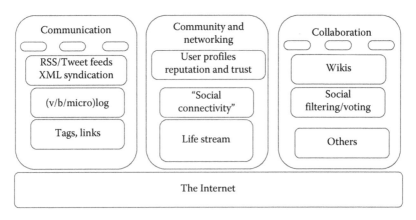

FIGURE 11.3   VO technologies included in ONE-ITS.

The ONE-ITS VO layer creates a social web within ITS stakeholders to support building a community of interest and collaborative knowledge generation. This web, being semantic also, is a means for building and strengthening knowledge diffusion practices. It draws on the expertise of relevant stakeholders documenting (for future data mining) the rationale and consequence of each decision through wikis, blogs, tweets, RSS feeds, and tag-based taxonomies as well as social networking components.

An example of ONE-ITS VO can be embodied into a virtual committee (e.g., ITS Policy group or Advanced Traffic Management group) of stakeholders with complementary knowledge and with a clear perspective of the short- and long-term impacts of a new project on the environmental, social, and economic sustainability. ONE-ITS VOs are provided with a secured workspace for an internal use of an agency to study, comment, propose, and evaluate creative solutions and design alternatives. Users can collaboratively author (e.g., using a wiki) to collectively build a consensus around a specific topic. This results in democratizing content management without loss of control (augmented with access control and rollback function). Each community can have a secure file repository where members of the community can access and contribute technical documents, tools and utilities, code samples, case studies, and other content.

Communities enable contextual delivery, findability, and filtering of knowledge based on community intelligence. Users can spontaneously and collaboratively (through, e.g., folksonomy) synthesize ranking, reputation, and other content items. A robust full-text search engine lets you search web contents (HTML, PDF, MS-Word, PowerPoint slides, etc.) based on your security access permissions. Advanced search features enable you to search the web content based on the metadata and keyword values. The ONE-ITS VO layer adopts a comprehensive security model based on a hierarchy of roles for access control (e.g., individuals, groups, projects, and team). This enables controlling access to information and content shared within the community.

Group members or users come from several remote locations and from different professional backgrounds. Users who are contributing to the community have personalized accounts and profiles. User accounts typically hold personal identifiers including a description of the user, the user's attributes, and photos (avatars), as well as their technical profile including technical industrial experience, education, and technical interests. Users can create a personal page with vanity URL, where they can post content. Users have personal file repositories for storing their contributed content.

11.3 ONE-ITS TECHNICAL ARCHITECTURE

To implement, deploy, and execute ONE-ITS, we used the software technical architecture shown in Figure 11.4. This architecture consists of these three layers described in detail in the following subsections. The ITS business/service layer aggregates software services that realize ITS business processes. Figure 11.4 shows different levels of abstractions of ITS services. The functionalities of the ITS services layer are augmented with a set of functionalities of the ITS data and presentation layers. ONE-ITS was implemented in compatibility with the Canadian ITS architecture [6].

11.3.1 PRESENTATION LAYER

ONE-ITS employs a portal server that receives and handles users' requests through a web browser. The portal server serves as a user interface, shown in Figure 11.5, to display results of executing ITS cyber services as well as VO services.

As a user logs on, through a single sign-on, to the portal and after authentication, a set of portlets are initiated and executed, and the output is displayed to the user. The ONE-ITS interface is personalized to the user's or VO's needs and objectives. Portlet execution involves consuming a set of services at the ITS service layer. Portlets use the presentation helper services to present results to end users. Presentation helper services include visualization and display services available on the Internet, such as Google Maps and Charts.

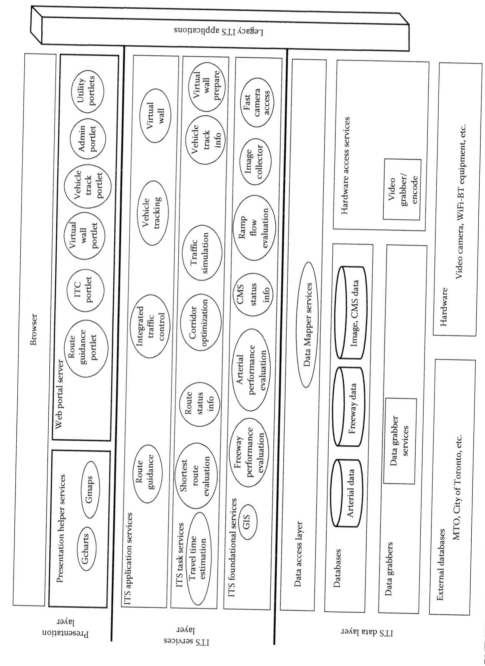

FIGURE 11.4 ONE-ITS software architecture.

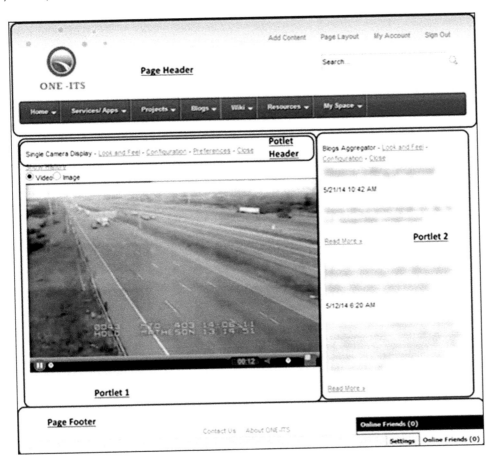

FIGURE 11.5   ONE-ITS user interface and portlets.

The portal offers uniform access through web browsers to ONE-ITS two types of portlets: social portlets and technical portlets. Examples to social portlets include VO support capabilities such as blogs, wikis, message boards, chats, and instant messages. Technical portlets include all transportation-related applications such as integrated traffic control (ITC), route guidance, virtual walls, and incidents/traffic reports.

To implement the VO collaboration features and interface for ONE-ITS, we leverage liferay open source components and utilities [7] as a backend for the presentation layers. Liferay is a Java-based portal that provides authentication, personalization, and content aggregation capabilities. It also offers a runtime environment for hosting a wide range of applications (portlets). Each page embeds several portlets that can be dynamically added by the page administrator. Administrators can also customize portlets by changing their locations, background colors, text styles, and so on. In addition, administrators can configure each portlet to allow only certain group(s) of users to use it. All ONE-ITS portlets are 100% compliant implementations of the Java Portlet Specification 2.0 [8], JSR-286, and hence they can be managed by any portlet container supporting this specification.

11.3.2   ITS SERVICE LAYER

The service layer consists of the implementation of web services that carry out selected ITS functionalities such as the visionary example described in Section 11.1. ONE-ITS task services are independent components that can be stitched together in an ITS workflow to build an ITS application service. These services provide the basic building blocks of ITS applications, as they access lower-level services in order to gather application-specific data and present them to ONE-ITS application services. Following are a few illustrative examples of how these services provide utility to upper software layers in the ONE-ITS architecture. Examples include the following:

1. An ITC software service that implements control strategies on urban corridors to be used by decision-making government agencies, local operators, and administrators of provincial highways and municipal arterial roads. Details of this service are described later in this chapter.
2. A traffic-responsive dynamic route guidance provisioned by private sector and/or media outlets can use physical data from provincial freeways and municipal arterials, data collected from connected vehicles [9–11], as well as global positioning systems data [12].
3. Emergency evacuation management services that respond to natural or man-made disasters to be used by researchers and emergency management organizations (police, fire, and emergency services) to control freeway ramps and/or traffic signals.
4. In addition, ONE-ITS service layer includes support services such as microscopic simulation using paramics [13] and mesoscopic simulation using dynusT [14].

11.3.3 Data/Resource Layer

Different traffic operations centers collect and provide a variety of physical traffic surveillance data in multiple formats. With around 5000 loop detectors measuring the pulse of the freeway system in Toronto every 20 s, this wealth of information furnishes significant research and development opportunities for national and international ITS researchers. Examples of traffic physical devices include changeable message signs (CMSs), freeway ramp meters, and arterial traffic lights. Examples of traffic data collected include road and highway data such as geometry and nonspatial metadata; loop detector data from departments of transportation including identification, functionality, real-time sensor readings, and location; CMS information from departments of transportation such as MTO and the City of Toronto including real-time message display, identification, and location; traffic incident reports; pavement temperature and ice formation data; and road weather information data from environmental departments.

In addition, these services manage the resources of high-performance computing infrastructure required to gather, transfer, and process these large amounts of data. These resources include grid compute resources and raw optical bandwidth. Communication between ITS task services and foundational services is carried in SOAP. Data are accessed and extracted from ITS databases through Java database connectivity.

This physical ONE-ITS layer provides uniform and efficient access to different physical hardware components and databases of ITS systems. Each of these components of the ITS infrastructure is represented by one or more ONE-ITS web services. The ramp metering web service, for example, is dependent on a service that collects loop detector data, on traffic flow models for highways and city streets, and on other services that use surveillance and control infrastructure. Other examples of this layer services include traffic data-plotting service, which is important for studying the evolution of traffic data over time. For example, ONE-ITS provides time series plots, shown in Figures 11.6 and 11.7, which are invaluable when studying the behavior of traffic. This layer's services involve processing data (e.g., aggregating, filtering, disaggregating) in order to present it to higher-level ITS services.

11.3.4 Example of a Technical ONE-ITS Application

This application illustrates the utility of ONE-ITS at the level of multiple traffic management centers, which represents different organizations grouped in VO. Figure 11.8 shows the ONE-ITS services

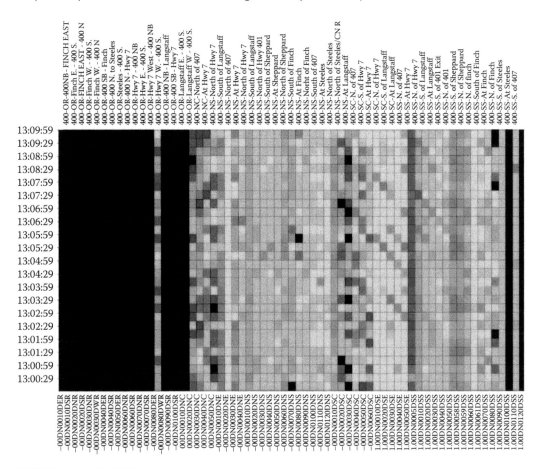

FIGURE 11.6   ONE-ITS: time series plots as data services.

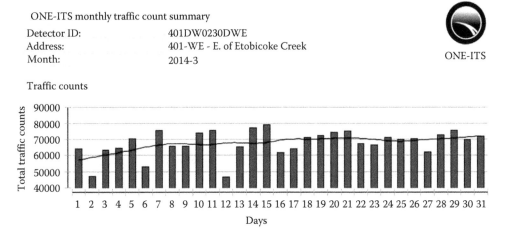

FIGURE 11.7   ONE-ITS: sensor data graph.

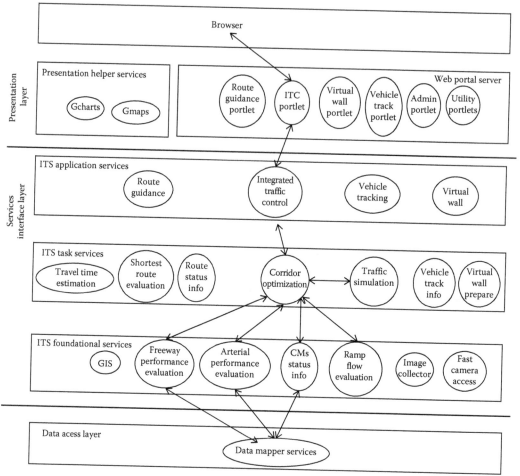

FIGURE 11.8 ONE-ITS: layered architecture of the ITC application services.

involved in this application. Within ONE-ITS, the highway administrator and the administrator of the city streets collaborate to provide the required integrated control of nonrecurring congestion on a freeway corridor. Nonrecurrent congestion is primarily due to capacity reduction caused by traffic incidents, such as crashes, disabled vehicles, adverse weather conditions, work zone, special events, and other temporary disruptions.

Congestion on highways affects traffic on city streets as well. The ITS infrastructure is used to control traffic on both the highway and the city streets. The main purpose is to control traffic flow on the freeway in order to improve efficiency of the overall corridor, including the freeway itself. The control infrastructure includes CMSs that divert the traffic from the congested highway to parallel city streets, metered ramps that limit the number of vehicles entering the freeway, and traffic lights that adapt to traffic flow fluctuations on the city streets. In addition, these control components may use other subcomponents such as traffic models or other software or hardware.

The steps of ITC application are as follows:

1. A traffic operator (human user) logs on username and password on the system login screen using an Internet browser. The customer is authenticated and is logged into the system.
2. The web portal server displays the personalized page of the user.

3. The user chooses the ITC portlet and enters the corridor ID that is to be controlled. Corridor IDs and descriptions can also be displayed to the user to select one of them.
4. The portlet checks the service registry for the requested operation via SOAP/HTTP. The service registry returns a handle of where the service is located in addition to the service interface.
5. The ITC application is invoked with the corridor ID to be controlled via SOAP/HTTP. The ITC application queries the service registry requesting a handler for the *optimizeCorridor* service.
6. The service *optimizeCorridor* decodes the SOAP message to extract input parameters and queries the registry for the location and interface of the *evaluateRampFlow, evaluateArterialPerformance, evaluate_FreewayPerformance, simulateCorridor* services.
7. The *optimizeCorridor* service gets a summary of traffic flows in the corridor using the capabilities of the *evaluateRampFlow* service. The *evaluateRampFlow* service gathers information about ramps using the corresponding DAO.
8. The *optimizeCorridor* service collects information about highway traffic in the corridor by using the capabilities of the *evaluateFreewayPerformance* service. The *evaluateFreewayPerformance* service gathers information about ramps using the corresponding DAO.
9. Then, the *optimizeCorridor* service collects information about the status of traffic in arterials in the corridor by using the capabilities of the *evaluateArterialPerformance* service. The *evaluateArterialPerformance* service gathers information about ramps using the corresponding DAO.
10. The *optimizeCorridor* service processes the information collected and invokes the *simulateCorridor* service. The *simulateCorridor* service returns a list of suggested corridor control actions. This list is returned to the *optimizeCorridor* service.
11. The *optimizeCorridor* service evaluates the suggested control actions and returns the list to the ITC application service, which returns the list to the ITC portlet for displaying them to the user.

11.4 CONCLUSIONS AND FUTURE OUTLOOK

ONE-ITS is a cyber–physical–social platform that *transparently* ties together elements of urban transportation infrastructure through ICT. ONE-ITS acquires real-time physical data about the urban transportation infrastructure. Such information is transcended, in the cloud (cyber-space), into knowledge of contextual urban dynamics. Such awareness is consequently applied to the management and control of urban infrastructure. In addition, it is applied to intelligently provide and deliver city applications, which affects travel patterns and social interactions.

We describe ONE-ITS as a double-sided coin with a VO side and an SOA side. On the VO side, ONE-ITS integrates human knowledge and expertise with technical resources allowing stakeholders to interact, collaborate, and share resources to achieve business objectives. On the SOA side, ONE-ITS depends on SOA to support VO management. SOA defines, builds, composes, and orchestrates services as required by operations in a dynamic manner. The core concept is to allow ITS stakeholders to establish and capture business logic in business models through the mother VO or, in the future, through dynamic dissolvable smaller VOs. These business models are realized in the form of cyber services that integrate ITS physical data. ONE-ITS allows VOs and stakeholders to perform these tasks in a collaborative manner. The collaborative social efforts of communities of practices and interests help build and maintain the development of ITS services, business processes, and data models. An important requirement for SOA is data engineering. SOA enables services to exchange data in a standard format. Defining data standard format can be done through xml policies and schemes. We describe here the high-level design and operational aspects of ONE-ITS, which are to be followed by technical experiences and implementation details in a following book chapter.

ACKNOWLEDGMENTS

The authors would like to acknowledge the contributions of the members of the ONE-ITS development team (a long list is provided at www.one-its.net) without whom this work was not possible and H.S. Ahmed for discussions of an earlier version of the architecture. The authors would like to thank CANARIE (www.canarie.ca) and the Ministry of Transportation Ontario (MTO) for funding and support.

REFERENCES

1. F. Qu, Wang, F., and Yang, L., Intelligent transportation spaces: Vehicles, traffic, communications, and beyond, *IEEE Communications Magazine*, 48(11), 136–142, 2010. doi:10.1109/mcom.2010.5621980.
2. B. Srivastava et al., A general approach to exploit available traffic data for a Smarter City, In *Proceedings of the 20th ITS World Congress Tokyo 2013*, Tokyo, Japan, 2013.
3. Z. Liu, Yang, D., Wen, D., Zhang, W., and Mao, W., Cyber-physical-social systems for command and control, *IEEE Intelligent Systems*, 26(4), 92–96, 2011. doi:10.1109/mis.2011.69.
4. J. Gubbi, Buyya, R., Marusic, S., and Palaniswami, M., Internet of Things (IoT): A vision, architectural elements, and future directions. *Future Generation Computer Systems*, 29(7), 1645–1660, 2013. doi:10.1016/j.future.2013.01.010.
5. J. Pitt et al., Transforming big data into collective awareness, *IEEE Computer*, 46(6), 40–45, 2013. doi:10.1109/mc.2013.153.
6. Tc.gc.ca, ITS Architecture for Canada- Release2.0—Transport Canada, 2013. Retrieved June 13, 2015, from https://www.tc.gc.ca/eng/innovation/its-architecture.htm.
7. Liferay.com, Liferay–Enterprise open source portal and collaboration software, 2015. Retrieved June 14, 2015, from http://www.liferay.com/.
8. S. Hepper and Nicklous, M., Java Portlet Specification Version 2.0 (JSR 286), 2008. Retrieved from https://jcp.org/ja/jsr/detail?id=286.
9. B. Hull et al., CarTel: A distributed mobile sensor computing system. In *SenSys '06 Proceedings of the Fourth International Conference on Embedded Networked Sensor Systems* (pp. 125–138), Boulder, CO, 2006.
10. L. Atzori, Lera, A., and Morabito, G., The Internet of Things: A survey. *Computer Networks*, 54(15), 2787–2805, 2010. doi:10.1016/j.comnet.2010.05.010.
11. A. Campbell et al., The rise of people-centric sensing. *IEEE Internet Computing*, 12(4), 12–21, 2008. doi:10.1109/mic.2008.90.
12. W. Khan, Xiang, Y., Aalsalem, M., and Arshad, Q., Mobile phone sensing systems: A survey. *IEEE Communications Surveys and Tutorials*, 15(1), 402–427, 2013. doi:10.1109/surv.2012.031412.00077.
13. Paramics-online.com, Quadstone Paramics|Traffic and Pedestrian Simulation, Analysis and Design Software, 2015. Retrieved June 14, 2015, from http://www.paramics-online.com/.
14. Dynust.net, DynusT-A simulation-based dynamic traffic assignment (DTA) software, 2015. Retrieved June 14, 2015, from http://dynust.net.

Section V

Architecture

12 ONE-ITS

Strategy, Architecture, and Software Implementation of a Cyber–Physical–Social Platform

Mohamed Elshenawy, Mohamed El-Darieby, and Baher Abdulhai

CONTENTS

12.1 INTRODUCTION

While Intelligent Transportation Systems (ITS) create new opportunities for enhancing transportation, those opportunities come with challenges that require rigorous R&D. Due to the complexity of the very nature of transportation problems and the required variety of expertise to address such problems, no research institution can practically maintain a critical mass of such diverse backgrounds. Many universities, public organizations, and societies attempt to cover as many aspects of transportation research as their resources allow. Inevitably, niches and specializations form and lead to narrower focuses on selected subsets of the overall problem. Efforts, no matter how comprehensive or ambitious, are limited by the capacity of the available researchers and resources. The nature of ITS R&D is complex due to the following:

- Current ITS solutions are mostly stand-alone systems that are used by a small group of users with little or no support for larger-scale collaboration and integration. A vivid example is the incompatible software platforms at different traffic operation centers.
- Advancing ITS applications is typically a labor-intensive and slow process because of using "older" technologies, architectures, and programming languages.

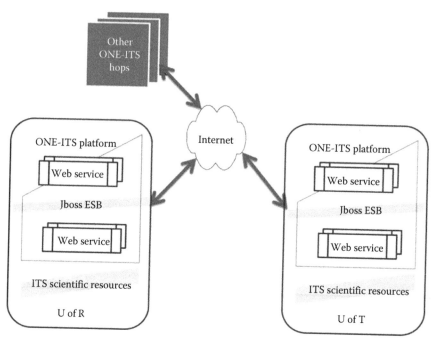

FIGURE 12.1   ESB bridging the different geographic locations of ONE-ITS.

- Currently, there are very limited efforts to enable remote sharing and "integration" of ITS infra- and info-structures. ITS applications and services often do not make use of high-performance computing technologies for processing and storage of real-time traffic information and services.

Bringing together national and international "centers" of excellence into virtual organizations (VOs) is the core objective of ONE-ITS. ONE-ITS adopts a service-oriented architecture (SOA) [1] to enable integration of ITS resources across the boundaries of different stakeholders. Proper SOA establishes and standardizes ability to encapsulate and expose, as services with standard interfaces, organizational resources such as legacy applications, databases, sensors, and computing resources already available at ITS stakeholders' infrastructure. Proper SOA allows such services interoperate with each other regardless of resource deployment platform. Intrinsic operability is provided by the underlying communication infrastructure. Loose-coupling is a fundamental requirement to enable building a VO. Figure 12.1 shows how ONE-ITS uses enterprise service bus (ESB) as a bridge between different ONE-ITS stakeholders; in this example, the universities of Toronto and Regina in Canada.

In this chapter, we describe how modern ICT furnishes a new opportunity to harness the integration of people, ideas, and tools in a collaborative environment. ONE-ITS, as a cyber–physical–social platform links organizational work processes, software services, physical databases, scientific instruments, and cyber infrastructure resources in various organizations. These elements are encoded and wrapped as services. ONE-ITS uses industry standards and proven methodologies and technologies to build this platform. The previous chapter described higher-level operational aspects of ONE-ITS. The purpose of this chapter is to introduce ONE-ITS implementation strategies and details.

12.2   MODEL-DRIVEN ONE-ITS ARCHITECTURE

ONE-ITS relies on a model-based architecture to enable the integration of social, cyber, and physical resources. In order to capture user needs, determine user requirements, and handle the ever-increasing complexity of ITS applications, ONE-ITS relies on business models for capturing the operational

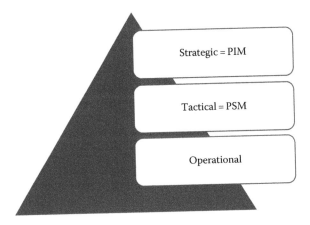

FIGURE 12.2   Model-based view of ONE-ITS.

expertise and resources of ITS stakeholders. The model-driven architecture (MDA) [2] defines three types of models, shown in Figure 12.2, for complex computing platform as follows:

The computation-independent model (CIM) captures high-level organizational structure and roles, the logic of business processes, and cooperation mechanisms. A CIM is essentially a visual representation of business processes. Relatively simple conceptual models such as UML diagram and/ or process definition diagrams are usually sufficient at early stages. These early stages must be strongly driven by people. In what follows, more details are provided on a CIM for ONE-ITS.

The platform-independent model (PIM) involves detailing major business processes, their interactions and roles of different stakeholders. This involves additional UML models such as class models, sequence diagrams, and state diagrams as well as more workflow and process definition models similar to Figure 12.1. PIMs typically include finer-grained use cases, component and interface specifications, and an integration strategy for different software processes. Later stages of the collaboration involve defining data requirements.

Finally, a platform-specific model (PSM) is a customized version of PIMs to support the software architecture of stakeholders involved. PSM are used to generate models for application code. Transformations from CIM to PIM and from PIM to PSM are carried out according to transformation rules that are also defined by the collaborating VO members.

12.2.1   MDA, NATIONAL ITS ARCHITECTURE, AND ONE-ITS

ONE-ITS adopts the Information Model of the Canadian ITS Architecture [3]. The ITS Architecture for Canada reflects the contributions of a wide range of the ITS experts including transportation specialists, systems engineers, consultants, practitioners, and so on. The architecture defines a unified framework for designing, planning, managing, and coordinating ITS deployments nationwide.

The architecture consists of the following four components: (1) user services and user subservices describe the rule of the system from the user's perspective; (2) the logical architecture defines the processes (activities or functions) that are required to satisfy the user services and dataflows exchanged among these processes; (3) the physical architecture defines the subsystems that make up an intelligent transportation system. It also defines the architecture flows that connect the various subsystems into an integrated system; and (4) market packages represent slices of the physical architecture that address specific services like, for instance, surface street control.

For example, the ONE-ITS integrated traffic control (ITC) application, described in Chapter 1, corresponds to user services 2.1 and 2.2. and to the following market packages: ATMS01, ATMS02, ATMS03, ATMS04, ATMS06, ATMS07, ATMS08, and, most importantly, to the ATMS01 market package, which is associated with the following set of input and output dataflows: {1.1.1.1— Process Traffic Sensor Data; 1.1.1.2—Collect and Process Sensor Fault Data; 1.1.2.1—Process

Traffic Data for Storage; 1.1.2.2—Process Traffic Data; 1.1.2.3—Update Data Source Static Data; 1.1.4.1—Retrieve Traffic Data; and 1.1.4.2—Provide Traffic Operations Personnel Traffic Data Interface; 1.1.4.4—Update Traffic Display Map Data; 1.2.8.1—Collect Indicator Fault Data; 1.2.8.2—Maintain Indicator Fault Data Store; 1.2.8.3—Provide Indicator Fault Interface; 1.2.8.4—Provide Traffic Operations Personnel Indicator Fault Interface; 1.3.1.3—Process Traffic Images}

The Canadian ITS Architecture information model (CIM) is designed as the coordinated semantic framework that enables meaningful data exchange between ITS services. The Canadian ITS architecture also defines a PIM that defines, among other things, ITS processes, data dictionaries, and dataflows among different business processes. Different ITS stakeholders have built PSM components that they use for their long-term and short-term operational planning as well as daily operations.

12.3 ONE-ITS PLATFORM ARCHITECTURE

Figure 12.3 illustrates the main components of the ONE-ITS architecture. As shown in figure, the architecture defines four main components: the client tier, the mediation tier, the service tier, and the back-end/database tier. This section describes each of these component in terms of its responsibility and functionality.

12.3.1 CLIENT TIER

ITS community members access, through a web browser, this tier (and the ONE-ITS platform) through a single sign-on using secure authentication. The ONE-ITS portal is where a set of personalized portlets is initiated and executed and the output is displayed to the user. Portlet execution involves consuming a set of services at the ITS Service tier as well as access to utilities and resources to support collaboration. This includes wiki, forums, ITS news services, ITS event calendar, and discussion groups. In addition, the portal serves as a hub for the ITS community to independently develop web services and data sets.

All ONE-ITS portlets are 100% compliant implementations of the Java Portlet Specification 2.0, JSR-286 [4], and hence can be managed by any portlet container supporting these specifications. According to the JSR-286, portlet container manages the life cycle of a portlet through four main

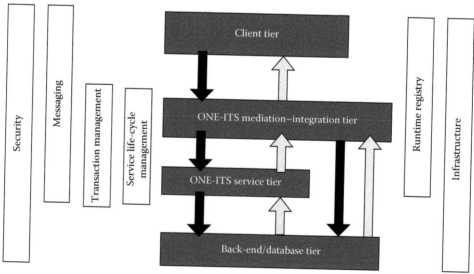

FIGURE 12.3 ONE-ITS implementation architecture.

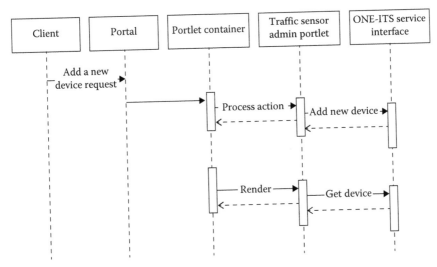

FIGURE 12.4 A sequence diagram for a ONE-ITS client tier.

methods as follows: *init, processAction, render, and destroy*. When the portlet is instantiated for the first time, *init*() method is invoked only once to perform all one-time activities. Client requests are then handled by invoking *processAction* and *render* methods. Client requests are typically triggered by URLs targeted to certain portlets, known as portlet URLs. Portlet URLs can be either action URLs or render URLs. Figure 12.4 shows an example of an action URL adding a new sensor to the system. Action URLs translate into one action request to the targeted portlet and many render requests to all portlets in the portal page. As shown in figure, portlet container invokes *processAction* method, which in turn invokes a ONE-ITS web service to add a device. The portlet container must wait until the action request finishes before invoking the render request. When a portlet container determines that a portlet is no longer required to be available, portlet container invokes *destroy*() method to release any resources held by the portlet and hence terminates the life cycle of a protlet.

12.3.2 Mediation Tier

ONE-ITS uses JBoss enterprise service bus (ESB) [5] as a middleware infrastructure layer to enable the delivery and integration of ONE-ITS services. ESB uses open technical standards such as SOAP [6], WSDL [7], XML, and XML schema to allow for messages to be self-contained. This limits the role of each ITS organization to only the implementation and hosting of application logic. In addition, JBoss ESB provides capabilities for secure and reliable request/response communication between loosely coupled ONE-ITS web services.

A second major component of the ONE-ITS ESB is the business rules management component. ONE-ITS relies on Drools business rules engine [8] to define message routing rules. These rules are used to enhance performance and address inter-operational aspects (e.g., service versioning). In addition, isolating business logic from source code allows web services to be connected differently in response to changing system dynamics.

A third component of the ONE-ITS ESB is the mediation component, which enables application connectivity. This is embodied in message transformation services that translate between different message transport protocols (e.g., JMS [9] to SOAP and vice versa). These services can also transform data between application consumers and providers to enable data interoperability. The transformations and translations are carried out in accordance to transformation rules that are defined by the system administrator. ONE-ITS will be using the JBoss transformation engine that is based on Smooks [10].

12.3.3 SERVICE TIER

The Service Layer contains the implementation of web services that correspond to the ITS work processes and applications. ONE-ITS services are contained within an application server and are described using the standard WSDL format. The application server keeps these services published on a community service registry via SOAP/HTTP. Portlets exchange SOAP messages to query the service registry and to interact with ITS services. ONE-ITS services gather ITS data, grid resources, and raw bandwidth required to carry out their functionality.

ONE-ITS employs Spring-WS [11] to build contract-first web services. In contract-first web services, developers start by designing WSDL contracts as a first implementation step. This is opposed to contract-last approach where developers start by writing the code and the WSDL contracts are generated accordingly. Contract-first approach facilitates cooperation between multiple stakeholders to implement new services. Staring by designing the interface ensures interoperability between stakeholders by allowing them to specify requirements for communication using language-independent XML schema. In addition, using contract-first approach allows developers to change the internal structure of services' code without affecting clients consuming these services. Figure 12.5 shows an example schema of a service data contract, the schema defines the structure of xml messages sent to and from a camera feed service. The second step in the spring-WS implementation is configure and write the service endpoint. Endpoints are created by using @Endpoint annotation, @Payload-Root annotation to map request payload elements to endpoint methods responsible for handling these requests.

ONE-ITS registry maintains a record of services including service description files (WSDL) and network references for services. This enables discovery, reuse, and management of services. In addition, the ONE-ITS registry enables users to add new ITS services to ONE-ITS over time. The registry is coupled with a UDDI service that resolves a logical service name to a specific software implementation at run time.

```xml
<?xml version="1.0" encoding="UTF-8"?>
<xs:schema id="CameraFeedInq"
    xmlns="http://oneits.net/schemas/messages"
    xmlns:xs="http://www.w3.org/2001/XMLSchema"
    xmlns:tns="http://oneits.net/schemas/messages"
    targetNamespace="http://oneits.net/schemas/messages"
    xmlns:types="http://oneits.net/schemas/types"
    elementFormDefault="qualified">
    <xs:include schemaLocation="PayloadTypes.xsd" />
    <xs:element name="CameraFeedGetRq">
        <xs:complexType>
            <xs:sequence>
                <xs:element name="CameraFeedId" type="xs:long"
                    minOccurs="0" maxOccurs="1" />
            </xs:sequence>
        </xs:complexType>
    </xs:element>

    <xs:element name="CameraFeedGetRs">
    <xs:complexType>
        <xs:sequence>
            <xs:element name="CameraFeed" type="CameraFeed"
                minOccurs="0" maxOccurs="1" />
        </xs:sequence>
    </xs:complexType>
    </xs:element>
</xs:schema>
```

FIGURE 12.5 ONE-ITS XML-based service schema for an ITS Camera.

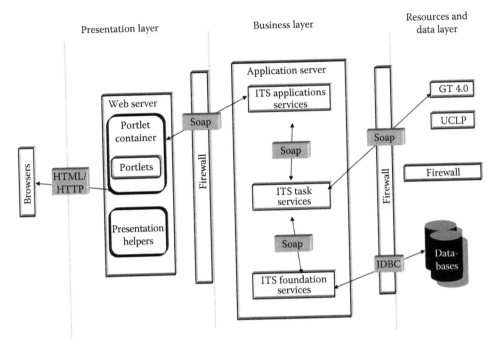

FIGURE 12.6 ONE-ITS deployment architecture.

12.3.4 BACKEND/DATABASE TIER

This layer defines mappings between different data models of different stakeholders. Data mapping involves defining standard data representation schemes at the conceptual and content levels and hence allows seamless flow of data and messages between ONE-ITS services. In addition, data mapping enables the inter-operation with legacy ITS data through data grabbers that capture data from other existing ITS databases. This results in decoupling consumers from databases' schemas and locations. In addition, various databases, whether relational, hierarchical, or object-oriented, can then be accessed through consistent interfaces and provides data in a standard format. The backend/database layer also encapsulates ITS data entities into Data Access Objects (DAO) with well-defined interfaces, which enables uniform access from higher-level ITS services.

In order to access a ONE-ITS service, a user issues a SOAP request for a ONE-ITS service. It is assumed that the user was authenticated earlier and was provided a security token used to build SAML [12] entries that are included in the SOAP-based request. The ONE-ITS portal server intercepts the user's request and forwards the request to the ESB to locate the requested service. The ESB receives the request and begins to verify SAML entries in the SOAP message. The ESB Routing module is invoked to locate a SAML security verify service. Then the ESB invokes the SAML service. If the SAML is verified, the ESB invokes its ESB routing service again to locate the requested ONE-ITS service. The ESB forwards the SOAP request to the provider service and, if any, returns results to the consumer service. Figure 12.6 shows the deployment diagram of ONE-ITS.

12.4 MAIN STEPS OF ONE-ITS SERVICE DEVELOPMENT

12.4.1 HIGH-LEVEL ONE-ITS SERVICE DEVELOPMENT SCENARIO

Figure 12.7 shows how services are integrated in ONE-ITS. Each service provider hosts an application server containing a group of web services that carry out selected ITS functionalities. A high-level service development scenario includes the following steps:

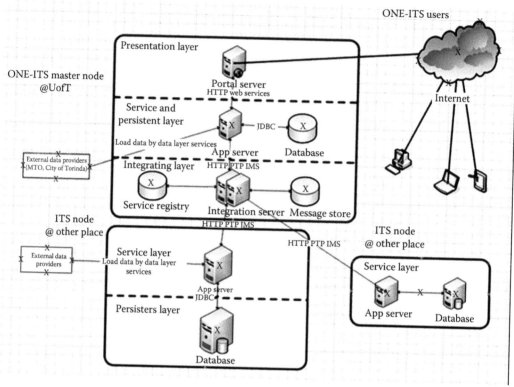

FIGURE 12.7   High-level deployment diagram.

```
<xs:schema id="AppConfigInq001" xmlns="http://www.w3.org/2001/XMLSchema"
    xmlns:xs="http://www.w3.org/2001/XMLSchema"
    xmlns:tns="http://oneits.net/schemas/messages"
    targetNamespace="http://oneits.net/schemas/messages"
    xmlns:types="http://oneits.net/schemas/types"
    elementFormDefault="qualified">

    <import namespace="http://oneits.net/schemas/types"  schemaLocation="PayloadTypes.xsd"/>
    <element name="AppConfigInqRq001">
        <xs:complexType>
          <xs:sequence>
            <xs:element name="configCode" type="xs:string"
                minOccurs="0" maxOccurs="1" />
          </xs:sequence>
        </xs:complexType>
    </element>
    <element name="AppConfigInqRs001">
      <xs:complexType>
        <xs:sequence>
            <xs:element name="configCode" type="xs:string"
                minOccurs="0" maxOccurs="1" />
            <xs:element name="configValue" type="xs:string"
                minOccurs="0" maxOccurs="1" />
        </xs:sequence>
      </xs:complexType>
    </element>
</xs:schema>
```

FIGURE 12.8   Example of a service interface.

1. Define service interface and the required data objects in a form of an xsd file (XML Schema Definition) as shown in Figure 12.8.
2. Generate WSDL from the xsd file.
3. Develop web service endpoint to create the request and response objects for the specific service.

4. Develop a business logic class to perform any pre- or postprocessing operations.
5. Develop an adaptor class to access database, JMS, or file system to get data.
6. Develop JUnit test cases that will be used to unit test the service implementation class and any other classes with significant logic (db adapters, utility classes, etc.). JUnit tests should contain a test case for each public method containing nontrivial logic. The JUnit tests will be run as part of the build, so unit tests must pass or the build will fail. Untested code is considered incomplete and unit test development must be included as part of the service design and development process.

12.5 CONCLUSION

In this chapter, we described how ONE-ITS links work processes, product, data, and decision makers in various organizations in order to synchronize and integrate elements of ITS. ONE-ITS depends on SOA to allow ITS stakeholders to establish and capture business logic in business models. These business models are realized through the execution of ITS software services running in Cyber online servers. These services, in turn, abstract and manage ITS infrastructure physical resources (e.g., CPUs, sensors) provided by different stakeholders.

ONE-ITS enables the exchange of not only information but also of knowledge across ITS organizations. We also describe how a layered service-oriented software architecture is used to organize ONE-ITS components. ONE-ITS uses enterprise service bus (ESB) as a mediation component in order to enable application connectivity. ESB capabilities enable the reuse of ONE-ITS services by a wider range of ITS stakeholders. This is embodied in message transformation services that translate between different message transport protocols (e.g., JMS to SOAP and vice versa). Open standards such as SOAP, WSDL, XML, and XML schema allow for messages to be self-contained. Using an open vendor-neutral communication limits the role of platform-specific technology to the implementation and hosting of application logic.

ACKNOWLEDGMENTS

The authors would like to acknowledge the contributions of the members of the ONE-ITS development team (a long list is provided at [13]) without whom this work would not have been possible. The authors would also like to thank CANARIE [14] and the Ministry of Transportation Ontario (MTO) [15] for their funding and support.

REFERENCES

1. M. Papazoglou, P. Traverso, S. Dustdar, and F. Leymann. Service-oriented computing: State of the art and research challenges. *Computer*, 40(11), 38–45 (2007). doi:10.1109/mc.2007.400.
2. Y. Singh and M. Sood. Model driven architecture: A perspective. In *IEEE International Advance Computing Conference, 2009. IACC 2009*, pp. 1644–1652, Patiala, India (2009).
3. Tc.gc.ca. ITS Architecture for Canada—Release 2.0—Transport Canada (2013). Retrieved June 13, 2015, from https://www.tc.gc.ca/eng/innovation/its-architecture.htm.
4. S. Hepper and M. Nicklous. Java Portlet Specification Version 2.0 (JSR 286) (2008). Retrieved from https://jcp.org/ja/jsr/detail?id=286.
5. Jboss.org. JBoss ESB—Reliable SOA infrastructure (2012). Retrieved June 13, 2015, from http://www.jboss.org/jbossesb/.
6. D. Box, D. Ehnebuske, G. Kakivaya, A. Layman, N. Mendelsohn, H. Nielsen et al. Simple Object Access Protocol (SOAP) 1.1 (2009). Retrieved from http://www.w3.org/TR/2000/NOTE-SOAP-20000508.
7. E. Christensen, F. Curbera, G. Meredith, and S. Weerawarana. Web Services Description Language (WSDL) 1.1 (2001). Retrieved from http://www.w3.org/TR/wsdl.
8. Drools.org. Drools—Drools—Business Rules Management System (2015). Retrieved June 13, 2015, from http://www.drools.org/.

9. M. Hapner, R. Burridge, R. Sharma, J. Fialli, and K. Stout. Java Message Service Specification—Version: 1.1 (2002). Sun Microsystems, Inc. Retrieved from http://download.oracle.com/otn-pub/jcp/7195-jms-1.1-fr-spec-oth-JSpec/jms-1_1-fr-spec.pdf.

10. Smooks.org. Smooks—Data Integration (n.d.). Retrieved June 13, 2015, from http://www.smooks.org/.

11. Static.springsource.org, Spring Web Services (2010). Retrieved June 13, 2015, from http://static.springsource.org/spring-ws/sites/1.5/.

12. S. Cantor, J. Kemp, R. Philpott, and E. Maler. Assertions and Protocols for the OASIS Security Assertion Markup Language (SAML) V2.0 (2015). Retrieved from https://docs.oasis-open.org/security/saml/v2.0/saml-core-2.0-os.pdf.

13. One-its.net. ONE-ITS- Funding and Supporting Agencies (n.d.). Retrieved June 13, 2015, from http://one-its.net.

14. Canarie.ca. Advancing Canada's knowledge and innovation infrastructure|CANARIE (2015). Retrieved June 13, 2015, from http://www.canarie.ca.

15. Mto.gov.on.ca. Ministry of Transportation (2010). Retrieved June 13, 2015, from http://www.mto.gov.on.ca.

13 Contracts for Specifying and Structuring Requirements on Cyber-Physical Systems

Jonas Westman and Mattias Nyberg

CONTENTS

13.1 INTRODUCTION

The notion of contracts was first introduced in Meyer [67] as a pair of pre- and postconditions [36,45,57] to be used in formal specification of software (SW) interfaces. In recent work [12,14], developed within the European research project SPEEDS [85], the use of contracts is extended from formal specification of software to serving as a central design philosophy in systems engineering to support the design of cyber-physical systems (CPS) [60,81]. One of the key challenges that triggered the extension of contracts is the increasingly complex development environment of cyber-physical systems, characterized by distributed original equipment manufacturer (OEM)/supplier chains [12,14].

In the context of an OEM/supplier chain, in order to achieve an overall set of intended properties of a system, the OEM needs to distribute the responsibilities between different suppliers that are to deliver components that are to be integrated with the rest of the system. Clearly defined interfaces and the separation of responsibilities between the different suppliers and the OEM are paramount in order to support seamless integration. A *contract* addresses such concerns by assigning the responsibility of achieving a certain property to a *component* under the form of a *guarantee*, given that certain constraints under the responsibility of the *environment* of the component, called *assumptions*, are fulfilled [12].

Although the aforementioned discussion focuses on the use of contracts for managing the complexity of OEM/supplier chains, the discussion can be generalized and is equally valid for any design context where clear separations of responsibilities are warranted. Another example of such a context is presented in Damm, Josko, and Peinkamp [28] and more explicitly in Westman, Nyberg, and Torngren [92] where contracts are shown to provide a suitable foundation for structuring safety requirements in the functional safety standard ISO 262626 [48]. The reason for this is that a contract explicitly declares what a component requires from its environment as assumptions, in order to guarantee that the safety requirements are met [92].

However, the achievements in Westman, Nyberg, and Torngren [92] were made possible only due to a modification of the work [12,81] in SPEEDS where the modification relaxes the constraint that a contract must be limited to the interface of its component. This modification has been further explored in Westman and Nyberg [90] thereby generalizing previous contract theories [7,10,12–15,17,22,27, 33,40–43,67,78,81,86,87] to fully support contracts that are not limited to component interfaces. As stated in Westman and Nyberg [90], the increased expressivity in which contracts can be specified, resulting from the generalization, is indeed needed in order to be able to support practical engineering and to make it possible to properly express safety requirements [90,92].

Building on the work in [12,14] and more specifically on [90,92], this chapter presents a general contract framework for specifying and structuring requirements on cyber-physical systems. The framework provides support for modeling both individual components and architectures of cyber-physical systems, at all levels of design, as well as the structuring and specification of requirements on the components using contracts.

At the core of the framework is a set of clearly separated conditions on a component and its environment where the conditions ensure that the component meets its responsibility expressed by the guarantee of a given contract. In the context of an OEM/supplier chain and also within a single company, the conditions specify design constraints that the developers of the component and its environment need to meet in order to ensure seamless integration. The conditions are general, which means that the conditions hold for any type of domain that is considered, for example, mechanical and SW, and further for the cases where a contract that is not limited to a component interface is used.

In general, requirements in industry are often of poor quality [3], and in order to therefore support the specification of requirements in practice, the framework includes a set of constraints on the variables, that is, *the scope*, over which a guarantee and the assumptions of a contract for a component can be specified. The constraints constitute necessary conditions in order for the component and its environment to meet their respective responsibilities in a given architecture, which means that the constraints serve as sanity checks of the contract.

To meet the high demands on requirement traceability from recent functional safety standards, for example, ISO 26262 [48] and IEC 61508 [47], and to support requirements traceability in general, a new graph, called a *contract structure*, is introduced. A contract structure supports both the structuring of the requirements on a cyber-physical systems and its components using contracts and, in particular, the tracing from low-level requirements on components to top-level requirements on the overall cyber-physical systems. The individual traceability of requirements is needed to comply with, for example, ISO 26262, where the assignment of *safety integrity levels* (SILs) to lower-level require- ments on components is determined based on the individual tracing from the lower-level requirements to top-level requirements on the system [91].

As a proof of concept, the framework is used for the structuring and specification of safety requirements in ISO 26262 [48] for a fuel level display (FLD) system, installed on all heavy trucks manufactured by Scania. A wide range of safety requirements are considered from a safety goal down to safety requirements on real industrially written C-code covering both application and basic SW, as well as requirements that specify the interaction between SW and hardware (HW) components. Thus, the case study does not only validate the framework but also serves as a reference example on the application of ISO 26262 in practice.

This chapter is organized as follows. Based on a theoretical framework in Section 13.2, contracts are introduced in Section 13.3 along with conditions that ensure that a component meets its respon- sibility expressed by the guarantee of a contract. Considering such conditions, Section 13.4 presents necessary constraints on the scopes of assumptions and guarantees. Section 13.5 introduces *contract structures* in order to structure requirements on a cyber-physical systems. Section 13.6 presents the industrial case study as a proof of concept and as a reference example on the application of ISO 26262 in practice. Section 13.7 compares this chapter with related work, and Section 13.8 summarizes the paper and draws conclusions.

13.2 ASSERTIONS, ELEMENTS, AND ARCHITECTURES

This section establishes a theoretic framework in order to model a cyber-physical systems and its parts and to describe the notion of a contract in Section 13.3. The framework mainly draws inspiration from the contract theory [12,14] developed in SPEEDS [85]. Similarities between the theoretic framework presented in this section and the one presented in Benveniste et al. and Codd [12,14], as well as with other frameworks and theories, are discussed briefly throughout the section and in more detail in Section 13.7.

13.2.1 ASSERTIONS AND RUNS

Let $X = \{x_1, \ldots, x_N\}$ be a set of variables. Consider a pair (x_i, ξ_i) consisting of a variable x_i and a trajectory ξ_i of values of x_i over a time window of possibly infinite length, starting at a certain time t_0, for example, as shown in Figure 13.1a. A set of such pairs, one for each variable in X, is called a *run* for X, denoted ω_X. For example, a run $\omega_{\{x_i, x_j\}}$ is shown in Figure 13.1b as a solid line, consisting of two pairs containing the trajectory shown in Figure 13.1a and another trajectory ξ_j of values of x_j, both represented as dashed lines.

Given a set of variables X' and a time window, an *assertion* W over X' is a possibly empty set of runs for X'. This notion corresponds to similar definitions in References [12,14,15,81,92].

Note that rather than specifying an assertion by explicitly declaring its sets of runs, assertions can be specified by a set of constraints, for example, by equations, inequalities, or logical formulas. For example, an assertion W' over $\{u, v\}$, specified by the equation $u = v$, is the set of all possible runs for $\{u, v\}$ where $u = v$ holds for all samples in the given time window.

As a second example, consider that W'' is an assertion over $\{x, y\}$, specified by the first-order differential equation

$$\frac{dy}{dt} = x(t), \quad \text{where } x(t) = t.$$

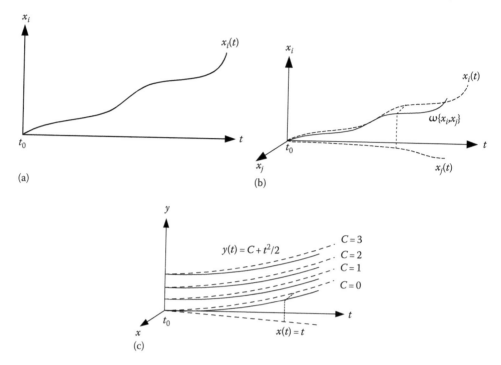

FIGURE 13.1 In (a), a trajectory of values of x_i is shown. In (b), a run $\omega_{\{x_i, x_j\}}$ is shown, consisting of two pairs containing the trajectory shown in (a) and another trajectory of values of x_j. In (c), a subset of the runs that are solutions to the differential equation $\frac{dy}{dt} = x(t)$, where $x(t) = t$, is shown.

The assertion W'' is hence the set of all possible runs that are solutions to the differential equation (see Figure 13.1c for a subset of these runs).

As a third example, consider that W''' is an assertion over $\{a, b\}$, specified by the logic formula $a = 0 \vee b = 0$, where both a and b take values from $\{0, 1\}$. The assertion W''' is hence the set of all possible runs for $\{a, b\}$ where, for each sample in the given time window, at least one of a and b has the value 0.

13.2.1.1 Projection of Assertions

Given an assertion W over $X = \{x_1, \ldots, x_N\}$, and another set of variables $X' \subseteq X$, the *projection* [13, 82,92] of W onto X', written $proj_{X'}(W)$, is the set of runs obtained when each pair that does contain an identifier $x \in X'$ is removed from each run ω_X in W, that is,

$$proj_{X'}(W) = \{\omega_{X'} | \omega_X \in W \text{ and } \omega_{X'} = \{(x, \xi) | (x, \xi) \in \omega_X \text{ and } x \in X'\}\}. \qquad (13.1)$$

Using notation of relational algebra [24], it holds that $proj_{X'}(W) = \prod_{X'}(W)$. Furthermore, the relation in Equation 13.1 corresponds to the definition of projection in Benveniste et al. and Westman, Nyberg, and Torngren [13,92], while that in Simko et al. [82] defines projection as an operation on a single run instead of on an assertion.

As an example, consider an assertion $\{\omega_{\{x_i, x_j\}}\}$ where $\omega_{\{x_i, x_j\}}$ is the run shown in Figure 13.1b. In accordance with the relation in Equation 13.1, $proj_{\{x_i\}}(\{\omega_{\{x_i, x_j\}}\})$ is hence the run containing only the pair with the trajectory of values of the variable x_i as shown in Figure 13.1a.

Given an assertion W' over X' and another set of variables X, $\widehat{proj}_X(W')$ is the set of runs where each run in W' is first extended with all possible runs for $X \setminus X'$, prior to applying the operation of projection. That is,

$$\widehat{proj}_X(W') = proj_X(\{\omega_{X \cup X'} | proj_{X'}(\{\omega_{X \cup X'}\}) \in W'\}). \qquad (13.2)$$

Note that if $X \subseteq X'$, then it holds that $\widehat{proj}_X(W') = proj_X(W')$. For the case where $X' \subseteq X$, the relation in Equation 13.2 corresponds to the definition of *inverse projection* in Benveniste et al. and Westman, Nyberg, and Torngren [13,92].

13.2.1.2 Dissimilar Sets of Variables

In the following, the symbols $\widehat{\cap}, \widehat{\cup}, \widehat{\subset}, \widehat{\subseteq}$, and so on, will be used to denote that prior to using operations and relations on assertions over *dissimilar sets of variables*, the assertions are first extended to the union of the sets of variables involved using the operator $\widehat{proj}$ in accordance with the relation in Equation 13.2. For example, given two assertions W and W' over the set of variables X and X', respectively,

$$W \widehat{\cap} W' = \widehat{proj}_{X \cup X'}(W) \cap \widehat{proj}_{X \cup X'}(W')$$

$$W \widehat{\subseteq} W' = \widehat{proj}_{X \cup X'}(W) \subseteq \widehat{proj}_{X \cup X'}(W').$$

Note that in the following and in the previous example, when using set operations and relations on assertions, it is assumed that all assertions are specified considering the same time window. That is, as stated in Benveniste, Cailaud, and Passerone [15], it is assumed that all assertions are specified with respect to *a universal time*.

As an example of how set operations and relations can be applied to assertions over dissimilar sets of variables, consider that $W_{0 \leq x \leq 1}$ and $W_{y=x}$ are assertions over $\{x\}$ and $\{x, y\}$, specified by the relations $0 \leq x \leq 1$ and $y = x$, respectively, where x and y both take values from $\mathcal{R}_{\geq 0}$. Considering that both the assertions $W_{0 \leq x \leq 1}$ and $W_{y=x}$ are independent of time, the set of runs of $W_{0 \leq x \leq 1}$ and $W_{y=x}$ at a single point in time are shown in Figure 13.2a and b, respectively. In accordance with the relation in Equation 13.2, $\widehat{proj}_{\{x,y\}}(W_{0 \leq x \leq 1})$ is the set of runs where each run in $W_{0 \leq x \leq 1}$ is extended with all possible runs for y, as shown in Figure 13.2c. Since $W_{y=x}$ is an assertion over $\{x, y\}$, in accordance with the relation in Equation 13.2, it holds that

$$W_{y=x} = proj_{\{x,y\}}(W_{y=x}) = \widehat{proj}_{\{x,y\}}(W_{y=x}).$$

As shown in Figure 13.2d, and in accordance with the notations presented earlier, it holds that

$$W_{0 \leq x \leq 1} \widehat{\cap} W_{y=x} = \widehat{proj}_{\{x,y\}}(W_{0 \leq x \leq 1}) \cap W_{y=x}.$$

13.2.1.3 Variables Constrained by Assertions

Let Ω_X denote the set of all possible runs for a set of variables X. An assertion W over X' *constrains* X if for each $x \in X$, it holds that

$$\begin{cases} W \widehat{\subset} proj_{X' \setminus \{x\}}(W) & X' \setminus \{x\} \neq \emptyset \\ \emptyset \neq W \subset \Omega_{\{x\}} & \text{otherwise,} \end{cases} \qquad (13.3)$$

where $\subset$ denotes a *proper* subset.

As an example, consider an assertion $W_{0 \leq x \leq 1}$ specified by the inequality $0 \leq x \leq 1$ as shown in Figure 13.2a at a single point in time. According to the relation in Equation 13.3, $W_{0 \leq x \leq 1}$ constrains $\{x\}$ since it holds that $W_{0 \leq x \leq 1} \subset \Omega_{\{x\}}$. However, $W_{0 \leq x \leq 1}$ does not constrain $\{y\}$ since $\{x\} = \{x\} \setminus \{y\}$ and $W_{0 \leq x \leq 1}$ is not a proper subset of itself.

As a second example, consider the assertion $W_{y=x}$, as shown in Figure 13.3a. In accordance with the relation in Equation 13.3, $W_{y=x}$ constrains $\{y\}$ if $W_{y=x} \widehat{\subset} proj_{\{x\}}(W_{y=x})$, that is, in accordance with point 2 in Section 13.2.1.2, $W_{y=x} \subset \widehat{proj}_{\{x,y\}}(proj_{\{x\}}(W_{y=x}))$. The assertion $proj_{\{x\}}(W_{y=x})$ is shown in Figure 13.3b. If this assertion is further extended with all possible runs for $\{y\}$ by projecting it onto $\{x, y\}$, the assertion obtained is the one shown in Figure 13.3c. Since $W_{y=x}$, shown in Figure 13.3a, is a proper subset of the assertion $\widehat{proj}_{\{x,y\}}(proj_{\{x\}}(W_{y=x}))$, shown in Figure 13.3c, it can hence be concluded that $W_{y=x}$ constrains $\{y\}$.

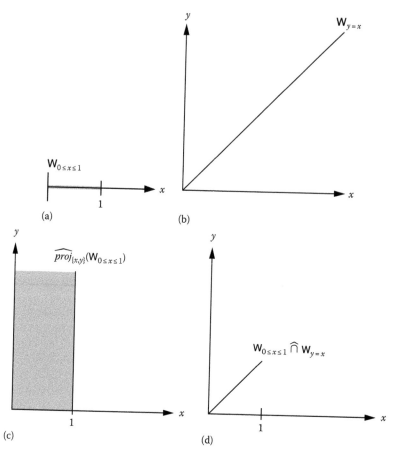

FIGURE 13.2 The assertions $W_{0 \le x \le 1}$ and $W_{y=x}$ are shown in (a) and (b), respectively. The assertions $\widehat{proj}_{\{x,y\}}(W_{0 \le x \le 1})$ and $W_{0 \le x \le 1} \widehat{\cap} W_{y=x}$ are shown in (c) and (d), respectively.

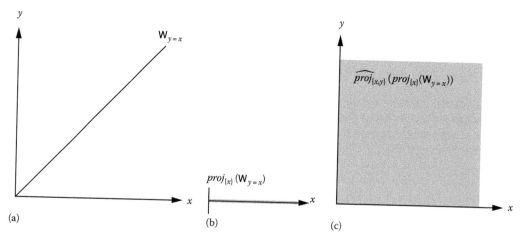

FIGURE 13.3 In a), an assertion $W_{y=x}$ is shown where $W_{y=x}$ constrains $\{y\}$. The assertions $proj_{\{x\}}(W_{y=x})$ and $\widehat{proj}_{\{x,y\}}(proj_{\{x\}}(W_{y=x}))$ are shown in b) and c), respectively.

The relation in Equation 13.3 can be compared to the notion of *receptiveness* as described in Benveniste et al. and Benveniste, Caillaud, and Passerone [12,14]. The notion of receptiveness is further discussed in Section 13.7.

13.2.2 ELEMENTS

In this section, the concept of an *element* is introduced in order to model any entity of a cyber-physical systems in general, such as an SW, HW, or physical entity, as well as to serve as a functional or logical design entity in general, for example, as a System Modeling Language block [39]. Elements essentially correspond to heterogeneous rich components (HRCs) [26,54], as used in the contract theory [12,14] of SPEEDS, but differs slightly in order to fit with the other concepts in this chapter.

Definition 13.1 (Element) *An element* $\mathbb{E}$ *is an ordered pair* (X, B) *where*

(a) *X is a nonempty set of variables, called the* interface *of* $\mathbb{E}$ *and where each* $x \in X$ *is called a*
    port variable
(b) B *is an assertion over X, called the* behavior *of* $\mathbb{E}$ □

An element is a model of a real-world object where the port variables model tangible quantities of the object from the perspective of an external observer to the object. The behavior of the element models the static and dynamic constraints that the object imposes on the quantities, independent of its surroundings.

As an illustrative example, consider an element $\mathbb{E}_{pot} = (X_{pot}, \mathsf{B}_{pot})$ modeling a potentiometer, as shown in Figure 13.4 where the rectangle filled with gray and the boxes on its edges symbolize the element and its port variables, respectively. The port variables v_{ref}, v_{branch}, and v_{gnd} model the reference, branch, and ground voltages, respectively. Furthermore, h models the position (0%–100%) of the "slider" that moves over the resistor and branches the circuit.

Given a model where it is assumed that the branched circuit is connected to a resistance that is significantly larger than the resistance of the potentiometer, the behavior B_{pot} can be specified by the equation

$$h = \frac{v_{branch} - v_{gnd}}{v_{ref} - v_{gnd}}.$$

13.2.3 ARCHITECTURE

This section describes how a set of elements can be structured in order to model a cyber-physical system, its parts, and its surroundings. Similar to, for example, [48,49], such a structure will be

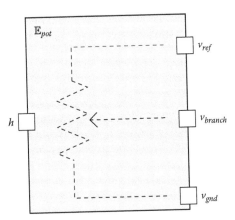

FIGURE 13.4   An element $\mathbb{E}_{pot} = (X_{pot}, \mathsf{B}_{pot})$ modeling a potentiometer.

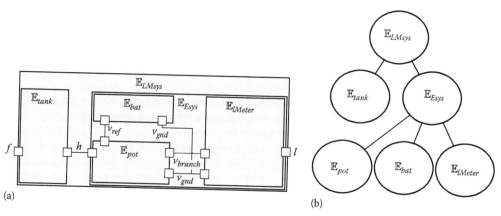

FIGURE 13.5 An architecture $\mathscr{A}_{LMsys}$ of a "level meter system." In (a) the architecture is represented as an hierarchy of rectangles with boxes on their edges where the boxes represent port variables. In (b) the architecture is represented as a rooted tree.

referred to as an *architecture*, which, in this chapter, will be denoted with the symbol $\mathscr{A}$. Although the definition of an architecture in this chapter is, in essence, in accordance with the definitions in References [48,49], the definition in this chapter is more specific since it is tailored to fit with the other concepts presented in the chapter.

The underlying principles are to structure elements in an hierarchical manner, modeling a system and the parts it consists of and in analogy with [12,14], use the *sharing of port variables* between element interfaces to model interactions between the parts. The sharing of port variables offers an expressive and simple way to model interactions, but requires the use of *connectors*, for example, as described in Modelica [69], to be explicitly represented as elements.

Prior to presenting the formal definition of an architecture, the concept is introduced informally by an example describing an architecture $\mathscr{A}_{LMsys}$ of a "level meter system" (LM system) $\mathbb{E}_{LMsys}$, as shown in Figure 13.5 where a set of elements is structured hierarchically as a graph as shown in Figure 13.5b. The sharing of port variables at the same hierarchical level is shown in Figure 13.5a visualized by connecting the port variables with a line. The sharing of port variables between different hierarchical levels is visualized by its appearance on several edges of rectangles.

As shown in Figure 13.5a, the LM system $\mathbb{E}_{LMsys}$ consists of a tank $\mathbb{E}_{tank}$ and an electric system $\mathbb{E}_{Esys}$. The electric system $\mathbb{E}_{Esys}$ consists of the potentiometer $\mathbb{E}_{pot}$ as shown in Figure 13.4, a battery $\mathbb{E}_{bat}$, and a level meter $\mathbb{E}_{lMeter}$ where the behaviors $\mathbb{B}_{bat}$ and $\mathbb{B}_{Lmeter}$ of $\mathbb{E}_{bat}$ and $\mathbb{E}_{lMeter}$ are specified by the equations $v_{ref} - v_{gnd} = 5V$ and $l = (v_{branch} - v_{gnd})/5$, respectively. The slider h is connected to a "floater," trailing the level f in the tank. In this way, the potentiometer $\mathbb{E}_{pot}$ is used as a level sensor to estimate the level in the tank. The estimated level is presented by the level meter $\mathbb{E}_{lMeter}$, where l denotes the presented level.

Notably, since each part of the electric system $\mathbb{E}_{Esys}$ will have quantities that may not be perceivable when the parts are integrated with each other, for example, the reference voltage v_{ref}, using standard terminology in *graph theory* [35], a port variable x of a child of $\mathbb{E}_{Esys}$ where $x \notin X_{Esys}$ cannot be a member of an interface of a nondescendant of $\mathbb{E}_{Esys}$*, for example, $\mathbb{E}_{tank}$. In order to further relate the individual behaviors of the children of $\mathbb{E}_{Esys}$ with the behavior of $\mathbb{E}_{Esys}$, the individual behaviors are first combined with each other using the intersection operator and subsequently restricted to the interface of $\mathbb{E}_{Esys}$ using the projection operator in accordance with the Section 13.2.1, that is,

$$B_{Esys} = \widehat{proj}_{X_{Esys}} \left(B_{pot} \cap B_{bat} \cap B_{Lmeter} \right).$$

---

* A descendant of $\mathbb{E}_{Esys}$ is a node reachable by iteratively proceeding from parent to child starting at $\mathbb{E}_{Esys}$.

Now that the concept of an architecture has been informally introduced, the formal definition follows.

Definition 13.2 (Architecture) *An architecture $\mathscr{A}$ is a set of elements organized into a rooted tree, such that:*

(a) for any nonleaf node $\mathbb{E} = (X, \mathsf{B})$, with children $\{(X_i, \mathsf{B}_i)\}_{i=1}^{N}$, it holds that $\mathsf{B} = \widehat{proj}_X(\widehat{\bigcap}_{i=1}^{N} \mathsf{B}_i)$ and $X \subseteq \bigcup_{i=1}^{N} X_i$; and

(b) if there is a child $\mathbb{E}' = (X', \mathsf{B}')$ and a nondescendent $\mathbb{E}'' = (X'', \mathsf{B}'')$ of $\mathbb{E} = (X, \mathsf{B})$, such that $x \in X'$ and $x \in X''$, then it holds that $x \in X$. □

Definition 13.2 is of a general type, which means that the conditions (a) and (b) of Definition 13.2 hold regardless of the domain, for example, mechanical and SW, that is considered. In some domains, however, for example, the SW domain, the interface of an element is typically organized into inputs and outputs, which means that additional constraints must be introduced that express which runs can be in each behavior of the elements. These constraints are further discussed in Section 13.7.

In the context of an architecture containing an element $\mathbb{E}$, the concept of *the environment of an element* $\mathbb{E}$ is introduced.

Definition 13.3 (Environment of Element) *Given an architecture $\mathscr{A}$, the environment of an element $\mathbb{E} = (X, \mathsf{B})$ in $\mathscr{A}$, denoted $Env_{\mathscr{A}}(\mathbb{E})$, is the set of elements $\{\mathbb{E}_i\}_{i=1}^{N}$ such that $\mathbb{E}_i = (X_i, \mathsf{B}_i)$ is either a sibling or a sibling of a proper ancestor of $\mathbb{E}$. Let $\mathsf{B}_{Env_{\mathscr{A}}(\mathbb{E})} = \widehat{\bigcap}_{i=1}^{N} \mathsf{B}_i$, denote the behavior of $Env_{\mathscr{A}}(\mathbb{E})$.* □

As an example, the elements $\mathbb{E}_{bat}$, $\mathbb{E}_{tank}$, and $\mathbb{E}_{lMeter}$ shown in Figure 13.5 are elements in the environment of the potentiometer $\mathbb{E}_{pot}$ in accordance with Definition 13.3.

13.2.3.1 Resolvable Architecture

Consider a lamp with a load resistance of $5\,\Omega$ as shown in Figure 13.6a. The lamp further has an internal fuse with a breaking capacity of $0.5\,A$, which means that $0.5\,A$ is the maximum current that can pass through the lamp without causing the fuse to melt. Now, consider that an element $\mathbb{E}_{lamp} = (\{V_+, V_-, i, R_L\}, \mathsf{B}_{lamp})$ is a model of the lamp. The behavior B_{lamp} of $\mathbb{E}_{lamp}$ is the set of runs that are solutions to the relation $i \leq 0.5\,A$ and the equation $i = (V_+ - V_-)/R_L$ where $R_L = 5\,\Omega$.

Suppose that there is a need to simulate a scenario where the fuse melts due to a passing of current larger than $0.5\,A$ through the lamp. The architecture shown in Figure 13.6b is hence considered where the element $\mathbb{E}_{lamp}$ and another element $\mathbb{E}'_{bat} = (\{V_+, V_-, i\}, B'_{bat})$, with a behavior B'_{bat} specified by the equation $V_+ - V_- = 5V$, are the leaf elements in the architecture. Since it holds that $i = 5\,V/5\,\Omega = 1\,A$ and the behavior B_{lamp} of $\mathbb{E}_{lamp}$ is only specified for values of i that are lower than or equal to $0.5\,A$, the scenario cannot be simulated since B'_{bat} and B_{lamp} have no common run, that is, it holds that $B'_{bat} \widehat{\cap} \mathsf{B}_{lamp} = \emptyset$. This is clearly an undesirable property of a model, and the concept *resolvable architecture* is introduced to characterize such cases.

Definition 13.4 (Resolvable Architecture) *An architecture $\mathscr{A}$ is resolvable if it holds that $\widehat{\bigcap}_{i=1}^{N} \mathsf{B}_i \neq \emptyset$, where $\{(X_i, \mathsf{B}_i)\}_{i=1}^{N}$ is the set of leaf nodes in $\mathscr{A}$.* □

The proposition that now follows presents a necessary and sufficient condition of resolvability. The proposition will be frequently used in the following, and especially in Section 13.3.

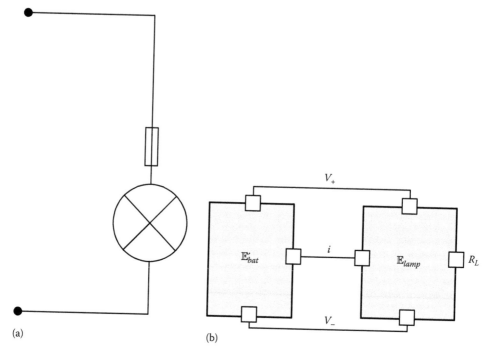

(a)   (b)

FIGURE 13.6 In (a), a lamp is shown with a load resistance of $5\,\Omega$ and an internal fuse with a breaking capacity of 0.5A. In (b), an architecture is shown where a model $\mathbb{E}_{lamp} = (\{V_+, V_-, i, R_L\}, \mathsf{B}_{lamp})$ of the lamp in (a) and an element $\mathbb{E}'_{bat} = (\{V_+, V_-, i\}, \mathsf{B}'_{bat})$ are the leaf elements in the architecture.

Proposition 13.1 *Given an architecture $\mathscr{A}$ containing an element $\mathbb{E} = (X, \mathsf{B})$, it holds that $\mathscr{A}$ is resolvable, if and only if $\mathsf{B}_{Env_{\mathscr{A}}(\mathbb{E})} \widehat{\bigcap}_{i=1}^{N} \mathsf{B}_i \neq \emptyset$.*

A similar proposition is presented in Westman and Nyberg [90], from which Proposition 13.1 can easily be derived.

As mentioned in Section 13.1, given an architecture $\mathscr{A}$, conditions that ensure that an element $\mathbb{E}$ meets its responsibility expressed by the guarantee of a contract will be presented in Section 13.3.1. For the cases where resolvability is not ensured by the overall context, to avoid the case highlighted in Figure 13.6, the conditions ensure that $\mathsf{B}_{Env_{\mathscr{A}}(\mathbb{E})} \widehat{\bigcap} \mathsf{B} \neq \emptyset$, which means that $\mathscr{A}$ is resolvable in accordance with Proposition 13.1.

Remark 13.1 *The architecture shown in Figure 13.6b is not resolvable since the behavior of the element $\mathbb{E}_{lamp}$ is not specified for a scenario where the circuit is broken due to the melting of the fuse. However, a more detailed model of the lamp can be specified by the relation $(i \leq 0.5\,A \land i = \frac{V_+ - V_-}{5\,\Omega}) \lor i = 0$, which captures the properties of the lamp in a context where the circuit is broken. Notably, if B_{lamp} is specified by this detailed model, then the architecture shown in Figure 13.6b is resolvable.*

13.3 CONTRACTS

Based on the theoretic framework presented in Section 13.2, this section introduces the concept of a contract, as well as conditions that, given an architecture, ensure that an element meets its responsibility expressed by the guarantee of a contract. In contrast to the contract theories [7,10,12–15,17,22,27,33,40–43,67,78,81,86,87], but in accordance with [90], the conditions are general, which

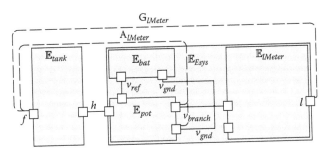

FIGURE 13.7   A contract $\mathcal{C}_{lMeter} = (\{A_{lMeter}\}, G_{lMeter})$.

means that the conditions hold for any type of architecture that is considered and further for the cases where the contract is not limited to the interface of the element.

As a start, a general definition of a contract is presented.

Definition 13.5 (Contract)   *A contract $\mathcal{C}$ is a pair $(\mathcal{A}, G)$, where*

(i) G is an assertion, called guarantee
(ii) $\mathcal{A}$ is a set of assertions $\{A_i\}_{i=1}^{N}$, where each A_i is called an assumption   □

In the context of an architecture, a guarantee of a contract for an element expresses an intended property under the responsibility of the element, given that the environment of the element fulfills the assumptions. The set of variables, over which an assumption $A_i \in \mathcal{A}$ or a guarantee G is expressed, is called the *scope* of A_i or G, denoted X_{A_i} and X_G, respectively. For the sake of readability, let $A_{\mathcal{A}} = \widehat{\bigcap}_{j=1}^{N} A_i$ and $X_{A_{\mathcal{A}}} = \bigcup_{i=1}^{N} X_{A_i}$.

As an illustrative example, consider the architecture $\mathcal{A}_{LMsys}$ shown in Figure 13.5 and a contract $\mathcal{C}_{lMeter} = (\{A_{lMeter}\}, G_{lMeter})$ for the level meter $\mathbb{E}_{lMeter}$, as shown in Figure 13.7 where the dashed lines represent the scopes of A_{lMeter} and G_{lMeter}, respectively. The guarantee G_{lMeter}, specified by the equation $l = f$, expresses that the indicated level, displayed by the meter, corresponds to the level in the tank. The guarantee G_{lMeter} is under the responsibility of $\mathbb{E}_{lMeter}$, but only if the voltage measured between v_{branch} and v_{gnd} maps to a specific level in the tank, that is, if the environment of $\mathbb{E}_{lMeter}$ fulfills the assumption A_{lMeter}, specified by the equation $f = (v_{branch} - v_{gnd})/5$.

The contract $\mathcal{C}_{lMeter}$ can, for example, be used for a scenario where an OEM develops the $\mathbb{E}_{lMeter}$ in-house, while the development of the elements $\mathbb{E}_{bat}$, $\mathbb{E}_{pot}$, and $\mathbb{E}_{tank}$ are outsourced to suppliers. The overall intended functionality of the LM system, as expressed by G_{lMeter}, is under the responsibility of $\mathbb{E}_{lMeter}$, meaning that the OEM does not only need to ensure the development of $\mathbb{E}_{lMeter}$, but also the successful integration of the elements $\mathbb{E}_{lMeter}$, $\mathbb{E}_{bat}$, $\mathbb{E}_{pot}$, and $\mathbb{E}_{tank}$ into the element $\mathbb{E}_{LMsys}$ with a behavior specified by $f = l$. A successful integration of the elements can be ensured by the OEM, given that the environment of $\mathbb{E}_{lMeter}$ in $\mathcal{A}_{LMsys}$ fulfills the assumption A_{lMeter}, which is a condition that is to be met by the suppliers.

13.3.1   Conditions on Element and Environment

This section presents conditions on an element $\mathbb{E}$ and its environment $Env_{\mathcal{A}}(\mathbb{E})$ in an architecture $\mathcal{A}$ in order to ensure that $\mathcal{A}$ is resolvable and that $\mathbb{E}$ *meets its responsibility expressed by the guarantee* G, that is, that $B_{Env_{\mathcal{A}}(\mathbb{E})} \widehat{\cap} B \widehat{\subseteq} G$. That is, in accordance with Proposition 13.1, the conditions ensure that

$$\emptyset \neq B_{Env_{\mathcal{A}}(\mathbb{E})} \widehat{\cap} B \widehat{\subseteq} G. \tag{13.4}$$

In order to get a better understanding of when these conditions are needed, a scenario in the context of an OEM/supplier chain as presented in Section 13.1 is examined. In the scenario, a contract

$C = (\mathcal{A}, G)$ is used to outsource the development of an element $\mathbb{E}$ with an interface X. Specifically, the scenario can be described in the following three phases:

1. A contract C and a set of variables X, that is, an *interface specification*, are handed from the OEM to a supplier.
2. The supplier develops an element $\mathbb{E} = (X, B)$ that is handed to the OEM.
3. The OEM integrates the element $\mathbb{E}$ with a set of elements $\{\mathbb{E}_i\}_{i=1}^{N}$ to form an architecture $\mathcal{A}$, where $\{\mathbb{E}_i\}_{i=1}^{N}$ is the environment $Env_{\mathcal{A}}(\mathbb{E})$ of $\mathbb{E}$.

As expressed in phases (1–2), the development of the element $\mathbb{E}$ is guided only by the information available in the contract C and the interface specification X, that is, without access to the elements in $\{\mathbb{E}_i\}_{i=1}^{N}$. Therefore, in order for the integration of $\mathbb{E}$ with the elements in $\{\mathbb{E}_i\}_{i=1}^{N}$ in phase (3) to result in an architecture $\mathcal{A}$, where the relation in Equation 13.4 holds with respect to C, conditions must be enforced on the element $\mathbb{E}$ and its environment $Env_{\mathcal{A}}(\mathbb{E})$. Conditions that ensure that the relation $B_{Env_{\mathcal{A}}(\mathbb{E})} \widehat{\cap} B \widehat{\subseteq} G$ holds are first presented, followed by additional conditions that ensure that also the relation $\emptyset \neq B_{Env_{\mathcal{A}}(\mathbb{E})} \widehat{\cap} B$ holds.

As previously mentioned, the guarantee G is under the responsibility of $\mathbb{E}$, given that *the environment of $\mathbb{E}$ fulfills the assumptions*. This means that it must hold that

$$B_{Env_{\mathcal{A}}(\mathbb{E})} \widehat{\subseteq} A_{\mathcal{A}}. \tag{13.5}$$

Supposing that the relation in Equation 13.5 holds, it follows that the relation $B_{Env_{\mathcal{A}}(\mathbb{E})} \widehat{\cap} B \widehat{\subseteq} G$ holds if

$$A_{\mathcal{A}} \widehat{\cap} B \widehat{\subseteq} G, \tag{13.6}$$

which is referred to as that $\mathbb{E}$ *satisfies* C in References [1,12–15,78,81]. Note that if $\mathcal{A} = \emptyset$, then the relation in Equation 13.6 simplifies to $B \widehat{\subseteq} G$.

To summarize the insights gathered so far, the conditions, in order for the relation $B_{Env_{\mathcal{A}}(\mathbb{E})} \widehat{\cap} B \widehat{\subseteq} G$ to hold, can be separated into the relation in Equation 13.5 that expresses a condition on the environment of $\mathbb{E}$ and the relation in Equation 13.6 is a condition on the element $\mathbb{E}$. However, in accordance with Proposition 13.1, the relation $B_{Env_{\mathcal{A}}(\mathbb{E})} \widehat{\cap} B \widehat{\subseteq} G$ trivially holds if $\mathcal{A}$ is not resolvable, which is a necessary condition of the relation in Equation 13.4. The relation in Equation 13.4 does not hence follow from the relations in Equations 13.5 and 13.6, so additional conditions must hence be imposed on the environment and on the element in order to ensure that the architecture is resolvable. In the following, these conditions will be examined.

13.3.1.1 Resolvability Conditions

Consider a contract $(\{A'\}, G')$ for the element $\mathbb{E}'_{bat}$ described in point 1 under Section 13.2.3. The guarantee G' is specified by the equation $i = 1$A, and A' is specified by the equation $i = (V_+ - V_-)/R_L$, where $R_L = 5\,\Omega$. The guarantee G', under the responsibility of $\mathbb{E}'_{bat}$, expresses that the current i through the element $\mathbb{E}'_{bat}$ shall be 1 A, given the assumption A' that $\mathbb{E}'_{bat}$ is connected to a circuit with a total resistance of $R_L = 5\,\Omega$.

Considering the *nonresolvable architecture* shown in Figure 13.6b, it holds that the environment of $\mathbb{E}'_{bat}$ fulfills the assumption A', that is, that $B_{lamp} \widehat{\subseteq} A'$ in accordance with the relation in Equation 13.5. Furthermore, the relation in Equation 13.6 also holds since the relation $A' \widehat{\cap} B'_{bat}$ is the set of runs over $\{V_+, V_-, i, R_L\}$ that are solutions to the equation $i = 1$ A. This means that resolvability is not ensured despite the fact that the relations in Equations 13.5 and 13.6 hold.

Notably, since G' is specified by the equation $i = 1$ A and B_{lamp} only contains runs where $i \leq 0.5$ A, it holds that $B_{lamp} \widehat{\cap} G' = \emptyset$. This implies that the architecture cannot be resolvable since it holds that $B_{lamp} \widehat{\cap} B'_{bat} \widehat{\subseteq} G'$ and if there does not exist a run in B_{lamp} that is also in G', then there cannot exist a run that is in both B_{lamp} and B'_{bat} that is also in G'. As previously shown in Westman and Nyberg [90], this leads to the following proposition:

Proposition 13.2   *Consider a contract* $\mathcal{C} = (\mathcal{A}, G)$ *and an element* $\mathbb{E} = (X, B)$. *If* $\mathcal{A}$ *is an architecture containing* $\mathbb{E}$ *where* $\emptyset \neq B_{Env_{\mathcal{A}}(\mathbb{E})} \cap B \subseteq G$, *then it holds that* $B_{Env_{\mathcal{A}}(\mathbb{E})} \cap G \neq \emptyset$.

Now that the necessary condition $B_{Env_{\mathcal{A}}(\mathbb{E})} \cap G \neq \emptyset$ on the environment of $\mathbb{E}$ has been identified in order for the relation in Equation 13.4 to hold, a complementary condition on the element $\mathbb{E}$ is examined in order to ensure that the relation in Equation 13.4 holds.

Consider that the relation $A_{\mathcal{A}} \cap B \subseteq G$ on the element $\mathbb{E}$ holds. As shown in the Wenn diagram in Figure 13.8a where all assertions have been extended to a common set of variables, if it holds that $B_{Env_{\mathcal{A}}(\mathbb{E})} \subseteq A_{\mathcal{A}}$ and $B_{Env_{\mathcal{A}}(\mathbb{E})} \cap G \neq \emptyset$, it is possible that the relation in Equation 13.4 holds. However, as shown in Figure 13.8b, this is not true for all cases. In fact, since $B_{Env_{\mathcal{A}}(\mathbb{E})} \cap G$ can simply consist of one run, and this run can possibly be any run in $A_{\mathcal{A}} \cap G$, in order to ensure that $B_{Env_{\mathcal{A}}(\mathbb{E})} \cap G \neq \emptyset$, it must hold that $A_{\mathcal{A}} \cap G \subseteq B$ as shown in Figure 13.8c.

These insights are now summarized in the following theorem [90].

Theorem 13.1   *Consider a contract* $\mathcal{C} = (\mathcal{A}, G)$ *and an element* $\mathbb{E} = (X, B)$ *where* $A_{\mathcal{A}} \cap B \subseteq G$. *It holds that*

$$A_{\mathcal{A}} \cap G \subseteq B,$$

if and only if $\emptyset \neq B \cap B_{Env_{\mathcal{A}}(\mathbb{E})} \subseteq G$ *for each architecture* $\mathcal{A}$ *containing* $\mathbb{E}$ *where* $B_{Env_{\mathcal{A}}(\mathbb{E})} \cap G \neq \emptyset$ *and* $B_{Env_{\mathcal{A}}(\mathbb{E})} \subseteq A_{\mathcal{A}}$.

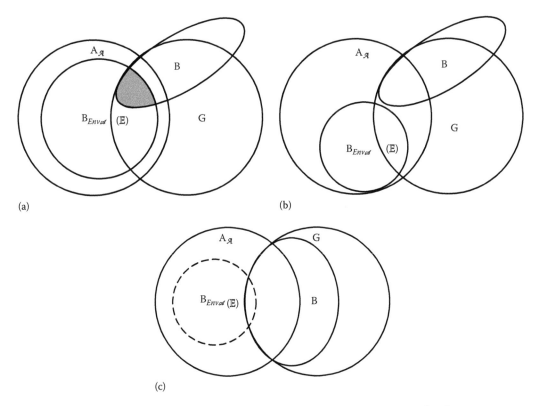

(a)   (b)

(c)

FIGURE 13.8   The Wenn diagram in (a) shows a case where it holds that $\emptyset \neq B_{Env_{\mathcal{A}}(\mathbb{E})} \cap B \subseteq G$ and (b) shows a case where this does not hold. In (c), a Wenn diagram is shown where it holds that $\emptyset \neq B_{Env_{\mathcal{A}}(\mathbb{E})} \cap B \subseteq G$ for each architecture $\mathcal{A}$ where $B_{Env_{\mathcal{A}}(\mathbb{E})} \subseteq A_{\mathcal{A}}$ and $B_{Env_{\mathcal{A}}(\mathbb{E})} \cap G \neq \emptyset$.

Given a contract $\mathcal{C} = (\mathsf{A}, \mathsf{G})$ and an element $\mathbb{E} = (\mathsf{X}, \mathsf{B})$ where $\mathsf{A}_\mathcal{A} \widehat{\cap} \mathsf{B} \widehat{\subseteq} \mathsf{G}$, using quantifiers, Theorem 13.1 can also be expressed as

$$\mathsf{A}_\mathcal{A} \widehat{\cap} \mathsf{G} \widehat{\subseteq} \mathsf{B} \Longleftrightarrow (\forall \mathcal{A} \ni \mathbb{E} : ((\mathsf{B}_{Env_\mathcal{A}(\mathbb{E})} \widehat{\subseteq} \mathsf{A}_\mathcal{A} \wedge \mathsf{B}_{Env_\mathcal{A}(\mathbb{E})} \widehat{\cap} \mathsf{G} \neq \emptyset) \Longrightarrow \emptyset \neq \mathsf{B} \widehat{\cap} \mathsf{B}_{Env_\mathcal{A}(\mathbb{E})} \widehat{\subseteq} \mathsf{G})).$$

Given that the relation $\mathsf{A}_\mathcal{A} \widehat{\cap} \mathsf{B} \widehat{\subseteq} \mathsf{G}$ on the element $\mathbb{E}$ holds, Theorem 13.1 expresses a necessary and sufficient condition on the element $\mathbb{E}$ such that if the element is integrated with a set of elements to form an architecture $\mathcal{A}$ where it holds that $\mathsf{B}_{Env_\mathcal{A}(\mathbb{E})} \widehat{\cap} \mathsf{G} \neq \emptyset$ and $\mathsf{B}_{Env_\mathcal{A}(\mathbb{E})} \widehat{\subseteq} \mathsf{A}_\mathcal{A}$, then the relation in Equation 13.4 holds. The condition expressed in Theorem 13.1 holds regardless of the type of architecture that is considered, for example, SW, HW, and mechanical.

For practical application, the following corollary clearly separates the respective conditions on the element $\mathbb{E}$ and its environment $Env_\mathcal{A}(\mathbb{E})$ in order for the relation in Equation 13.4 to hold with respect to $\mathcal{C}$.

Corollary 13.1   *Given a contract $\mathcal{C} = (\mathsf{A}, \mathsf{G})$ and an architecture $\mathcal{A}$ containing an element $\mathbb{E} = (\mathsf{X}, \mathsf{B})$, it holds that $\emptyset \neq \mathsf{B} \widehat{\cap} \mathsf{B}_{Env_\mathcal{A}(\mathbb{E})} \widehat{\subseteq} \mathsf{G}$ if both the following conditions hold:*

(i) The element $\mathbb{E}$ is such that

$$\mathsf{A}_\mathcal{A} \widehat{\cap} \mathsf{B} \widehat{\subseteq} \mathsf{G}, \text{ and} \tag{13.7}$$

$$\mathsf{A}_\mathcal{A} \widehat{\cap} \mathsf{G} \widehat{\subseteq} \mathsf{B}. \tag{13.8}$$

(ii) The environment $Env_\mathcal{A}(\mathbb{E})$ is such that

$$\mathsf{B}_{Env_\mathcal{A}(\mathbb{E})} \widehat{\subseteq} \mathsf{A}_\mathcal{A}, \text{ and} \tag{13.9}$$

$$\mathsf{B}_{Env_\mathcal{A}(\mathbb{E})} \widehat{\cap} \mathsf{G} \neq \emptyset. \tag{13.10}$$

In the context of the scenario presented in the beginning of this section, Corollary 13.1 specifies the conditions that the OEM and the supplier need to meet in order to ensure that the integration of $\mathbb{E}$ with the elements in $\{\mathbb{E}_i\}_{i=1}^N$ in phase (3) results in a *resolvable* architecture $\mathcal{A}$ where the element $\mathbb{E}$ meets its responsibility expressed by the guarantee G, that is, that the relation in Equation 13.4 holds.

Given that the relation in Equation 13.9 holds, the relations in Equations 13.8 and 13.10 are *sufficient* conditions to ensure resolvability. The relations in Equations 13.8 and 13.10 are also *necessary* to ensure resolvability in a general context such as the OEM/supplier chain, but in fact not necessary for the case when $\mathbb{E}$ and the elements in $\{\mathbb{E}_i\}_{i=1}^N$ are not developed in isolation from each other. An example of such a case is when both the element $\mathbb{E}$ and the elements in $\{\mathbb{E}_i\}_{i=1}^N$ are developed within the same company. Since the team that develops $\mathbb{E}$ has full access to the elements in $\{\mathbb{E}_i\}_{i=1}^N$, the integration of $\mathbb{E}$ with $\{\mathbb{E}_i\}_{i=1}^N$ into a *resolvable* architecture $\mathcal{A}$ is done by trial and error and by relying on the expertise of the in-house development teams. Therefore, in order for the relation in Equation 13.4 to hold in such a context, it is sufficient that the respective relations in Equations 13.7 and 13.9 on the element and the environment hold.

As a conclusion to this section, it is examined whether the element $\mathbb{E}_{lMeter}$ and the environment of $\mathbb{E}_{lMeter}$ in the architecture $\mathcal{A}_{LMsys}$ shown in Figure 13.5 are such that the respective conditions (i) and (ii) of Corollary 13.1 holds with respect to the contract $\mathcal{C}_{lMeter}$ shown in Figure 13.7. As expressed in the condition (i) of Corollary 13.1, it must hold that

$$\mathsf{A}_{lMeter} \widehat{\cap} \mathsf{B}_{lMeter} \widehat{\subseteq} \mathsf{G}_{lMeter}, \text{ and} \tag{13.11}$$

$$\mathsf{A}_{lMeter} \widehat{\cap} \mathsf{G}_{lMeter} \widehat{\subseteq} \mathsf{B}_{lMeter}. \tag{13.12}$$

Furthermore, as expressed in the condition (ii) of Corollary 13.1, it must hold that

$$B_{Env_{\mathscr{A}_{LMsys}}(\mathbb{E}_{lMeter})} \, \widehat{\subseteq} \, A_{lMeter}, \text{ and} \tag{13.13}$$

$$B_{Env_{\mathscr{A}_{LMsys}}(\mathbb{E}_{lMeter})} \, \widehat{\cap} \, G_{lMeter} \neq \emptyset. \tag{13.14}$$

By applying the operation of intersection, $A_{lMeter} \, \widehat{\cap} \, B_{Lmeter}$ yields an assertion specified by the equation $f = l$. Since this is equal to the equation that specifies G_{lMeter}, the relation in Equation 13.11 holds. Furthermore, since $A_{lMeter} \, \widehat{\cap} \, G_{lMeter}$ is an assertion specified by the equation $l = (v_{branch} - v_{gnd})/5$, which is equal to the equation that specifies B_{lMeter}, the relation in Equation 13.12 also holds.

Suppose that the behavior of $\mathbb{E}_{tank}$ is specified by the equation $f = h$. Since $Env_{\mathscr{A}_{LMsys}}(\mathbb{E}_{lMeter}) = \{\mathbb{E}_{tank}, \mathbb{E}_{Esys}, \mathbb{E}_{pot}\}$, it holds that $B_{Env_{\mathscr{A}_{LMsys}}(\mathbb{E}_{lMeter})}$ is specified by the equation $f = h = (v_{branch} - v_{gnd})/5$. This means that the relation in Equation 13.13 holds. Furthermore, since $B_{Env_{\mathscr{A}_{LMsys}}} \, \widehat{\cap} \, G_{lMeter}$ is specified by $f = h = l = (v_{branch} - v_{gnd})/5$, the relation in Equation 13.14 also holds.

Since the relations in Equations 13.11 through 13.14 hold, Corollary 13.1 implies that the architecture $\mathscr{A}_{LMsys}$ is resolvable and that the guarantee G_{lMeter}, under the responsibility of $\mathbb{E}_{lMeter}$, is met.

13.4 SCOPING CONSTRAINTS

Section 13.3 presented conditions that ensure that an element and its environment in an architecture meet their respective responsibilities with respect to a contract. In contrast to the contract theories [7,10,12–15,17,22,27,33,40–43,67,78,81,86,87], the conditions are applicable for the cases where contracts that are not limited to element interfaces are used, which means that the framework in this chapter is strictly more expressive with respect to how a contract for an element can be specified. However, the increased expressivity is not unlimited since there are necessary constraints on the scopes of the assumptions and the guarantee of a contract for an element in order for the relations in Equations 13.7 and 13.9 to hold in a resolvable architecture. Therefore, to facilitate the application of the framework, this section introduces such constraints to serve as sanity checks of the contract.

Prior to presenting such constraints in a formal manner, two relevant propositions will first be presented, followed by two representative examples of when the scopes of either an assumption or the guarantee of a contract for $\mathbb{E}_{Esys}$ leads to the violation of the relations in Equations 13.7 and 13.9 in the architecture $\mathscr{A}_{LMsys}$ shown in Figure 13.5. Note that the architecture $\mathscr{A}_{LMsys}$ is resolvable, as was shown in point 1 under Section 13.3.1.

Proposition 13.3 *An assertion* W *over* X *can only constrain subsets of* X.

Proposition 13.4 *Given two assertions* W *and* W' *where* $\emptyset \neq W \, \widehat{\subseteq} \, W'$, *if* W' *constrains a set of variables* X'', *then* X'' *is also constrained by* W.

Consider the following examples (a) and (b) of two contracts $C'_{Esys} = (\{A'_{Esys}\}, G'_{Esys})$ and $C''_{Esys} = (\{A''_{Esys}\}, G''_{Esys})$, respectively:

(a) As shown in Figure 13.9a, the variable v_{branch} is in the scope of A'_{Esys}, but not in the interface of the tank $\mathbb{E}_{tank}$, that is, the environment of $\mathbb{E}_{Esys}$. In the generic case, that is, when A'_{Esys} constrains $\{v_{branch}\}$, B_{tank} can only constrain port variables on the interface X_{tank} in accordance with Proposition 13.3. This means that it does not hold that $X_{A'_{Esys}} \subseteq X_{tank}$, which means that it does not hold that $\emptyset \neq B_{tank} \, \widehat{\subseteq} \, A'_{Esys}$ in accordance with Proposition 13.4. Hence, $\mathbb{E}_{tank}$ is not such that the relation in Equation 13.9 holds with respect to C'_{Esys}.

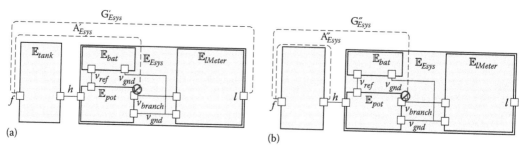

FIGURE 13.9 Two contracts C'_{Esys} and C''_{Esys} for $\mathbb{E}_{Esys}$ where $\mathbb{E}_{Esys}$ and are shown in (a) and (b), respectively, where $\mathbb{E}_{Esys}$ and its environment in the architecture $\mathscr{A}_{LMsys}$ shown in Figure 13.5 do not meet the relations in Equations 13.7 and 13.9.

(b) As can be seen in Figure 13.9b, the variable v_{branch} is in the scope of G''_{Esys}, but neither in the interface of $\mathbb{E}_{Esys}$ nor in the scope of A''_{Esys}. This means that in the generic case, that is, when G''_{Esys} constrains $\{v_{branch}\}$ and when $\mathsf{A}''_{Esys} \widehat{\cap} \mathsf{B}_{Esys} \neq \varnothing$, since $\mathsf{A}''_{Esys} \widehat{\cap} \mathsf{B}_{Esys}$ does not constrain v_{branch} in accordance with Proposition 13.3, the relation $\varnothing \neq \mathsf{A}''_{Esys} \widehat{\cap} \mathsf{B}_{Esys} \widehat{\subseteq} \mathsf{G}''_{Esys}$ does not hold in accordance with Proposition 13.4. Therefore, the element $\mathbb{E}_{Esys}$ is not such that the relation in Equation 13.7 holds, that is, $\mathbb{E}_{Esys}$ does not satisfy C''_{Esys}.

As illustrated in the examples (a) and (b), if either an assumption in $\mathcal{A}$ or the guarantee G of a contract for an element $\mathbb{E}$ has a scope that includes variables that are neither part of the interface of any element in the environment of $\mathbb{E}$ nor on the interface of $\mathbb{E}$, then at least one of the relations in Equations 13.7 and 13.9 does not hold in the generic case.

Theorem 13.2 *Given a contract $C = (\mathcal{A}, \mathsf{G})$, if $\mathscr{A}$ is a resolvable architecture containing an element $\mathbb{E} = (X, \mathsf{B})$ where*

(i) $\mathsf{A}_\mathcal{A} \widehat{\cap} \mathsf{B} \widehat{\subseteq} \mathsf{G}$
(ii) $\mathsf{B}_{Env_{\mathscr{A}}(\mathbb{E})} \widehat{\subseteq} \mathsf{A}_\mathcal{A}$,

then it holds that

(a) $X'_{\mathsf{A}_\mathcal{A}} \subseteq X_{Env_{\mathscr{A}}(\mathbb{E})}$,
(b) $X'_\mathsf{G} \subseteq X_{Env_{\mathscr{A}}(\mathbb{E})} \cup X$,

where $X'_{\mathsf{A}_\mathcal{A}} \subseteq X_{\mathsf{A}_\mathcal{A}}$ and $X'_\mathsf{G} \subseteq X_\mathsf{G}$ are the sets of variables constrained by $\mathsf{A}_\mathcal{A}$ and B, respectively, and $X_{Env_{\mathscr{A}}(\mathbb{E})}$ denotes the union of the interfaces of the elements in the environment of $\mathbb{E}$ in $\mathscr{A}$.

Theorem 13.2 is a slight modification from a theorem in Westman and Nyberg [90] and can easily be derived from Propositions 13.1, 13.3, and 13.4.

The relations (a) and (b) of Theorem 13.2 express necessary conditions in order for an element and its environment in a resolvable architecture to be such that the relations in Equations 13.7 and 13.9 hold with respect to a contract. Notably, the relations (a) and (b) of Theorem 13.2 only express constraints on the contract. To characterize a contract that complies with such constraints, the concept of a *scope-compliant contract* is introduced.

Definition 13.6 (Scope-Compliant Contract) *A contract $(\mathcal{A}, \mathsf{G})$ is scope-compliant with respect to two interface specifications X and $X_{Env_{\mathscr{A}}(\mathbb{E})}$ if*

(a) $X_{\mathsf{A}_\mathscr{A}} \subseteq X_{Env_\mathscr{A}(\mathbb{E})}.$
(b) $X_\mathsf{G} \subseteq X_{Env_\mathscr{A}(\mathbb{E})} \cup X.$

□

Given an architecture $\mathscr{A}$ containing an element $\mathbb{E} = (X, \mathsf{B})$ where $X_{Env_\mathscr{A}(\mathbb{E})}$ is the union of the interfaces of the elements in the environment of $\mathbb{E}$, assuming that the scopes of $\mathsf{A}_\mathscr{A}$ and G are equal to the set of variables that they constrain, the relations (a) and (b) of Definition 13.6 hold for all cases where the relations (i) and (ii) of Theorem 13.2 also hold. This includes all of the practical cases since the relations (i) and (ii) of Theorem 13.2 would still hold even if all variables that are not constrained by $\mathsf{A}_\mathscr{A}$ and G are removed from $X_{\mathsf{A}_\mathscr{A}}$ and X_G, respectively. Hence, the inclusion of such variables simply does not make sense, therefore such cases are excluded from Definition 13.6.

Regarding the examples shown in Figure 13.9, since $X_{\mathsf{A}'_{Esys}} \not\subseteq X_{tank}$, in accordance with relation (a) of Definition 13.6, the contract $\mathcal{C}'_{Esys}$ is not scope-compliant with respect to X_{Esys} and $X_{Env_\mathscr{A}(E_{Esys})}$. Furthermore, since $X_{\mathsf{G}''_{Esys}} \not\subseteq X_{tank} \cup X_{Esys}$, in accordance with relation (b) of Definition 13.6, the contract $\mathcal{C}''_{Esys}$ is not scope-compliant with respect to X_{Esys} and $X_{Env_\mathscr{A}(E_{Esys})}$. Definition 13.6 hence provides means to detect that the scopes of the assumptions and the guarantee of each contract in Figure 13.9 lead to the violation of the relations in Equations (13.7) and (13.9) in $\mathscr{A}_{LMsys}$.

13.5   HIERARCHICAL STRUCTURING OF REQUIREMENTS USING CONTRACTS

Consider a scenario where it is infeasible to verify that the element $\mathbb{E}$ in a resolvable architecture $\mathscr{A}$ satisfies its contract $\mathcal{C}$, due to the complexity of $\mathbb{E}$. A solution to such an issue is to *iteratively* split the contract $\mathcal{C}$ into a set of contracts for proper descendants of $\mathbb{E}$ until it is possible to verify that a descendant $\mathbb{E}_i$ of $\mathbb{E}$ satisfies $\mathcal{C}_i$ with the intent that

$$\text{if each } \mathbb{E}_i \text{ satisfies } \mathcal{C}_i, \text{ then } \mathbb{E} \text{ satisfies } \mathcal{C}. \qquad (13.15)$$

If the property given in Equation 13.15 holds and the environment of $\mathbb{E}$ is such that the relation in Equation 13.9 holds, in accordance with point 1 under Section 13.3.1, the relation in Equation 13.4 holds with respect to $\mathcal{C}$ since the architecture $\mathscr{A}$ is resolvable. If an architecture only consists of two hierarchical levels, then the property given in Equation 13.15 corresponds to *dominance/refinement* in References [10,12–14,78,81] and the basic idea of *compositional verification* [1,2,19,20,25,29,32, 46,52,53,63,65,66,68,76,79,94], as well as, in essence, the notion of *completeness* in ISO 26262 [48].

Despite the fact that the aforementioned scenario describes a case where the element $\mathbb{E}$ has been developed prior to verifying that each $\mathbb{E}_i$ satisfies $\mathcal{C}_i$, the verification can be done prior to the integration of the proper descendants of $\mathbb{E}$ into $\mathbb{E}$, as long as resolvability of $\mathscr{A}$ is ensured. As presented in point 1 under Section 13.3.1, resolvability of $\mathscr{A}$ can either be ensured by the context of which the elements in $\mathscr{A}$ are developed in or by enforcing additional conditions on the elements in $\mathscr{A}$.

This section introduces the concept of a graph, called a *contract structure* $\mathfrak{C}$, that organizes a set of contracts for elements in an architecture with the intent of having the property given in Equation 13.15. As will be shown, contract structures support the structuring of requirements on a cyber-physical systems and the individual tracing of lower-level requirements to their top-level goals. This claim will be further validated in Section 13.6 where contract structures will be used in an industrial case study to structure safety requirements for the FLD system mentioned in Section 13.1. Prior to presenting the formal definition of a contract structure in Section 13.5.2, an underlying concept of using a contract to express a relation between requirements is described in Section 13.5.1.

13.5.1   CONTRACTS AS REQUIREMENT RELATIONS

As mentioned in Section 13.3, in the contract shown in Figure 13.7, the guarantee G_{lMeter} is under the responsibility of $\mathbb{E}_{lMeter}$, given that the environment of $\mathbb{E}_{lMeter}$ in an architecture fulfills the assumption A_{lMeter}. Formulated differently, the guarantee G_{lMeter} is a *requirement on* $\mathbb{E}_{lMeter}$, given an architecture where the environment fulfills the assumption A_{lMeter}. This was also observed in Westman, Nyberg, and Torngren [92] where guarantees are used to express safety requirements on elements.

Consider a scenario where the environment that the element $\mathbb{E}_{Lmeter}$ is to be deployed in is unknown, for example, when developing $\mathbb{E}_{Lmeter}$ *out of context* [48]. The assumption A_{Lmeter} of the contract shown in Figure 13.7 hence expresses the conditions that the environment of $\mathbb{E}_{Lmeter}$ is to fulfill in *an arbitrary architecture containing* $\mathbb{E}_{Lmeter}$. However, in a *specific architecture*, such as the architecture $\mathscr{A}_{LMsys}$ in Figure 13.5, *the assumption* A_{Lmeter} *can rather be seen as a requirement* that is to be guaranteed by the potentiometer $\mathbb{E}_{pot}$. This was also observed in Westman, Nyberg, and Torngren [92] where, in the context of an architecture, assumptions are in fact references to other guarantees.

Therefore, in the definition of a contract structure for a *specific architecture* in Section 13.5.2, an assumption of a contract for an element $\mathbb{E}$ will correspond to a guarantee of a contract for an element in the environment of $\mathbb{E}$. Formulated differently, *the assumption of a contract for* $\mathbb{E}$ *is a requirement on an element in the environment of* $\mathbb{E}$, *while the guarantee is a requirement on* $\mathbb{E}$. The cases, where the use of explicit assumptions are needed, for example, when it is necessary to express that an assumption is to be fulfilled by two or more guarantees, are further discussed in Remark 13.2.

13.5.2 CONTRACT STRUCTURE

Consider that each guarantee G_i of a contract $C_i = (A_i, G_i)$ for an element $\mathbb{E}_i$ in the architecture $\mathscr{A}_{LMsys}$ is represented by a node in an edge-labeled directed graph as an overlay onto the hierarchical structure of the elements of $\mathscr{A}_{LMsys}$ as shown in Figure 13.10. A guarantee G_i is an assumption of another contract C_j, if there exists an arc labeled "assumption of" from G_i to G_j, visualized as a line with a circle filled with black at the end. For example, the arc from G_{pot} to G_{lMeter} represents that G_{pot} is an assumption of the contract C_{lMeter}.

As also shown in Figure 13.10, an incoming arc labeled "Fulfills," visualized as an arrow, to a guarantee G_i of a contract for an element $\mathbb{E}_i$ from a guarantee G_j of a contract for a child $\mathbb{E}_j$ of $\mathbb{E}_i$, represents the intent of $G_j \sqsubseteq G_i$. For example, the arc from the guarantee G_{Esys} to G_{LMsys} represents the intent of $G_{Esys} \sqsubseteq G_{LMsys}$.

Now that the concept of a contract structure has been introduced informally, the formal definition of a contract structure follows.

Definition 13.7 (Contract Structure for Architecture) *Given an architecture* $\mathscr{A}$ *and a set* $\{(A_{i,j}, G_{i,j})\}_{j=1}^{N_i}$ *where* $(A_{i,j}, G_{i,j})$ *is a contract for an element* $\mathbb{E}_i$ *of* $\mathscr{A}$ *and where each assumption in each set* $A_{i,j}$ *is either of the following:*

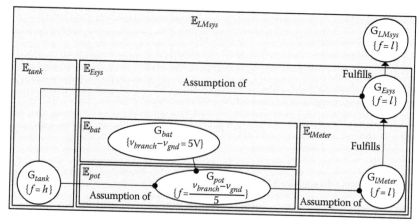

FIGURE 13.10 A contract structure $\mathfrak{C}_{LMsys}$ for the architecture $\mathscr{A}_{LM-sys}$.

(a) *A guarantee of a contract for a sibling of* $\mathbb{E}_i$
(b) *An assumption of a contract for a proper ancestor of* $\mathbb{E}_i$
(c) *An assumption of the contract for the root element in* $\mathscr{A}$

then a contract structure $\mathfrak{C}$ *for* $\mathscr{A}$ *is an arc labeled directed acyclic graph (DAG), such that*

(i) *The guarantees* $\mathsf{G}_{i,j}$ *and the assumptions of the contract for the root element in* $\mathscr{A}$ *are the nodes in* $\mathfrak{C}$
(ii) *Each arc is uniquely labeled either "assumption of" or "fulfills"*
(iii) *There is an arc labeled "assumption of" from a node* W *to* $\mathsf{G}_{i,j}$, *if and only if* W *is in* $\mathcal{A}_{i,j}$
(iv) *If there is an arc labeled "fulfills" from* $\mathsf{G}_{i,j}$ *to* $\mathsf{G}_{k,l}$, *then* $\mathsf{G}_{k,l}$ *is a guarantee of a contract for a proper ancestor of* $\mathbb{E}_i$
(v) *If a guarantee* $\mathsf{G}_{i,j}$ *is reachable from an assumption* A *of a contract for a proper ancestor* $\mathbb{E}_m$ *of* $\mathbb{E}_i$, *then* A *is also an assumption of any contract* $(\mathcal{A}_{k,l}, \mathsf{G}_{k,l})$ *where* $\mathbb{E}_k$ *is a proper ancestor of* $\mathbb{E}_i$ *and a descendant of* $\mathbb{E}_m$ *(including itself) and where* $\mathsf{G}_{k,l}$ *is reachable from* $\mathsf{G}_{i,j}$ □

As discussed in Section 13.5.1 and as also shown in Figure 13.10, the conditions (a) and (b) of Definition 13.7 express that an assumption of a contract for an element $\mathbb{E}$ correspond to a *guarantee of a contract for an element in the environment of* $\mathbb{E}$, that is, an assumption is either a guarantee of a contract for a sibling of $\mathbb{E}$ or an assumption of a contract for a proper ancestor of $\mathbb{E}$. The only exception is when the assumption is part of a contract for the root element in the architecture as expressed in condition (c) of Definition 13.7.

Furthermore, as expressed in the conditions (i–iv) of Definition 13.7 and shown in Figure 13.10, each node in a contract structure corresponds to a guarantee of a contract or an assumption of a contract for the root node and each arc either expresses that a node is an assumption of a contract or that the intent is that a guarantee is to fulfill another guarantee.

Consider the contract structure shown in Figure 13.11 that is intended to clarify why the graph would not be a contract structure if the arc marked with "(*)" is added to the graph, as expressed in the condition (v) of Definition 13.7. In Figure 13.11, the intent is that *the guarantee* G_2 *is under the (full) responsibility of* $\mathbb{E}$, *without relying on any assumptions*. However, if the arc labeled "(*)" is added, then the aforementioned statement is contradicted, since the responsibility of the guarantee G_2 is split between the children of $\mathbb{E}$, *given that* G' *is fulfilled by the environment of* $\mathbb{E}$.

Prior to presenting a theorem based on a contract structure where the theorem expresses sufficient conditions of the property given in Equation 13.15, two definitions that will be used in the theorem are introduced.

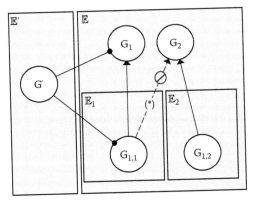

FIGURE 13.11 A contract structure that would not be a contract structure if the arc marked with "(*)" is added to the graph.

Definition 13.8 (Goal in Contract Structure) *A node in a contract structure for an architecture is a goal if the node does not have any successors.* □

As an example, the guarantee G_{LMsys} is a goal in the contract structure shown in Figure 13.10.

Definition 13.9 (Atomic Guarantee of Goal in Contract Structure) *Given a goal* G *in a contract structure* $\mathfrak{C}$ *for an architecture* $\mathscr{A}$, *a guarantee* G' *is an atomic guarantee of* G *if* G' *does not have any predecessors with an outgoing "fulfills" arc to* G' *and* G *is reachable from* G'. □

As an example, the guarantees G_{tank}, G_{bat}, G_{pot}, and G_{lMeter} are the atomic guarantees of G_{LMsys} in the contract structure shown in Figure 13.10. The contract structure in Figure 13.10 hence supports the identification of the lowest-level requirements G_{tank}, G_{bat}, G_{pot}, and G_{lMeter} on the elements in $\mathscr{A}_{LMsys}$ where the requirements can be traced to the top-level requirement G_{LMsys} on the element $\mathbb{E}_{LMsys}$.

Theorem 13.3 *Consider a contract structure* $\mathfrak{C}$ *for an architecture* $\mathscr{A}$ *where the guarantee* G *of a contract* $(\mathcal{A}, G)$ *for an element* $\mathbb{E} = (X, B)$ *in* $\mathscr{A}$ *is a goal in* $\mathfrak{C}$. *It holds that* $A_{\mathcal{A}} \widehat{\cap} B \widehat{\subseteq} G$ *if*

(a) *For each atomic guarantee* $G_{i,j}$ *of* G, *it holds that* $A_{\mathcal{A}_{i,j}} \widehat{\cap} B_i \widehat{\subseteq} G_{i,j}$ *where* $(\mathcal{A}_{i,j}, G_{i,j})$ *is a contract for* $\mathbb{E}_i = (X_i, B_i)$ *in* $\mathscr{A}$

(b) $X_G \subseteq X_{Env_{\mathscr{A}}(\mathbb{E})} \cup X$ *where* $X_{Env_{\mathscr{A}}(\mathbb{E})}$ *is the union of the interfaces of the elements in the environment of* $\mathbb{E}$

(c) *For each node* G' *that is reachable from at least one atomic guarantee* $G_{i,j}$ *of* G, *it holds that* $\widehat{\bigcap}_{k=1}^{N} G_k \widehat{\subseteq} G'$ *where* $\{G_1, \ldots, G_N\}$ *is the set of direct predecessors of* G' *with "fulfills" arcs to* G'

Proof: Consider the contracts $\{(\mathcal{A}_{i,j}, G_{i,j})\}_{i,j=1}^{M,M_i}$ containing the atomic guarantees of G. In accordance with Definition 13.7, for each contract $(\mathcal{A}', G')$ where G' is reachable from at least one guarantee $G_{i,j}$, it holds that each assumption in $\mathcal{A}'$ is either an atomic guarantee $G_{i,j}$ of G, a guarantee G' where each descendant of G' is an atomic guarantee $G_{i,j}$ of G if the descendant does not have any predecessors, or an assumption in $\mathcal{A}$. This and the conditions (a) and (c) imply that $A_{\mathcal{A}} \widehat{\cap} \widehat{\bigcap}_{i=1}^{M} B_i \widehat{\subseteq} G'$, and further that $A_{\mathcal{A}} \widehat{\cap} \widehat{\bigcap}_{i=1}^{M} B_i \widehat{\subseteq} \widehat{\bigcap}_{i,j=1}^{M,M_i} G_{i,j}$, and finally that $\widehat{\bigcap}_{i,j=1}^{M,M_i} G_{i,j} \widehat{\subseteq} G$ in accordance with Definitions 13.8 and 13.9. This and since the condition (b) and Proposition 13.3 imply that G can only constrain subsets of $X_{Env_{\mathscr{A}}(\mathbb{E})} \cup X$, it follows that $A_{\mathcal{A}} \widehat{\cap} \widehat{proj}_X(\widehat{\bigcap}_{i=1}^{M} B_i) \widehat{\subseteq} G$ in accordance with Section 13.2.1. Since $\{\mathbb{E}_i\}_{i=1}^{M}$ is a subset of the proper ancestors of $\mathbb{E}$ in accordance with Definition 13.7, it can easily be realized that $A_{\mathcal{A}} \widehat{\cap} B \widehat{\subseteq} G$ in accordance with Definition 13.2.

The conditions (b) and (c) of Theorem 13.3 expresses sufficient conditions of the property given in Equation 13.15 while the condition (c) ensures that the antecedent, that is, the if part, of the property given in Equation 13.15 holds. The condition (b) ensures that the scope of the goal G of the contract for $\mathbb{E}$ does not include a port variable x of a proper descendant of $\mathbb{E}$ where x is not part of the interface of $\mathbb{E}$. Since the behavior of $\mathbb{E}$ cannot constrain $\{x\}$ and since $A_{\mathcal{A}}$ is intended to be fulfilled by the environment of $\mathbb{E}$, the violation of the condition (b) either implies that the constraint (a) of Definition 13.6 does not hold or that the element $\mathbb{E}$ cannot satisfy $\mathcal{C}$. The condition (c) ensures that all of the "fulfills" arcs between guarantees that can be reached from the atomic guarantees of G do in fact hold.

Considering the contract structure for $\mathscr{A}_{LMsys}$ shown in Figure 13.10, since the scope of G_{LMsys} is a subset of X_{LMsys} and since the relation $G_{lMeter} \widehat{\subseteq} G_{Esys} \widehat{\subseteq} G_{LMsys}$ holds, it can be inferred, through the use of Theorem 13.3, that if the leaf elements of $\mathscr{A}_{LMsys}$ satisfy their contracts, then $\mathbb{E}_{LMsys}$ satisfies $\mathcal{C}_{LMsys}$.

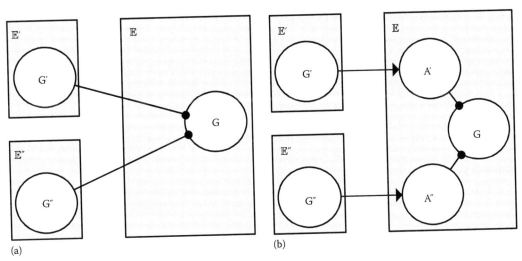

(a)

(b)

FIGURE 13.12   A possible extension to Definition 13.7 is shown in (b) where, in contrast to (a), assumptions are used as intermediate nodes.

Remark 13.2 (Intermediate Assumption Nodes)   *In the general case, the use of explicit assumptions of a contract C for an element $\mathbb{E}$ can be captured in a contract structure by introducing intermediate assumption nodes with outgoing "assumption of" arcs to the guarantee of C and incoming "fulfills" arcs from both guarantees of contracts for siblings of $\mathbb{E}$ and assumptions of contracts for proper ancestors of $\mathbb{E}$. The use of intermediate assumption nodes should only be used when absolutely necessary since it introduces an extra verification step than those presented in Theorem 13.3 in order to ensure that the property given in Equation 13.15 holds. The extra verification step is to check that the direct predecessors of an assumption node fulfill the assumption (Figure 13.12).*

Remark 13.3 (Circular Reasoning)   *Since the assumptions and guarantees of a contract structure are organized into a directed acyclic graph, the use of circular argumentation is avoided. Note that circularity can be resolved in other ways, for example, by introducing assumptions about the computational model [1] or the timing model [65]. See also [64,70] for more discussions on such matters.*

13.6   INDUSTRIAL CASE—THE FUEL LEVEL DISPLAY SYSTEM

This section shows an explicit use of the concepts presented in Sections 13.2 through 13.5 by structuring and specifying *safety requirements*, that is, requirements that have been assigned an SIL [91], on an industrial system. The safety requirements will be structured as proposed in ISO 26262— a functional safety standard for the automotive industry. The industrial system is an FLD system, installed on all heavy trucks manufactured by Scania. In accordance with ISO 26262, a top-level safety requirement, that is, a *safety goal*, for the FLD system is broken down all the way into SW and HW safety requirements on the C-code implementation and associated HW.

Specifically, this is done by modeling the FLD system and its environment as an architecture $\mathscr{A}_{FLDSys}$ in accordance with Section 13.2. Contracts are formed for elements in $\mathscr{A}_{FLDSys}$ in accordance with Section 13.3 and the guarantee and the assumptions are specified by considering the scoping constraints presented in Section 13.4. The contracts, expressing relations between safety requirements on the elements in $\mathscr{A}_{FLDSys}$ in accordance with Section 13.5.1, are structured using a contract structure as presented in Section 13.5.

13.6.1 Architecture of the FLD System

The FLD system is a safety-critical system, and this section will consider the actual C-code that is compiled and flashed onto the produced vehicles. The basic functionality of the FLD system is to provide an estimate of the fuel volume in the fuel tank to the driver. The functionality is distributed across three electronic control unit (ECU) systems, that is, an ECU with sensors and actuators, in the E/E system of the vehicle $\mathbb{E}_{Vehicle}$: engine management system (EMS) $\mathbb{E}_{EMS}$, instrument cluster (ICL) $\mathbb{E}_{ICL}$, and Coordinator (COO) $\mathbb{E}_{COO}$. The ECU systems also interact with the fuel tank $\mathbb{E}_{FuelTank}$ and the parking brake system $\mathbb{E}_{pBrakeSys}$ that both are outside of the E/E system of the truck. The environment of the vehicle $\mathbb{E}_{Vehicle}$ in $\mathscr{A}_{FLDSys}$ solely consists of a "driver" $\mathbb{E}_{Driver}$ with an interface identical to the vehicle $\mathbb{E}_{Vehicle}$.

There are several architectural variants of the FLD system, for example, variability in fuel tanks and types of sensors. Due to space restrictions, only one type of architecture is considered here. The considered variant is shown in Figure 13.13 where all port variables and elements that are not relevant have been removed for reasons of readability. The sharing of port variables is either visualized by connecting port variables with a line or by the appearance of port variables on several edges of rectangles.

COO $\mathbb{E}_{COO}$ estimates the fuel volume $actualFuelVolume[\%]$ in the tank $\mathbb{E}_{FuelTank}$ by a Kalman filter. The input signals to $\mathbb{E}_{COO}$ are the position $sensedFuelLevel[\%]$ of a floater in the fuel tank $\mathbb{E}_{FuelTank}$, as sensed by the fuel sensor $\mathbb{E}_{fuelSensor}$, and the controller area network (CAN) signal $FuelRate[L/h]$ in the message $FuelEconomy$, transmitted on $CAN1$ from EMS $\mathbb{E}_{EMS}$. The CAN signal $FuelRate[L/h]$ is an estimate of the current fuel consumption. The estimated fuel volume is transmitted on $CAN2$ as the CAN signal $FuelLevel[\%]$ in the CAN message $DashDisplay$. The CAN message is received by ICL $\mathbb{E}_{ICL}$ where a fuel gauge $indicatedFuelVolume[\%]$ in the display presents the information to the driver.

A development according to ISO 26262 revolves around an *item*, that is, "a system that implements a function at a vehicle level" [48]. For the analysis in this section, COO $\mathbb{E}_{COO}$ is chosen to be the item. In Figure 13.13, the architecture of the SW $\mathbb{E}_{SW}$ of the ECU $\mathbb{E}_{ECU}$ of $\mathbb{E}_{COO}$ is shown, as well as the ECU-HW $\mathbb{E}_{HW}$, the fuel sensor $\mathbb{E}_{fuelSensor}$, and the relevant elements in the environment of the item in $\mathscr{A}_{FLDSys}$.

The SW architecture of COO is structured as follows: the application (APPL) layer consists of SW components that implement the high-level functionality of the ECU $\mathbb{E}_{ECU}$; the middleware (MIDD) layer contains the SW components in charge of controlling the I/O, for example, sensors and actuators, connected to the ECU and the encoding/decoding of CAN messages; and the basic input/output system (BIOS) layer contains the SW components that manages the low-level interaction with the executing platform, that is, the ECU HW $\mathbb{E}_{HW}$.

For further understanding, functions inside of the different elements in the SW architecture are also shown in Figure 13.13, as well as data and control flow between the functions. Each function that is shown in Figure 13.13 is part of an element and is associated with a *function trigger*, that is, a Boolean port variable of the element where the port variable shares the same identifier as the function. Due to space restrictions, the identifiers of the function triggers are not shown in Figure 13.13, and each function trigger that is associated with a main function, identified as `funcID_20ms` in Figure 13.13, is not shown at all.

13.6.2 Specification and Structuring of Safety Requirements

In this section, the safety goal and safety requirements are specified and structured in parallel to the architecture $\mathscr{A}_{FLDSys}$ of the FLD system using a contract structure for $\mathscr{A}_{FLDSys}$ as shown in Figure 13.14. The specification of a guarantee of a contract for an element in $\mathscr{A}_{FLDSys}$ is made by considering the intended responsibility of the element in the context as a whole. This means that the guarantee, that is, a safety requirement on the element, and also the assumptions, will not necessarily be limited to the interface of the element.

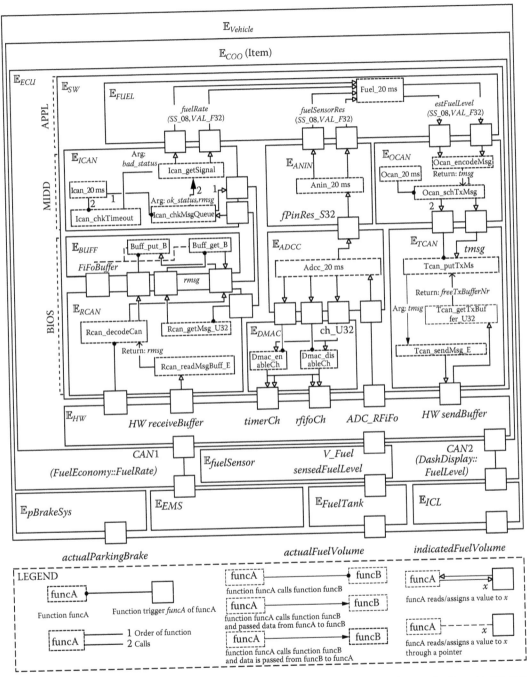

FIGURE 13.13 The architecture $\mathscr{A}_{FLDSys}$ of the fuel level display system where only the relevant port variables and elements are shown. Different relations, for example, data and control flow, between different SW functions and port variables are explained in the legend at the bottom of the figure. *Note:* The architecture and the naming are modified to protect the integrity of the implementation. The original complexity and intent of the architecture is, however, sustained.

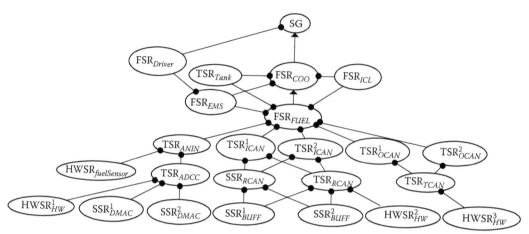

FIGURE 13.14 Contract structure for the architecture $\mathscr{A}_{FLDSys}$ where the structure of the elements are not shown for reasons of readability.

Regarding the representation of requirements, port variables are referred to using the format "*name*[*unit*]" or simply "*name*," such as *actualFuelVolume*[%] and *FifoBuffer*. To pair a requirement with assumptions, a reference is made to one or a set of requirement on elements in the environment, by writing "{SR#}" besides the requirement.

Different hierarchical levels of requirements as described in ISO 26262, that is, safety goals, functional safety requirements (FSRs), technical safety requirements (TSRs), and HW and SW safety requirements (HWSRs/SSRs) are mapped to the type of port variable referred to in the requirements. That is, if a requirement refers to port variables that model properties at a vehicle level or those shared between ECU systems, it is considered to be an FSR. If both port variables with HW and SW properties are referred to in the requirements, it is considered to be a TSR and if only port variables with, for example, only SW properties, are referenced, it is considered to be a SSR and so on.

As a limitation due to space restrictions, only requirements that are applicable when the ignition is on are considered. In SW, that corresponds to only including requirements applicable during run time. It is assumed that every main function, identified as `funcID_20ms` in Figure 13.13, is run every 20 ms and that the dataflow is exclusive to what is shown in Figure 13.13, that is, no other element can read or write to a variable other than what is shown. Furthermore, upon triggering a function by setting its function trigger from false to true, the function is assumed to terminate (which means that the function trigger is again set to false) within a negliable time. The term "*corresponds to*" is used when two variables/values are approximately equal, for example, when they only differ in type or deviate due to small delays.

13.6.2.1 Specification of the Safety Goal

As shown in Table 13.1, the safety goal SG, under the responsibility of the vehicle $\mathbb{E}_{Vehicle}$, expresses that *the indicated fuel volume, shown by the gauge, shall not be greater than the fuel volume in the tank, while the parking brake is not applied.* The state when the vehicle $\mathbb{E}_{Vehicle}$ is parked is hence considered as a safe state of the vehicle. The safety goal SG is also shown as a node in the contract structure shown in Figure 13.14.

13.6.2.2 Safety Requirements on COO (the Item) and Its Environment

As shown in Figure 13.14, the safety goal SG has an incoming "assumptions of" arc from the FSR FSR_{Driver}, which means that FSR_{Driver} is an *assumption* of the contract ({FSR_{Driver}}, SG) for $\mathbb{E}_{Vehicle}$ in accordance with Section 13.5.2. This means that the safety goal is under the responsibility of the

TABLE 13.1

Safety Requirements on Coordinator $\mathbb{E}_{COO}$ (the Item) and Safety Goal

SG	If *actualParkingBrake[Bool]* is false, then *indicatedFuelVolume[%]*, shown by the fuel gauge, is less than or equal to *actualFuelVolume[%]* {FSR$_{Driver}$}
FSR$_{COO}$	If *actualParkingBrake[Bool]* is false, then *indicatedFuelVolume[%]*, shown by the fuel gauge, is less than or equal to *actualFuelVolume[%]* {TSR$_{Tank}$, FSR$_{EMS}$, FSR$_{ICL}$}.

TABLE 13.2

Safety Requirements on the Environment of the Item (Instrument Cluster $\mathbb{E}_{ICL}$, Tank $\mathbb{E}_{Fueltank}$, Driver $\mathbb{E}_{Driver}$, and Engine Management System $\mathbb{E}_{EMS}$)

FSR$_{Driver}$	If *actualParkingBrake[Bool]* is false, Then the derivative of *actualFuelVolume[%]* is less than or equal to 0.
TSR$_{Tank}$	The position of the floater *sensedFuelLevel[%]*, sensed by the fuel sensor, does not deviate more than $\pm10\%$ from *actualFuelVolume[%]* in the fuel tank.
FSR$_{EMS}$	If *actualParkingBrake[Bool]* is false AND *CAN1* is equal to CAN message *FuelEconomy* and it has not passed more than $0.3s^{a}$ since the last time *CAN1* was equal to *FuelEconomy*, then CAN signal *FuelRate[L/h]* in *FuelEconomy* does not deviate more than $\pm1\%$ from the derivative of *actualFuelVolume[%]*; OR *FuelRate[L/h]* is equal to *0xFE* (error) {FSR$_{Driver}$}.
FSR$_{ICL}$	If it has not passed more than $1s^{a}$ since the last time *CAN2* was equal to CAN message *DashDisplay* containing CAN signal *FuelLevel[%]* that is not equal to *0xFE* (error), Then *indicatedFuelVolume[%]*, shown by the fuel gauge, corresponds to *FuelLevel[%]*. Otherwise, *indicatedFuelVolume[%]* is equal to 0.

[a] For confidentiality reasons, the values are either modified or not provided.

vehicle $\mathbb{E}_{Vehicle}$, but only if the driver $\mathbb{E}_{Driver}$ fulfills the assumption FSR$_{Driver}$, that is, that *the driver does not refuel the vehicle, while the parking brake is not applied*, as expressed in Table 13.2.

In accordance with Section 13.5.2, the arc from the FSR FSR$_{coo}$ to the safety goal SG in Figure 13.14 means that the intent is that FSR$_{coo}$ fulfills SG. As can be seen in Table 13.1, these two requirements are the same, which means that the responsibility of SG is fully delegated from the vehicle $\mathbb{E}_{Vehicle}$ to COO $\mathbb{E}_{coo}$. However, the FSR FSR$_{coo}$ is only under the responsibility of $\mathbb{E}_{coo}$, if the environment of $\mathbb{E}_{coo}$ fulfills the requirements TSR$_{Tank}$, FSR$_{EMS}$, and FSR$_{ICL}$, as shown in Figure 13.14.

As shown in Table 13.2, the requirements TSR$_{Tank}$ and FSR$_{EMS}$ express the following in the nominal case, respectively: *the fuel sensor $\mathbb{E}_{fuelSensor}$ is correctly installed in the tank $\mathbb{E}_{Fueltank}$*; and *EMS $\mathbb{E}_{EMS}$ provides an accurate estimate of the fuel consumption, when driving*. The FSR FSR$_{ICL}$, under the responsibility of ICL $\mathbb{E}_{ICL}$, expresses in the nominal case that *the indicated fuel volume, shown by the gauge, shall correspond to the estimated fuel volume, transmitted onto CAN2 by $\mathbb{E}_{coo}$*. As further shown in Figure 13.14, the assumption FSR$_{Driver}$ of the contract for the vehicle is also an assumption of the contract containing FSR$_{EMS}$, under the responsibility of $\mathbb{E}_{EMS}$.

13.6.2.3 Safety Requirements on Application and Middleware SW

In Table 13.3, the safety requirements FSR$_{FUEL}$, TSR$_{ANIN}$, TSR$_{ICAN}^{1-2}$, and TSR$_{OCAN}^{1-2}$ on the SW components FUEL $\mathbb{E}_{FUEL}$, ANIN $\mathbb{E}_{ANIN}$, ICAN $\mathbb{E}_{ICAN}$, and OCAN $\mathbb{E}_{OCAN}$, respectively, are presented. The SW components in the MIDD SW provide the APPL SW with SW signals that correspond to readings from sensors and CAN signals and also encode SW signals from the APPL SW into CAN messages.

As can be seen in Table 13.1 and Table 13.3, the FSR FSR$_{FUEL}$ on fuel is the same as FSR$_{COO}$, which means that the responsibility of FSR$_{COO}$ is delegated to fuel, given that the environment of

TABLE 13.3

Safety Requirements on Application Software and Middleware Software Components

FSR_{FUEL}	If *actualParkingBrake[Bool]* is false, then *indicatedFuelVolume[%]*, shown by the fuel gauge, is less than or equal to *actualFuelVolume[%]* {TSR_{Tank}, FSR_{EMS}, FSR_{ICL}, TSR_{ANIN}, TSR_{ICAN}^{1-2}, TSR_{OCAN}^{1-2}}.
TSR_{ANIN}	A *fuelSensorRes_Val_F32[%]* corresponds to the floater position *sensedFuelLevel[%]*, sensed by the fuel sensor; or *fuelSensorRes_SS_U08[Enum]* has the value *ERR* {TSR_{ADCC}, $HWSR_{fuelSensor}$}.
TSR_{ICAN}^{1}	If it has not passed more than $0.3s^{a}$ since the last time *CAN1* was equal to CAN message *FuelEconomy* containing CAN signal *FuelRate[L/h]* that is not equal to *0xFE* (error), Then *fuelRate_Val_F32[L/h]* corresponds to *FuelRate[L/h]* {SSR_{RCAN}, FSR_{RCAN}}.
TSR_{ICAN}^{2}	If CAN signal *FuelRate[L/h]* in CAN message *FuelEconomy* was equal to *0xFE* (error) the last time *CAN1* was equal to *FuelEconomy*, or it has passed more than $0.3s^{a}$ since the last time *CAN1* was equal to *FuelEconomy*, then *fuelRate_SS_U08[Enum]* is equal to *ERR*. {SSR_{RCAN}, TSR_{RCAN}}.
TSR_{OCAN}^{1}	If *CAN2* is equal to CAN message *DashDisplay* containing CAN signal *FuelLevel[%]* that is equal to *0xFE* (error) and it has not passed more than $1s^{a}$ since the last time *CAN2* was equal to *DashDisplay*, then *estFuelLevel_SS_U08[Enum]* corresponds to *ERR* {TSR_{TCAN}}.
TSR_{OCAN}^{2}	IF *CAN2* is equal to CAN message *DashDisplay* containing CAN signal *FuelLevel[%]* that is not equal to *0xFE* (error) and it has not passed more than $1s^{a}$ since the last time *CAN2* was equal to *DashDisplay*, then *FuelLevel[%]* corresponds to *estFuelLevel_Val_F32[%]*. {TSR_{TCAN}}.

^{a} For confidentiality reasons, the values are either modified or not provided.

$\mathbb{E}_{FUEL}$ fulfills the requirements TSR_{Tank}, FSR_{EMS}, and FSR_{ICL}, and furthermore TSR_{ANIN}, TSR_{ICAN}^{1-2}, and TSR_{OCAN}^{1-2} as shown in Figure 13.14.

The TSR TSR_{ANIN} expresses that the input signal *fuelSensorRes_Val_F32[%]* corresponds to the position of the floater *sensedFuelLevel[%]*, or the status signal *fuelSensorRes_SS_U08[Enum]* has the value *ERR*. The TSR TSR_{ICAN}^{1-2} expresses that the input signal *fuelRate_Val_F32[L/h]* to fuel corresponds to CAN signal *FuelRate[L/h]* in CAN message *FuelEconomy* in case *FuelRate[L/h]* is not equal to *0xFE* (error). In case *FuelRate[L/h]* is equal to *0xFE* (error) or if the signal was expected sooner, then *fuelRate_SS_U08[Enum]* is equal to *ERR*. The TSR TSR_{OCAN}^{1-2} expresses that the output signal *estFuelLevel_Val_F32[%]* to fuel is transmitted on *CAN2* as the signal *FuelLevel[%]* in CAN message *DashDisplay* if *estFuelLevel_SS_U08[Enum]* is not equal to *ERR*. In case *estFuelLevel_SS_U08[Enum]* is equal to *ERR*, then *FuelLevel[%]* is equal to *0xFE*.

13.6.2.4 Safety Requirements on Basic Input/Output System SW and HW Components

In Table 13.4, the safety requirements TSR_{TCAN}, SSR_{RCAN}, TSR_{ADCC}, SSR_{DMAC}^{1-2}, and SSR_{BUFF}^{1-2} on the SW components TCAN $\mathbb{E}_{TCAN}$, RCAN $\mathbb{E}_{RCAN}$, ADCC $\mathbb{E}_{ADCC}$, direct memory access channel (DMAC) $\mathbb{E}_{DMAC}$, and BUFF $\mathbb{E}_{BUFF}$, respectively, are presented. The SW components in the BIOS layer provide the MIDD layer with SW signals that correspond to voltage values at the input pins for analogue sensors and manages the HW/SW interaction.

As shown in Figure 13.14, the TSR TSR_{ADCC} and the HWSR $HWSR_{fuelSensor}$ are assumptions of the contract containing the guarantee TSR_{ANIN}, under the responsibility of $\mathbb{E}_{ANIN}$. The TSR TSR_{ADCC} expresses that the SW signal *fPinRes_s32[mV]* corresponds to a voltage value *V_Fuel[mV]* at one of the input pins of the ECU $\mathbb{E}_{ECU}$, and the HWSR $HWSR_{fuelSensor}$ expresses that the fuel sensor $\mathbb{E}_{fuelSensor}$ either provides the intended values or values that are out of range, as shown in Table 13.5.

However, the TSR TSR_{ADCC} is only under the responsibility of $\mathbb{E}_{ADCC}$, given that the environment of $\mathbb{E}_{ADCC}$ fulfills the assumptions $HWSR_{HW}^{1}$ and SSR_{DMAC}^{1-2} as shown in Figure 13.14. The assumption $HWSR_{HW}^{1}$ expresses a requirement on the ECU HW $\mathbb{E}_{HW}$, that is, that a raw analog-to-digital converter (ADC) value is available if the ADC is allowed to sample for 20 ms, that is, an execution tick, as shown in Table 13.5. In order for ADCC $\mathbb{E}_{ADCC}$ to control the ADC, it has to enable/disable direct memory access channels by calling functions in DMAC $\mathbb{E}_{DMAC}$. Hence, as stated in the assumptions

TABLE 13.4
Safety Requirements on Basic Input/Output System Software Components

TSR_{ADCC}	A $fPinRes\_s32[mV]$ corresponds to the voltage value $V\_Fuel[mV]$ {$HWSR^1_{HW}$, SSR^1_{DMAC}, SSR^2_{DMAC}}.
TSR_{RCAN}	If $CAN1$ is equal to a CAN message $rmsg$, then $rmsg$ is in $FiFoBuffer$ within 20 ms {$HWSR^2_{HW}$, SSR^{1-2}_{BUFF}}.
SSR_{RCAN}	If the oldest message in $FiFoBuffer$ has PGN $0xFEF2$ when $Rcan\_getRxMsg\_U32$ transitions from false to true, then $rmsg$ corresponds to $FuelEconomy$ when $Rcan\_getRxMsg\_U32$ transitions from true to false {SSR^{1-2}_{BUFF}}.
TSR_{TCAN}	If $tmsg$ has PGN $0xFEFC$ when $Tcan\_putTxMsg\_E$ transitions from false to true, then $DashDisplay$ corresponds to $tmsg$ the next time $CAN2$ is equal to $DashDisplay$. {$HWSR^3_{HW}$}.
SSR^1_{DMAC}	The direct memory access channels that corresponds[a] to ch_U32 when $Dmac\_enableCh$ transitions from false to true is enabled when $Dmac\_enableCh$ transitions from true to false.
SSR^2_{DMAC}	The direct memory access channels that corresponds[a] to ch_U32 when $Dmac\_disableCh$ transitions from false to true is disabled when $Dmac\_disableCh$ transitions from true to false.
SSR^1_{BUFF}	The value of $rmsg$ when $Buff\_put\_B$ transitions from false to true is added to $FiFoBuffer$ before $Buff\_put\_B$ transitions from true to false.
SSR^2_{BUFF}	When $Buff\_get\_B$ transitions from true to false, $rmsg$ is the oldest message in $FiFoBuffer$.

[a] $TimerCh\_U32$ and $rfifoCh\_U32$ corresponds to the direct memory access channels $timerCh$ and $timerCh$, respectively.

TABLE 13.5
Safety Requirements on Coordinator Electric Control Unit HW or HW Components

$HWSR_{fuelSensor}$	The fuel sensor converts the floater position $sensedFuelLevel[\%]$ into a voltage value $V\_Fuel[mV]$ according to table Y[a] or $3000 < V\_Fuel[[mV]$ OR $V\_Fuel[[mV] < 200$[a].
$HWSR^1_{HW}$	If the DMA Cs $timerCh$ and $rfifoCh$ are enabled for approximately 20 ms, then a raw value of $V\_Fuel[mV]$ is available in ADC_RFIFO.
$HWSR^2_{HW}$	If $CAN1$ is equal to a CAN message $rmsg$, then $rmsg$ is in $HWreceiveBuffer$, and $Rcan\_decodeCan$ is set to true.
$HWSR^3_{HW}$	If a CAN message $tmsg$ is the oldest message in $HWsendBuffer$, then $CAN2$ is equal to $tmsg$, the next time $CAN2$ is equal to a message in $HWsendBuffer$.

[a] For confidentiality reasons, the values are either modified or not provided.

SSR^{1-2}_{DMAC}, under the responsibility of DMAC $\mathbb{E}_{DMAC}$, the DMACs are enabled/disabled by calling the functions $Dmac\_disableCh(ch\_U32)$ and $Dmac\_enableCh(ch\_U32)$, respectively, where appropriate values of the argument ch_U32 match specific DMACs.

In Figure 13.14, the TSR TSR_{RCAN} and the SSR SSR_{RCAN} are assumptions of both the contracts containing the guarantees TSR_{RCAN} and SSR_{RCAN}. The SSR SSR_{RCAN} expresses that if the oldest message $rmsg$ in the queue $FiFoBuffer$ has PGN number $0xFEF2$ when the function $Rcan\_getRxMsg\_U32$ is called, then $rmsg$ corresponds to $FuelEconomy$. The TSR TSR_{RCAN} expresses that if $FuelEconomy$ has arrived within 20 ms, that is, since the latest execution tick, then it has been placed in the queue $FiFoBuffer$.

As shown in Figure 13.14, the SSRs SSR^{1-2}_{BUFF} and $HWSR^2_{HW}$ are the assumptions of the contract containing the guarantee TSR_{RCAN}, and SSR^{1-2}_{BUFF} is the assumption of the contract containing the guarantee SSR_{RCAN}. The SSRs SSR^{1-2}_{BUFF} express that the responsibility of $\mathbb{E}_{BUFF}$ is to manage the queue $FiFoBuffer$, and $HWSR^2_{HW}$ expresses the requirement on the ECU HW $\mathbb{E}_{HW}$ that RCAN $\mathbb{E}_{RCAN}$ is notified whenever a new CAN message has arrived in $HWreceiveBuffer$.

The TSRs TSR^{1-2}_{OCAN} are under the responsibility of $\mathbb{E}_{OCAN}$ as shown in Figure 13.14, given that the environment of $\mathbb{E}_{OCAN}$ fulfills the TSR TSR_{TCAN}, under the responsibility of $\mathbb{E}_{TCAN}$. The TSR TSR_{TCAN} expresses that if the function $Tcan\_putTxMsg\_E$ is called with an argument $tmsg$ with a PGN 0xFEFC, then *DashDisplay*, corresponding to *tmsg*, is to be transmitted on *CAN2*, given that the ECU HW $\mathbb{E}_{HW}$ fulfills the HWSR $HWSR^3_{HW}$. The HWSR $HWSR^3_{HW}$ expresses that the messages placed in *HWsendBuffer* are to be transmitted on *CAN2* in the order in which they are placed in *HWsendBuffer*.

13.6.3 DISCUSSION

As can be observed in Section 13.6.2, each contract in Tables 13.1 through 13.5 allows a clear separation of responsibilities between an element in $\mathscr{A}_{FLDSys}$ and its environment by explicitly declaring what the element requires from its environment as assumptions, in order to ensure that the requirement is met. That is, if, for example, the fuel sensor $\mathbb{E}_{fuelSensor}$ is installed incorrectly, that is, if the environment of $\mathbb{E}_{COO}$ does not fulfill the assumption TSR_{Tank} of the contract for $\mathbb{E}_{COO}$, then it cannot be ensured that $\mathbb{E}_{COO}$ meets its responsibility expressed by the FSR FSR_{COO}.

Regarding the specification of the requirements, references to port variables were made in compliance with the conditions (a) and (b) of Definition 13.6. Considering the contract ($\{\mathsf{FSR}_{Driver}\}$, SG) for $\mathbb{E}_{Vehicle}$ as an example, since the scope $\{actualParkingBrake[Bool], actualFuelVolume[\%]\}$ of the assumption FSR_{Driver} is a subset of the interface of $\mathbb{E}_{Driver}$, that is, the environment of the vehicle $\mathbb{E}_{Vehicle}$ as mentioned in Section 13.6.1, the condition (a) of Definition 13.6 holds. Furthermore, since the scope of SG is a subset of the interface of $\mathbb{E}_{Vehicle}$, the condition (b) of Definition 13.6 holds.

In accordance with Theorem 13.2, this means that necessary constraints on the scopes of the assumptions and the guarantee of the contract ($\{\mathsf{FSR}_{Driver}\}$, SG) hold in order for the environment $\mathbb{E}_{Driver}$ and the element $\mathbb{E}_{Vehicle}$ to meet their responsibilities, expressed by FSR_{Driver} and SG, respectively. Therefore, the conditions (a) and (b) of Definition 13.6 serve as a sanity check of the contract ($\{\mathsf{FSR}_{Driver}\}$, SG) and also of all the other contracts presented in Tables 13.1 through 13.5.

The contract structure for $\mathscr{A}_{FLDSys}$ shown in Figure 13.14 organizes the safety requirements in Tables 13.1 through 13.5 on elements in $\mathscr{A}_{FLDSys}$. The contract structure gives an overview of all the relations between the safety requirements and the possibility of tracing atomic safety requirements to the safety goal. The intent of the relations are well defined in Theorem 13.3, which means that the exact conditions that need to be verified are given, in order to ensure that the vehicle meets its responsibility expressed by the safety goal SG.

Furthermore, since an assumption of a contract in the contract structure in Figure 13.14 corresponds to a guarantee of another contract in accordance with Section 13.5.1, the contract structure offers a compact representation of the dependencies between the requirements on the elements in $\mathscr{A}_{FLDSys}$. That is, if intermediate assumption nodes were instead used as discussed in Remark 13.2, the contract structure in Figure 13.14 would have almost twice as many nodes. Moreover, as also mentioned in Remark 13.2, the use of intermediate assumption nodes introduces an extra verification step than those presented in Theorem 13.3 in order to ensure that the property given in Equations 13.15 holds.

13.7 RELATED WORK

As mentioned in Section 13.3, the notion of contracts was first introduced in Meyer [67] as a pair of pre- and postconditions [36,45,57] to be used as formal specification in object-oriented programming. Since then, the work in Meyer [67] has been extended to component-based design [40] and analog systems [86,87] and introduced in the formalisms *Behavior Interaction Priority* [16] and *refinement calculus* [8], [7,42,78], respectively. Furthermore, in the European research project SPEEDS [85], a contract theory [12,14] was introduced as a means to meet the challenges in the design of cyber-physical systems [60,81]. Similar work to [12,14] is also presented in Derler et al. and Torngren et al. [34,88]. Recent work include metatheories of contracts in Bauer et al. and Benveniste et al. [10,13] and also in Cimatti and Tonetta [22] with tool support [21], as well as extensions to modalities [59] in Damm et al. and Goessler and Raclet [27,41] and to a stochastic setting [15,33,43]. The use of contracts has been proposed for analyses integration [80] and as means to achieve functional safety in References [5,9,11,28] and also in Soderberg and Johansson [83] with tool support [84].

The notion of contracts is also closely related to the idea of compositional verification/ reasoning [32,46] that has its origins in two independent theories [52,53,68], inspired by ideas on proof methods for *sequential* [36,38,45] and *concurrent* [6,56,73] programs. Since the conception of [52,53,68], several frameworks that extend the ideas References [52,53,68], have emerged, such as [1,2,76], respectively. Furthermore, techniques to automate compositional verification have been proposed, see, for example, [77] or [23] for a survey. Moreover, compositional verification has been proposed for model checking, see, for example, [4] or [55] for an overview. Given that the two approaches [52,53,68] are the same in principle, meta theories [19,63,94] have been introduced to unify [52,53,68]. Recently, with inspiration from [44,65,66], the use of compositional verification in SW was extended to system verification in Cofer et al. [25] and compositional theories that extend interface automa [30,62] have been proposed in References [20,29,79].

Considering the approaches for compositional reasoning presented earlier, none has any explicit support for the specification or structuring of requirements that are not limited to element interfaces. This means that, in contrast to this chapter, the organization of contracts presented in Section 13.6.2 is not supported.

In accordance with [12,14] and also with [15,81,90,92], contracts and behaviors of elements are in this chapter both defined by relying on the concept of a set of runs, also referred to as *traces* [1]. The concept of runs is, in turn, largely inspired by the works in References [18,71,75] that generalize the concept of *trace structures* [37,93] to represent behaviors that are independent of a particular model of computation.

In Section 13.2.2, the concept of an element that essentially corresponds to a HRC [26,54] as used in Benveniste et al. Benveniste, Caillaud, and Passerone [12,14] was introduced. The main difference is that an HRC can have several *implementations*, that is, behaviors, which means that an element corresponds to an implementation of a HRC, rather than to a HRC itself. An element in this chapter is in that sense more similar to a component as defined in Simko et al. [82] that is inspired by the *tagged signal model* [61] and *interface theory* [31].

Section 13.2.3 introduced a *general* definition of an architecture [48,49], which means that the definition is applicable regardless of the domain, for example, mechanical, and sw, that is considered. To capture additional conditions on an architecture $\mathscr{A}$ where the interface X of each element $\mathbb{E} = (X, \mathbf{B})$ in $\mathscr{A}$ is split into two disjoint sets of *inputs* X_{in} and *outputs* X_{out}, the approach in the contract framework [12,14] of SPEEDS can be used. Specifically, the following conditions can be introduced for each element $\mathbb{E} = (X, \mathbf{B})$ in $\mathscr{A}$: (a) $\widehat{proj}_{X_{in}}(\mathbf{B}) = \Omega_{X_{in}}$ where $\Omega_{X_{in}}$ is the set of all runs for X_{in}, which means that the behavior is *receptive* to X_{in}; and (b) $X_{out} \cap X_{out}^{Env_{\mathscr{A}}(\mathbb{E})} = \emptyset$ where $X_{out}^{Env_{\mathscr{A}}(\mathbb{E})}$ is the union of the set of output ports of elements in the environment $Env_{\mathscr{A}}(\mathbb{E})$ of $\mathbb{E}$ in $\mathscr{A}$. This is essentially the approach in Benveniste et al. Benveniste, Caillaud, and Passerone [12,14] but where the constraint (b) is enforced in Benveniste et al. Benveniste, Caillaud, and Passerone [12,14] by a composition operator on HRCs rather than by relying on the notion of an architecture.

Concerning the notion of contract structures presented in Section 13.5.2, out of an extensive litera- ture search of contract approaches [7,12–15,27,33,40,41,43,67,78,78,81,86,87,90,92], only the work in Westman and Nyberg [90] has a graph-based approach for structuring contracts. However, in con- trast to contract structures as defined in this chapter, the organization of contracts in Westman and Nyberg [90] is limited to express an hierarchy of contracts with only two levels and is further limited to only allow one contract for an element. However, the work by Westman and Nyberg [90] explic- itly supports the use of explicit assumption nodes, which means that the organization of contracts in Westman and Nyberg [90] can be seen as a complementary approach to contract structures.

Contract structures also have a lot in common with goal oriented requirements engineering (GORE) models, see, for example, I* [95] or KAOS [89] or [58] for a survey, which draws on ideas presented in References [50,51,74]. The main difference is that while a contract structure repre- sents an hierarchy of contracts in general, the use of assumptions, also called *expectations*, in GORE models are strictly limited to top-level specifications that split the responsibilities between an *SW system* and its environment. Furthermore, a similar concept to a contract structure is presented in Nyberg [72], based on Bayesian networks.

Regarding properties of contracts, the reader is referred to [90] that presents definitions of the contract properties *consistency* [12–14], *compatibility* [12–14], and *dominance* [10,78] where the definitions are not restricted to the cases where a contract is limited to interfaces of elements and are further not confined to a certain domain in particular. For other properties of contracts, such as *parallel composition* [12–15,33,81], the conditions in Corollary 13.1 can be used in combination with the metatheories of contracts in Bauer et al. and Benveniste et al. [10,13] to instantiate definitions of properties of contracts. The conditions in Corollary 13.1 are needed since instantiation conditions for the cases where contracts that are not limited to element interfaces are not presented in either [10] or [13].

13.8 CONCLUSION

In Section 13.1, recent works [12,14,90,92] were described where contracts are proposed as means to meet the challenges in the design of cyber-physical systems. Building on the works in References [12,14,90,92], Sections 13.2 and 13.3 presented a theoretic foundation capable of modeling cyber-physical systems at all levels of design and conditions that ensure that an element and its environment in an architecture meet their respective responsibilities with respect to a contract. The conditions are general, which means that the conditions hold for any type of architecture that is considered and further for the cases where the contract is not limited to the interface of the element.

Based on the concepts established in Sections 13.2 and 13.3, Section 13.4 introduced necessary constraints on the scopes of the assumptions and the guarantee of a contract in order for the conditions presented in Section 13.3 to hold for an element in a given architecture. Furthermore, a new graph, called a *contract structure*, was introduced in Section 13.5 in order to support the structuring and tracing of requirements on a cyber-physical systems using contracts.

The explicit use of the concepts presented in Sections 13.4 and 13.5 was shown in the industrial case study in Section 13.6 where safety requirements were specified and structured in parallel to the architecture of the FLD system, as proposed by ISO 26262. A contract structure was used to organize the safety requirements and enabled the individual tracing of safety requirements on HW and SW components back to the safety goal. The scoping constraints were used as sanity checks of the contracts and hence facilitated the specification of the safety requirements in practice.

Consider the generality of the theoretic framework and the conditions presented in Sections 13.2 and 13.3, and the adequate support for the specification and structuring of requirements presented in Sections 13.4 and 13.5. Taking into account these two aspects, this chapter constitutes a general contract framework for structuring and specifying requirements on cyber-physical systems. As mentioned in Section 13.7, the framework can also be refined for a specific domain, for example, SW, and serve as a complement to other frameworks with a slightly different focus.

REFERENCES

1. M. Abadi and L. Lamport, Composing specifications, *ACM Trans. Program. Lang. Syst.*, 15(1), 73–132, January 1993. [Online]. Available: http://doi.acm.org/10.1145/151646.151649.
2. M. Abadi and L. Lamport, Conjoining specifications, *ACM Trans. Program. Lang. Syst.*, 17(3), 507–535, May 1995. [Online]. Available: http://doi.acm.org/10.1145/203095.201069.
3. J.-R. Abrial, M. Butler, S. Hallerstede, T. S. Hoang, F. Mehta, and L. Voisin, Rodin: An open toolset for modelling and reasoning in Event-B, *STTT*, 12(6), 447–466, 2010.
4. R. Alur, T. Henzinger, F. Mang, S. Qadeer, S. Rajamani, and S. Tasiran, Mocha: Modularity in model checking, in *Computer Aided Verification*, ser. Lecture Notes in Computer Science, A. Hu and M. Vardi, Eds. Berlin, Germany: Springer, 1998, vol. 1427, pp. 521–525. [Online]. Available: http://dx.doi.org/10.1007/BFb0028774.

5. T. Arts, M. Dorigatti, and S. Tonetta, Making implicit safety requirements explicit, in *Computer Safety, Reliability, and Security*, ser. Lecture Notes in Computer Science, A. Bondavalli and F. Di Giandomenico, Eds. Springer International Publishing, 2014, vol. 8666, pp. 81–92. [Online]. Available: http://dx.doi.org/10.1007/978-3-319-10506-2_6.

6. E. A. Ashcroft, Proving assertions about parallel programs, *J. Comput. Syst. Sci.*, 10(1), 110–135, February 1975. [Online]. Available: http://dx.doi.org/10.1016/S0022-0000(75)80018-3.

7. R.-J. Back and J. von Wright, Contracts, games, and refinement, *Inf. Comput.*, 156(1–2), 25–45, January 2000. [Online]. Available: http://dx.doi.org/10.1006/inco.1999.2820.

8. R.-J. J. Back, A. Akademi, and J. V. Wright, *Refinement Calculus: A Systematic Introduction*, 1st edn., F. B. Schneider and D. Gries, Eds. Secaucus, NJ Springer-Verlag New York, Inc., 1998.

9. I. Bate, R. Hawkins, and J. McDermid, A contract-based approach to designing safe systems, in *Proceedings of the Eighth Australian Workshop on Safety Critical Systems and SW*, vol. 33 ser. SCS '03. Australian Computer Society, Inc., Canberra, Australia, 2003, pp. 25–36.

10. S. Bauer, A. David, R. Hennicker, K. Guldstrand Larsen, A. Legay, U. Nyman, and A. Wsowski, Moving from specifications to contracts in component-based design, in *Fundamental Approaches to Software Engineering*, ser. Lec. Notes in Computer Science, J. Lara and A. Zisman, Eds. Berlin, Germany: Springer, 2012, vol. 7212, pp. 43–58. [Online]. Available: http://dx.doi.org/10.1007/978-3-642-28872-2_3.

11. A. Baumgart, P. Reinkemeier, A. Rettberg, I. Stierand, E. Thaden, and R. Weber, A model-based design methodology with contracts to enhance the development process of safety-critical systems, in (Eds. Min, SangLyul and Pettit, Robert and Puschner, Peter and Ungerer, Theo.) *Software Technologies for Embedded and Ubiquitous Systems*, ser. Lecture Notes in Computer Science. Berlin, Germany: Springer, 2011, vol. 6399, pp. 59–70. [Online]. Available: http://dx.doi.org/10.1007/978-3-642-16256-5_8.

12. A. Benveniste, B. Caillaud, A. Ferrari, L. Mangeruca, R. Passerone, and C. Sofronis, Multiple viewpoint contract-based specification and design, in *Formal Methods for Components and Object*, F. S. Boer, M. M. Bonsangue, S. Graf, and W.-P. Roever, Eds. Berlin, Germany: Springer-Verlag, 2008, pp. 200–225. [Online]. Available: http://dx.doi.org/1multiviewPoint0.1007/978-3-540-92188-2_9.

13. A. Benveniste, B. Caillaud, D. Nickovic, R. Passerone, J.-B. Raclet, P. Reinkemeier, A. Sangiovanni-Vincentelli, W. Damm, T. Henzinger, and K. G. Larsen, Contracts for system design, INRIA, Rapport de recherche RR-8147, November 2012. [Online]. Available: http://hal.inria.fr/hal-00757488.

14. A. Benveniste, B. Caillaud, and R. Passerone, A generic model of contracts for embedded systems, INRIA, Research Report RR-6214, 2007. [Online]. Available: http://hal.inria.fr/inria-00153477.

15. A. Benveniste, B. Cailaud, and R. Passerone, Multiple viewpoint contract-based specification and design, in *Model-Based Design for Embedded Systems*, G. Nicolescu and P. Mosterman, Eds. Taylor & Francis, Amsterdam, The Netherlands, 2009, pp. 487–518. [Online]. Available: http://www.google.se/books?id=8Cjg2mM-m1MC.

16. S. Bliudze and J. Sifakis, The algebra of connectors: Structuring interaction in bip, in *Proceedings of the seventh ACM &Amp; IEEE International Conference on Embedded Software*, ser. EMSOFT '07. New York: ACM, 2007, pp. 11–20. [Online]. Available: http://doi.acm.org/10.1145/1289927.1289935.

17. M. Broy, Towards a theory of architectural contracts: Schemes and patterns of assumption/promise based system specification, in *Software and Systems Safety — Specification and Verification*, ser. NATO Science for Peace and Security Series D: Information and Communication Security, M. Broy, C. Leuxner, and T. Hoare, Eds. IOS Press, Amsterdam, The Netherlands, 2011, vol. 30, pp. 33–87. [Online]. Available: http://dblp.uni-trier.de/db/series/natosec/natosec30.html#Broy11.

18. J. Burch, R. Passerone, and A. Sangiovanni-Vincentelli, Overcoming heterophobia: Modeling concurrency in heterogeneous systems, in *Proceedings, 2001 International Conference on Application of Concurrency to System Design*, Newcastle, UK, *2001*, 2001, pp. 13–32.

19. A. Cau and P. Collette, Parallel composition of assumption-commitment specifications, *Acta Inform.*, 32(2), 153–176, 1996. [Online]. Available: http://dx.doi.org/10.1007/s002360050039.

20. T. Chen, C. Chilton, B. Jonsson, and M. Kwiatkowska, A compositional specification theory for component behaviours, in *Programming Languages and Systems*, ser. Lecture Notes in Computer Science, H. Seidl, Ed. Berlin, Germany: Springer, 2012, vol. 7211, pp. 148–168. [Online]. Available: http://dx.doi.org/10.1007/978-3-642-28869-2_8.

21. A. Cimatti, M. Dorigatti, and S. Tonetta, Ocra: A tool for checking the refinement of temporal contracts, in *2013 IEEE/ACM 28th International Conference on Automated Software Engineering (ASE)*, Palo Alto, CA, November 2013, pp. 702–705.

22. A. Cimatti and S. Tonetta, Contracts-refinement proof system for component-based embedded systems, *Sci. Comput. Prog.*, 97(3), 333–348, 2015, Object-Oriented Programming and Systems (OOPS 2010) Modeling and Analysis of Compositional Software (papers from {EUROMICRO} SEAA–12). [Online]. Available: http://www.sciencedirect.com/science/article/pii/S0167642314002901.

23. J. M. Cobleigh, G. S. Avrunin, and L. A. Clarke, Breaking up is hard to do: An evaluation of automated assume-guarantee reasoning, *ACM Trans. Softw. Eng. Methodol.*, 17(2), 7:1–7:52, May 2008. [Online]. Available: http://doi.acm.org/10.1145/1348250.1348253.

24. E. F. Codd, A relational model of data for large shared data banks, *Commun. ACM*, 13(6), 377–387, June 1970. [Online]. Available: http://doi.acm.org/10.1145/362384.362685.

25. D. Cofer, A. Gacek, S. Miller, M. W. Whalen, B. LaValley, and L. Sha, Compositional verification of architectural models, in *Proceedings of the Fourth International Conference on NASA Formal Methods*, ser. NFM'12. Berlin, Germany: Springer-Verlag, 2012, pp. 126–140. [Online]. Available: http://dx.doi.org/10.1007/978-3-642-28891-3_13.

26. W. Damm, Controlling speculative design processes using rich component models, in *Fifth International Conference on Application of Concurrency to System Design, 2005. ACSD 2005*, Rennes, France, June 2005, pp. 118–119.

27. W. Damm, H. Hungar, B. Josko, T. Peikenkamp, and I. Stierand, Using contract-based component specifications for virtual integration testing and architecture design, in *Design, Automation Test in Europe Conference Exhibition (DATE), 2011*, Grenoble, France, March 2011, pp. 1–6.

28. W. Damm, B. Josko, and T. Peinkamp, Contract based ISO CD 26262 safety analysis, in *Safety-Critical Systems, 2009*. SAE, Detroit, MI, 2009.

29. A. David, K. G. Larsen, A. Legay, U. Nyman, and A. Wasowski, Timed I/O automata: A complete specification theory for real-time systems, in *Proceedings of the 13th ACM International Conference on Hybrid Systems: Computation and Control*, ser. HSCC '10. New York: ACM, 2010, pp. 91–100. [Online]. Available: http://doi.acm.org/10.1145/1755952.1755967.

30. L. de Alfaro and T. A. Henzinger, Interface automata, *SIGSOFT Softw. Eng. Notes*, 26(5), 109–120, September 2001. [Online]. Available: http://doi.acm.org/10.1145/503271.503226.

31. L. de Alfaro and T. A. Henzinger, Interface theories for component-based design, in *Embedded Software*, ser. Lecture Notes in Computer Science, T. A. Henzinger and C. M. Kirsch, Eds. Berlin, Germany: Springer, 2001, vol. 2211, pp. 148–165. [Online]. Available: http://dx.doi.org/10.1007/3-540-45449-7_11.

32. W.-P. de Roever, The need for compositional proof systems: A survey, in *Compositionality: The Significant Difference*, ser. Lecture Notes in Computer Science, W.-P. de Roever, H. Langmaack, and A. Pnueli, Eds. Berlin, Germany: Springer, 1998, vol. 1536, pp. 1–22. [Online]. Available: http://dx.doi.org/10.1007/3-540-49213-5_1.

33. B. Delahaye, B. Caillaud, and A. Legay, Probabilistic contracts: A compositional reasoning methodology for the design of systems with stochastic and/or non-deterministic aspects, *Form. Methods Syst. Des.*, 38(1), 1–32, Feb. 2011. [Online]. Available: http://dx.doi.org/10.1007/s10703-010-0107-8.

34. P. Derler, E. A. Lee, M. Törngren, and S. Tripakis, Cyber-physical system design contracts, in *ICCPS '13: ACM/IEEE Fourth International Conference on Cyber-Physical Systems*, Philidelphia, 2013. [Online]. Available: http://chess.eecs.berkeley.edu/pubs/959.html.

35. R. Diestel, *Graph Theory*, 4th edn., ser. Graduate Texts in Mathematics. Springer, Berlin, Germany, 2012, vol. 173.

36. E. W. Dijkstra, Guarded commands, nondeterminacy and formal derivation of programs, *Commun. ACM*, 18(8), 453–457, August1975. [Online]. Available: http://doi.acm.org/10.1145/360933.360975.

37. D. L. Dill, Trace theory for automatic hierarchical verification of speed-independent circuits, in *Proceedings of the fifth MIT conference on Advanced Research in VLSI*. Cambridge, MA: MIT Press, 1988, pp. 51–65. [Online]. Available: http://dl.acm.org/citation.cfm?id=88056.88061.

38. R. W. Floyd, Assigning meanings to programs, in *Mathematical Aspects of Computer Science*, ser. Proceedings of Symposia in Applied Mathematics, J. T. Schwartz, Ed., vol. 19. Providence, RI: American Mathematical Society, 1967, pp. 19–32.

39. S. Friedenthal, A. Moore, and R. Steiner, *A Practical Guide to SysML: Systems Modeling Language*. San Francisco, CA: Morgan Kaufmann Publishers Inc., 2008.

40. H. Giese, Contract-based component system design, in *33rd Annual Hawaii International Conference on System Sciences (HICSS-33)* Maui, HI: IEEE Press, 2000.

41. G. Goessler and J.-B. Raclet, Modal contracts for component-based design, in *Proceedings of the 2009 Seventh IEEE ernational Conference on Software Engineering and Formal Methods*, ser.

SEFM '09. Washington, DC: IEEE Computer Society, 2009, pp. 295–303. [Online]. Available: http://dx.doi.org/10.1109/SEFM.2009.26.

42. S. Graf and S. Quinton, Contracts for bip: Hierarchical interaction models for compositional verification, in *Proceedings of the 27th IFIP WG 6.1 International Conference on Formal Techniques for Networked and Distributed Systems*, ser. FORTE '07. Berlin, Germany: Springer-Verlag, 2007, pp. 1–18. [Online]. Available: http://dx.doi.org/10.1007/978-3-540-73196-2_1.

43. G. GÃ¶ssler, D. Xu, and A. Girault, Probabilistic contracts for component-based design, *Formal Methods System Design*, 41(2), 211–231, 2012. [Online]. Available: http://dx.doi.org/10.1007/s10703-012-0162-4.

44. J. Hammond, R. Rawlings, and A. Hall, Will it work? [requirements engineering], in *Proceedings of the Fifth IEEE International Symposium on Requirements Engineering, 2001*, Trento, Italy, 2001, pp. 102–109.

45. C. A. R. Hoare, An axiomatic basis for computer programming, *Commun. ACM*, 12(10), 576–580, October 1969. [Online]. Available: http://doi.acm.org/10.1145/363235.363259.

46. J. Hooman and W. P. de Roever, The quest goes on: A survey of proofsystems for partial correctness of CSP, in *Current Trends in Concurrency, Overviews and Tutorials*, Berlin, Germany, 1986, pp. 343–395. [Online]. Available: http://dx.doi.org/10.1007/BFb0027044.

47. IEC 61508, Functional safety of electrical/electronic/programmable electronic safety-related systems, Geneva, Switzerland, 2010.

48. ISO 26262, Road vehicles-Functional safety, Geneva, Switzerland, 2011.

49. ISO/IEC/IEEE 42010, System and software Engineering—Architecture description, Geneva, Switzerland, 2011.

50. M. Jackson, *Software Requirements &Amp; Specifications: A Lexicon of Practice, Principles and Prejudices*. New York: ACM Press, 1995.

51. M. Jackson, The world and the machine, in *Proceedings of the 17th International Conference on Software Engineering*, ser. ICSE '95. New York: ACM, 1995, pp. 283–292. [Online]. Available: http://doi.acm.org/10.1145/225014.225041.

52. C. B. Jones, Tentative steps toward a development method for interfering programs, *ACM Trans. Program. Lang. Syst.*, 5(4), 596–619, Utrecht, The Netherlands, Oct. 1983. [Online]. Available: http://doi.acm.org/10.1145/69575.69577.

53. C. B. Jones, Specification and design of (parallel) programs, in *Information Processing 83*, ser. IFIP Congress Series, R. E. A. Mason, Ed., vol. 9, IFIP. Paris, France: North-Holland, September 1983, pp. 321–332.

54. B. Josko, Q. Ma, and A. Metzner, Designing embedded systems using heterogeneous rich components, in *Proceedings of the INCOSE International Symposium 2008*, 01 2008.

55. O. Kupferman and M. Y. Vardi, An automata-theoretic approach to modular model checking, *ACM Trans. Program. Lang. Syst.*, vol. 22, no. 1, pp. 87–128, Jan. 2000. [Online]. Available: http://doi.acm.org/10.1145/345099.345104.

56. L. Lamport, Proving the correctness of multiprocess programs, *IEEE Trans. Softw. Eng.*, 3(2), 125–143, March 1977. [Online]. Available: http://dx.doi.org/10.1109/TSE.1977.229904.

57. Win and sin: Predicate transformers for concurrency, *ACM Trans. Program. Languages Syst.*, 12, 396–428, 1990.

58. A. Lapouchnian, Goal-oriented requirements engineering: An overview of the current research, University of Toronto, Toronto, Ontario, Canada, 2005.

59. K. Larsen, Modal specifications, in *Automatic Verification Methods for Finite State Systems*, ser. Lecture Notes in Computer Science, J. Sifakis, Ed. Berlin, Germany: Springer, 1990, vol. 407, pp. 232–246. [Online]. Available: http://dx.doi.org/10.1007/3-540-52148-8_19.

60. E. Lee, Cyber physical systems: design challenges, in *11th IEEE International Symposium on Object Oriented Real-Time Distributed Computing (ISORC)*, Orlando, FL, 2008, pp. 363–369.

61. E. Lee and A. Sangiovanni-Vincentelli, A framework for comparing models of computation, *IEEE Trans. Computer-Aided Design Int. Circuits Syst.*, 17(12), 1217–1229, December 1998.

62. N. A. Lynch and M. R. Tuttle, An introduction to input/output automata, *CWI Quar.* 2, 219–246, 1989.

63. P. Maier, A set-theoretic framework for assume-guarantee reasoning, in *Automata, Languages and Programming*, ser. Lecture Notes in Computer Science, F. Orejas, P. Spirakis, and J. van Leeuwen, Eds. Berlin, Germany: Springer, 2001, vol. 2076, pp. 821–834. [Online]. Available: http://dx.doi.org/10.1007/3-540-48224-5_67.

64. P. Maier, Compositional circular assume-guarantee rules cannot be sound and complete, in *Foundations of Software Science and Computation Structures*, ser. Lecture Notes in Computer Science, A. Gordon, Ed.

Berlin, Germany: Springer, 2003, vol. 2620, pp. 343–357. [Online]. Available: http://dx.doi.org/10.1007/3-540-36576-1_22.

65. K. L. Mcmillan, Circular compositional reasoning about liveness, in *Advances in Hardware Design and Verification: IFIP WG10.5 International Conference on Correct Hardware Design and Verification Methods (CHARME 99),* volume 1703 of Lecture Notes in Computer Science. Springer-Verlag, Bad Herrenalb, Germany, 1999, pp. 342–345.

66. K. McMillan, Verification of infinite state systems by compositional model checking, in *Correct Hardware Design and Verification Methods,* ser. Lecture Notes in Computer Science, L. Pierre and T. Kropf, Eds. Berlin, Germany: Springer, 1999, vol. 1703, pp. 219–237. [Online]. Available: http://dx.doi.org/10.1007/3-540-48153-2_17.

67. B. Meyer, Applying Design by Contract, *IEEE Comput.,* 25, 40–51, 1992.

68. J. Misra and K. Chandy, Proofs of networks of processes, *IEEE Trans. Softw. Eng.,* SE-7(4), 417–426, 1981.

69. Modelica, A unified object-oriented language for physical systems modeling, language specification, 2012. [Online]. Available: http://www.modelica.org.

70. K. S. Namjoshi and R. J. Trefler, On the completeness of compositional reasoning methods, *ACM Trans. Comput. Logic,* 11(3), 16:1–16:22, May 2010. [Online]. Available: http://doi.acm.org/10.1145/1740582.1740584.

71. R. Negulescu, Process spaces, in *Proceedings of the 11th International Conference on Concurrency Theory,* ser. CONCUR '00. London, UK: Springer-Verlag, 2000, pp. 199–213. [Online]. Available: http://dl.acm.org/citation.cfm?id=646735.701627.

72. M. Nyberg, Failure propagation modeling for safety analysis using causal bayesian networks, in *2013 Conference on Control and Fault-Tolerant Systems (SysTol),* Nice, France, October 2013, pp. 91–97.

73. S. Owicki and D. Gries, An axiomatic proof technique for parallel programs i, *Acta Infor.,* 6(4), 319–340, 1976. [Online]. Available: http://dx.doi.org/10.1007/BF00268134.

74. D. L. Parnas, Functional documents for computer systems, *Sci. Comp. Program.,* 25, 41–61, 1995.

75. R. Passerone, Semantic foundations for heterogeneous systems, PhD dissertation, University of California, Berkeley, CA, 2004, aAI3146975.

76. A. Pnueli, In transition from global to modular temporal reasoning about programs, in *Logics and Models of Concurrent Systems,* ser. NATO ASI Series, K. Apt, Ed. Berlin, Germany: Springer, 1985, vol. 13, pp. 123–144. [Online]. Available: http://dx.doi.org/10.1007/978-3-642-82453-1_5.

77. C. S. Păsăreanu, D. Giannakopoulou, M. G. Bobaru, J. M. Cobleigh, and H. Barringer, Learning to divide and conquer: Applying the l* algorithm to automate assume-guarantee reasoning, *Form. Methods Syst. Des.,* 32(3), 175–205, Jun. 2008. [Online]. Available: http://dx.doi.org/10.1007/s10703-008-0049-6.

78. S. Quinton and S. Graf, Contract-based verification of hierarchical systems of components, in *Sixth IEEE International Conference on, Software Engineering and Formal Methods, 2008. SEFM '08.,* Cape Town, South Africa, November 2008, pp. 377–381.

79. J.-B. Raclet, E. Badouel, A. Benveniste, B. Caillaud, A. Legay, and R. Passerone, A modal interface theory for component-based design, *Fundam. Inf.,* 108(1–2), 119–149, January 2011. [Online]. Available: http://dl.acm.org/citation.cfm?id=2362088.2362095.

80. I. Ruchkin, D. De Niz, D. Garlan, and S. Chaki, Contract-based integration of cyber-physical analyses, in *Proceedings of the 14th International Conference on Embedded Software,* ser. EMSOFT '14. New York: ACM, 2014, pp. 23:1–23:10. [Online]. Available: http://doi.acm.org/10.1145/2656045.2656052.

81. A. L. Sangiovanni-Vincentelli, W. Damm, and R. Passerone, Taming Dr. Frankenstein: Contract-based design for cyber-physical systems, *Eur. J. Control,* 18(3), 217–238, 2012. [Online]. Available: http://dblp.uni-trier.de/db/journals/ejcon/ejcon18.html#Sangiovanni-VincentelliDP12.

82. G. Simko, T. Levendovszky, M. Maroti, and J. Sztipanovits, Towards a theory for cyber-physical systems modeling, in *Proceedings of the Fourth ACM SIGBED International Workshop on Design, Modeling, and Evaluation of Cyber-Physical Systems,* ser. CyPhy '14. New York: ACM, 2014, pp. 56–61. [Online]. Available: http://doi.acm.org/10.1145/2593458.2593463.

83. A. Soderberg and R. Johansson, Safety contract based design of software components, in *2013 IEEE International Symposium on Software Reliability Engineering Workshops (ISSREW),* Pasadena, CA, November 2013, pp. 365–370.

84. A. Soderberg and B. Vedder, Composable safety-critical systems based on pre-certified software components, in *2012 IEEE 23rd International Symposium on Software Reliability Engineering Workshops (ISSREW),* Dallas, TX, November 2012, pp. 343–348.

85. SPEEDS, SPEculative and exploratory design in systems engineering, 2006-2009. [Online]. Available: http://www.speeds.eu.com/.

86. X. Sun, Compositional design of analog systems using contracts, PhD dissertation, EECS Department, University of California, Berkeley, CA, May 2011. [Online]. Available: http://www.eecs.berkeley.edu/Pubs/TechRpts/2011/EECS-2011-49.html.

87. X. Sun, P. Nuzzo, C.-C. Wu, and A. Sangiovanni-Vincentelli, Contract-based system-level composition of analog circuits, in *46th ACM/IEEE Design Automation Conference, 2009. DAC '09*, July 2009, pp. 605–610.

88. M. Törngren, S. Tripakis, P. Derler, and E. A. Lee, Design contracts for cyber-physical systems: Making timing assumptions explicit, EECS Department, University of California, Berkeley, CA, Technical Report UCB/EECS-2012-191, August 2012. [Online]. Available: http://www.eecs.berkeley.edu/Pubs/TechRpts/2012/EECS-2012-191.html.

89. A. van Lamsweerde and E. Letier, From object orientation to goal orientation: A paradigm shift for requirements engineering, in *Radical Innovations of Software and Systems Engineering in the Future*, ser. Lecture Notes in Computer Science, M. Wirsing, A. Knapp, and S. Balsamo, Eds. Berlin, Germany: Springer, 2004, vol. 2941, pp. 325–340. [Online]. Available: http://dx.doi.org/10.1007/978-3-540-24626-8_23.

90. J. Westman and M. Nyberg, Environment-centric contracts for design of cyber-physical systems, in *Model-Driven Engineering Languages and Systems*, ser. Lecture Notes in Computer Science, J. Dingel, W. Schulte, I. Ramos, S. Abraho, and E. Insfran, Eds. Springer International Publishing, Valencia, Spain, 2014, vol. 8767, pp. 218–234. [Online]. Available: http://dx.doi.org/10.1007/978-3-319-11653-2_14.

91. J. Westman and M. Nyberg, Extending contract theory with safety integrity levels, in *To be published at 2014 IEEE 15th International Symposium on High-Assurance Systems Engineering (HASE)*. Springer International Publishing, Daytona Beach, FL, January 2015.

92. J. Westman, M. Nyberg, and M. Törngren, Structuring safety requirements in ISO 26262 using contract theory, in *Computer Safety, Reliability, and Security*, ser. Lecture Notes in Computer Science, F. Bitsch, J. Guiochet, and M. Kaniche, Eds. Berlin, Germany: Springer, 2013, vol. 8153, pp. 166–177. [Online]. Available: http://dx.doi.org/10.1007/978-3-642-40793-2_16.

93. E. S. Wolf, Hierarchical models of synchronous circuits for formal verification and substitution, PhD. dissertation, Stanford University, Stanford, CA, 1996, uMI Order No. GAX96-12052.

94. Q. Xu, A. Cau, and P. Collette, On unifying assumptioncommitment style proof rules for concurrency, in *CONCUR 94: Concurrency Theory*, ser. Lecture Notes in Computer Science, B. Jonsson and J. Parrow, Eds. Berlin, Germany: Springer, 1994, vol. 836, pp. 267–282. [Online]. Available: http://dx.doi.org/10.1007/978-3-540-48654-1_22.

95. E. Yu, Towards modelling and reasoning support for early-phase requirements engineering, in *Proceedings of the Third IEEE International Symposium on Requirements Engineering, 1997*, Annapolis, MD, January 1997, pp. 226–235.

14 The Internet of Interlaced Cyber-Physical Things

Claudio Estevez, César Azurdia, and Sandra Céspedes

CONTENTS

14.1 INTRODUCTION

The way most people view a computer is as a device used to perform simulations, prepare presentations, edit documents, and even play video games, hence in the form of a desktop or notebook. Nevertheless, most computers go unperceived, working in the background of our daily lives, including in vehicles, house appliances, office equipment, manufacturing, and even the ubiquitous cell phone. These less perceived computers are known as embedded systems, and in lame terms, it is a computer that is designed specifically with one (or few) purpose to reduce processing time, costs, and/or size. Today, embedded systems are no longer restraint by processing power, the capability to maintain strict task deadlines, high costs, or size. Since most embedded systems are connected to the local network and can communicate with other embedded systems, these have evolved to what is known today as cyber-physical systems or CPS [1]. These are systems that still interact with the physical world and perform specific tasks, as embedded systems do, but are much more versatile and powerful in terms of processing capabilities. Cyber-physical systems are able to communicate within an intranet, but are not necessarily connected to the Internet. On the other hand, we have the Internet of Things (IoT), in which we have embedded systems connected to the Internet periodically feeding it with data. A server or receiving device somewhere in the Internet, commonly referred to as the cloud, turns these data into information, knowledge, or optimistically wisdom, terminology used in the context of information science, more specifically, the data, information, knowledge, and wisdom (DIKW) pyramid. The IoT, in contrast to cyber-physical systems, can also interact with data, which even though is stored in a physical media is not considered interacting with the physical world. Also, the IoT attempts to climb the DIKW pyramid as much as possible to use the data acquired in the most useful way possible, while cyber-physical systems are meant to mainly control a specific aspect in the physical domain, and to achieve this, the quickest route is the optimal (hence keeping the system simple, reducing the amount of DIKW layers climbed). There may be more ways to explain these terms and their differences; if these are not clear, we encourage the reader to pursue further readings into these interesting topics. Some good sources to start from are the books [1,2] and the website [3]. When the IoT is combined with cyber-physical systems, the resulting system is an interlaced mesh of subsystems that interact not only in the virtual domain but also in the physical domain. The control decisions performed in one discipline can potentially affect (positively or negatively) the decisions of other disciplines; therefore, the affected discipline takes control over a certain action that itself might affect another discipline creating a single entity that is constantly optimizing resources while maintaining balance. We call this system the Internet of interlaced cyber-physical things (IoICPT). Figure 14.1 attempts to portray this concept. In this system, the cyber-physical subsystems interact not only through the communication links but also through the physical world. A simple example could be an unmanned system that is manufacturing a particular product and, in real-time, it is been informed that the demand of this product is increasing; hence, the system enters an overdrive mode. This decision caused the system to generate more pollution, so another cyber-physical subsystem which is attempting to filter the air informs all devices, which are generating air pollutants, to decrease their use when possible. The system may also: alert hospitals that more polysomnograms will be required and to prepare the rooms necessary for the estimated amount of patients admitted, and so forth. The possibilities are endless. Basically, this system would be the brain of a smart city and just like the Internet does not reside in any particular place nor does the brain of a smart city; rather, it is spread throughout the city. To portray the interlacing of various applications in the different disciplines, we have chosen six fields: e-health, energy, transportation, building/home, mobile networks, and environment. The interleaved system is shown in Figure 14.2.

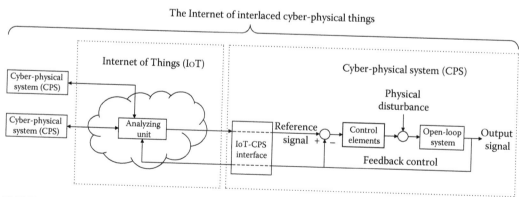

FIGURE 14.1 System portraying the interlacing of cyber-physical systems through the Internet of Things architecture.

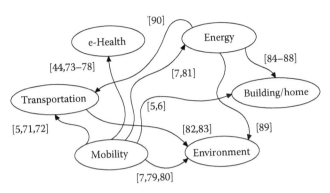

FIGURE 14.2 Mesh of disciplines and work that show the interaction between these domains.

This chapter is organized in the following manner. The first section discusses fundamental attributes of IoT systems. These attributes are found in IoT end systems and are an organized way to view the individual components of these devices, which are identification, sensing, communication, and purpose. The following section discusses communication-related work that has been researched in the field of the IoT. It is focused on the bottom portion of the OSI model and organized by layers. The third section describes work done in different disciplines, such as e-health, energy, transportation, building/home, mobile networks, and environment. There are other interesting disciplines that can be tied to this mesh of systems, such as economy, but the chosen disciplines have a strong interaction with the physical world, which is the focus of this book. The final section discusses the conclusions that are drawn from the work presented.

14.2 BUILDING BLOCKS

There are various traits that the IoT systems have in common and therefore can be used to define the components necessary to build this fairly complex system. Without any of these attributes, it would not be possible to have an IoT. These traits are identification, sensing, communication, and purpose. *Identification* is necessary to distinguish the different nodes. To have identity is to be unique, and this enables various applications such as determining the location of the information and what exact mechanisms to control. *Sensing* is the means used to extract information. This can be any type of sensor, active or passive, for example, pressure, temperature, light, and RFID. The amount of sensors or amount of information generated by the sensors is not a requisite of the system, though, if too much data are produced, then usually some techniques to reduce the data are employed, such as data mining or information fusion. *Communication* is the attribute that enables the sharing of data or information. The sharing can be performed in real time or it can be stored for later use. This attribute serves as an interface between the sensing and the purpose. It is not limited to a single communication technology, but actually quite on the contrary; the objective is to cover as much physical area as possible to broaden the amount of nodes (or subsystems) in the system, and to do this, it can use as many communication technologies as necessary. Finally, all these steps are done with a *purpose*. The purpose can be to control, inform, or monitor and has a wide range of other tasks.

14.3 ADVANCES IN Io-TCPS RESEARCH

This section discusses research advances achieved in various fields. The research advances discussed are organized by OSI layer, more specifically the bottom three layers, which are physical, data link, and network layers. In each layer, there are various published works discussed in detail, including methodology of the work, summaries, and standards.

14.3.1 Physical Layer Technologies for IoT: 5G Systems

New research directions will lead to essential changes in the design of future fifth-generation (5G) cellular mobile networks [4]. In general, there are five technologies that could lead to both architectural and component complex design changes. The five key technologies that are being considered at the moment are the following: device-centric architectures [5–7], millimeter-wave (mmWave) bands [8,9], massive multiple input, multiple output (MIMO) [9], smarter devices, and native support for machine-to-machine (M2M) communications [10,11]. Therefore, the device-centric architecture of the actual cellular mobile networks needs to be redesign for 5G systems. Hence, it is time to reconsider the concepts of uplink and downlink, as well as control and data channels, to better route information flows with different priorities and purposes toward different sets of nodes within the network. While spectrum has become scarce at microwave frequencies, it is plentiful in the mmWave frequency bands. Such a free and unlicensed bandwidth spectrum has led to researchers with diverse backgrounds to study the different aspects of mmWave signal transmission. Although far from being fully understood, technologies operating at mmWave bands have already been standardized for short-range services (IEEE 802.11ad) and are being implemented in some applications such as small-cell backhauls. Massive MIMO proposes utilizing a very high number of antennas to multiplex messages for several devices on each time–frequency resource, focusing the radiated energy toward the intended directions, while minimizing intercell interference. On the other hand, massive MIMO requires major architectural changes, particularly in the design of macro base stations (BS), and it may also lead to new types of deployments, but will be an efficient solution to achieve higher error-free data rates.

Cellular generation networks 2G–3G–4G were built under the design assumption of having complete control at the infrastructure without considering intelligence at the device side. But 5G systems should drop this design assumption and exploit intelligence at the device side within different layers of the protocol layer, for example, by allowing device-to-device (D2D) connectivity or employing smart cache by the mobile user. While this design structure mainly requires a change at the node level (component change), it also has implications at the architectural level. For M2M communication in 5G systems, it involves satisfying three different requirements associated with different classes of low-data-rate services: support of an extensive number of low-rate devices, sustaining a minimal data rate in virtually all circumstances, and very-low-latency data transfer. Addressing these requirements in 5G requires new methods and ideas at both the component and architectural levels, as will be discussed in the next subsections.

14.3.1.1 Massive MIMO

Massive MIMO is also referred to as large-scale antenna systems or as large-scale MIMO. Massive MIMO is an emerging technology that scales up conventional MIMO systems. In general, massive MIMO is a technology similar to multiuser MIMO in which the number of antennas at the BS is much larger than the number of devices per signaling resource [4,12]. Having more base station antennas than devices makes the channels of the different devices almost orthogonal; therefore, simple spatial multiplexing/demultiplexing techniques can be applied. By adding multiple antennas, in addition to frequency and time dimensions, extra degrees of freedom can be added into the system. Further, a compelling improvement can be obtained in terms of energy efficiency, reliability, and spectral efficiency.

Massive MIMO in general depends on spatial multiplexing, which assumes that the BS has good knowledge of the channel in the uplink and downlink [9]. In the uplink, the user equipments (UEs) send pilot signals, and the BS sends the channel responses of each UE, whereas the downlink is more complex. In traditional MIMO systems, the BS sends out pilot signals, and the UE estimates the channel responses and feeds back the information back to the BS. But this will not be a feasible solution in massive MIMO systems. Initially, optimal downlink pilot signals should be orthogonal between antennas. Therefore, the amount of time–frequency resources needed to implement the orthogonal pilot signal increases proportionally to the number of antennas used by the system. Normally, a massive MIMO system would require up to 100 times more resources than a conventional MIMO system [9]. Additionally, the number of channel responses that each UE must estimate is proportional to the number of antennas implemented at the BS. In a similar manner, the complexity of the system would be up to 100 larger than the conventional system. A possible solution for the downlink would be to implement the time division multiplexing mode and focus on the correlation between the downlink and uplink channels.

Systems that operate with massive MIMO technology are equipped with a large number of antennas at the transmitter and/or receiver sides. These systems are characterized by having tens or hundreds of antennas. The antennas mounted at the transmitter can be distributed or co-located according to the application. In a similar way, the antennas at the receiver can be mounted into one single device or distributed into several devices. Many different configurations and deployment scenarios for the antenna array used by massive MIMO systems can be implemented. Some of the possible antenna configurations and deployment scenarios for a massive MIMO BS would be a distributed array of antennas, linear array, rectangular array, and cylindrical array [9]. Each antenna would be a small and active device, preferably supplied by an electric or optical digital bus.

Besides providing energy and spectral efficiency, massive MIMO systems are characterized by reducing the effects of noise and fast fading. Further, intracell interference can be mitigated using simple linear precoding and detection methods. To achieve this, MIMO technology relies on phase coherence using very powerful signal processing algorithms. Some specific benefits of a massive MIMO system include the following:

a. Massive MIMO is capable of increasing the capacity of the system up to 10 times or even more [9]. The larger capacity results from the extensive spatial multiplexing used in massive MIMO. This is an extremely important result because with the amount of devices that will be connected to the IoT, we need to increase the channel capacity to meet the system's parameters. Further, the radiated energy efficiency of the system can be improved on the order of 100 times. The improvement in terms of energy efficiency is possible because with a large number of antennas, beamforming (coherent superposition) algorithms can be implemented to concentrate energy into small regions of space. According to [13], massive MIMO combined with maximum ratio combining is a feasible option to improve the uplink spectral efficiency 10 times and simultaneously increase the radiated power 100 times.

b. Massive MIMO simplifies the multiple access layer. In massive MIMO combined with orthogonal frequency-division multiplexing (OFDM)-based systems, each subcarrier will basically have the same channel gain for the entire bandwidth.

c. Massive MIMO decreases latency produced by multipath channels. Wireless communication channels are characterized by multipath fading, which reduces the energy of the received signal at a given time. Massive MIMO depends on beamforming and the law of large numbers to avoid fading dips and diminish the effects of latency.

d. Massive MIMO systems can be built with inexpensive low-power components. Hundreds of low-cost amplifiers with output power in the milliwatt range can replace a very expensive ultralinear 50 W amplifier, which is the one implemented in conventional systems [9].

e. Massive MIMO increases the robustness against intentional jamming and interference intentionally created by men.

Despite the benefits and opportunities presented by massive MIMO systems [14], there are certain drawbacks that need to be addressed. One of these issues is that time-division duplexing applied in massive MIMO systems requires channel reciprocity. Another issue that needs to be considered is that every terminal in a massive MIMO system is assigned an orthogonal uplink pilot sequence. But the maximum number of orthogonal pilot sequences that can exist is limited by the channel delay spread. The effect of reusing pilots from one cell to another and the negative consequences it produces is known as pilot contamination [15]. Pilot contamination is not only specific to massive

MIMO, but its effect on massive MIMO appears to be much more radical than in classical MIMO systems [12]. Other challenges that need to be addressed are the realization of distributed processing systems, synchronization of the antenna units, and diminishing the computational complexity required to implement massive MIMO systems.

14.3.1.2 Millimeter-Wave Technology:

While spectrum has become limited at microwave frequencies, it is abundant in the mmWave frequency bands. Microwave cellular systems are characterized by having extremely valuable but limited spectrum. Around 600 MHz of bandwidth are currently in use for mobile communications and are divided among several operators [16]. Two ways to gain access to more microwave spectrum have been proposed: to reuse the terrestrial TV spectrum for applications such as broadband access and to share spectrum by implementing cognitive radio techniques. By implementing the previous techniques, it appears that doubling the current cellular bandwidth is the best case scenario at microwave frequencies [4], which is not enough to achieve the capacities needed by 5G systems and the devices connected to the IoT.

There is an enormous amount of spectrum available at the mmWave frequency bands, ranging from 3 to 300 GHZ. Several bands have been considered and are being studied at the moment, such as the following frequency bands [17]:

a. Multipoint distribution service band at 28–30 GHz
b. License-free band at 60 GHz
c. E-band at 71–76 GHz, 81–86 GHz, and 92–95 GHz

If all of these bands were to be implemented, several tens of gigahertz would become available for 5G cellular systems, offering much more bandwidth than what actual systems implement. With a larger bandwidth, improvements in cell capacity and higher user data rates could be easily achieved. It is expected that 5G systems will offer a data rate with a minimum of 1 Gb/s rate to all users and data rates up to 5 and 50 Gb/s for high mobility and pedestrian users, respectively [8].

Despite all of the available spectrum available at the mmWave frequency bands, there are certain issues that need to be addressed to implement mmWave technology in 5G cellular networks. Traditionally, due to high propagation loss and the absence of cost-effective components (distinct hardware constraints), these frequency bands have mostly been utilized for outdoor point-to-point line of sight (LOS) backhaul links. An example of this is the transmission of high-resolution multimedia streams for indoor applications by using mmWaves. To fully exploit these unused frequency bands, two key issues need to be solved: sufficiently large geographical coverage and support mobility in non-line-of-sight (NLOS) environments. Supporting NLOS communication links is essential in cellular networks with high mobility, as well as higher error-free data rates.

Certain concerns exist regarding using mmWave frequency bands for mobile cellular communications. Some of these concerns rely on the propagation characteristics at higher frequencies, which behave as an optical system, and it is characterized by a high propagation loss and often modelled as an LOS link. In Roh et al. [8], researchers at Samsung Electronics revealed that the key parameters that characterize the propagation properties of the mmWave frequency bands, such as the path loss exponent, are comparable to those of traditional cellular frequency bands when the antennas at the transmitter and receiver are used to produce beamforming gains. The previous discovery is extremely important because beamforming gains can be achieved by implementing a MIMO antenna array system; hence, NLOS links can be achieved at mmWave frequency bands. A patch antenna at 3 GHz and an array antenna at 30 GHz, of the same physical size, were designed and tested. The results showed the same amount of propagation loss disregarding of the operation frequency.

As previously stated, an appropriate beamforming scheme that is needed to focus the transmitted and/or received power in a desired direction to overcome the undesired path loss is a key element for feasible cellular communications in the mmWave frequency bands. When combined with OFDM systems, digital beamforming is carried out on a subcarrier basis before the inverse fast Fourier transform (IFFT) operation at the transmitter side is performed and after the FFT operation at the receiver side. Therefore, OFDM-based systems combined with MIMO techniques modulated in the mmWave frequency bands are being suggested as the key technologies to be used at the physical layer of 5G cellular networks [18]. This is due to the high data rate transmission capability, high bandwidth efficiency, the robustness to multipath fading given by the ability to convert a frequency selective fading channel into several nearly flat fading channels, and also the way OFDM-based systems deal with delay spread by using a guard interval [19,20].

14.3.1.3 Device-Centric Architectures:

To address the challenges and meet the requirements of 5G cellular systems, a startling change in the design of the cellular architecture is needed. Typically, cellular designs have relied on the fundamental concept of cells as the essential metric within the radio access network. We know that wireless users normally stay indoors about 80% of the time, while they stay outdoors only about 20% of the time [21]. But the current cellular architecture normally uses an outdoor base station in the middle of a cell to communicate the UEs within the cell, no matter if the users stay indoors or outdoors. Indoor UEs require more energy to communicate efficiently with the BS because the signal has to go through building walls and the effect multipath is more harmful, decreasing the data rate, reducing the spectral efficiency, and producing high penetration loss. Therefore, over the last few years, different trends have been pointing out to a disruption of the cell-centric structure.

One of the key ideas in the design of 5G cellular architectures is to separate indoor and outdoor scenarios; this is to avoid penetration loss through buildings in indoor environments. A distributed antenna system and massive MIMO technology shall be combined with BSs with different transmit powers and coverage areas [5]. Outdoor BSs will be equipped with large antenna arrays with some antenna elements distributed around the cell and connected to the BS via optical fiber. Because outdoor UEs are normally equipped with a limited number of antennas, they will be capable of collaborating with each other to form a virtual large antenna array. Further, large antenna arrays shall also be installed outside of every building to communicate with outdoor base stations or distributed antenna elements of the base stations. This way, we obtain a heterogeneous cellular architecture formed by macrocells, microcells, small cells, and relays. By using this architecture, indoor UEs only need to communicate with indoor access points (APs) with large antenna arrays installed outside the buildings. Therefore, some existing technologies suitable for short-range and high error-free data rates could be implemented, such as Wi-Fi, femtocells, and visible light communications [22–24].

To accommodate high mobility, users in vehicles, and high-speed trains, the concept of mobile femtocells (MFemtocell) has been proposed [5]. The MFemtocell combines the concepts of a femtocell and mobile relays. A MFemtocell is a small cell that can move around and can dynamically adjust its connection to an operator's core network. This cell can be deployed on public transportation buses and trains and inside vehicles. For example, MFemtocells could be located inside vehicles to communicate users within the cars, while large antenna arrays are located outside the vehicle to communicate with outdoor BSs.

14.3.1.4 Device-to-Device D2D Connectivity:

Initially, cellular systems were built on the concept of having total control at the infrastructure side, whereas 5G systems will allow the devices to play a more active role in the system. Therefore, the design of a 5G network should take into account that the devices will be smarter, establishing D2D connectivity. Without D2D connectivity, multiple wireless hops are needed to achieve what normally requires a single hop, therefore making it an energy and spectral inefficient system.

14.3.1.5 Machine-to-Machine Communication:

Wireless communications systems are becoming a normal asset these days; therefore, a large variety of emerging services with new types of requirements are in demand. One of these services is

M2M communication. M2M communication in 5G systems involves satisfying different requirements associated with different types of low-data-rate services, such as supporting a massive number of low-rate devices, maintaining a minimal data rate in basically all circumstances, and achieving a very-low-latency data transfer.

M2M communication systems expect an enormous number of connected devices. Current systems typically operate with a few hundred devices per BS. But some M2M services will require much more connected devices for sensing, smart grid components, metering, and other type of services with wide area coverage. To be able to interact with many devices, we need a very high link accuracy. Systems were initially dominated by wireline connectivity because wireless links did not offer the same degree of confidence. Therefore, we need wireless links to be reliably operational all of the time and in all places.

Probably the most difficult requirement needed for the proper operation of M2M communication systems is a very-low-latency data transfer. M2M communication requires that data be transmitted accurately within a given time interval. A clear example of low latency and real-time operation can be found in vehicle-to-device connectivity, where traffic safety can be improved if the data are transmitted with low latency.

Massive M2M communication is classified into two categories [4]. The first one refers to massive M2M communication where each connected sensor or device transmits small data blocks intermittently, such as temperature, pollution levels, humidity, and atmospheric pressure, among others. Current systems are not designed to simultaneously serve data traffic from a large number of devices. For example, a current system could serve 5 sensors at 2 Mb/s each, but cannot serve 1000 devices each requiring 1kb/s. The second type of massive M2M communication refers to the operation of systems that require low latency and high reliability, but with a low average rate per device.

In general, M2M communication is characterized by devices that transmit small data blocks. This changes the actual tendency for the following reasons:

a. Existing coding algorithms normally rely on long code words; therefore, existing codes are not applicable to very small data blocks.
b. Short data blocks are inefficient with channel estimation and control overheads. In current systems, the control and channel estimation overhead represents only a chunk of the payload data. But with short data blocks, this will not be the case. Hence, an optimized design needs to be implemented to achieve a balance between the data and control planes.

14.3.2 MAC LAYER PROTOCOLS AND STANDARDS

The IoT era has already started and it is growing rapidly. With 12.5 billion devices in 2010 and doubling every 5 years, it is estimated that there will be approximately 50 billion devices connected to the Internet by 2020 [25], hence a worldwide average of 6.5 devices per person. With such a large device density, accessibility will be, and already is, an important research topic. This section briefly discusses recent work related to the media access control (MAC) layer, which is a sublayer of the data link layer. Because the logical link control sublayer plays a much smaller role than the MAC layer in the topic of IoT, it will not be discussed.

14.3.2.1 Millimeter-Wave-Oriented Protocols:

There are many interesting publications related to MAC layer protocols published in recent years. The first two works that will be discussed are oriented to 60 GHz wireless personal area networks (WPAN). In Estevez et al. [26], a FD–time division multiple access hybrid protocol is proposed in which the AP uses a frequency–time 2D grid that organizes the service order. This mechanism provides a collision-free environment in which the sensors can achieve energy independence. To provide services to a larger number of users, this work suggests to create a FD-TDMA grid, similar to orthogonal frequency division multiple access (OFDMA), as shown in Figure 14.3. The focus of this work is in the self-sustainability feature of this MAC layer protocol. The key to obtain this

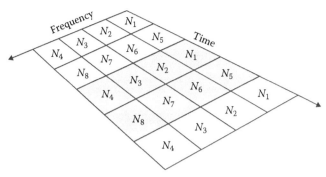

FIGURE 14.3 FD–time division multiple access time–frequency grid proposed in Estevez et al. [26] for 60 GHz WPANs.

energy independence is to have an energy harvesting device and a smart scheduler. The scheduler must measure the energy consumed by the frame transmission to compute the sleep time necessary to balance the energy consumed with the energy harvested. Since the power consumed during transmission is assumed to be greater than the rate of charge, the scheduler will need to decrease its overall transmission rate significantly to achieve energy self-sustainability. For example, using FD-TDMA, the results show that the transmission rate will decrease by a factor of 100 with respect to the link rate. This is a small price to pay for energy independence and very useful in CPS.

In Estevez and Kailas [27], a process-stacking multiplexing access algorithm is designed for single-channel operation. The concept is simple and intuitive, but its implementation is not trivial. The key to stacking single-channel events is to operate while simultaneously obtaining and handling a posteriori time frame information of scheduled events. This information is used to shift a global time pointer that the wireless AP manages and uses to synchronize all serviced nodes. This scheme does not work under a Time Division Multiple Access-Frequency Division Multiple Access (TDMA-FDMA) hybrid technique as its predecessor technique [26], so it can only be compared with TDMA. In this scheme, the slot duration is variable and the BS is given the capability to interlace data with processes. This allows for the same algorithm that is scheduling the frames to schedule idle time to mimic the same energy self-sustainability capability of its predeceasing work. This eliminates the need for an external process to interrupt frame transmission. The packet encapsulator has a crucial role in the Process-stacking Multiple Access (PSMA) scheme. It gathers the network layer packets (typically IP) coming from the queue and encapsulates these into a frame. Once the frame is built, its size is translated into a synchronization time using the transmission bit rate. The time, which is used to synchronize this frame, is inserted in a previously stored frame. The previous frame is sent to the main processor for transmission, while the recently encapsulated frame is temporarily stored. Since the recently built frame triggers the transmission of the previously stored frame, this process is referred to as a push (i.e., the arriving frame pushes the stored one). This continues until the queue is empty. To better illustrated this process, see Figure 14.4.

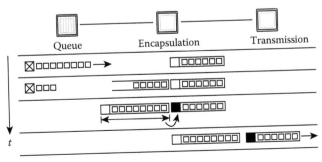

FIGURE 14.4 Node process used by process-stacking multiplexing access proposed in [27].

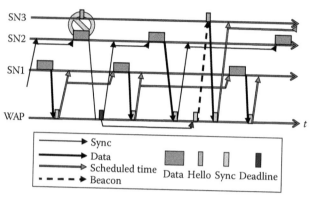

FIGURE 14.5 Scheduling diagram that portrays the behavior of process-stacking multiplexing access proposed in [27].

Inserting nodes is an interesting subject in a scheme that does not have any idle time. The task of inserting the first Sensor Node (SN) is trivial, since there are no active transmissions. If various nodes are sharing the channel and an inactive node turns active, such that it must enter the sharing cycle, it will send a hello frame. The processes are stacked such that there is insufficient time to transmit in between frames; this will cause collisions. When a collision occurs, the Wireless Access Point (WAP) sends a new schedule time to the expected SN with the maximum allocation time (assuming the sync field could not be recovered). Since this is the last event, a beacon signal is scheduled after this, allowing enough time for a hello frame to be retransmitted. Because there is a scheduled SN, the unlinked SN has to wait until the beacon signal establishes a link. Once the WAP has serviced and rescheduled the linked SNs, it sends the beacon signal. The unlinked SN detects that this is a broadcast message (not a hello frame response) and retransmits the hello frame and holds on to the data until the next cycle. The WAP links the SN, and from there on, the frame exchange returns to its routine, but with an additional SN. This node insertion process is portrayed in Figure 14.5.

14.3.2.2 Internet of Things–Oriented Protocols:

Two recent MAC layer protocols proposed for IoT in dense large-scale networks are the Robust Asynchronous Resource Estimation (RARE) protocol [28] and the X-MAC protocol [29]. RARE is a MAC layer protocol that manages densely deployed single-hop clusters in a self-organized fashion. This protocol is a useful contribution to the IoT, as transceivers can be costly. Replacing them with transmitters, and the capability to manage these, saves cost on the overall system. How the scheduling works is that the nodes that are equipped with only transmitters will also transmit a seed used to generate a pseudorandom number generator (PRNG). With this seed, the BS is able to accurately compute the transmission time of these nodes, so it can easily schedule transceiver-equipped nodes during the idle times. This scheme is collision free between transmitter- and transceiver-equipped nodes, but it is not immune to collisions between transmitter nodes. It uses a PRNG to schedule the transmitter nodes, opposed to using periodic scheduling, because if the transmission interval was periodic and happens to be synchronized with another node, all frames would collide.

X-MAC is a low-power listening mechanism in which nodes decide their schedule independently from their neighbors. X-MAC implementations help the serviced nodes to reduce their energy consumption and latency. It does this by embedding the address information of the target in the preamble so that nontarget receivers can quickly go to sleep, which addresses the overhearing problem. The second part of this scheme is to use a strobed preamble to allow the target receiver to interrupt the long preamble as soon as it wakes up and determines that it is the target receiver. This short strobed preamble approach reduces the time and energy wasted waiting for the entire preamble to complete. In Beaudaux et al. [30], a large-scale test bed is assembled that showed the scalability capabilities of X-MAC. It performed satisfactory in a 3D grid composed of 240 nodes. In this work, the authors point out the weaknesses and suggest improvements.

There is also work that focuses on smaller-scale networks [31,32] and rather attempts to improve the existing protocols. In Tang et al. [31], two opportunistic random access mechanisms are proposed: overlapped contention and segmented contention, to favor the node of the best channel condition. These algorithms are based on Carrier Sense Multiple Access with Collision Avoidance (CSMA/CA) and polished to provide temporal fairness and avoid starving the nodes with poor channel conditions. In Liu et al. [32], a hybrid between CSMA and TDMA is proposed. It consists of a contention period and a transmission period and is designed for heterogeneous M2M networks. In this protocol, different devices with preset priorities first contend the transmission opportunities following the convention-based-persistent CSMA mechanism. Only the successful devices will be assigned a time slot for transmission following the reservation-based TDMA mechanism. If the devices failed in contention at previous frame, to ensure the fairness among all devices, their contending priorities will be raised by increasing their contending probabilities at the next frame. There is also MAC layer work oriented to industry applications, such as [33]; this work studies the effect of varying superframe size of different MAC layer industry standards (WirelessHART, ISA100.11a, and WIA-PA) and the latency effects these cause. The motivation is the access control of machinery. The last work discussed in this section is oriented to video streaming. In Xu et al. [34], the transmission delay deadlines of each sender's video packets are translated into a monotonically decreasing weight distribution within the considered time horizon. Higher weights are assigned to the slots that have higher probability for deadline-abiding delivery. Given the sets of weights of the sender's video streams, a low-complexity delay aware resource allocation approach to compute the optimal slot allocation policy that maximizes the deadline-abiding delivery of all the senders is proposed. As it can be observed, the work in the MAC layer is quite diverse, covering monitoring, management, control, and informing applications, in a wide range of network architectures, such as WPAN, wireless local area network (WLAN), and intravehicular, wireless sensor (WSN), M2M, industrial, and access networks.

14.3.2.3   Standards:

There are numerous standards dedicated just to the MAC layer that are related to the IoT. To start, some IEEE standards will be briefly described: IEEE 802.1Q-2011 discusses MAC bridges and virtual bridged LANs. IEEE 802.11-2012 mentions about enhancements to the existing MAC layer functions in WLANs. IEEE 802.11ad-2012 defines modifications to IEEE 802.11 to enable operation in the 60 GHz range. IEEE 802.15.1-2005 gives specifications for MAC and physical (PHY) layers in WPANs. IEEE 802.15.3-2003 gives specifications for high-rate WPANs. IEEE 802.15.4e-2012 gives specifications for low-rate WPANs (LR-WPANs). IEEE 802.15.4j-2013 defines modifications to support medical body area network services in the 2.4 GHz range for LR-WPANs. IEEE 802.15.6-2012 gives specifications for wireless body area networks. Some standards that are still under development are as follows: IEEE P802.15.4k/g for low-energy critical infrastructure monitoring networks, IEEE P802.11ai for fast initial link setup and IEEE P802.11ah for sub 1 GHz license-exempt operation, IEEE P1609 for wireless access in vehicular networks (WAVE), IEEE P1856 for prognostics and health management of electronic systems, and IEEE P1901.2 for low-frequency (less than 500 kHz) narrowband power line communication for smart grid applications. There are just too many to summarize all, but these are some key standards. Some non-IEEE standards worth mentioning are IETF RFC 6282/4944 (6LoWPAN) for IPv6 datagrams over IEEE 802.15.4-based networks, IETF RFC 2464 for IPv6 packet transmission over Ethernet, IETF RFC 5072 that specifies IPv6 over point-to-point protocol, and IETF RFC 5121 that discusses the transmission of IPv6 via the IPv6 convergence sublayer over IEEE 802.16 networks.

14.3.3   Network Layer Technologies

The key for a CPS to become part of the IoT corresponds to its ability to connect to the global network. As in any Internet-connected system, the network layer is expected to provide two main services: host identification and routing/forwarding of packets. Nevertheless, a number of additional protocols are required to complement the main services; for example, host identification also requires a mechanism for address configuration/detection in a local network or a way to provide seamless connectivity under mobile host circumstances. With the appearance of Internet-connected CPS, a number of challenging conditions are now present: resource-constrained devices, low-rate connectivity, and unstable links, among others. Networks that involve some or all of the challenging conditions are referred to as low-power and lossy networks. These networks have a different communication model compared to traditional networks, which triggered new research work and the redefinition of what otherwise would be well-studied and standardized network layer protocols. In this section, we discuss such new standardization and research works related to making low-power and lossy networks into IP-compliance networks.

14.3.3.1 Host Identification and IPv6-over-foo:

In the IoT, all the sensing and embedded computing devices are expected to be equipped with the IP protocol for connection to the outside world. Accordingly, IPv6 is the proper protocol to achieve network layer identification for a large number of devices, which are expected to be nearly three devices per capita in 2018 and, as mentioned earlier, around a total of 50 billion Internet-connected devices by 2020 [25]. The number of devices immediately rules out the possibility of employing IPv4. Although IPv6 is a mature technology that has been standardized since 1998 [35], it is well known that deployment of native IPv6 networks and applications has been delayed until recent years. Furthermore, the idea of IP-based CPS was not present at the time when low-cost proprietary sensor networks allowed for a rapid proliferation of non-IP solutions. The most representative solution, ZigBee, employs the IEEE 802.15.4 standard for PHY and MAC layers, but for a long time, it had a proprietary ZigBee specification for upper layers of the protocol stack. Just recently, ZigBee has standardized an IP-compliant protocol stack to support IP connectivity in home automation, smart energy, and other application areas [36]. Other popular non-IP examples of WSNs are Z-WAVE, INSTEON, and WAVENIS [37].

Ultimately, the academic and industrial communities converged to the idea that the IoT should be based in standardized communication approaches. In addition to interoperability, high reliability and low-power consumption are also among the core requirements expected from the IoT communication stack [38]. Naturally, IP is conceived as the universal language that will allow devices, machines, and sensors in CPS to talk to each other, regardless of the access technology each one is employing. Figure 14.6a illustrates a realization of the initial proposal for a standardized communication stack. Two new aspects are observed in the figure: an intermediate layer named 6LoWPAN and a specific routing protocol designed with the intricacies of low-power and lossy networks in mind.

To achieve full IP compliance, two main aspects to be solved for IoT's resource-constrained devices are the transmission of IPv6 packets over limited–maximum transfer unit (MTU) access technologies and the support of multihop transmission. This is tackled by means of an adaptation layer, such as the one depicted in Figure 14.6a between MAC and IP layers. In the particular case of 802.15.4-based networks, the MTU is limited to 127 bytes, whereas an IPv6 packet holds 1280 bytes or more. The problem has been partially solved by means of a standard defined by the IETF, the so-called 6LoWPAN [39,40], which defines an intermediate layer that allows for the adaptation of IP packets before transmission over 802.15.4-based networks. The standard provides mechanisms for fragmenting IP packets, so as to make them fit in the available MTU; it also defines compression of headers and reassembly. Nevertheless, fragmentation has an impact on the performance of constrained network; hence, a body of research has been devoted to study fragmentation [41,42], as well as other issues such as the general performance of 6LoWPAN networks [43] and practical demonstrations of 6LoWPAN implementations [44–47].

However, 6LoWPAN is not the only protocol proposed at the adaptation layer. Upon the definition of the IEEE802.15.4e amendment for medium access control following time division access, namely, the Timeslotted Channel Hopping (TSCH) mode (instead of the CSMA/CA employed in

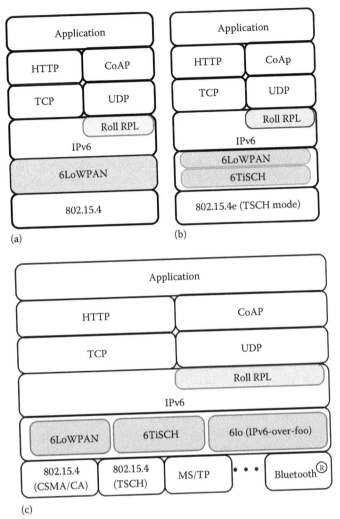

FIGURE 14.6 Evolution of standardized communication stacks for the Internet of Things. (a) Initial standardized communication stack; (b) 6TiSCH communication stack; and (c) general IPv6-enabled communication stack.

the original standard), it became necessary to define 6TiSCH. The 6TiSCH IETF working group specifies the mechanisms for enabling IPv6 over the TSCH mode of the IEEE802.15.4e standard [48]. There is ongoing work that defines the complete architecture for 6TiSCH integrated to 6LoW-PAN [49,50], as well as 6TiSCH demonstrations of operative implementations of this network stack [51]. We illustrate the 6TiSCH communication stack in Figure 14.6b. In parallel, a more general framework for facilitating IPv6 connectivity over constrained node networks is also under discussion within the Internet's community experts [52]. This working group tackles the specifications for using 6LoWPAN in link layer technologies of interest in constrained node networks. In discussion are the integration of IPv6 over MS/TP networks, Bluetooth® low energy, and DECT ultra low energy. The evolved communication stack for a general IP-enabled IoT is illustrated in Figure 14.6c.

14.3.3.2 Address Configuration and Neighbor Discovery(ND):

The low-power and lossy networks connected to the IoT are, as the name indicates, networks with unpredictable link conditions: connections go down due to nodes going to sleep mode to save energy consumption, links may be asymmetric and have time-varying connectivities and qualities, availability varies when nodes operate in mobile environments, and so forth. Therefore, the IP-addressing model that governs traditional networks is hardly applicable in such unpredictable conditions. As a result, addressing autoconfiguration schemes with heavy signaling and multicast operation showed not to be practical when employed for configuring devices in a CPS. Differences of the addressing model in traditional local area networks and ad hoc networks (which are also applicable to low-power and lossy networks) have been identified in Baccelli and Townsley [53].

In terms of protocols for addressing configuration, two general categories are defined: stateful and stateless. In the former, an allocation table is employed for addressing assignment; in the latter, nodes autogenerate their own addresses. Distributed approaches have been proposed for maintaining the allocation table in low-power and lossy networks, since traditional centralized approaches (e.g., DHCP) have scalability and reliability issues [54]. On the standardization side, the ND protocol, which defines addressing mechanisms for IPv6 networks, has been redefined based on the new addressing model identified in Baccelli and Townsley [53]. A lighter version of the protocol eliminates the need for multicast signaling, which is not supported in sensor networks where nodes are in sleep mode for a good part of the time. It also defines the distribution of network prefixes and 6LoW-PAN compressing context information through multihop paths [55]. Furthermore, a secure version of the ND protocol for constrained networks is currently under discussions at the IETF [56].

Together with addressing configuration, ND is also fundamental for setting up communications. Connectivity maintenance and contextual awareness are achieved through the ND process. A perfect balance between low-frequency signaling for discovery (i.e., the duty cycle) and low time to discover newcomers in the network (i.e., the discovery latency) would be the best-case scenario. However, these two aspects have a trade-off between each other [57]. To improve the balance, a number of ND protocols have been proposed to minimize the duty cycle based on probabilities, so even when nodes go to sleep, there is still a high chance to discover newcomers with a low duty cycle [58]. Other ND protocols employ a deterministic approach, so that nodes are awake exactly half of the time. To ensure a reduced energy consumption, the active slots are distributed across multiple cycles [59]. The standard ND protocol for constrained networks has defined the format of messages and the mandatory behavior, such as route-over and mesh-under support and the prohibition to employ multicast messages [55]. Nevertheless, optimizations to find the aforementioned balance are left to the implementers.

14.3.3.3 IP Mobility:

Many IoT environments require mobility from the CPS, for example, moving nodes that are part of Intelligent Transportation Systems or mobile health scenarios. Furthermore, the mobility pattern of nodes is different depending on the context in which the CPS has been deployed. In a transportation scenario such as the one depicted in Figure 14.7, sensor and control nodes attached to vehicles may move at very high speeds and at the same time connect to the Internet through multihop communications. In a mobile health scenario, nodes usually move at low speeds but may also experience vehicular speeds when there are communications from on-the-move ambulances; in that scenario, seamless Internet connectivity of body area networks is important to guarantee a continuous reception/transmission of information while the patient receives on-the-move medical treatment or when he or she moves around health care facilities.

Mobility management is then one of the most important issues in 6LoWPAN research. Same as with other standard protocols for IP support, network layer mobility protocols need to be adapted to fit in the multihop sensor/control network with a reduced signaling overhead. Protocols such as network mobility support, proxy mobile IPv6 (PMIP), and distributed network mobility are studied to better adapt to the characteristics of the constrained sensor network. Typically, adaptations to the standard protocols are designed according to the mobility patterns of nodes. In Céspedes, Shen, and Lazo [60], the aforementioned network mobility protocols are analyzed in the context of smart transportation systems. An adapted version of the standard PMIP is proposed in Cespedes, Lu, and

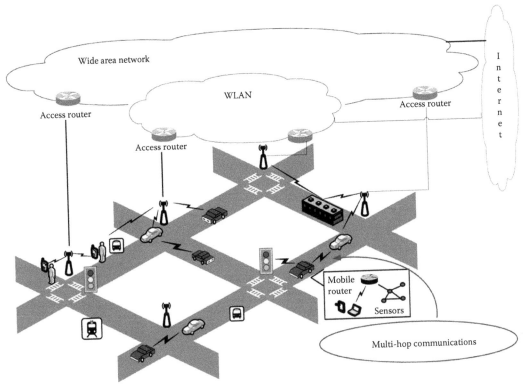

FIGURE 14.7 Mobile cyber-physical systems in a transportation scenario.

Shen [61] to enable network-based IP mobility in a vehicular network with single-hop and multihop connectivity. Along the same lines, the requirements of constrained wireless networks for mobile health are described in Hahm, Pfeiffer, and Schiller [51]. To cover part of these requirements, Kim et al. propose a framework to support network-based mobility for a 6LoWPAN network in a health care system in Kim, Haw, and Hong [62] and for general low-power and lossy networks in Kim et al. and Bag et al. [44,63].

14.3.3.4 Routing and Forwarding:

CPS typically perform data acquisition, that is, the spatially distributed sensor networks transmit data resulting from monitoring processes. The data are usually transmitted toward a data concentrator across a multihop network formed by nodes that are sensors themselves. Hence, the proper routing and forwarding of information are key aspects to guarantee the collection of information. Furthermore, the CPS may be not only monitors but also controllers. Hence, data traffic direction will be multipoint to point (i.e., from sensors to the data concentrator), but also point to multipoint (i.e., from a central controller to control nodes).

A routing protocol (RPL) for low-power and lossy networks is standardized in RFC 6550 [64]. It is a distance-vector routing protocol for which urban, industrial, home, and building automation low-power and lossy network requirements were considered at the time of designing the protocol. Some of these requirements include (1) scalability, (2) dynamicity (e.g., self-organizing and self-healing features), (3) latency, and (4) parameter-constrained routing (e.g., identification of nodes' capabilities and adaptation of forwarding mechanisms according to those capabilities). In the protocol, a set of metrics are available depending on the nature of traffic that is being routed. Different metrics can be combined according to an objective function, which contributes to guarantee quality of service and to handle real-time traffic. The network status is created and maintained based on one or more directed acyclic graphs (DAGs). A DAG is a directed graph in which all edges are oriented in such a way that no cycles exist. A root node is selected for each DAG created in RPL, and the nodes that belong to a given DAG have a rank associated with them. By employing the rank information, RPL builds a destination-oriented DAG with several routes from each node, which is intended to enhance the performance and robustness of the network [65]. Performance evaluations of RPL demonstrate there is still room for improvement in terms of support to bidirectional traffic and loop avoidance [66,67].

Another RPL is the hybrid routing protocol [68]. It also employs DAGs to provide multiple reliable paths to a border router. Each node builds a default route table by adding the neighboring nodes toward a border router. Entries in the default table have information about the link layer packet success rates, so evaluations of the quality of the links are available to rank the best entries in the table. This feature is especially important to fulfil with reliability, as multiple routes are provided to a given destination. Nodes send topology reports according to the top-ranked entries in the default table; such reports are piggybacked with periodic messages for collection traffic, and they allow border routers to build and maintain a global view of the topology. HYDRO is considered both a centralized and a distributed routing mechanism: on the one hand, the low-power nodes maintain a distributed DAG that provides the set of default routes for communicating with border routers; on the other hand, the border routers maintain a global view of the network topology through the topology reports sent by each node.

As for the forwarding plane, there is also experimental standardization work that intends to improve the reliability of the forwarding process in the highly unstable environment of lossy networks. In Herberg et al. [69], a depth-first forwarding (DFF) protocol is proposed for increasing the reliability of data delivery in networks with dynamic topologies and/or lossy links. The base algorithm for the forwarding of data is the depth-first search over a network graph. While data are being forwarded, the information about link failures and loops is employed to redirect packets through successive neighbors. DFF may select new forwarders for a packet, even if the routing table has not yet been updated due to rapid changes in network topology or links' statuses. In a later work, DFF protocol has been tested when combined with reactive on-demand routing protocols also employed in lossy networks [70].

14.4   INTERNET OF INTERLACED CYBER-PHYSICAL THINGS

In this section, the advances achieved in various disciplines are explained. In each discipline, we choose a few published research proposals and focus on the IoT applications and how it interacts with the cloud or other cyber-physical systems. The intention is to demonstrate how the IoICPT forms a symbiotic environment.

14.4.1   e-Health

e-Health is the research field related to health in which electronic devices, mainly communication based, play the main role. In the field of e-health, most applications fall within the following topics: monitor and control of devices that collect data from living beings, remote control of instruments related to health care, and analysis of real-time or off-line health data.

The work of [77] discusses the use of satellite-based networks for telemedicine purposes. These networks enable tools for health care services. Among these tools, we can find various technologies such as high-resolution cameras with stereoscopic visualization, interactive real-time video communication with remote control of medical devices, and virtual reality simulations with tracked visualization and haptic feedback, which are all digital media based and allow the interaction with the Internet. This work mentions various works of remote-controlled systems oriented to health care. One example is to perform intraoperative telepathology, such that a tissue has been taken and a slice has been prepared for biopsy; this sample is put under a remote-controlled, camera-equipped microscope, and the diagnosis is then formulated by a remote expert.

In the work of [78], a new framework called the virtual remote nursing (VRN) is proposed that provides a virtual-nurse agent installed on the client's personal computer or smartphone to help manage the client's health condition continuously. With this approach, medical practitioners can assign different tasks to the virtual-nurse using a generic task definition mechanism, where a task is defined as a combination of medical workflow, operational guidelines, and associated data. VRN is guided by the practitioners who decide on the patient's treatment. This allows a VRN to act as a personalized full-time nurse for its patient, by performing practitioner-support tasks, using the data collected on the patient's health.

A compendium of *monitoring technology literature* proposed by health-oriented projects is provided in Gatzoulis and Iakovidis [74]. Nine personal health management system examples are provided, each addressing a specific issue. The AMON, MOBIHEALTH, WEALTHY, MERMOTH, and MYHEART projects all have wireless connection to a medical center. The AMON project designed a wrist-worn device that monitors high-risk cardiac/respiratory patients. The MOBIHEALTH project is working on a body area network that monitors chronically ill patients. The WEALTHY and MERMOTH projects study biomedical clothes with textile electrodes and electronic sensors oriented to prevent and monitor cardiovascular diseases. The VTAMN project uses biomedical clothes with integrated sensors and electronics to monitor diseases. The Mamagoose Pyjama, Verhaert project uses a biomedical pajama for infants to study the detection of sudden death syndrome. LifeShirt, Vivometrics Inc. is an organization dedicated to the study of biomedical shirt with integrated sensors for sleep diagnostics and disease monitoring. The SmartShirt, Sensatex Inc. organization works on a textile platform with embedded electronic sensors and a conductive fiber grid to transmit data used for health monitoring. Finally, the GlucoWatch G2 Biographer from Cygnus Inc. is a wearable device to monitor noninvasive glucose for diabetics.

Another example is the work in Niyato, Hossain, and Camorlinga [76] that presents a general system model and optimization formulation that can be used for many e-health services. One example is the monitoring of ECG signals for patients with cardiac diseases. The ECG signal is sampled by the ECG device. The data are then packetized and stored in a queue as *normal* data waiting for transmission over wireless connections. However, if an abnormal event is detected (e.g., low heart rate), a special message with the detailed information of the event (e.g., location, time, and condition of the patient) will be generated and stored in a separate queue labeled as *critical* data. In this case, priority will be given to these data, so that the patient can be treated promptly.

14.4.2 Energy

Energy, in the context of IoT and CPS, focuses on smart grid and microgrid topics, though it is not limited to these topics. Energy, in general, is an important field as it addresses the systems that power all the electrical devices, directly or indirectly. Smart grid and microgrid deal more specifically with the optimization of resources and by reducing the impact that users have on the environment.

In Hamzeh, Karimi, and Mokhtari [89], a new control strategy for the islanded operation of a multibus medium voltage (MV) microgrid is discussed. The problem it addresses is related to the distributed generation (DG) units. The microgrid is composed of dispatchable electronically coupled DG units, and each of these units supplies a local load that can be unbalanced due to the insertion of single-phase loads. In the proposed scheme, each DG unit is controlled using several components, which include a proportional resonance controller with an adjustable resonance frequency that is used to regulate the load voltage, a droop control system used to share the average power components among all DG units, and a negative-sequence impedance controller that is used to compensate the negative-sequence currents of the unbalanced loads and minimize the negative-sequence currents in the MV lines improving the power quality of the microgrid. Results show that the proposed scheme is able to regulate the voltage and frequency of the microgrid.

Building energy saving, communication, and automated control are interesting topics related to CPS and IoT. In the work of Wei and Li [87], the information technology infrastructure of in-building energy monitoring is addressed. The authors propose a system framework that monitors the energy consumption of the building and analyzes the data to provide control over an architecture that resembles that of the IoT. Using this scheme, the authors achieve a real-time monitoring and control system that improves the energy savings of the smart building.

Energy management is an important topic in the context of CPS and IoT. In Wandhare, Thale, and Agarwal [88], a multilevel reconfigurable hierarchical control system is implemented on a laboratory prototype microgrid, which is composed of photovoltaic cells, wind turbine, micro hydro generator, and a fuel cell. Notice that these are all renewable energy sources. A transient and momentary power backup is provided by means of an ultracapacitor. Because these renewable energy sources produce raw power under varying weather conditions, disruptive effect such as voltage fluctuations, voltage dips, frequency variation, and harmonic distortion can arise on the grid if these are not properly managed. The hierarchical control scheme proposed in this work is an attempt to create a hybrid control system between the centralized and decentralized architecture. The hierarchies are divided as follows: The highest level is the advisory control level, which does not really control the individual microgrid components but acts as an advisor, hence the name. The second level is a global microgrid controller, which is in charge of optimization, synchronization, islanding, stability, protection, and so on. The third level is a resource manager, which is responsible for determining individual source contribution (generation of the reference operating points for the sources) based on the reference values acquired from the global microgrid control layer and the feedback received from the local control layer. The local controller makes the final decisions and directly acts on the power converters associated with sources and energy storage devices. It attempts to optimize the energy efficiency of the local system using centralized information. Results show that this scheme performs better than both centralized and decentralized systems by providing high flexibility and high fault tolerance margin to the microgrid.

14.4.3 Environment

Among the different applications considered for the IoT, the environmental/earth monitoring is receiving a growing interest as environmental technology is becoming an essential aspect in our daily life. Further, environmental monitoring techniques are becoming a crucial component in sustainable growth worldwide. One of the key elements of an environmental monitoring system is a WSN platform. WSN environmental monitoring includes indoor and outdoor applications. Regarding outdoor applications, there are several deployment categories, such as city environments and open nature environments. For the case of city environments, environmental monitoring can be done for pollution monitoring, traffic, and lighting, among others, whereas for open nature environments, environmental monitoring can be performed for flooding detection, volcano and habitat monitoring, weather forecasting, earthquakes, chemical hazard, and crop and agriculture optimization, among others.

The open nature environmental monitoring is extremely challenging: technologically and logistically because it normally is done under severely harsh conditions [79]. Physical access to the operation field for maintenance and deployment is one of the toughest and most difficult tasks to perform. But a typical IoT application requires for low cost, low maintenance, and high service availability; therefore, there are many challenges in designing an IoT-based WSN environmental monitoring system.

A cost-effective WSN operating with many sensor nodes normally functions with energy restrictions. Therefore, limited energy can critically limit the services offered by the network and is a topic to be addressed considering the requirements of the IoT environmental monitoring system [79]. Normally, a WSN data acquisition system for IoT environmental applications is challenging. The nodes can be exposed to variable and acute climatic conditions in a deployment field difficult to reach.

In all environmental monitoring IoT-based techniques, each sensor node performs periodic measurements of the environmental variable to be monitored, such as temperature, humidity, and pollution levels, among others. Then, the sensor nodes process the measured variable on board to reduce energy consumption and send the data (field communication) to the gateway repeaters.

The gateway repeaters will transmit the data (out-of-field communication) to the application servers, whereas the application servers will generate back-end alerts and have the data available via the IoT.

Several researchers have proposed different applications and concepts regarding environmental monitoring and its challenges using WSNs. One of these contributions is the one proposed by Delphine et al. [80]. In this manuscript, the authors discuss the existing challenges in implementing WSNs applied to the IoT. The authors indicate that WSNs are finding a wide range of applications in various domains, including health care, industrial and production monitoring, control networks, and assisted and enhanced living scenarios, among other fields. And in the near future, WSNs are expected to be integrated into the "Internet of Things," where sensor nodes will join the Internet dynamically to collaborate and accomplish certain tasks, such as environmental monitoring. However, certain issues must be analyzed and investigated to integrate WSNs into the IoT. Delphine et al. [80] indicate that connecting WSNs to the Internet is possible via three approaches: security, network configuration, and quality of service. Further, it can be seen that designing an environmental monitoring system needs to consider aspects related to mobility and energy harvesting. In References [7,79–81,87,91], other topics regarding environmental monitoring can be found. Besides environmental monitoring, WSNs can be implemented for e-health systems [75], as well as green communication systems [7], among others.

14.4.4 MOBILE NETWORKS

Fourth-generation (4G) wireless communication systems have been deployed or are soon to be deployed in many countries around the world. But with the explosion of wireless and mobile devices and services, there are many challenges that cannot be solved with 4G systems, such as high energy consumption and limited frequency spectrum [5]. Designers of wireless communication system designers have been facing the harsh demand for higher error-free data rates and the mobility required by new wireless applications; therefore, research on the fifth-generation wireless systems is an ongoing topic.

Wireless communication networks are perhaps the most crucial component of an information and communication technology system [5]. Further, wireless communication networks are one of the fastest growing and most dynamic sectors in the world. Additionally, the development of wireless technologies has improved people's ability to stay connected anytime and anywhere. The exceptional success of wireless mobile communication systems is because of a rapid pace of technological innovation. Since second-generation (2G) systems came out in 1991, to the first third-generation (3G) system launched in 2001, the wireless mobile network has evolved from a traditional telephony system to a network that can transmit rich multimedia contents. The demand of rich multimedia contents has created the necessity to develop new technologies capable of transmitting higher error-free data rates in a band-limited channel. Therefore, 4G systems were designed to meet the requirements of the International Mobile Telecommunications-Advanced standard, which is characterized by using the IP in all types of services [5,92]. By implementing IP in all services, a dramatic increase in devices connected to mobile broadband systems is taking place. Subscribers desire faster Internet access to their mobile network provider all of the time. This has resulted in an explosion of wireless mobile services and devices. The Wireless World Research Forum has estimated that 7 trillion wireless devices will serve 7 billion people by 2017 [93]. Hence, the number of network-connected wireless devices will reach 1000 times the population of the world. It can be seen that as more and more devices are connected to the Internet, current technologies will not be sufficient to manage highly dense networks. For the devices to be connected to the IoT, mobile providers will have to migrate from IPv4 to the IPv6 protocol to permit more devices to be connected to the mobile network. Therefore, many research challenges need to be addressed, and this is why research and development of 5G networks is a hot topic at this moment [4–7,9–11,21].

As mentioned previously, several research challenges need to be addressed to allow more network-connected wireless devices interact via the Internet. One of the most crucial challenges for achieving the system capacity expected in 5G systems is physical scarcity of the RF spectrum allocated for mobile communications at this moment. In Roh et al. [8], the authors introduce a mmWave beam-forming technique as an enabling technology for 5G cellular systems. During the past few years, academic and industrial research has been carried out to develop and optimize 4G systems to provide higher data rates to end users by deploying more BS and improving spectral efficiency. Despite the efforts that have been done, none of them are seen as a viable solution to support the traffic demands expected in 2020 and beyond. It is expected that 5G systems will offer a minimum data rate of 1 Gb/s anytime and anywhere to all end users [8]. To achieve these high data rates, Roh et al. [8] proposed to use large frequency bands of underutilized spectrum. Most of the underutilized spectrum is concentrated in the mmWave bands. Traditionally, these frequency bands have only been used for outdoor point-to-point backhaul links. This is because mmWave bands are characterized by high propagation loss and shortage of cost-effective components. Further, to use the underutilized mmWave bands for outdoor applications, two things need to be addressed. The cell needs to cover a large geographical area and support mobility even in NLOS environments.

One of the key discoveries done Roh et al. [8] affirms that the crucial parameters characterizing the propagation properties of the mmWave bands, such as the path loss exponent, are comparable to those of typical cellular bands, which normally operate from several hundred megahertz to several gigahertz [5], when transmit and receive antennas are used to produce beamforming gains. The beamforming gains were obtained by implementing an array antenna operating at 30 GHz. The initial test results provided in Wang et al. [5] give valuable insights and understanding of the mmWave bands for future use in mobile networks and urge academic and industrial research to focus in this sizzling topic.

Besides the work that has been done by the researchers from Samsung Electronics [5], research and development regarding 5G mobile networks is a research topic that is being explored by a vast number of researchers around the world [4,7,9,79–81,87,91]. Topics such as massive MIMO for the next-generation mobile networks are being studied [9], as well as the technologies [4] and cellular architectures that are to be implemented [5].

14.4.5 TRANSPORTATION

In transportation scenarios, the cyber-physical systems are present in the form of onboard sensors, cameras, and radar applications, among others. With safety and navigation in mind, the initial proposals for providing communications in a transportation scenario were oriented to enable vehicle-to-vehicle (V2V) and vehicle-to-roadside (V2R) communications. In this manner, crash avoidance, safety warnings, navigation control, and platoon coordination became feasible applications that are making the deployment of Intelligent Transportation Systems a reality [94]. However, information coming from the vehicular network may be of interest not only to nearby vehicles but also to a remote node located in the Internet. For example, a vehicle can be monitored from a remote mechanical workshop or directly by the auto maker, to track the performance of the different vehicle's components. Hence, transportation systems are also an area of application in the IoT. Although vehicular networks look similar to WSNs, the former employs different wireless communications systems, which changes the communications paradigms from those in sensor networks. Besides, special security and privacy approaches also need to be specified and implemented [95].

Currently, governmental entities such as the U.S. Department of Transportation are evaluating the readiness for the application of V2V communications in real scenarios [94]. When this type of communication becomes available, a vast amount of information will be ready to feed remote systems located in the Internet, such as congestion control systems, fleet management and tracking, content-distribution systems for onboard entertainment, urban and environmental monitoring systems, and many more. In fact, many of these functionalities are already being provided by auto makers with connected car products, where a connection may be provided by a data plan from a cellular phone that is integrated with the vehicle's computer or by an Internet connection provided by a mobile router integrated with the car.

An interesting analysis of the future of vehicular networking and the mobile Internet is provided in Gerla and Kleinrock [96]. Networks formed by actors involved in a transportation system, namely, vehicles, pedestrians, and public transportation, form a large urban sensing network with different objectives (e.g., congestion, pollution, and forensics) [97–99]; the transferring of information may not only involve direct connections from sensors to the Internet, but it may also involve Peer-to-peer (P2P) exchanges with other vehicles [96]. Although the advancements in intelligent transportation are evident and many working prototypes are already running, the following areas have been identified as the forthcoming research challenges in vehicular networking: (1) scientific foundations of inter-vehicle communications, (2) field operational tests, (3) vehicular networking applications, and (4) heterogeneous vehicular networks [95].

14.4.6 SMART HOME/BUILDING

With the aim of improving the quality of life, the concept of smart homes/buildings made its appearance more than a decade ago. In a formal definition, smart homes/buildings involve the integration of technology and services together with networking devices and communications. Most of the smart home applications are oriented to controlling different aspects of our daily life, whether it be the energy consumption of home appliances or the automation of daily activities such as setting clock alarms or preparing a cup of coffee at a certain time. Although the initial smart home networks were formed by isolated sensor and control networks [100], nowadays IoT-based smart home systems are considered the solution to overcome previous problems like poor portability, weak updating capabilities, and personal computer dependency [101].

Along the line of exploiting Internet connections to benefit the effectiveness of smart homes, architectures based on cloud computing and web services are proposed in Soliman et al. and Jian et al. [102,103]. Jian et al. [103], nodes in the smart home exchange information with the cloud server by means of P2P communications, with the objective of providing stable smart home services. Furthermore, Park et al. [104] propose to manage control and sensor nodes in smart homes through a global network by means of smart phones. In general, the IoT-based smart home systems provide a centralized control through the Internet. In this way, information, telecommunications, entertainment, and local control and sensor devices are combined to create a personalized home system platform [101].

As a future direction for this area's research, smart homes are proposed to be connected to each other in what has been identified as another paradigmatic class of CPS: the smart community [105]. The purpose of the smart community will be to cooperate and distribute monitoring information about the community environment. The homeowners agree to participate according to local geographic, terrain, and zoning conditions.

14.5 SUMMARY AND FUTURE DISCUSSION

The Internet of Things, as a cyber-physical system, is a very powerful technology that is arising. It should be no surprise that the interlacing of this technologies is, and will continue, causing a big impact on our society. Big advances have been made in the physical, media access control (MAC), and networking layers that enable more advantages. At the physical layer, we have work that aims at increasing the transmission throughput, and in the MAC layer, there is work that increases the efficiency and capacity of node multiplexing, and at the network layer, there is much work in supporting largely populated networks. The tools attained from this work allows a greater interaction between interdisciplinary fields and more complex control systems. The work discussed here were mainly from the e-health, energy, environment, mobile network, transportation, smart home, and smart building disciplines (Table 14.1). The Internet of interlaced cyber-physical things is a step forward to smart cities. A few ingredients are missing in the symbiosis of IoT and CPS to become a smart city, and these are nonphysical, mainly the interaction with the economy and possibly politics.

TABLE 14.1

Summary of State-of-the-Art Research in Internet of Things–Based Cyber-Physical Systems

e-Health	[77] Discusses the use of satellite-based networks for telemedicine purposes.
	[78] New framework that provides a virtual-nurse agent for the client.
	[74] Compendium of nine personal health management and monitoring technologies.
	[76] General system model and optimization formulation for many e-health services.
Energy	[89] Discusses a control strategy for the islanded operation of a MV microgrid.
	[87] Information technology infrastructure of in-building energy monitoring.
	[88] Prototype microgrid implementation of a reconfigurable hierarchical control system.
Smart home/building	[103] Smart home nodes exchange information with a cloud server by means of P2P communications.
	[104] An IPv6-based global network for controlling sensor nodes through smartphones.
	[105] A proposal for smart community networks: smart homes that talk to each other to cooperate and inform about community environment.
Environment	[79] Addresses the basic requirements of an IoT environmental monitoring system.
	[80] The authors discuss the existing challenges in implementing a WSN applied to the IoT.
	[81] Autonomous monitoring wireless sensors.
	[87] Energy consumption monitoring and energy-saving management system of intelligent building based on IoT.
	[87] Design of WSNs applied to the IoT.
Mobile networks	[4] Indicates the five disruptive technologies directions to accomplish 5G networks.
	[5] Discusses the cellular architecture and key technologies for 5G wireless communication networks.
	[8] A mmWave beamforming technique is introduced for 5G cellular systems.
	[9] Proposes using massive MIMO in next-generation wireless systems.
	[10] Exploiting content caching and delivery techniques for 5G systems.
Transportation	[97] Harvesting of information for urban streets through sensor nodes in vehicles.
	[94] Evaluation of the readiness of V2V communications in real scenarios.
	[98] The vehicular network as a large surveillance and monitoring system.
	[99] An IoT vehicular network employed for traffic estimation based on traffic monitoring reports sent by vehicles.

REFERENCES

1. E. A. Lee and S. A. Seshia, *Introduction to Embedded Systems: A Cyber-Physical Systems Approach.* Lee & Seshia, Berkeley, CA, 2011.
2. J. Holler, V. Tsiatsis, C. Mulligan, S. Avesand, S. Karnouskos, and D. Boyle, *From Machine-to-Machine to the Internet of Things: Introduction to a New Age of Intelligence.* Academic Press, Waltham, MA, 2014.
3. Berkeley CPS Team. (2009) Cyber-physical systems. [Online]. Available: http://cyberphysicalsystems. org/. Accessed October 25, 2014.
4. F. Boccardi, R. W. Heath Jr., A. Lozano, T. L. Marzetta, and P. Popovski, Five disruptive technology directions for 5G, *IEEE Commun. Mag.*, 52(2), 74–80, February 2014.
5. C. X. Wang, F. Haider, X. Gao, X. H. You, Y. Sang, D. Yuan, H. M. Aggoune, H. Herald, S. Fletcher, and E. Hepsaydir, Cellular architecture and key technologies for 5G wireless communication networks, *IEEE Commun. Mag.*, 52(2), 122–130, February 2014.
6. B. Bangerter, S. Talwar, R. Arefi, and K. Stewart, Networks and devices for the 5G era, *IEEE Commun. Mag.*, 52(2), 90–96, February 2014.
7. I. Chi-Lin, C. Rowell, S. Han, G. Li, and Z. Pan, Toward green and soft: A 5G perspective, *IEEE Commun. Mag.*, 52(2), 66–73, February 2014.

8. W. Roh, J. Y. Seol, J. H. Park, B. Lee, L. Lee, Y. S. Kim, J. Cho, K. Cheun, and F. Arfanyar, Millimeter-wave beamforming as an enabling technology for 5G cellular communications: Theoretical feasibility and prototype results, *IEEE Commun. Mag.*, 52(2), 106–113, February 2014.
9. E. Larsson, O. Edfords, F. Tufvesson, and T. L. Marzetta, Massive MIMO for next generation wireless systems, *IEEE Commun. Mag.*, 52(2), 186–195, February 2014.
10. X. Wang, M. Chen, T. Taleb, A. Ksentini, and V. Leung, Cache in the air: Exploiting content caching and delivery techniques for 5G systems, *IEEE Commun. Mag.*, 52(2), 131–139, February 2014.
11. G. Fettweis and S. Alamouti, 5G: Personal mobile Internet beyond what cellular did to telephony, *IEEE Commun. Mag.*, 52(2), 140–145, February 2014.
12. T. L. Marzetta, Noncooperative cellular wireless with unlimited numbers of base station antennas, *IEEE Trans. Commun.*, 9(11), 3590–3600, November 2010.
13. H. Q. Ngo, E. G. Larsson, and T. L. Marzetta, Energy and spectral efficiency of very large multiuser MIMO systems, *IEEE Trans. Commun.*, 61, 1436–1449, April 2013.
14. E. G. Larsson, O. Edfors, F. Tufvesson, and T. L. Marzetta, Massive MIMO for next generation wireless systems, *IEEE Commun. Mag.*, 17(6), 48–56, November 2013.
15. H. Yin, D. Gesbert, M. C. Filippou, and Y. Liu, Decontaminating pilots in massive MIMO systems, in *2013 IEEE International Conference on Communications (ICC)*. IEEE, Budapest, Hungary, 2013, pp. 3170–3175.
16. Z. Pi and F. Khan, An introduction to millimeter-wave mobile broadband systems, *IEEE Commun. Mag.*, 49(6), 101–107, June 2011.
17. International Telecommunication Union. (2014) Annex 6 to working party 5c chairmans report. [Online]. Available: http://http://stakeholders.ofcom.org.uk/binaries/spectrum/spectrum-industry-groups/FWILF/2014/documents/FWILF14_005.pdf.
18. T. Rappaport, S. Sun, R. Mayzus, H. Zhao, Y. Azar, K. Wang, G. Wong, J. Schulz, M. Samimi, and F. Gutierrez, Millimeter wave mobile communications for 5g cellular: It will work! *Access, IEEE*, 1, 335–349, 2013.
19. R. Nee and R. Prasad, *OFDM Wireless Multimedia Communications*. Artech House, Boston, Massachusetts, 2000.
20. T. Hwang, C. Yang, G. Wu, S. Li, and G. Y. Li, OFDM and its wireless applications: A survey, *IEEE Trans. Veh. Technol.*, 58(4), 1673–1694, May 2009.
21. V. Chandrasekhar, J. G. Andrews, and A. Gatherer, Femtocell networks: A survey, *IEEE Commun. Mag.*, 46(9), 59–67, September 2008.
22. C. A. Azurdia-Meza, Z. Li, and C. Li, Analysis of the power control error in closed-loop power control algorithms applied to visible light communication systems, in *IEEE Latin America Conference on Communications*, Santiago, Chile, 2013.
23. P. Li, Y. Li, and C. Azurdia, Optical interference cancellation in visible light identification system based on wireless mesh network topology, in *Fifth International Conference on Machine Vision*. International Society for Optics and Photonics, Wuhan, China, 2013, pp. 87 830N–87 830N–8.
24. D. Iturralde, C. Azurdia-Meza, N. Krommenacker, I. Soto, Z. Ghassemlooy, and N. Becerra, A new location system for an underground mining environment using visible light communications, in *Ninth International Symposium on Communication Systems, Networks and Digital Signal Processing*. IEEE, Manchester, UK, July 2014, pp. 1165–1169.
25. D. Evans, The Internet of Things: How the next evolution of the Internet is changing everything, *CISCO White Paper*, vol. 1, San José, California, 2011.
26. C. Estevez, J. Wei, A. Kailas, D. Fuentealba, and G.-K. Chang, Very-high-throughput millimeter-wave system oriented for health monitoring applications, in *2011 13th IEEE International Conference on e-Health Networking Applications and Services (Healthcom)*, Columbia, MO, June 2011, pp. 229–232.
27. C. Estevez and A. Kailas, Energy-efficient process-stacking multiplexing access for 60-ghz mm-wave wireless personal area networks, in *2012 Annual International Conference of the IEEE Engineering in Medicine and Biology Society (EMBC)*, San Diego, CA, Auguest 2012, pp. 2084–2087.
28. J. Zhao, C. Qiao, R. Sudhaakar, and S. Yoon, Improve efficiency and reliability in single-hop wsns with transmit-only nodes, *IEEE Trans. Parallel Distrib. Syst.*, 24(3), 520–534, March 2013.
29. M. Buettner, G. V. Yee, E. Anderson, and R. Han, X-mac: A short preamble mac protocol for duty-cycled wireless sensor networks, in *Proceedings of the Fourth International Conference on Embedded Networked Sensor Systems*. ACM, Boulder, CO, 2006, pp. 307–320.

30. J. Beaudaux, A. Gallais, J. Montavont, T. Noel, D. Roth, and E. Valentin, Thorough empirical analysis of x-mac over a large-scale Internet of Things testbed, *Sensors J., IEEE*, 14(2), 383–392, February 2014.

31. C. Tang, L. Song, J. Balasubramani, S. Wu, S. Biaz, Q. Yang, and H. Wang, Comparative investigation on csma/ca-based opportunistic random access for Internet of Things, *Internet Things J., IEEE*, 1(2), pp. 171–179, April 2014.

32. Y. Liu, C. Yuen, X. Cao, N. Hassan, and J. Chen, Design of a scalable hybrid Mac protocol for heterogeneous m2m networks, *Internet Things J., IEEE*, 1(1), 99–111, February 2014.

33. H. Yan, Y. Zhang, Z. Pang, and L. D. Xu, Superframe planning and access latency of slotted MAC for industrial WSN in IOT environment, *IEEE Trans. Ind. Inform.*, 10(2), 1242–1251, May 2014.

34. J. Xu, Y. Andrepoulos, Y. Xiao, and M. van der Schaar, Non-stationary resource allocation policies for delay-constrained video streaming: Application to video over Internet of Things-enabled networks, *IEEE J., Sel. Areas Commun.*, 32, 782–794, April 2014.

35. S. Deering and R. Hinden, Internet protocol, version 6 (IPv6) Specification, *IETF RFC 2460*, December 1998, http://www.rfc-editor.org/info/rfc2460.

36. Zig Bee IP Specification, *ZigBee Public Document 13-002r00*, pp. 1–72, February 2013. http://www.zigbee.org/zigbee-for-developers/network-specifications/zigbeeip/.

37. L. Mainetti, L. Patrono, and A. Vilei, Evolution of wireless sensor networks towards the Internet of Things: A survey, in *19th International Conference on Software, Telecommunications and Computer Networks (SoftCOM), 2011*, Split, Croatia, 2011, pp. 1–6.

38. M. R. Palattella, N. Accettura, X. Vilajosana, T. Watteyne, L. A. Grieco, G. Boggia, and M. Dohler, Standardized protocol stack for the Internet of (Important) Things, *IEEE Commun. Surveys Tutorials*, 15(3), 1389–1406, January 2013.

39. G. Montenegro, N. Kushalnagar, J. Hui, and D. Culler, Transmission of IPv6 packets over IEEE 802.15.4 networks, *IETF RFC 4944*, September 2007, http://www.rfc-editor.org/info/rfc4944.

40. J. Ko, A. Terzis, S. Dawson-Haggerty, D. E. Culler, J. W. Hui, and P. Levis, Connecting low-power and lossy networks to the Internet, *IEEE Commun. Mag.*, 49(4), 96–101, April 2011.

41. D. F. Ramirez Hincapie and S. Cespedes, Evaluation of mesh-under and route-over routing strategies in AMI systems, in *2012 IEEE Colombian Communications Conference (COLCOM)*, Cali, Colombia, May 2012, pp. 1–6.

42. F. Touati, R. Tabish, and A. Ben Mnaouer, Towards u-health: An indoor 6LoWPAN based platform for real-time health care monitoring, in *Sixth Joint IFIP Wireless and Mobile Networking Conference (WMNC)*, Dubai, United Arab Emirates, April 2013, pp. 1–4.

43. X. Wang and H. Qian, Constructing a 6LoWPAN wireless sensor network based on a cluster tree, *IEEE Trans. Veh. Technol.*, 61(3), 1398–1405, March 2012.

44. J. Kim, R. Haw, E. J. Cho, C. S. Hong, and S. Lee, A 6LoWPAN sensor node mobility scheme based on proxy mobile IPv6, *IEEE Trans. Mobile Comput.*, 11(12), 2060–2072, December 2012.

45. G. Moritz, F. Golatowski, C. Lerche, and D. Timmermann, Beyond 6LoWPAN: Web services in wireless sensor networks, *IEEE Trans. Ind. Inform.*, 9(4), 1795–1805, November 2013.

46. D. H. Le and W. Pora, Implementation of smart meter working as IEEE1888-6LoWPAN gateway for the building energy management systems, in *2014 11th International Conference on Electrical Engineering/Electronics, Computer, Telecommunications and Information Technology (ECTI-CON)*, Nakhon Ratchasima, Thailand, May 2014, pp. 1–5.

47. C. Yibo, K.-M. Hou, H. Zhou, H.-l. Shi, X. Liu, X. Diao, H. Ding, J.-J. Li, and C. de Vaulx, 6LoWPAN stacks: A survey, in *2011 Seventh International Conference on Wireless Communications, Networking and Mobile Computing*, Wuhan, China, September 2011, pp. 1–4.

48. P. Thubert, T. Watteyne, and R. Assimiti, An architecture for IPv6 over the TSCH mode of IEEE 802.15.4e, *Internet Draft (work in progress)*, 2014. [Online]. Available: http://www.ietf.org/id/draft-ietf-6tisch-architecture-03.txt. Accessed October 31, 2014.

49. N. Accettura, M. R. Palattella, M. Dohler, L. A. Grieco, and G. Boggia, Standardized power-efficient and Internet-enabled communication stack for capillary M2M networks, in *2012 IEEE Wireless Communications and Networking Conference Workshops (WCNCW)*, Paris, France, April 2012, pp. 226–231.

50. N. Accettura and G. Piro, Optimal and secure protocols in the IETF 6TiSCH communication stack, in *2014 IEEE 23rd International Symposium on Industrial Electronics (ISIE)*, Istanbul, Turkey, June 2014, pp. 1469–1474.

51. O. Hahm, S. Pfeiffer, and J. Schiller, On real-time requirements in constrained wireless networks for mobile health, in *Proceedings of the Fourth ACM MobiHoc Workshop on Pervasive Wireless Health Care—MobileHealth '14*, August 2014, pp. 1–6.

52. C. Bormann, 6LoWPAN generic compression of headers and header-like payloads, *Internet Draft (work in progress)*, 2014. [Online]. Available: http://tools.ietf.org/html/draft-ietf-6lo-ghc-03. Accessed October 15, 2014.

53. E. Baccelli and M. Townsley, IP addressing model in ad hoc networks, *IETF RFC 5889*, p. 7, September 2010, http://www.rfc-editor.org/info/rfc5889.

54. E. Ancillotti, R. Bruno, and M. Conti, On the interplay between RPL and address autoconfiguration protocols in LLNs, in *2013 Ninth International Wireless Communications and Mobile Computing Conference (IWCMC)*. IEEE, July 2013, pp. 1275–1282. [Online]. Available: http://ieeexplore.ieee.org/lpdocs/epic03/wrapper.htm?arnumber=6583740.

55. Z. Shelby, S. Chakrabarti, E. Nordmark, and C. Bormann, Neighbor discovery optimization for IPv6 over low-power wireless personal area networks (6LoWPANs), *IETF RFC 6775*, 2012.

56. B. Sarikaya and F. Xia, Lightweight and secure neighbor discovery for low-power and lossy networks, *Internet Draft (work in progress)*, 2014. [Online]. Available: http://www.ietf.org/id/draft-sarikaya-6lo-cga-nd-00.txt.

57. W. Sun, Z. Yang, X. Zhang, and Y. Liu, Energy-efficient neighbor discovery in mobile ad hoc and wireless sensor networks: A survey, *IEEE Commun. Surv. Tutorials*, 16(3), 1448–1459, January 2014.

58. S. Fang, S. M. Berber, and A. K. Swain, Analysis of neighbor discovery protocols for energy distribution estimations in wireless sensor networks, in *2008 IEEE International Conference on Communications*, 2008, Beijing, China, pp. 4386–4390.

59. M. Bakht, M. Trower, and R. H. Kravets, Searchlight: Won't you be my neighbor? in *Proceedings of the 18th Annual International Conference on Mobile Computing and Networking*, Istanbul, Turkey, 2012, pp. 185–196.

60. S. Céspedes, X. Shen, and C. Lazo, IP mobility management for vehicular communication networks: Challenges and solutions, *IEEE Commun. Mag.*, 49(5), 187–194, May 2011.

61. S. Cespedes, N. Lu, and X. Shen, VIP-WAVE: On the feasibility of IP communications in 802.11p vehicular networks, *IEEE Trans. Intell. Transp. Syst.*, 14(1), 82–97, March 2013.

62. J. H. Kim, R. Haw, and C. S. Hong, Development of a framework to support network-based mobility of 6LoWPAN sensor device for mobile health care system, *2010 Digest of Technical Papers International Conference on Consumer Electronics (ICCE)*, pp. 359–360, Las Vegas, Nevada, January 2010.

63. G. Bag, S. M. S. Shams, A. H. Akbar, H. M. M. T. Raza, K.-H. Kim, and S.-W. Yoo, Network assisted mobility dupport for 6LoWPAN, in *Sixth IEEE Consumer Communications and Networking Conference*, Las Vegas, NV, January 2009, pp. 1–5.

64. T. Winter, P. Thubert, A. Brandt, J. Hui, R. Kelsey, P. Levis, K. Pister, R. Struik, J. Vasseur, and R. Alexander, RPL: IPv6 routing protocol for low-power and lossy networks, *IETF RFC 6550*, March 2012, http://www.rfc-editor.org/info/rfc6550.

65. B. Pavković, F. Theoleyre, and A. Duda, Multipath opportunistic RPL routing over IEEE 802.15.4, in *Proceedings of the 14th ACM International Conference on Modeling, Analysis and Simulation of Wireless and Mobile Systems—MSWiM '11*. New York, ACM Press, October 2011, p. 179. [Online]. Available: http://dl.acm.org/citation.cfm?id=2068897.2068929.

66. T. Clausen, U. Herberg, and M. Philipp, A critical evaluation of the IPv6 routing protocol for low power and lossy networks (RPL), in *2011 IEEE Seventh International Conference on Wireless and Mobile Computing, Networking and Communications (WiMob)*, Shanghai, China, October 2011, pp. 365–372.

67. U. Herberg and T. Clausen, A comparative performance study of the routing protocols LOAD and RPL with bi-directional traffic in low-power and lossy networks (LLN), in *Proceedings of the Eighth ACM Symposium on Performance Evaluation of Wireless Ad Hoc, Sensor, and Ubiquitous Networks—PE-WASUN '11*, New York, USA, November 2011, p. 73.

68. S. Dawson-Haggerty, A. Tavakoli, and D. Culler, Hydro: A hybrid routing protocol for low-power and lossy networks, in *2010 First IEEE International Conference on Smart Grid Communications*. IEEE, June 2013, http://www.rfc-editor.org/info/rfc6971. October 2010, pp. 268–273. [Online]. Available: http://ieeexplore.ieee.org/lpdocs/epic03/wrapper.htm?arnumber=5622053.

69. U. Herberg, A. Cardenas, T. Iwao, M. Dow, and S. Cespedes, Depth-first forwarding (DFF) in unreliable networks, *IETF RFC 6971*, p. 40, 2013.

70. T. Clausen, J. Yi, A. Bas, and U. Herberg, A Depth First Forwarding (DFF) extension for the LOADng Routing Protocol, in *2013 First International Symposium on Computing and Networking*, December 2013, pp. 404–408.

71. V. Kostakos, T. Ojala, and T. Juntunen, Traffic in the Smart City: Exploring City-Wide Sensing for Traffic Control Center Augmentation, *IEEE Internet Comput.*, 17(6), 22–29, November 2013.

72. M. Gerla, E.-K. Lee, G. Pau, and U. Lee, Internet of vehicles: From intelligent grid to autonomous cars and vehicular clouds, in *2014 IEEE World Forum on Internet of Things (WF-IoT)*, Seoul, Korea, March 2014, pp. 241–246.

73. F. Hu, D. Xie, and S. Shen, On the application of the Internet of Things in the field of medical and health care, in *IEEE International Conference on Green Computing and Communications (GreenCom), 2013 IEEE and Internet of Things (iThings/CPSCom), and IEEE Cyber, Physical and Social Computing*, August 2013, pp. 2053–2058.

74. L. Gatzoulis and I. Iakovidis, Wearable and portable ehealth systems, *Eng. Med. Biol. Mag., IEEE*, 26(5), 51–56, 2007.

75. J. Caldeira, J. Rodrigues, and P. Lorenz, Toward ubiquitous mobility solutions for body sensor networks on health care, *IEEE Commun. Mag.*, 50(5), 108–115, May 2012.

76. D. Niyato, E. Hossain, and S. Camorlinga, Remote patient monitoring service using heterogeneous wireless access networks: Architecture and optimization, *IEEE Journal on Selected Areas in Communications*, 27(4), 412–423, 2009.

77. G. Graschew, S. Rakowsky, T. Roelofs, and P. Schlag, New medical technologies of the future, in *2013 International Conference on Digital Technologies (DT)*, Zilina, Slovakia, May 2013, pp. 84–89.

78. M. Najafi, S. Aghtar, K. Sartipi, and N. Archer, Virtual remote nursing system, in *Consumer Communications and Networking Conference (CCNC), 2011 IEEE*, Las Vegas, Nevada, January 2011, pp. 13–17.

79. M. T. Lazarescu and S. C. Mukhopadhyay, Internet of Things Low-Cost Long-Term Environmental Monitoring with Reusable Wireless Sensor Network Platform, in *Internet of Things Challenges and Opportunities*, Ed. Springer International Publishing, New York, 2014.

80. C. Delphine, A. Reinhardt, P. S. Mogre, and R. Steinmetz, Wireless sensor networks and the Internet of Things: Selected challenges, in *Proceedings of the Eighth GI/ITG KuVS Fachgespräch "Drahtlose Sensornetze"*, Hamburg, Germany, August 2009, pp. 31–34.

81. D. Y. and G. Andia Vera, Towards Autonomous Wireless Sensors: RFID and Energy Harvesting Solutions, in book, *Internet of Things Challenges and Opportunities*, S. C. Mukhopadhyay, Ed. Springer International Publishing, Springer International Publishing, New York, 2014.

82. M. Tesanovic and S. Vadgama, Short paper: Vehicle emission control in smart cities, in *2014 IEEE World Forum on Internet of Things (WF-IoT)*, Seoul, South Korea, March 2014, pp. 163–164.

83. B. Hull, V. Bychkovsky, Y. Zhang, K. Chen, M. Goraczko, A. Miu, E. Shih, H. Balakrishnan, and S. Madden, CarTel, in *Proceedings of the Fourth International Conference on Embedded Networked Sensor Systems—SenSys '06*, Boulder, CO, October 2006, p. 125.

84. E.-K. Lee, P. Chu, and R. Gadh, Fine-grained access to smart building energy resources, *Internet Comput., IEEE*, 17(6), 48–56, November 2013.

85. X. Guan, Z. Xu, and Q.-S. Jia, Energy-efficient buildings facilitated by microgrid, *IEEE Trans. Smart Grid*, 1(3), 243–252, December 2010.

86. K. Hua, Intelligent controls for cyber physical energy buildings, in *2013 IEEE/CIC International Conference on Communications in China—Workshops (CIC/ICCC)*, Xian, China, August 2013, pp. 120–124.

87. C. Wei and Y. Li, Design of energy consumption monitoring and energy-saving management system of intelligent building based on the Internet of Things, in *2011 International Conference on Electronics, Communications and Control (ICECC)*, Ningbo, China, September 2011, pp. 3650–3652.

88. R. Wandhare, S. Thale, and V. Agarwal, Reconfigurable hierarchical control of a microgrid developed with pv, wind, micro-hydro, fuel cell and ultra-capacitor, in *2013 28th Annual IEEE Applied Power Electronics Conference and Exposition (APEC)*, Long Beach, California, March 2013, pp. 2799–2806.

89. M. Hamzeh, H. Karimi, and H. Mokhtari, A new control strategy for a multi-bus mv microgrid under unbalanced conditions, *IEEE Trans. Power Syst.*, 27(4), 2225–2232, November 2012.

90. R. Vicini, O. Micheloud, H. Kumar, and A. Kwasinski, Transformer and home energy management systems to lessen electrical vehicle impact on the grid, *Gen., Transm. Distrib., IET*, 6(12), 1202–1208, December 2012.

91. Y. Zhan, L. Liu, L. Wang, and Y. Shen, Wireless sensor networks for the Internet of Things, *Int. J. Distrib. Sensor Netw.*, vol. 2013, Article ID 717125, 2 pages, 2013.
92. A. Hashimoto, H. Yorshino, and H. Atarashi, Roadmap of IMT-advanced development, *IEEE Microw. Mag.*, 9(4), 80–88, August 2008.
93. L. Sorensen and K. E. Skouby, User scenarios 2020, Wireless World Research Forum (WWRF), Technical Report, July 2009.
94. N. Naylor, U.S. Department of Transportation Announces Decision to Move Forward with Vehicle-to-Vehicle Communication Technology for Light Vehicles, May 2014. [Online]. Available: http://www.nhtsa.gov/About+NHTSA/Press+Releases/2014/USDOT+to+Move+Forward+with+Vehicle-to-Vehicle+Communication+Technology+for+Light+Vehicles. Accessed on October 16, 2014.
95. F. Dressler, H. Hartenstein, O. Altintas, and O. K. Tonguz, Inter-Vehicle communication—Quo Vadis, *IEEE Commun. Mag.* 52(6), 170–177, 2014.
96. M. Gerla and L. Kleinrock, Vehicular networks and the future of the mobile Internet, *Comput. Netw.*, 55(2), 457–469, February 2011. [Online]. Available: http://linkinghub.elsevier.com/retrieve/pii/S1389128610003324
97. E. Magistretti, M. Gerla, P. Bellavista, and A. Corradi, Dissemination and harvesting of urban Data using vehicular sensing platforms, *IEEE Trans. Veh. Technol.*, 58(2), 882–901, February 2009. [Online]. Available: http://ieeexplore.ieee.org/lpdocs/epic03/wrapper.htm?arnumber=4573261.
98. X. Yu, Y. Liu, Y. Zhu, W. Feng, L. Zhang, H. Rashvand, and V. Li, Efficient sampling and compressive sensing for urban monitoring vehicular sensor networks, *IET Wireless Sensor Syst.*, 2(3), 214, 2012. [Online]. Available: http://digital-library.theiet.org/content/journals/10.1049/iet-wss.2011.0121.
99. R. Du, C. Chen, B. Yang, N. Lu, X. Guan, and X. Shen, Effective urban traffic monitoring by vehicular sensor networks, *IEEE Trans. Veh. Technol.*, 2014. [Online]. Available: http://ieeexplore.ieee.org/lpdocs/epic03/wrapper.htm?arnumber=6807778. 64(1), 273–286, 2015.
100. S. Chemishkian, Building smart services for smart home, in *Proceedings 2002 IEEE Fourth International Workshop on Networked Appliances (Cat. No.02EX525)*, Gaithersburg, MD, 2002, pp. 215–224.
101. K. Bing, L. Fu, Y. Zhuo, and L. Yanlei, Design of an Internet of Things-based smart home system, in *2011 Second International Conference on Intelligent Control and Information Processing*, vol. 2, Harbin, China, July 2011, pp. 921–924.
102. M. Soliman, T. Abiodun, T. Hamouda, J. Zhou, and C.-H. Lung, Smart home: Integrating Internet of Things with web services and cloud computing, in *2013 IEEE Fifth International Conference on Cloud Computing Technology and Science*, vol. 2, Bristol, UK, December 2013, pp. 317–320.
103. W. Jian, C. Estevez, A. Chowdhury, Z. Jia, and G.-K. Chang, A hybrid mac protocol design for energy-efficient very-high-throughput millimeter wave wireless sensor communication networks, in *Communications and Photonics Conference and Exhibition (ACP), 2010 Asia*, Shanghai, China, December 2010, pp. 631–632.
104. S. O. Park, T. H. Do, and S. J. Kim, A hierarchical middleware architecture for efficient home control through an IPv6-based global network, in *2011 IEEE Ninth International Symposium on Parallel and Distributed Processing with Applications*, Busan, South Korea, May 2011, pp. 167–170.
105. X. Li, R. Lu, X. Liang, X. Shen, J. Chen, and X. Lin, Smart community: An Internet of Things application, *IEEE Commun. Mag.*, 49(11), 68–75, November 2011.

Section VI

Security Issues

15 Security in Delay-Tolerant Mobile Cyber-Physical Applications

Naércio Magaia, Paulo Pereira, and Miguel Correia

CONTENTS

15.1 INTRODUCTION

Unlike conventional embedded systems where the emphasis tends to be on the computational elements, a *cyber-physical system* [1] is typically designed as a network of physically distributed embedded sensor and actuator devices equipped with computing and communicating capabilities to process and react to stimuli from the physical world and make decisions that also impact the physical world. Example applications of CPS include high-confidence medical devices and systems, assisted living, traffic control and safety, advanced automotive systems, process control, energy conservation, environmental control, avionics, instrumentation, critical infrastructure control (e.g., electric power, water resources, and communications systems), distributed robotics (e.g., telepresence, telemedicine), defense systems, manufacturing, and smart structures [2].

The broad dissemination of mobile devices with substantial computation resources (e.g., processing and storage capacity), a variety of sensors (e.g., cameras, GPS, speakers, microphone, and light and proximity sensors), and multiple communication mechanisms (e.g., cellular, Wi-Fi, Bluetooth) allowing interconnection to the Internet as well as to other devices made possible the rise of a noticeable subcategory of CPS, known as *mobile CPS* [3]. In mobile CPS, since cyber-physical systems are combined with mobile devices with Internet access, tasks requiring more resources than those locally available can now be executed using the mobile link, to access either a server or cloud environment, fostering a new range of applications such as augmented reality, interaction with social networks, and health monitoring [3].

Delay-tolerant mobile cyber-physical applications are mobile cyber-physical applications that use the *delay-tolerant network* (DTN) [4] paradigm to communicate. DTNs are networks in which end-to-end connectivity between a source and target nodes might never exist, so nodes have to store (or buffer) data packets and forward them to others until they reach their target. This lack of end-to-end connectivity is the main difference in relation to mobile ad hoc networks (MANETs), where continuous end-to-end connectivity has to be established prior to the forwarding of messages. Due to nodes' mobility and/or network's dynamics (nodes joining or leaving the network, e.g., due to devices turned off or run out of battery), end-to-end connectivity is not guaranteed, even in real MANETs. However, in the DTN routing paradigm (a store-carry-and-forward approach), mobility-related issues are no longer seen as obstacles, since nodes can carry messages with them while moving until an appropriate forwarder is found. The DTN routing strategy allows messages to be relayed among nodes until the destination is reached, or they are discarded.

Nevertheless, DTN routing involves the challenging task of finding suitable nodes to forward messages. A variety of network information is used to address this problem, namely, (1) dynamic network information, for example, location information, traffic information, and encounter information, and (2) static network information, for example, social relations among nodes. Static network information, like social ties and behaviors between nodes, tends to be more stable over time, and when used, they facilitate message transmission [5]. A considerable number of social-aware or social-based routing protocols have been proposed, which use static information, that is, they use social relations among nodes to determine the most appropriate node to forward messages.

In CPS, due to the increase in the interaction between the physical systems and cyber systems, physical systems are increasingly exposed to cybersecurity threats. Similarly, in decentralized and self-organized networks, *security* aspects like confidentiality, integrity, authentication, privacy, trust, and cooperation enforcement arise. To deal with confidentiality, integrity, and privacy, nodes should encrypt or somehow protect information, as forwarding decisions can be made based on the packet's content and/or context. Due to nodes' resource scarcity and to the fact that they are controlled by rational entities, nodes might misbehave. Node misbehavior, malicious or selfish, can significantly impact network performance [6,7]. Decision making in various fields, for example, commerce or trade, becomes much simpler with the use of reputation and trust. On the one hand, trust can be seen as the belief a node has in the peer's qualities; on the other hand, reputation can be seen as a peer's perception about a node [8].

The authors of [5,9] proposed a taxonomy to classify mechanisms to stimulate cooperation among nodes in DTNs, that is, cooperation enforcement. Cooperation enforcement schemes are categorized as reputation based, remuneration based (also called credit based), and game theory based (hereafter called game based). In *reputation-based schemes*, nodes use others' reputation records to make forwarding decisions. Good or bad reputation is gained by forwarding or not messages from other nodes, respectively. Nodes with bad reputation (i.e., misbehaving) are often excluded from the network. Some form of credit (or reward) is used to control message forwarding in *remuneration-based schemes*. Nodes that forward others' messages are rewarded. The earned credit is used to obtain forwarding services from other nodes. In *game-based schemes*, forwarding decisions are modeled by game theory, where each node follows a strategy aiming at maximizing its benefits and minimizing its resource consumption. This can be accomplished, for example, by maximizing its delivery probability or perhaps by minimizing its end-to-end delay. A well-known strategy is tit for tat (TFT), in which a node forwards as many messages for a neighbor as the neighbor has forwarded its.

Besides selfishness, which is common in self-organized environments, CPS are prone to the following attacks: [10] eavesdropping, compromised key attacks, man-in-the-middle attacks, and denial-of-service (DoS) attacks. Eavesdropping occurs if an attacker intercepts any information communicated by the system. For example, in medical CPS applications, privacy issues arise if patient's health data are disclosed. If an attacker obtains a secret code (key), used to decrypt secure information, the attacker gains access to secured communications without the perception of the sender or receiver, by means of the compromised key. In a man-in-the-middle attack, falsified messages can be sent to the system that may cause it to go into an incorrect state. An example can be a false message sent to the system control saying that the system is fine, while some actions are required. A DoS attack consists in making the system unavailable.

This chapter is organized as follows. Section 15.2 presents security threats of cyber systems and DTNs. Section 15.3 presents security requirements of such systems. Section 15.4 presents mechanisms to enhance trust and cooperation enforcement in DTNs. Section 15.5 surveys security mechanisms for delay-tolerant mobile cyber-physical applications. Section 15.6 presents some examples of delay-tolerant mobile cyber-physical applications. Finally, Section 15.7 presents concluding remarks.

15.2 SECURITY THREATS

CPS consist of sensing, processing, and communication platforms tightly coupled with physical processes. A CPS is composed of a physical process and a cyber system, in which the physical system is controlled and monitored by the cyber system. CPS often aim to monitor the behavior of a physical process and to trigger actions in order to change its behavior, resulting in a correct and better physical environment. However, due to the interaction between physical systems and cyber systems, physical systems are exposed to a cyber system's security threats. The following are some examples of real attacks to CPS or existing vulnerabilities: hackers have broken into the air force control mission-support systems of the U.S. Federal Aviation Administration (FAA) several times; [11] hackers can hack wirelessly networked medical devices implanted in the human body; [12] hackers have penetrated power systems in several regions outside the United States; [13] and CarShark [14]—a software tool—that can kill a car engine remotely turns off its brakes, makes its instruments give false readings, and inserts fake data packets to carry out attacks.

Mobile CPS are those in which CPS are combined with mobile devices, such as smartphones and tablets, allowing interconnection to the Internet as well as to other devices in a decentralized manner. In some scenarios, these interconnections are made in a delay-tolerant manner, which fostered the appearance of delay-tolerant mobile cyber-physical applications.

In decentralized networks, like DTNs, forwarding decisions are made individually by each node (or entity), which may incur in the consumption of nodes' limited resources, for example, battery power, bandwidth, processing, and memory, all over the network. These networks share a tricky notion of being self-organized and/or self-managed, but they require for their correct operation that each node gives its own contribution.* Some concerns arise in these networks (1) in the establishment of trust between entities, (2) in the stimulation of their cooperation, or even (3) due to the fairness of their contributions.

There are two possible types of node's misbehavior: selfish and malicious. When a node manifests a *selfish* behavior, it aims to maximize its benefits by using the network while saving its own resources (e.g., battery power). As such, cooperation enforcement schemes can be leveraged to foster cooperation. If a node manifests a *malicious* behavior, it tends to maximize the damage caused to the network for its own benefit. A way to deal with such misbehavior is by detecting and isolating those nodes from the network.

* It is assumed that nodes have equivalent privileges and responsibilities.

Selfish behaviors can be classified as individual or social selfishness [5,9]. A node presents individual selfishness if it only aims at maximizing its own utility, hence disregarding a system-wide criterion. Social selfishness is manifested when nodes only forward messages of others to whom they have social ties with.

Besides selfishness, entities on a self-organized environment are also prone to other forms of attacks such as flooding and cheating. A flooding attack consists in exhaustively using others' resources, for example, by initiating an enormous amount of requests, in order to render the network useless. A cheating (or retention) attack consists in gaining an unfair advantage over other nodes by holding essential system's data.

Yet, another way of classifying attacks is based on the nature of the attacks and the type of attackers. As such, they can be classified as passive or active [15]. Passive attacks occur if an unauthorized party gains access to a message without modifying its contents. There are two types of passive attacks, namely, (1) the release of message contents [16], which occur if message contents are made available or disclosed to unauthorized parties, and (2) traffic analysis [17], which allows an attacker to infer communication patterns, thus guessing the nature of the communication that was taking place. Active attacks are characterized by an unauthorized party modifying the contents of the message.

Table 15.1 summarizes potential attacks to delay-tolerant mobile cyber-physical applications. As an example, Figure 15.1, as in Magaia, Pereira, and Correia [18], shows the average delivery probability (i.e., the fraction of successfully received packets over all packets sent) for five different percentages of nodes performing a black-hole attack on a map-based mobility model [19] of the Helsinki City, for 8 DTN routing protocols, namely, Direct Delivery (DD), First Contact (FC), Epidemic, PRoPHET, MaxProp, RAPID, and Spray and Wait (SnW). DD is not affected by black-hole attacks, as it can only deliver a message if it meets the final destination. SnW delivers more messages than DD as it sprays multiple copies to intermediate nodes that may come in contact with the destination node sooner. Naturally, it is vulnerable to attacks by relying on other nodes for forwarding messages. As FC forwards only one copy of the message to the first node met, it is the most affected routing protocol by black-hole attacks. It was also observed that black-hole misbehaving nodes reduce congestion (e.g., PRoPHET with less than 60% of misbehaving nodes), that is, misbehaving nodes cause a reduction of the number of message copies circulating in the network as they drop them. Epidemic and PRoPHET are less affected in comparison with other routing protocols by black-hole misbehaving nodes, due to their unlimited copy algorithms. But MaxProp and Rapid also suffer from black-hole attacks because of the contact pattern (i.e., shorter contact durations) since longer messages require more contacts to be transmitted.

15.3 SECURITY REQUIREMENTS

As a wireless medium is accessible (or open) for everyone within communication range, misbehaving nodes might attempt to compromise message contents, justifying the existence of security requirements [15] such as authentication, confidentiality, integrity, availability, and privacy, in such environments.

15.3.1 AUTHENTICATION

Authentication assures that the communication is genuine. There are two possible cases: (1) the single message case, for example, a warning message, where the objective is to assure the source's legitimacy to the message's destination, that is, the message is from the source that it claims to be from, and (2) the ongoing interaction case, for example, the connection of an entity to another, where the objective is twofold: first, it is necessary to assure the entities' authenticity, that is, each entity is who it claims to be, and second, it is also necessary to assure that a third party cannot interfere with the connection, even if masquerading as one of the legitimate parties. In a cyber-physical system, authentication assures the authenticity of all the related processes such as sensing, communication, and actuation.

TABLE 15.1
Summary of Potential Attacks to Delay-Tolerant Mobile Cyber-Physical Applications

Attack Type	Attack Description
Individual selfishness [5,9]	A selfish node only aims at maximizing its own utility, disregarding the system-wide criteria.
Social selfishness [5,9]	Nodes only forward messages of others to whom they have social ties with.
Eavesdropping [15,16]	Message content is made available or disclosed to unauthorized parties.
Compromised-key attack [10]	An attacker obtains a cryptographic key and uses it to read an encrypted communication.
Traffic analysis [15,17]	Extracting unauthorized information by analyzing communication patterns (but not their content).
Routing loop attacks [22]	Modifying routing packets so they do not reach their destination.
Wormhole attacks [22]	A set of malicious nodes creates a worm link to connect distant network points with low latency, causing disruption in normal traffic load and end flow.
Black-hole attacks [22]	A malicious node drops all packets forwarded to it and responds positively to incoming route requests despite the fact of not having proper routing information.
Gray-hole attacks [22]	Special case of a black-hole attack where a malicious node selectively drops packets.
DoS [22]	Attacks that obstruct the normal use or management of a service.
False information or false recommendation [22]	Colluding and providing false recommendation/information in order to isolate good nodes while keeping bad ones connected.
Incomplete information	Consists in not cooperating to provide proper or complete information.
Packet modification/insertion	Consists in the modification or malicious insertion of packets.
Newcomer attacks [22]	A malicious node registers as a new user to discard its bad reputation or distrust.
Sybil attacks [22]	Consists in using multiple network identities.
Blackmailing [22]	Consists in using a majority-voting scheme trying to cause routing topology change.
Replay attacks [22]	Consists in maliciously or fraudulently repeating or delaying valid transmitted packets.
Selective misbehaving attacks [22]	Consists in selectively misbehave to other nodes.
On–off attacks [22]	Disrupting services by behaving correctly/incorrectly in alternation.
Conflicting behavior attacks [22]	Behave differently to different nodes to cause contradictory opinions.
Credit forgery attack (or layer injection attack) [50]	Forge valid credit in order to reward itself for work it did not do or for more than it has done.
Nodular tontine attack (or layer removal attack) [50]	Remove one or more layers of a multilayer credit generated by previous forwarding nodes.
Submission refusal attack [50]	If the source and last intermediate node collude, the latter may refuse to submit the received credit from a VB and receive another compensation from the source node.

15.3.2 CONFIDENTIALITY

Confidentiality assures the protection of the transmitted data from passive attacks, since, and as stated previously, the wireless medium is accessible for everyone within communication range. Regarding the content of the data transmission, several levels of protection can be identified, in which the broadest level assures the protection of all user data transmitted between two nodes over a period of time [15].

Confidentiality may also aim to assure protection against traffic analysis. The idea is that an attacker shall not be able to perceive traffic flow information such as the source and destination, frequency, length, or even other traffic's characteristics.

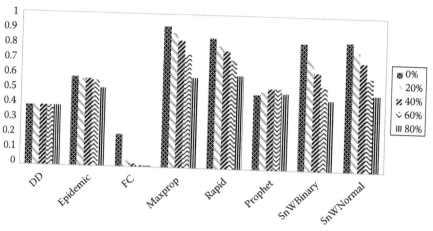

FIGURE 15.1 Average delivery probability as a function of the percentage of nodes performing a black-hole attack.

As an example, consider a delay-tolerant mobile cyber-physical application, such as health monitoring, in which the personal health records need to be transmitted from the personal health record system to the doctor or the front medical devices. By means of confidentiality, the system prevents third parties from inferring data about the patients' personal health records either by reading their content or by understanding to what kind of doctor it is being sent to.

15.3.3 INTEGRITY

Integrity assures that the transmitted data are received as sent, without duplication, insertion, modification, reordering, or replay. Similar to confidentiality, integrity can apply to (1) a message stream, (2) a single message, or even (3) certain fields of the message. And as earlier, the complete stream protection is the best approach.

There are two types of integrity [15]: (1) connection oriented and (2) connectionless. The connection-oriented integrity type addresses both message stream modification and DoS. The destruction of data is also covered under this type. The connectionless integrity type deals with individual messages without considering any larger context. It usually only offers protection against the modification of messages.

Integrity is compromised in a CPS when, for instance, false data are received by the system control stating that the system is fine, whereas urgent actions are required.

15.3.4 AVAILABILITY

Availability refers to a system's property of being accessible and usable when requested by an authorized entity, in accordance with the system's design. Loss or reduction of availability can be caused by a variety of attacks. An example of such attacks is DoS (see Table 15.1). With automated countermeasures, for example, authentication and encryption, or some sort of physical action, it is possible to prevent/recover from the loss of availability resulting from such attacks. For example, consider a smart grid infrastructure where information, such as price information, control commands, and meter data, flows to/from a smart meter. In this scenario, the availability of price information and control commands is critical: the former, because of the serious financial and possible legal implications, and the latter, as it is necessary to turn the meter back on after completing the payment of the electric bill. Instead, the availability of the meter data may not be critical as the data can be accessed later on.

15.3.5 PRIVACY

Privacy can be understood as the users' willingness to disclose or not its information to others (family, friends, or even the general public). As a property, privacy is the confidentiality of personal information.

The lack of infrastructure in decentralized and self-managed environments leverages nodes' forwarding decisions, allowing the opportunist exploitation of any other nodes in their vicinity to help messages reach their intended destinations. In such environments, message forwarding relies on the nodes' participation in the network.

If some nodes are deemed untrusted, privacy issues may rise due to passive attacks, which are in the nature of eavesdropping on transmissions.

15.4 TRUST AND COOPERATION ENFORCEMENT SCHEMES

In this section, a survey of mechanisms to foster cooperation in decentralized and self-managed networks, such as DTNs, is presented.

15.4.1 TRUST

According to the *Oxford Dictionary of English* [20], trust is a "firm belief in the reliability, truth, or ability of someone or something." In the literature, many other definitions of trust can be found according to various disciplines. For example, in line with *social sciences* [21], trust can be defined as "as the degree of subjective belief about the behavior of a particular entity"; in line with *economics*— as shown in the prisoner's dilemma—trust is based on the assumption that humans are rational and strict utility maximizers of their own interest, therefore selfish. Nevertheless the emergence of altruistic behavior can be seen in initially purely selfish mechanisms. In line with *communication and networking* [22], trust relationships among participating nodes are important, as with them cooperative and collaborative environments can be built, which improve system objectives in terms of scalability, reconfigurability, reliability, dependability, or security.

Trust management is necessary in order for participating nodes without previous interactions to form a decentralized and self-organized network (e.g., MANET or DTN) with an acceptable level of trust relationships between them [22]. As stated in Blaze, Feigenbaum, and Lacy [23], "trust management provides a unified approach for specifying and interpreting security policies, credentials, and relationships." Some applications of trust management in decision-making situations are intrusion detection, authentication, access control, key management, and the isolation of misbehaving nodes for effective routing [22].

Trust management is composed of trust establishment, trust update, and trust revocation. A *trust establishment* process consists of the representation, evaluation, maintenance, and distribution of trust among nodes. A *trust update* process consists in modifying a node's trust values as a consequence of their collaboration with others, thus favoring trustworthy nodes and penalizing untrustworthy ones. *Trust revocation* consists in dropping/cancelling trust relationships among nodes.

The main properties of trust in decentralized and self-managed environments [22] are (1) *dynamicity*, since trust establishment is based on temporally and spatially local information; (2) *subjectivity*, due to network's dynamics, wherein a trustee node may be assigned to different levels of trust as a result of different experiences; (3) *incomplete transitivity*, in which trust's transitivity among two entities (trustor and trustee) and a third party is guaranteed if a trustor trusts the trustee and the trustee's recommendation of the third party; (4) *asymmetry*, wherein entities with different capacities (e.g., energy or computation power) may not trust each other; and (5) *context dependency*, in which trust types depend on the foreseen task [22].

The properties presented earlier should be taken into account during the design of a trust-based framework. Here are other important aspects to consider for decentralized and self-managed environments: (1) an entity decision procedure of trust should be *fully distributed* and based on a cooperative evaluation with uncertainty and incomplete evidence, since a trusted third party (TTP) is unreliable in such environments; (2) trust's determination should be flexible to membership changes and to deployment scenarios, therefore in a *highly customizable manner*; (3) *selfishness* should be taken into account by a trust decision framework, and hence no assumption that all nodes are cooperative should be made; (4) also because of the network's dynamics, trust should be established in a *self-organized and reconfigurable manner*.

Additional care should be taken to ensure that a trust management system is not easily subverted, attacked, or compromised. It is important to mention that trust management schemes are devised to detect misbehaving nodes (i.e., selfish nodes along with malicious ones). In addition, if the available information or evidence does not provide a certain level of trust, the trust engine should be robust enough to gracefully degrade.

15.4.2 COOPERATION ENFORCEMENT

Cooperation enforcement [24] (incentive schemes/mechanisms) can be leveraged to manage and organize decentralized and self-managed systems, therefore compensating for the nonexistence of a central or dedicated entity. As a consequence, it is possible to deal with the security challenges previously mentioned. Even though a cooperative behavior may result in an increase of the nodes' resource consumption, for example, forwarding nodes may incur in additional energy and bandwidth usage during packets' transmissions and receptions. It was demonstrated in the context of MANETs that cooperation can succeed over competition [25]. So, the idea is to guarantee that a cooperative behavior is overall more beneficial than a passive or malicious uncooperative behavior. It is often desirable that these mechanisms distinguish uncooperative behaviors due to valid reasons such as energy shortage and crashing, from malicious uncooperative ones.

The use of cooperation enforcement schemes in MANETs has been exhaustively researched [26–33]. Cooperation enforcement schemes have been envisaged for many application domains, namely, infrastructure-based P2P applications (e.g., file sharing, distributed processing, and data backup), wireless networks (e.g., DTNs, wireless sensor networks, MANETs, wireless ad hoc backup, and nomadic computing), and web commerce (e.g., auction sites, review and recommendation sites) [24].

Cooperation enforcement schemes can be categorized as reputation-based, remuneration-based, and game-based schemes. The following sections discuss in detail each of these categories.

15.4.2.1 Reputation-Based Schemes

Reputation-based schemes are those in which the decision to interact depends on the other node's reputation.

Architecture: The architecture of the reputation management system, an integral part of the reputation mechanism, can be centralized, decentralized, or hybrid [24]. In the centralized approach, a central authority collects nodes' information and derives and provides the scores for all participants. In the distributed approach, since no central authority is available, nodes' ratings are stored in a distributed fashion. The evaluation of reputation is commonly based on subsets of information (e.g., neighbor nodes' information), which may be prone to inconsistencies if compared to the centralized approach. Nevertheless, a distributed management system scales better and is more common in self-organized networks than the centralized one. A hybrid approach is a combination of both.

Operations: Reputation-based mechanisms are composed of three phases: collection of evidence, cooperation decision, and cooperation evaluation [24]. In the first phase, a node collects reputation information by observing and experiencing and/or by means of recommendations of third parties. In the second phase, a node evaluates the collected information in order to decide if it should cooperate or not, based on the other node's reputation. Some evaluation methods used in this phase are voting schemes, average ratings, Bayesian-based computation, and flow mode [24]. In the last phase, the degree of cooperation between nodes is evaluated. It is done after their interaction and consists in rewarding nodes that presented a good behavior by adequately increasing their local reputation. Consequently, nodes with bad reputation are isolated, hence not receiving others' services.

Attacks and countermeasures: Misbehaving nodes can cause a variety of attacks such as individual and social selfishness, DoS, functionality attacks (e.g., subversion attacks), and single or group attacks to the reputation system (e.g., a liar or collusion,* respectively). For example, the CORE [34] mechanism can be used to guard against the impact of liars. In Marti et al. [35], the Watchdog and Pathrater are used to identify misbehaving nodes and selecting paths to avoid them, respectively. For a more detailed discussion of attacks and countermeasures, please refer to Sen [36].

15.4.2.2 Remuneration-Based Schemes

Remuneration-based schemes are those in which cooperating nodes should receive an equivalent complement (remuneration), and misconduct is punished with a penalty. The exchange of services, for some sort of payment, calls for a TTP (e.g., bank) to manage the process.

Architecture: A remuneration-based mechanism is composed of four operations: negotiation, cooperation decision, cooperation evaluation, and remuneration [24]. During negotiation, nodes decide on the terms of their interaction. This can be done only between them, or between them and a TTP. The cooperation decision, that is, if a node can or cannot cooperate, is taken based on the outcome of the negotiation. The cooperation evaluation is twofold: on the one side, the service requesting party decides based on the acceptability of the service to the request; on the other side, the service providing party decides based on the acceptability of the remuneration. At last, the collaborating node is remunerated. There are three types of remuneration envisaged in this scheme: virtual currency units, real money, and bartering units [24].

Fair exchange: A fair exchange protocol offers ways to guarantee that at the end of the exchange held by two or more nodes, either all of them have received what they were expecting or none of them has received anything. The correctness of the exchange depends on the availability of a neutral TTP. There are two types of protocols: online and off-line fair exchange protocols [24]. In the former, the TTP constitutes a bottleneck as it mediates every interaction between the nodes. In the latter, the TTP is used as an intermediary if and only if one of the nodes has doubts about the fairness of the exchange.

15.4.2.3 Game-Based Schemes

Game-based schemes are those in which forwarding decisions are modeled using game theory. Game theory provides guidelines on how to model situations such as social dilemmas (e.g., the prisoner's dilemma). It also provides insights on how individual node-to-node interactions, without a centralized entity, can still spawn cooperation toward a more efficient outcome. Classical (rational) game theory assumes that nodes (i.e., players) have well-defined and consistent goals, which can be described by a utility function, which can be seen as a measure of the players' satisfaction resulting from a certain game outcome. So, in these schemes, each node follows a strategy aiming at maximizing its benefits and minimizing its resource consumption. For example, a node decides to forward a message if the (direct or indirect) result of that action maximizes its delivery probability, or possibly it minimizes its end-to-end delay.

Some definitions are presented here:

Definition 15.1 *Game. A game constitutes a formal description of a strategic interaction between players.*

Definition 15.2 *Player. A player is an entity entitled to its own decisions and subsequent actions. It can also be interpreted as a node or as a group of nodes making decisions.*

Definition 15.3 *Action. An action is the act of performing a move in the game.*

---

* According to the *Oxford Dictionary of English* [20], collusion is a secret or illegal cooperation or conspiracy in order to deceive others.

Definition 15.4   *Payoff. A payoff (or utility), typically represented by a (positive or negative) number, reflects the desirability of an outcome to a player. As a consequence, it incorporates the player's attitude toward the risk.*

Definition 15.5   *Strategy. A strategy represents a set of actions that can be performed by the player during the game.*

Definition 15.6   *Payoff matrix. A payoff matrix is a matrix that represents the players, their strategies, and the payoffs for each player with every possible combination of actions.*

A plethora of ways to classify games is available nowadays [37,38]. Some classification examples are according to (1) the level of cooperation, (2) the symmetry of the payoff matrix (or the dependency between the strategy and the player); (3) the sum of the players' payoffs; (4) the amount of information known in advance; (5) how the plan of action is chosen; and (6) the number of players.

In relation to (1), games can be classified as cooperative or noncooperative. In noncooperative games, the game describes (or focuses on) the strategy of the node, where a node has to make a decision if it is going to cooperate or not with another random node. If the cooperation decisions are taken by a group of nodes, these types of games are called cooperative games. In relation to (2), games can be classified as symmetric or asymmetric. Symmetric games (also called matrix games) are those in which the strategy options and payoffs do not depend on the players, but only of the other strategies employed. All players have the same strategy set [39]. The games where the strategies of the players are not identical are known as asymmetric ones (also called bimatrix games). In relation to (3), games can be classified as zero-sum or non-zero-sum games. In zero-sum games, the sum of all players' payoffs is zero, for any possible outcome. Thus, a player's benefit is equal to the loss of other players. However, if, for any outcome, the sum of all players' payoffs is greater or less than zero, they are called non-zero-sum games. In relation to (4), games are classified as of perfect or imperfect information. If all previous players' moves are known, the game is of perfect information. A concept similar to a perfect information game is a complete information game, in which all players' strategies and payoffs are known, excluding their actions. In relation to (5), games can be classified as strategic or extensive. If each player chooses (once and for all) its action's plan and all players' decisions are simultaneously made, it is called a strategic game. In extensive games, each player can choose its action's plan while taking decisions, which does not have to happen at the beginning of the game. In relation to (6), games can be classified as two-player or multiplayer game. As the name suggests, the two-player games are those played by exactly two players. Instead, if there are more than two players, the game is called a multiplayer one.

15.4.2.3.1   Social Dilemmas

Classical game theory is based on the following key assumptions: (1) player's perfect rationality, that is, players have well-defined payoff functions, being fully aware of their own and their opponents' strategy options and payoff values; (2) that this is common knowledge, meaning that all players are aware of their own rationality and of the rationality of other players; and that all players are aware that all are rational ad infinitum. A strategy profile of a game is said to be a Nash equilibrium (NE), if and only if no player has an unilateral incentive to deviate and play another strategy, since there is no way it could be better off given the others' choices.

Social dilemmas occur in certain situations where the game has a single NE, which is not the Pareto efficient* so that the sum of individual utilities (social welfare) is not maximized in equilibrium. One of game theory's main tasks is to provide guidelines on how to solve social dilemmas and to provide

---

* The Pareto efficiency is an NE refinement concept used to provide equilibrium selection in cases where the NE concept alone could provide multiple solutions to the game.

insights on how player-to-player interactions (excluding the intervention of a central entity) may still generate an aggregate cooperation toward a more efficient outcome in many real-life situations.

The most popular dilemmas of cooperation are the snowdrift game, the stag-hunt game, and the prisoner's dilemma [40].

Stag Hunt: In Rousseau's A Discourse on Inequality in 1755, he described the stag-hunt game's story. In it, each hunter prefers stag S over hare H and hare over nothing. According to game theory, the highest income is achieved if each hunter chooses hunting stag, but the chance of a successful stag hunt increases with the number of hunters. So, there is nearly no chance of catching a stag alone, but the odds of getting a hare does not depend on others. Hence, the payoff matrix for two hunters is given by

		Hunter$_2$	
		H	S
Hunter$_1$	H	(1,1)	(2,0)
	S	(0,2)	(3,3)

Prisoner's Dilemma: In 1950, Tucker aiming at showing the difficulty of analyzing certain kinds of games previously studied by Dresher and Flood came up with the prisoner's dilemma story:

Two burglars (B_1 and B_2) are arrested after their joint burglary and held separately by the police. However, the police does not have sufficient proof in order to have them convicted; therefore, the prosecutor visits each of them and offers the same deal: if one confesses (called *defection D* in the context of game theory) and the other remains silent (called *cooperation C*—with the other prisoner), the silent accomplice receives a 3-year sentence and the confessor goes free. If both stay silent then the police can only give both burglars 1-year sentence for minor charges. If both confess, each burglar receives a 2-year sentence.

Let P be the punishment for mutual deflection, T the temptation to defect, S Sucker's payoff, and R the reward for mutual cooperation, such that the matrix payoff is

		B_2	
		D	C
B_1	D	(P,P)	(T,S)
	C	(S,T)	(R,R)

$S < P < R < T$ is the ranking ordering satisfied by the matrix elements.

Snowdrift: Two drivers trapped on opposite sides of a snowdrift have the following options: (1) cooperation, by getting out and shoveling, and (2) defection, by remaining in their cars. If both drivers decide to shovel, each of them gets the benefit b of getting home and both share the work's cost c. Therefore, each receives a mutual cooperation reward $R = b - c/2$. If they both choose to defect, none of them gets home and they both obtain no benefit ($P = 0$). If only one of them shovels, then both get home but the defector's income becomes $T = b$, as it was not reduced by shoveling, while the cooperator gets $S = b - c$.

Snowdrift is mathematically equivalent to the hawk–dove and the chicken games.

15.4.2.3.2 Strategies

Tit for Tat. Rapoport proposed the *tit for tat* strategy for the Axelrod computer tournament in 1984. This strategy, for the Iterated Prisoner's Dilemma, starts by cooperating in the first step and subsequently repeating the opponent's previous action. In the long run, the tit-for-tat strategy cannot be exploited, since it retaliates defection (never being also the first to defect) by playing defection until the coplayer decides on cooperating again. Thus, the extra income gained by the opponent during its first defection is returned to tit for tat. It is also a forgiving strategy as it is willing to cooperate again, defecting only if the opponent defects. This strategy effectively helps to maintain cooperative behaviors in multiplayer evolutionary prisoner's dilemma games [40]. Tit-for-tat modified versions were proposed to overcome the drawbacks of the strategy's determinism (e.g., in noise environments) such as the tit for two tats, which only defects if its opponents have defected twice in a row, and the generous (forgiving) tit-for-tat, which cooperates with some probability even if the coplayer has previously defected.

Win–stay, lose–shift (WSLS): The concept WSLS was introduced by Thorndike in 1911. WSLS strategies make use of a heuristic update rule depending on a direct payoff criterion (aspiration level) that dictates when the player should change its planned action. The player maintains its original action, if the recent rounds' average payoff is above the aspiration level, changing to a new one if not. By doing so, the aspiration level differentiates between winning and losing situations. A random choice can be performed, if there are multiple changing alternatives. An example of WSLS strategy is Pavlov [41], which was demonstrated to be able to defeat tit for tat in noisy environments (e.g., for the Iterated Prisoner's Dilemma), due to its ability to correct mistakes and exploit unconditional cooperators.

15.5 SECURITY IN DELAY-TOLERANT MOBILE CYBER-PHYSICAL APPLICATIONS

Security in delay-tolerant mobile cyber-physical applications has to be handled at two levels. First, *protection mechanisms* have to be used for ensuring authentication, confidentiality, integrity, and privacy. Second, *cooperation enforcement and trust mechanisms* have to be used to guarantee that misbehaving nodes do not impair communication. The following two sections survey work in these areas.

15.5.1 PROTECTION

Protection mechanisms for ensuring authentication, confidentiality, integrity, and privacy have been widely studied for decades and can today be better understood by resorting to textbooks in network security. Here, we are not going to focus on these mechanisms in general, but just on specific requirements of CPS, following Wang et al. [10]:

- *Sensing*: The integrity of the readings from environmental parameters and settings has to be assured, as they are often critical for CPS applications. Well-known mechanisms, such as physical shielding and access control, are extremely important to enforce this requirement. Moreover, more recent, hardware-based mechanisms, such as the Trusted Computing Group's Trusted Platform Module [42] and ARM's TrustZone [43], can be used. The former is a chip in x86 motherboards that allow attesting the integrity of the software of a node. The latter is a functionality of recent ARM processors, which allows running software components in an isolated environment.
- *Actuation*: CPS not only sense but also actuate on physical processes, which is even more critical, as it may have costs in terms of human health [44] or even the destruction of important devices [45]. This requirement ensures that actuation takes place under appropriate authorization. Besides that, mechanisms used to ensure the integrity of the sensing requirement? may also be used.
- *Communication*: The networked nature of some CPS applications allows the establishment of a network for data fusion, the delivery of data to back-end servers (e.g., cloud-based ones), or taking of coordinated response actions. This requirement imposes the use of secure (inter and intra) CPS communication mechanisms for protection from both active and passive adversaries. It encompasses a confidentiality and privacy protection scheme (involving encryption) to prevent eavesdropping and stealing of user's private information; a mutual authentication protocol, to address authenticity issues; an access control and authorization scheme, to address unauthorized access issues; a key management scheme, to generate

and distribute cryptographic keys; and an intrusion detection and prevention mechanism, to detect intrusions and block DoS attacks.

- *Computing*: Data collected and processed in a CPS platform have to be secured against physical and cyber tampering and invalid access. This can be accomplished using encryption, digital signatures, and access control.
- *Feedback*: Any CPS uses control loops that involve feedback data. This requirement is about the protection of the CPS's control system providing feedback for performing actuation. Most mechanisms already mentioned also apply for this requirement.

In DTNs, security mechanisms can be applied on a hop-by-hop (between two nodes that meet) or end-to-end (between sender and receiver) basis, depending on specific goals [46]. For instance, these mechanisms are important for nodes to verify if they are not relaying/forwarding data that were modified by another node in the path. Specific requirements of DTNs are as follows:

- *Authentication*: It has to be possible to distinguish legitimate nodes (i.e., nodes that belong to the DTN) from unauthorized ones.
- *Confidentiality*: Sensitive information cannot be disclosed to unauthorized third parties during its propagation through the DTN.
- *Integrity*: Transmitted data cannot be modified while in transit through the DTN.
- *Privacy*: It is more related to specific DTN application requirements. Sensitive information of an entity controlling/owning a DTN node cannot be disclosed to other entities.

All these requirements can be enforced using mostly cryptographic methods (availability being the exception). Confidentiality, integrity, and privacy can be mostly obtained using *symmetric cryptography* (e.g., AES) and *message authentication codes* (e.g., HMACs based on SHA-3). However, these methods require key distribution, which typically requires *public cryptography*, also required for authentication.

Public cryptography is usually based on a public key infrastructure (PKI) and a set of certification authorities, as in the World Wide Web. However, this is not an optimal approach due to the disconnected nature of DTN environments, where there is no end-to-end connectivity. Decentralized mechanisms, such as identity-based cryptography (IBC) [16], in which each node's identity acts as a key, are more suitable to DTNs in alternative to a PKI. Yet, IBC may not be feasible in certain DTN environments, as it calls for a global TTP for private key generation, to guarantee that new nodes can enter the network. Some alternatives have been proposed for such environments such as the use of trusted social contacts [47] or obfuscating routing information [48].

15.5.2 COOPERATION ENFORCEMENT AND TRUST

In a DTN, nodes can be controlled by rational users or entities. Due to resource scarcity, these entities may attempt to maximize their utilities and preserve their resources, that is, behaving selfishly by only forwarding messages for nodes with whom they have social ties with, which can significantly impact network performance [6,7]. In other words, selfish entities (or nodes) may be faithful to the nodes from the same group (based on a common interest) and uncooperative to outsiders. However, despite the fact of selfish behavior being harmful, selfish nodes can also be used to control message overhead in resource-limited networks.

Selfishness [49] measures the level of interaction or cooperation among nodes. However, the lack of connectivity or large transmission delays, which are typical in DTNs, makes the task of designing cooperation enforcement schemes for DTNs more challenging. Consequently, a selfish behavior is difficult to identify and measure, which is aggravated by the delayed feedback information. Still, recently, some DTN routing solutions have been proposed to address this problem [50–55].

As stated earlier, cooperation enforcement schemes for DTNs are categorized as reputation based, remuneration based, and game based. Next, cooperation enforcement schemes to handle selfishness in DTNs are reviewed.

15.5.2.1 Reputation-Based Mechanisms

RCAR: The authors of [51] proposed a reputation-based extension to the context aware routing [56] protocol, called RCAR, to address the problem of black holes in DTNs.

In RCAR, every node keeps a local notion of reputation, therefore avoiding the overhead and technical complication associated with a centralized reputation management system in decentralized networks. Upon message forwarding, each node estimates the likelihood of selecting forwarding nodes based on the node's reputation, that is, based on past interactions with possible forwarding nodes.

The reputation management system employs both data and acknowledgment messages. The data messages' format incorporates the nodes' list (*nlist*), that is, the list of nodes a message has passed through, as well as a list of digital signatures (*slist*) that is used to prove the integrity and authenticity of every node in *nlist*. The update mechanism, employed by the proposed reputation management system, does not disseminate updates based on broadcast/multicast mechanisms, which are expensive.

Reputation is maintained by means of three mechanisms: acknowledgments, nodes' list, and aging. Since each message contains a list of forwarding nodes that the message has traversed, upon message reception, nodes should update the reputation of the forwarding nodes in *nlist*. Despite that, the sender waits for an acknowledgment from the destination node and only increases the reputation of the forwarding node upon the reception of the acknowledgment. At last, the aging mechanism is used to decrease the reputation of all nodes. As messages can get lost, there is no way a node can know the reasons behind it, for example, message drop due to a black-hole node (which node misbehaved?), or due to buffer overflow, or even due to time-to-live expiration. To fulfill requirements of DTNs, the aging mechanism's decrease period is dynamically updated using the Kalman filters [57].

Give2Get (G2G): Mei and Stefa [52] proposed two strategy proof forwarding protocols for pocket-switched networks [58] of selfish individuals, namely, G2G Epidemic and G2G Delegation. The proposed protocols are strategy proof, which means that the strategies of following the protocol are NEs. So, no individual has any incentive to deviate.

In Epidemic Forwarding [59], nodes use every contact opportunity to forward messages. When a node is in contact with another one and it has a message that the other does not have, the message is relayed to the other node. If selfish nodes that simply drop messages are considered in the network (also called message droppers), Epidemic Forwarding performance degrades [7].

In Delegation Forwarding [60], nodes have associated to them a forwarding quality, which may depend on the message's destination. A message, upon its generation, is associated with the forwarding quality of the sender. Then, when a relay node gets in contact with another node, it checks whether the forwarding quality of the other relay node is higher than that of the forwarding quality of the message. If so, it creates a replica of the message, labels both messages with the forwarding quality of the other relay node, keeps one of them, and forwards the other message to the other relay node. If not, the message is not forwarded.

The idea behind G2G Epidemic Forwarding is that the protocol works correctly even if all nodes in the network are selfish. That can be accomplished if no selfish node has a better choice than following the protocol truthfully, thus being a NE. G2G Epidemic Forwarding consists of three phases: message generation, relay, and test. In the message generation phase, the message is modified, that is, the message sender is hidden from every possible relay except the destination, so that the relay candidate has no interest in not accepting it.* In the relay phase, nodes collect proof of relay (POR) to show to the source and/or previous relays during the test phase. In the test phase, nodes present evidence of their correct behavior, making it impossible for relays to drop messages. If no evidence is presented by the relay, a proof of misbehavior is generated and the relay is removed from the network.

---

* The authors made the following assumptions: (1) every node is selfish and accepts messages destined to it; (2) there are no Byzantine nodes in the network, that is, nodes that behave arbitrarily; (3) selfish nodes do not collude; and (4) nodes are capable of making use of public key cryptography.

G2G Delegation Forwarding makes use of G2G Epidemic Forwarding techniques with the intention of stopping message droppers since G2G Epidemic Forwarding techniques are not enough for the algorithms to be NE, due to the fact of selfish nodes (1) being able to lie about the forwarding quality (i.e., being liars) and (2) being able to change the forwarding quality of messages, for example, set it to zero, therefore getting rid of the message sooner (i.e., being cheaters). Two approaches have been devised for G2G Delegation Forwarding: delegation destination frequency and delegation destination last contact. In the former, a node forwards a message to another node if the other node has contacted the destination of the message more frequently than any other node in the nodes' list inside the message. In the latter, a node forwards a message to another node if the other node has contacted the destination of the message more recently than any other node in the nodes' list inside the message.

15.5.2.2 Remuneration-Based Mechanisms

Secure multilayer credit-based incentive (*SMART*). The authors of [50] proposed a SMART scheme to stimulate bundle forwarding cooperation among selfish nodes, which can be implemented in a distributed manner without relying on any tamperproof hardware in a DTN environment. In SMART, intermediate nodes, without the involvement of the sender, can transfer/distribute credit since a forwarding path cannot be predicted by the sender (unlike in MANETs where forwarding paths are known *a priori*), and due to nodes' mobility, intermediate nodes and the sender may be disconnected.

SMART assumes the existence of two main entities: an off-line security manager (OSM), which is responsible for key distribution, and a virtual bank (VB), that is, a special network component, such as a roadside unit in vehicular networks [61] or the information publisher in social networks [62], which takes charge of credit clearance. DTN nodes submit collected coins to the VB, by exploiting opportunistic links to these special network components. Upon joining the network, every node should register with the OSM. During the clearance phase, nodes should submit the collected layered coins to the VB in order to receive their rewards.

SMART is based on the concept of a layered coin—composed of multiple layers, where each layer is generated by the source/destination or an intermediate node—providing virtual electronic credits to charge for and reward the provision of data forwarding in DTN environments. The first layer, also known as the base layer, is created by the source node, and it indicates the payment rate (credit value), the remuneration conditions, and the class of service (CoS) requirements, besides other reward policies. In the following propagation process, a new layer, also known as the endorsed layer, which is built on the previous one, is created by each intermediate node by appending a nonforgeable digital signature. It implies that the forwarding node agrees in providing forwarding services underneath the predefined CoS requirement, thus being accordingly remunerated in accordance with the reward policy. By checking the signature at each endorsed layer, it is easy to check the propagation path and determine each intermediate node. If the provided forwarding service fulfills remuneration conditions defined in the predefined reward policy, in the rewarding and charging phase, each forwarding node along a single/multiple path(s) will share the credit defined in this coin depending on single/multicopy data-forwarding algorithms and its forwarding results.

Four aspects were considered in SMART's design, namely, effectiveness, by stimulating cooperation among selfish nodes; security, by being robust to various attacks; efficiency, by not introducing extra communication and transmission overhead; and generality, by being compatible with most existing DTN routing schemes. A trade-off had to be taken into account between security and performance. Security, as intermediate nodes, manages all security issues related to a coin; during the forwarding process, $a(n)$ (individual or social) selfish node(s) may attempt to maximize its expected benefit by cheating the system. The final aspect is performance, due to the extra computation and transmission overhead as a result of any security functionality, during the design of a secure credit-based incentive scheme.

Mobicent: Chen and Chan [53] proposed a credit-based system to support Internet access service for heterogeneous wireless network environments. Two modes of operation are supported by mobile devices in such environments, namely, (1) a long-range low-bandwidth link, for example, cellular interface, to maintain an always-on connection, used in particular by the source and destination nodes, and (2) a short-range high-bandwidth link, for example, Wi-Fi, to opportunistically exchange large amount of data with neighboring nodes (which is used by all nodes), since due to node mobility these links tend to be intermittent.

The Mobicent network architecture consists of three components: (1) a TTP, which stores key information for all nodes, providing also verification and remuneration services; (2) helpers, which are mobile or static nodes that help in data relaying using the short-range high-bandwidth link; and (3) mobile clients, which are the destination nodes.

The payment mechanism works as follows: the data payload is encrypted by each relay with a one-time symmetric key before being forwarded. When a client receives the encrypted data and intends to access the decrypted data, it must make a payment to the TTP in exchange for the encrypted keys. This happens since the key is sent along with the data in encrypted form and the TTP is the only means to recover them. Only relays involved in data forwarding receive payment.

Chen and Chan [53] dealt with two forms of selfish actions (or attacks), namely, edge insertion attacks and edge hiding attacks. Let $G = (V, E)$ be a contact graph. Each vertex (or node) $v \in V$ can be identified by an integer value $i = 1, 2, \ldots, |V|$. Each edge $e \in E$, identified by a pair $\{v_1, v_2\}$, denotes the opportunistic contact between two nodes at time $t(e)$. Thus, $(\{v_1, v_2\}, t_1)$ means that v_1 meets v_2 at time t_1. The former, that is, an *edge insertion attack* of a node v, consists in creating a Sybil v' so that $G \to G' = (V', E')$, where $V' = V \cup \{v'\}$ and $E' = E^{v \to (v,v')} \cup \{(v, v', t)\}$. The latter, that is, an *edge hiding attack* for a node v, consists in modifying $G \to G' = (V, E - e)$, where $e \in E(v)$.

According to the authors, due to network's dynamics, edge insertion and hiding attacks are extremely difficult to be detected in DTNs. As the proposed scheme provides incentives for self-ish nodes to honestly behave (i.e., by setting the client's payments and relay's rewards so that nodes behave truthfully), mechanisms to detect selfish actions are not required. And also, by working on top of the DTN routing layer, this scheme ensures that selfish actions do not result in a large reward, not requiring predetermined routing paths either.

15.5.2.3 Game-Based Mechanisms

TFT. Shevade and Zhang [54] propose a pairwise TFT—a simple, robust, and practical coop-eration enforcement scheme for DTNs—which incorporates generosity and contrition to tackle bootstrapping or exploitation problems common in basic TFT mechanisms [29,63,64].

Upon the first encounter between two nodes, as they have not previously relayed packets suc-cessfully among them, the basic TFT mechanisms would prevent relaying. Generosity enables bootstrapping by allowing an initial cooperation between the nodes up to ε, that is, a node is allowed to send ε packets more than it should according to what it had previously relayed. It also handles asymmetric traffic demands by absorbing traffic imbalance up to a ε amount. But any imbalance exceeding ε could lead to lengthy retaliation among neighbors. As a result, generosity is insufficient by itself. Contrition addresses the previous situation by refraining from reacting to a valid retaliation to its own mistake, that is, preventing mistakes from causing endless retaliation. With contrition, a node realizes that the other node's action in the current interval was due to its own action in the pre-vious interval and so does not lower service in the future interval. Similarly, contrition cannot work by itself, since it only provides a way to return to stability after perturbation, not providing a way to reach stability.

The authors also proposed an incentive-aware routing protocol in which selfish nodes are allowed to maximize their individual utilities taking into account the TFT constraints. For a given pair of nodes, the TFT constraints state that the total amount of traffic through a link is equal to the total amount of traffic in the opposite direction.

The proposed routing protocol consists of the following components: (1) link state (i.e., link capacity, mean, and variance of the waiting time on links) is periodically exchanged by every node, similarly to many link state protocols (e.g., OSPF)*; (2) with link state, each source node computes forwarding paths and uses source routing to send traffic; (3) each destination node sends an acknowledgement via flooding, upon receiving a packet. Then, the source node uses it to update its TFT constraints for the subsequent interval.

Barter: The authors of [55] proposed a mechanism, which is modeled using game theory, to discourage selfish behavior based on the principles of barter.† In the context of the proposed mechanism, exchanges are made in messages. Hence, when two mobile nodes are in communication range with each other, (1) they send messages' descriptions that they currently store to each other, (2) and then they reach a consensus on the subset of messages they want to exchange. Fairness is ensured (1) by guaranteeing that the selected subsets have the same size and (2) by using a preference order, in which messages are exchanged in a message-by-message manner. Notice that the number of exchanged messages depends on the length of the shorter list or the duration of the connection. The exchange can be interrupted, if any party cheats. So, any major disadvantage is not experienced by the honest party, since it at most downloaded one less message in comparison to the misbehaving party.

In this scheme, the mobile nodes decide which messages they want to download from each other. The nodes behave selfishly, that is, they only download messages that are of primary interest (or destined) to them, in the long run, and according to the principle of barter, they will not have other messages to exchange for the ones they are interested in. Therefore, messages that are secondary (or not destined) for a given mobile node may still have a barter value for the mobile node. Thus, it can be perceived as an investment to acquire new primary messages.

Two assumptions were made by the authors: (1) mobile nodes offer all their valid and only valid messages to download and (2) two mechanisms are present in the system to prevent the injection of fake messages, specifically, digital signatures, where only nodes with a digital signature, supplied by an authority, can exchange messages among themselves, and a reputation mechanism, which is based on the quality of the message contents.

Table 15.2 presents a summary and comparison of cooperation enforcement schemes applied to DTN routing protocols based on the security mechanism used, the main idea of the routing protocol, the attacks considered, and the overhead.

15.6 DELAY-TOLERANT MOBILE CYBER-PHYSICAL APPLICATIONS

In this section, some examples of delay-tolerant mobile cyber-physical applications prone to security issues are presented.

15.6.1 HEALTH MONITORING

With relatively affordable onboard sensors or connected external sensors, mobile cyber-physical applications can be leveraged to *monitor patient's health*. By means of wireless connectivity (e.g., cellular, Wi-Fi, Bluetooth), standard IP networking can be used to send data back to Internet services that aggregate information for doctors. Doctors can use this service to more precisely adjust medication dosages, based on trends in symptoms over the course of a day. Onboard smartphone sensor data collection is relatively easy for a mobile cyber-physical application, but processing and disseminating data is much more challenging for applications that use multiple external sensors networked through USB, Bluetooth, or other means. Therefore, appropriate architectures to buffer data when cellular connections are unavailable are necessary (thus involving delay tolerance), hence not overrunning the device's memory. Indeed, through onboard phone processing, the amount of data to be transmitted from the phone to the Internet service or buffered can be reduced.

An example of a health monitoring system is the mobile electrocardiogram (ECG) system [65] that uses smartphones as base stations for ECG measurement and analysis. The main idea behind mobile

---

* It is assumed Shevade and Zhang [54] that link state is disseminated faithfully. Thus, the authors focused on making the data-plane incentive compatible. Security of the control plane was left for future work.
† According to Stevenson [20], barter means "exchange (goods or services) for other goods or services without using money."

TABLE 15.2

Summary and Comparison of Incentive Mechanisms Applied to DTNs

Publication	Security Mechanism	Main Idea	Attack(s) Considered	Overhead
RCAR [51]	Reputation based	Every node keeps locally the reputation of every forwarding node it comes in contact with. Nodes with the highest reputation are selected as message forwarders.	Black-hole attack	The overhead is reduced, since RCAR reputation mechanism is integrated in the routing protocol (data and ACK messages). Thus, broadcast/multicast is not used by RCAR.
G2G [52]	Reputation based	Every node keeps locally evidences of their correct behavior as a relay. If no evidences are presented in future encounters, nodes are removed from the network.	Individual selfishness (message droppers, liars, and cheaters) Social selfishness (collusion[a])	Storage and communication overhead due to the use of PORs. The probabilities of detection of selfish behavior are of more than 90% and 60% for G2G Epidemic and Delegation Forwarding, respectively.
SMART [50]	Remuneration based	SMART scheme to stimulate bundle forwarding cooperation among selfish nodes in a DTN environment.	Credit forgery attack, nodular tontine attack, submission refusal attack	Additional cryptographic (computational and communication) overhead due to the use of layered coins. The performances of the underlying routing protocol, with or without SMART, are very close.
Mobicent [50]	Remuneration based	Runs on top of DTN routing layer and provides incentives for selfish nodes to behave honestly, thus not requiring a selfish action detection mechanisms.	Edge insertion attack, edge hiding attack	No overhead analysis mentioned in this chapter.
TFT [54]	Game based	Simple, robust, and practical incentive mechanism for DTNs that incorporate generosity and contrition.	Individual selfishness	Additional control overhead caused by the dissemination of link state and ACK messages.
Barter [55]	Game based	Incentive mechanism to discourage selfish behavior based on the principles of barter.	Individual selfishness	No overhead analysis mentioned in this chapter.

[a] Despite not being the target attack addressed in Mei and Stefa [52], the authors proposed two mechanisms, namely, random checks of conformity and rewarding traitors, to mitigate the presence of colluding nodes or to limit their possible harm.

ECG was, on the one hand, to reduce the workload of the medical personnel and, on the other hand, to accelerate the measurement to analysis cycle. The smartphone that received the transmitted data from the measurement device stored them in a memory card. So, under normal circumstances, the analysis is performed automatically near the patient, and only in critical cases (i.e., if some abnormities are detected), the data are sent to medical personnel for further analysis. In previous mobile measurement systems like the one proposed in Chen [66], smartphones, which were used as gateways, were used to send measurement data continuously to a separate server, where data analysis occurred, incurring in high financial costs as continuous data communication on mobile networks is charged by the amount of data sent [67]. Another feature of this system was the possibility of the patient sending an alarm in the case of occurrence of one of the following symptoms: anxiety, faintness, or other distress.

As stated previously, mobile measurement systems that rely on a mobile network for data communication to a server- or cloud-based infrastructure might incur in a financial penalty because of the amount of collected data by wearable and wireless medical sensors, which monitors, for example, heart rate, oxygen level, blood flow, respiratory rate, muscle activities, movement pattern, body inclination, and oxygen uptake. Since the measurement data are stored locally on the smartphone, a possible approach could be the use of data mules to carry the data to the hospital provided that encryption schemes were in place to safeguard the data.

In Syed-Abdul [68], a study of the potential of DTNs to support health care services, such as e-mail access, notification of lab results, backup of electronic health records, and teleconsultation in a low-resource setting, is presented. As each service has its own requirements, the frequency of data delivery is context dependent. For example, for services like notification of lab results and/or ordering of medical supplies, the physical transportation of digital data at a frequency of less than once per week is tolerable.

15.6.2 CARTEL

The CarTel project [69] designed a multipurpose distributed sensor computing system to collect, process, deliver, and visualize data from remote and intermittently connected sensors located on mobile units (e.g., cars), which in comparison to static sensor networks can sense at much better fidelity and higher scale the environment, in particular over large areas. It is a multipurpose project as its application areas are diverse such as traffic monitoring, route planning, environmental monitoring, civil infrastructure monitoring, automotive diagnostics, augmented reality, and data muling.

A CarTel node consists of a mobile embedded computer with a set of sensors, which collects and processes sensor readings locally before sending them to a central portal, for further processing.

CarTel is composed of three main components:

1. The portal, which is a central location that hosts CarTel applications and functions, being the point of control and configuration for the distributed system. All data collected by the mobile nodes are sent to the portal.
2. The intermittently connected database (ICEDB), which is a delay-tolerant continuous query processor. ICEDB distributes query execution and results between the ICEDB server running on the portal and the remote nodes. It supports heterogeneous data types and its queries are written in SQL with several extensions for continuous queries and prioritization.
3. The carry-and-forward network (CafNet), which is a general-purpose network stack for delay-tolerant communication, which can be used by applications to send messages across an intermittently connected network. Two kinds of intermittency are envisioned:
   a. The first one, through opportunistic networking (e.g., Wi-Fi, Bluetooth), since end-to-end connectivity is available but intermittent. Mobile nodes can access Wi-Fi access points to communicate with the portal.
   b. The second one, through a best-effort approach, by using other mobile node storage devices, such as USB keys and flash memory as data mules, relying on them to deliver data to the portal.

ICEDB and CafNet specify how nodes collect, process, and deliver sensor data. Generally, CarTel applications use three main components of the portal environment, namely, (1) the portal framework, (2) the ICEDB server used to retrieve sensor data, and (3) a data visualization library used to display geocoded attributes. The portal framework provides a platform for building applications that share a mutual user authentication mechanism and look and feel. And in order to lessen privacy concerns, users are only allowed to view collected sensor data from remote nodes hosted by them.

For example, in the traffic monitoring CarTel application, each car is instrumented with a GPS sensor to opportunistically gather information, such as traffic delays observed as cars move, allowing to infer congestion hot spots. Furthermore, equipping cars with cameras allows building applications that can help users to navigate unfamiliar territories.

15.7 CONCLUSION

A discussion of the main security concerns in delay-tolerant cyber-physical applications was presented. Precisely, issues related to authentication, confidentiality, integrity, availability, privacy, trust, and cooperation enforcement were analyzed. A survey of cooperation enforcement schemes available for DTNs was also presented.

The design of delay-tolerant mobile cyber-physical applications poses significant security challenges. The increased interaction between the physical systems and cyber systems exposes physical systems to cyber-system security threats. Mobility, self-organization, and delay tolerance require finding a suitable node to forward messages, which may incur in the consumption of nodes' limited resources, along with additional concerns such as trust establishment and fairness of contributions among the nodes.

Using cooperation enforcement (reputation based, remuneration based, or game based) and encryption schemes, security in delay-tolerant mobile cyber-physical applications can be leveraged, providing confidentiality, integrity, and availability. Indeed, such schemes may incur in additional computational costs and/or communication overhead. For example, additional data may be transferred between nodes to establish trust, or the sensitive application's data traversing the network need to be encrypted, in a hop-by-hop or end-to-end manner, thus incurring in additional computation costs. Additionally, a trusted third party may be used to help in providing such services, which also increases the complexity of the architecture and protocols.

REFERENCES

1. J. Sztipanovits, Composition of cyber-physical systems, in *Engineering of Computer-Based Systems*, pp. 3–6, IEEE Computer Society, Los Alamitos, CA, 2007.
2. E.A. Lee, Cyber physical systems: Design challenges, in *IEEE International Symposium on Object Oriented Real-Time Distributed Computing (ISORC)*, Orlando, FL, pp. 363–369, 2008.
3. J. White et al., R&D challenges and solutions for mobile cyber-physical applications and supporting Internet services, *Internet Services and Applications*, 1(1), 45–56, 2010.
4. M.J. Khabbaz, C.M. Assi, and W.F. Fawaz, Disruption-tolerant networking: A comprehensive survey on recent developments and persisting challenges, *IEEE Communications Surveys and Tutorials*, 14(2), 607–640, 2012.
5. K. Wei, X. Liang, and K. Xu, A survey of social-aware routing protocols in delay tolerant networks: Applications, taxonomy and design-related issues, *IEEE Communications Surveys and Tutorials*, 16(1), 556–578, 2014.
6. P. Hui et al., Selfishness, altruism and message spreading in mobile social networks, in *IEEE INFOCOM*, Rio de Janeiro, Brazil, pp. 1–6, 2009.
7. N. Magaia, P.R. Pereira, and M.P. Correia, Selfish and malicious behavior in Delay-Tolerant Networks, in *Future Network and Mobile Summit*, Lisboa, Portugal, pp. 1–10, 2013.
8. Z. Liu, A.W. Joy, and R.A. Thompson, A dynamic trust model for mobile ad hoc networks, in *Future Trends of Distributed Computing Systems*, Suzhou, China, pp. 80–85, 2004.
9. Y. Zhu, B. Xu, X. Shi, and Y. Wang, A survey of social-based routing in delay tolerant networks: Positive and negative social effects, *IEEE Communications Surveys & Tutorials*, 15(1), 387–401, 2013.
10. E.K. Wang et al., Security issues and challenges for cyber physical system, in *IEEE/ACM Int'l Conference on Green Computing and Communications and International Conference on Cyber, Physical and Social Computing*, Washington, DC, pp. 733–738, 2010.
11. E. Mills, Hackers broke into FAA air traffic networks, *The Wall Street Journal*, A6, May 2009. [Online]. http://online.wsj.com/articles/SB124165272826193727.

12. N. Leavitt, Researchers fight to keep implanted medical devices safe from hackers, *Computer*, 43(8), 11–14, 2010.
13. K. O'Connell, CIA report: Cyber extortionists attacked foreign power grid, disrupting delivery, *Internet Business Law Services*, 2008. [Online]. http://www.ibls.com/internet_law_news_portal_view.aspx?id=1963&s=latestnews. Accessed on October 31, 2014.
14. K. Koscher et al., Experimental security analysis of a modern automobile, in *Security and Privacy*, pp. 447–462, 2010.
15. W. Stallings, *Network Security Essentials: Applications and Standards*, Pearson Education Limited, India, 2007.
16. A. Shikfa, M. Onen, and R. Molva, Privacy and confidentiality in context-based and epidemic forwarding, *Computer Communications*, 33(13), 1493–1504, 2010.
17. Z. Le, G. Vakde, and M. Wright, PEON: Privacy-enhanced opportunistic networks with applications in assistive environments, in *Pervasive Technologies Related to Assistive Environments*, p. 76, 2009.
18. N. Magaia, P.R. Pereira, and M.P. Correia, Nodes' misbehavior in vehicular delay-tolerant networks, in *Conference on Future Internet Communications (CFIC)*, Lisbon, Portugal, pp. 1–9, 2013.
19. A. Keränen, J. Ott, and T. Kärkkäinen, The ONE simulator for DTN protocol evaluation, in *Proceedings of the Second International Conference on Simulation Tools and Techniques*, Brussels, Belgium, 2009, p. 55.
20. A. Stevenson, *Oxford Dictionary of English*, Oxford University Press, New York, 2010.
21. K. Cook, *Trust in Society*, Russell Sage Foundation, New York, 2001.
22. J.-H. Cho, A. Swami, and R. Chen, A survey on trust management for mobile ad hoc networks, *IEEE Communications Surveys and Tutorials*, 13(4), 562–583, 2011.
23. M. Blaze, J. Feigenbaum, and J. Lacy, Decentralized trust management, in *Security and Privacy*, pp. 164–173, 1996.
24. L. Strigini et al., Resilience-building technologies: State of knowledge, 2007.
25. L. Buttyán and J.-P. Hubaux, Stimulating cooperation in self-organizing mobile ad hoc networks, *Mobile Networks and Applications*, 8(5), 579–592, 2003.
26. M. Jakobsson, J.-P. Hubaux, and L. Buttyán, A micro-payment scheme encouraging collaboration in multi-hop cellular networks, in *Financial Cryptography*, Guadeloupe, French West Indies, pp. 15–33, 2003.
27. L. Blazevic et al., Self organization in mobile ad hoc networks: The approach of Terminodes, *IEEE Communications Magazine*, 39(6), 166–174, 2001.
28. L. Buttyán and J.-P. Hubaux, Enforcing service availability in mobile ad-hoc WANs, in *Mobile Ad Hoc Networking & Computing*, pp. 87–96, 2000.
29. V. Srinivasan, P. Nuggehalli, C.-F. Chiasserini, and R.R. Rao, Cooperation in wireless ad hoc networks, in *IEEE INFOCOM*, San Francisco, CA, vol. 2, pp. 808–817, 2003.
30. L. Anderegg and S. Eidenbenz, Ad hoc-VCG: A truthful and cost-efficient routing protocol for mobile ad hoc networks with selfish agents, in *Mobile Computing and Networking*, 2003, pp. 245–259.
31. W. Wang, S. Eidenbenz, Y. Wang, and X.-Y. Li, OURS: Optimal unicast routing systems in non-cooperative wireless networks, in *Mobile Computing and Networking*, pp. 402–413, 2006.
32. W.Z Wang, X.-Y. Li, and Y. Wang, Truthful multicast routing in selfish wireless networks, in *Mobile Computing and Networking*, pp. 245–259, 2004.
33. S. Zhong, L.E. Li, Y.G. Liu, and Y.R. Yang, On designing incentive-compatible routing and forwarding protocols in wireless ad-hoc networks, *Wireless Networks*, 13(6), 799–816, 2007.
34. P. Michiardi and R. Molva, Core: A collaborative reputation mechanism to enforce node cooperation in mobile ad hoc networks, in *Advanced Communications and Multimedia Security*, Springer, pp. 107–121, 2002.
35. S. Marti, T.J. Giuli, K. Lai, and M. Baker, Mitigating routing misbehavior in mobile ad hoc networks, in *Mobile Computing and Networking*, pp. 255–265, 2000.
36. J. Sen, A survey on reputation and trust-based systems for wireless communication networks, *arXiv preprint arXiv:1012.2529*, 2010.
37. M.J. Osborne and A. Rubinstein, *A Course in Game Theory*, MIT Press, Cambridge, MA, 1994.
38. T.L. Turocy, Texas a&m university, Bernhard von Stengel, London School of Economics "Game Theory" *CDAM Research Report (October 2001)*, Elsevier, New York, 2001.
39. S.-F. Cheng, D.M. Reeves, Y. Vorobeychik, and M.P. Wellman, Notes on equilibria in symmetric games, in *Game Theory and Decision Theory*, 2004.
40. G. Szabó and G. Fath, Evolutionary games on graphs, *Physics Reports*, 446(4), 97–216, 2007.
41. D. Kraines and V. Kraines, Pavlov and the prisoner's dilemma, *Theory and Decision*, 26(1), 47–79, 1989.

42. Trusted Computing Group, Trusted Platform Module Library Specification, Family "2.0"", Level 00, Revision 01.07, March 2004.
43. T. Alves and D. Felton, TrustZone: Integrated hardware and software security, *ARM White Paper*, 3(4).
44. J. Robertson, The trials of a diabetic hacker, *BusinessWeek*, February 2012. [Online]. http://www.businessweek.com/articles/2012–02–23/the-trials-of-a-diabetic-hacker.
45. J. Meserve, Sources: Staged cyber attack reveals vulnerability in power grid, *CNN*, September 2007. [Online]. http://edition.cnn.com/2007/US/09/26/power.at.risk/.
46. S. Farrell, H. Weiss, S. Symington, and P. Lovell, Bundle security protocol specification, *RFC*, 6257, 2011. [Online]. https://tools.ietf.org/html/rfc6257.
47. K.E. Defrawy, J. Solis, and G. Tsudik, Leveraging social contacts for message confidentiality in delay tolerant networks, *IEEE International Computer Software and Applications Conference*, 1, 271–279, 2009.
48. I. Parris and T. Henderson, Privacy-enhanced social-network routing, *Computer Communications*, 35(1), 62–74, 2012.
49. Q. Li, S. Zhu, and G. Cao, Routing in socially selfish delay tolerant networks, in *IEEE INFOCOM*, pp. 1–9, 2010.
50. H. Zhu, X. Lin, R. Lu, Y. Fan, and X. Shen, Smart: A secure multilayer credit-based incentive scheme for delay-tolerant networks, *IEEE Transactions on Vehicular Technology*, 58(8), 4628–4639, 2009.
51. G. Dini and A.L. Duca, Towards a reputation-based routing protocol to contrast blackholes in a delay tolerant network, *Ad Hoc Networks*, 10(7), 1167–1178, September 2012.
52. A. Mei and J. Stefa, Give2Get: Forwarding in social mobile wireless networks of selfish individuals, *IEEE Transactions on Dependable and Secure Computing*, 9(4), 569–582, July 2012.
53. B.B. Chen and M.C. Chan, MobiCent: A credit-based incentive system for disruption tolerant network, in *IEEE INFOCOM*, San Diego, CA, pp. 1–9, 2010.
54. U. Shevade and Y. Zhang, Incentive-aware routing in DTNs, in *IEEE International Conference on Network Protocols*, Orlando, FL, pp. 238–247, 2008.
55. L. Buttyán, L. Dóra, M. Félegyházi, and I. Vajda, Barter trade improves message delivery in opportunistic networks, *Ad Hoc Networks*, 8(1), 1–14, January 2010.
56. M. Musolesi and C. Mascolo, CAR: Context-aware adaptive routing for delay-tolerant mobile networks, *IEEE Transactions on Mobile Computing*, 8(2), 246–260, February 2009.
57. R. Kalman, A new approach to linear filtering and prediction problems, *Basic Engineering*, 82(1), 35, March 1960.
58. P. Hui et al., Pocket switched networks and human mobility in conference environments, in *ACM SIGCOMM Workshop on Delay-Tolerant Networking—WDTN*, New York, pp. 244–251, 2005.
59. A. Vahdat and D. Becker, Epidemic routing for partially connected ad hoc networks, Duke University, Durham, NC, Technical report 2000.
60. V. Erramilli, A. Chaintreau, M. Crovella, and C. Diot, Delegation forwarding, in *Mobile Ad Hoc Networking and Computing—MobiHoc*, New York, p. 251, 2008.
61. S.-B. Lee, G. Pan, J.-S. Park, M. Gerla, and S. Lu, Secure incentives for commercial ad dissemination in vehicular networks, in *Mobile Ad Hoc Networking and Computing—MobiHoc*, New York, p. 150, 2007.
62. A. Garyfalos and K.C. Almeroth., Coupons: A multilevel incentive scheme for information dissemination in mobile networks, *IEEE Transactions on Mobile Computing*, 7(6), 792–804, June 2008.
63. J.J. Jaramillo and R. Srikant, DARWIN, in *Mobile Computing and Networking—MobiCom*, New York, p. 87, 2007.
64. F. Milan, J.J. Jaramillo, and R. Srikant, Achieving cooperation in multihop wireless networks of selfish nodes, in *Game Theory for Communications and Networks—GameNets*, New York, p. 3, 2006.
65. H. Kailanto, E. Hyvarinen, and J. Hyttinen, Mobile ECG measurement and analysis system using mobile phone as the base station, in *Pervasive Computing Technologies for Health Care*, pp. 12–14, January 2008.
66. W. Chen et al., A mobile phone-based wearable vital signs monitoring system, in *Computer and Information Technology*, pp. 950–955, 2005.
67. J. Rodriguez, L. Dranca, A. Goñi, and A. Illarramendi, Web access to data in a mobile ECG monitoring system, in *Transformation of Health Care with Information Technologies.*: IOS Press, pp. 100–111, 2004, ch. Web access to data in a mobile ECG monitoring system.
68. S. Syed-Abdul et al., Study on the potential for delay tolerant networks by health workers in low resource settings, *Computer Methods and Programs in Biomedicine*, 107(3), 557–64, September 2012.
69. B. Hull et al., CarTel: A distributed mobile sensor computing system, in *Embedded Networked Sensor Systems*, pp. 125–138, 2006.

16 Towards Secure Network Coding–Enabled Wireless Sensor Networks in Cyber-Physical Systems

*Alireza Esfahani, Georgios Mantas, Du Yang,
Alberto Nascimento, Jonathan Rodriguez,
and Jose Carlos Neves*

CONTENTS

16.1 INTRODUCTION

Cyber-physical systems are complex systems that integrate computation and networking with physical processes. Computing devices and networks are used to monitor and control the physical processes with feedback loops where physical processes affect computations and vice versa. In other words, cyber-physical systems target at monitoring the behavior of the physical processes and actuating, when it is required, in order to control and make them work properly. Today, cyber-physical systems are found in a wide spectrum of important daily life domains such as physical rehabilitation and telemedicine, elderly assisted living, critical infrastructure control (e.g., electric power, water resources), traffic management, environmental control, public safety, smart cities, smart building automation, automobiles, avionics, planet exploration, social networking, and video games [1–4].

Since the monitoring of physical processes constitutes a fundamental function in cyber-physical systems, a lot of research effort has been placed on modern cyber-physical systems in order for their monitoring operation to be based on wireless sensor networks (WSNs) and benefit the advantages derived from the wireless communication in terms of connectivity, flexibility, and scalability. Hence, modern cyber-physical systems integrate multiple WSNs consisting of distributed wireless sensor nodes within the environment of the monitored physical processes. The wireless sensors are responsible to gather data related to the state of the physical processes and send them to the controller entity through single-hop or multihop communications over wireless channels. The number and scale of WSNs integrated by a cyber-physical systems depend highly on the target CPS application [4,5].

However, WSNs are characterized by low communication bandwidth, packet loss, and power consumption constraints that are all inherited by cyber-physical systems that integrate WSNs [6]. For this reason, the success and prosperity of cyber-physical systems are affected significantly by advances against these limitations of WSNs. In this sense, Network Coding–enabled WSNs, where Network Coding (NC) technology is applied to WSNs, can be adopted by cyber-physical systems in order to benefit from the improvements of the application of NC technology in WSNs. Particularly, NC can provide network capacity improvement [7], robustness to packet losses [8], and lower energy consumption [9]. Nevertheless, from the security point of view, NC offers both benefits and security vulnerabilities [8]. Based on that and the fact that security in cyber-physical systems is of utmost importance, since most of the CPS applications are safety critical, we steer our focus on the mitigation of those security attacks against NC-enabled WSNs that exploit the vulnerabilities of the NC technology and compromise the security of the whole CPS. Any disruption or outage of cyber-physical systems can severely affect not only the physical processes being controlled but also the people who depend on them [2].

Therefore, in this chapter, we intend to provide a foundation for organizing research efforts toward the deployment of the appropriate countermeasures against attacks targeting NC-enabled WSNs integrated in cyber-physical systems. In this sense, we provide a categorization of the current existing NC-enabled WSN attacks that can have a negative impact on the security of the whole CPS. Furthermore, we present mitigation techniques, derived from the literature, for the most common included in the attack categories that we have defined. It is our hope that this work will help researchers to design and develop more reliable and efficient defense mechanisms against known attacks for NC-enabled WSNs in cyber-physical systems or against attacks that have not yet appeared but can be potential threats in the future.

Following the introduction, this chapter is organized as follows. In Section 16.2, we give an overview of the CPS architecture and provide examples of current CPS applications in order to demonstrate efficiently the main CPS functionalities according to the CPS architecture. In Section 16.3, the adoption of NC-enabled WSNs in CPS applications is discussed. In Section 16.4, a comprehensive review of the security requirements for NC-enabled WSNs deployed in cyber-physical systems is provided. In Section 16.5, the main categories of attacks against NC-enabled WSNs that can affect the security of the whole CPS are presented. In Section 16.6, countermeasures for attacks included in the defined categories in Section 16.5 are discussed. Finally, Section 16.7 concludes this chapter.

16.2 OVERVIEW OF CYBER-PHYSICAL SYSTEMS ARCHITECTURE

A CPS consists of a central processing unit (i.e., controller) interacting with the physical world (i.e., physical process) through multiple sensors and actuators in a feedback loop as shown in Figure 16.1. This feedback architecture is also called a *closed loop* and implies the integration of the cyber world with the physical world. Furthermore, a CPS may include many different types of networks spanning from WSNs to the Internet interconnecting its many different communicating entities [2,4,5]. The sensors of the CPS are used to sense the operation of the monitored physical process and send the collected data to the controller. Upon receipt of the gathered data, the controller processes and analyzes it in order to take decisions regarding the required actions that have to be performed onto the monitored physical process so as to ensure the desired behavior of the physical process according to its original objectives. Afterward, based on the defined actions, the appropriate control commands are determined and sent to the actuators, which are responsible to perform the corresponding actions and change the behavior of the physical process [3,5].

Since a lot of CPS applications have been deployed recently adding more intelligence to the social life, some of them are reviewed to demonstrate efficiently the main functionalities of cyber-physical systems taking into consideration the closed-loop architecture. For instance, a lot of examples of CPS applications are derived from the health care sector targeting at the provision of high-quality health care services to patients or elderly people [10–12]. In a typical health care CPS application, vital signs are gathered from the human body through medical sensors worn by patients or elderly people. Afterward, the obtained data are sent to a central processing unit located at a hospital or a health care center, where the obtained data are processed and analyzed in order to take decisions and determine the required actions that have to be performed. Additionally, the data can be stored for further processing and analysis [13,14]. For example, as a health care CPS application can be considered the health care monitoring architecture presented by Huang et al. [15]. This architecture integrates wearable sensor systems and an environmental sensor network for monitoring elderly or

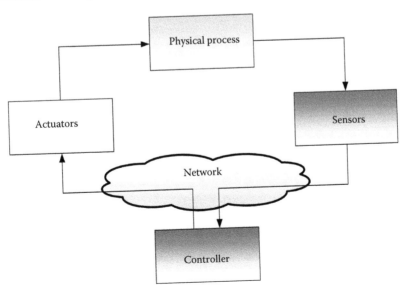

FIGURE 16.1 General CPS architecture.

chronic patients in the residence. The wearable sensor system consists of various medical sensors that collect a timely set of vital signs transmitted via low-energy wireless communication to mobile computing devices.

Furthermore, the environmental monitoring and control application is an additional example of CPS applications. The environmental monitoring can help to enhance the human ability to realize the real world [16]. Environmental monitoring can be classified into two categories: indoor and outdoor monitoring [17]. Indoor environmental monitoring applications may include power monitoring, prod-uct address monitoring, and civil structures deformations monitoring. On the other hand, the outdoor environmental monitoring applications may consist of habitat monitoring, weather forecasting, and earthquake detection.

Finally, vehicular sensor networks (VSNs) have been receiving a lot of attention in smart trans-portation applications, which can be considered as CPS applications too. Smart transportation applications can improve road safety [16].

16.3 NETWORK CODING–ENABLED WSNs IN CYBER-PHYSICAL SYSTEMS

WSNs have been recently used in many CPS applications, such as environment monitoring, traffic control, health care applications, elder assistance, and so forth. Due to their wide range of applica-tions and the fact that they offer a lot of new possibilities, WSNs have already become considerably attractive to research community. However, low communication bandwidth, packet loss, and power consumption constraints are the main challenges that the research community has to face in order for WSNs to reach their full potential and contribute significantly to the success and advancement of cyber-physical systems [6]. Toward this direction, NC can be a promising solution for WSNs. Compared to the classical commodity flow, where the information is only routed or replicated, in an NC-enabled network the information flow can also be mixed, manipulated, encoded, and decoded at the nodes. NC does not only preserve most of the store-and-forward paradigm features but also provides significant benefits to the network on which network capacity improvement, robustness to packet losses, and lower energy consumption are applied.

Figure 16.2 demonstrates how the capacity of a network is improved when the NC method is applied instead of the traditional store-forward method. In this demonstration, there are two nodes A and B which transmit their packets a and b to each other through an intermediate node S, respectively.

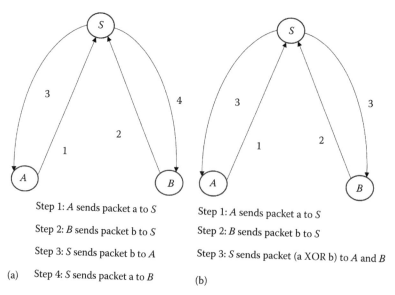

Step 1: A sends packet a to S

Step 2: B sends packet b to S

Step 3: S sends packet b to A

(a)   Step 4: S sends packet a to B

Step 1: A sends packet a to S

Step 2: B sends packet b to S

Step 3: S sends packet (a XOR b) to A and B

(b)

FIGURE 16.2 (a) Traditional store-forward method and (b) Network Coding examples of three nodes.

In the traditional store-forward method, A transmits its packet to S and after that B transmits its packet to S. Afterward, S transmits the received packet from A to B and the received packet from B to A. On the other hand, in NC method A and B transmit their packets a and b in a similar way to the traditional store-forward method. However, S transmits only the XOR result of the two packets, $a \oplus b$, to both A and B simultaneously. This capability provides one transmission step less for the NC method compared to the traditional store-forward method.

16.3.1 NETWORK CODING PRELIMINARY

To study Network Coding, we consider single-source multicast transmission on a network, which is defined by a triple (G,S,I) consisting of the following components:

Acyclic and Directed Multi-Graph G: We consider a pair of (V, E) as a directed acyclic graph where V represents the set of all nodes of G and E represents the set of all edges of G.

Source Node S: In our network model, we have a source S that transmits messages in multicast mode. The source S is one of the nodes of G.

Nonsource Node I : In our network model, we consider that the intermediate and sink nodes are included in a set of nodes, which is defined as $I = \{\forall x \in V - \{S\}\}$.

More precisely, the source S wants to send packets $u_1, \ldots, u_m$ to multiple sinks. Each message u_i is divided into n packets and each packet includes independent random symbols from the finite field $\mathbb{F}_p$. In NC model, each node has the ability to encode its incoming data in order to generate its outgoing packets. Hence, for each node, NC uses an encoding function f_v to generate the output packets by taking as input the packets arriving at the node. We demonstrate the NC operation at the node $v \in I, S$ as

$$\left(y_1^v, \ldots, y_l^v\right) = f_v\left(x_1^v, \ldots, x_d^v\right) \tag{16.1}$$

where parameters l and d are the number of packets departing from and arriving at the node v, respectively. For the source node S, the input packets in Equation 16.2 will be the original message packets. On the other hand, the sink nodes can recover the original message packets from the output packets in Equation 16.3. In other words, we assume without loss of generality that the sink nodes do not have any output edge.

16.3.2 NETWORK CODING BACKGROUND

According to the max-flow min-cut theorem [18], the first advent of the NC scheme was proposed by Ahlswede et al. [7] in 2000. After that, the authors in Cai and Yeung [19] demonstrated a model for a collection of subsets of wiretap channels for an NC system called *wiretap network*. Subsequently, a solution for error correction was presented in Cai and Yeung [20] further. Furthermore, a different approach to NC was proposed Koetter and Medard [21], where an algebraic framework is derived to verify the feasibility of the multicast problem.

Thus, according to this definition, the max-flow min-cut theorem was modified to the problem of finding a point on an algebraic set. Subsequently, Ho et al. [22] proposed the multicast problem in a linear NC scenario that improved the algebraic framework presented in previous work [21]. Moreover, [22] investigated an upper and lower bound required for the alphabet size of the network codes. The authors explained the relation between a linear dispersion and a generic NC. Additionally, they found a relation on the sizes of the base fields of the code. Next, Jaggi et al. [23] provided deterministic polynomial time algorithms and randomized algorithms for linear NC in directed acyclic graphs. They also realized a connection between linear NC and linear system theory, especially with the theory of codes on graphs.

Finally, in a random NC approach presented in Ho et al. [24], the source tries to multicast an original information and the intermediate nodes transmit the linear combinations of the incoming information on the outgoing channels, using independently and randomly chosen code coefficients from some finite fields. Nevertheless, in order for the sink to recover the original information, a decoding process is required. Thanks to this approach, there are possibilities that there will be NC schemes such as the stateless NC scheme that can be suitable for networks with unknown or changing topologies.

16.3.3 NETWORK CODING PROTOCOLS

According to Lima et al. [25], NC protocols can be categorized into the following two main types of protocols: stateless and state-aware NC-based protocols. This categorization is based on whether the NC protocol makes use or does not make use of the network state information that includes information related to the state of the network such as network topology, link cost, and node location.

16.3.3.1 Stateless NC Protocols

Stateless NC-based protocols do not rely on network state information, as mentioned earlier, to perform coding operations such as the mixing of data packets. In stateless NC, coding operations can work properly even in the presence of dynamically changing topologies, such as in mobile ad hoc networks (MANETs). For instance, Random Linear Network Coding (RLNC) [24] is a distributed stateless NC protocol in which a node without any network state information can randomly and independently choose a set of coefficients to perform a linear combination of two or more received data packets and then send the resulted packet to its output links.

Stateless NC protocols are more immune against different types of attacks compared to the state-aware NC-based protocols because of their independence of the network state information [26]. However, due to this independence, these protocols not only need more sophisticated coding operations, but also the network codes are required to be selected from a sufficiently large field in order to guarantee that encoding and decoding operations can perform correctly in the source and sink nodes. This results in increasing the data overhead in the network.

16.3.3.2 State-Aware NC Protocols

State-aware NC protocols require that each node obtains partial or full network state information. A very famous state-aware NC-based protocol is COPE, which is presented in Katti et al. [27]. This protocol uses local information and network state information to achieve improvements in terms of throughput and robustness. More specifically, COPE proposes an opportunistic NC in order to reduce the number of transmissions. In state-aware NC protocols, the fact that each node has knowledge of network state information such as the other nodes' location, network topology, and link state leads to security vulnerabilities.

16.3.4 SECURE NETWORK CODING

In one of the very earliest attempts, Krohn, Freedman, and Mazieres [28] proposed a security hashing scheme based on homomorphic functions to validate blocks of rate-less codes only at source and receiver nodes. After that in 2006, Gkantsidis and Rodriguez [29] analyzed the disruptive and epidemic effect of packet corruption on NC systems. Moreover, they proposed the use of signatures and hashes for addressing this issue instead of using traditional cryptographic schemes.

The proposed work by Han et al. [30] overviewed the security capabilities of the NC against Byzantine attacks in P2P scenarios. Ho et al. [24] proposed an information-theoretic approach for detecting Byzantine attacks in multicast networks by using RLNC and augmenting each source packet with a flexible number of hash symbols calculated as a polynomial function of the data symbols. This model assumes that the attacker does not have full access to all random coding coefficients of all the packets sent to the sink nodes. Siavoshani et al. [31] proposed a mechanism for locating the exact Byzantine attackers. Yu et al. [32] defined a homomorphic signature scheme for both linear and XOR NC to prevent pollution attacks. In this scheme, they exploit probabilistic key pre-distribution as well as message authentication codes (MACs), and it can filter polluted messages in a few hops with a high probability.

In the context of CPS, two NC schemes have been proposed. Su and Zhang [33] used NC in order to support a QoS-provisioning MAC protocol. Moreover, in Prior et al. [34] NC strategies are presented to be implemented in smart grid applications. However, the main challenge here is to deal with the coding type and the instruction that brings some constraints. Therefore, the authors proposed two protocols for applying NC in smart grid communications, for example, advanced metering infrastructures. The proposed protocols are based on tunable sparse NC.

16.4 SECURITY REQUIREMENTS FOR NETWORK CODING–ENABLED WSNs IN CYBER-PHYSICAL SYSTEMS

Ensuring security in NC-enabled WSNs is a critical factor for the success of CPS applications since attacks exploiting the vulnerabilities of NC-enabled WSNs can affect the security of the whole CPS application. However, NC-enabled WSNs can be the target of many known and unknown security threats since many security breaches can raise in its environment due to its heterogeneous and dynamic nature. Hence, the following security properties are required to be assured in NC-enabled WSNs integrated by cyber-physical systems: authentication, data confidentiality, data integrity, availability, and nonrepudiation [3,4,6].

Authentication: Ensures that the origin of a packet delivered over a wireless communication channel is traceable. Thus, a receiving node is able to verify that the received message has indeed come from the legitimate sender node and not from a malicious actor. Furthermore, authentication guarantees that two nodes entering into a communication authenticate each other. In other words, authentication ensures that each communicating node is the one that it claims to be [3,6].

Data confidentiality: Prevents an unauthorized entity from accessing a message shared between entities authorized to access it. For example, this property allows only the legitimate sensor nodes to access the content of the transmitted messages (e.g., sensor's readings) between them. Moreover, sensitive information located on the sensor nodes such as private keys and identities must be protected from unauthorized access [3,6,35].

Data integrity: Data integrity methods attempt at preventing modification of communicated/stored data by an unauthorized entity. Integrity is compromised when a malicious actor accidentally or purposely modifies or deletes data [3,6].

Availability: Ensures that any resource of a given NC-enabled WSN is always available for any legitimate entity. Availability guarantees that the provided services of the NC-enabled WSN are always available even in presence of internal or external attacks [3,6].

Nonrepudiation: Prevents an entity from denying previous activities. For example, after completing the process of sending and receiving data, neither the sender later can deny that he sent the information nor the receiver can deny the reception [36].

16.5 SECURITY ATTACKS FOR NETWORK CODING–ENABLED WSNs IN CYBER-PHYSICAL SYSTEMS

Taking into consideration the broadcast nature of the wireless channel and the security vulnerabilities of the NC technology, the satisfaction of the aforementioned security requirements in NC-enabled WSNs integrated by cyber-physical systems is a challenging task. A known classification of security attacks, which is defined both in X.800 and RFC 2828, includes passive and active attacks [37].

Passive attacks only observe the communication without interfering in any way; in other words, they do not disturb the normal operation of the network. Malicious nodes that perform a passive attack only snoop (or read) the exchanged information without modifying it. As they do not compromise the operation of the network, detection of this type of attacks is difficult. On the other side, active attacks try to disrupt the normal network operation and may also alter, corrupt, or delete the data packets being exchanged over the network. Furthermore, the most active attackers can threaten

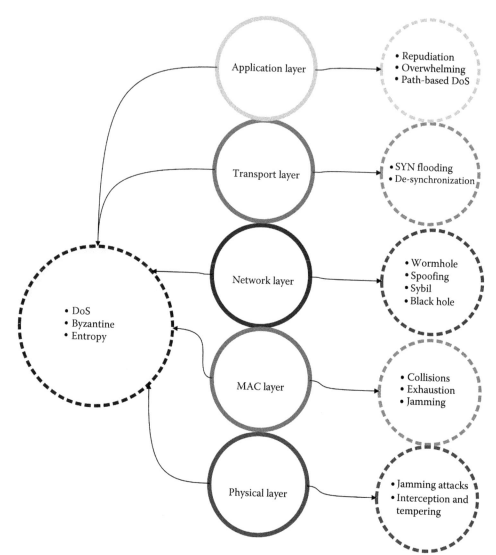

FIGURE 16.3 Attacks against NC-enabled WSNs in different network layers. On the right side, we show the specific attacks for each layer, and on the left side, we show the common attacks targeting all the network layers.

different network layers (i.e., MAC layer, network layer, transport layer, and application layer). Attacks targeting the NC-enabled WSNs in different network layers are presented in Figure 16.3.

Some of the most destructive network attacks such as pollution attacks, Byzantine fabrication, and modification attacks are caused by active attackers inside the network. Despite its benefits, NC-enabled networks are more susceptible to attacks than traditional store-and-forward one since even a single number of corrupted messages can infect a large number of downstream nodes because the corrupted messages propagate via recoding.

We consider two types of attackers: internal and external attackers. As internal attackers, we consider the nodes that have been compromised or the legitimate nodes that have been turned into malicious ones. Moreover, they have access to network resources and key material. However, the external attackers are limited to adversaries without having access to network resources; this includes eavesdropping, modification, and replay of packets. However, our main purpose in this chapter is to study the common attacks that are possible for NC-enabled WSNs in cyber-physical systems.

Specifically, our main idea here is to show how the specific characteristics of NC can be exploited by attackers.

16.5.1 Eavesdropping

The main goal of an eavesdropper is to read data traffic and obtain sensitive information (e.g., native packets, public keys, private keys, location, or passwords of other nodes). In this case, the attacker not only acts as a normal node but also tries to attain some sensitive information that should be kept secret during the communication (see Figure 16.4). In order to obtain this information, the eavesdropper node only listens to the message transmission in the broadcasting wireless medium [38]. Eavesdropper targets at degrading the confidentiality of the transmitted data.

16.5.2 Impersonation

This attack is a kind of active eavesdropping, where the attacker nodes in order to obtain the information send queries to the victim nodes by concealing themselves as friendly nodes. By this bogus authenticity, the attacker can introduce conflicting routes or routing loops, cause waste of sensors' power, and portion the network (see Figure 16.5). State-aware NC protocols rely on network nodes that can be effected by this type of attack [25]. Impersonation targets to degrade the authenticity property in NC-enabled WSNs.

16.5.3 Byzantine Attacks

Byzantine attacks include a wide range of attacks such as selectively dropping packets, forwarding data packets through nonoptimal or even invalid routes, generating routing loops, modifying or altering the message in transit, changing the header of packets, and so on. The main objective of these attacks is to decrease the network throughput. In state-aware NC protocols, packet headers normally contain topology states and routing information, and in stateless NC protocols, headers normally contain the required decoding vectors [39]. Therefore, the Byzantine attacks are not only disruptive on state-aware NC protocols but also impose a negative impact on stateless NC. In NC-enabled

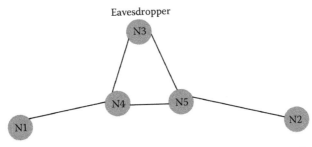

FIGURE 16.4 A simple eavesdropping attack behavior, where he tries to listen to the data that are available at the channel.

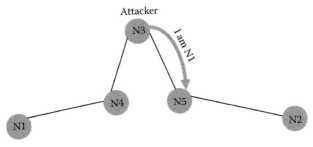

FIGURE 16.5 A simple impersonation attack behavior.

WSNs integrated in CPS applications, the main categories of Byzantine attacks that may occur are the following.

16.5.3.1 Byzantine Modification

In this type of attack, the adversary aims at modification of the transmitted data in transit. For instance, the attacker can send a lot of unnecessary messages to a victim node by changing the destination node native messages. More precisely, the adversary causes a destination node to decode the native packets incorrectly by changing the coded packets in transit in stateless NC protocols. State-aware NC protocols can be affected by changes on either coded data packets [25].

16.5.3.2 Byzantine Fabrication (Pollution Attacks)

Based on the nature of packet mixing in NC schemes, these schemes are vulnerable to a severe security threat also known as pollution attacks, where an adversary node injects corrupted packets into the network. So, by propagating a corrupted packet that an intermediate node uses during the coding process, all the packets that are coded and forwarded by the node will be corrupted as well. Moreover, Byzantine fabrication attacks can disrupt the routing operation of network in different ways such as forwarding data packets through nonoptimal or even invalid routes and generating routing loops. In state-aware NC protocols, packet headers normally contain topology states and routing information. Also, in stateless NC protocols, headers normally contain required decoding vectors. Therefore, the Byzantine fabrication attacks are disruptive for both stateless and state-aware NC protocols. We present a pollution attack in Figure 16.6. Pollution attack targets to compromise the data integrity property in NC-enabled WSNs.

16.5.3.3 Byzantine Replay Attacks

A node sends the old authenticated previously transmitted messages to the network. Sending these old messages causes waste network resources and eventually degrades the overall throughput rate [40]. Stateless NC protocols can be affected in terms of NC gain and processing time by the injection of packets that are repeated into the information flow. State-aware NC protocols can be affected by sending the old messages to the specific victims in order to waste its resource as well. An example of a Byzantine replay attack is demonstrated in Figure 16.7.

16.5.4 Wormholes

Wormholes can constitute a serious threat in WSNs, where two or more malicious nodes can collaborate and create a tunnel between two nodes. Afterward, these malicious nodes record packets and

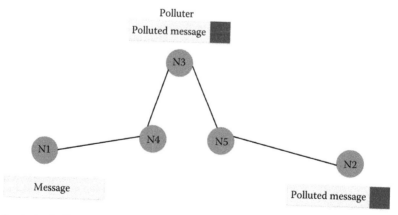

FIGURE 16.6   A simple Byzantine fabrication attack.

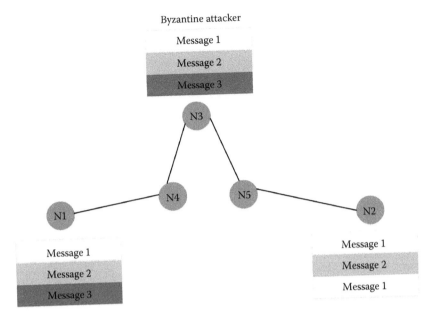

FIGURE 16.7   Byzantine replay attack.

retransmit them into the network through the tunnel [41]. The adversaries target at convincing the neighbor nodes that the two sides of a tunnel are in the same range and very close to each other. By using this strategy, an adversary may be able to send the messages to the other part of the tunnel in the sensor network without being in the same range. The wormhole attack can impose more severe impact on the state-aware NC protocols, such as disrupting the route discovery process and taking the control of flows by leading them through the tunnel, in comparison with the impact on the stateless protocols. Figure 16.8 depicts the wormhole attack.

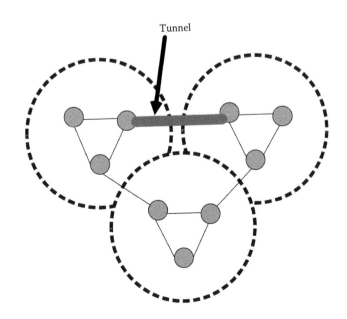

FIGURE 16.8   Wormhole attack representation.

16.5.5 Entropy Attacks

An adversary intends to generate valid but noninnovative packets that are valid linear combination that are stored or overheard at an earlier time by the adversary. In other words, the adversary node creates and distributes a noninnovative coded packet that is a nonrandom linear combination of coded packets such that the generated coded packet is linearly dependent on the previous coded packets stored at a downstream node. These valid but noninnovative packets decrease the decoding opportunities at sinks in both stateless and state-aware NC systems, cause the waste of network resources, and degrade the overall throughput rate. In a pollution attack, a node creates and generates a coded packet that is not valid linear combinations of the native data. However, entropy attackers create and distribute coded packets that are valid linear combinations of the native data. As a result, a pollution defense is not helpful against an entropy attack because all of the defenses against pollution attacks rely on the fact that an invalid coded packet is not a valid linear combination [40].

16.5.6 Denial of Service Attacks

A denial of service (DoS) attack is an attempt to make the resources of a system unavailable to the legitimate users. More specifically, an adversary aims to prevent some services from functioning efficiently. Hardware failures, software bugs, or resource exhaustion can lead to DoS attacks. Actually, the adversary in DoS attacks targets availability, which ensures that an authorized node can access all the services of the network. In this attack, the malicious node may inject a high volume of junk packets into the network and this leads to the waste of network resources like nodes' battery power, memory, and wired or wireless channels. Various types of DoS attacks work at different layers and affect differently the network such as jamming and tampering at the physical layer, collision and exhaustion at the link layer, black holes and routing table overflow at the network layer, SYN flooding and de-synchronization at the transport layer, and finally failure in the web servers at the application layer [42].

In NC state-aware schemes, a malicious node can easily perform a DoS attack by flooding its neighbors via injecting lots of corrupted blocks (pollution attack) or even clean but old and reparative packets (Byzantine replay attack) into the downstream victim nodes.

16.6 SECURITY COUNTERMEASURES FOR NETWORK CODING–ENABLED WSNs IN CYBER-PHYSICAL SYSTEMS

The fact that NC allows intermediate nodes to manipulate and mix incoming data packets from neighbor nodes as well as transmits data packets to neighbor nodes leads to open security issues in NC-enabled WSNs. Therefore, the NC-enabled WSNs are vulnerable to various types of attacks that can compromise the security of the cyber-physical systems that integrate these WSNs. These types of attacks, which are presented in Section 16.5, are summarized in Table 16.1.

TABLE 16.1

Attacks against NC-Enabled WSNs in Cyber-Physical Systems

Type of Attack	State-Aware/Stateless NC	Target
Eavesdropping	Both	Confidentiality
Impersonation	State-aware NC	Authenticity
Byzantine	Both	Integrity, availability
Wormhole	Both	Confidentiality
Entropy	Both	Integrity
DoS	Both	Availability

TABLE 16.2

Categories of the Security Schemes against Byzantine Fabrication, Eavesdropping, and Entropy Attacks

Type of Attack	Security Schemes		Related Works
Byzantine fabrication	Cryptographic and key management	Homomorphic hash functions	[29,44]
		Homomorphic digital signatures	[32,45]
		Homomorphic MAC	[46–48]
	Network codes		[49–51]
	Cooperative schemes		[29,52–56]
Eavesdropping	Cryptographic and key management		[57,58]
	Network codes		[59]
	Cooperative schemes		[25]
Entropy	Only a limited number of defense mechanisms—Not any specific security scheme		[40,60]

In this section, we particularly focus on the mitigation of Byzantine fabrication, eavesdropping, and entropy attacks, which are among the most common attacks that can be used against NC-enabled WSNs and have a negative impact on the security of the whole cyber-physical systems. These attacks can result in the degradation of the performance gain of coded networks or even worse in the complete disruption of the whole network operation. However, these attacks have to be addressed in order for the NC-enabled WSNs to reach their full potential [43].

Toward this direction, several security schemes, focused on NC-enabled WSNs, have been recently proposed to handle these attacks since the traditional security mechanisms are infeasible solutions for the resource-constrained NC-enabled WSNs due to the large bandwidth for communication that they require. We classify the security schemes against Byzantine fabrication, eavesdropping, and entropy attacks into three main categories as follows: (1) cryptographic and key management–based schemes that are used by NC as an extra line of defense through assigning proper keys or cryptographic functions; (2) schemes that allow legitimate nodes to add redundancy to their information flows by applying an appropriate secure network codes algorithm; and (3) cooperative schemes that are able to detect compromised nodes and isolate them from the other legitimate nodes. This categorization of the security schemes against Byzantine fabrication, eavesdropping, and entropy attacks is also demonstrated in Table 16.2.

16.6.1 COUNTERMEASURES TO BYZANTINE FABRICATION ATTACKS

16.6.1.1 Cryptographic and Key Management–Based Schemes

For secure NC-enabled WSNs in cyber-physical systems, the ability to distribute secret keys in a secure manner is an obvious fundamental requirement. The most attention of researchers is absorbed to pollution attacks that are a type of Byzantine fabrication attacks. To address pollution attacks, a number of cryptographic-based schemes have been proposed in References [28,32,40,61–65]. Specifically, the proposed cryptographic schemes include homomorphic hash functions [29,44], homomorphic digital signatures [32,45], and homomorphic MAC schemes [46–48] for NC-enabled systems. However, these solutions have several drawbacks, when they are applied to NC-enabled WSNs, due to the fact that they are computationally expensive and they cannot be supported by the resource-constrained NC-enabled WSNs.

Moreover, trusted party schemes and public key infrastructures (PKIs) are used as solutions for ensuring security in NC-enabled WSNs. In addition, the key pre-distribution schemes emerge as appropriate schemes for NC-enabled WSNs that can afford less computation processes. Furthermore, according to [25] it is possible to design a secret key distribution scheme that requires only a small number of pre-stored keys by exploiting the benefits of NC. Nevertheless, the proposed scheme ensures that shared-key connectivity is established with probability one and that the mobile node is probably oblivious to the distributed keys. In the rest of this section, we present current specific cryptographic and key management–based schemes that can be applied to NC-enabled WSNs.

Zhao et al. [66] designed a signature scheme to maintain NC properties for file distribution. In this scheme, the source calculates a signature from the spanned subspace and broadcasts this signature to all nodes. They use a standard public-key method and the verification is accomplished by checking the membership of encoded block in the spanned subspace. The authors proved that breaking this scheme is as hard as solving a Diffie–Hellman problem.

Another signature-based scheme was proposed by Yu et al. [32]. This idea is a kind of homomorphic signature function that resists against pollution attacks. In this scheme, the source signs its messages using its private key, and the forwarders verify the messages by using the public key. The authors believe that a hash collision is equivalent to solving a discrete logarithm problem.

Boneh et al. [67] also present two signature schemes for NC to prevent pollution attacks. The proof of their first scheme relies on the co-computational Diffie–Hellman problem, which is slightly stronger when compared to the Diffie–Hellman problem. Both two approaches are based on a single public key that can be used to sign many linear spaces.

An efficient scheme for securing XOR NC was presented for the first time in Yu et al. [68]. The authors studied the integrity of XOR-coded packets by proposing a MAC scheme; however, this scheme is vulnerable to tag pollution attacks. In particular, they assume that each node picks a fixed number of keys randomly from a large global key pool and a public key. Through meticulously managing the key pool size and the number of keys that each node picks, the authors assure that any two nodes have certain probability to find some shared keys. The source uses its keys to generate MACs for its messages. Since the forwarders have the shared keys, they verify the MACs of received messages by using their shared keys.

A practical defense scheme, called DART, against pollution attacks in NC for wireless mesh networks was proposed in Dong, Curtmola, and Nita-Rotaru [69]. It is based on checksums that are very efficient to create and verify. Each node uses a checksum to verify only those packets that were received before the checksum itself was created. However, the delayed key distribution that is used in this scheme requires a clock synchronization of all the nodes.

RIPPLE, which was presented by Li et al. [46], is based on a symmetric key scheme. It allows a node to detect corrupted packets and encode only the authenticated ones. This scheme is collusion resistant as well as resistant against tag pollution attacks; however, similar to [69], this scheme requires time synchronization among all the nodes.

The hybrid-key cryptography approach was explored by Zhang et al. [70]. The authors proposed a scheme called MacSig that resists not only against data pollution but also against tag pollution attacks. The basic idea is to pad each source packet with an extra symbol to make it orthogonal to a given vector. Moreover, a signature is also appended at the end for resisting against tag pollution attacks.

KEPTE [21] presented by Wu et al. is based on a key pre-distribution tag encoding scheme and it enables all intermediate and sink nodes to verify the correctness of the received data messages. The authors believe that KEPTE can resist against tag pollution. The authors showed that this approach can be used as a practical NC scheme.

16.6.1.2 Security via Network Codes

By exploiting appropriate network codes, it is possible to detect and correct corrupted packets, and there is opportunity to mitigate a significant devastating effect of Byzantine attacks (pollution attack) in NC-enabled WSNs in cyber-physical systems. However, the mechanisms [49–51] that provide error detection and error correction may lead to some undesired problems. An error detection scheme creates overhead monitoring and there is a possibility for the sensors to correct the errors only after pollution attacks occur, which may bring about epidemic disruptive problems for an NC system.

16.6.1.3 Security via Cooperative Mechanisms

Cooperative schemes that arise from NC can improve the verification efficiency. In these schemes, users cooperate not only to distribute content but also to protect the rest sensors against adversary users in NC-enabled WSNs. Moreover, the cooperative schemes for NC-enabled WSNs are able to detect compromised nodes and isolate the adversary nodes [29,52,53].

In Gkantsidis and Rodriguez [29], one of the most famous cooperative schemes is presented. It is an efficient scheme to prevent Byzantine attacks against bogus alert messages based on secure random checksums. Another cooperative mechanism is a watchdog, which detects and removes adversaries by dynamically adjusting the routing paths [54]. It consists of the following three phases: monitoring, detecting, and isolating the malicious nodes. After monitoring and detection phases, nodes inform each other about these malicious nodes and finally they run the isolating phase. However, due to the significant overhead produced, this mechanism is not efficient for WSNs.

In the algebraic watchdog presented by Kim et al. [55] a network was proposed in which the participants check their neighborhood locally to enable a secure global network. According to the homomorphism property of the used hash functions, all the intermediate nodes in the pollution attack model are able to verify the validity of the encoded packets on the fly prior to mixing them algebraically. As a result, the intermediate nodes can collaborate to verify the transit packets and the polluted packets will be dropped early and malicious nodes can be detected and isolated.

Le and Markopoulou introduced a homomorphic MAC scheme, called SpaceMac [56]. SpaceMac allows a node to verify the received messages if they belong to a specific subspace. Moreover, by using SpaceMac, the authors have proposed another cooperative scheme to detect and locate the adversary nodes.

16.6.2 Countermeasures to Eavesdropping Attacks

16.6.2.1 Cryptographic and Key Management–Based Schemes

The compromise of even a single sensor node in WSNs can reveal the secret keys and for that reason all its messages can be decrypted easily. Therefore, key distribution is a vital requirement of WSNs. Nevertheless, due to the limited computational power of current sensor devices, the generation and distribution of keys are too expensive in terms of system overhead. Moreover, the basic communication patterns of WSNs differ from traditional networks' patterns since key-distribution techniques need to scale to networks with hundreds or thousands of nodes.

A simple approach for distributing the keys in WSNs is to pre-configure the network with a shared unique symmetric key for each node. However, we need to establish a lot of keys and distribute them to all sensor nodes of the network. On the other hand, public-key cryptography (e.g., Diffie–Hellman key scheme) is another option that is useful in WSNs. Besides, bootstrapping keys that are based on a trusted base station is another solution for key distribution [57]. In this case, all the processes related to the establishment and distribution of the keys are done by the base station and each node only needs to share a key with the base station. However, this mechanism makes the base station a single point of failure and for that reason this security mechanism is considered as weak one. Moreover, random-key pre-distribution protocols are given in Eschenauer and Gligor [58].

16.6.2.2 Security via Network Codes

By exploiting appropriate network codes, it is possible to prevent eavesdropper from obtaining meaningful information from the eavesdropped data. Based on the use of appropriate network codes, there is the opportunity to mitigate a significant devastating effect of eavesdropping attack in NC-enabled WSNs in CPS. By specific definition of an eavesdropper model, which is described Zhang et al. [59], in NC-enabled WSNs in CPS there are the following three different security levels for NC systems: Shannon security level, weak security level, and computational security level. Preparing a NC system, which supports the Shannon security level, is not practical as it is feasible only under ideal assumption. Moreover, for the weak security level, the authors show that it requires a sufficiently large filed size. Finally, by utilizing cryptographic approaches, the computational security can be achieved in practical NC-enabled WSNs in CPS.

16.6.2.3 Security via Cooperative Mechanisms

In case of eavesdropping attacks, Lima et al. [25] studied a cooperative scheme with special emphasis on three different nodes in the NC-enabled WSNs. First, they consider nice but curious nodes, which they do not ignore the data for which they are not the intended receivers. In the second type, the eavesdropper is able to wiretap a subset of network links. Finally, the third type of attacker is a worst-case eavesdropper who is given full access to all the traffic in the network.

16.6.3 COUNTERMEASURES TO ENTROPY ATTACKS

In contrast to the other threats and attacks that have been mentioned earlier, the entropy attacks have been emerged recently. Hence, only a limited number of defense mechanisms have been proposed for entropy attacks. In the context of entropy attacks, an attacker creates packets that contain information, which has been already known by the other sensors, and send them to the next nodes. The result is a severe degradation of the system performance.

An efficient symmetric-key-based authentication scheme with network coding, called PMAC, was proposed by Cheng et al. [60]. In particular, their approach requires that any node can construct a valid MAC tag for any legitimate block it wants to upload. This idea can resist against entropy attacks.

The main research in this topic was done by [40], where the authors they consider two local and global entropy attacks. The attack that produces coded packets that are noninnovative to local neighbor nodes is called local entropy attack. A main goal of attacker in global entropy attacks is producing coded packets that seems innovative to local neighbor nodes but they are noninnovative to at least an outlying downstream node. The simulations in this work show the impact of their solution to the entropy attacks. Moreover, they present the difficulties in defending against a global entropy attack. However, developing an extremely efficient countermeasure mechanism against entropy attacks is still required and it is an ongoing research.

16.7 CONCLUSION AND OPEN ISSUES

The emerging cyber-physical systems are a promising technology that can contribute significantly toward the improvement of safety, convenience, and comfort in the daily life of modern societies. Cyber-physical systems are complex systems integrating physical processes with cyber systems. The main objective of cyber-physical systems is to monitor and control efficiently physical process. In the context of cyber-physical systems, monitoring of the physical processes is a fundamental function that is currently based on WSNs due to the advantages that they offer in terms of connectivity, flexibility, and scalability. In other words, WSNs are one of the main components of cyber-physical systems for collecting physical information. However, WSNs are characterized by limitations that can affect the advancement of cyber-physical systems. These limitations are related to low communication bandwidth, packet loss, and power consumption. Therefore, it is obvious that the success and prosperity of cyber-physical systems are dependent on advances of WSNs in these limitations.

Toward this direction, the application of NC technology in WSN, leading to NC-enabled WSNs, can be a promising solution. Recent research efforts show that NC technology can provide significant benefits to WSNs in terms of improvement in network capacity and robustness among others. However, due to the fact that NC technology enables the intermediate nodes to manipulate and mix incoming data packets from neighbor nodes as well as transmit data packets to neighbor nodes, security issues are raised in NC-enabled WSNs. Additionally, these security issues can be exploited by attackers to compromise the security of the cyber-physical systems, which is a critical factor for their success.

Therefore, in this chapter, we have focused on the provision of a categorization of the current existing NC-enabled WSN attacks that can have a negative impact on the security of the whole CPS.

Furthermore, we have presented the current mitigation techniques, derived from the literature, against the most common attacks included in the attack categories that we have defined. Our objective is the provision of a foundation for organizing research efforts toward the design and deployment of more reliable and efficient defense mechanisms against known attacks for NC-enabled WSNs in cyber-physical systems or against attacks that have not yet appeared but can be potential threats in the future.

ACKNOWLEDGMENTS

The research leading to these results has received funding from the European Community's Seventh Framework Programme (FP7/2007–2013) under Grant Agreement No. 285969 (CODELANCE) and from the Fundação para a Ciência e Tecnologia (PTDC/EEA-TEL/119228/2010 – SMARTVISION). The first author would like to thank the University of Madeira and the MAP-TELE PhD Program that have contributed to these works.

REFERENCES

1. E. A. Lee, Cyber physical systems: Design challenges, in *2008 11th IEEE International Symposium on Object Oriented Real-Time Distributed Computing (ISORC)*, pp. 363–369, 2008.
2. A. A. Cardenas, S. Amin, and S. Sastry, Secure control: Towards survivable cyber-physical systems, *System*, 1, a3, 2008.
3. E. K. Wang, Y. Ye, X. Xu, S. Yiu, L. Hui, and K. Chow, Security issues and challenges for cyber physical system, in *Proceedings of the 2010 IEEE/ACM Int'l Conference on Green Computing and Communications & Int'l Conference on Cyber, Physical and Social Computing*, pp. 733–738, 2010.
4. F.-J. Wu, Y.-F. Kao, and Y.-C. Tseng, From wireless sensor networks towards cyber physical systems, *Pervasive and Mobile Computing*, 7, 397–413, 2011.
5. F. Xia, X. Kong, and Z. Xu, Cyber-physical control over wireless sensor and actuator networks with packet loss, in *Wireless Networking Based Control*, Springer, New York, pp. 85–102, 2011.
6. J. Sen, A survey on wireless sensor network security, *International Journal of Communication Networks and Information Security (IJCNIS)*, 1(2), pp. 55–78, 2009.
7. R. Ahlswede, N. Cai, S. Y. R. Li, and R. W. Yeung, Network information flow, *IEEE Transactions on Information Theory*, 46, 1204–1216, July 2000.
8. T. Ho and D. S. Lun, *Network Coding: an Introduction*, vol. 6, Cambridge University Press, Cambridge, UK 2008.
9. Y. Wu, P. A. Chou, and K. Sun-Yuan, Minimum-energy multicast in mobile ad hoc networks using network coding, *IEEE Transactions on Communications*, 53, 1906–1918, 2005.
10. G. Mantas, D. Lymberopoulos, and N. Komninos, Integrity mechanism for ehealth tele-monitoring system in smart home environment, in *Annual International Conference of the IEEE Engineering in Medicine and Biology Society, 2009. EMBC 2009*, pp. 3509–3512, 2009.
11. G. Brettlecker, C. Cáceres, A. Fernández, N. Fröhlich, A. Kinnunen, S. Ossowski, H. Schuldt, and M. Vasirani, Technology in health care, in *CASCOM: Intelligent Service Coordination in the Semantic Web*, Springer, Science & Business Media, pp. 125–139, 2008.
12. M. Fengou, G. Mantas, D. Lymberopoulos, N. Komninos, S. Fengos, and N. Lazarou, A new framework architecture for next generation e-health services, *IEEE Journal of Biomedical and Health Informatics*, 17, 9–18, 2013.
13. C.-Y. Lin, S. Zeadally, T.-S. Chen, and C.-Y. Chang, Enabling cyber physical systems with wireless sensor networking technologies, *International Journal of Distributed Sensor Networks*, 1–21, 2012.
14. L. Xu, L. Rongxing, L. Xiaohui, S. Xuemin, C. Jiming, and L. Xiaodong, Smart community: An Internet of Things application, *IEEE Communications Magazine*, 49, 68–75, 2011.
15. Y.-M. Huang, M.-Y. Hsieh, H.-C. Chao, S.-H. Hung, and J. H. Park, Pervasive, secure access to a hierarchical sensor-based health care monitoring architecture in wireless heterogeneous networks, *IEEE Journal on Selected Areas in Communications*, 27, 400–411, 2009.
16. N. Gershenfeld, R. Krikorian, and D. Cohen, The Internet of Things, *Scientific American*, 291, 76, 2004.

17. T. Arampatzis, J. Lygeros, and S. Manesis, A survey of applications of wireless sensors and wireless sensor networks, in *Intelligent Control, 2005. Proceedings of the 2005 IEEE International Symposium on, Mediterrean Conference on Control and Automation*, pp. 719–724, 2005.
18. B. Bollobás, *Graph Theory, An Introductory Course*, Springer-Verlag, New York, 1979.
19. N. Cai and R. W. Yeung, Secure network coding on a wiretap, *IEEE Transactions on Information Theory*, 1, 424–435, 2002.
20. N. Cai and R. W. Yeung, Network coding and error correction, in *Proceedings of the 2002 IEEE Information Theory Workshop, 2002*, pp. 119–122, 2002.
21. R. Koetter and M. Medard, Beyond routing: an algebraic approach to network coding, in *INFOCOM 2002. Proceedings of the IEEE Twenty-First Annual Joint Conference of the IEEE Computer and Communications Societies*, vol. 1, pp. 122–130, 2002.
22. T. Ho, D. R. Karger, M. Medard, and R. Koetter, Network coding from a network flow perspective, in *Proceedings of the IEEE International Symposium on Information Theory, 2003*, p. 441, 2003.
23. S. Jaggi, P. Sanders, P. A. Chou, Michelle Effros, S. Egner, K. Jain, and L. M. G. Tolhuizen, Polynomial time algorithms for multicast network code construction, *IEEE Transactions on Information Theory*, 51, 1973–1982, 2005.
24. T. Ho, M. Medard, R. Koetter, D. R. Karger, M. Effros, S. Jun, and B. Leong, A random linear network coding approach to multicast, *IEEE Transactions on Information Theory*, 52, 4413–4430, 2006.
25. L. Lima, J. Vilela, P. Oliveira, and J. Barros, Network coding security: Attacks and countermeasures, *CoRR*, vol. abs/0809.1366, 2008.
26. M. Bloch and J. Barros, *Physical-Layer Security: From Information Theory to Security Engineering*, Cambridge University Press, Cambridge, UK, 2011.
27. S. Katti, D. Katabi, W. Hu, H. S. Rahul, and M. Médard, The importance of being opportunistic: Practical network coding for wireless environments, *IEEE Transaction on Networking*, 16, 497–510, 2006.
28. M. N. Krohn, M. J. Freedman, and D. Mazieres, On-the-fly verification of rateless erasure codes for efficient content distribution, in *IEEE Symposium on Security and Privacy*, pp. 226–240, 2004.
29. C. Gkantsidis and P. R. Rodriguez, Cooperative security for network coding file distribution, in *INFOCOM, 25th IEEE International Conference on Computer Communications*, 2006.
30. K. Han, T. Ho, R. Koetter, M. Medard, and F. Zhao, On network coding for security, in *IEEE Military Communications Conference, 2007. MILCOM 2007*, 2007, pp. 1–6.
31. M. J. Siavoshani, C. Fragouli, and S. Diggavi, On locating byzantine attackers, in *Fourth Workshop on Network Coding, Theory and Applications, 2008. NetCod 2008*, pp. 1–6, 2008.
32. Z. Yu, Y. Wei, B. Ramkumar, and Y. Guan, An efficient signature-based scheme for securing network coding against pollution attacks, in *IEEE INFOCOM*, 2008.
33. H. Su and X. Zhang, Network coding based qos-provisioning mac for wireless smart metering networks, in *Quality, Reliability, Security and Robustness in Heterogeneous Networks*, Springer, Berlin, Heidelberg, pp. 161–171, 2012.
34. R. Prior, D. E. Lucani, Y. Phulpin, M. Nistor, and J. Barros, Network coding protocols for smart grid communications, *IEEE Transactions on Smart Grid*, 5, 1523–1531, 2014.
35. W. Stallings, *Cryptography and Network Security: Principles and Practice*, Prentice Hall Press, Upper Saddle River, NJ, 2010.
36. A. J. Menezes, P. C. Van Oorschot, and S. A. Vanstone, *Handbook of Applied Cryptography*, CRC press, Boca Raton, FL, 2010.
37. G. Padmavathi and S. D., A survey of attacks, security mechanisms and challenges in wireless sensor networks, *(IJCSIS) International Journal of Computer Science and Information Security*, 4(1,2), 1–9, 2009.
38. M. Anand, Z. Ives, and I. Lee, Quantifying eavesdropping vulnerability in sensor networks, in *Proceedings of the Second International Workshop on Data Management for Sensor Networks*, pp. 3–9, 2005.
39. S. Jaggi, M. Langberg, S. Katti, T. Ho, D. Katabi, M. Médard, and M. Effros, Resilient network coding in the presence of byzantine adversaries, *IEEE Transactions on Information Theory*, 54, 2596–2603, 2008.
40. A. J. Newell, R. Curtmola, and C. Nita-Rotaru, Entropy attacks and countermeasures in wireless network coding, *Presented at the Proceedings of the fifth ACM Conference on Security and Privacy in Wireless and Mobile Networks*, Tucson, AZ, 2012.
41. C. Hon Sun and L. King-Shan, DelPHI: Wormhole detection mechanism for ad hoc wireless networks, in *2006 1st International Symposium on Wireless Pervasive Computing*, p. 6, 2006.

42. M. Illyas, Security in wireless ad hoc networks, Chapter 30, (Eds. Mishra, A., & Nadkarni, K. M.) in *The Handbook of Ad Hoc Wireless Networks*, CRC Press, Boca Raton, FL, pp. 499–549, 2003.
43. C. Fragouli and J. W. Jean-Yves Le Boudec, Network coding: An instant primer, *ACM SIGCOMM Computer Communication*, 36 63–68, 2006.
44. M. Adeli and L. Huaping, Secure network coding with minimum overhead based on hash functions, *IEEE Communications Letters*, 13, 956–958, 2009.
45. Z. Fang, T. Kalker, M. Medard, and K. J. Han, Signatures for content distribution with network coding, in *IEEE International Symposium on Information Theory, 2007. ISIT 2007*, pp. 556–560, 2007.
46. L. Yaping, Y. Hongyi, C. Minghua, S. Jaggi, and A. Rosen, RIPPLE authentication for network coding, in *INFOCOM, 2010 Proceedings IEEE*, pp. 1–9, 2010.
47. E. Kehdi and L. Baochun, Null keys: Limiting malicious attacks via null space properties of network coding, in *IEEE INFOCOM 2009*, pp. 1224–1232, 2009.
48. S. Agrawal and D. Boneh, Homomorphic MACs: MAC-based integrity for network Codinc, *Presented at the Proceedings of the Seventh International Conference on Applied Cryptography and Network Security*, Paris-Rocquencourt, France, 2009.
49. N. Cai and R. W. Yeung, Network error correction, II: Lower bounds, *Communication in Information and Systems*, 6, 37–54, 2006.
50. R. W. Yeung and N. Cai, Network error correction, I: Basic concepts and upper bounds, *Communication in Information and Systems*, 6, 19–36, 2006.
51. R. Koetter and F. R. Kschischang, Coding for errors and erasures in random network coding, *IEEE Transactions on Information Theory*, 54, 3579–3591, 2008.
52. Q. Wenbo, L. Jian, and R. Jian, An efficient error-detection and error-correction (EDEC) scheme for network coding, in *Global Telecommunications Conference (GLOBECOM 2011), 2011 IEEE*, pp. 1–5, 2011.
53. L. Yongkun and J. C. S. Lui, Identifying pollution attackers in network-coding enabled wireless mesh networks, in *2011 Proceedings of 20th International Conference on Computer Communications and Networks (ICCCN)*, pp. 1–6, 2011.
54. S. Marti, T. J. Giuli, K. Lai, and M. Baker, Mitigating routing misbehavior in mobile ad hoc networks, *Presented at the Proceedings of the Sixth Annual International Conference on Mobile Computing and Networking*, Boston, MA, 2000.
55. M. Kim, M. Medard, and J. Barros, Algebraic watchdog: Mitigating misbehavior in wireless network coding, *Journal of Selected Areas in Communications*, 29, 1916–1925, 2011.
56. L. Anh and A. Markopoulou, Cooperative defense against pollution attacks in network coding using SpaceMac, *IEEE Journal on Selected Areas in Communications*, 30, 442–449, 2012.
57. A. Perrig, R. Szewczyk, J. Tygar, V. Wen, and D. E. Culler, SPINS: Security protocols for sensor networks, *Wireless Networks*, 8, 521–534, 2002.
58. L. Eschenauer and V. D. Gligor, A key-management scheme for distributed sensor networks, in *Proceedings of the Ninth ACM Conference on Computer and Communications Security*, pp. 41–47, 2002.
59. P. Zhang, Y. Jiang, C. Lin, Y. Fan, and X. Shen, P-coding: Secure network coding against eavesdropping attacks, in *INFOCOM, 2010 Proceedings IEEE*, pp. 1–9, 2010.
60. C. Cheng, T. Jiang, and Q. Zhang, TESLA-based Homomorphic MAC for authentication in P2P system for live streaming with network coding, *IEEE Journal on Selected Areas in Communications*, 31, 291–298, 2013.
61. J. Dong, R. Curtmola, and C. Nita-Rotaru, Practical defenses against pollution attacks in intra-flow network coding for wireless mesh networks, *Presented at the Proceedings of the Second ACM Conference on Wireless Network Security*, Zurich, Switzerland, 2009.
62. T. Ho, L. Ben, R. Koetter, M. Medard, M. Effros, and D. R. Karger, Byzantine modification detection in multicast networks with random network coding," *IEEE Transactions on Information Theory*, 54, 2798–2803, 2008.
63. M. Kim, L. Lima, Z. Fang, J. Barros, M. Medard, R. Koetter, T. Kalker, and K. J. Han, On counteracting Byzantine attacks in network coded peer-to-peer networks, *IEEE Journal on Selected Areas in Communications*, 28, 692–702, 2010.
64. L. Anh and A. Markopoulou, On detecting pollution attacks in inter-session network coding, in *2012 Proceedings IEEE INFOCOM*, pp. 343–351, 2012.
65. L. Lima, S. Gheorghiu, J. Barros, M. Médard, and A. L. Toledo, Secure network coding for multi-resolution wireless video streaming, *Journal of Selected Areas in Communications*, 28, 377–388, 2010.

66. F. Zhao, T. Kalker, M. Médard, and K. J. Han, Signatures for content distribution with network coding, in *In Proceedings of International Symposium on Information Theory (ISIT)*, 2007.

67. D. Boneh, D. Freeman, J. Katz, and B. Waters, Signing a linear subspace: Signature schemes for network coding, in *12th International Conference on Practice and Theory in Public Key Cryptography, PKC '09*, pp. 68–87, 2009.

68. Z. Yu, W. Yawen, B. Ramkumar, and G. Yong, An efficient scheme for securing XOR network coding against pollution attacks, in *INFOCOM 2009, IEEE*, pp. 406–414, 2009.

69. J. Dong, R. Curtmola, and C. Nita-Rotaru, Practical defenses against pollution attacks in wireless network coding, *ACM Transactions on Information and System Security (TISSEC)*, 14, 7, 2011.

70. Z. Peng, J. Yixin, L. Chuang, Y. Hongyi, A. Wasef, and S. Xuemin, Padding for orthogonality: Efficient subspace authentication for network coding, in *INFOCOM, 2011 Proceedings IEEE*, pp. 1026–1034, 2011.

Section VII

Sensors and Applications

17 Unified Framework toward Solving Sensor Localization, Coverage, and Operation Lifetime Estimation Problems in Cyber-Physical Systems

Zhanyang Zhang

CONTENTS

17.1 INTRODUCTION

Recent technological advancement in the areas of sensing, computing, and wireless communication has made it possible to design and develop wireless sensors that are small in size and low in cost. These sensors have data collecting, processing, and communicating capabilities. They can form a wireless sensor network that contains hundreds of tiny sensors. Wireless sensor networks (WSNs) extend our capability to explore, monitor, and control the physical world at large [1]. Today, many wireless sensors are being embedded in electrical and mechanical components that are integral parts of physical systems widely used in medicine, military, environmental monitoring, transportation, and industry process control. After decades of rapid growth of Internet and wireless networks, countries have invested tremendous resources (from both public and private sectors) to build cyber infrastructures that provide ubiquitous services to users with from anywhere and at anytime style of connection and access. WSNs provide the opportunities for a wide range of applications in all domains to bridge the cyberspace and the physical world by forming cyber-physical systems (CPS).

The primary concept of cyber-physical systems is to integrate computing (sensing, analyzing, predicting, understanding), communication (interaction, intervene, interface management), and control (inter-operate, evolve, evidence-based certification) together to make intelligent and autonomous systems.

Since wireless sensors have limited data processing, storage, and communication capabilities, and most critically limited energy resources, these limitations present a set of challenging problems in cyber-physical system application design and operation. In this chapter, we investigate these problems and provide a unified framework to address them with the focus on a particular class of cyber-physical system applications for homeland security, environment monitoring, and control. Here are some of the most common issues in design and operating such kinds of cyber-physical system applications:

- Due to the nature of such applications, it is required to deploy a large number of homogeneous wireless sensors randomly over a targeted area without knowing the sensors' locations. To determine the locations of these sensors is essential to correlate the data and their source.
- In order to ensure high quality of sensing data, sometimes, it is required that there must be a minimum of K-sensors to cover any given point in the targeted area (known as the K-coverage problem). For example, it needs three or more sensor coverage to determine the speed and direction of an intrusion object in a targeted area.
- In addition, many cyber-physical system applications have to operate under harsh environments with high failure rates and nonreplenishable power sources. As many sensors come out of commission over a period of time due to failure or power depletion, a CPS application is no longer able to provide the required services. A sound CPS application life span estimation model is vital for us to answer fundamental questions, such as how long a CPS application can operate without compromising required quality of services (QoS).

Each of these issues presents a complex problem on its own. They are also closely linked and are all essential to the success of design and operation of CPS applications. For example, the location information of sensors can be used to verify K-coverage for a targeted area in a CPS application. It is often the case where extra numbers of sensors are deployed in the same area beyond required K-coverage. A CPS application can take advantage of these redundant sensors by having a duty-cycle schedule to let some sensors sleep in order to save energy while others perform the duty of service. A sleep sensor can be called to duty once an on-duty sensor fails or is scheduled to sleep. Therefore, the duty-cycle schedule can increase reliability and prolong the life span of a CPA application. Our survey shows that there are a number of published works addressing these issues individually. Many researchers provided unique and novel solutions to individual problems with their own merit. But they often take a narrow approach to gain optimization in one dimension while making trade-offs of others that, sometimes, even compromise the required QoS and overall system performance. To our best knowledge, today we are still lack of a unified approach or framework to address all of the aforementioned issues and provide an integrated solution for all of these problems while achieving overall optimization and without compromising the required QoS.

The goal of this chapter is to investigate the complexity of these problems and their intricate relationships through a survey of current published work. Then we present several key results of our research work in CPS during the past few years. These results lead to a unified framework to solve the three critical problems in CPS design and operation, namely, localization, K-coverage, and CPS operation life span estimation.

The rest of the chapter is organized as follows. We present a survey of previous work related to the problems of our investigation in Section 17.2. In Section 17.3, we describe a signal stimulation model (SSM) for solving both sensor localization and K-coverage problems. The sensor cluster localization algorithm (SCLA) and its performance analysis are also presented in Section 17.3. In Section 17.4, we present the sensor cluster maintenance algorithm (SCMA) and discuss the behaviors of different

419

Unified Framework toward Solving Sensor Localization

cluster components. State transition diagrams are used to demonstrate the interactions between different cluster components. In Section 17.5, we build up a theoretical foundation with definitions and theorems that are necessary to prove that the SCLA and SCMA algorithms, when they work together, can solve the K-coverage problem during the life of CPS operations. In Section 17.6, we introduce a cluster-based Markov process as a closed-form mathematical model for the estimation of a CPS application life span. We present a set of analytical results to depict relationships between CPS lifetime and other factors, such as sensor population, sensor failure rate, and QoS constraints. Finally, in Section 17.7, we present the results of our simulation study about SCLA/SCMA algorithms and the proposed Markov process model for the purpose of validation and performance analysis. We conclude the chapter in Section 17.8 with a discussion on a list of open problems for future research.

17.2   RELATED WORK

Given the resource limitations of these sensors within a CPS application and their operating environments, it is not always feasible to equip each sensor with a GPS receiver (such as wild fire monitoring systems where GPS does not work due to dense forest). This has motivated many researchers to seek GPS-less solutions for locating sensors. Patwari et al. and Savvides et al. [2,3] designed distributed algorithms that yield sensor node locations without addressing the K-coverage problem. In addition, their algorithms require the existence of anchor nodes (sensor nodes with known locations) and time synchronization of the whole network. Wang et al. [4] address the connected K-coverage problem with a localized heuristic algorithm for the problem, but their algorithm does not provide a guaranteed K-coverage for the entire life span of the applications. Charkrabarty et al. [5] investigate linear programming techniques to optimally place a set of sensors on a sensor field (3D grids) for a complete coverage of the field. In many applications, such precision deployment of sensor nodes may not be feasible (such as in harmful environment or rough terrain). Our approach to the sensor location problem does not assume the existence of anchor nodes and time synchronization. We provide an integrated solution to sensor localization and the K-coverage problem. In addition, due to the short battery life that powers individual sensors, the lifetimes of sensor networks are severely restricted. Significant research has been done to improve energy efficiency at both the individual sensor level and the sensor network as a whole [6]. By observing societies formed by certain natural species, such as an ant colony, we learn that a colony can sustain its functions and life for a long period of time despite the fact that each individual ant has a short life span. This is achieved through member collaborations and reproduction. Inspired by such observations, our research addresses the K-coverage problem while maximizing the lifetime of sensor networks and supporting robust CPS operations through self-organized sensors.

The key contributions of our research for localization and K-coverage problems are a cluster formation/localization algorithm and a cluster maintenance algorithm [7]. We developed a signal stimulation model (SSM) that uses location-guided laser beams to trigger sensors within one hop of communication range to form geographically bound sensor clusters. Based on the SSM model, we proposed an SCLA that can form a cluster and elect a cluster head based on their responses to the stimulation signal projected at the center of a grid cell. Once the cluster is formed, all the cluster members are aware of their locations. The SCLA algorithm can verify the required K-coverage as long as the number of members in each grid cell is greater than or equal to K.

In reality, sensor failures occur randomly due to hardware/software malfunction or power depletion. We proposed an SCMA that can prolong the CPS lifetime by maintaining the smallest subset of cluster members in active mode necessary to perform required tasks while permitting the remaining cluster members to enter sleep mode in order to conserve energy. The sensors in sleep mode turn off their communication and sensing functions while sleeping. Periodically, they awaken and may replace one or more active nodes that are deceased (either because they have depleted their power or failed prematurely) or when their energy levels fall below a minimum threshold. Unlike an ant colony that can maintain its population through reproduction, we assume that a CPS does not have sensor replacement capability. To achieve the simultaneous goals of energy conservation and providing

K-coverage with fault tolerance operations, it is obvious that SCMA requires a sensor population with sufficient density and redundancy. Our approach is particularly suitable to open-space environment monitoring CPS applications that are subject to long network operation time and harsh operating conditions with a high degree of random node failures.

In addition to sensor localization and K-coverage problems, sensor failures and disconnections are part of reality in many CPS operations. They present a new set of challenging problems for CPS applications, such as

- *Network connectivity and routing management*—Making sure that all sensor nodes are connected and data can be routed efficiently in the network.
- *CPS life expectancy management*—Planning and managing the network to meet the required network operation life expectancy without compromising QoS requirements.

To address these challenges, our research focuses on how to estimate CPS life expectancy with consideration of QoS constraints that are defined as a set of requirements, such as data rate, delay time, and degree of sensing coverage (K-coverage). Based on our previous work on sensor localization and K-coverage problems, we introduce a Markov process model with QoS constraints and derive a closed-form mathematical model for CPS life span estimation [8].

17.3   SIGNAL STIMULATION MODEL (SSM)

We introduce SSM as a basic model that defines the parameters and constraints within which a CPS application is deployed and operated. The SSM model is defined by the following set of assumptions:

1. Sensors—Each sensor has sensing, data storage, data processing, and wireless communication capability that is equivalent to an MICA Mote developed at UC Berkeley [9,10]. Each sensor covers a communication cell and a sensing cell defined by radius Rc and Rs, respectively. All of the sensors are nonmobile and they can sense optical signals (delivered by laser beams), in addition to other sensing capabilities.
2. Sensor network—A sensor network consists of a set of homogeneous sensors. Sensors can communicate with each other via wireless channels in single or multiple hops, thus they form an ad hoc network. There are one or more base stations located outside the sensor region but near the border of the sensor network with wired or long-range wireless communication links to the Internet for collecting data or disseminating queries and control commands to the sensor network.
3. Deployed region—Sensors are deployed over an open-space area. A virtual grid marks this area. Each cell in the grid is a D-by-D square.
4. There is a *lightweight location guided laser* designator system that can project a laser beam to a given location (x, y) with acceptable accuracy, such as the system produced by Northrop Grumman that has target range up to 19 km with accuracy of 5 m [11]. Obviously, this assumption requires a line-of-path for laser beams to reach the sensors at ground level. However, for many open-space environment monitoring applications, this is not a major problem.
5. To ensure the coverage and connectivity of a sensor network, the model requires that D, Rc, and Rs satisfy the following constraints:
   - To ensure that a sensor anywhere in a cell can cover the cell, it must satisfy the condition, $Rs^2 \geq 2D^2$. In our model, we assume $Rs^2 = 2D^2$.
   - To ensure a sensor anywhere in a cell can communicate with a sensor anywhere in a neighboring cell, it must satisfy the condition, $Rc \geq 2Rs$. In our model, we assume $Rc = 2Rs$.

Figure 17.1 shows the parameters that define a virtual grid with four neighboring cells.

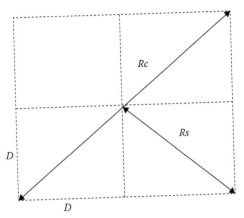

FIGURE 17.1 Parameters of a virtual grid.

Every point in the region can be represented by a pair of (x, y) coordinate values. A sensor has three possible states, U (unknown), H (cluster head), and M (cluster member). Initially, all sensor states are set to U. During the postdeployment phase, an object flies over the deployed region and projects a laser beam to the center of a grid cell (Xc, Yc). The sensors nearby will sense the signal. The sensor readings are stronger if they are closer to the projected laser beam. The sensor with the strongest reading is identified as the cluster head (state = H). All sensors that have a reading greater than λ (λ-cut) and are one hop away from the cluster head become members of the cluster (state = M). Ideally, λ should maximize the possibility of including a sensor in the cluster if it is within the cell, and minimize the possibility of including a sensor in the cluster if it is outside the cell. An optimal value of λ can be obtained through experimentation and simulation. Since an accurate light energy propagation model is extremely difficult to obtain and light wave is a form of electromagnetic wave, we believe it is a reasonable assumption that the light signal decay model should be similar to the attenuation of radio waves between antenna and wireless nodes close to the ground for our study and simulation. Radio engineers typically use a model that attenuates the power of a signal as $1/r^2$ at short distances (where r is the distance between the nodes), and as $1/r^4$ at longer distances [12]. In our model, the size of the virtual grid is small, and thus, we assume the light signal attenuation follows the short distance model.

Figure 17.2 shows a cluster formed in grid cell 5. The black dot indicates that the sensor node is a cluster head in the cluster.

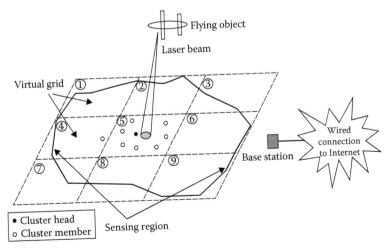

FIGURE 17.2 Sensor cluster localization and formation.

In this chapter, we assume that a laser beam is projected to one cell at a time with a cluster forming time interval, T, for each cell. T should be just long enough to allow the sensors in a cell to form a cluster, but not so long as to cause unnecessary delay for the operations between cells. In general, T is a function of sensor density—n (the number of sensor nodes in a cell), radio propagation delay τ and IEEE 802.11 MAC layer back off delay β in its CMSA/CA protocol with p as a probability of package collision in the form of

$$T = n^2\beta p + n\tau + c \tag{17.1}$$

where c is a constant for initial delay from sensing the light signal to transmitting the data. With this assumption, a sensor node can only belong to one cluster, since once it joins a cluster it will not respond to the laser signals projected to other cells. This works even for the sensors located on the border of grid cells.

Based on the SSM model described earlier, we present the SCLA algorithm with the following steps:

1. Let t_0 be the time when the laser beam is projected to a cell and let T be the time interval for cluster formation. Let $t = t_0$ when starting the algorithm.
2. While $t > t_0$ and $t < (t_0 + T)$ repeat steps 3 and 4.
3. For each sensor with unknown status that has detected the signal, if the sensor reading is greater than λ, then it will broadcast a message with the sensor id and sensor reading (SID, value) to its neighbors within one hop of communication. Otherwise, it keeps silent.
4. When receiving a message, a sensor with unknown status acts according to the following rules:
   a. Rule 1—If the reading value of the received message is greater than its own reading and its own reading is greater than λ, then it will set the state = M (a cluster member) and reset its local memory.
   b. Rule 2—If the reading value of received message is less than its own reading and its own reading is greater than λ, then save the message (SID, value) in its local memory.
5. For a sensor that still has unknown status, if its own reading is greater than λ it will set the state = H (a cluster head).
6. The cluster head sends the cluster membership information (SID, value) pairs, which were saved at step 4, to the closest base station.
7. Project laser beam to the next cell and repeat steps 1–6 until all the cells are visited.

Our analysis shows that the SCLA algorithm performs and scales well when sensor network size increases in terms of both the number of cells in a grid and the total number of sensors in a cell. Let L be the largest number of communication hops from a cluster head to the closest base station. Let M be the total number of cells in the grid. Let N be the total number of sensors deployed. Let $n(i)$ be the number of sensors in cell(i). The cost of SCLA algorithm in terms of the number of messages transmitted is given as

$$\text{Cost}\,(L, M, N) \leq \sum_{i=1}^{M} n(i) + M^*L \tag{17.2}$$

where $\sum_{i=1}^{M} n(i) = N$. If we assume sensors are uniformly distributed, then we have

$$\text{Cost}\,(L, M, N) \leq M^*\,(N/M) + M^*L = N + M^*L \tag{17.3}$$

Formula (17.3) is equivalent to the notation of $O(N)$, when M and L are significantly smaller than N, which is true in most high-density sensor networks. These results show that it is feasible to deploy large amounts of redundant sensor nodes in order to compensate for high rates of sensor failure with a linear overhead cost growth over the sensor population.

17.4 SENSOR CLUSTER MAINTENANCE ALGORITHM (SCMA)

The SCMA algorithm consists of three components, namely, cluster member component, cluster head component, and base station component. These components work together to achieve the following functions:

1. Coordinating a subset of cluster members as active nodes that perform required tasks while tagging the remaining cluster members as sleep nodes to conserve energy.
2. Replacing the nodes that failed unexpectedly or whose energy level falls below a certain threshold to guarantee required coverage.
3. Warning the network operators if the required coverage is going to be compromised.

After the cluster is formed according to the SCLA algorithm, all cluster members are in active mode $(m = A)$ by default. A member listens to the initialization message $(I, \text{sensor\_list})$, from the cluster head. A member turns into sleep mode $(m = S)$ unless its SID is on the sensor_list in the initialization message. We use finite state automata to precisely describe the behavior of each component in a simple canonical form. Particularly, we employ a special type of finite state automata, called Mealey machine, which is formally defined by Hopcropt and Ullman [13] as

Let finite state automata, $F = (Q, \Sigma, \Delta, \delta, \gamma, q_0)$ where

q_0 is the initial state $(q_0 \in Q)$
Σ is a set of inputs
Q is a set of states
Δ is a set of outputs
δ is a state transition function: $Q \times \Sigma \to Q$
γ is an output function: $Q \times \Sigma \to \Delta$

To help understand the state transition diagrams, we list all the message definitions in the table given below (Table 17.1).

TABLE 17.1
Message Definitions for an SCMA Algorithm

Message	From	To	Description
(A, SID)	Head	Member	Activate a member (SID)
(Ack, SID)	Member	Head	Acknowledgement: a member (SID) is activated
(Ack, SID)	Member	Base	Acknowledgement: a member (SID) is a new head
(E, HID)	Head	Base	Emergency message to the base station
(H, HID)	Head	All members	Broadcast "hello" message
(H, SID)	Member	Head	Reply to "hello" message by a member (SID)
(I, Sensor_list)	Head	All members	Broadcast initial active sensor list
(N, SID)	Base	Member	Appoint a member (SID) to be the new head
(R, SID)	Member	Base	A member requests the cluster status
(Status, HID)	Head	Base	Send cluster status to the base station
(Status, SID)	Base	Member	Send cluster status to a member (SID)
(W, HID)	Head	All members	Broadcast "wakeup" message
(W, SID)	Member	Head	Reply to "wakeup" message by a member (SID)
Tx/e			Time trigger—while in state x for time interval Tx then return to previous state without output. $x \in (c, r, s, w, p, u, n, g)$.
Pa/e			Event trigger—power level below a.

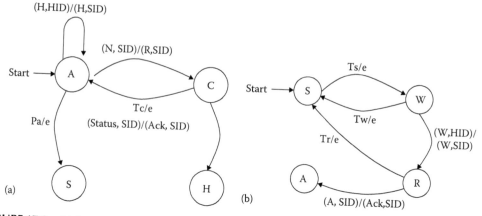

FIGURE 17.3   (a) State transition diagram for active members. (b) State transition diagram for sleep members.

We use two state transition diagrams, one represents cluster members in an active mode and the other represents cluster members in a sleep mode. Figure 17.3a defines the behavior of an active member node within the SCMA algorithm and how it can change to sleep mode or become a cluster head when a cluster head has failed. For example, when an active member receives a "Hello" message from the cluster head (H, HID), it replies with message (H, SID) and goes back to the same state. An arrow indicates such a transition with a text label on top in the form of (input message)/(output

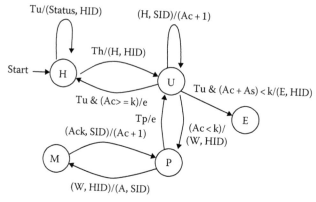

FIGURE 17.4   State transition diagram for cluster heads.

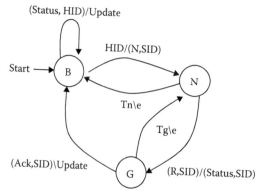

FIGURE 17.5   State transition diagram for base stations.

message). The same notation applies to all the state transition diagrams in this section. Figure 17.3b describes the behavior of a sleep member node and how it can change to active mode.

The cluster head component coordinates members between active and sleep modes. It updates cluster status and synchronizes the cluster status with the base station. It uses an active node counter (Ac) and a sleep node counter (As) to track active members and sleep members. Figure 17.4 describes how a cluster head works. The cluster head issues an emergency message, (E, HID) to base station if the required coverage is going to be compromised within this cluster. In the SCMA algorithm, the function of a base station component is to oversee the status of clusters and the condition of networks. Figure 17.5 shows how it works.

17.5 *K*-COVERAGE THEOREMS

We introduce the following definitions and theorems to formally prove that the proposed SCLA and SCMA algorithms together can solve the *K*-coverage problem.

Definition 17.1 *K-coverage problem: a CPS application is required to guarantee there are at least k sensors whose sensing range can cover any given point in the deployed region.*

Definition 17.2 *The membership degree of a cluster is the number of sensors in a cluster including the header.*

Theorem 17.1 *For a cluster(i) formed at a cell(i) in accordance to the SCLA algorithm, if the membership degree of the cluster(i) is no less then K during the lifetime of operation, then the area within the cell(i) is K-covered.*

Proof: Based on the assumption 5.1 in the model definition section, a sensor located anywhere within a cell can cover any points in the cell including the border of the cell. If the *cluster(i)* has no less than K member sensors during the lifetime of the operation then, by definitions 1 and 2, any points within the *cell(i)* are covered by no less then K sensors during the lifetime of the operation. Therefore, *cell(i)* is *K*-covered. □

Theorem 17.2 *Given a deployed region R that is enclosed in a virtual grid G, if there is one cluster formed in each cell and its membership degree is no less then K during the lifetime of the operation, then the deployed region R is K-covered.*

Proof: By Theorem 17.1, every cell in G is K-covered and the virtual grid G encloses the entire region R; therefore, the deployed region R is *K*-covered. □

Based on Theorem 17.2, it becomes obvious that SCMA algorithm can satisfy a CPS application K-coverage requirement as long as it can maintain at least K active member sensors in the cluster at each cell during the *life of CPS* operations. Theorem 17.2 also forms the basis that is used to develop a closed-form mathematical model utilizing a Markov process in the next section.

17.6 MARKOV PROCESS MODEL AND ANALYSIS

Many CPS applications have to operate under harsh environments with high failure rates. A good CPS life estimation model will enable us to address a fundamental question, such as how long a CPS application can operate without compromising required QoS. In this section, we make certain assumptions about sensor network operations that allow us to use a modified Markov process to produce a closed-form model to answer this question and we then validate the model by running various simulations in Section 17.5.

17.6.1 CPS Life Span Problem Statement

First, let us consider these fundamental questions about CPS applications,

1. How long can a CPS application operate while meeting the required QoS?
2. How many sensor nodes need to be deployed for a CPS application to reach the expected network life without compromising the required QoS?

There are no simple answers to these questions. The life span of a CPS application depends on many variables, such as sensor population, battery capacity, sensor energy consumption rate, QoS constraints, and sensor random failure rate. Power depletions and sensor failures are the two major factors that impact CPS life [14]. For many environment monitoring CPS applications, sensor nodes are powered by nonrefreshable power sources, such as commercially used nonrechargeable batteries with certain capacities. Power depletion due to energy consumption is mainly determined by how often a sensor needs to transmit data (data transmission rate) that is highly dependent on the nature of applications. For example, a long-term environment monitoring application may only need to collect data every few hours. In comparison, an intrusion detection application may need to collect data every few seconds.

Our goal is to develop a generic model without limitation to specific applications. Therefore, we limited our study in this chapter to relationships among CPS life, sensor population, sensor failure rate, and QoS constraints. Sensor failure rate is independent of the power level of the sensors. Sensor population is a factor of CPS life due to the fact, in many CPS applications, that sensors are densely deployed (more sensors than necessary for the operation requirements). Some sensors are put in "sleep" mode to conserve energy and they can "wake" up to replace the sensors that have failed. This is a common practice called "duty-cycle" [15]. Sensor failure rate is mainly determined by both qualities of manufactured sensors and deployment environment conditions. One can obtain the specific sensor node failure rate from its manufacturer or from lab tests with similar deployment conditions. K-coverage requirement as a QoS constraint is a factor of *CPS application life* due to the fact that as soon as a CPS application can no longer provide K-coverage, the usefulness of the CPS application is ended. In general, the model of a CPS application life can be expressed as follows:

$M = (L, N, E_0, e, p, Q)$ is the model for CPS life, where
L is the network lifetime
N is the sensor population
E_0 is the initial battery capacity
e is the energy consumption rate
p is the sensor failure rate
Q is the QoS constraints

Our goal is to answer the two questions stated earlier in a closed mathematical form by deriving these two functions:

- Let $L = f(N, E_0, e, p, Q)$ be the lifetime function
- Let $N = g(L, E_0, e, p, Q)$ be the sensor population function

The first function is useful to project the lifetime of a CPS application. The second function is useful for deployment planning and cost budgeting, a question often asked by your boss.

17.6.2 MARKOV PROCESS MODEL FOR CPS LIFE SPAN ESTIMATION

In a CPS application, let us consider the question of how long a sensor cluster can last with required K-coverage. In other words, how long it takes to reduce the sensor population from n sensors to k sensors (from state S_n to state S_k) due to random sensor failures.

Let Δt be the time interval between two observations of a cluster member degree. The measurement of Δt can be in hours or days based on the quality of sensor. We set Δt small enough so that it is only possible for one sensor to fail within this time window. Let X_i, denote the number of observations made while the cluster remains at state S_i, where the cluster contains exactly i sensor nodes. This forms a sequence of Bernoulli trials since the failure of each individual sensor is independent of others. Thus, each observation can be considered as a Bernoulli trial with two outcomes, sensor failed or not failed. The random variable X_i follows geometric distribution. It has the character of a "memory less" process, and it can be modeled by a discrete Markov process. Then according to a Markov chain (Figure 17.6), the expected number of observations made while a cluster stays at S_i can be expressed as

$$E[X_i] = \frac{1}{p} \tag{17.4}$$

where p is the probability of failure (failure rate) during Δt.

Let T_i be the time duration that a cluster stays at S_i. Thus, the expected time of T_i is

$$E[T_i] = \frac{\Delta t}{p} \tag{17.5}$$

Therefore, the life expectancy of a cluster(i) with consideration of K-coverage requirement is

$$LC_i(K, n_{i,}) = \Sigma E(T_i), \quad i = n_i \text{ to } K \quad \text{and} \quad n_i \geq K \tag{17.6}$$

or

$$LC_i(K, n_{i,}) = \Sigma E(T_i) == \Delta t(n_i - K + 1)/p \tag{17.7}$$

where

K stands for K-coverage

n_i is the initial number of sensors in the cluster(i)

According to Theorem 17.3, the CPS life expectancy is determined by the shortest cluster life expectancy:

$$L(K, N) = \min\{LC_i(K, n_i)\}, \quad i = 1 \text{ to } M \tag{17.8}$$

$$N = \Sigma(n_i)$$

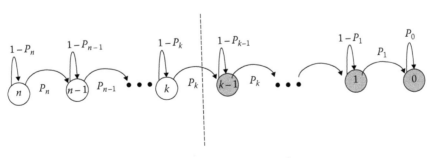

FIGURE 17.6 A modified Markov process model with QoS constraints.

where

N is the sensor population in the network
M is the total number of clusters or cells in the network

Assuming a perfect uniform distribution, then in theory each cluster should have the same number of sensors $n(n = N/M)$. From Equation 17.8, we have

$$L(K,N) = LC(K,n) = \Delta t\,(n - K + 1)/p \tag{17.9}$$

Finally, with Equation 17.9, we established a closed-form mathematical function of CPS application life expectancy in terms of sensor population N, sensor failure rate, and QoS constraints in the form of K-coverage requirement. Thus, we find an answer to the first fundamental question proposed in Section 17.4.1.

From Equation 17.9, with a required CPS lifetime (L), we can derive

$$n = N/M = K - 1 + pL/\Delta t \tag{17.10}$$

Equation 17.10 enables us to plan and budget the total number of sensors (N) needed to ensure the CPS can last the required lifetime (L) with certain QoS constraints (K). This provides an answer to the second fundamental question proposed in Section 17.4.2.

17.7 SIMULATION STUDY AND PERFORMANCE ANALYSIS

We conducted two set of simulations, one is for the SCLA and SCMA algorithms and the other is for the Markov process model to estimate a CPS lifespan.

17.7.1 SIMULATION STUDY OF SCLA AND SCMA ALGORITHMS

In order to validate the model and algorithms presented in this chapter and to gain insights into how the algorithm works, we conducted simulation studies using the NS2 simulator with Monarch Extensions to ns for SCLA algorithm. We extended our simulation for SCMA algorithm with C++ modules to study the energy conservation and the impact on CPS lifetimes. Our simulations are implemented with two scenarios. The first scenario involves simulating a single cell grid with different sensor densities (number of sensors in the grid).

The focus of this simulation is to study the performance and scalability of our model against sensor density. In the second scenario, we take the same measurements from a multicell grid simulation with considerations of both sensor density and the size of the deployment area in terms of the number of cells. The purpose of this simulation study is to understand the performance and scalability of our model in a multicell grid.

We set the cell dimension to 10 m for all the simulations presented in this chapter. Our simulation tests indicate that the outcomes are not as sensitive to the cell dimension as they are to sensor density. We let the number of sensor nodes vary from 10 to 80 in increments of 10. In our simulation model, we set the propagation delay between two nodes as 10 ms. We use multicast in UDP protocol to simulate sensor node broadcast in one-hop distance. We set p as the probability for a node to receive the broadcast message successfully (p in the range 0–1).

The message package size is set to 128 bytes and the bandwidth between two nodes is set to 2 mbps. To simulate IEEE 802.11 MAC layer CMSA/CA protocol, we introduce a back-off time delay, a random number between 50 and 100 ms, which is assigned to a node when it detects that a channel is busy. The node will back off for a delay interval before it tries to broadcast again. The simulation results capture two key measurements, the number of messages being transmitted and the time interval for cluster formation in the grid. All of the simulation results presented are the average of five simulation runs.

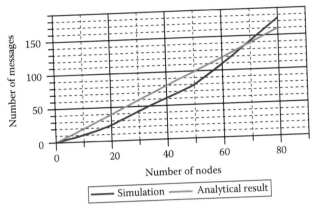

FIGURE 17.7 Messages transmitted in single cells.

Figure 17.7 shows the number of messages being transmitted in a single-cell grid. It compares the analytical result with simulation results. It indicates that the cost of message transmissions is close to a linear function of n, the number of sensor nodes in the cell. To better understand cluster membership distributions and study the impact of λ values (the λ-cut defined in SSM model) on member selections, we used a single grid cell of 10 m by 10 m with λ values in the range [0.02, 0.04]. The simulation result in Figure 17.8 shows the percentage of sensors that are dropped from the cluster for the cell as the value of λ changes. It shows the higher λ value leads to more sensor nodes being excluded from the cluster. The simulation results below are for a multicell grid scenario with the same key measurement as we presented for a single-cell grid. Figure 17.9 shows that with a fixed number of sensor nodes, the number of messages being transmitted actually drops as we expected as the cost function defined in Equation 17.3. *Because there are fewer collisions as sensor density decreases.* This simulation result indicates that the number of messages being sent is more sensitive to the density in each cell than the number of cells in a grid. Figure 17.10 presents an interesting measurement, the percentage of sensors wrongly claimed by clusters. It is closely correlated with the γ values. The ratio of disputed sensors drops or stays at the same level after the number of cells reaches 16 due to the decrease in sensor density. Figure 17.11 shows the percentage of sensors that are unclaimed in a multicell grid with 80 sensors total. This more likely happens to sensors at the cell borders and is sensitive to λ values.

Our simulation study of the SCMA algorithm is designed to understand the relationships between the network lifetime, sensor node density, and the overhead cost with K-coverage constraints. The CPS lifetime is defined as the time interval that a CPS network can sustain its operations and services

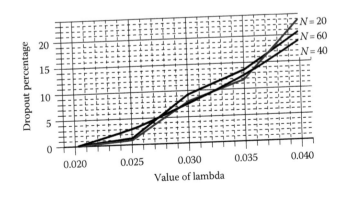

FIGURE 17.8 Membership distribution over λ values.

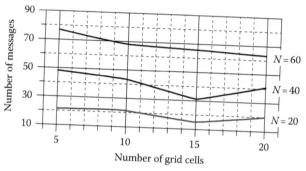

FIGURE 17.9 Performance and scalability in multicells.

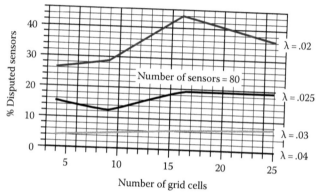

FIGURE 17.10 Number of disputed sensors with λ values.

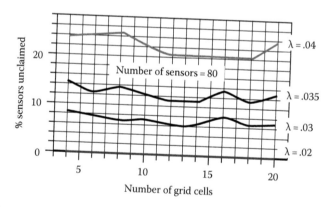

FIGURE 17.11 Percentage of unclaimed sensors.

with respect to K-coverage requirement. In other words, the service quality, such as sensing and communication coverage requirements, cannot be compromised during the lifetime of CPS operations.

Given the parameters defined in the SSM model, the communication coverage of a CPS can be reduced to the sensing coverage (assumption 5). The sensor network lifetime is defined by the shortest cluster lifetime. If a cluster is unable to provide the required coverage in one cell, then the whole network is compromised. In the notation of SCMA, K (K-coverage requirement) is defined as an input parameter. It is obvious that different CPS applications may require different K-coverage.

In our initial simulation, we baseline our study by setting $K = 1$. We assume sensors are uniformly distributed over grid cells during deployment for simulation study purposes. We use the sensor

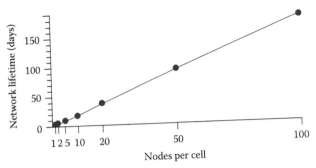

FIGURE 17.12 Network lifetime over number of nodes in a cluster.

node power consumption characteristics published by Crossbow [16] for our energy consumption computations.

Figure 17.12 shows a cluster lifetime versus the number of nodes in a cluster, which has a linear growth in terms of sensor population. But we expect this linear growth at lower slopes as K-coverage increases (more active nodes required). It also incurs overhead costs for the cluster head to probe active nodes with "hello" messages. Further simulation can be done to study this impact and how to control the head node's probing cycle to balance the trade-off of required coverage and overhead costs. Given the cell size in our simulation (10 m by 10 m), 100 nodes/cell is far beyond the most population density of CPS applications.

It is interesting to observe the relationship between the average node life span and node density. If we let an active node moves into sleep mode when its power level falls below a certain threshold, it might increase the average member node lifetime as node density increases since the workload is more evenly distributed among a larger node population. But the simulation result in Figure 17.13 shows the average node lifetime is slightly below linear growth in terms of node density. Particularly, as the node density goes beyond 50 nodes. This result is not in total agreement with our theoretical analysis on SCMA. We contribute this discrepancy to the additional overhead cost of switching an active node to sleep node on a volunteer basis when the active node's power level is below a certain threshold.

We are aware that the average lifetime of a cluster head is expected to be shorter than the member nodes since the head node assumes more duties than the members. We could introduce the same logic to the SCMA algorithm by permitting a head node to sleep when its power level falls below a certain threshold. We might expect a greater overhead cost in doing so. Currently, we are conducting more extensive simulation studies to understand this issue better.

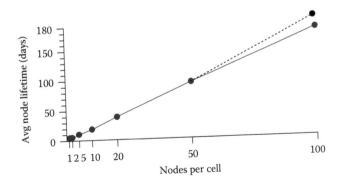

FIGURE 17.13 Average node lifetime versus node density.

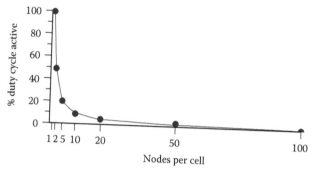

FIGURE 17.14   Percentage of active duty time versus number of nodes in a cluster.

Figure 17.14 shows that as the percentage of active duty cycle (the percentage of active time over total node lifetime) is decreased the node density increases with $K = 1$ coverage. We expect this percentage to follow the same trend but at less decline rate percentage as the K value increases since more nodes must stay active.

Based on the simulation results shown in Figure 17.14, we expect the cluster head probing cycle (T_h is the time interval before sending next "hello" message) of "hello" messages to impact the overhead cost of the SCMA algorithm. As interval T_h increases, the overhead cost of SCMA will decrease. We conducted simulation tests under three T_h values.

The simulation results shown in Figure 17.15 validate our expectations. It shows the overhead cost as the percentage of SCMA message transmitted over the total potential bandwidth capacity of a sensor cluster. For three different T_h values, we can see *the impact of T_h remains significant* only before the node population reaches a "critical mass" where nodes/cell = 5 for $K = 1$ coverage. As the node population increases beyond the critical mass, the total message capacity of the cluster grows at a much faster than the overhead cost. This result tells us that SCMA scales well as node population increases measured by the percentage of overhead cost. This result also suggests that it is most cost effective to increase the T_h value for the clusters with a smaller population (less than 5 sensor nodes per cell).

17.7.2   SIMULATION STUDY OF MARKOV PROCESS MODEL FOR WNS LIFE SPAN ESTIMATION

We performed a simulation study to validate the analytical model under different deployment scales and QoS constraints (K-coverage requirements) for a general environment monitoring application. The simulation program is implemented in C++ using Microsoft Visual Studio 2005. What follows is a brief description of conditions and scenarios for our simulation study.

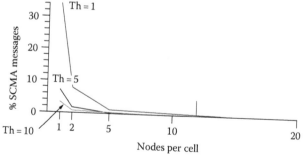

FIGURE 17.15   Percentage of SCMA messages versus node density per cell.

Following the cluster-based operation model defined in Section 17.4, we defined a virtual grid of 500 m by 500 m that encloses the targeted CPS deployment region. There are 100 cells in the grid. Each cell is a square of 50 m by 50 m. To mimic a "real-world" CPS deployment and the cluster-based operation model, the simulation program performs these tasks:

1. It randomly deploys all sensors (sensor population N) over the target area (500 m by 500 m).
2. It tracks the number of sensors in each cell. We assume all sensors in a cell belong to a cluster under an optimal condition. It will find the cell with the minimum number of sensors, C_{min}. under an optimal condition. It will find the cell with the minimum number of sensors, C_{min}.
3. It simulates the Markov process with given parameters for the cell. It records the time for the cell to go from n sensors to k sensors. This will be the lifetime of the CPS application since C_{min} has the shortest lifetime among all the cells.
4. In order to minimize the variance and bias of the simulation results, we execute 100 runs for each set of parameters. The results presented in this chapter are average values over 100 runs.

We designed two simulation scenarios to study the impacts on L (CPS lifetime) from vantage points of different values of N (total sensor population), K (QoS constraint), and p (sensor failure rate). In this simulation study, the sensor failure rate, p, is measured as the number of sensors failing per day. The observation interval is set to a half day, $\Delta t = 0.5$.

In scenario 1, we assume a fixed failure rate p, where an individual sensor has a 0.01 chance of failing within the observation interval. Thus in scenario 1, we study the relationship of between L and N with a fixed value of $p = 0.01$. We let N change in the range of 100–2000 and set $K = 1, 3$, and 5 respectively. Figure 17.15 shows the results for simulation scenario 1. We should explain the data items and terms on the chart to help readers understand the data. It shows the CPS life (L—unit in days) over sensor population (N—number of sensors). There are three sets of data in each chart. Simulated CPS life (S-Life) is the estimation of CPS life by simulation program. M-Life is the projected CPS life-based Markov mathematical mode using Equations 17.7 and 17.8. M-Life computation is done with simulation of random deployment and the minimum cluster size. In theory, if the deployment is truly uniform, then each cell should have the same number of sensors. But this is rarely the case in real-world deployment. We introduce a theoretical measurement called Theoretical CPS life (T-Life), which is computed using Equation 17.9. T-Life provides a useful means for CPS application engineers and operators to do pre-deployment planning and budgeting since no one can exactly pre-determine the sensor footprints of the deployment due to the nondeterministic nature of the randomness.

Figure 17.16a,b,c show that S-Life and M-Life are very close regardless of the K-coverage requirement. As a matter of fact, at first glance it might be hard to distinguish between the M-life (square-dot line) and S-life (circle-dot line) on the graph since they often coincide. These suggest that the Markov process model is well supported by simulation and suitable for modeling CPS life expectancy under these scenarios. Major characteristics of CPS life expectancy are all in agreement between the model and simulation and they behave well in line with our following predictions:

1. CPS life increases as sensor population grows under the same K and p values due to the fact of "duty-cycle" of sensor nodes.
2. CPS life decreases as K-coverage goes up.

It is interesting to point out that there is a gap between the T-Life and S-Life/M-Life due to the fact that S-Life and M-Life are based on minimal cluster size while T-Life is based on average cell size under ideal uniform distribution. There are 100 clusters in the grid. But average cluster size and minimal cluster size seem to converge as the sensor population reaches a critical mass. As indicated in Figure 17.17, the critical mass of sensor population for a 100 cells grid is about 1500 sensor nodes. This behavior is well explained by the "large number theorem" in probability theory [17].

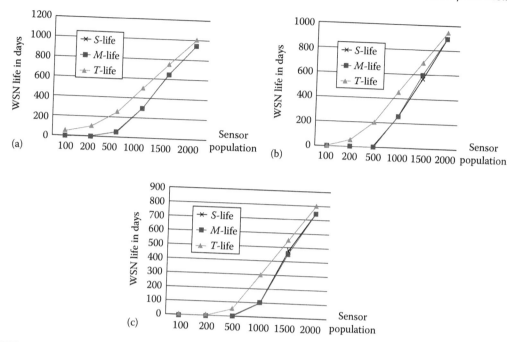

FIGURE 17.16 CPS life expectation over sensor population. (a) When $K = 1$ and $p = 0.01$. (b) When $K = 2$ and $p = 0.01$. (c) When $K = 5$ and $p = 0.01$.

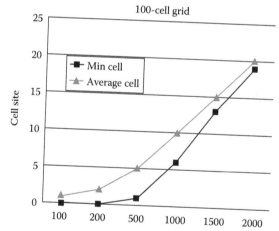

FIGURE 17.17 CPS cell size convergence (a 100-cell grid).

The impact of the "large number theorem" is clearly illustrated in the sensor deployment foot print charts in Figure 17.18. Figure 17.18a shows a deployment of 100 sensor nodes where some cells have no sensor coverage at all (coverage holes in the networks). Figure 17.18b shows a deployment of 1000 sensor nodes where all cells seemingly have more even coverage.

In scenario 2, we let the failure rate increase from 0.01, to 0.05, to 0.1 to study the impact of L while keeping a fixed K-coverage value (Figure 17.19).

It clearly shows that CPS life (L) drops sharply as sensor failure rate p value increases. CPS life seems more sensitive to the change of p values than K value. The same is true with higher K values, $K = 2$ or $K = 5$. One of the insights we gain from this simulation is how a harsh deployment environment deprecates the performance of a CPS application.

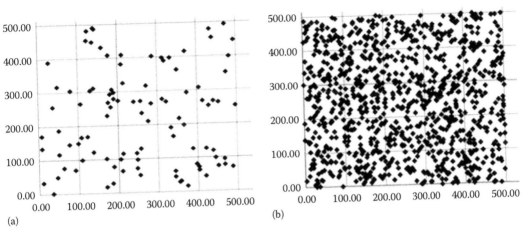

(a)   (b)

FIGURE 17.18   CPS sensor deployment footprint charts. (a) Deployment of 100 sensors. (b) Deployment of 1000 sensors.

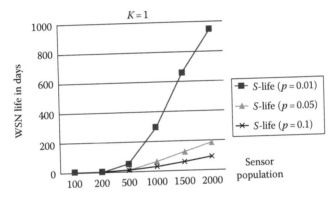

FIGURE 17.19   CPS life expectation with impacts of different sensor failure rates.

17.8   CONCLUSION

We propose a unique solution to address both sensor localization and K-coverage problems that can conserve energy while supporting robust CPS operations. Simulation results show that both algorithms, SCLA and SCMA, perform and scale well with overheads in linear proportions to the deployed sensor population in a variety of deployment densities. SCMA can guarantee the K-coverage requirements of CPS applications over their network lifetimes and it can warn network operators should coverage requirements be compromised.

The work reported in this chapter leads to several interesting topics for future research. We plan to investigate the possibility of using this model for differential K-coverage problems [18] where different cells may require different K-coverage within the same CPS application. This will help us to refine our model and to achieve even greater energy efficiency. We are aware that we did not take into consideration the energy consumed by sensors for providing normal operational tasks. Our analysis and simulation results only explore the overhead portion of energy consumed by SCLA and SCMA. The energy credited to CPS normal operational tasks are highly application dependent; therefore, it is out of the scope of our study.

In this chapter, we also present our research on using a Markov process to model CPS application life span. We are able to establish a closed-form mathematical function to express a CPS application's life expectancy in terms of sensor population, sensor failure rate and QoS constraints, and then validate the model by running various simulations. With this work, we demonstrate that the fundamental questions about CPS applications can be answered with reasonable certainty. We are currently extending our study to include the impact to CPS life due to power consumption and to develop an integrated model that takes into consideration both sensor failure and power depletion. We are also investigating the possible ways to integrate some of the well-known "duty-cycle" algorithms and study their overall impacts on CPS lifetime.

ACKNOWLEDGMENTS

This work was supported, in part, by the College of Staten Island, City University of New York Provost Research Award. This research was also supported, in part, under National Science Foundation Grants CNS-0958379 and CNS-0855217 and the City University of New York High Performance Computer Center at the College of Staten Island.

REFERENCES

1. I. F. Akyildiz et al., A survey on sensor networks, *IEEE Communications Magazine*, 40(8), 102–114, August 2002.
2. N. Patwari et al., Locating the nodes: Cooperative localization in wireless sensor networks, *IEEE Signal Processing Magazine*, 22, 54–69, July 2005.
3. A. Savvides et al., Dynamic fine-grained localization in ad-hoc networks of sensors, in *Proceeding of ACM SIGMOBILE*, Rome, Italy, July 2001.
4. X. Wang et al., Integrated coverage and connectivity configuration in wireless sensor networks, in *Proceedings of the ACM SenSys*, Los Angeles, November, 2003.
5. K. Charkrabarty et al., Grid coverage for surveillance and target location in distributed sensor networks, *IEEE Transaction on Computers*, 51, 1448, 2002.
6. C. Jone et al., A survey of energy efficient network protocols for wireless networks, *Journal of Wireless Networks*, Kluwer Academic Publishers, 7, 343–358, 2001.
7. Z. Zhang et al., A cluster based approach toward sensor localization and K-coverage problems, *Ubiquitous Computing and Communication Journal*, 2(4), 15–23, 2007.
8. Z. Zhang, and M. Tausner, Markov process modeling and simulation for wireless sensor network life estimation with QoS constraints, in *Proceeding of 2010 Summer Computer Simulation Conference*, Ottawa, Ontario, Canada, July, 2010.
9. XBow Inc. *MICA Wireless Measurement System*, http://www.xbow.com/Products/. March 18, 2005.
10. Berkeley WEBS (Wireless Embedded Systems). *University of Berkeley Mote*, http://webs.cs.berkeley.edu/. December 8, 2004.
11. Northrop Grumman, *Lightweight Laser Designator Rangefinder Data Sheet*, http://www.northropgrumman.com/Capabilities/ANPED1LLDR/Documents/lldr.pdf. June 12, 2015.
12. J. Broch et al., A performance comparison of multi-hop wireless ad hoc network routing protocols, in *Proceeding of the Fourth Annual ACM/IEEE International Conference on Mobile Computing and Networking*, Dallas, TX, October, 1998.
13. J. Hopcroft and J. Ullman, *Introduction to Automata Theory, Language, and Computation*, Addison Wesley, Reading, MA, 1979.
14. G. Giridhar and P. R. Kumar, Maximizing the function lifetime of sensor networks, in *Proceedings of the Fourth International Symposium on Information Processing in Sensor Networks, IPSN*, Los Angeles, April 2005.
15. H. Zhang and J. C. Hou, Maximizing α—Lifetime for wireless sensor networks, *International Journal of Sensor Networks*, 1, 64–71, 2006.
16. XBow Inc. *MICA2 AA Battery Pack Service Life Test* http://www.xbow.com/Support/Support_pdf_files/MICA2_BatteryLifeTest.pdf. March 18, 2005.
17. K. S. Trivedi, *Probability & Statistics with Reliability, Queuing and Computer Science Applications*, Prentice-Hall, Englewood Cliffs, NJ, 1982.
18. X. Du and F. Lin, Maintaining differentiated coverage in heterogeneous sensor networks, *EURASIP Journal on Wireless Communications and Networking*, 4, 269210, 2005.

18 Application Development Framework for Developing Cyber-Physical Systems

Pankesh Patel, Tie Luo, Umesh Bellur, and Sanjay Chaudhary

CONTENTS

18.1 INTRODUCTION

Cyber-physical systems have recently moved closer to reality, thanks to the proliferation of smart objects such as smoke detectors, air pollution monitors, and communication-enabled cars. These smart objects sense the physical world via sensors, affect the physical world via actuators, and interact with users when necessary. In effect, they capture data from the physical world and communicate it to the cyber space for further processing. Cyber-physical systems are pervasive and can be found in many application domains. For example, a building interacts with its residents and surrounding buildings in case of fire for safety and security of residents [1], offices adjust themselves automatically to user preferences while minimizing energy consumption [2], traffic signals control the influx of vehicles according to the current highway status [3,4], or an industrial control system monitors the local environment for alarm conditions and controls local operations such as opening and closing valves and breakers [5].

18.1.1 CASE STUDY: A SMART BUILDING APPLICATION

To illustrate the characteristics of cyber-physical systems, we consider the building automation [6, p. 361]. A building cluster might consist of several buildings, with each building in turn consisting of one or more floors, each floor is a cluster of several rooms. Each room may consist of a large number of heterogeneous devices equipped with sensors to sense environment, actuators to influence the environment, and storage that stores some persistent information, as illustrated in Figure 18.1. Many applications can be developed on top of these devices, one of which is discussed in the next paragraph.

  To accommodate mobile workers' preference in a room, a database is used to keep the profile of each worker, including his preferred temperature level. A badge reader in the room detects the worker's entry and queries the database for the worker's preference. Based on this, the thresholds used by the room's devices are updated. To reduce electricity waste when a person leaves the room, detected by the badge disappear event, the heating level is automatically set to the lowest level. All the relevant parameters can be set according to the building's policy. Moreover, the system generates the current status (e.g., temperature, energy consumption) of each room, which is then aggregated to determine the current status of each floor and, in turn, the entire building. A monitor installed at the building entrance presents the information to the building operator for situational awareness.

18.1.2 CHALLENGES

Our main objective is to enable the rapid development of cyber-physical systems with minimal effort from various stakeholders* involved in the process. To achieve this, we focus on the following challenges:

  Complexity: Cyber-physical systems application development is a multi disciplined process. As a result, conflicts among various sets of skills required during the process arise, pertaining to domain expertise (e.g., smart building applications are characterized in terms of rooms and floors, and smart city applications are expressed in terms of sectors), platform-specific knowledge (e.g., Android-specific *APIs* to get data from sensors, vendor-specific databases such as MySQL),

---

* Throughout this chapter, we use the term *stakeholders* as is used in software engineering [7] to indicate people who are involved in application development. Examples of stakeholders defined in Taylor, Medvidovic, and Dashofy [7] include software designer, developer, domain expert, and technologist.

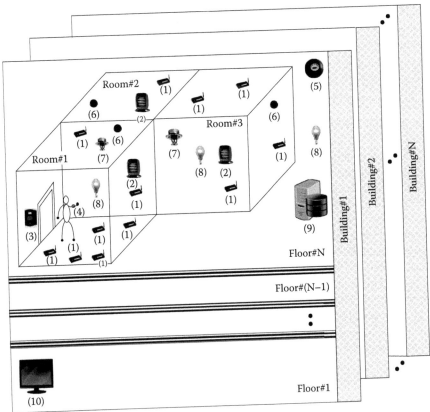

FIGURE 18.1 A cluster of multifloor buildings with deployed devices with a (1) temperature sensor, (2) heater, (3) badge reader, (4) badge, (5) alarm, (6) smoke detector, (7) sprinkler, (8) light, (9) data storage, and (10) monitor.

application-specific features (e.g., regulating temperature, detecting fire), and deployment-specific knowledge (e.g., understanding of the specific target area where the application is to be deployed).

Heterogeneity of target devices and protocols: A cyber-physical systems executes on a network consisting of heterogeneous devices in terms of types (e.g., sensor, actuator, and storage devices), interaction modes (e.g., publish/subscribe, request/response, command [8]), and different platforms (e.g., microcontrollers with no OS, Android mobile OS, Java SE on laptops). The heterogeneity largely spreads into the application code and makes it difficult to port code between different deployment environments.

Large number of devices: A cyber-physical systems typically executes on a distributed system that consists of hundreds to thousands of devices, involving the coordination of their activities. It is impractical in general to require the ability of reasoning at such levels of scale, as has been largely the view in the Wireless Sensor Network (WSN) community [9]. Consequently, there is a need of adequate abstractions that allow stakeholders to express their requirements in a compact manner regardless of the scale.

Different life-cycle phases: Stakeholders have to address issues that are attributed to different life-cycle phases, including *development*, *deployment*, and *evolution*. At the *development phase*, the application logic has to be analyzed and separated into a set of distributed tasks for the underlying network consisting of a large number of heterogeneous entities. Then, these tasks have to be implemented for the specific platform of a device. At the *deployment phase*, the application logic has to be deployed onto a large network of devices. At the *evolution phase*, stakeholders often need to be readdressed issues such as changes in application requirements and deployed devices. Manual effort in these phases on a large number of heterogeneous devices is a time-consuming and error-prone process.

The challenges discussed earlier are not completely new. In fact, they have been investigated at length in the domains of software engineering and model-driven design and more specifically in wireless sensor network macroprogramming. In the next section, we summarize the application development techniques from those domains, which our work is based on.

18.1.3 Application Development Approaches

Existing application development approaches, available to stakeholders for cyber-physical systems development, are summarized here:

Node-centric programming: Currently, the development of cyber-physical systems is performed at the *node level*, by experts of embedded and distributed systems, who are directly concerned with operations of each individual device, as illustrated in Figure 18.2. They think in terms of activities of individual devices and explicitly encode interactions with other devices. For example, they write a program that reads sensing data from appropriate sensor devices; aggregates data pertaining to some external events; decides where to send it, addressed by ID or location; and communicates with actuators if needed. Stakeholders in WSN, for example, use general-purpose programming languages (such as nesC [10], galsC [11], or Java) and target a particular middleware API or node-level service [12–14]. The Gaia [15] distributed middleware infrastructure for pervasive environments supports application development using C++.

Although node-centric programming allows for the development of extremely efficient systems based on complete control over individual devices, it is unwieldy for cyber-physical systems due to the large size and heterogeneity of systems.

Macroprogramming: This approach aims to aid stakeholders by providing the ability to specify their applications at a global level rather than individual nodes as illustrated in Figure 18.2. In macroprogramming systems, abstractions are provided to specify high-level collaborative behaviors, while

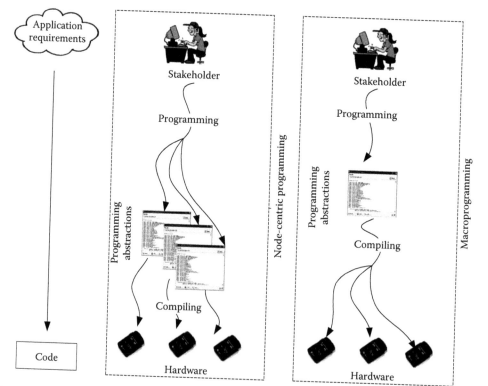

FIGURE 18.2   Comparing node-centric programming and macroprogramming.

hiding low-level details such as message passing or state maintenance from stakeholders. Stakeholders describe their application using these abstractions, which is then compiled to node-level code [16].

Macroprogramming is a viable approach compared to the node-centric programming. However, most of the macroprogramming systems largely focus on the development phase while ignoring the fact that it represents a tiny fraction of the application development life cycle [17]. The lack of a software engineering methodology to support the entire application development life cycle often results in big difficulty to maintain and reuse, as well as platform-dependent designs, which can be tackled by the model-driven approach.

Model-driven development MDD: It applies the basic separation of concerns principle both vertically and horizontally. *Vertical separation* principle reduces the application development complexity by separating the specification of the system functionality from its specification on a specific platform such as programming languages and middleware [18]. The former is defined as a platform-independent model, while the latter as platform-specific model. *Horizontal separation* principle reduces it by describing the system using different system views, each view describing a certain facet of the system.

By following the MDD guidelines, many benefits can be achieved [19]. These benefits come from the basic idea that by separating different concerns of a system at a certain level of abstraction and by providing transformation engines to convert these abstractions to a target code, productivity (e.g., reusability, maintainability) in the application development process can be improved.

18.1.4 AIM AND OBJECTIVES

Our overall aim is to ease cyber-physical systems development for stakeholders, as is the case in software engineering, by taking inspiration from the MDD approach and building upon sensor network macroprogramming. Specifically, we set to achieve the following objectives:

- To isolate cyber-physical systems development into mutually independent concerns so that stakeholders can deal with them individually, both in time (evolution) and in space (reuse across applications).
- To provide high-level modeling languages addressing characteristics of cyber-physical systems. This helps stakeholders to reduce both complexity and development effort associated with cyber-physical systems.
- To automate cyber-physical systems development wherever possible to reduce development efforts.

18.1.5 CONTRIBUTIONS

We can achieve the aforementioned aim by proposing an integrated *development framework*, which includes support programs, code libraries, high-level languages, and other software that help stakeholders to develop and glue different components of a software product. This framework separates an application development process into different concerns and provides a set of *modeling languages* to specify them. Moreover, it offers *automation techniques* at different phases of cyber-physical systems development.

Modeling languages: The development framework integrates three modeling languages: (1) vocabulary language (VL) to describe domain-specific features of a cyber-physical systems, (2) architecture language (AL) to describe application-specific functionalities of a cyber-physical systems, and (3) deployment language (DL) to describe deployment-specific features consisting information about a physical environment where devices are deployed.

Automation techniques: Our development framework is supported by code generation, task-mapping, and linking techniques, which together provide automation at various phases of cyber-physical systems development. Code generation supports the application development phase by generating a programming framework that reduces the effort in specifying the details of the components of a cyber-physical systems. Task-mapping and linking techniques together support the deployment phase by producing device-specific code that runs on a distributed system collaboratively hosted by individual devices.

Our work is supported by a middleware that enables delivery of messages across physical regions, thereby enabling a distributed system.

18.1.6 OUTLINE

The remainder of this chapter is organized as follows: Section 18.2 reviews related works. Section 18.3 presents the proposed development framework. Section 18.4 presents an implementation of our development framework. Section 18.5 evaluates the development framework quantitatively. Finally, Section 18.6 summarizes this chapter and Section 18.7 discusses some of possible future directions.

18.2 RELATED WORKS

This section focuses on existing works in literature that would address the research challenges discussed in Section 18.1.2. As stated earlier, while the application development life cycle has been discussed in general in the software engineering domain, a similar structured approach is largely lacking in the cyber-physical systems. Consequently, in this section, we present not only the existing approaches geared toward the cyber-physical systems but also its precursor fields of pervasive computing and wireless sensor networking. These are mature fields, with several excellent surveys available on programming models [9,20] and middleware [21].

We organize this section based on the perspective of the system provided to the stakeholders by the various approaches. Section 18.2.1 presents the node-level programming approaches. Section 18.2.2 summarizes approaches that aim to abstract the entire (sensing) system as a database on which one can run queries. Section 18.2.3 presents the evolution of these approaches to macroprogramming. Section 18.2.4 then describes the macroprogramming approaches more grounded in MDD techniques.

18.2.1 NODE-CENTRIC PROGRAMMING

In the following, we present systems that adopt the node-centric approach.

In the pervasive computing domain, *Olympus* [22] is a programming model on top of *Gaia* [15]—a distributed middleware infrastructure for pervasive environments. Stakeholders write a C++ program that consists of a high-level description about active space entities (including service, applications, devices, physical objects, and locations) and common active operations (e.g., switching devices on/starting/stopping applications). The Olympus framework takes care of resolving high-level description based on properties specified by stakeholders. While this approach certainly simplifies the IoT application development involving heterogeneous devices, stakeholders have to write a lot of code to interface hardware and software components, as well as to interface software components and its interactions with a distributed system. This makes it tedious to develop applications involving a large number of devices.

The *Context toolkit* [23,24] simplifies the context-aware application development on top of heterogeneous data sources by providing three architectural components, namely, widgets, interpreters, and aggregators. These components separate application semantics from platform-specific code. For example, an application does not have to be modified if an Android-specific sensor is used rather than a Sun SPOT sensor. It means stakeholders can treat a widget in a similar fashion and do not have to deal with differences among platform-specific codes. Although context toolkit provides support for acquiring the context data from the heterogeneous sensors, it does not support actuation that is an essential part of cyber-physical systems.

Henricksen et al. [25,26] propose a middleware and a programming framework to gather, manage, and disseminate context to applications. This work introduces context modeling concepts, namely, context modeling languages, situation abstraction, and preference and branching models. This work presents a software engineering process that can be used in conjunction with the specified concepts. However, the clear separation of roles among the various stakeholders is missing. Moreover, this framework limits itself to context gathering applications, thus not providing the actuation support that is important for cyber-physical systems development.

Physical-virtual mash-up: as indicated by its name, connects web services from both the physical and virtual worlds through visual constructs directly from web browsers. The embedded device runs tiny web servers [27] to answer HTTP queries from users for checking or changing the state of a device. For instance, users may want to see temperature of different places on map. Under such requirements, stakeholders can use the mash-up to connect physical services such as temperature sensors and virtual services such as Google Maps. Many mash-up prototypes have been developed that include both the physical and virtual services [28–32]. The mash-up editor usually provides visual components representing web service and operations (such as add, filter) that stakeholders need to connect together to program an application. The framework takes care of resolving these visual components based on properties specified by stakeholders and produces code to interface software components and distributed systems. The main advantage of this mash-up approach is that any service, either physical or virtual, can be mashed up if they follow the standards (e.g., Representational State Transfer [REST]). The physical-virtual mash-up significantly lowers the barrier of the application development. However, stakeholders have to manage a potentially large graph for an application involving a large number of entities. This makes it difficult to develop applications containing a large numbers of entities.

18.2.2 DATABASE APPROACH

Since a large part of these systems consists of devices that sense the environment, a natural abstraction for thinking of their functionality at a high level is that of a *database*, providing access to sensed data. This approach views the whole sensor network as a virtual database system. It provides an easy-to-use interface that lets stakeholders issue queries to a sensor network to extract the data of interest.

In *TinyDB* [33] and Cougar [34] Systems, an (structured query language) SQL-like query is submitted to a WSN. On receiving a query, the system collects data from the individual device, filters it, and sends it to the base station. They provide a suitable interface for data collection in a network with a large number of devices. However, they do not offer much flexibility for introducing the application logic. For example, stakeholders require extensive modifications in the TinyDB parser and query engine to implement new query operators.

The work on sensor information networking architecture (SINA) [35] overcomes this limitation on specification of custom operators by introducing an imperative language with an SQL query. In SINA, stakeholders can embed a script written in sensor querying and tasking language [36] in the SQL query. By this hybrid approach, stakeholders can perform more collaborative tasks than what SQL in TinyDB and Cougar can describe.

The TinyDB, Cougar, and SINA systems are largely limited to homogeneous devices. The Internet-Scale Resource-Intensive Sensor Network(*IrisNet*) [37] allows stakeholders to query a large number of distributed heterogeneous devices. For example, Internet-connected PCs source sensor feeds and cooperate to answer queries. Similar to the other database approaches, stakeholders view the sensing network as a single unit that supports a high-level query in XML. This system provides a suitable interface for data collection from a large number of different types of devices. However, it does not offer flexibility for introducing the application logic, similar to TinyDB and Cougar.

Semantic streams [38]: Allow stakeholders to pose a declarative query over semantic interpretations of sensor data. For example, instead of querying raw magnetometer data, stakeholders query whether a vehicle is a car or truck. The system infers this query and decides sensor data to use to infer the type of vehicle. The main benefit of using this system is that it allows people, with less technical background to query the network with heterogeneous devices. However, it presents a centralized approach for sensor data collection that limits its applicability for handling a network with a large number of devices.

Standardized protocols: A number of systems have been proposed to expose functionality of devices accessible through standardized protocols without having worry about the heterogeneity of underlying infrastructure [39]. They logically view sensing devices (e.g., motion sensor, temperature sensor, door and window sensor) as service providers for applications and provide abstractions usually through a set of services. We discuss these examples as follows:

TinySOA [40] is a service-oriented architecture that provides a high-level abstraction for WSN application development. It allows stakeholders to access WSNs using service-oriented APIs provided by a gateway. The APIs provide functions to obtain information about the network, listing devices, and sensing parameters. The gateway component acts as a bridge between a WSN and an application. The gateway consists of a WSN infrastructure registry and discovery components. Through these components, the gateway allows stakeholders to access data from a WSN without dealing with low-level details. This system provides suitable interfaces from a large number of different types of devices for data collection. However, every device in TinySOA is pre configured and sends data to the base station only. Thus, the flexibility for in-network processing is limited, similar to TinyDB, Cougar, and IrisNet.

Priyantha et al. [41] present an approach based on Simple Object Access Protocol (SOAP) [42] to enable an evolutionary WSN where additional devices may be added after the initial deployment. To support such a system, this approach has adopted two features: (1) structured data (the data generated by sensing devices are represented in a XML format for that may be understood by any application) and (2) structured functionality (the functionality of a sensing device is exposed by Web Service Description Language) [43]. While this system addresses the evolution issue in a target deployment, the authors do not demonstrate the evolution scenarios such as a change in functionality of an application and technological advances in deployment devices.

A number of approaches based on REST [44] have been proposed to overcome the resource needs and complexity of SOAP-based web services for sensing and actuating devices. *TinyREST* [45] is one of the first attempts to overcome these limitations. It uses the HTTP-based REST architecture to access a state of sensing and actuating devices. The TinyREST gateway maps the HTTP request to TinyOS messages and allows stakeholders to access sensing and actuating devices from their applications. The aim of this system is to make services available through standardized REST without having to worry about the heterogeneity of the underlying infrastructure; that said, it suffers from a centralized structure similar to TinySOA.

18.2.3 MACROPROGRAMMING LANGUAGES

In the following, we present macroprogramming languages for cyber-physical systems development, which are grounded in traditional general-purpose programming languages (whether imperative or functional) in order to provide developers with familiar abstractions.

Kairos [46] allows stakeholders to program an application in a Python-based language. The Kairos developers write a centralized program of a whole application. Then, the preprocessor divides the program into subprograms, and later its compiler compiles it into binary code containing code for accessing local and remote variables. Thus, this binary code allows stakeholders to program distributed sensor network applications. Although Kairos makes the development task easier for stakeholders, it targets homogeneous network where each device executes the same application.

Regiment [47] provides a high-level programming language based on Haskell to describe an application as a set of spatially distributed data streams. This system provides primitives that facilitate processing data, manipulating regions, and aggregating data across regions. The written program is compiled down to an intermediate token machine language that passes information over a spanning tree constructed across the WSN. In contrast to the database approaches, this approach provides greater flexibility to stakeholders when it comes to the application logic. However, the regiment program collects data to a single base station. It means that the flexibility for any-to-any device collaboration for reducing scale is difficult.

MacroLab [48] offers a vector programming abstraction similar to MATLAB for applications involving both sensing and actuation. Stakeholders write a single program for an entire application using MATLAB like operations such as `addition`, `find`, and `max`. The written macroprogram is passed to the MacroLab decomposer that generates multiple decompositions of the program. Each decomposition is analyzed by the cost analyzer that calculates the cost of each decomposition with respect to a cost profile (provided by stakeholders) of a target deployment. After choosing a best decomposition by the cost analyzer, it is passed to the compiler that converts the decomposition into a binary executable. The main benefit is that it offers flexibility of decomposing code according to cost profiles of the target platform. While this system certainly separates the deployment aspect and functionality of an application, this approach remains general purpose and provides little guidance to stakeholders about the application domain.

18.2.4 MODEL-BASED MACROPROGRAMMING

As an evolution of the approaches discussed earlier, which are based on general-purpose programming languages, model-driven approaches to macroprogramming aim to provide greater coverage of the software development life cycle. In the following, we present model-driven approaches that have been proposed to make cyber-physical systems development easy.

PervML [49] allows stakeholders to specify pervasive applications at a high level of abstraction using a set of models. This system raises the level of abstraction in program specification, and code generators produce code from these specifications. Nevertheless, it adopts generic UML notations to describe them, thus providing little guidance to stakeholders about the specific application domain. In addition to this, the main focus of this work is to address the heterogeneity associated with pervasive computing applications, and the consideration of a large number of devices in an application is missing. PervML integrates the mapping process at the deployment phase. However, stakeholders have to link the application code and configure device drivers manually. This manual work in the deployment phase is not suitable for cyber-physical systems involving a large number of devices. Moreover, the separation between deployment and domain-specific features are missing. These limitations would restrict PervML to a certain level.

Our work takes inspiration from *DiaSuite* [8], which is a suite of tools to develop pervasive computing applications. It combines design languages and covers application development life cycle. The design language defines both a taxonomy of an application domain and an application architecture. Stakeholders define entities in a high-level manner to abstract heterogeneity. However, the consideration of a large number of devices in an application is largely missing. Moreover, the application deployment for a large number of heterogeneous devices using this approach is difficult because stakeholders require manual effort (e.g., mapping of computational services to devices).

ATaG [50] is a WSN, that is, a macroprogramming framework to develop SCC applications. The notion of abstract task, abstract data item, and abstract channel is the core of this framework. An abstract task encapsulates the processing of data items. The flow of information among abstract tasks is specified in terms of input and output data items. Abstract channels connect abstract tasks to data items. ATaG presents a compilation framework that translates a program, containing abstract notations, into executable node-level programs. Moreover, it tackles the issue of scale reasonably well. The ATaG linker and mapper modules support the application deployment phase by producing device-specific code to result in a distributed software system collaboratively hosted by individual devices, thus providing automation at deployment phase. Nevertheless, the clear separation of roles among the various stakeholders in the application development, as well as the focus on heterogeneity among the constituent devices, is largely missing. Moreover, the ATaG program notations remain general purpose and provide little guidance to stakeholders about the application domain.

RuleCaster [17,51] introduces an engineering method to provide support for SCC applications, as well as evolutionary changes in the application development. The RuleCaster programming model is based on a logical partitioning of the network into spatial regions. Each region is in one discrete state. A rule-based language is used to specify a transition of the state, where each rule consists of two parts: (1) body and (2) head. The body part specifies a fire condition, and the head part specifies one or more actions when the condition becomes true. The RuleCaster compiler takes as input the application program containing rules and a network model that describes device locations and its capabilities. Then, it maps processing tasks to devices. Similar to ATaG, this system handles the scale issue reasonably well by partitioning the network into several spatial regions. Moreover, it supports automation at the deployment phase by mapping computational components to devices. However, the clear separation of roles among the various stakeholders, support for application domain, and the focus on heterogeneity among the constituent devices are missing.

Pantagruel [2] is a visual approach dedicated to the development of home automation applications. The Pantagruel application development consists of three steps: (1) specification of taxonomy to define entities of the home automation domain (e.g., temperature sensor, alarm, door, smoke detector), (2) specification of rules to orchestrate these entities using the Pantagruel visual language, and (3) compilation of the taxonomy and orchestration rules to generate a programming framework. The novelty of this approach is that the orchestration rules are customized with respect to entities defined in the taxonomy. While this system reduces the requirement of having domain-specific knowledge for other stakeholders, the clear separation of different development concerns, support for large scale, automation both at the development and deployment phases are largely missing. These limitations make it difficult to use for cyber-physical systems development.

This related work section has investigated existing works for cyber-physical systems development. Our investigation has revealed that many ideas have been proposed for addressing the research challenges individually. However, none of the examined approach addresses all of the identified challenges to a sufficient extent. So there is a need of designing a new approach that utilizes advantages and promising features of the existing works to develop a comprehensive integrated approach, discussed in the next section.

18.3 OUR APPROACH

Applying software engineering design principles *separation of concerns* [52], we partition the identified concepts and associations among them into different concerns, represented in a conceptual model [53] described in Section 18.3.1. The identified concepts are linked together into a concrete development framework [54–56] presented in Section 18.3.2.

18.3.1 CONCEPTUAL MODEL

A conceptual model often serves as a base of knowledge about a problem area [57]. It represents the concepts as well as the associations among them and also attempts to clarify the meaning of various terms. Figure 18.3 illustrates the concepts and their associations along with these four separate concerns: (1) domain-specific concepts, (2) functionality-specific concepts, (3) deployment-specific concepts, and (4) platform-specific concepts.

18.3.1.1 Domain-Specific Concepts

These concepts are specific to a target application domain (e.g., building automation, transport). For example, the building automation domain is expressed in terms of rooms and floors, while the transport domain is expressed in terms of highway sectors. Furthermore, each domain has a set of entities of interest (e.g., average temperature of a building, smoke presence in a room), which are observed and controlled by sensors and actuators, respectively, with a storage storing the information about the entities of interest. We describe these concepts in detail:

- An *entity of interest (EoI)* is an object (e.g., room, book, plant), including attributes that describe it and its state that is relevant from a user or an application perspective [58, p. 1]. The entity of interest has an observable property called phenomenon. Typical examples are the temperature value of a room and a tag ID.
- A *resource* is a conceptual representation of a sensor, an actuator, or a storage. We consider the following types of resources:

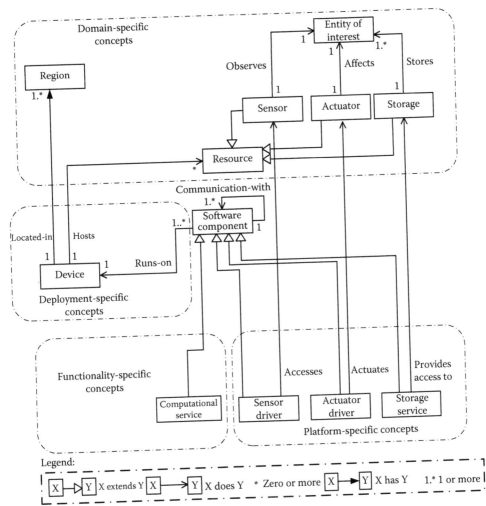

FIGURE 18.3 A conceptual model for cyber-physical systems.

- A *sensor* has the ability to detect changes in the environment. Thermometer and tag readers are examples of sensors. The sensor *observes* a phenomenon of an EoI. For instance, a temperature sensor observes the temperature phenomenon of a room.
- An *actuator* makes changes in the environment through an action. Heating or cooling elements, speakers, and lights are examples of actuators. The actuator *affects* a phenomenon of an EoI by performing actions. For instance, a heater is set to control a temperature level of a room.
- *Storage* has the ability of storing data in a persistent manner. The storage *stores* information about a phenomenon of an EoI. For instance, a database server stores information about an employee's temperature preference.
- A device is located in a *region* (or location). The region specifies the location of a device. In the building automation domain, a region of a device can be expressed in terms of building, room, and floor IDs.

18.3.1.2 Functionality-Specific Concepts

These concepts describe the computational elements of an application and interactions among them. A computational element is a type of software component, which is an architectural entity that (1) encapsulates a subset of the system's functionality and/or data, and (2) restricts access to that subset via an explicitly defined interface [7, p. 69]. We use the term *application logic* to refer a functionality of a software component. An example of the application logic is to open a window when the average temperature value of a room is greater than 30°C.

The conceptual model contains the following functionality-specific software component, a *computational service*, which is a type of software component that consumes one or more units of information as inputs, processes it, and generates an output. An output could be a data message that is consumed by others or a command message that triggers an action of an actuator. A computational service is a representation of the processing element in an application.

A software component *communicates-with* other software components to exchange data or control. These interactions might contain instances of various interaction modes such as request-response, publish-subscribe, and command. Note that this is in principle an instance of the component-port-connector architecture used in software engineering.

18.3.1.3 Deployment-Specific Concepts

These concepts describe information about devices. Each device *hosts* zero or more resources. For example, a device could host resources such as a temperature sensor to sense, a heater to control a temperature level, a monitor to display a temperature value, and a storage to store temperature readings. Each device is *located in* regions. For instance, a device is located in `room#1` of `floor#12` in `building#14`. We consider the following definition of a device:

- A *device* is an entity that provides resources the ability of interacting with other devices. Mobile phones and personal computers are examples of devices.

18.3.1.4 Platform-Specific Concepts

These concepts are computer programs that act as a (operating system specific) translator between a hardware device and an application. We identify the following platform-specific concepts:

- A *sensor driver* is a type of software component that operates on a sensor attached to a device. It *accesses* data observed by the sensor and generates the meaningful data that can be used by other software components. For instance, a temperature sensor driver generates temperature values and its metadata such as unit of measurement and time of sensing. Another software component takes these temperature data as input and calculates the average temperature of the room.
- An *actuator driver* is a type of software component that controls an actuator attached to a device. It translates a command from other software components and *actuates* the actuator appropriately. For instance, a heater driver translates a command "turn the heater on" to regulate the temperature level.
- A *storage service* is a type of software component that provides a read and write access to a storage. A storage service *provides access to* the storage. Other software components access data from the storage by requesting the storage service. For instance, SensApp [59] provides storage service to access to a no-SQL MongoDB database server.

The next section presents a development framework that links the four aforementioned concerns and provides a conceptual framework to develop cyber-physical systems.

18.3.2 DEVELOPMENT FRAMEWORK

This section presents our proposed development framework that provides a conceptual framework for cyber-physical systems development. It provides a set of modeling languages and offers automation techniques at different phases of cyber-physical systems development for the respective concerns.

18.3.2.1 Domain Concern

This concern is related to domain-specific concepts of a cyber-physical systems. It consists of two steps:

1. *Specifying domain vocabulary*: The domain expert specifies a domain vocabulary using the VL (detail in Section 18.3.3). The vocabulary includes specification of resources, which are responsible for interacting with entities of interest. In the vocabulary, resources are specified in a high-level manner to abstract low-level details from the domain expert. Moreover, the vocabulary includes definitions of regions that define spatial partitions (e.g., room, floor, building) of a system.
2. *Compiling vocabulary specification*: Leveraging the vocabulary, the development framework generates: (1) a vocabulary framework to aid the device developer, (2) a customized architecture grammar according to the vocabulary to aid the software designer, and (3) a customized deployment grammar according to the vocabulary to aid the network manager. The key advantage of this customization is that the domain-specific concepts defined in the vocabulary are made available to other stakeholders and can be reused across applications of the same application domain.

18.3.2.2 Functional Concern

This concern is related to functionality-specific concepts of a cyber-physical systems (Figure 18.4). It consists of the following steps:

1. *Specifying application architecture*: Using a customized architecture grammar, the software designer specifies an application architecture using the AL (detail in Section 18.3.4). It is an architecture description language to specify computational services and their interactions with other software components. To facilitate scalable operations within cyber-physical systems, AL offers scope constructs. These constructs allow the software designer to group devices based on their spatial relationship to form a cluster (e.g., "devices are in `room#1`") and to place a cluster head to receive and process data from that cluster. The grouping and cluster head mechanism can be recursively applied to form a hierarchical clustering that facilitates the scalable operations within cyber-physical systems.
2. *Compiling architecture specification*: The development framework leverages an architecture specification to support the application developer. To describe the application logic of each computational service, the application developer is provided an architecture framework, preconfigured according to the architecture specification of an application [1].
3. *Implementing application logic*: To describe the application logic of each computational service, the application developer leverages a generated architecture framework (detail in Section 18.3.5). It contains abstract classes*, corresponding to each computational service, that hide interaction details with other software components and allow the application developer to focus only on application logic. The application developer implements only the abstract methods of generated abstract classes.

18.3.2.3 Deployment Concern

This concern is related to deployment-specific concepts of a cyber-physical systems. It consists of the following steps:

1. *Specifying target deployment:* Using a customized deployment grammar, the network manager describes a deployment specification using the DL (detail in Section 18.3.7). The deployment specification includes the details of each device, including its regions (in terms

---

* We assume that the application developer uses an object-oriented language.

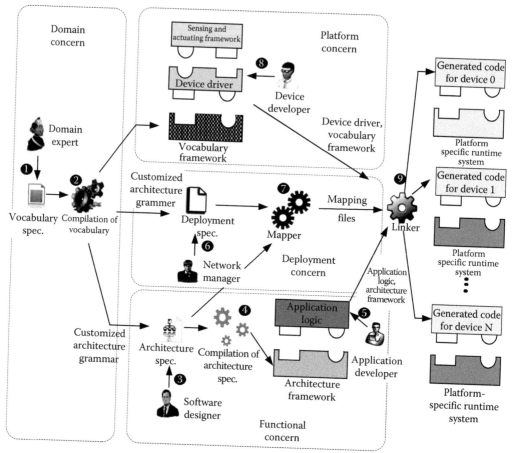

FIGURE 18.4 Cyber-physical system development: the overall process.

of values of the regions defined in the vocabulary), resources hosted by devices (a subset of those defined in the vocabulary), and the type of the device. Ideally, the same cyber-physical systems could be deployed on different target deployments (e.g., the same inventory tracking application can be deployed in different warehouses). This requirement is dictated by separating a deployment specification from other specifications.

2. *Mapping*: The mapper produces a mapping from a set of computational services to a set of devices (detail in Section 18.3.6). It takes as input a set of placement rules of computational services from an architecture specification and a set of devices defined in a deployment specification. The mapper decides where each computational service will be deployed. The current version of this algorithm selects devices randomly and allocates computational services to the selected devices. A mapping algorithm aware of heterogeneity, associated with devices of a target deployment, is a part of our future work.

18.3.2.4 Platform Concern

This concern is related to platform-specific concepts of a cyber-physical systems. It consists of the following step:

1. *Implementing device drivers*: Leveraging the vocabulary, our system generates a vocabulary framework to aid the device developer (detail in Section 18.3.8). The vocabulary framework contains *interfaces* and *concrete classes* corresponding to resources defined in the

vocabulary. A concrete class contains concrete methods to interact with other software components and platform-specific device drivers. The interfaces are implemented by the device developer to write platform-specific device drivers.

18.3.2.5   Linking

The linker combines and packs code generated by various stages into packages that can be deployed on devices. It merges a generated architecture framework, application logic, mapping files, device drivers, and vocabulary framework. This stage supports the application deployment phase by producing device-specific code to result in a distributed software system collaboratively hosted by individual devices, thus providing automation at the deployment phase.[*]

18.3.2.6   Handling Evolution

Evolution is an important aspect in cyber-physical systems development where new resources and computational services are added, removed, or extended during application development or execution (detail in Section 18.3.9). To deal with these changes, our development framework separates cyber-physical systems development into different concerns and allows an iterative development [60] for these concerns.

18.3.3   Specifying Domain Vocabulary

VL is designed to enable the domain expert to describe a domain vocabulary domain. It offers constructs to specify concepts that interact with entities of interest. These concepts can be described as $V = (P, D, R)$. P represents the set of regions, D represents the set of data structure, and R represents the set of resources. We describe these concepts in detail as follows:

Regions (P): It represents the set of regions that are used to specify locations of devices. A region definition includes a region label and region type, for example, rooms and floors (considered as region labels) in the building automation domain and highway sectors in the transportation domain. Each room or floor in a building may be annotated with an integer value (e.g., `room:1` interprets as room number 1) considered as region type. This construct is declared using the `regions` keyword. Listing 18.1 (lines 1–4) shows region definitions for the building automation domain.

Data structures (D): Each resource is characterized by types of information it generates or consumes. A set of information is defined using the `structs` keyword (Listing 18.1, line 5). For instance, a temperature sensor may generate a temperature value and unit of measurement (e.g., Celsius or Fahrenheit). This information is defined as `TempStruct` and its two fields (Listing 18.1, lines 9–11).

Resources (R): It defines resources that might be attached with devices, including sensors, actuators, or storage. It is defined as $R = (R_{sensor}, R_{actuator}, R_{storage})$. R_{sensor} represents a set of sensors, $R_{actuator}$ represents a set of actuators, and $R_{storage}$ represents a set of storage. We describe them in detail as follows:

- *Sensors* (R_{sensor}). It defines a set of various types of sensors. A set of sensors is declared using the `sensors` keyword (Listing 18.1, line 13). $S_{generate}$ is a set of sensor measurements produced by R_{sensor}. Each sensor ($S \in R_{sensor}$) produces one or more sensor measurements ($op \in S_{generate}$) along with the data types specified in the data structure (D). A sensor measurement of each sensor is declared using the `generate` keyword (Listing 18.1, line 17). For instance, a temperature sensor generates a temperature measurement of `Tempstruct` type (lines 16–17) defined in data structures (lines 9–11).
- *Actuators* ($R_{actuator}$). It defines a set of various types of actuator. A set of actuators is declared using the `actuators` keyword (Listing 18.1, line 18). A_{action} is a set of actions performed

---

[*] We assume that a middleware is already installed on the deployed devices. The installed middleware enables inter-device communication among devices.

by $R_{actuator}$. Each actuator ($A \in R_{actuator}$) has one or more actions ($a \in A_{action}$) that is declared using the `action` keyword. An action of an actuator may take inputs specified as parameters of an action (Listing 18.1, line 21). For instance, a heater may have two actions. One is to switch off the heater and second is to set the heater according to a user's temperature preference illustrated in Listing 18.1, lines 19–21. The `SetTemp` action takes a user's temperature preference shown in line 21.

- *Storage* ($R_{storage}$). It defines a set of storage (e.g., user's profile storage) that might be attached to a device. A set of storage is declared using the `storage` keyword (Listing 18.1, line 22). $ST_{generate}$ represents a set of retrievals of $R_{storage}$. A retrieval ($rq \in ST_{generate}$) from the storage ($ST \in R_{storage}$) requires a parameter. Such a parameter is specified using the `accessed-by` keyword (Listing 18.1, line 24). For instance, a user's profile is accessed from profile storage by his unique badge identification illustrated in Listing 18.1, lines 23–24.

```
1  regions:
2      Building: integer;
3      Floor: integer;
4      Room: integer;
5  structs:
6      BadgeDetectedStruct
7          badgeID: string;
8          timeStamp: long;
9      TempStruct
10         tempValue: double;
11         unitOfMeasurement: string;
12 resources:
13     sensors:
14         BadgeReader
15             generate badgeDetected: BadgeDetectedStruct;
16         TemperatureSensor
17             generate tempMeasurement: TempStruct;
18     actuators:
19         Heater
20             action Off();
21             action SetTemp(setTemp: TempStruct);
22     storage:
23         ProfileDB
24             generate  profile: TempStruct accessed-by
                       badgeID: string;
```

Listing 18.1 Code snippet of the building automation domain using VL. The keywords are in **blue**.

The regions (P), data structures (D), and resources (R) defined using VL in the vocabulary are referred in the functional concern to write an architecture specification, described in the next section.

18.3.4 Specifying Application Architecture

Based on concepts defined in the vocabulary specification, the software designer specifies an architecture specification to design an application. Specifically, sensors (R_{sensor}), actuators ($R_{actuator}$), storage ($R_{storage}$), and regions (P) defined in the vocabulary are referred to write the architecture specification. The architecture specification can be described as $A_v = (C)$. C represents a set of computational services. It is described as $C = (C_{generate}, C_{consume}, C_{request}, C_{command}, C_{in-region}, C_{hops})$. $C_{generate}$ represents a set of outputs produced by computational services. $C_{consume}$ is a set of inputs consumed by computational services. The inputs could be data produced by other computational services or sensors (R_{sensor}). $C_{request}$ represents a set of request by computational services to retrieve data from the storage ($R_{storage}$). $C_{command}$ represents a set of commands to invoke actuators ($R_{actuator}$). $C_{in-region}$ is a set of regions (R_{region}) where computational services can be placed. C_{hops} is a set of regions (R_{region}) where computational services receive data. In the following, we describe these concepts in detail:

Consume ($C_{consume}$) and *generate* ($C_{generate}$) These two concepts together define publish/subscribe interaction mode that provides subscribers with the ability to express their interest in an event, generated by a publisher, that matches their registered interest. A computational service represents the publish and subscribe using *generate* and *consume* concept, respectively. We describe these two concepts in details as follows:

1. *Consume.* This represents a set of subscriptions (or consumes) expressed by computational services to get event notifications generated by sensors or other computational services. A consumption ($c \in C_{consume}$) of a computational service is expressed using the `consume` keyword. A computational service expresses its interest by an event name. For instance, a computational service `RoomAvgTemp`, which calculates an average temperature of a room, subscribes its interest by expressing event name `tempMeasurement` illustrated in Listing 18.2, line 9.

2. *generate.* This represents a set of publications (or generates) that are produced by computational services. A generation ($g \in C_{generate}$) of a computational service is expressed using the `generate` keyword. The computational service transforms data to be consumed by other computational services in accordance with the application needs. For instance, the computational service `RoomAvgTemp` consumes temperature measurements (i.e., `tempMeasurement`), calculates an average temperature of a room, and generates `roomAvgTempMeasurement` (Listing 18.2, lines 7–9) that is used by `RoomController` service (Listing 18.2, lines 11–12).

Request ($C_{request}$). This is a set of requests, issued by computational services, to retrieve data from storage ($R_{storage}$). A request is a one-to-one synchronous interaction with a return values. In order to fetch data, a requester sends a request message containing an access parameter to responder. The responder receives and processes the request message and ultimately returns an appropriate message as a response. An access ($rq \in C_{request}$) of the computational service is specified using `request` keyword. For instance, a computational service `Proximity` (Listing 18.2, line 5), which wants to access user's profile data, sends a request message containing profile information as an access parameter to a storage `ProfileDB` (Listing 18.1, line 24).

Command ($C_{command}$). This is a set of commands, issued by a computational service to trigger actions provided by actuators ($R_{actuator}$). The software designer can pass arguments to a command depending on action signature provided by actuators. Moreover, he specifies a scope of command, which depends on a region where commands are issued. A command is specified using the `command` keyword. An example of command invocation is given in line 14 of Listing 18.2. The room controller service (i.e., `roomController`), which regulates temperature, issues a `SetTemp` command with a preferred temperature as an argument (i.e., `settemp`) to heaters (Listing 18.1, line 21).

In-region ($C_{in-region}$) and *hops* (C_{hops}). To facilitate the scalable operations within an IoT application, devices should be grouped to form a cluster based on their spatial relationship [35] (e.g.,"devices are in room#1"). The grouping could be recursively applied to form a hierarchy of clusters. Within a cluster, a computational service is placed to receive and process data from its cluster of interest. Figure 18.5 shows this concept for more clarity. The temperature data are first routed to a local average temperature service (i.e., `RoomAvgTemp`), deployed in per room, then later per floor (i.e., `FloorAvgTemp`), and then ultimately routed to building average temperature service (i.e., `BuildingAvgTemp`).

AL offers *scope* constructs to define both the service placement ($C_{in-region}$) and its data interest (C_{hops}). The service placement (defined using the `in-region` keyword) is used to govern a

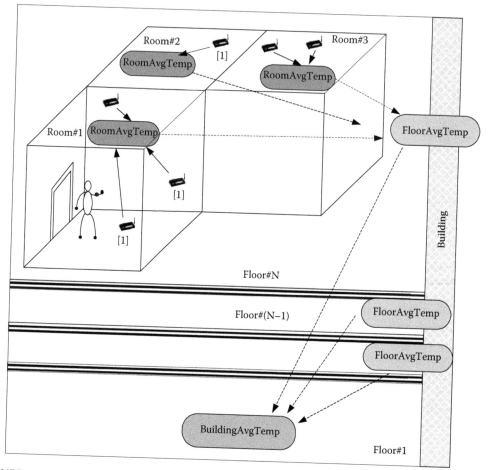

FIGURE 18.5 Clustering in a smart building application. Temperature sensing device is numbered as [1].

placement of computational service in a cluster. The service placement can be in regions defined in a vocabulary specification. So, it is a subset of P.

The data interest of a computational service is used to define a cluster from which the computational service wants to receive data. The data interest can be in regions defined in the vocabulary specification. So, it is a subset of P. It is defined using the `hops` keyword. The syntax of this keyword is `hops: radius: unit of radius`. Radius is an integer value. The unit of radius is a cluster value. For example, if a computational service `FloorAvgTemp` deployed on floor number 12 has a data interest `hops: i: Floor`, then it wants data from all floors starting from 12th floor to `(12+i)` th floor and all floors starting from 12th floor to `(12-i)` th floor.

Figure 18.6 shows the layered architecture of the smart building application. Computational services are fueled by sensing components. They process inputs data and take appropriate decisions by triggering actuators. We illustrate AL by examining a code snippet in Listing 18.2, which describes a part of Figure 18.6. This code snippet revolves around the actions of the `Proximity` service (Listing 18.2, lines 2–6), which coordinates events from the `BadgeReader` with the content of `ProfileDB` storage service. To do so, the `Proximity` composes information from two sources: one for badge events (i.e., badge detection) and one for requesting the user's temperature profile from `ProfileDB`, expressed using the `request` keyword (Listing 18.2, line 5). Input data are declared using the `consume` keyword that takes source name and data interest of a computational service from logical region (Listing 18.2, line 4). The declaration of `hops: 0: room` indicates that the computational service is interested in consuming badge events of the current room. The `Proximity` service

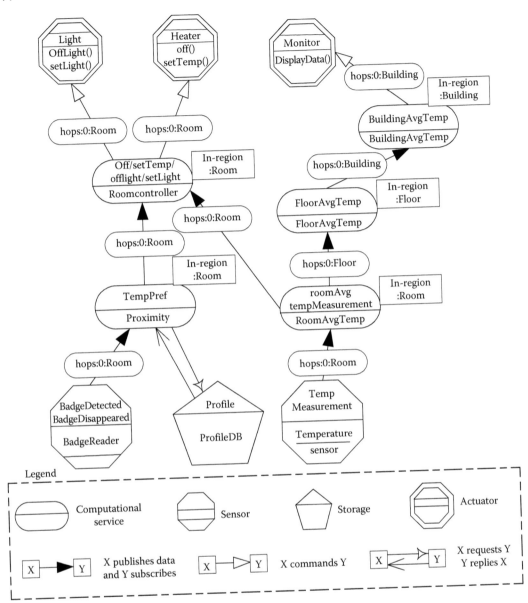

FIGURE 18.6   The layered architecture of a smart building application.

is in charge of managing badge events of room. Therefore, we need `Proximity` service to be partitioned per room using `in-region: room` (Listing 18.2, line 6). The outputs of the `Proximity` and `RoomAvgTemp` are consumed by the `RoomController` service (Listing 18.2, lines 11–15). This service is responsible for taking decisions that are carried out by invoking commands declared using the `command` keyword (Listing 18.2, line 14).

```
1  computationalServices:
2      Proximity
3          generate  tempPref: UserTempPrefStruct;
4          consume   badgeDetected from hops:0: Room;
5          request   profile( badgeID);
6          in-region: Room;
```

```
7    RoomAvgTemp
8      generate roomAvgTempMeasurement:TempStruct;
9      consume tempMeasurement from hops:0: Room ;
10     in-region: Room;
11   RoomController
12     consume roomAvgTempMeasurement from hops:0: Room;
13     consume tempPref from hops:0: Room;
14     command SetTemp( setTemp) to hops:0: Room;
15     in-region: Room;
```

Listing 18.2　A code snippet of the architecture specification for the smart building application using AL. The language keywords are in **blue**, while the keywords referred from vocabulary are underlined.

18.3.5　Implementing Application Logic

Leveraging the architecture specification, we generate a framework to aid the application developer. The generated framework contains abstract classes corresponding to the architecture specification. The abstract classes include two types of methods: (1) *concrete methods* to interact with other components transparently through the middleware and (2) *abstract methods* that allow the application developer to program the application logic. The application developer implements each abstract method of generated abstract class. The key advantage of this framework is that a framework structure remains uniform. Therefore, the application developer has to know only abstract methods where they have to specify the application logic.

Abstract methods: For each input declared by a computational service, an abstract method is generated for receiving data. This abstract method is then implemented by the application developer in the subclass of the abstract class. Listing 18.3 illustrates this concept. From the badgeDetected input of the Proximity declaration in the architecture specification (Listing 18.2, lines 2–6), the onNewbadgeDetected() abstract method is generated (Listing 18.3, line 16). This method is implemented by the application developer.

```
1  public abstract class Proximity {
2    private String partitionAttribute = "Room";
3    public void notifyReceived(String eventName, Object arg
       ) {
4      if (eventName.equals("badgeDetected")) {
5        onNewbadgeDetected((BadgeDetectedStruct) arg);
6      }
7    }
8    public void subscribebadgeDetected() {
9      Region regionInfo = getSubscriptionRequest(
10       partitionAttribute, getRegionLabels(),
         getRegionIDs());
11     PubSubMiddleware.subscribe(this, "badgeDetected",
         regionInfo);
12   }
13   protected TempStruct getprofile(String arg) {
14     return (TempStruct) PubSubMiddleware.sendCommand("
         getprofile", arg, myDeviceInfo);
15   }
16   protected abstract void onNewbadgeDetected(
       BadgeDetectedStruct arg);
```

```
17    protected void settempPref(UserTempPrefStruct newValue)
       {
18       if (tempPref != newValue) {
19          tempPref = newValue;
20          PubSubMiddleware.publish("tempPref", newValue,
             myDeviceInfo);
21       }
22    }
23  }
```

Listing 18.3 The Java abstract class `Proximity` generated from the declaration `Proximity` in the architecture specification.

Concrete methods: The compilation of an architecture specification generates concrete methods that take care of heterogeneous interactions with other software components transparently. For instance, a computational service processes input data and produces refined data to its consumers. The input data are either notified by other component (i.e., publish/subscribe) or requested (i.e., request/response) by the service itself. The concrete methods for these interaction modes are generated in an architecture framework. The lines 2–6 of Listing 18.2 illustrate these two heterogeneous interactions:

1. *Publish/subscribe.* The `Proximity` service receives `badgeDetect` event (Listing 18.2, line 4). Our framework generates the `subscribebadgeDetected()` method to subscribe `badgeDetected` event (Listing 18.3, lines 8–12). This method defines the data interest of a service from where it receives data. The value of `partitionAttribute` (Listing 18.3, line 2), which comes from the architecture specification (Listing 18.2, line 6), defines the scope of receiving data. The aforementioned constructs are empowered by our choice of middleware, which is a variation of the one presented in Mottola et al. [3] and enables delivery of data across logical scopes. Moreover, it generates the implementation of `notifyReceived()` method to receive the published events (Listing 18.3, lines 3–7).
2. *Request/response.* The `Proximity` service requests `profile` data (Listing 18.2, line 5). A `sendcommand()` method is generated to request data from other components (Listing 18.3, lines 13–15).

18.3.6 MAPPING

This section presents our mapping algorithm that decides devices for a placement of computational services. It takes inputs as (1) a list of devices D defined in a deployment specification (see Listing 18.4) and (2) a list of computational services C defined in an architecture specification (see Listing 18.2). It produces a mapping of computational services to a set of devices.

We presents the mapping algorithm (see Algorithm 18.1) that comprises two steps. The first step (lines 4–9) constructs the two key-value data structures from a deployment specification. These two data structures are used in the second step. The second step (lines 10–20) selects devices randomly and allocates computational services to the selected devices*. In order to give more clarity to readers, we describes these two steps in detail below.

The first step (Algorithm 18.1, lines 4–9) constructs two key-value data structures *regionMap* and *deviceListBy RegionValue* from D. The *regionMap* (line 6) is a key-value data structure where *regionName* (e.g., `Building`, `Floor`, `Room` in Listing 18.4) is a key and

---

* A mapping algorithm cognizant of heterogeneity, associated with devices of a target deployment, is a part of our future work. See Section 18.7 for details.

Algorithm 18.1 Mapping Algorithm

Input: List D of m numbers of devices, List C of k numbers computational services

Output: List *mappingOutput* of m numbers that contains assignment of C to D

1: Initialize *regionMap* key-value pair data structure
2: Initialize *deviceListByRegionValue* key-value pair data structure
3: Initialize *mappingOutput* key-value pair data structure
4: **for all** *device* in D **do**
5:   **for all** pairs (*regionName, regionValue*) in *device* **do**
6:     *regionMap*[*regionName*] ← *regionValue* // construct *regionMap* with *regionName* as key and assign *regionValue* as Value
7:     *deviceListByRegionValue*[*regionValue*] ← *device*
8:   **end for**
9: **end for**
10: **for all** *regionName* in *regionMap.getKeySet()* **do**
11:   **for all** *computationalservice* in C **do**
12:     **if** *computationalservice.partitionValue()* = *regionName* **then**
13:       **for all** *regionValue* in *regionMap.getValueSet(regionName)* **do**
14:         *deviceList* ← *deviceListByRegionValue.getValueSet(regionValue)*
15:         *selectedDevice* ← *selectRandomDeviceFromList(deviceList)*
16:         *mappingOutput*[*selectedDevice*] ← *computationalservice*
17:       **end for**
18:     **end if**
19:   **end for**
20: **end for**
21: **return** *mappingOutput*

---

regionValue (e.g., `15`, `11`, `1` in Listing 18.4) is a value. The *deviceListByRegionValue* (line 7) is a key-value data structure where *regionValue* is a key and *device* (e.g., `TemperatureMgmt-Device-1` in Listing 18.4) is a value. Once these two data structures are constructed, we use them for the second step (lines 10–20).

The second step (Algorithm 18.1, lines 10–20) selects a device and allocates computational services to the selected device. To perform this task, the line 10 retrieves all keys (in our example `Building`, `Floor`, `Room`) of *regionMap* using *getKeySet()* function. For each computational service (e.g., `Proximity`, `RoomAvgTemp`, `RoomController` in Listing 18.2), the selected key from the *regionMap* is compared with a partition value of a computational component (line 12). If the value match, the next step (lines 13–17) selects a device randomly and allocates a computational service to the selected device.

Computational complexity: The first step (Algorithm 18.1, lines 4–9) takes $O(mr)$ time, where m is a number of devices and r is a number of region pairs in each device specification. The second step (Algorithm 18.1, lines 10–20) takes $O(nks)$ time, where n is a number of region names (e.g., building, floor, room for the building automation domain) defined in a vocabulary, k is a number of computational services defined in an architecture specification, and s is a number of region values specified in a deployment specification. Thus, the total computational complexity of the mapping algorithm is $O(mr + nks)$.

18.3.7 SPECIFYING THE TARGET DEPLOYMENT

The deployment specification includes device properties, regions where devices are placed, and resources that are hosted by devices. The resources (R) and regions (P) defined in a vocabulary become a set of values for certain attributes in DL. DL can be described as $T_v = (D)$. D represents a set of devices. A device ($d \in D$) can be defined as ($D_{region}, D_{resource}, D_{type}$). D_{region} represents a set of device placements in terms of regions defined in a vocabulary. $D_{resource}$ is a subset of resources defined in a vocabulary. D_{type} represents a set of device type (e.g., Java SE device, Android device) that is used to pick an appropriate device driver from a device driver repository. Listing 18.4 illustrates a deployment specification of the smart building application. This snippet describes a device called `TemperatureMgmt-Device-1` with an attached `TemperatureSensor` and `Heater`, situated in `building` 15, `floor` 11, `room` 1: it is Java S E enabled and a nonmobile device.

Note that although individual listing of each device's attributes appears tedious, (1) we envision that this information can be extracted from inventory logs that are maintained for devices purchased and installed in systems, and (2) thanks to the separation between the deployment and functional concern in our approach, the same deployment specification can be reused across IoT applications of a given application domain.

```
1  devices:
2     TemperatureMgmt-Device-1:
3        region:
4           Building: 15 ;
5           Floor: 11;
6           Room: 1;
7        resources: TemperatureSensor, Heater;
8        type: JavaSE;
9        mobile: false;
10    . . .
```

Listing 18.4 Code snippet of a deployment specification for the building automation domain using SDL. The language keywords are in **blue**, while the keywords derived from a vocabulary are underlined.

18.3.8 IMPLEMENTING DEVICE DRIVERS

Leveraging the vocabulary specification, our system generates a vocabulary framework to aid the device developer. The vocabulary framework contains *concrete classes* and *interfaces* corresponding to resources defined in a vocabulary. A concrete class contains concrete methods for interacting with other software components and platform-specific device drivers. The interfaces are implemented by the device developer to write platform-specific device drivers.

In order to enable interactions between concrete class and platform-specific device driver, we adopt the factory design pattern [57]. This pattern provides an interface for a concrete class to obtain an instance of different platform-specific device driver implementations without having to know what implementation the concrete class obtains. Since the platform-specific device driver implementation can be updated without changing the code of concrete class, the factory pattern has advantages of encapsulation and code reuse. We illustrate this concept with a `BadgeReader` example. The class diagram in Figure 18.7 illustrates the concrete class `BadgeReader`, the interface `IBadgeReader`, and the associations between them through the factory class `BadgeReaderFactory`. The two abstract methods of the `IBadgeReader` interface are implemented in the `AndroidBadgeReader` class. The platform-specific implementation is accessed through the `BadgeReaderFactory` class. The `BadgeReaderFactory` class returns an instance of platform-specific implementations according to request by the concrete method `registerBadgeReader()` in the `BadgeReader` class.

18.3.9 HANDLING EVOLUTION

Evolution is an important aspect in cyber-physical systems development where resources and computational services are added, removed, or extended. To deal with these changes, we separate cyber-physical systems development into different concerns and allow an iterative development for these concerns. This iterative development requires only a change in evolved specification and

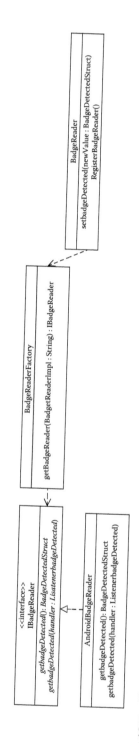

FIGURE 18.7   Class diagram representing (1) the interface `IBadgeReader` and the implementation of two abstract methods in the `AndroidBadgeReader` class and (2) the concrete class `BadgeReader` that refers the `AndroidBadgeReader` through the `BadgeReaderFactory` factory class.

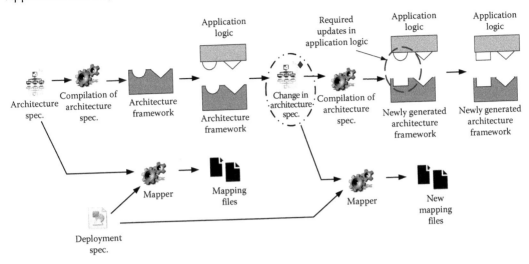

FIGURE 18.8   Handling evolution in the functional concern.

reusing dependent specifications/implementation in compilation process, thus reducing effort to handle evolution, similar to the work by Cassou et al. [8].

Figure 18.8 illustrates evolution in the functional concern. It could be addition, removal, or extension of computational services. A change in an architecture specification requires recompilation of it. The recompilation generates a new architecture framework and preserves the previously written application logic. This requires changes in the existing application logic implementations manually and recompilation of the architecture specification to generate new mapping files that replace old mapping files.

18.4   TOOLSUITE: AN IMPLEMENTATION OF OUR APPROACH

We believe that our development framework, discussed in Section 18.3, should be supported by tools for it to be applicable in an effective way. This section presents an implementation of the development framework,* realized as a suite of tools; we call it *ToolSuite*. ToolSuite is composed of different components, mentioned as follows, at each phase of application development that stakeholders can use.

Editor: It helps stakeholders to write high-level specifications, including vocabulary, architecture, and deployment specification with the facilities of syntax coloring and syntax error reporting. We use Xtext[†] for a full-fledged editor support, similar to the work by Bertran et al. [61]. Xtext is a framework for the development of domain-specific languages and provides an editor with syntax coloring by writing Xtext grammar.

Compiler: The compiler parses high-level specifications and translates them into code that can be used by other components in the system. The parser module of compiler is implemented using ANTLR[‡], a well-known parser generator that creates parser files from grammar descriptions. The code generator module of compiler manages a repository of plug-ins. Each plug-in, defined as template files, is specific to a target implementation language (e.g., Java, Python). The key advantage of it is that it simplifies an implementation of a new code generator for a target implementation. The plug-ins are implemented using StringTemplate Engine,[§] a Java template engine for generating source code

---

* An open source version, targeting on Android- enabled and Java SE-enabled devices and MQTT middleware, is available
  on: https://github.com/pankeshlinux/IoTSuite/wiki.
† http://www.eclipse.org/Xtext/.
‡ http://www.antlr.org/.
§ http://www.stringtemplate.org/.

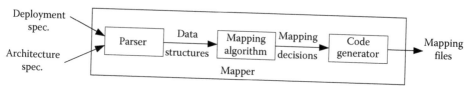

FIGURE 18.9   Architecture of the mapper component in ToolSuite.

or any other formatted text output. We build two compilers to aid stakeholders shown in Figure 18.4:
(1) compiler for a vocabulary specification and (2) compiler for an architecture specification. The
current version of these compilers generates, frameworks, compatible with Eclipse IDE.

Mapper: The mapper produces a mapping from a set of computational services to a set of devices.
Figure 18.9 illustrates the architecture of the mapper component. The parser converts high-level
specifications into appropriate data structures that can be used by a mapping algorithm. The mapping
algorithm produces mapping decisions into appropriate data structures. The code generator consumes
the data structures and generates mapping files. Our current implementation of the mapper randomly
maps computational services to a set of devices. However, due to generality of our architecture, more
sophisticated mapping algorithm can be plugged into the mapper.

Linker: This combines and packs code generated by various stages of compilation into packages
that can be deployed on devices. The current version of the linker generates Java source packages
for Android and Java SE platform. In order to execute code, these packages still need to be compiled
by a device-level compiler designed for a target platform. The current version of the linker generates
Java source packages for Android and Java SE platform, which can be imported into Eclipse IDE. In
order to execute code, these packages still need to be compiled by a device-level compiler designed
for a target platform.

Runtime system: The main responsibility of the runtime system is a distributed execution of cyber-
physical systems. It is composed of three parts: (1) *middleware* (it runs on each individual device and
provides a support for executing distributed tasks). (2) *wrapper*: (it plugs packages, generated by the
linker module, and middleware), and (3) *support library* (it separates packages, produced by the
linker component, and underlying middleware by providing interfaces that are implemented by each
wrapper). The integration of a new middleware into ToolSuite consists of an implementation of interfaces, specified by the support library, in the wrapper. The current implementation of ToolSuite uses
the MQTT* middleware, which enables interactions among Android devices and Java SE-enabled
devices.

18.4.1   Eclipse Plug-In

We have integrated the system components as an Eclipse plug-in to provide end-to-end support for
cyber-physical systems development. Figure 18.10 illustrates the use of our plug-in at various phases
of cyber-physical systems development: *IoT domain project* (① in Figure 18.10)—, using which the
domain expert can describe and compile a vocabulary specification of an application domain; *IoT
architecture project* (② in Figure 18.10)—, using which the software designer can describe and compile an architecture specification of an application; *IoT deployment project* (③ in Figure 18.10)—,
using which the network manager can describe a deployment specification of a target domain and
invoke the mapping component; and *IoT Linking project* (④ in Figure 18.10)—, using which the network manager can combine and pack code generated by various stages of compilation into packages
that can be deployed on devices.

---

* http://mqtt.org/.

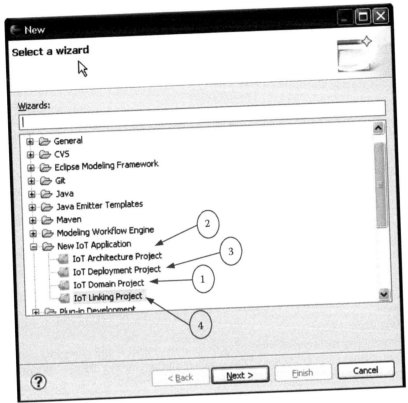

FIGURE 18.10 Eclipse plug-in for cyber-physical systems development.

18.5 EVALUATION

This section evaluates our approach in terms of how much it reduces effort of cyber-physical systems development. We measure development effort through the number of lines of code (LoC) written by the authors. To evaluate our approach we consider two representative cyber-physical systems: (1) the smart building application described in Section 18.1.1 and (2) a fire detection application that aims to detect fire by analyzing data from smoke and temperature sensors. When fire occurs, residents are notified via their smart phones by an installed mobile application. In addition, residents of the building and neighborhood are also informed through a set of alarms. A fire state is computed based on a current average temperature value and smoke presence, by a local fire state service deployed per room. This state is then sent to a service deployed per floor, and finally a computational service decides whether alarms should be activated to notify users. Table 18.1 summarizes the components used by each application.

Evaluating automation: The primary aim is to evaluate automation provided by our approach. Specifically, we answer this question: *How complete is the code automatically generated by our approach?* To answer this question, we have implemented two cyber-physical systems, as discussed earlier, using our approach. These applications are implemented independently. We did not reuse specifications and implementations of one application in the other. We deployed the 2 applications on 10 simulated devices running on top of a middleware that simulates a network on a single PC dedicated to the evaluation.

We measured development effort using Eclipse EclEmma 2.2.1 plug-in.* This tool counts actual Java statement as lines of code and does not consider blank lines or lines with comments.

---

* http://www.eclemma.org/.

TABLE 18.1

List of Components of Smart Building and Fire Detection Applications

Component Type	Smart Building	Fire Detection
Sensing	TemperatureSensor	TemperatureSensor
	BadgeReader	SmokeDetector
Actuating	Heater	Alarm
	Light	Phone
	Monitor	—
Storage	ProfileDB	None
Computational	RoomAvgTemp	RoomAvgTemp
	Proximity	RoomFireState
	FloorAvgTemp	FloorFireState
	BuildingAvgTemp	BuildingFireController
	RoomController	—

Our measurements reveal that more than 81% of the total number of lines of code is generated in two applications (see Table 18.2).

Code coverage: The measure of lines of code is only useful if the generated code is actually executed. We measured code coverage of the generated programming frameworks of two applications (see Table 18.2) using the EclEmma Eclipse plug-in. Our measures show that more than 90% of generated code is actually executed, the remaining portion being error-handling code for errors that did not happened during the experiment. This high value indicates that most of the execution takes place in the generated code.

Evaluating development effort for a large number of devices: The aforementioned experiment was conducted for 10 simulated devices. In this experiment, the primary goal is to evaluate the effort of developing cyber-physical systems involving a large number of devices.

We developed the smart building application on a set of simulated devices running on top of the middleware dedicated to the evaluation. The assessments were conducted over an increasing number of devices. The first development effort assessment was conducted on 10 devices instrumented with heterogeneous sensors, actuators, and storage. In the subsequent assessments, we kept increasing the number of devices equipped with sensors and actuators up to 500. In each assessment, we measured lines of code to specify vocabulary, architecture, and deployment, application logic, and device drivers. Table 18.3 illustrates the assessment results containing a number of devices involved in the experiment and handwritten lines of code to develop the smart building application.

In Table 18.3, we make the following two observations:

1. As the number of devices increases, lines of code for vocabulary and architecture specification, device drivers, and application logic remain constant for a deployment consisting a large number of devices. The reason is that our approach provides the ability to specify an application at a global level rather than individual nodes.
2. As the number of devices increases, lines of code for a deployment specification increase. The reason is that the network manager specifies each device individually in the deployment specification. This is a limitation of DL. Our future work will be to investigate how a deployment specification can be expressed in a concise and flexible way for a network with a large number of devices. We believe that the use of regular expressions is a possible technique to address this problem.

TABLE 18.2
Lines of Code and Code Coverage in Smart Building and Fire Detection Applications.

Application Name	Handwritten (Lines of Code)					Generated (Lines of Code)			Generated Handwritten+Generated	Code Coverage of Generated Code
	Vocab Spec.	Arch. Spec.	Deploy. Spec. (Devices = 10)	Device Driver	App. Logic	Mapping Code	Archi. Fram.	Vocab. Fram.		
Smart building	41	28	81	98	131	561	408	757	81.99%	92.22%
Fire detection	27	21	81	53	72	528	292	476	83.61%	90.38%

TABLE 18.3

Number of Devices Involved in the Development Effort Assessment and Handwritten Lines of Code to Develop the Smart Building Application

Number of Devices	Handwritten (Lines of Code)				
	Vocab Spec.	Arch. Spec.	Deploy. Spec.	Device Driver	App. Logic
10	41	28	81	98	131
34	41	28	273	98	131
50	41	28	401	98	131
62	41	28	497	98	131
86	41	28	689	98	131
110	41	28	881	98	131
200	41	28	1601	98	131
300	41	28	2401	98	131
350	41	28	2801	98	131
500	41	28	4001	98	131

18.6   SUMMARY

Application development in the cyber-physical systems involves dealing with a wide range of related issues such as lack of separation of concerns and lack of high-level of abstractions to address both the large scale and heterogeneity. Moreover, stakeholders involved in application development have to address issues that can be attributed to different life-cycle phases when developing applications. First, the application logic has to be analyzed and then separated into a set of distributed tasks for an underlying network. Then, the tasks have to be implemented for the specific hardware. Apart from handling these issues, they have to deal with other aspects of life cycle such as changes in application requirements and deployed devices.

We note that although the software engineering, MDD, WSN, and pervasive computing community have discussed and analyzed similar challenges in the general case, this has not been applied to the case of cyber-physical systems in particular. Therefore, this work proposes a new approach that utilizes advantages and promising features of the existing works to develop a comprehensive integrated approach while focusing on ease of cyber-physical systems development.

This chapter presents a development framework for cyber-physical systems development, based on techniques presented in the domains of sensor network macroprogramming and MDD. It separates cyber-physical systems development into different concerns and integrates a set of high-level languages to specify them. This methodology is supported by automation techniques at different phases of cyber-physical systems development. Our evaluation based on two realistic cyber-physical systems shows that our approach generates a significant percentage of the total application code and drastically reduces development effort for cyber-physical systems involving a large number of devices. Thus, our approach improves the productivity of stakeholders involved in cyber-physical systems application development.

18.7   FUTURE RESEARCH DIRECTIONS

This chapter addresses the challenges, presented by the steps involved in cyber-physical systems development, and prepares a foundation for our future research work. Our future work will proceed in the following complementary directions, discussed as follows.

Mapping algorithms cognizant of heterogeneity: While the notion of region labels is able to reasonably tackle the issue of scale at an abstraction level, the problem of heterogeneity among the devices still remains. We will provide rich abstractions to express both the properties of the devices (e.g., processing and storage capacity, networks it is attached to, as well as monetary cost of hosting a computational service) and the requirements from stakeholders regarding the preferred placement of the computational services of the applications. These will then be used to guide the design of algorithms for efficient mapping (and possibly migration) of computational services on devices.

Developing concise notion for DL: In the current version of DL, the network manager is forced to specify the detail of each device individually. This approach works reasonably well in a target deployment with a small number of devices. However, it may be time consuming and error prone for a target deployment consisting of hundreds to thousands of devices. Our future work will be to investigate how the deployment specification can be expressed in a concise and flexible way for a network with a large number of devices. We believe that the use of regular expressions is a possible technique to address this problem.

Testing support for cyber-physical systems development: Our near-term future work will be to provide support for the testing phase. A key advantage of testing is that it emulates the execution of an application before deployment so as to identify possible conflicts, thus reducing application debugging effort. The support will be provided by integrating an open source simulator. This simulator will enable transparent testing of cyber-physical systems in a simulated physical environment. Moreover, we expect to enable the simulation of a hybrid environment, combining both real and physical entities. Currently, we are investigating open source simulators. We see Siafu* as a possible candidate due to its open source and thorough documentation.

Runtime adaptation in cyber-physical systems: Even though our approach addresses the challenges posed by evolutionary changes in target deployments and application requirements, stakeholders have to still recompile the updated code. This is common practice in a single PC-based development environment, where recompilation is generally necessary to integrate changes. However, it would be very interesting to investigate how changes can be injected into the running application that would adapt itself accordingly. For instance, when a new device is added into the target deployment, a cyber-physical systems can autonomously include a new device and assign a task that contributes to the execution of the currently running application.

REFERENCES

1. D. Cassou, B. Bertran, N. Loriant, C. Consel et al., A generative programming approach to developing pervasive computing systems, in *GPCE'09: Proceedings of the Eighth International Conference on Generative Programming and Component Engineering*, pp. 137–146, Denver, Colorado, 2009.
2. Z. Drey, J. Mercadal, and C. Consel, A taxonomy-driven approach to visually prototyping pervasive computing applications, in *Domain-Specific Languages*, pp. 78–99, Springer, Oxford, UK, 2009.
3. L. Mottola, A. Pathak, A. Bakshi, V. Prasanna, and G. Picco, Enabling scope-based interactions in sensor network macroprogramming, in *IEEE International Conference on Mobile Ad hoc and Sensor Systems*, pp. 1–9, IEEE, Pisa, Italy, 2007.
4. J.-S. Lau, C.-K. Tham, and T. Luo, Participatory cyber physical system in public transport application, in *2011 Fourth IEEE International Conference on, Utility and Cloud Computing (UCC)*, pp. 355–360, Melbourne, Australia, December 2011.
5. K. Stouffer, J. Falco, and K. Scarfone, Guide to industrial control systems (ICS) security, *NIST Special Publication*, 800(82), 16–16, 2008.
6. J.-P. Vasseur and A. Dunkels, *Interconnecting Smart Objects with IP: The Next Internet*. Morgan Kaufmann, San Francisco, CA, 2010.
7. R. N. Taylor, N. Medvidovic, and E. M. Dashofy, *Software Architecture: Foundations, Theory, and Practice*. Wiley Publishing, 2009.
8. D. Cassou, J. Bruneau, C. Consel, and E. Balland, Toward a tool-based development methodology for pervasive computing applications, *IEEE Transactions on Software Engineering*, 38(6), 1445–1463, 2012.
9. L. Mottola and G. Picco, Programming wireless sensor networks: Fundamental concepts and state of the art, *ACM Computing Surveys (CSUR)*, 43(3), 19, 2011.

---

* http://siafusimulator.org/.

10. D. Gay, P. Levis, R. Von Behren, M. Welsh, E. Brewer, and D. Culler, The nesC language: A holistic approach to networked embedded systems, in *Proceedings of the ACM SIGPLAN 2003 Conference on Programming Language Design and Implementation*, vol. 38, pp. 1–11, San Diego, California, 2003.

11. E. Cheong and J. Liu, galsC: A language for event-driven embedded systems, in *DATE*, pp. 1050–1055, IEEE Computer Society, Munich, Germany, 2005.

12. K. Whitehouse, C. Sharp, E. Brewer, and D. Culler, Hood: A neighborhood abstraction for sensor networks, in *Proceedings of the Second International Conference on Mobile Systems, Applications, and Services*, pp. 99–110, ACM, Boston, 2004.

13. C. Frank and K. Römer, Algorithms for generic role assignment in wireless sensor networks, in *Proceedings of the Third International Conference on Embedded Networked Sensor Systems*, pp. 230–242, ACM, San Diego, 2005.

14. P. Costa, L. Mottola, A. Murphy, and G. Picco, Programming wireless sensor networks with the teeny lime middleware, *Middleware 2007*, pp. 429–449, Newport Beach, CA, 2007.

15. M. Román, C. Hess, R. Cerqueira, A. Ranganathan, R. H. Campbell, and K. Nahrstedt, Gaia: A middle-ware platform for active spaces, *ACM SIGMOBILE Mobile Computing and Communications Review*, 6(4), 65–67, 2002.

16. A. Pathak, L. Mottola, A. Bakshi, V. Prasanna, and G. Picco, A compilation framework for macroprogram-ming networked sensors, *Distributed Computing in Sensor Systems*, 4549, 189–204, 2007.

17. U. Bischoff and G. Kortuem, Life cycle support for sensor network applications, in *Proceedings of the Second International Workshop on Middleware for Sensor Networks*, pp. 1–6, ACM, Newport Beach, CA, 2007.

18. D. C. Schmidt, Guest editor's introduction: Model-driven engineering, *IEEE Computer Society Available*: http://www.cs.wustl.edu/~schmidt/GEI.pdf, 39, 25–31, February 2006.

19. R. Picek and V. Strahonja, Model driven development-future or failure of software development, in *IIS*, vol. 7, pp. 407–413, Venice, Italy, 2007.

20. R. Sugihara and R. Gupta, Programming models for sensor networks: A survey, *ACM Transactions on Sensor Networks (TOSN)*, 4(2), p. 8, 2008.

21. K. Henricksen and R. Robinson, A survey of middleware for sensor networks: State-of-the-art and future directions, in *Proceedings of the International Workshop on Middleware for Sensor Networks*, pp. 60–65, ACM, Melbourne, Australia, 2006.

22. A. Ranganathan, S. Chetan, J. Al-Muhtadi, R. Campbell, and M. Mickunas, Olympus: A high-level pro-gramming model for pervasive computing environments, in *Third IEEE International Conference on Pervasive Computing and Communications (PerCom)*, pp. 7–16, Hawaii, 2005.

23. A. Dey, G. Abowd, and D. Salber, A conceptual framework and a toolkit for supporting the rapid prototyping of context-aware applications, *Human–Computer Interaction*, 16, 97–166, 2001.

24. D. Salber, A. K. Dey, and G. D. Abowd, The context toolkit: Aiding the development of context-enabled applications, in *Proceedings of the SIGCHI Conference on Human Factors in Computing Systems: The CHI is the Limit*, (New York), pp. 434–441, ACM, 1999.

25. K. Henricksen and J. Indulska, Developing context-aware pervasive computing applications: Models and approach, *Pervasive and Mobile Computing*, 2(1), 37–64, 2006.

26. C. Bettini, O. Brdiczka, K. Henricksen, J. Indulska, D. Nicklas, A. Ranganathan, and D. Riboni, A survey of context modelling and reasoning techniques, *Pervasive and Mobile Computing*, 6, 161–180, April 2010.

27. S. Duquennoy, G. Grimaud, and J.-J. Vandewalle, The Web of Things: Interconnecting devices with high usability and performance, in *International Conference on Embedded Software and Systems (ICESS)*, pp. 323–330, Zhejiang, China, 2009.

28. M. Blackstock and R. Lea, WoTKit: A lightweight toolkit for the web of things, in *Proceedings of the Third International Workshop on the Web of Things*, p. 3, ACM, Newcastle, UK, 2012.

29. D. Guinard, V. Trifa, and E. Wilde, A resource oriented architecture for the Web of Things, in *Internet of Things (IOT), 2010*, pp. 1–8, IEEE, Tokyo, Japan, November 2010.

30. A. P. Castellani, M. Dissegna, N. Bui, and M. Zorzi, WebIoT: A web application framework for the Internet of Things, in *WCNC Workshops*, pp. 202–207, IEEE, Paris, France, 2012.

31. G. Ghidini, S. K. Das, and V. Gupta, Fuseviz: A framework for web-based data fusion and visualization in smart environments, in *IEEE Nineth International Conference on Mobile Ad hoc and Sensor Systems (MASS)*, pp. 468–472, IEEE, Las Vegas, 2012.

32. V. Gupta, P. Udupi, and A. Poursohi, Early lessons from building sensor. network: An open data exchange for the web of things, in *Eighth IEEE International Conference on Pervasive Computing and Communications Workshops*, pp. 738–744, Mannheim, Germany, 2010.

33. S. Madden, M. Franklin, J. Hellerstein, and W. Hong, TinyDB: An acquisitional query processing system for sensor networks, *ACM Transactions on Database Systems (TODS)*, 30(1), 122–173, 2005.

34. Y. Yao and J. Gehrke, The cougar approach to in-network query processing in sensor networks, *SIGMOD Record*, 31, 9–18, September 2002.

35. C. Shen, C. Srisathapornphat, and C. Jaikaeo, Sensor information networking architecture and applications, *Personal Communications, IEEE*, 8(4), 52–59, 2001.

36. C. Jaikaeo, C. Srisathapornphat, and C.-C. Shen, Querying and tasking in sensor networks, in *AeroSense 2000*, pp. 184–194, International Society for Optics and Photonics, 2000.

37. P. B. Gibbons, B. Karp, Y. Ke, S. Nath, and S. Seshan, Irisnet: An architecture for a worldwide sensor web, *Pervasive Computing, IEEE*, 2(4), 22–33, 2003.

38. K. Whitehouse, F. Zhao, and J. Liu, Semantic streams: A framework for composable semantic interpretation of sensor data, in *Proceedings of the Third European Conference on Wireless Sensor Networks, EWSN'06*, (Berlin, Germany), pp. 5–20, Springer-Verlag, 2006.

39. N. Mohamed and J. Al-Jaroodi, A survey on service-oriented middleware for wireless sensor networks, *Service Oriented Computer Applications*, 5, 71–85, June 2001.

40. E. Avilés-López and J. García-Macías, TinySOA: A service-oriented architecture for wireless sensor networks, *Service Oriented Computing and Applications*, 3, 99–108, June 2009.

41. N. B. Priyantha, A. Kansal, M. Goraczko, and F. Zhao, Tiny web services: Design and implementation of interoperable and evolvable sensor networks, in *Proceedings of the Sixth ACM Conference on Embedded Network Sensor Systems, SenSys '08*, (New York), pp. 253–266, ACM, 2008.

42. D. Box, D. Ehnebuske, G. Kakivaya, A. Layman, N. Mendelsohn, H. F. Nielsen, S. Thatte, and D. Winer, Simple object access protocol (SOAP) Google Patents, 1.1, 2000.

43. R. Chinnici, J.-J. Moreau, A. Ryman, and S. Weerawarana, Web services description language (WSDL) version 2.0 part 1: Core language, *W3C Recommendation*, 26, 2007.

44. R. T. Fielding, Architectural styles and the design of network-based software architectures. PhD thesis, University of California, Oakland, CA, 2000.

45. T. Luckenbach, P. Gober, S. Arbanowski, A. Kotsopoulos, and K. Kim, TinyREST—A protocol for integrating sensor networks into the Internet, in *Proceedings of REALWSN*, Citeseer, Stockholm, Sweden, 2005.

46. R. Gummadi, O. Gnawali, and R. Govindan, Macro-programming wireless sensor networks using kairos, *Distributed Computing in Sensor Systems*, 3560, 466–466, 2005.

47. R. Newton, G. Morrisett, and M. Welsh, The regiment macroprogramming system, in *Proceedings of the Sixth International Conference on Information Processing in Sensor Networks*, pp. 489–498, ACM, Cambridge, 2007.

48. T. W. Hnat, T. I. Sookoor, P. Hooimeijer, W. Weimer, and K. Whitehouse, Macrolab: A vector-based macroprogramming framework for cyber-physical systems, in *Proceedings of the Sixth ACM Conference on Embedded Network Sensor Systems*, pp. 225–238, ACM, Raleigh, NC, 2008.

49. E. Serral, P. Valderas, and V. Pelechano, Towards the model driven development of context-aware pervasive systems, *Pervasive and Mobile Computing*, 6(2), 254–280, 2010.

50. A. Pathak and V. Prasanna, High-level application development for sensor networks: Data-driven approach, *Theoretical Aspects of Distributed Computing in Sensor Networks*, 5431, 865–891, 2011.

51. U. Bischoff and G. Kortuem, Rulecaster: A macroprogramming system for sensor networks, in *Proceedings OOPSLA Workshop on Building Software for Sensor Networks*, Portland, Oregon, 2006.

52. W. L. Hürsch and C. V. Lopes, *Proceedings of the 21st international conference on Software engineering*, N degrees of separation: multi-dimensional separation of concerns, Separation of concerns, ACM, pp. 107–119, 1995.

53. P. Patel, A. Pathak, T. Teixeira, and V. Issarny, Towards application development for the Internet of Things, in *Proceedings of the Eighth Middleware Doctoral Symposium*, p. 5, ACM, Lisbon, Portugal, 2011.

54. P. Patel, A. Pathak, D. Cassou, and V. Issarny, Enabling high-level application development in the Internet of Things, in *Sensor Systems and Software* (M. Zuniga and G. Dini, eds.), vol. 122 of *Lecture Notes of the Institute for Computer Sciences, Social Informatics and Telecommunications Engineering*, pp. 111–126, Springer International Publishing, Lucca, Italy, 2013.

55. P. Patel, B. Morin, and S. Chaudhary, A model-driven development framework for developing sense-compute-control applications, in *Proceedings of the First International Workshop on Modern Software Engineering Methods for Industrial Automation, MoSEMInA 2014*, (New York), pp. 52–61, ACM, 2014.

56. P. Patel, *Enabling High-Level Application Development for the Internet of Things*. These, Université Pierre et Marie Curie - Paris, France, VI, November 2013.

57. M. Fowler, *Analysis Patterns: Reusable Object Models*. Addison-Wesley Longman, Amsterdam, the Netherlands, 1996.

58. S. Haller, The things in the Internet of Things, *Poster at the (IoT 2010)*. Tokyo, Japan, November, 2010.

59. S. Mosser, F. Fleurey, B. Morin, F. Chauvel, A. Solberg, and I. Goutier, Sensapp as a reference platform to support cloud experiments: From the Internet of Things to the Internet of services, in *2012 14th International Symposium on Symbolic and Numeric Algorithms for Scientific Computing (SYNASC)*, pp. 400–406, Timisoara romania, September 2012.

60. I. Sommerville, *Software Engineering*. Addison-Wesley, Harlow, UK, 9 edn., 2010.

61. B. Bertran, J. Bruneau, D. Cassou, N. Loriant, E. Balland, and C. Consel, DiaSuite: A tool suite to develop sense/compute/control applications, *Science of Computer Programming, Fourth Special Issue on Experimental Software and Toolkits*, 2012.

Section VIII

Computing Issues

19 Integration of the Cloud for Cyber-Physical Systems

Ajay Rawat and Abhirup Khanna

CONTENTS

19.1 CYBER-PHYSICAL SYSTEMS

19.1.1 OVERVIEW OF CYBER-PHYSICAL SYSTEMS

In recent times, it can be very well seen that computation and communication abilities of devices have significantly increased. The reach of such kind of devices to a large number of people has increased due to the availability of computation and communication capabilities in all kinds of entities in the physical world. The systems that bridge the gap between the cyber world (computation and communication) and physical world are known as *cyber-physical systems*. A cyber-physical systems integrates computing, communication, and storage capabilities along with monitoring and controlling the entities of the physical world. It is a convergence of computation, communication, information, and control. The term cyber-physical systems refers to a new age of systems that integrate computational and physical components of devices to facilitate better interaction between humans and their nearby

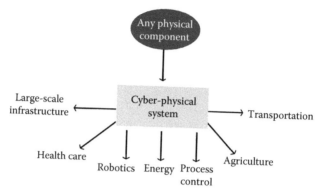

FIGURE 19.1 Cyber-physical systems.

devices. Cyber-physical systems are those systems that are monitored, controlled, and managed by the use of sensors and actuators. The work of the sensors is to provide input for the cyber-physical systems by monitoring the physical environment. Based on these inputs, cyber-physical systems perform certain computations and exhibits the results by the help of actuators in the physical world. It can also be considered as the convergence of many systems such as embedded systems, distributed sensor systems, and real-time systems. Due to the diverse characteristics of CPS, it allows the users to add new capabilities to the physical word. It provides a way in which the precision of the cyber world meets the uncertainties of the physical environment. As with the advancement of the Internet humans could communicate among themselves in revolutionary ways, cyber-physical systems would also give rise to new innovative ways in which humans could interact and communicate with the physical world around them. The rise in the field of cyber-physical systems could be seen as a result of many factors including revolution in wireless communication, high Internet bandwidth, availability of low-cost sensors, and low-power-consuming computing devices. Figure 19.1 exhibits how exactly a cyber-physical systems looks like.

19.1.2 ESSENCE OF CYBER-PHYSICAL SYSTEMS

Cyber-physical system are the result of advancement in the field of embedded systems and the need for human–computer interaction. Cyber-physical system have gained widespread acceptance across all walks of life and finds its application in almost every domain. The question arises is what makes cyber-physical system so popular among users and how it is able to fit in every domain. The answer to this is the following characteristics that form the essence of cyber-physical system:

- Provides direct association between the physical and real world
- It is heterogeneous as it can accommodate varied computational entities
- Comprises a well-networked and -coordinated system
- Involves fusion of data and information
- Systems are autonomic and adaptive
- Supports multifunctionality
- Out-and-out user interfaces
- Deployment possible under adverse physical conditions
- Supports mission critical deployment

19.1.3 COMPOSITION FOR CYBER-PHYSICAL SYSTEM

As discussed earlier, cyber-physical system are a next-generation network-integrated systems that involve the grouping of loosely coupled distributed cyber and physical systems that are monitored and controlled according to the requirements of the user. In this context, the cyber systems comprise sensors and logical compositions while the physical systems consist of actuator units. The different layers of this architecture are as follows:

- *Device layer*: This layer forms the interface for the cyber-physical system. It creates a resource construct and merges the management of various devices. User applications reside over this layer.
- *Computation layer*: The role of this layer is to prove compute capabilities of a cyber-physical system. It identifies user requirements and allocates resources accordingly. Scheduling and management of resources also takes place at this layer.
- *Context layer*: This layer acts like a decision-making unit where all the activities of CPS applications are monitored and controlled.
- *Network layer*: The work of this layer is to manage communication between all the components of a CPS. It focuses on the network behavior such as network latency, bandwidth, and network topologies.

19.1.4 CHALLENGES OF CYBER-PHYSICAL SYSTEM

Some of the prominent challenges that prevent the widespread use of cyber-physical system are

- Lack of security and privacy
- Inflexible and nonscalable
- Absence of virtualization
- Dearth of reconfiguration and robustness
- Absence of in-house fault tolerance
- Limited storage capabilities
- Improper use of resources
- Absence of high compute capabilities, that is, *big data*
- Lack of proper standards
- Location-dependent access

19.2 CONCEPT OF CLOUD COMPUTING AND ITS ARCHITECTURE

19.2.1 CLOUD COMPUTING AT A GLANCE

As rightly said by Alan J. Perlis, "One can only display complex information in the mind, like seeing, movement or flow or alteration of view is more important than the static picture, no matter how lovely."

Cloud computing (CC) is a new paradigm that aims at delivering IT as a service to business on an on-demand basis. It is an amalgamation of many technologies like distributed computing, web services, SOA, virtualization, and so forth. There are many characteristics that CC provides in order to face the challenges related to traditional IT systems. Some of these are like scalability, optimization, provisioning, and heterogeneity. The cloud architecture consists of various components comprising front-end platform (mobile devices, thin clients), back-end platforms (servers, data centers), a cloud-based delivery model, and a network topology. Based on the type of service being delivered by the cloud, CC can be elaborated on the basis of four delivery models:

- Infrastructure as a service (IaaS)
- Platform as a service (PaaS)
- Software as a service (SaaS)
- Business process as a service (BPaaS)

With the help of these delivery models, the end user is able to access cloud services in a confined manner as every model tries to focus on a specific kind of a service. IaaS provides infrastructure as a service to the end user on a subscription basis. This type of a cloud on the whole focuses on hardware-related resources and services. The services provided by the cloud are usually customized virtual machines, storage space, backup support, and so forth. PaaS provides platform as a service to the end user on a subscription basis. The services provided by the PaaS cloud includes web server stack, middleware, facilities for application design and development, application versioning, marshaling, and other similar kinds of platforms. SaaS provides software as a service to the end user on an on-demand basis. This type of a cloud primarily focuses on software resources and services. It is not required for the end user to purchase any kind of a license related to the software provided by the cloud. All updates and maintenance-related activities of the software are handled by the cloud service provider. In case of BPaaS, which is a new delivery model that arises from the changing needs of today's business world, it provides a complete business process along with the manpower required to operate the process. People from the business community are very keen to adopt this kind of a delivery model as they only need to invest their money and the rest is done by the cloud provider.

19.2.2 DELIVERY AND DEPLOYMENT MODELS FOR CLOUD

Based on how and where the cloud is set up and deployed, the cloud may be classified into three major deployment models:

1. *Private Cloud*
   a. An organization may choose to build a cloud within their datacenter.
   b. The organization purchases own hardware and software to set up the cloud.
   c. The main intention behind this kind of cloud is to deliver cloud service to internal departments within the organization.
   d. Security could be a major factor contributing to the decision to set up a cloud in-house.
2. *Public Cloud*
   a. This is a more general form of cloud. It is deployed to provide cloud services to the general population irrespective of their organization affiliation.
   b. The services are generally available through a website using an on-demand payment and subscription mechanism.
   c. Public cloud is considered less secure than private clouds.
   d. From an end user perspective, there is no capital expenditure involved in setting up a public cloud. The end user pays only a monthly subscription fee based on the usage.
3. *Hybrid Cloud*
   a. The cloud is set up to handle a fraction of the workload on private cloud and a fraction of the workload on the public cloud.
   b. Typically, a customer would normally place their production workload on the private cloud and use public cloud for development and test environments.
   c. The workload can be moved between the private and public section of the hybrid cloud based on demand.
   d. The nonproduction workload on the public cloud moves back to the private cloud when the private cloud is less loaded.
   e. Hybrid clouds combine the benefits of public and private cloud and help further optimize the capital and operational expenses of running a workload.

Based on the layer at which a cloud service is delivered, the three delivery models are

1. *IaaS (Infrastructure as Service)*
   a. The cloud provides infrastructure services to the end user on a subscription basis.
   b. The infrastructure services may include custom-designed virtual machines, storage and backup infrastructure, tape backup as service, etc.
   c. This type of cloud deals primarily with the hardware resources and services. Some common examples are Amazon EC2, Rackspace, etc.

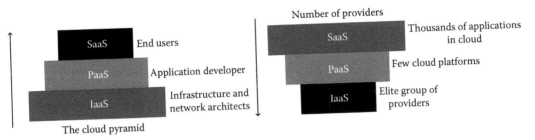

FIGURE 19.2 Delivery and deployment models of cloud computing.

2. *PaaS (Platform as a Service)*
    a. This type of cloud provides platform services to end users on pay-as-use basis.
    b. The platform services may include the web server stacks, middleware, or other similar platforms.
    c. Some common examples are the Apache-PHP-Perl webserver platform, IBM Web-sphere as a service, Tomcat servlet container as a service, Moodle LMS, Microsoft Azure, etc.
3. *SaaS (Software as a Service)*
    a. This type of cloud provides software services to end users on a subscription basis.
    b. The end user customer is not required to maintain any license for using the cloud provided software.
    c. All updates to the software is taken care of by the cloud service provider.
    d. For example: Google Apps, etc.
4. *BPaaS (Business Process as a Service)*
    a. This type of cloud provides a particular business process as a service along with the staff that is required to run the process activities.
    b. The end user is not required to hold any license or hire any staff for using the Business Process services.
    c. This kind of service provides an end-to-end business process coverage for a business on an on-demand subscription basis.
    d. For example: Providing business analytics as a service to end customers, tailor-made for the business (Figure 19.2).

19.2.3 ARCHITECTURE FOR CLOUD COMPUTING

As discussed earlier, there are three deployment models and four delivery models for cloud computing. But the question still arises is what makes these models different from one another and how a customer would select any one from them. To have a better understanding of cloud and differentiate among its models, one needs to have some insight into the cloud architecture. The architecture for cloud describes the various components required to set up cloud and their interaction with one another. It also helps to portray the requirements, considerations, and specifications enlisted by the consumer. Some of the standard cloud architectures are NIST Cloud Computing Reference Architecture, IBM Cloud Architecture, and Oracle Cloud Computing Reference Architecture. The various actors that constitute the cloud architecture are as follows:

1. *Cloud provider*: Is a person or an organization whose work is to provide services to its customers.
2. *Cloud auditor*: The work of the cloud auditor is to evaluate the services rendered by the cloud provider. These services are evaluated on the basis of certain parameters such as performance, security and privacy measures, and cost. In case of security auditing, a cloud auditor would ensure that the security requirements of the system are maintained. In order

to do so, it would create test case scenarios to assess the security controls of the system and would also determine the extent to which the controls are implemented.

3. *Cloud broker*: Is one of the most important entities of the cloud architecture. As a result of continuous growth in the field of CC, it becomes difficult for the consumer to keep track of all the different services being rendered to it. Here comes the role of the cloud broker that overcomes this problem by integrating all the services into one and keeping the consumer away from the complexities of service allocation. The major services provided by a cloud broker include the following:

   a. *Service arbitrage*: Service arbitrage is very much similar to service aggregation, with the only difference is that the services being integrated are not fixed. Service arbitrage allows varied choices for the broker.

   b. *Service intermediation:* A cloud broker enhances a given service by adding some more features to it and thus providing a value-added service to the consumer.

   c. *Service aggregation:* In case of service aggregation, new services are rendered to the consumers by combining one or more services. The broker tends to provide data integration and also ensures secure data travel between consumer and multiple cloud providers.

4. *Cloud carrier*: The work of the carrier is to provide access of services to the cloud consumers through a network or telecommunication channels. For example, network access devices include computers, laptops, mobile phones, or any other personal digital assistant (PDA), etc. It maintains various service-level agreements (SLAs) with different cloud providers to ensure quality of service to the cloud consumers. The cloud carrier also ensures distribution of services with the help of certain telecom or network providers.

5. *Cloud consumer*: A person or an organization that uses services provided from a cloud provider and also maintains an affiliation with it.

19.2.4 CHARACTERISTICS AND BENEFITS OF CLOUD COMPUTING

Cloud provides tangible benefits to business. The business saves costs thus improving the bottom line. It also adds value to the existing business processes by incorporating new functions for increasing efficiency, flexibility, manageability, and improved transparency. Some of the key benefits of cloud computing are as follows:

- *Provisioning* is automated and on-demand and can be done on a self-service basis. The provisioning typically takes from a few seconds to a few hours. Also, the demand for resources can be estimated well in advance to plan for procuring hardware.
- *Utilization* typically is around 60%–70%. The freed-up resources goes into a pool that can be assigned transparently to other users. It is also possible to downscale and upscale based on demand.
- *Storage capacity*: The cloud gives you almost unlimited storage capacity to store your data along with its remote access. Hence, the end users need not worry about running out of storage space or frequently increasing its current storage space availability.
- *Computation*: Huge processing is allowed to be done on the cloud. Techniques such as BigData analytics along with the use of dynamic databases make computation efficient and faster at the cloud end.
- *Scalability is on-demand*: Capacity can be planned and operational expenses can be fine-tuned to meet the current demand.
- *Availability*: Typically, the VM instances are not tied to any particular hardware. Rather they are designed to run over a range of hardware. Hence, it is possible to restart the instance on secondary hardware if the primary hardware fails. This happens automatically and transparently.
- *Elasticity* allows scaling up and down on-demand.

- *Capital expenditure*: Depending on the cloud deployment, a customer saves about 40% in upfront capital expenditure required to procure hardware. The operational expense can be further fine-tuned based on demand, thus resulting in much higher savings. Chargeback allows for more granular monitoring of usage of resources in terms of cost. This opens up avenues for further optimization.
- *Monitoring* in cloud allows for further optimization of resources for maximum utilization and reduced wastage. This results in higher savings over time. Based on the cloud implementation, the savings in cloud may vary. There are multiple factors that may impact the features of cloud and the savings and benefits inherent in cloud.

19.3 CYBER-PHYSICAL CLOUD COMPUTING

19.3.1 VISION FOR CYBER-PHYSICAL CLOUD COMPUTING

Internet is the most rapid and ever-changing entity that exists. With every passing minute, a new device gets connected to it and becomes an integral part of our lives. These devices change the way we interact with the world around us. New technologies and concepts arise every day but only some of them could make it in creating a revolution. Two such technologies are cyber-physical system and cloud computing. Both technologies work on different compute paradigms, each of which having its own benefits and constraints. As discussed previously, cyber-physical system involve the fusion of the cyber and the physical world whereas cloud computing provides a new way to compute and store information. CPS has many disadvantages like security, big data, scalability, and in-house fault tolerance that can easily be avoided by the use of cloud computing. In order to enhance the interaction between different devices and to foster the mission of ubiquitous computing, the amalgamation of these two technologies took place, which was termed as cyber-physical cloud computing (CPCC), sometimes also addressed as cyber-physical cloud (CPC). According to NSIT, it is defined as "a System environment that can rapidly build, modify and provision cyber-physical systems composed of a set of cloud computing-based sensor, processing, control, and data services." CPCC is basically having CPS with cloud integration. It has cloud as its backbone for compute and storage, whereas CPS as its sense organs to interact and communicate with the real world. Information would be gathered by the sensors, processed, and stored on the cloud sent back to the cyber-physical system and exhibited through the actuators. The driving force behind CPCC was the ever-increasing digital data and a need to enhance the interaction between the cyber and physical world in a robust manner on a real-time basis. It allows users to have access to services any time any place through the use of appropriate devices. CPCC constitutes a dynamic cyber-physical system environment by allowing autonomic service creation and virtualization of physical devices. The physical components associated with CPS are virtualized and represented as VMs on the cloud, where they can perform their normal functionalities. The software components are directly offloaded to the cloud, resulting in processed information that is and sent back to the CPS through communication protocols. In short, CPCC tends to open new doors for better communication between the cyber and physical world.

19.3.2 ARCHITECTURAL OVERLAY OF CPCC

As discussed before, the main aim of CPCC is to gather data from a variety of sources, processes, and convert them into information, understand, and then transform them in the form of actions in the physical world. In order to have a deep insight about CPCC, one needs to understand its architecture. The architecture of CPCC can be broadly classified into three components. Figure 19.3 depicts the architecture of a CPCC environment. Three components of CPCC are as follows:

1. *Sensor*: Components that gather data from the physical world and pass it on to the cyber space.
2. *Cloud servers*: Components that process sensor's data into valuable information.
3. *Actuator*: Components that exhibit this information in form of actions in the physical world.

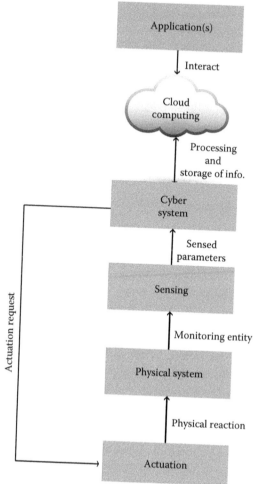

FIGURE 19.3 CPCC architecture.

These three components have various actors involved that work in coordination to support the proper working of a CPCC system. The different actors that constitute the architecture of CPCC are as follows:

- *Sensors*: The work of the sensors is to monitor and perceive events or phenomena that take place in the physical world. Two types of sensors are involved in the system, namely physical sensors and social sensors. *Physical sensors* gather data from physical entities such as temperature, pressure, direction of the wind, and so on. *Social sensors* extract data from human beings in form of social engineering. They are also known as virtual sensors. These sensors need to communicate with the provisioner depending upon the type of information they want to sense. A collection of virtual sensors constitute a cloud of sensors (CoS). The intelligence level of sensors varies from sensor to sensor. A lot of work has been done on the working model and interpretation of physical sensors that makes it mature enough as compared to social sensors. Standard interfaces need to be adopted in case of working with physical sensors.
- *Processors*: They are the entities that process data collected by the sensors into valuable information. In CPCC, the processing units are the different servers that reside on the cloud

and the entire processing takes place at the cloud end. Data are processed in accordance with certain algorithms. The processed data then act as input for the actuators or other processors.

- *Actuators*: These are physical entities that convert digital information into actions in the physical world. There are two kinds of actuators, active actuators and passive actuators. Active actuators directly exhibit information into physical actions, for example, barometer, thermostat. Passive actuators notify humans to take certain actions depending upon the information it has received. It does not incur any direct action into the physical world, for example, smartphones and smart watch.
- *Communication links*: They structure a pathway for information to travel between the cloud and CPS. They can be in the form of standard network protocols like TCP/IP, IPv6, E6, etc.
- *Provisioner*: The work of the provisioner is to allocate resources according to the needs of the VMs that run on a cloud server. These VMs are actually an abstraction of a physical world entity. Provisioner is an entity that caters to the requirements of all the other actors like sensors, processors, and actuators. Services are scheduled by the provisioner depending upon the system requirements. The concept of self-provisioning that allows change in the system requirements on a real-time basis can only be incurred by the help of an efficient provisioner.
- *Coordinator*: The work of this component is to orchestrate all the processes that take place in a CPCC environment. The coordinator resides on the cloud from where it maintains a proper work flow for all the activities that occur in a system.

19.3.3 Benefits of CPCC

As rightly said by Jeffrey Fry, "Unless you can demonstrate your benefits, all you are left to compete with is price." Thus, it is very important to highlight the benefits of CPCC as they are the ones which make it popular among clients and users. Some of the benefits of CPCC are as follows:

- *Resource utilization*: Optimum resource utilization is a key advantage of cloud computing and the same is inherited by CPCC. Resources are well used whenever we talk about the CPCC environment. The presence of sensors provides an approximate estimation for the actual amount of resources required for a physical device. The provisioner at the cloud would only allocate the required amount of resources preventing any kind of resource fragmentation. On numerous occasions, the actuators and sensors are shared by different processes thus utilizing them to the most of their abilities. Better utilization of resources also leads to increased efficiency of the system.
- *Modular composition*: As described in the architecture itself, the model of CPCC is divided into different components or modules. This feature of CPCC makes it much more flexible and agile. The modules can be arranged in different manners creating customized systems according to the requirements of the customer.
- *Scalability*: The CPCC architecture allows it to be developed and deployed in a scalable manner. All the different components are orchestrated and work in sync with one another. This results in the formation of resource and service pools that give rise to smart systems whenever needed. After the purpose has been fulfilled, the resources are sent back to the resource pool.
- *Processing capabilities*: The processing capabilities of CPCC are much higher than that of a cyber-physical system. Large amounts of knowledge can be generated from huge chunks of heterogeneous data. Due to its integration with BigData, countless calculations can be performed instantly.
- *Reliability*: The CPCC framework is very reliable and can even operate in places where natural disasters have occurred or are occurring. The robustness of the system continues to be there even when some components stop functioning.

- *Social interaction*: CPCC facilitates social interaction that in itself is a very rare thing for many of the computing domains. Due to its vast network of sensors, it allows data to be collected on a real-time basis and that too involving human interaction. The system is also able to handle unreliable data from dynamic sources of information.

All of the aforementioned benefits help to improve the quality of life and bring about a positive change in the world.

19.3.4 SIMILARITIES AND DIFFERENCES BETWEEN CPCC AND INTERNET OF THINGS (IoT)

The Internet of Things (IoT) is one of the most talked about technologies in present-day computing world. It intends to connect all the things that surround us form mobile phones to refrigerators through the Internet. It is estimated that by 2015 more than six billion things will be connected to the Internet. The basic concept behind IoT is to establish a certain means of communication between different things (electronic devices). The two fundamental components of IoT are the sensors and the Internet. The sensors gather information relating to things, which is communicated to other things by the use of Internet. It has also been seen that IoT is being amalgamated with cloud in order to have better interaction between devices and foster better storage and communication of information. This is the reason why it is sometimes considered as an alternative to CPCC, which is not true. With considerable differences between the two, there are even some similarities that prevail. The following tables depict the similarities and differences between the two technologies (Tables 19.1 and 19.2).

19.3.5 DISTINCTIVE TRAITS OF CPCC

Every computing model requires certain characteristics that would ensure proper functioning of the model. The same also applies to CPCC, where certain characteristics are mentioned that would ensure

TABLE 19.1
Similarities between CPCC and IoT

Similarities	CPCC	IoT
Cloud	Present	Present but optional
Sensors	Yes	Yes
Data sharing	Yes	Yes
Resource sharing	Yes	Yes
Scalability	Scalable	Highly scalable
Protocols	WiFi, WiMax, 3G	RFID, WiFi, WiMax
Data generated	Large	Very large
Security and privacy concerns	High	Very high
Real-time system	Yes	Yes

TABLE 19.2
Differences between CPCC and IoT

Differences	CPCC	IoT
Coupling between physical and cyber world	High	High
Domain specific	Yes	No
Fault tolerance	Moderate	No
On-demand accessibility	No	No
Cost	High	Very high
Storage space	Very large	Moderate

its proper performance and also illustrate the requirements that would lead to its proper functioning. The desirable characteristics of a CPCC system are as follows:

- *Agile*: The system should be agile enough so that its configuration can change depending on different situations. The agile nature of the system makes it adaptable and customer oriented for any kind of health care organization. The system could withstand scenarios involving high emergency and low availability of resources, thus resulting in higher throughput.
- *Real-time operations*: The CPCC model should always work on a real-time basis. It should be able to adapt to the changing situations that take place in the real world. Efficient data streaming should be there between the cloud and the sensors. Quick actions need to be taken by the actuators on the basis of processed information. Resources to the VMs should be allocated dynamically on a real-time basis.
- *BigData support*: A CPCC model should ensure storage, processing of large amounts of heterogeneous data. This is one of the key factors that make it different from a CPS model. CPCC should be able to aggregate and manipulate large chunks of data in order to derive valuable information from it.
- *Standard interfaces*: In order to bear efficiency and robustness of a system, standardized interfaces need to be implemented. They are very essential to ensure proper coordination and communication between different components of a CPCC environment. The interfaces should be interoperable, interpretable, and derived from the concepts of ubiquitous computing. Provisioning of different kinds of services will also be fostered by the use of standardized interfaces. Management of resources also becomes convenient in the case of standard interfaces.
- *Security and privacy*: A CPCC system has to ensure certain levels of security in order to protect the processed information. The current security measures may not be beneficial for a CPCC system. A CPCC environment ought to use private cloud as a deployment model to maintain the privacy of the processed information.

19.3.6 INTEGRATING DOMAINS

19.3.6.1 CPCC and Health Care

In the present-day scenario, it is well evident that health care organizations lag behind many industries when it comes to adoption of new technologies. This can be seen as IT organizations generate a lesser amount of revenue from the health care sector as compared to other industries. In every health care organization, machines are deployed that interact with one another, process raw data into valuable information, and exhibit this information in form of results. But as rightly said "rolling stone gathers no mass," similarly the health care sector is looking for ways in order to enhance its capabilities and productivity. Cyber-physical systems were seen as alternatives to improve the working of health care organizations by enhancing the interaction between different devices. Input would be taken from the biosensors, computed into information that would lead to efficient decision-making. The next step toward perfection was the involvement of CC that can be seen as the biggest change after introduction to computers in the health care industry. Embracing CC along with cyber-physical systems would help health organizations to render better clinical services and enhance a patient's experience. This indeed would result in the practice of a new technology called CPCC in the health care sector. The cloud part will act as a center for storage and computation, whereas cyber-physical systems would enhance the inter-machine interaction.

To adopt a new technology, one needs to have a complete blueprint of how the technology works and what are the various components that constitute that technology. Here describe the complete architecture for the use of CPCC in a health care organization. The various components of CPCC architecture in health care sector are as follows:

1. *Sensors*: They are the sense organs of a CPCC system. Talking about health care organizations, the main work of a sensor is to gather physiological data from varied sources that are

later processed into valuable information. The data collected from the sensors act as input for the entire system. Some of the parameters sensed by medical sensors are blood pressure, insulin levels, heartbeat, essential minerals, and vitamins. Every sensor can be categorized on the basis of three components: sensor type, methodology, and sensing parameters. *Sensor type* defines what kind of a sensor it is. Whether it is homogeneous or heterogeneous, single-dimensional, or multidimensional sensor. *Methodology* defines the method it follows in order to accumulate physiological data. The sensors may collect data from a hospital or from a patient's home. The sensor can also be part of a wireless sensor network (WSN). The data can also be collected in an active or a passive manner like MRI data can be collected in an active manner, whereas blood pressure and heart rate can be passively inferred from this data. The set of attributes a sensor can sense constitutes its parameters. A sensor might sense a single parameter like blood pressure or multiple parameters such as in an ECG. The various kinds of sensors incorporated by a CPCC system in a health care organization are pressure sensors, temperature sensors, biosensors, accelerometers, SQUIDs (superconducting quantum interference devices), and implantable sensors.

2. *Computation unit*: In a CPCC environment, computation takes place at the cloud end. The computation unit performs various activities on the data received from the sensors. Some of these activities are data integration, data processing, and data modeling. Data integration involves gathering of data from various kinds of sensors in order to attain enhanced knowledge. After successful data integration, the data are then sent for further processing. Huge chunks of data are processed in a distributed manner at the cloud end. Different techniques such as BigData analytics and data mining are used to process large amounts of heterogeneous data. Next comes the modeling of CPCC components and applications on the basis of the processed data. Important trends can be predicted from the processed data that can help in modeling of efficient and scalable systems.

3. *Monitoring unit*: The complexity of a CPCC system is defined by the level of monitoring it incurs. Different algorithms need to be deployed depending upon the nature of monitoring. Monitoring can either be static or dynamic. Static monitoring occurs at a give place for certain amount of time, whereas for dynamic monitoring the patient is monitored and observed from anywhere and anytime. Resource allocation and provisioning at the cloud is done in accordance with the level of monitoring adopted by a system.

4. *Data storage*: Large amounts of data collected from the sensors and monitoring unit is stored in the cloud storage. Data are stored in an encrypted form as it is of high importance to the patient and the health care organization. A private cloud is mostly preferred for storing such kinds of valuable data. Use of dynamic and graph databases should be incurred to ensure fact access of data. Archival copies of data are also maintained as a precautionary measure for any kind of data loss.

5. *Communication links*: They form the medium of communication between different devices and the cloud environment. Without proper communication links, the CPCC system can never become functional. Communication links in CPCC for health care are established in three parts: collection of data from sensors, transfer of this data to the cloud, and transfer of results to the actuators. Efficient routing algorithms and network topologies need to be deployed in order to ensure real-time data transfer. Some of the widely used communication protocols are WiFi, WiMax, and TCP/IP.

6. *Actuation unit:* The process of actuation can either be manual or automatic. Manual actuators require human intervention to work on the basis of the suggestions made by the compute unit. Automatic actuators work on their own in sync with the results provided to them. Useful insights from the processing unit are reflected in the physical world by the help of actuators. In a health care organization, actuators are usually in the form of threshold alarms. These alarms get activated when certain essential physiological parameters cross there threshold value. They may also make phone calls or book appointments with the doctor depending upon the condition of the patient.

7. *Health care applications*: CPCC offers various applications for end users as well as for health care organizations. These applications help users to have a better interaction with a health care system of a hospital, diagnostic lab or other health care organizations. Customized apps may also be designed in accordance to the needs of an organization. Health care applications can be divided into two areas, namely assisted and controlled.

   a. The *assisted apps* are used for monitoring the health parameters of a patient without restricting its everyday activities. These apps render medical advice to patents on the basis of the data collected by the biosensors on a real-time basis. Such apps monitor the in house activities of a patient and would immediately inform the hospital authorities in case of any kind of mishap. These kinds of applications foster remote care for patients and are very much beneficial for elderly persons who live alone.

   b. The second class of applications is *controlled apps*. These applications are present in a controlled environment of a hospital such as the ICU or the operation theatre. The level of observation of these apps is much higher and intense than assisted applications. The information collected by these apps is from varied sources like clinical observations, biosensors, monitors, etc.

19.3.6.2   CPCC and Smart Homes

A smart home is just on the horizon. The rise of new technologies coupled with varied web applications and their easy accessibility are enabling the creation of intelligent products and services. A smart home acts as a cohesive system involving the use of such products and services. Smart home technology is gaining widespread acceptance among users as it integrates different technologies to enhance the quality of living. The evolution of smart home took place in order to overcome the drawbacks of the traditional home setups. These age old setups had many problems like, lack of home automation, high-energy consumption, inflexibility, reduced manageability, and nonscalable. The two prominent technologies that constitute the concept of smart home are cloud computing and cyber-physical systems. In a smart home environment, CC acts as a data repository where large amounts of data can be stored and also a place for application development and deployment. It provides flexibility and scalability to the home automation system. On the other hand, cyber-physical systems act as a facilitator for human interaction between users and home appliances. But the question still arises that what makes the smart home different from the current day homes. The answer to it is the following characteristics of a smart home that make it different and exclusive form present-day homes:

- *Intelligent*: Every smart home has the facility to make decisions based on the data collected by the sensors. The implementations of intelligent applications that manage the household activities result in optimization of resources. Intelligent services help to foster better understanding of the home environment and improve interaction between different users. It is very essential for every home automation system to make use of appliances and services that are intelligent enough to cope with the real-time scenarios.
- *Interconnected*: One main feature of a smart home is its ability to communicate and interact with all the home appliances along with the end users. A proper network connection is established in a smart home that involves interconnectivity of different sensors and their respective devices. A communication link is also established between cloud and other physical devices. This allows users to access home applications and appliances remotely through the Internet. The service also provides benefit from it as it helps them to analyze customer's characteristics and its usage patterns. This interconnected setup provides uptime flexibility and scalability to a smart home.
- *Instrumented*: It is the ability to sense and monitor changes in an environment. Instrumented devices provide high-quality detailed information of their surroundings. They also describe about the functionalities of the system in which they operate. For instance, devices would monitor their electricity consumption and would provide a detailed report upon it. This

report will help the user to reduce its electricity expenditure by effectively using its appliances. Data-driven product manufacturing may also be achieved by properly analyzing the data collected from these devices.

The smart home architecture comprises of many different components of cloud, CPS, and the home environment. Each of them is loosely coupled and can any time join or leave the home network. What follows is the list of various components that constitute a CPCC integrated smart home:

1. *Sensors*: The home automation system talks about making a home smart. Just as humans smart homes also require information to work in an intelligent manner. This information is provided to them by the use of sensors. Sensors that are an integral part of a CPCC system may be operated in an automated or command driven fashion. The quality and range of the sensors used determine the extent of home automation. Some of the commonly used sensors in a smart home are light sensors, temperature and humidity sensors, fire alarm sensors, leakage sensors, microphones, and proximity sensors.

2. *Home network*: It forms a common link that connects all the components of a smart home. A home network may comprise of both wired and wireless actors. The wired components can be fiber optics or twisted wires. The wired actors are used to foster communication among different home appliances. Whereas on the other hand, wireless actors establish communication between the control unit and the cloud. They also help the control unit to interact with all the home appliances. Some of the examples for wireless system are radio frequency, WiFi, Bluetooth Z-wave, microwaves, and so on.

3. *Control unit*: It is an entity that orchestrates all the functionalities of various appliances of a smart home. It can either be a physical or even a logical entity. In case of a logical entity, it would be residing on the cloud in form of a virtual machine. The instructions provided by the user are transmitted to the control unit that further directs different appliances in order to fulfill the instructions. The control unit acts like an intermediary between the cloud and different home appliances.

4. *Cloud*: It is the breeding ground for all the home applications. Applications are developed and deployed on the cloud. It also acts as a storage unit from which the user can remotely access its data.

5. *Smart grid*: It is used to facilitate efficient energy management in the house. It allocates only the required amount of power required for an appliance to work that indeed saves power and prevents damage due to high power input. The smart grid works in accordance with the control unit to ensure proper functioning of all the appliances. The smart grid can turn on or off the power of a certain appliance depending upon the requirements of the user.

6. *Home appliances*: They are the entities that form the last part of the hierarchy. They can be seen as the actuators of a CPCC system as they are the ones that exhibit user instructions in the real world. For a smart home, the appliances can be divided into three categories, namely kitchen appliances, living room, and bedroom appliances. Kitchen appliances consist of: microwaves, refrigerators, dish washers, and coffee makers. Living room appliances are television, video games, audio video systems, home calendar, and video phone. Lastly, the bedroom appliances comprise of: lighting, health monitoring, heating, air conditioning, and security systems.

7. *Smart home applications*: They form the interface between user and the smart home system. Smart home apps help the user to monitor and control the activities of various home appliances from a remote location. The applications can be classified into two categories namely, user centric, and system centric. User centric applications involve direct involvement of the user. The input as well as the output of such apps depends upon the state of the user. These apps may either be used to store user's data on to the cloud or act as voice recognition systems. They can also be used to monitor the health parameters of a user or allow the user to play games on a cloud console whose output can be directly streamed to the television sets.

System centric applications are those that help to orchestrate all the components of a smart home system. They may be used to divert or manage network traffic within the hose and on the cloud servers.

As every coin has two sides similarly smart homes have certain disadvantages to it. Some of the disadvantages of smart homes are as follows:

- *Cost*: High provisioning and maintenance costs are incurred by a smart home. Not all users can afford to setup a fully functional smart home.
- *Adoption*: Only technically sound users are able to operate in a smart home environment, thus constraining its adoption to only a certain class of people. Traditional appliances may not do thus new kind of appliances also needs to be deployed in case of a smart home system. This also limits people from changing over to a smart home environment.
- *Security*: A smart home also comes with security concerns that are inherited from its cloud part. Valuable user data saved on the cloud might be at the risk of unauthorized access. The security system of a smart home may also be tampered due to certain leaks in its network channel.

19.3.6.3 CPCC for Smatter Roads

Ever since the commencement of globalization different means of transportation have contributed to social as well as economic aspects of the society. Continuous work has been done in this field leading to better roads and traffic management techniques. Earlier embedded systems where used for monitoring and controlling the movement of traffic, followed by the use of cyber-physical systems. The advancement of embedded systems and large-scale adoption of sensors gave rise to the use of cyber-physical systems. It provides a complete set of new solutions for traffic management that were very different from the traditional techniques both in nature and functionalities. But with changing times the needs do change and so does the technology. Since the last couple of years, cloud computing has evolved that led to the rise of new possibilities for web services. These new advancements in CC have extended its reach to almost all walks of life and one of them being the transportation system. Cloud computing amalgamated with cyber-physical systems gave rise to CPCC. Today, CPCC is being looked upon as a technology for achieving the goal of smarter roads and dynamic traffic management systems. It aims at integrating the benefits of CC and CPS to foster efficient and robust traffic management systems.

To understand the working of CPCC in case of traffic management and how it would lead to smatter roads, one needs to have an insight of all the components that compose the systems. The architecture of CPCC for traffic management needs to be studied in order to investigate its different feasible possibilities. Many new applications such as interactive navigation system can be created after analyzing the architecture of CPCC. The various actors that constitute the architecture are as follows:

- *Sensors*: They are used to monitor the activities of various automobiles running on the roads. Sensors act as data collectors and collect data from different roads and automobiles. These data are then processed into information from which road traffic patterns can be identified. Some of the widely used sensors in cars are oxygen sensor, temperature sensor, speedometer, car alarm, automatic locks, antilock brakes, and air bags.
- *Traffic control center (TCC)*: Data sensed by the sensors is sent to the TCC whose work is to maintain the trajectory for all the vehicles on road. The working of traffic signals is also monitored by the TCC. It acts like a management entity for the data collected and transfers this data to other actors for further processing.
- *Control unit*: The data received from the TCC is stored and processed by the control unit that resides on the cloud. Its work is to obtain useful information from this raw data. Traffic patterns can be predicted from the information generated by the control unit. The processed

data also acts as input for traffic management techniques that would control the movement of vehicles and working of traffic lamps. Effective traffic management algorithms can be created from this processed information.

- *Traffic lamps*: They act as actuators for the system. They basically exhibit the processed information into the real world. Instructions given by the control unit are demonstrated by these traffic lamps. The flow of traffic is controlled by them according to the orders of the compute unit. All the traffic lamps situated in an area are interconnected and share their information with the TCC.

- *Vehicles*: They are the end user entities of the system. Vehicles are constantly connected to the cloud either from an inbuilt compute system or through a user device. Information relating to the flow of traffic and occurrence of traffic jams is provided to the vehicles. Nearest route to its destination is also given to it depending upon its in house or outdoor sensed information.

19.3.6.4 CPCC and Library Management Systems

In recent times, fast growth of web-based applications along with technologies such as CC and CPS have led to the creation of efficient and robust management systems. Both the technologies can be used differently in different domains but when put together can create magic. CPCC is the amalgamation of these two technologies that can be applied in real-time systems to enhance their productivity and manageability. One such system is the Library Management System, popularly known as the LMS. The present-day LMS are very obsolete, and they still tend to use age-old technologies. Serious technological enhancement is required in this domain in order to provide useful experience to the students. There are many flaws in the current day LMS such as inflexible, nonscalable, and inconsistent. CPCC tends to overcome all these limitations by the use of cloud computing that provides flexibility and scalability to a system. The main aim of implementing CPCC for a LMS is to transform it to a behavior-driven system rather than a functionality driven system. CPCC tends to inherit the concepts of ubiquitous computing into a LMS, thus enhancing its capabilities and extending its reach to a large number of students.

The LMS that we discuss consists of RFID technology in integration with cloud computing. LMS provides information to the students on the basis of their location, preferences, and their previous search results. Normally, students are able to access a LMS within a closed network whereas the proposed LMS could be accessed remotely. The use of RFID provides robustness and reliability to the LMS, whereas the cloud provides scalability and remote access to it. To have a deep insight about the LMS, one needs to focus at all the different components of the system. The following are the various components that constitute the proposed architecture for a library management system:

- *RFID tag*: The RFID tag holds information related to a book or student. The information can be a student's unique ID, its name, and the course in which it is enrolled. Every tag has a unique tag serial number. RFID tags can be of two types, namely active and passive. The passive tag is cheaper as compared to active tag as it is small in size and does not comprise of an in house battery. Passive tags get activated by the signals (RF energy) received from the RFID reader. The active tag periodically transmits information in form of signals, which is read by the RFID reader. Each RFID tag comprises of a volatile memory. The storage capacity of RFID tags is much more than that of traditionally used bar code tags. The tags are placed on every book and student ID card that is enrolled in the database of the system.

- *RFID reader*: The work of the RFID reader is to read information from the RFID tags and send it for further processing. RFID readers need not be placed in direct contact with the tags in order to read information that was the case when bar code readers where used. The readers act as sensors for a CPCC system. Every RFID reader is connected to the cloud where all the records are maintained. A burglary alarm would ring if an unissued book is taken out of the library. This would happen when a student ID card was not read by the reader soon after reading the book tag.

- *Cloud*: It is the place where all the information and records relating to different students and books are stored. It uses dynamic and graph databases to fetch quick results in response to the information sensed by the reader. Proper level of abstraction is maintained by the cloud that allows prevented access to students and complete access to professors. Only authenticated students and professors are allowed to access the cloud. The librarian acts as an admin for the entire system and can have a check on the number of books issued. A number of eBooks and research papers in PDF formats are also made available from the cloud. In general, a public cloud is deployed for a LMS.
- *Communication channel*: Proper communication channel needs to be established to foster transfer of information from the tags to the cloud database. Popular communication protocols used in a smart LMS are TCP/IP, ZigBee, DNP3, and IEC 62351.
- *Library applications*: They provide the interface between the cloud and the user. Valid credentials are entered from an application that is then authenticated by the cloud. These applications allow remote access to the cloud and allow downloading of eBooks and other documents in PDF format.

What follows is an algorithm that shows the workings of a smart LMS:

1. Read data from a book tag and student ID.
2. Send this information to the cloud.
3. Match this information with the records present in the database.
4. Update the database.
5. Input credentials from an application.
6. Authenticate these credentials in the cloud.
7. If authenticated allow the user to access its database.
8. If not re-enter the credentials.

19.3.7 RELATED WORK AND ISSUES

To comprehend the architecture for CPCC and work toward building a smatter planet, the following issues need to be addressed:

- *CPCC semantics and ontology*: There is a need to explore different programming paradigms and languages that can be adopted in CPCC.
- *CPCC architecture*: A better understanding and investigation of all the different components of the system should be done that could result to varied possible implementations. Construct architecture for the implementation of CPCC systems. The architecture should provide basic building blocks of a system that can be implemented by different stakeholders according to their desired requirements.
- *Knowledge generation from social sensors*: Construct systems that are able to understand and work upon human language as input parameters. This would involve the use of Natural Language Processing.
- *Service discovery, provisioning and conducting*: A deep insight of how the CPCC systems work and how its different components interact with one another. Algorithms need to be proposed that could orchestrate all the different components of the system.
- *There is a need to develop interoperable, portable*, and related standards.
- Developing both technology and standards that would ensure security and privacy.
- Categorize business goals and needs according to the principles proposed by a CPCC system.

REFERENCES

1. T.W. Wlodarczyk and C. Rong, On the sustainability impacts of cloud enabled cyber physical space, in *Second IEEE International Conference on Cloud Computing Technology and Science*, 2010.
2. L.T.T. Phuong et al., Energy efficiency based on quality of data for cyber physical system, in *IEEE International Conferences on Internet of Things, and Cyber, Physical and Social Computing*, 2011.
3. J. Jaehoon Paul, and E. Lee., Vehicular cyber-physical systems for smart road networks. *KICS Information and Communications Magazine* 31(3), 103–116, 2014.
4. W. Jin et al., A secured health care application architecture for cyber-physical systems. arXiv preprint arXiv:1201.0213, 2011.
5. A. Koubaa and B. Andersson, A vision of cyber-physical Internet.
6. K. Kim, Use cases of cyber-physical data cloud computing, in Ontology Summit, 2012.
7. H. Jung La and S. Dong Kim, A service-based approach to designing cyber physical systems, in *9th IEEE/ACIS, 2010, International Conference on Computer and Information Science*, pp. 895–900, August 2010.
8. D. Tracey and C. Sreenan, A holistic architecture for the Internet of Things, sensing services and big data, in *13th IEEE/ACM International Symposium on Cluster, Cloud, and Grid Computing*, 2013.
9. Z. Yang et al., Study and application on the architecture and key technologies for IOT, in *International Conference on Multimedia Technology (ICMT)*, 2011.
10. N., Zubair, and A. Alvi, Clome: The practical implications of a cloud-based smart home. arXiv preprint arXiv:1405.0047, 2014.
11. J. Gubbi et al., Internet of Things (IoT): A vision, architectural elements, and future directions, in *Future Generation Computer Systems*, 29(7), 1645–1660, September 2013; Dillon, T., Web-of-things framework for cyber-physical systems, in *The Sixth International Conference on Semantics, Knowledge & Grids (SKG)*, Ningbo, China, 2010 (Keynote).
12. Talcott, C., Cyber-physical systems and events, in Wirsing, M., Banâtre, J.-P., Hölzl, M., Rauschmayer, A. (eds.) *Soft-Ware Intensive Systems*. LNCS, vol. 5380, pp. 101–115. Springer, Heidelberg, Germany, 2008.
13. L. Fang et al., NIST cloud computing reference architecture. NIST special publication 500, p. 292, 2011.
14. Y. Tan, M.C. Vuran, and S. Goddard, Spatio-temporal event model for cyber-physical systems, in *29th IEEE International Conference on Distributed Computing Systems Workshops*, pp. 44–50, 2009.
15. T. Prashant, K. Deshmukh, and M. Shrivastava, Cloud computing for intelligent transportation system. *International Journal of Soft Computing and Engineering IJSCE*, 2012.
16. Y. Tan, M.C. Vuran, S. Goddard, Y. Yu, M. Song, and S. Ren, A concept lattice-based event model for cyber-physical systems, in presented at the *Proceedings of the 1st ACM/IEEE International Conference on Cyber-Physical Systems*, Stockholm, Sweden, 2010.
17. K. Yue, L. Wang, S. Ren, X. Mao, and X. Li, An adaptive discrete event model for cyber-physical system, in *Analytic Virtual Integration of Cyber-Physical Systems Workshop*, pp. 9–15, 2010; T.S. Dillon, J. Singh, O. Hussain, and E. J. Chang. Semantics of cyber physical systems, in *Keynote IFIP Intelligent Information Systems*, Guilin, China, 2012.
18. T. Dillon, V. Potdar, J. Singh, and A. Talevski. Cyber-physical systems providing quality of service (QoS) in a heterogeneous systems-of-systems environment, Keynote, *IEEE DEST*, Korea, 2011.
19. L. Xu et al., Human factors-aware service scheduling in vehicular cyber-physical systems, in *INFOCOM, 2011 Proceedings IEEE*. IEEE, 2011.
20. P. Mell and T. Grance, Draft NIST working definition of cloud computing, Referenced on June. 3rd, 15, 32 v15, August 21, 2009.
21. M. Armbrust, A. Fox, R. Griffith, A. Joseph, R. Katz, A. Konwinski, G. Lee, D. Patterson, A. Rabkin, and I. Stoica, Above the clouds: A Berkeley view of cloud computing," EECS Department, University of California, Berkeley, CA, Technical Report UCB/EECS-2009-28, 2009 National Science Foundation, Cyber-physical systems (CPS) workshop series, http://varma.ece.cmu.edu/Summit/Workshops.html.
22. R.A. Shaw, A. B. Kostinski, and M. L. Larsen, Towards quantifying droplet clustering in clouds. *Quarterly Journal of the Royal Meteorological Society,* 128(582), 1043–1057, 2002.
23. E. Lee, Cyber physical systems: Design challenges, in *IEEE Object Oriented Real-Time Distributed Computing*, pp. 363–369, 2008.
24. P. Suraj et al., An autonomic cloud environment for hosting ECG data analysis services. *Future Generation Computer Systems* 28(1), 147–154, 2012.

25. National Science Foundation, Cyber-physical systems executive summary, 2008, http://varma.ece.cmu.edu/Summit/CPS_Summit_Report.pdf.
26. T.S. Dillon, A. Talevski, V. Potdar, and E. Chang, Web of things as a framework for ubiquitous intelligence and computing, in Zhang, D., Portmann, M., Tan, A.-H., and Indulska, J. (eds.) UIC 2009. LNCS, vol. 5585, pp. 2–13. Springer, Heidelberg, Germany, 2009.
27. A. David, Natural disasters: a framework for research and teaching. *Disasters* 15(3), 209–226, 1991.
28. S. Eric et al., A vision of cyber-physical cloud computing for smart networked systems. NIST Interagency/Internal Report (NISTIR) 7951 http://www2. nict. go. jp/univ-com/isp/doc/NIST. IR. 7951. pdf, August 2013.
29. T. Ying, S. Goddard, and L.C. Perez, A prototype architecture for cyber-physical systems. *ACM Sigbed Review* 5(1), 26, 2008.

20 Big Data Analysis for Cyber-Physical Systems

Mainak Adhikari, Sukhendu Kar, Sourav Banerjee, and Utpal Biswas

CONTENTS

20.1 INTRODUCTION

Cyber-physical systems (CPS) interconnect the cyber world and the physical world by embedding sensors and computational nodes. Cyber-physical systems have the potential to transform the way a person lives and works. Smart cities, smart power grids, intelligent homes with a network of appliances, robot-assisted living, environmental monitoring, and transportation systems are examples of some complex systems and applications. In a CPS, the physical world is integrated with sensing, communication, and computing components. These four components have complex interactions.

The complexity of cyber-physical systems has resulted in model-based designs, and the development of CPS plays a key role in engineering. Naturally, the sensing component of a CPS is critical to its modeling and management because it provides real operational data. The goal of sensing is to provide high-quality data with good coverage of all components at low cost. However, these goals may not always be achievable, for example, the usage of high-fidelity sensors such as loop detectors and radars to detect traffic on a road. These sensors are expensive and hence cannot be used to cover a large metropolitan area. Smart devices with GPSs can provide good coverage, but the quality of their data is low.

Traditionally, the design of the sensing component of a cyber-physical systems focuses on how to deploy a limited number of sophisticated, reliable, and expensive sensors and how to optimize the coverage of an environment or physical phenomenon. However, advances in sensing and communication technologies over the past few years have disrupted the traditional method. The sensors have become smaller and cheaper, and with the maturity of wireless networking, now a large number of sensors can be deployed to collect massive amount of data at a low cost.

The availability of big monitoring data on cyber-physical system creates challenges for traditional CPS modeling by breaking some of the assumptions, but also provides opportunities to simplify the CPS model identification task, achieve proactive maintenance, and build novel applications by combining cyber-physical system modeling with data driven by learning and mining techniques.

Combining cyber-physical system modeling with data analytics cannot be accomplished by simply transplanting approaches from the general purpose of the computing domain—for example, techniques for performance modeling and analytics in services like Facebook and Google's web services. Though there are obvious similarities between CPS modeling and software engineering, few important differences also exist. In the IT system, the execution time of a task comes under performance issues, and it is not incorrect to take longer time to perform a task, unlike in a cyber-physical system where a task's execution time can be critical than its correct functioning.

20.2 INTRODUCTION OF CYBER-PHYSICAL SYSTEMS

A cyber-physical system is the technology that enables numerous innovative technologies. It can be defined as a computational system integrated with physical processes and networking. It refers to engineered systems of the new generation along with their integrated physical capabilities and computations. It can interact with humans through modalities [1]. In cyber-physical system, strong coupling is required to achieve stability, performance, robustness, efficiency, and reliability to deal with physical systems of various application domains [2] (Figure 20.1).

20.2.1 WHAT IS A CYBER-PHYSICAL SYSTEM?

Embedded system: An embedded system is an electronic device that is incorporated into a computer for that device to be implemented. A computer has software, embedded into its hardware, that makes it a system dedicated for specific application(s) or part of an application or a product. Embedded systems are electronic systems that contain microprocessor(s) or microcontroller(s) to

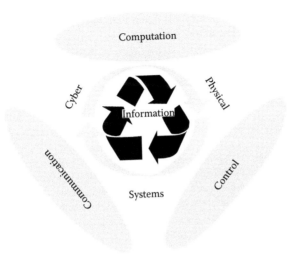

FIGURE 20.1 Basic components of a cyber-physical system.

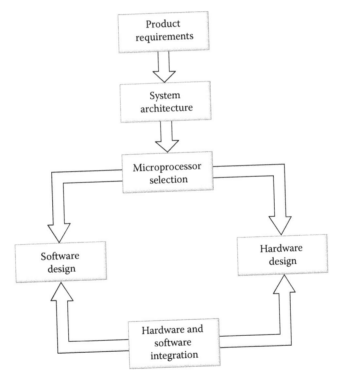

FIGURE 20.2 Diagrammatical representation of an embedded system.

perform specific dedicated application(s) or task(s). An embedded system is preprogrammed to do a specific function, while a general-purpose system can be used to run any program. To complete its functions, an embedded processor must cooperate with the rest of the system components (Figure 20.2).

Nowadays, CPS have drawn the attention of many great researchers from both academia and industry. They are considered a revolution in next-generation computing systems. CPS can transform our world from a physical system to a cybersystem with faster response, accuracy, and higher qualities of life. In a cyber-physical systems, each of the computing elements consists of a low-weight computational system. In cyber-physical systems, many embedded devices like sensors and actuators are connected to control, sense, and monitor the physical world. These two embedded devices bridge the gap between the physical and cyberworlds.

Here is a list of enablers [4], and their functions, that are used in CPS:

Enablers	Function
Software	Enables high performance of computing and real-time embedded systems
Hardware	Enables the fabrication of techniques and designs of microscale and nanoscale.
Computing and control	Enables decision aids, optimization, and verifications
Networking	Enables Internet, wireless techniques, and cloud-based techniques, controlled by a developer.

20.2.2 CYBER-PHYSICAL SYSTEM MODELS

Here, modeling is an approach toward designing CPS. The analytics refer to the learning and mining approaches by extracting knowledge from monitoring data. Big data analytics complement cyber-physical systems models [4,10]. These models can have several properties, for example, determinism can be used to capture a system's evolution and synthesis implementation. It enables analysis, stimulation, and detection of the defects in a design. The academia and industry have built tools for CPS modeling, for example, MATLAB and Simulink, which support the designing of cyber-physical systems:

- Signal processing tools are used to design automated handling of specific scenarios.
- Integrated communication modeling analyzes channel, protocol, and operation logic interactions.
- Qualification and certification deploy high-integrity cyber-physical systems.

System model: In a physical environment, a large class of CPS are modeled as a finite collection of operational modes. A CPS switches between "m" different modes and "n" dimensional systems. Its state is represented as a combination of continuous and discrete components, for example, a moving car—a combination of gear (discrete component), and contiguous metrics (torque) (Figure 20.3).

20.2.3 CHARACTERISTICS OF CYBER-PHYSICAL SYSTEMS

A CPS is constituted by sensors, actuators, networks, and desktop/laptop computers, and embedded or real-time systems [5]. But none of these can be considered a cyber-physical systems in itself. Cyber-physical systems dynamically recognize and reconfigure control systems with a high degree of automation; complexity at multiple, spatial, and temporal scale; and control loops that are closed at all scales. Several parameters, such as a high degree of complexity, strong coupling, and coordination between a system's computational parts and physical entities through networked communication, characterize a CPS.

Characteristics of cyber-physical systems are described as follows

Life cycle: In most cases, cyber-physical security features last for two to three decades, which is the life span of the systems. However, technologies may change during this period.

Accessibility: Administrative users, technology vendors, and manufacturers cannot replace security mechanisms in CPS.

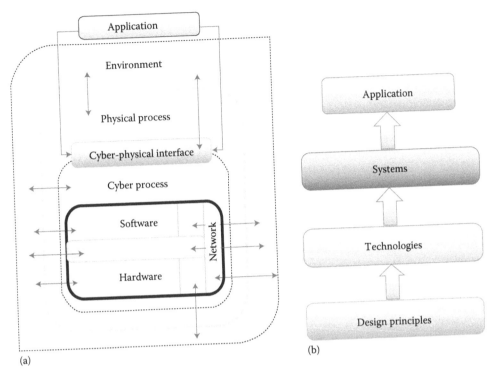

FIGURE 20.3 (a) Design model of a CPS. (b) Working model of a CPS.

Invisibility: It is one of the most important characteristics. The *cyber* portion is often invisible. For example, attacks on smart pointers with software defined radio (SDR) can potentially reach beyond the power grid.

Verification and validation (V&V): System-level evaluations of safety and trustworthiness can be performed with V&V methods. But they do not follow standardized requirements. The V&V methods are customized and expensive. System reliability increases, and cost-effective methods for verifying and validating CPS can help to decrease the cost of system integration.

Probabilistic verification to ensure that software behaves as required.

Functional verification maximizes the likelihood of a CPS to meet its desired goals.

Autonomy: Autonomous systems optimize scalability, extensibility, and performance by creating an open-source, portable, decentralized operating environment.

Adaptive: An adaptive system is able to change its behavior, structure, and modify itself dynamically with no, or little, human intervention.

Certification: The certification processes attest to assure system performance. Basically, certification is a judgment on whether a system is adequately safe, secure, and can meet other criteria for a given application.

20.2.4   ARCHITECTURE OF CYBER-PHYSICAL SYSTEMS

A CPS comprises of control logics and sensor units and, on the other hand, physical systems are a collection of actuator units. The prototype architecture [6–8,10] of a CPS includes global reference time, semantic control laws, networking techniques, event/information, quantified confidence, and subscription schemes.

Global reference time: Next-generation networks provide global reference time. In the architecture, all system components (physical devices, cyberworlds, and humans) should accept this.

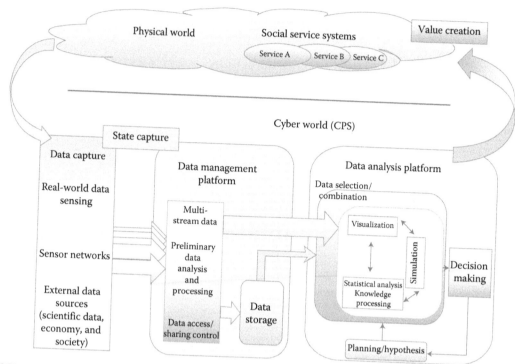

FIGURE 20.4 Architecture of a CPS.

Semantic control laws: It is defined as an event–conditio–action-type law, and it forms the core of each CPS control unit.

Networking techniques: Data management schemes are provided by next-generation networks.

Event or information driven: Events are either *raw facts* that are reported by sensor units/humans (called sensor events) or *actions* made by actuator units/humans (called actuator events). Information is the abstraction of physical world made either by CPS control units or by humans.

Quantified confidence: In this architecture, any event/information should contain built-in properties.

Confidence: This specifies the event/information confidence level.

Digital signature and authentication code: This specifies who published the event/information.

Trustworthiness: This specifies the level of trust a subscriber has on a particular publisher.

Dependability: This specifies the subscriber's dependence on event/information provided by a publisher.

Criticalness: This specifies the critical urgency of each event/information.

Subscription scheme: Each of the CPS control unit behaves as a human being by using this scheme (Figure 20.4).

CPS [6,11] focus on a strong interaction between computational and physical elements in a system. At present, CPS find application in the fields of agriculture, automotives, critical infrastructure, energy, health care, transportation and mobility, and consumer appliances.

20.2.5 Applications of Cyber-Physical Systems

Cyber-physical technology [7] can be applied in many government and private sectors, offering numerous applications in products we use in our daily life.

Sector	Products	Cyber-Physical Systems	Benefits
Defense	Soldier equipment Weapons and weapon platforms Autonomous and smart underwater sensors	Smart weapons Wearable computing/sensing uniforms Intelligent vehicles Supply chain and logistics systems Energy-efficient technologies	Increased war fighter effectiveness, security and agility Decreased exposure for human war fighters and greater capability for remote warfare
Agriculture	Resource optimization Environmental impact optimization		Increased automation Closed-loop bioengineering processes Improved safety of food products
Heath and biomedical	Medical devices Disease diagnosis Personal care equipment	Implantable devices Assistive health care systems Wearable sensors Wireless body area networks	Cost-effective health care Timely disease diagnosis and prevention Improved outcomes and quality of life
Energy and industrial automation	Electricity systems Oil and gas production Renewable energy supply	Smart electric power grid Smart oil and gas distribution grid, Plug-in vehicle charging systems	Increased energy efficiency Greater reliability, security, and diversity of energy supply
Transportation	Autonomous or smart vehicles (surface, air, water, and space) Vehicle-to-vehicle communication Vehicle-to-infrastructure communication	Plug-in and smart cars Next-generation air transport control Interactive traffic control systems	Accident prevention and congestion, reduction (zero-fatality highways or freeways)Greater safety and convenience of travel
Critical infrastructure	Commercial buildings Net-zero energy buildings High-performance residential Bridges and dams Municipal water and wastewater treatment	Building automation systems Smart HVAC equipment Smart grids for water and wastewater Active monitoring and control system Networked appliance systems Early warning systems	More safe, secure, and reliable infrastructure Increased building efficiency, comfort and convenience Assurance of water quality and supply Improved occupant health and safety Accident warning and prevention Control of indoor air quality
Emergency Response	First responder equipment Firefighting equipment Communications equipment	Integrated emergency response systems Resilient communications networks Detection and surveillance systems	Increased emergency responder effectiveness, safety, efficiency, and agility Rapid ability to respond to natural and other disasters

20.2.6 CHALLENGES IN CYBER-PHYSICAL SYSTEMS

The application of CPS in terms of technical, methodological, and functional research and development is quite complex, and so is its effects on the economy and society [11,12]. The underlying technical challenges also have a great deal of commonality reflecting a range of fundamental scientific, engineering, institutional, and societal issues. Economically, the new technology of CPS will allow designing systems by sharing both abstract knowledge and concrete tools and allow designing more dependable on CPS.

A number of challenges have been identified that prevent advancements in CPS education, workforce training, and technology transition.

Multidepartment degrees and resources: Basically, university systems are divided into more traditional disciplines (mechanical engineering, electrical engineering, civil engineering, computer science, etc.). The challenge is to incorporate a multidisciplinary CPS program within the existing university structure. In CPS, a more formal teaching and training approach will be provided; hence, the training cycles will be reduced by the resulting program.

Dynamic training and certification: CPS is a dynamic field for continuous education and retraining and it is suitable for certification and accreditation programs.

Value proposition: CPS need simplified and enhanced descriptions of research goals, benefits, and risks. Quicker and less expensive industry adoption of research can be facilitated by this type of collaboration and the understanding of the benefits and applications of CPS research can be improved.

20.2.6.1 Scientific and Technical Challenges of Cyber-Physical Systems

A new system science is required to encompass both physical and computational aspects of any system. Electronics and software are created by CPS and networked structures by the physical systems. Systems can be developed by designing new models and methods. A key challenge for computer scientists is to find a way with precise real-time requirements to work via communication networks whose behavior is randomly presented, that is, is based on probabilities.

Due to its complexity and interdisciplinary behavior, a CPS presents new requirements on the controllability of engineering and operation. The architectural design for CPS includes several issues, such as communication topology, reference architectures, open architecture, and modular service architecture. So challenges in the fields of *usability and reliability, security and safety*, future proof (capacity to evolve), and usage (human–machine interaction, acceptance, ergonomics) are of vital importance. In addition, there are numerous issues related to technical implementation through hardware and mechanics (sensors, actuators, energy provision, end devices, mechanics, middleware and platforms, wired and wireless communication). A number of processes and methods are required for *managing development and engineering*, such as distributed development, user involvement, integrated methods and models for physical components, electronics, and software. CPS require various applications to be quickly and easily networked with one another during the development period and dynamically during operation. They require distinctive *interoperability* on all levels of abstraction. In many sectors, the functions of the system are independent of materials, locations, and devices and are detached from physical restrictions, thus creating an image of reality; this is known as *virtuality* of CPS. One of the technical challenges is the interplay of the physical linking of components and their virtual networking.

System design: There are many factors to protect or impede system-level design. For example, lack of formalized high-fidelity models for large systems and inadequate scientific foundations. Compositionality is the strong interdependence of the software impact in CPS design. Both in system design and verification space, the CPS designers aspire of this modular and compositional approach. This approach makes it possible to design the system in a narrow domain with restricted properties. In the system design space, the scientific and technical challenges include evaluation techniques, lack of mathematical and system foundations, formalized metrics, and methods to achieve compositionality for dealing with cross-cutting properties, and so on.

Human–system interaction: The challenge here is to model and measure situational awareness, which is critical in decision-making, such as human perception of the system and its environment and the changes in parameters. These are used in power plant operations, air traffic control, military command and control, and emergency services.

Integrating complex, heterogeneous large-scale systems: Across heterogeneous systems, a fundamental issue is the lack of common terminology, modeling languages, and rigorous semantics for describing the interconnection between the physical and computational worlds. Another major challenge is in achieving the interoperability and compositionality of various components constructed in different engineering sectors. Integration problems also arise due to the lack of ownership of interfaces between systems.

Measuring and verifying system performance: In CPS, it is very difficult to verify performance, accuracy, reliability, security, and various other requirements, which impedes development and investment. The methods of V&V of CPS are limited, cumbersome, and expensive. One of the most important challenges is to develop test beds and data sets to enable a straightforward approach to the validation of complex CPS. Another major challenge is the creation of methodologies to further extend the capabilities of V&V methods for complex systems. Testing processes require less time if the design phase is more reliable. Modeling privacy requirements also has challenges. Against the requirements, a system can be validated henceforth.

Dealing with uncertainty: In new and uncertain environments, CPS need to evolve and operate reliably. During cyber-physical system design and development stages, innovative methods will be required to quantify uncertainty. In software, the limits of reliability and accuracy of physical components, the validity of models characterizing them, network connections, and potential design errors exacerbate the uncertainty.

20.2.6.2 Institutional, Societal, and Other Challenges of Cyber-Physical Systems

Economic challenges: CPS should be marketed, operated, and distributed because of their technical development. The changes in our economic systems are supported and accelerated by CPS. As a result of CPS, new business models are being developed that help companies and customers to find and contact each other. Existing relationships (between subcontractors, production companies, and customers) are developed through corporate networks. New company roles and functions (service aggregators, who collect individual services from suppliers and market them as whole solutions via shared platforms) are created by the CPS.

Social challenges: Acceptance by users is a crucial prerequisite for the use of CPS. Acceptance is highly dependent on well-designed human–machine interaction. More robust social discourse deals with a series of fundamental issues relating to CPS.

Political challenges: To protect safety-critical infrastructures and to clarify the issues of liability, it is very much essential to create legal policies. The rules of open system need to be created. A high-level of investments in technical infrastructure is needed to implement CPS.

Trust, security, and privacy: Cybersecurity is a critical aspect in CPS, including the protection of national infrastructure, privacy of individuals, system integrity, and intellectual property. Cybersecurity is a strong national priority, and much progress has been made to ensure protection from cyberattacks. Industries require protection of intellectual property and sensitive business and demographic information.

Effective models of governance: Good governance can provide structured control and regulation and reduce liabilities. Good governance is a part of many organizations, ranging from expert forums to treaty based.

Skilled workforce: A CPS is a sophisticated and advanced technology system that requires knowledge and training to be designed, developed, and implemented. It requires new workforce and new skills.

20.3 INTRODUCTION OF BIG DATA

Data: Data are a collection of facts and statistics. In other words, data can be defined as a set of quantitative and qualitative variables. They can be numbers or words on a piece of paper, measurement, observation, as bits and bytes in electronic memory, or as facts stored in a person's mind or description of things. It can be visualized using graphs and images. It is usually used to distinguish binary machine-readable information and human-textual information. For example, data files (binary data) and text files (ASCII data) are distinct in some applications.

20.3.1 WHAT ARE BIG DATA?

The term "big data" is used to describe structured, semistructured, and unstructured data; availability of data; and exponential growth of data. Big data is used to describe a massive volume of data sets that are very large and complex. This makes processing of data with standard methods or by using traditional database and software techniques difficult. The term may also refer to the technology (including tools and processes) that an organization requires to handle the large amounts of data and storage facilities, when it is used by vendors.

Big data has the following characteristics:

Volume: Big data suggests that its volume can take up terabytes and petabytes of storage space. Increasingly, an enterprise demands to use and analyze more types of structured and unstructured data that do not fit into existing operational and analytic business systems.

Variety: The next aspect of big data is its variety. This means that the category to which big data belongs to is also an essential fact that the data analysts should be aware of. This helps them to use the data effectively to their advantage, thus upholding the importance of the big data.

Velocity: The term "velocity" in this context refers to how fast data are generated and processed to meet the demands and the challenges of growth and development (Figure 20.5).

Variability: This factor can pose a problem for data analysts. This refers to the inconsistency that can be seen in the data at times, which hampers the process of handling and managing the data effectively.

Complexity: Data management can become a very complex process, especially when large volumes of data come from multiple sources. These data need to be linked, connected, and correlated in order to grasp the information that they are supposed to convey, and aptly termed as the "complexity" of big data.

20.3.2 TYPES OF DATA

There are basically two types of data.

1. Qualitative data
2. Quantitative data

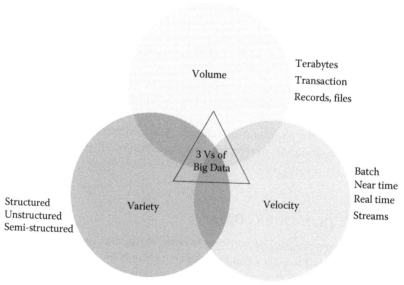

FIGURE 20.5 Basic components of big data.

Qualitative data: If data have no numerical value, that is, if they cannot be measured in numbers, then this type of data are called qualitative data.

Qualitative data can be classified into two:

Nominal data: "Nominal" comes from the Latin word *nomen*, which means "name." The data that are differentiated by a naming system are called nominal data. Nominal data usually belong to a definable category. Example: a set of groups.

Ordinal data: The data that show sequential symbols or letters are called ordinal data. Arithmetic operations cannot be presented as ordinal data because they show only the sequence. This type of data is also categorical because it belongs to a definable category. Example: first, fourth, and sixth participants in a race.

Quantitative data: This type of data are measured with only numeric values.

Quantitative data are also of two types:

1. *Discrete quantitative data*: These types of data are measured with only specific values.
   Examples: Shoe size, number of cars in a car park, number of students.
   Number of students is an example of discrete quantitative data though it is a numeric value; it cannot be a negative or a fractional number.
2. *Continuous quantitative data*: These types of data can be measured by any numeric value.
   Examples: length, width, mass.
   Mass is an example of continuous quantitative data because it can be measured in numerical values.

20.3.3 ARCHITECTURE OF BIG DATA

Nowadays, multilayered architecture is used to deal with big data. Big data can be stored, processed, managed, and analyzed in several ways. Many additional areas such as education, industry, government, and defense utilize big data processing. In big data systems, many machines work in parallel to store and process data [13,14]. Big data architecture is used to take advantage of clustered hardware. Big data architecture describes a scalable, easy-to-understand approach. A big data system can be built and run by a small team. The architecture of big data is not a fixed, one-size-fits-all approach. For a particular workload, each of the components has several alternatives with their respective advantages and disadvantages.

Big data is horizontally scaled and is for general-purpose storage and a computing fabric for ordered data designed to operate on a cluster of commodity hardware. Many different types of data are incorporated by big data and its analytics architecture, such as operational data, authoritative data, external data, system-generated data, analytical data, and historical data.

Operational data: Data residing in operational systems (ERP, warehouse management systems, etc.) are very well structured.

Authoritative data: These are high-quality data used to provide context in the operation.

External data: Other common sources of big data tend to originate from outside of the organization. These include social media feeds, blogs, and independent product and service ratings.

Analytical data: These are structured data that provide easy access to and solve analytical queries using analytical tools. They take the form of dimensional data models and OLAP cubes. Some types of analytical data can be processed as historical data.

Historical data: These data are is organized and structured to accommodate large volumes and business changes without the necessity for schema revisions.

System-generated data: These data originate from an organization and have historically been overlooked in terms of business analytics value. System logs, RFID tags, and sensor output are the forms of system-generated big data that must be captured and analyzed (Figure 20.6).

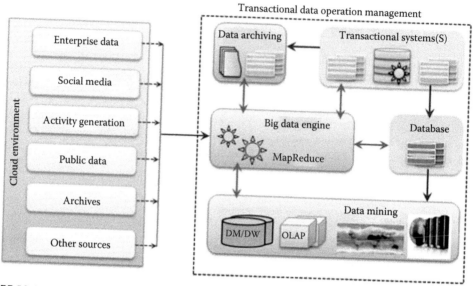

FIGURE 20.6 Architecture of big data (data processing and management).

20.3.3.1 Big Data Architecture Capabilities

- Database capability
  Oracle NoSQL
  - Efficiently processes data without a row-and-column structure
  - Highly scalable multinode, multiple datacenters
  - Not limited to SQL
  Apache Cassandra
  - Fault-tolerance capability
  - Replicates across multiple datacenters
- Statistical analysis capability
  Open-source project R
  - Programming language for statistical analysis
- Processing capability
  MapReduce
  - Problems are broken down into some several smaller problems.
  - Workloads are distributed across thousands of nodes.
  - Can be exposed via SQL.
- Storage and management capability
  Hadoop distributed management systems (HDFS)
  - An open-source distributed system
  - Automatic data replication
- Data integration capability
  Oracle Big Data Connectors
  - Connects Hadoop to relational databases.
  - Has the ability to optimize data processing with parallel data importation.

In modern architecture, basically, structured data are involved and used only for reporting and analyzing purposes. There are one or more unstructured sources present that contribute to a very small portion of the overall data. In big data architectures, there are different types of sources involved; each of the sources provides data in different formats, at different intervals, and with different volumes.

Source system: Various sources are involved in the architecture of big data, such as enterprise, social media data, activity-generated data, public data, archive files, structured and unstructured sources, and so on.

Transactional system: Usually, there is one transactional system used for mission-critical applications as a backend database.

Big data engine: It plays a major role in the data processing and management architecture. The engine can process large volume of data, ranging from a few megabytes to hundreds of terabytes. The engine consists of a Hadoop framework that allows distributed processing of heterogeneous data sets across clusters of computers.

20.3.4 APPLICATIONS OF BIG DATA

Big data environments are growing fast and are becoming more dynamic and complex, resulting in poor visibility, performance, and reduced control over applications. Big data are used to describe multiterabytes of data sets [15,16]. Social media channels, websites, automatic censors at the workplace, and robotics produce a plethora of structured, semistructured, and unstructured data. There are some application areas where big data is already making a real difference today:

1. *Targeting customers*: It is one of the most publicized areas of big data today. Big data is used to understand customer behavior and preferences. Companies look for intelligent ways to expand their traditional data sets into social media data to get a more complete picture of their customers.
2. *Personal quantification*: Big data is not only for companies and government but also for general, individual users.
3. *Optimizing business process*: Increasingly, big data is used to optimize business processes. Retailers are able to optimize their stock based on predictions generated from social media data, web search trends, and weather forecasts. For example, geographic positioning and radio frequency identification sensors are used to track goods or delivery vehicles. This improves the HR business process.
4. *Improving health care*: Over the last hundred years, advances in the diagnosis, treatment, and prevention of disease have dramatically improved the quality of patient care. Today, health care has become increasingly digitized. Big data analytics helps to reduce hospitalizations by providing physicians with better technology and more detailed information about each of their patients. Using these resources, physicians are able to treat their patients appropriately to ensure best outcomes. So big data is drastically humanizing the quality of health care.
5. *Improving sports performance*: Big data analytics have aided sports teams and athletes to enhance performance and develop relationships with their fans. The match analytics tool collects and analyzes the massive amount of players' performance data, captured during trainings through the cameras installed on the field or sensors placed in players' clothes and cleats, which is then run in the SAP Hana platform. The SAP tools allow the analysis of a range of factors concerning players' performance such as speed, kick power, the number of successful passes of a footballer.
6. *Big data service refinery*: "Big data refinery" is a new system capable of storing, aggregating, and transforming a wide range of multistructured raw data into usable formats that help in providing new insights for the business. Big data refinery provides a cost-effective platform for unlocking the potential value of the data and discovering business questions that are worth answering with the help of these data.
7. Optimizing machine: Big data analytics help machines and devices become smarter and more autonomous. Big data tools are used to operate Google's self-driving car. They are also used to optimize energy grids using data from smart meters.

8. *Improving security*: Big data is applied to improve security and enable law enforcement. Police forces use big data tools to catch criminals and even predict criminal activity, and credit card companies use big data to detect fraudulent transactions.
9. *Financial trading*: Big data algorithms are used to make trading decisions. The majority of equity trading now takes place via data algorithms that increasingly take into account signals from social media networks and news websites, and it is done to make, buy, and sell decisions within seconds.
10. *Data warehouse optimization*: The term how an organization optimizes its data storage, processing for cost, and performance while data volumes continue to grow from an ever-increasing variety of data sources. Data warehouses reach their maximum capacity too quickly as the demand for more data and more types of data forces IT organizations to invest in expensive upgrades.

20.3.5 ADVANTAGES OF BIG DATA

Data are now woven into every sector. The use of big data is becoming a crucial factor in leading companies. Big data can add value to almost any industry (government and private) sector, enabling new insights and enhancing decision support for a wide variety of projects. Some examples of industry sectors that can benefit from big data: social media and search engines, IT operations, cloud computing, insurance, banking, and finance, telecommunication, Internet service providers, mobile service providers, health care, research and development, marketing, retail and point-of-sale analytics, manufacturing optimization, utility and energy, and transportation [16–19]. Customers can benefit by understanding the utility of big data that can be used for industry-specific purposes to extract actionable value from analyzed data sets. The application of big data is an emerging field. Use cases of big data in some industries are described as follows:

* *Research and development (R&D)*: To comb through enormous volumes of text-based research and other historical data, many companies (such as pharmaceutical manufacturers) are using Hadoop.
* *System usage*: Lightweight Directory Access protocol that generates large amounts of data to illuminate trends and enable enterprises to plan better. Using these data, operations staff can monitor how the subsystems are behaving and establish rules and policies to respond to usage thresholds. Operations staff can optimize system utilization by tracking spikes.
* *Data accumulation and archiving*: Data archiving is the process of identifying and moving inactive data out of current production systems into specialized long-term archival storage systems. It is essential for organizations that accumulate new information but still need to retain older information.
* *Sentiment analysis and customer churn management*: Companies can use voice analytics and text analytics based on voice modulation and keyword analysis to understand customer sentiments better.
* *Improve delinquent collections*: Prepaid phone services are able to gain popularity. Basically, the services are delivered on a post-pay basis, which means that with widely varying credit histories, providers can extend credit to tens of millions of customers. High-performance analytics from SAS have transformed companys' delinquent collections process to increase collection rates, decrease churn, and increase lift from *next-best offers*.
* *Network optimization to prevent failure*: Big data technologies are used to analyze any type of networks.
* *Root-cause analysis*: If there is a large system failure, the root cause can be unclear. Big data is collected from numerous sources, including process control machinery, supply chain management systems, and performance monitoring programs, that track products already sold. It optimizes the supply chain using real-time actionable insights about customers and

suppliers. It prevents failures by using predictive root-cause analysis on machine data. It detects fraud before incurring losses by applying predictive analytics to expose patterns and anomalies.

- *Data preservation*: This helps in searching the right information and discovering trends. It basically focuses on collecting ("ingestion"), maintaining ("curating"), and providing access (building open-access "archives") to large time-based data sets, for example, data sets consisting of trace data of large distributed systems and applications.

- *Information security*: This includes malware and fraud detection. Big data security can also be seen as an opportunity. While multiple distributed systems and data sets may provide smaller, less appealing targets, organizations are often inconsistent in the management and application of security to these data sets. A more centralized, consolidated model of big data offers organizations a chance to revisit data security and improve on their legacy policies.

- *Recommendation engine*: App store and e-commerce. Recommendation engines can be used anywhere. Users are looking for products/services or people. Recommendation engines require vast amount of data from different data sets that may be useful.

- *Predictive analysis*: Predictive models and analysis are typically used to forecast future probabilities. They are also used to analyze current data and historical facts in order to understand customers, products, and partners better and to identify potential risks and opportunities for a company.

- *Real-time decision-making and scenario tuning*

- *Perform risk analysis*: Social and economic factors are crucial for the success of any business. In predictive analytics, big data allows scanning and analyzing newspaper reports or social media feeds. Detailed data on health tests of suppliers and customers are another benefit of using big data. It helps in taking actions when one predicts any issues with suppliers or customers.

- *Customize your website in real time*: Big data analytics allow personalizing the content or look of a website in real time to suit each customer visiting the website. A well-known example is Amazon's use of real-time, item-based, collaborative filtering to fuel its "Frequently bought together" and "Customers who bought this item also bought" features; Amazon generates about 20% more revenue via this method.

- *Keep the data safe*: Customers will be able to detect sensitive information that is not protected in an appropriate manner. With real-time big data analytics, anyone can flag up any situation where 16-digit numbers (credit card data) are stored or e-mailed out and investigated accordingly.

- *Tailored health care*: We now live in a hyper-personalized world, but the health care sector still seems to be lagging behind in using generalized approaches. When someone is diagnosed with cancer, the doctor follows one therapy to treat the patient, and if it does not work, they try another. With big data tools, it will soon be boon for everyone to have their genes mapped as part of their medical record, which will bring the medicine closer than ever and help find the genetic determinants that can cause a disease and develop a personalized medicine.

- *Make smarter city*: An increasing number of smart cities are indeed leveraging big data tools for the benefit of their citizens and the environment. For example, in Norway, the city of Oslo reduced street lighting energy consumption by 62% with a smart solution. To optimize the timing of its traffic signals, Portland, Oregon, used the smart technology and was able to eliminate more than 157,000 metric tons of CO_2 emissions within 6 years.

- *Speed, capacity, and scalability of cloud storage*: Internally, companies are analyzing massive data sets without making a significant capital investment in hardware to host. The IT industry has realized that new skills and training are required for big data hosting platforms.

- *Redevelop products*: Big data can also help understand how others sense products. Analysis of unstructured social media text reveals the sentiments of customers and segments them based on their geographical locations.
- *To develop the next generation of products and services*: Manufacturers use data obtained from sensors embedded in products to create innovative after-sales service offers.
- *New strategies of competition*: With real-time big data analytics, one can stay a step ahead of the competition.
- *Cost savings*: Real-time big data analytics tools may be expensive but, eventually, it will save a lot of money.
- *Unlock significant value by making information transparent*: There is a significant amount of information that is not captured in digital form. This represents a significant source of inefficiency.
- Sophisticated analytics can substantially improve decision-making and minimize risks.

20.4 BIG DATA ANALYSIS AND CYBER-PHYSICAL SYSTEMS

Control and analyses of cyber-physical and robotics systems is increasingly becoming a big data challenge. Big data in CPS are divided into two types: sensor data and derived data. The data issued from the sensors are known as sensor data and those that are computed from the sensor data are derived data [20]. A big data driven by a CPS is designed to specify, model requirements, implement these requirements on big data platform, and guarantee quality of service (QoS) in a highly scalable manner.

20.4.1 OVERVIEW

The collection and subsequent analysis of a significantly large collection of data can be referred to as big data, which may contain hidden insights or intelligence (user data, sensor data, and machine data). Big data supports novel extensions for durable, named solution sets, efficient storage and querying of reified statement models, and scalable graph analytics. Big data enables new business insights that help with diversification and opening new markets, creating competitive advantages. Big data database supports multitenancy. The data can be deployed as a stand-alone server, a highly available replication cluster, an embedded database, and a horizontally shared federation of services similar to Google's Bigtable and Apache Cassandra [21].

A CPS is controlled by embedded software [22,23]. CPS are at the heart of our critical infrastructure and form the basis of future smart services. They use computations and communication that are deeply embedded in and interact with physical processes to add new capabilities to physical systems. The system increases efficiency and interaction between computer networks and the physical world, enabling advances that improve the quality of life. The system is also improved to enable personalized health care, traffic flow management, emergency response, and the electric power generation and delivery. CPS range from miniscule (pacemakers) to large scale (the national power grid). In a CPS, the physical world integrates with sensing, communication, and computing components.

Computing systems are being increasingly used to gather data from physical infrastructures related to individuals, enterprise, and communities, and to make such infrastructures intelligent through powerful analytics of the data. One type of ultra-large system (ULS) is a CPS. An ULS includes ultra-large scale along many dimensions. It includes interdependent webs of software-intensive systems, people, policies, cultures, and economics.

Naturally, the sensing component of a CPS is critical for the modeling and management of the CPS because it provides real operational data. The goal of sensing is to provide high-quality data with good coverage of all components at a low cost. Today, low-cost ubiquitous sensing is causing a paradigm shift from resource-constrained sensing toward using big data analytics to extract information and actionable intelligence from massive amounts of sensor data. The availability of big monitoring data on CPS creates challenges for traditional CPS modeling by breaking some of the assumptions and also provides opportunities to simplify the CPS model identification task, achieve proactive maintenance, and build novel applications by combining CPS modeling with data-driven learning and mining techniques.

20.4.2 ARCHITECTURE OF CYBER-PHYSICAL SYSTEMS WITH BIG DATA

Device architectures are highly proprietary, noncommunicable, and reliable in providing inputs and assessing outputs. Usually, complex instruction set computer and reduced instruction set computer architectures are used. Multicore, system-on-a-chip architectures and flexible reconfigurable architectures are becoming more common in device designs. Nowadays, emerging medical device architectures provide wired and wireless interfaces to facilitate networked communication (patient). Participants' reports suggest that all the medical data within devices be aggregated, but traditional devices are designed to operate separately. Multiple cyber-physical devices enable interoperation to optimize patient care, with a very high level of confidence. It improves patient care and treatment and facilitates and accelerates aggregation of patient information.

CPS architecture in health care is essential for the quality and performance of the system. In order to expand CPS application in medical devices, the architecture is to be designed based on the application domain, user data requirement, and system integration. The architecture of medical CPS (MCPS) can be categorized into three elements: (1) infrastructure, (2) composition, and (3) data requirement.

Infrastructure: The architecture for the medical device should be designed from the perspective of the infrastructure (it may be server based or cloud based). For small architectures, server-based infrastructure is appropriate, and it requires individual maintenance. On the other hand, recently, cloud-based architecture is utilized for scalability, cost-effectiveness, and accessibility.

Data requirement: To manage the MCPS, different types (input data, output data, and historic data) of data are required. Size of the data used depends on the type of biosensors used. Both simple data (temperature) and large image data (magnetic resonance imaging) are required for processing. Examples of low data rate application are blood pressure and temperature, and for high data rate application high-density surface electromyogram.

Composition: In MCPS, both computation and communication processes are conducted in parallel. According to the application, the system architecture of CPS in health care is able to identify the system composition. The system supports the interconnection between the devices and their software applications (Figure 20.7).

Medical records are generally crucial and must be protected by highly secure physical facilities, multi-layered software or hardware network security systems and data encryption and redundancy. Operation of in-facility server rooms is expensive but less secure to process and store medical data. Some medical imaging devices generate extremely massive data. Many uncertainties during transferring data between various networks pose serious issues with using public cloud solutions for MCPS.

20.4.3 CHALLENGES OF CYBER-PHYSICAL SYSTEMS WITH BIG DATA ANALYSIS

Advancements in CPS continue to be challenged by a variety of technical (i.e., scientific and engineering), institutional, and societal issues. The objective is to develop a new system of science and engineering methods to build highly confidential systems in which cyber and physical designs are compatible, synergistic, and integrated at all scales. A number of barriers hinder the progress of designing, developing, and deploying CPS in health care and in other application domains. The big data paradigm shift, enabled by low-cost, ubiquitous sensing, presents new challenges in CPS.

1. *Knowledge of input–output requirements*: Sensor data from a CPS is in the form of time series, and today, it is not uncommon to encounter data sets with thousands of metrics.

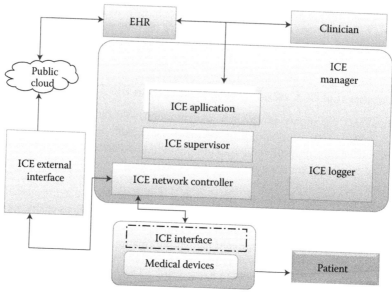

FIGURE 20.7 Architecture of a medical cyber-physical system.

In distributed computing systems, several factors make it impossible for humans to understand the entire system: large scale, complex interaction across components, the demand to deploy a heterogeneous class of applications using limited resources while still maintaining high QoS, etc.

2. *Online processing of high-dimensional and low-quality data*: With the rise of industrial big data and smart infrastructure (cities, transportation, homes, etc.), CPS modeling across multiple domains has become a challenge. An interdisciplinary approach that includes designing novel algorithms to model and build systems is required to tackle this challenge.

3. *Biomedical and health care systems*: Numerous challenges in medicine and biomedical engineering are accepted and resolved by CPS research. These include intelligent operating rooms and hospitals, image-guided surgery and therapy, fluid flow control for medicine, and biological assays. CPS are also applied in operating room devices such as infusion pumps for sedation, ventilators and oxygen delivery systems for respiratory support, and a variety of sensors for monitoring patient condition.

4. *Medical device interoperability*: Every medical device has a different communication interface. To integrate heterogeneous medical devices in a safe, secured, and certified manner, there must be a well-maintained system that manages their interoperability.

5. *Security and privacy*: The System security is critical as individual patient data is communicated over CPS networks and illegal use of patient data may cause mental unrest and abuse leading to further physical illness and may hamper reputation and lead to loss of privacy.

6. *Autonomy*: The Medical Cyber-Physical System (MCPS) can be used for increasing the autonomy of the system. The autonomy of the system can be increased by enabling actuation of therapies, and it can be done based on the patient's current health state. Closing the loop in this manner must be done safely and effectively.

7. *Software reliability*: In medical devices, software plays an important role. Traditionally, many functions, including safety interlocks, are implemented in hardware. Device functionality and multiple functions are ensured via software. Software ensures proper cooperation between medical devices and patients. Therefore, safety and efficiency of the system depend on proper software design, development, and management.

8. *Certifiability*: A cost-effective approach is required to handle the complex and safety-critical nature of MCPS. It is required to prove medical device software dependability. Therefore, certifiability is an essential for eventual viability of MCPS.
9. *Data extraction*: Medical devices collect multiple physiological parameters from patients. These parameters vary widely and can provide general information about patients. They can predicts illness, possible nullification of the emergency situation of a patient. Design this type of system that can seamlessly extract complex physiological parameters from patients is considered a challenge.
10. *Lack of prototype architecture in CPS for health care*: Nowadays, lack of secure and trustworthy prototype architecture is absent for testing, evaluation, and system developments that include health care devices. For this, in uncertain environmental scenarios, there is inability to ensure correctness of CPS in health care architecture.

20.4.4 ADVANTAGES OF CYBER-PHYSICAL SYSTEMS

In the physical world, CPS are computers that perform functions in real time as a result of stimuli. They are prevalent in various parts of everyday life. Cyber-physical technology can be applied in a wide range of domains, offering numerous advantages in products. Some of the advantages are as follows:

1. *Interaction between human and system*: For decision-making, considering environmental changes in fixing parameters is critical. This is an absolute necessity for complex and dynamic systems. Some CPS include humans as an integral part of the system. This makes the interaction (between human and system) easier, because, basically, humans pose difficulties in modeling using stand-alone systems.
2. *Better system performance*: CPS are capable of providing better system performance in terms of feedback and automatic redesign for the close interaction between sensors and cyber-infrastructure. In a CPS, better computational resources and cyber-subsystems ensure the presence of multiple sensing entities, multiple communication mechanisms, high-level programming language, and end-user maintenance that also ensures better system performance by the CPS.
3. *Faster response time*: Due to the fast communication capability of sensors and cloud infrastructure, a CPS can provide faster response times, which in turn facilitates early detection of remote failure and proper utilization of shared resources. Example: bandwidth.
4. *National competitiveness*: An example of the advantages is increased energy efficiency that the CPS can bring to society. The value of both the physical and the *cyber* or IT portions will be increased by the product. The design and manufacturing base and the talent base are critical for national competitiveness.
5. *Uncertainty*: Uncertainty is a key ingredient with a high degree of interconnectivity of *emergent* behavior. It also provides motivation for a large degree of coupling in CPS. The motivation is to react to customer needs and desires or to environmental issues.
6. *Scalability*: CPS are capable of scaling the system according to the demand utilizing the properties of cloud computing. Users can acquire the necessary infrastructure without having to invest in additional resources. They combines physical dynamics with computational processes. Networking infrastructure, programming tools, and software modeling is combined by the cyber-domain. Mechanical motion control, chemical processes, biological processes, and human involvement are combined by the physical domain.
7. *Dealing with certainty*: Certainty is that which provides proof that a design is valid and trustworthy. A CPS is capable of demonstrating unknown system behavior that it can study and evolve into a better system.

8. *Capability*: A CPS allows us to add capabilities to physical systems. Complex systems with new capabilities can be built with the help of a CPS.
9. *Computing and communication with physical processes*: It is the safest and most efficient system for computing and communication that reduces the cost of building a separate operating system to perform such functions.
10. *Technological and economic drivers*: Computers and communication are ubiquitous and enable national or global-scale CPS. Social and economic forces require more efficient use of the national infrastructure.

20.4.5 REAL-LIFE APPLICATIONS OF CYBER PHYSICAL SYSTEMS WITH BIG DATA ANALYSIS

The technology in medical devices evolved in tandem with the overall technology innovation of various other industries and the establishment of engineering best practices, like transitioning from vacuum tube electronics to transistor-based electronics and from metal to plastics. Over the last 20 years, designing of medical devices has evolved from analog to digital systems [22,23]. Today, software, microprocessor, sensor, and actuator technologies are ubiquitous in these devices. Most devices contain embedded systems that depend on some components, such as a combination of proprietary, commercial-off-the-shelf (COTS), and custom software or software of unknown pedigree (SOUP). An embedded system can be thought of as a special-purpose computer system often designed to perform dedicated functions and to have resource-limitation constraints, which form part of a mechanical device. Embedded system programs are also implemented in read-only memory and generally are not intended to be reprogrammable (Figure 20.8).

Globally, intelligent transportation plays an important role in reducing risks, accident rates, traffic congestion, carbon emissions, and air pollution and alternatively in increasing safety, reliability, traveling speed, and traffic flow. In metropolitan deployments, intelligent transportations comprise of freeway management systems, freight management systems, incident and emergency management systems, traveler information systems, and information management systems. Many applications in an Intelligent Transportation System (ITS) play a significant role, for example, electronic toll

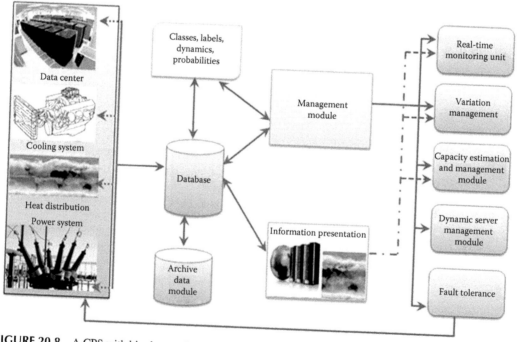

FIGURE 20.8 A CPS with big data analysis.

collection, vehicle data collection, highway data collection, and transmit signal priority. Different types of transmissions are employed in these applications. Some applications are for long-distance transmission, some for short-distance, and others for radio modem transmission for the collection of computerized information. An example of a CPS in this area is the air transportation network that is coordinated over long distances. Air traffic density is expected to increase so it requires a hierarchical suite of network management tool to maximize efficiency and maintain safety on earth and in air.

20.4.5.1 Medical Cyber-Physical Systems (MCPS)

A MCPS integrates the notion of combining the aspects of both the physical world and the cyberspace. It refers to the systems that have a combination of embedded devices, software for controlling those devices, and a communication channel for interaction. Systems work with a set of embedded devices and those devices are controlled by the cyberspace with the help of a set of commands and control statements. Any change in the physical world directly affects the cyberspace. MCPS are commonly used in health care-related applications. New design, verification, and evaluation techniques are required to develop safe and effective MCPS due to increase in size and complexity [24,25] of the system. Lots of devices are used in MCPS. Based on their primary functionality, such devices can be categorized into two broad categories:

1. *Monitoring devices*: These devices provide different kinds of clinic-relevant information about patients, such as bedside heart rate and oxygen level monitors and sensors.
2. *Delivery devices*: These devices provide information on the actuate therapy, such as infusion pumps and ventilators, that is capable of changing the patient's physiological state (Figure 20.9).

Administrators control the information in electronic health records (EHRs), the availability of medicines in the pharmacy stores, and the records of the patients and other accessories, and so on. Decision support entities can process the data collected and generate alarms for various medical emergencies. Alarms are required to inform the clinicians about the patient's condition. However, smart alarm systems have been developed that go beyond the current threshold with methods to provide more accuracy so that the information can be analyzed by the caregivers. Delivery devices can be used to initiate treatment, thus bringing the caregiver into the control loop. In other words, decision support entities can utilize a smart controller to analyze the data received from such monitoring devices, estimate the state of the patient's health, and automatically initiate treatment (e.g., drug infusion) by issuing commands to delivery devices, thereby closing the loop.

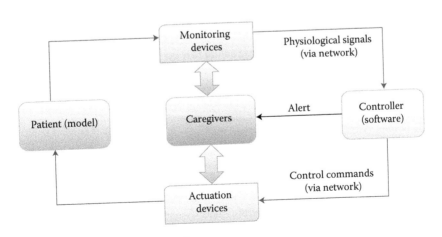

FIGURE 20.9 Working principle of an MCPS.

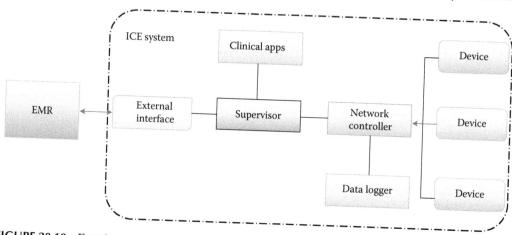

FIGURE 20.10 Functional architecture of an MCPS.

Medical device interoperability has the potential to reduce health care costs, and improve treatment outcome and patient safety. Medical devices (software, applications) and other equipments share the same information model and communication protocol that is required to achieve interoperability. They can work with any compatible data regardless of its source, such as the manufacturer or specific device type. If a system uses different types of medical devices, then the system is known as a distributed device. Technical requirements are used for verification ("the system was built right"), and user requirements are used for validation ("you built the right system"). Intelligent operating rooms and hospitals, image-guided surgery and therapy, fluid flow control for medicine and biological assays, and the development of physical and neural prostheses are included in MCPS or health care systems (Figure 20.10).

Clinicians use applications that interact with medical devices and run on a Supervisor. Patient-connected medical devices and other equipments are connected to the system via adapters or a built-in ICE equipment interfaces. The network controller ties the devices to the Supervisor and sends data to the data logger, which functions analogously to the black box on airplanes. The ICE system communicates with outside resources like an EHR system, physician order entry system, or pharmacy system through external interfaces. The patient and clinicians are key elements to the system, and this is intended explicitly to improve patient's safety and treatment outcomes. Middleware can be used to implement this functional architecture by taking over the responsibilities of the network controller.

20.4.5.1.1 Medical Cyber-Physical Systems for Big Data Analysis

Medical systems are integrated with different classes of devices so that various kinds of functions can be performed in real time. A good infrastructure is required to store and process the data, which becomes complex in computation. It is necessary to improve the service functionality of today's health care system [26]. To perform various functions such as filtering, listening, processing, and accelerating, systems require different types of subsystems. In health care systems, the body of a patient is connected to various sensors to measure different physiological data such as ECG and oxygen and hemoglobin levels. These data are then sent to the remote application server without any loss of information at the receiving center. It would help the doctor to analyze the data and take a decision for the patient before the condition becomes severe. Storage systems can store a huge amount of data (both significant and nonsignificant data) daily. With increased processing time, the system requires a large amount of storage capacity that will reduce the performance of the system. The system lacks flexibility and cannot differentiate the cyberworld and the real world. In health monitoring, the system is designed to operate in a controlled environment with a set of predefined rules and semantics. When video-monitoring a patient, the system will be able to track objects.

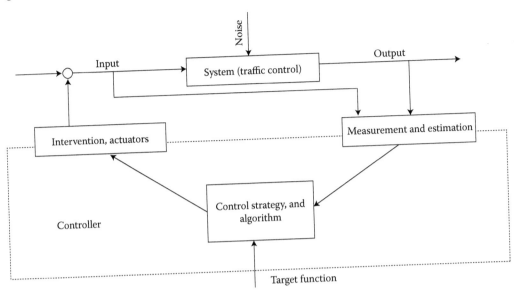

FIGURE 20.11 Intelligent transportation system.

20.4.5.2 Intelligent Transportation Systems

Another important real-life application of CPS with big data analytics is the ITS. The developments in transportation have drastically changed the way it works. It now utilizes new technologies like positioning systems, sensor technologies, telecommunication, virtual operation, planning techniques, and data processing. An ITS provides solutions for the corporation and reliable platform. It is not only for highway traffic but also for services in navigation systems, railway systems, water transport systems, and air transport systems. ITSs are divided into generations [27–29] (Figure 20.11):

Periods	Generation	Technology
2000	First generation (ITS 1.0)	One-way infrastructure
2000–2003	Second generation (ITS 2.0)	Two-way communication
2004–2005	Third generation (ITS 3.0)	Automated vehicle operations and interactive system operations
2006–2011	ITS (ITS 4.0)	Multimodel incorporating personal mobile devices, vehicle, infrastructure, and information networks for system operations

20.4.5.2.1 Freeway Management Systems

Globally, traffic jams have been one of the most serious problems due to increased population, poor infrastructure, and defective traffic systems. In many developed countries, freeway management systems are used to solve the problem. An example of this system is the California Freeway Management System. Millions of unprocessed data are collected and stored by the system, including traffic flow data and speed records of vehicles. The system works with the help of hardware and different technologies like ramp meters, circuit televisions (CCTV), and traffic signal control systems [30–32]. It also provides navigational services. Traditionally, freeway systems use fixed sensors such as loop detectors and television cameras. There are many problems like high cost and maintenance. Nowadays, increasingly, GPS, mobile phones, and mobile sensors are used in freeway management systems (Figure 20.12).

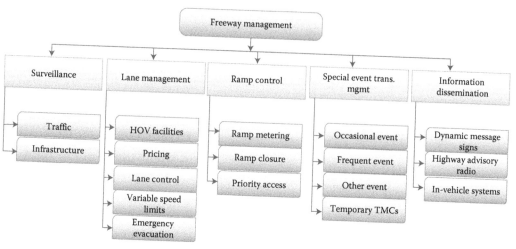

FIGURE 20.12   Freeway management system.

20.4.5.2.2 Transmit Management Systems

A transmit management systems provide accurate information about the satisfaction safety, position, and security of the traveler. It is used to increase ridership and enhance service reliability and operating efficiency [34–37]. The user or customer has the right to use the information, and so on. The information may be operational, schedule related, maps, and other general data. Information about arrival or departure times can also obtained by the customers.

For example, automatic vehicle location provides a foundation for vehicle tracking (Figure 20.13).

20.4.5.3 Incident Management Systems

Incident management systems play a major role disaster management, such as the 2008 Sichuan earthquake, the September 11 attack, and road accidents. In these situations, the main objective of the system is to provide safety to the victims with the help of fire vehicles and ambulances. It has many subsystems such as incident detection, traffic management, and roadway incident management (Figure 20.14).

20.4.5.4 Emergency Management Systems

An emergency management system deals with risk and its remedies. The emergency management information system monitors and computes the information obtained from the environment. Decision support systems help in making decisions during crises in the transportation systems (Figure 20.15).

20.4.6 APPLICATION OF INTELLIGENT MANAGEMENT SYSTEMS

Intelligent management systems are used for transportation safety, efficiency, and user services. Some of the applications of an intelligent management systems are described in the following:

Electronic toll collection: One of the most popular applications of intelligent management systems is electronic toll collection. It eliminates delays in the transportation and enhances the mechanism of collecting tolls. An electronic toll collection systems can be classified into two:

1. *Dedicated short-range communication*: In this system, and onboard unit is placed in the vehicle.
2. *Roadside units*: These are installed on the road, and communicate with each other within a range of 30 miles (Figure 20.16).

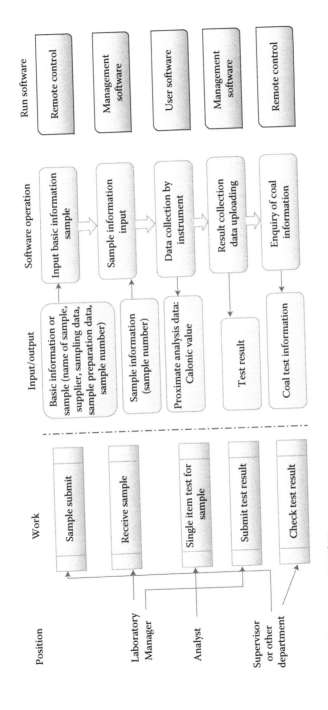

FIGURE 20.13 Transmit management system.

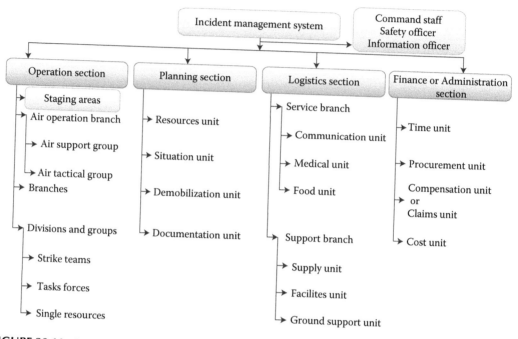

FIGURE 20.14   Incident management system.

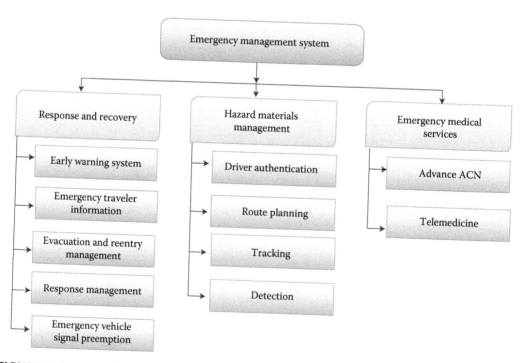

FIGURE 20.15   Emergency management system.

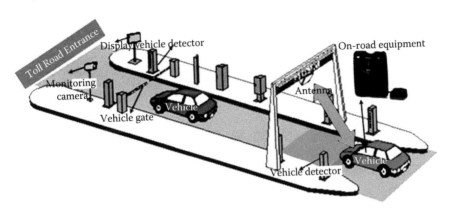

FIGURE 20.16 Electronic toll collection.

Some of the devices used in electronic toll collection include:

Toll lanes

- Tag sensors
- Automated coin baskets
- Toll booth attendants
- Loop sensors
- Axle sensors
- Weigh-in-motion scales
- Video cameras
- Traffic signals

Vehicles

- Active tags
- Passive tags

Highway data collection: In intelligent management systems, the highway data collection technology enables recording of road traffic data, such as weather, wind velocity data, complex terrain, and reliable safety of vehicles. Positioning and communication systems work with the help of sensors on the road.

Vehicle data collection: A vehicle data collection system collects data related to the quality and performance of vehicles for processing, analyzing, and remote monitoring. Application of such systems provides support for vehicle tracking, military, engineering, and ground durability testing. See, for example, the Georgia Tech Trip Data Collector in Figure 20.17.

Transmit signal priority: A transmit signal priority system makes transit services faster, reliable, inexpensive, and cost effective. The objective of the system includes efficient schedule, optimized travel time, and uninterrupted traffic movement through the control of traffic signal intersections.

Traffic management systems: Traffic management plays a major role in transportation management systems. The overall efficiency of the traffic management system is to improve safety and better mobility, flow, and economic productivity. It collects data from various hardware devices (camera, sensor, etc.), and feeds them into a traffic management centers where the data are processed and analyzed [38,39] (Figure 20.18).

FIGURE 20.17   Vehicle data collection.

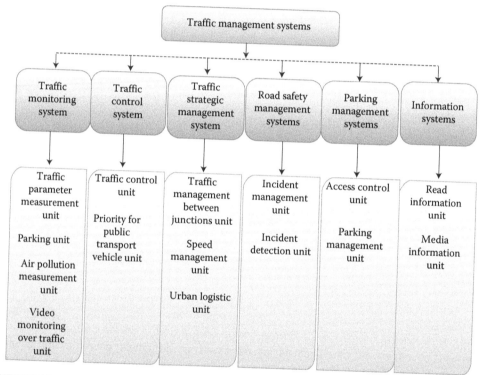

FIGURE 20.18   Traffic management system.

A traffic management system includes:

- Traveler information system
- Traffic control system
- Video control system
- Event management system

Examples: Railway traffic management, road traffic management in cities, and aircraft traffic management.

20.4.7   TECHNOLOGIES USED IN INTELLIGENT MANAGEMENT SYSTEMS

Current and growing communication technologies are integrated by ITS.

FIGURE 20.19 (a) Traffic jams. (b) Traffic management.

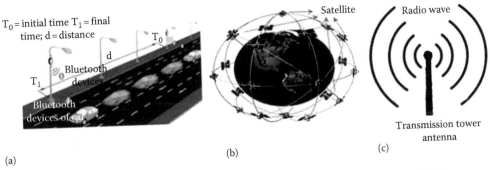

FIGURE 20.20 (a) Bluetooth detection. (b) Satellite communication. (c) Wireless communication.

The transportation system is capable of improving transportation conditions that include safety and services (Figures 20.19 and 20.20):

- Wireless communications (use radio waves, microwaves as a medium)
- Floating car data/floating cellular data (works on different types of networks such as UMTS, GRPS, CDMA, GSM)
- Sensing technologies (microchip, RFID, etc.)
- Inductive loop detection (three components: loop, detector, loop extension cable)
- Computational technologies
- Video vehicle detection
- Bluetooth detection

20.5 CONCLUSION

CPS play a major role in the design and development of future engineering systems with new capabilities that far exceed today's levels of autonomy, functionality, usability, reliability, and cybersecurity. The potential of CPS to change every aspect of life is enormous. Concepts such as autonomous cars, robotic surgery, intelligent buildings, smart electric grid, smart manufacturing, and implanted medical devices are some of the practical examples. The *big* monitoring data collected from CPS using ubiquitous sensing with respect to CPS modeling and analytics have a great impact on the environment. We discussed new challenges for CPS modeling and opportunities related to CPS management and novel applications arising from big data. The challenges facing MCPS are formidable, yet they give vast opportunities for research and immediate practical impact. We identified major challenges in MCPS development and discussed promising research directions that will help to overcome these challenges.

KEY TERMS AND DEFINITION

1. *Cyber*: In science, cyber is a prefix that means "computer" or "computer network" in *cyberspace*, which is an electronic medium for online communication. Common usages of cyber are cyber-culture, cyberpunk, and cyberspace. Cyber is a combining form that represents computer network (cybertalk, cyberart, cybercafe).

   To protect networks, computers, programs, and data from attack, damage, or unauthorized access, cybersecurity is implemented. Cybersecurity is the body of technologies, processes, and practices designed to protect computer/online networks. The elements of cybersecurity systems are information security, application security, network security, and end-user education. The evolving nature of security risk is one of the problematic issues facing cybersecurity. A proactive and adaptive approach is promoted to deal with the current environment and advisory organizations.

   Cyberphobia: An irrational fear of computers.

   Cyberpunk: A genre of science fiction that draws heavily on computer science ideas.

   Cyberspace: The nonphysical terrain that is created by computer systems. Anything related to the Internet also falls under the cyber category.

   Cyber-bullying: The bullying that takes place using electronic technology. Communication devices such as cell phones, computers, and tablets and social media like text messages, chat, and websites are included in electronic technology. Examples of cybersecurity are text messages; e-mails; rumors sent by e-mail or posted on social networking sites; inappropriate pictures, videos, and websites; and fake profiles. Social media sites can be used for positive activities such as connecting with friends and family, helping students with education, and for good entertainment.

2. *Google Bigtable*: Google Bigtable is a high-performance, compressed, distributed, and proprietary data storage system built on the Google file system, Chubby lock service, and a few other Google technologies. It is used by over 60 Google products. It is designed for massive information databases such as billions of URLs that are frequently visited and that have more than 100 TB of satellite image data.

   Characteristics of Bigtable:

   - Map: A map is a set associative arrays.
   - Persistence: Data are persistently stored on disks.
   - Distribution: Bigtable is distributed among many independent machines.
   - Storage: Bigtable stores data by keys, usually in the same machine, to keep related data close together.

   Features of Google Bigtable:

   - A distributed storage system for managing structured data
   - Scalable
     - Terabytes of in-memory data
     - Petabyte of disk-based data
   - Self-managing
     - Servers can be added/removed dynamically
   - The map is indexed by
     - A row key
     - A column name
     - A timestamp
   - Column oriented
   - Dynamic control over data layout and format
   - Data in uninterrupted strings
   - Dynamic control over serving data from memory or disk

3. *Virtual medical device (VMD)*: A set of medical devices coordinating over a network in a clinical scenario. The Medical Device Coordination Framework (MDCF) is a prototype

middleware that manages the correct composition of medical devices and transfers them into the VMD.

Research issues in VMD:
- Real-time support
  - Leverage current hospital networks
- Noninterference
  - Assume guarantee interface
- Development environment for VMD Apps
  - Support for programming clinical algorithms with timing constraints
- MDCF platform implementation
  - Device connection and configuration protocols
  - VMD setup/tear-down algorithm
- Safety analysis of the platform
  - Guarantees of communication

4. *Requirements for VMD and medical device capabilities.*
- Automatic device
  - Ensures correct devices used in any given VMD
  - Reduces scope of standardization efforts to manageable size
- Precise VMD development artifact
  - Feed into VMD simulation (i.e., testing)
  - Feed into verification (i.e., model checking)
- Formal semantics
  - May, must, at least one of transitions
  - Refinement relations between specification and implementation

5. *Vasospasm decision caddy*: Vasospasm decision caddy is a postbrain surgery risk. It is very difficult to diagnose and if not treated, can be perilous. It provides some information such as contexts for alarms.

Its feature include
- Three-pronged approach
  - Guideline driven
  - Physician driven
  - Data driven
- Current deployment barriers
  - Few real-time data stream feeds
  - No interfacing of streams to the systems
- Analyze data in new ways
  - New device sources
  - Trending
  - Clinician provide data
  - Interpolate missing data

REFERENCES

1. R. Baheti and H. Gill, Cyber-physical systems, Retrieved from http://nsf12520.htm, 2011.
2. R. Rajkumar, I. Lee, L. Sha, and J. Stankovic, Cyber-physical systems: The next computing revolution, in *Proceedings of the 47th Design Automation Conference*. ACM, pp. 731–736, 2010.
3. G. Santhosh Kumar, Sureshkumar, Research Direction in Cyber Physical Systems, pp. 494.
4. Dr. Meeko Oishi, Introduction to Cyberphysical Systems. January 21, 2014.
5. Y. Tan, S. Goddard, L.C. P´erez, A prototype architecture for cyber-physical systems.
6. Cyber-physical Systems, Retrieved from http://cyberphysicalsystems.html.
7. R. Rajkumar, Cyber physical systems: A natural convergence of engineering and computer science. Distinguished Lecturer Series- A lecture videocast from the computer science department at UNC, September 28, 2009. Available: Retrieved from http://www.youtube.com/watch?v=7wvq7pVjdJA.

8. Radhakisan Baheti and Helen Gill, 2011. *NSF Workshop on Cyber-Physical Systems*, Retrieved from http://varma.ece.cmu.edu/cps/, October 2006.
9. W. Wolf, Cyber-physical systems, *IEEE Computer*, 42, 8–89, March 2009.
10. P. Derler, E.A. Lee, A.S. Vincentelli, Modeling cyber-physical systems, *Proceedings of the IEEE*, 100(1), 1–28, 2012.
11. I.F. Akyildiz, I.H. Kasimoglu, Wireless sensor and actor networks: Research challenges, *Ad Hoc Networks*, 2(4), 251–267, 2004.
12. Cyber-Physical Systems, Retrieved from http://cyberphysicalsystems.org/.
13. An oracle white paper in enterprise architecture, *Oracle Information Architecture: An Architect's Guide to Big Data*, August 2012.
14. An Oracle White Paper, Oracle enterprise transformation solutions series, *Big Data & Analytics Reference Architecture*, September 2013.
15. *The Advantages and Disadvantages of Real-Time Big Data Analytics*, February 8, 2013. Retrieved from http://the-power-of-real-time-big-data.
16. T, McGuire, J. Manyika, and M. Chui, Why Big Data is the new competitive advantage, August 2012.
17. The Rise of Industrial Big Data. GE whitepaper. Retrieved from http://www.geip.com/library/detail/13170/.
18. Ten Practical Big Data Benefits, Retrieved from http://ten-practical-big-data-benefits.htm.
19. R. Chouffani, May 21, 2013, 5 Reasons to Move to Big Data (and 1 Reason Why It Won't Be Easy), Retrieved from http://5-reasons-to-move-to-big-data–and-1-reason-why-it-won-t-be-easy-.html.
20. F. Lauer and G. Bloch. Switched and PieceWise Nonlinear Hybrid System Identification, in *Proceedings of HSCC*, 2008.
21. V. L. Le, F. Lauer, L. Bako, and G. Bloch. Learning nonlinear hybrid systems: From sparse optimization to support vector regression, in *Proceedings of HSCC*, 2013.
22. S. Paoletti, A. L. Juloski, G. Ferrari-trecate, and R. Vidal. Identification of hybrid systems: A tutorial.
23. The Ptolemy Project. Retrieved from http://ptolemy.eecs.berkeley.edu/.
24. Y.W. Ahn, A.M.K. Cheng, Automatic resource scaling for medical cyber-physical systems running in private cloud computing architecture.
25. High confidence software and systems coordinating group of the networking and information technology research and development program, *High-Confidence Medical Devices: Cyber-Physical Systems for 21st Century health care*, February 2009.
26. Technical whitepaper, The big data RDF database, systap, "bigdata".
27. I. Lee, *Fellow, IEEE*, O. Sokolsky, *Member, IEEE*, S. Chen, *Student Member, IEEE*, J. Hatcliff, E. Jee, *Member, IEEE*, B. Kim, Andrew King, Margaret Mullen-Fortino, Soojin Park, Challenges and Research Directions in Medical Cyber-Physical Systems.
28. A. Roederer, K. Venkatasubramanian, *Member, IEEE*.
29. L. Sha, G. Gopalakrishnan, X. Liu, Q. Wang. Cyber-physical systems: A new frontier, in *Proceedings of the IEEE International Conference on Sensor Networks, Ubiquitous, and Trustworthy Computing (SUTC)*, 2008.
30. T. Hunter, T. Das, M. Zaharia, P. Addeel, A.M. Bayen, Large scale estimation in cyber physical systems using streaming data: A case study with smartphone traces. *arXiv.org*, 1212.3393v1, 2012.
31. Chen, S. et al., Mobile mapping technology of wind velocity data along highway for traffic safety evaluation, *Transportation Research Part C: Emerging Technologies*, 18(4), 507–518, 2010.
32. E. Davey, Rail traffic management systems (TMS), in *IET Professional Development Course on Railway Signalling and Control Systems (RSCS 2012)*.
33. J. Jun, R. Guensler, J. Ogle, Differences in observed speed patterns between crash-involved and crash-not involved drivers: Application of in-vehicle monitoring technology, *Transportation Research Part C: Emerging Technologies*, 19(4), 569–578, 2011.
34. S. Das, B.L. Matthews, A.N. Srivastava, N.C. Oza. Multiple kernel learning for heterogeneous anomaly detection: Algorithm and aviation safety case study, in *Proceedings of the KDD*, 2010.
35. C. Bachmann, Fusing a bluetooth traffic monitoring system with loop detector data for improved freeway traffic speed estimation, *Journal of Intelligent Transportation Systems*, 17, 1–4.
36. H. Dia, K. Thomas, Development and evaluation of arterial incident detection models using fusion of simulated probe vehicle and loop detector data. *Information Fusion*, 12(1), 20–27, 2011.
37. J. Jun, R. Guensler, J. Ogle, Differences in observed speed patterns between crash-involved and crash-not involved drivers: Application of in-vehicle monitoring technology. *Transportation Research Part C: Emerging Technologies*, 19(4), 569–578, 2011.

38. G. Haddow, J. Bullock, D.P. Coppola, *Introduction to Emergency Management*, Boston, MA, Butterworth-Heinemann, 2010.

39. Lambert, J.H. et al., Understanding and managing disaster evacuation on a transportation network, *Accident Analysis & Prevention*, 50, 645–658, 2012.

40. E.E. Ozguven, K. Ozbay, A secure and efficient inventory management system for disasters, *Transportation Research Part C: Emerging Technologies*, 2011.

41. Yoon, S. et al., Transportation security decision support system for emergency response: A training prototype. *Decision Support Systems*, 46(1), 139–148, 2008.

42. D. Chunlin, Y. Liu, Sample average approximation method for chance constrained stochastic programming in transportation model of emergency management, *Systems Engineering Procedia*, 5, 137–143, 2012.

43. S. Messelodi, S., Intelligent extended floating car data collection, *Expert Systems with Applications*, 36(3, Part 1), 4213–4227, 2009.

44. I. Lee, Assuring the safety, security, and reliability of Medical-Device Cyber Physical Systems (MDCPS), November 16, 2012.

Sotiris Nikoletseas and Theofanis P. Raptis

CONTENTS

21.1 INTRODUCTION

Cyber-physical systems (CPS) bridge the cyber world (e.g., information, communication, and intelligence) to the physical world through sensors and actuators. A CPS may consist of multiple static/mobile sensor and actuator networks integrated under an intelligent decision system [25]. Characteristic paradigms of such systems rely on an underlying communication network to transport data packets between sensors, computational units, and actuators [11]. Through closing the loop involving both the cyber world and the physical world, *wireless ad hoc networking* is one of the most critical concepts for building future CPS, which promises to revolutionize the way we interact with the physical world [26]. In particular, wireless ad hoc networks exploit the methodology of feedback, which has been recognized as the central element of control systems. Their advent has the potential to revolutionarily promote existing control applications by enabling an unprecedented degree of distributed cyber-physical control.

Wireless ad hoc networking allows CPS nodes to communicate with each other over the wireless medium, possibly by using other nodes as relays or cooperating in other more information theoretic ways. By attaching sensors to nodes and providing them with computational capability, one obtains wireless sensor networks (WSNs). They can be deployed to monitor their environment, for example, monitoring facilities for anomalies or monitoring wildlife [14,19,24], or to conduct physics in the large by offering scientists the means to deploy large number of sensors in the field and wirelessly gather information from them, as at the Center for Embedded Networked Sensing [1].

Lack of continuous energy supply is an obvious limitation in this type of networks. Usually, wireless nodes are deployed with batteries that will not be recharged or replaced. In this scenario, the network is considered disposable and power conservation is paramount. Energy-efficient design techniques have been studied for ad hoc networks at all levels from hardware design to protocols for medium access control, routing, data gathering, topology control, and so on [32]. However, in some applications, the network is not treated as disposable. It is possible to sustain the sensors by recharging or replacing batteries when needed. Some have taken a further step to extract energy directly from the deployment environment. These harvesting techniques power sensors via solar power, kinetic energy, floor vibration, acoustic noise, and so forth [10]. However, harvesting techniques are only becoming capable of generating the level of power required to sustain some of current applications. In some cases, they cannot generate sufficient energy for sustained operation, since the application situation of harvesting techniques is limited by dynamic environments. As there is generally a lack of *a priori* knowledge of energy profiles, such dynamics impose much difficulty on the design of protocols that keep network nodes from running out of energy. *Wireless power transfer* can be a key solution for overcoming these barriers.

The vision of transferring power wirelessly can be dated back to the early twentieth century (earlier than electric power grids). After the first electrical signal was sent across the Atlantic, Nikola Tesla, a pioneering electrical engineer, experimented with large-scale wireless power distribution by building the world's first power station in Long Island, New York [22]. He planned to use the power station, called Wardenclyffe Tower, to transmit not only signals but also wireless electricity. Unfortunately, due to its large electric fields, which significantly diminished the power transfer efficiency, Tesla's invention was never put into practical use. In the late twentieth century, the need for wireless power transfer reemerged when mobile electronic devices (e.g., laptops, cell phones, PDAs, tablets) became popular. Further, the rapid development of electric and plug-in hybrid vehicles in the auto industry also contributed to the need for wireless power transfer. Due to these demands, there have been many active efforts to develop efficient technologies for wireless power transfer [29]. Recently, the Wireless Power Consortium [2] was established to set the international standards for interoperable wireless charging. Member companies in the consortium include integrated circuit manufacturers, mobile phone makers, and mobile telecom operators. These standardization efforts are helping accelerate the pace of bringing wireless power transfer technologies to cross-domain applications, like CPS.

Wireless power transfer technologies offer new possibilities for managing the available energy in CPS and lead the way toward a new paradigm for wireless ad hoc networking. Wireless power transfer-enabled networks consist of nodes that may be either stationary or mobile, as well as few mobile nodes with high energy supplies. The latter, by using wireless power transfer technologies, are capable of fast recharging the network nodes. This way, the highly constrained resource of energy can be managed in great detail and more efficiently. Another important aspect is the fact that energy management can be performed passively from the perspective of nodes and without the computational and communicational overhead introduced by complex energy management algorithms. Finally, energy management can be studied and designed independently of the underlying routing mechanism used for data propagation.

Due to their concrete implementation of all basic CPS components, wireless power transfer-enabled networks lead the way toward a new application domain for CPS. CPS are consisted of three main parts [12]: (1) the physical part, (2) the computational part, and (3) the networking/communication part. Together, the computational and the networking parts form the *cyber* component of the CPS.

1. The physical is the part of the system that is not realized with computers or digital networks. It can include mechanical parts, biological or chemical processes, or human operators. In the context of wireless power transfer-enabled networks, there is a wide use of such processes ranging from wireless battery charging to control of mobile robots carrying wireless chargers.
2. The computational is the part that comprises one or more computational platforms of sensors, actuators, one or more computers, and (possibly) one or more operating systems. Ad hoc networks are fundamentally structured on platforms of this type.
3. The networking is the part where a network fabric exists, which provides the mechanisms for the communication among the computational platforms. According to the type of the networking application, different networking mechanisms may be implemented. A characteristic example is the routing mechanisms implemented in WSNs where communication among network nodes can be implemented via different methods (hop by hop, direct transmissions, energy balanced, etc.).

There are considerable challenges in making such wireless power-enabled CPS work. Some indicative are the following: First of all, controlling (stationary or mobile) wireless chargers is not trivial. Assuming a finite initial energy, those devices have a limited lifetime and their available energy supplies should be injected in the network wisely. Second, the wireless power transfer process itself consists of a challenging task. For example, the extent to which a network node should be charged, in order that the global network lifetime is prolonged, is not obvious. Finally, other issues like the amount of energy given to the chargers, the trajectory that they should follow inside the network their behavior w.r.t. the communication pattern, and energy dissipation inside the network further complicate the design and implementation of a CPS of this kind.

In this chapter, we present some state-of-the-art protocols designed for improving several critical aspects of the mobile charging node configuration in order to improve energy efficiency, prolong the lifetime of the network, and also improve important network properties [3,4,17,18]. An underlying routing protocol is taking care of the data propagation. Unlike other methods in the state of the art, we do not couple the charging process and the data propagation; actually, we wish to perform efficient wireless power transfer in a way that is agnostic to the routing protocol, via adaptive techniques that (without knowing the routing protocol) implicitly adapt to any routing protocol. While interesting research has been contributed to wireless power transfer problems and particularly to the scheduling of the mobile charging nodes, most methods so far necessitate significant (in many cases even global) network knowledge (e.g., it is assumed that the charger knows the energy levels of all nodes in the network), and the solutions are centralized. On the contrary, our methods are distributed and adaptive and use only local (or limited) network information. Also, unlike many state-of-the-art approaches that opt for integration and coupling of the charging and routing problems, our methods can be used together with any underlying routing protocol (since they adapt on it implicitly). Furthermore, our protocols dynamically and distributively adapt to network diversities, for example, they cope well with heterogeneous node placement.

21.1.1 BACKGROUND

21.1.1.1 Stationary Chargers

In this line of research, the charging nodes are considered as stationary and the network nodes as either mobile or stationary. This static setting, although realistic, limits the chargers from fully exploiting their energy management capabilities.

Xiang et al. [27] formulate a set of power flow problems and propose algorithms to solve them based on a detailed analysis on the problem structure. Moreover, they further investigate the joint data and power flow problems. Applying the proposed algorithms, they obtain extensive numerical results on the optimal configurations of power flow and joint data/power flows, which yield instructive insights on practical system constructions.

Rault, Bouabdallah, and Challal [21] propose an optimization model to determine the minimum number of stationary chargers needed to recharge the elements of a network in a multihop scenario, taking into account the energy demand of the nodes, the energy loss that occurs during a transfer, and the capacity of the chargers.

Dai et al. [5] study the problem of scheduling stationary chargers so that more energy can be received while no location in the field has electromagnetic radiation (EMR) exceeding a given threshold. They design a method that transfers the problem to two traditional problems, namely, a multidimensional 0/1 knapsack problem and a Fermat–Weber problem. The method includes constraint conversion and reduction, bounded EMR function approximation, area discretization and expansion, and a tailored Fermat–Weber algorithm. In order to evaluate the performance of their method, they build a testbed composed of eight chargers.

Dai et al. [6] consider the problem of scheduling stationary chargers with adjustable power, namely, how to adjust the power of chargers so as to maximize the charging utility of the devices, while assuring that EMR intensity at any location in the field does not exceed a given threshold. They present an area discretization technique to help in reformulating the problem into a traditional linear programming problem. Further, they propose a distributed redundant constraint reduction scheme to cut down the number of constraints and thus reduce the computational efforts of the problem.

Dai et al. [8] propose a new metric that characterizes the expected portion of time that a mobile node can sustain normal operation in networks with stationary chargers, under the constraints of node speed and battery capacity. They investigate the upper and lower bounds of the metric in 1D networks with one single source and multiple sources, respectively, and they give tight lower bounds in both 2D and 3D cases. They also perform extensive simulations to verify their findings.

21.1.1.2 Centralized Charging/Global Knowledge

In this line of research, the charging process and the charging nodes' coordination is performed centrally and not distributively. Also, the amount of knowledge that is required is in most cases global with respect to various network parameters, like the residual energy level or the location of each individual node at any given time. This approach may not be realistic for large-scale network deployments as it introduces high communication overhead (i.e., every charger has to propagate its status over large distances) and does not scale well with network size.

Peng et al. [20] build a proof-of-concept prototype by using a wireless power charger installed on a robot and nodes equipped with wireless power receivers, carry out experiments on the prototype to evaluate its performance in small-scale networks of up to 10 nodes, and conduct simulations to study its performance in larger networks of up to a 100 nodes. They also conduct simulations of the proposed heuristics, limited to a small number of nodes in the network.

Dai et al. [7] investigate the minimum changing nodes problem for 2D networks, that is, how to find the minimum number of energy-constrained charging nodes and design their charging routes in a 2D network such that each node in the network maintains continuous work, assuming that the energy consumption rates for all nodes are identical. They prove the hardness of the problem by reduction from the distance-constrained vehicle routing problem. Then they propose approximation algorithms for this problem and conduct simulations to validate the effectiveness of the algorithms.

Wang et al. [23] leverage concepts and mechanisms from named data networking (NDN) to design energy monitoring protocols that deliver energy status information to mobile charging nodes in an efficient manner. They study how to minimize the total traveling cost, ensuring no node failure, and derive theoretical results on the minimum number of mobile vehicles required for perpetual network operations.

Xie et al. [28] introduce the concept of renewable energy cycle and offer both necessary and sufficient conditions. They study an optimization problem, with the objective of maximizing the ratio of the wireless charging node's vacation time over the cycle time. For this problem, they prove that the optimal traveling path for the charging node is the shortest Hamiltonian cycle and provide a number of important properties. Subsequently, they develop a near-optimal solution by a piecewise linear approximation technique and prove its performance guarantee.

Xie et al. [30] propose a cellular structure that partitions the 2D plane into adjacent hexagonal cells and pursue a formal optimization framework by jointly optimizing traveling path, flow routing, and charging time. By employing discretization and a reformulation-linearization technique, they develop a provably near-optimal solution for any desired level of accuracy. Through numerical results, they demonstrate that the solution can address the charging scalability problem.

21.1.1.3 Joint Charging and Data Gathering:

According to the following works, the charging problem can be coupled together with the routing problem. The mobile charging node is also serving as a data collector or a routing relay intermediate. Approaches of this kind, although realistic, prevent the protocols from implicitly adapting to any underlying routing policy. For this reason, the routing protocol should be *a priori* known to the designer, something that may be impossible in networks of high dynamics.

Xie et al. [31] minimize energy consumption of the entire system while ensuring none of the nodes runs out of energy. They develop a mathematical model for this complex problem. Instead of studying the general problem formulation that is time dependent, they show that it is sufficient to study a special subproblem that only involves space-dependent variables.

Zhao, Li, and Yang [34] give a two-stage approach for the joint design. They formulate the problem into a network utility maximization problem and propose a distributed algorithm to adjust data rates, link scheduling, and flow routing so as to adapt to the up-to-date energy replenishing status of nodes. The effectiveness of the approach is validated by extensive numerical results. Comparing with solar harvesting networks, their solution can improve the network utility by 48% on average.

Li, Zhao, and Yang [13] propose a wireless energy replenishment and mobile data gathering mechanism that charges nodes effectively and collects data from the network using a mobile charging node. They provide an efficient algorithm that maximizes network utility. The numerical results demonstrate the performance advantage of the mechanism and provide a guidance on parameter selection for system design.

Li et al. [16] propose a practical and efficient joint routing and charging scheme. Through proactively guiding the routing activities in the network and delivering energy to where it is needed, the scheme effectively improves the network energy utilization, thus prolonging the network lifetime. They conduct experiments in a small-scale testbed and simulations and demonstrate that the scheme significantly elongates the network lifetime.

Li et al. [15] formulate the problem of maximizing the network lifetime via codetermining routing and charging, prove the hardness of the problem, derive an upper bound of the maximum network lifetime, present a set of heuristics to determine the wireless charging strategies under various routing schemes, and demonstrate their effectiveness via in-depth simulation.

21.2   THE MODEL

Our model features three types of devices: stationary nodes, mobile chargers, and one stationary control center. The control center acts as a sink for the data that are routed throughout the network. We assume that there are N nodes of wireless communication range r distributed at random in a circular area of radius R. We virtually divide the network into slices, the number of which is equal to the number of the mobile chargers. K mobile chargers initially deployed at coordinates $(x, y) = \left(\frac{R}{2} \cos(\frac{\pi}{K}(2j-1)), \frac{R}{2} \sin(\frac{\pi}{K}(2j-1))\right)$ of the circular area, where $j = 1, 2, \ldots, K-1$ (one mobile charger per slice). In the case where $K = 1$, the mobile charger is initially deployed at the center of the circular area. The control center lies at the center of the circular area. In our model, we assume that the mobile chargers do not perform any data gathering process.

We denote by E_{total} the total available energy in the network. Initially,

$$E_{total} = E_{nodes} + E_{MC}(t_{init}),$$

where

E_{nodes} is the amount of energy shared among the nodes
$E_{MC}(t_{init})$ is the total amount of energy that the mobile chargers have and may deliver to the network by charging nodes

The maximum amount of energy that a single node and a single charger may store is E_{node}^{max} and E_{MC}^{max}, respectively. Energy is split among the nodes and the chargers as follows:

$$E_{node}^{max} = \frac{E_{nodes}}{N} \text{ and } E_{MC}^{max} = \frac{E_{MC}(t_{init})}{K}.$$

We denote E_i and E_j as the residual energies of node i and mobile charger j, respectively.

In our model, the charging is performed point to point, that is, only one node may be charged at a time from a mobile charger by approaching it at a very close distance so that the charging process has maximum efficiency. The time that elapses while the mobile charger moves from one node to another is considered to be very small when compared to the charging time; still the trajectory followed (and particularly its length) is of interest to us, since it may capture diverse cost aspects, like gas or electric power needed for charger movement. We assume that the charging time is equal for every node and independent of its battery status.

We assume a quite heterogeneous data generation model. Each node chooses independently a relative data generation rate $\lambda_i \in [a, b]$ (where a, b are constant values) according to the uniform distribution $\mathcal{U}[a, b]$. Values of λ_i close to a imply low data generation rate and values close to b imply high data generation rate. The routing protocol operates at the network layer, so we are assuming appropriate underlying data-link, MAC, and physical layers.

21.2.1   CHARGER DISPATCH DECISION PROBLEMS

In the following, we give a formal definition for the decision version of the problem, when using a single mobile charger. Note that the problem is difficult even when we consider only a single mobile charger.

Definition 21.1 (CDDP [4]). *Suppose that we are given a set S of nodes, each one capable to store E units of energy, and for each node $s \in S$ a list L_s of pairs (t_s^j, e_s^j), $j \geq 1$ in which t_s^j corresponds to the time that the jth message of s was generated and e_s^j the energy that the node used to transmit it. We are also given an $|S| \times |S|$ matrix D, where $D_{i,j}$ is the distance between nodes i and j and a mobile charger M that can charge a node in one time unit to its initial energy (notice that if we assume that the charger moves with constant speed v, then the time needed to travel between i and j is $\frac{D_{i,j}}{v}$).*

The charger dispatch decision problem (CDDP) is to determine whether there is a feasible schedule for M to visit the nodes so that no message is lost due to insufficient energy.

Notice that we neglect to include energy needed by nodes in order to receive messages. Moreover, the messages a node s might receive by other nodes are included in each L_s. Thus, we suppose that these messages are generated by the node itself. This allows the consideration of different routing policies in a unified manner.

The general version of the CDDP is $\mathcal{NP}$-complete.

Theorem 21.1 ([4]) *CDDP is $\mathcal{NP}$-complete.*

Proof: We first note that, given a certain walk W of the charger visiting nodes in S, we can verify whether this walk is sufficient so that no message is lost, that is, no message x is generated on a node s such that x is the jth message of s and s has less than e_s^j available energy at time t_s^j. In particular, this can be done in $O(T \cdot |W|)$ time, where T is the total number of events generated in the network. Therefore, CDDP $\in \mathcal{NP}$.

For the hardness part, we use the geometric traveling salesman problem (G-TSP in short, see Garey and Johnson [9], page 212). Let $P \subseteq \mathbb{Z} \times \mathbb{Z}$ and $B \in \mathbb{N}$ be the input of G-TSP. We now transform this into an input for CDDP as follows: we use a set S of $|S| = |P|$ nodes and set $D_{i,j}$ equal to the Euclidean distance between the ith and jth points, in P. Furthermore, for each node $s \in S$, we define its event list to be

$$L_s = \left\{ (0, E), \left(\frac{B}{v}, 1\right) \right\},$$

where v is the charger's speed. That is, two events occur in each node s, namely, one at time 0 depleting all the energy available in s and one at time $\frac{B}{v}$ requiring energy 1. Notice that a solution to this instance of the CDDP problem would provide an answer to G-TSP, which means that G-TSP $\leq_m$ CDDP. This completes the proof. $\square$

21.3 TRADE-OFFS OF THE CHARGING PROCESS

Theorem 21.1 states that the problem is hard to solve, even with global knowledge of the energy dissipation. The protocols that we present encounter an even harder problem, since the knowledge is limited to a local level. In order to properly understand and investigate the charging process, we first present two of its aspects and specify their inherent trade-offs.

21.3.1 ENERGY PERCENTAGE AVAILABLE TO CHARGERS

In order to be fair in our evaluation, we assume that the total available energy to the network E_{total} is finite and same in all cases. This way, we will be able to investigate whether the energy efficiency is increased (and to what extent) with and without the introduction of the mobile charger and the charging process to the network. This particular trade-off consists in how much energy (in terms of the total energy available) should the mobile charger be initially equipped with. On the other hand, more energy to the mobile charger leads to better online management of energy in the network. However, since $E_{total} = E_{nodes} + E_{MC}^{init}$, more energy to the mobile charger also means that the nodes will initially be only partially charged. Therefore, it will be more likely that they run out of energy before the mobile charger charges them, leading to possible network disconnection and low coverage of the network area. In order to determine the optimal energy amount available to charger as a percentage of E_{total}, we conduct experiments for various ratios between E_{nodes} and E_{MC}^{init}.

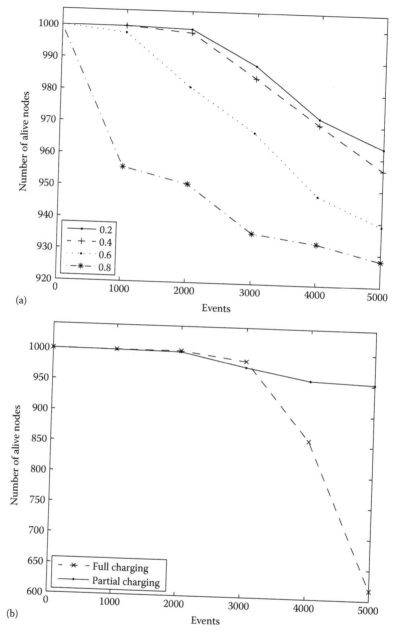

FIGURE 21.1   Alive nodes over time. (a) Various initial energy percentages and (b) full versus partial charging.

Indicatively, we present results about the single-charger case. The protocol used is described in Section 21.4.2. A key conclusion is that a rather modest percentage of energy at the mobile charger is a wise strategy (see Figure 21.1a), with a 20% percentage value being the best choice.

21.3.2   FULL VERSUS PARTIAL CHARGING

Each time the mobile charger visits a node, a straightforward strategy would be to fully charge that node. This way the mobile charger would maximize the time interval of revisiting the node before it runs out of energy. However, as the network operates, energy is dissipated in nodes due to data propagation and in the mobile charger due to the charging process. Therefore, the mobile charger will have increasingly less energy to distribute to more and more nodes.

Another approach is to judiciously spread the precious available energy to as many nodes as possible in order to extend the network lifetime. Following this rationale, the amount of energy the mobile charger delivers to a node i is proportional to the residual charging energy of the mobile charger. More formally, the mobile charger charges a node until its energy becomes

$$E_i \approx \frac{E_j}{E_{MC}(t_{init})} \cdot E_{node}^{max}.$$

In order to determine the best strategy, we conduct detailed experiments comparing the full charging strategy against the adaptive, partial charging strategy. We use again the protocol described in Section 21.4.2. The basic result is that partial charging is more efficient than the full charging case (see Figure 21.1b).

21.4 PROTOCOLS USING A SINGLE MOBILE CHARGER

21.4.1 ENERGY/FLOW CRITICALITY [3]

In order to develop efficient protocols for the mobile charger and address the corresponding trade-offs, we introduce *an attribute that captures a node's importance* in the network, under any given routing protocol. This attribute relies on two factors: (1) the *traffic served* by the node and (2) the *energy consumed* by the node.

The need for combining these two factors emerges from the fact that the traffic served by a node captures different aspects than its energy consumption rate. A node may consume a large amount of energy either because it serves a high network flow, protocols using a single mobile charger, or because its transmissions have high cost (e.g., long-range transmissions) (or both). The purpose of the attribute is to indirectly prioritize the nodes according to their flow rate and energy consumption; a node serving high traffic and/or having low residual energy should be charged at higher energy level.

We denote as $c_i(t)$ the *energy/flow criticality* (also referred to as criticality for simplicity) of node i at time t, with $c_i(t) = f_i(t) \cdot \rho_i(t)$. Given the time t_{MC} when the last charging of the node occurred,

$$f_i(t) = 1 - \frac{\text{generation rate of node } i}{\text{traffic rate of } i \text{ since } t_{MC}} = 1 - \frac{\lambda_i}{\lambda_i + \frac{m_i(t)}{t - t_{MC}}}$$

is the *normalized traffic flow* served by node i, where $m_i(t)$ is the amount of traffic (number of messages) that i has processed (received and forwarded) toward the control center by time t since time t_{MC} and

$$\rho_i(t) = \frac{\text{energy consumed since last charging}}{\text{max node energy since } t_{MC}} = \frac{E_i(t_{MC}) - E_i(t)}{E_i(t_{MC})} = 1 - \frac{E_i(t)}{E_i(t_{MC})}$$

is the *normalized energy consumption* by time t, since the last charging. The criticality is thus a number in [0, 1] that captures the importance of a given node by taking into account its flow rate, its energy consumption, its possible special role in the network, and its influence to the routing protocol; nodes serving high traffic (large $m_i(t)$) and/or having consumed a lot of energy (low $E_i(t)$) have high criticality $c_i(t)$ at time t and *are prioritized* by the mobile charger.

21.4.2 LOCAL KNOWLEDGE PROTOCOL [4]

Given the symmetric geometry, uniform density, and uniform data generation rate of the network, we propose that the mobile charger follows a circular trajectory around the control center. The radius of the trajectory varies and adapts to the energy depletion rates of each subregion of the network. Starting from the control center, the mobile charger traverses a path that forms a set of concentric circles, centered around the control center with varying (increasing or decreasing) radii. In particular, the mobile charger charges the nodes inside the ring that contains the corresponding trajectory. The width d of each ring is prespecified and constant. For example, if we denote by S the set of the nodes to be charged at the kth ring, then

$$|S| = \frac{N}{R^2} \left((kd)^2 - (k-1)^2 d^2\right) = \frac{N}{R^2} \left(2kd - d^2\right).$$

While at a given distance from the control center, the mobile charger records the mean value of the energy of the nodes lying on the corresponding circular trajectory; we denote this value by $\overline{E}_i$. Accordingly, the mobile charger keeps record of the mean value of the energy of the nodes lying on its previous circular trajectory; we denote this value by $\overline{E}_{i'}$. Based on these two values, the mobile charger tries to optimize its trajectory in terms of charging the nodes that deplete their energy faster. The protocol is shown here:

```
1:  while E_j > 0 do
2:      E_tmp = 0
3:      for every i ∈ S do
4:          E_tmp += E_i
5:          Charge until E_i ≈ (E_j / E_j(t_init)) · E_node^max
6:      end for
7:      Ē_i = E_tmp / |S|
8:      if Ē_i ≈ Ē_init then
9:          if Ē_i' ≥ Ē_i then
10:             Keep direction
11:         else
12:             Change direction
13:         end if
14:     end if
15: end while
```

The mobile charger starts traversing the network at first from the center of the network area by setting, and it first visits the nodes that lie one hop away from the control center. Once all corresponding nodes are charged, the mobile charger increases its radius by visiting nodes that are two hops away from the control center. By comparing the values of $\overline{E}_i$ and $\overline{E}_{i'}$ the mobile charger is able to figure out whether it moves toward areas that are stressed by the routing protocol or not. More specifically, if $\overline{E}_i < \overline{E}_{i'}$, then the mobile charger assumes it is moving toward a stressed area of the network. Otherwise, if $\overline{E}_i > \overline{E}_{i'}$, then it assumes it is moving away from the stressed areas and therefore should change direction.

21.4.3 GLOBAL KNOWLEDGE PROTOCOL [3]

The global knowledge charger we suggest is an online method that uses criticality as a ranking function. In each round, the charger moves to the node that minimizes the product of the negation of each node's criticality times its distance from the current position of the mobile charger. More specifically, in each moving step, the global knowledge protocol (GKP) minimizes the product

$$\min_i \left\{(2 - c_i(t)) \cdot \left(1 + \frac{dist_i}{2R}\right)\right\},$$

where $dist_i$ is the distance of node i from the mobile charger, with the minimum taken over all nodes in the network (or at least a large part of it). In other words, *this protocol prioritizes nodes with high criticality and small distance to the mobile charger*. Since this protocol requires *a global knowledge of the state of the network*, it is expected to outperform all other strategies that use only local or limited network information, thus somehow representing an online centralized performance upper bound. However, it would not be suitable for large-scale networks as it introduces great communication overhead (i.e., every node has to propagate its criticality to the mobile charger) and does not scale well with network size.

21.4.4 LIMITED REPORTING PROTOCOL [3]

The control center is informed about the status of some representative nodes scattered throughout the network and is able to provide the mobile charger with some guidance. In other words, this protocol distributively and efficiently *simulates* the GKP. We assume that the control center can transmit to the mobile charger wherever in the network the latter might be. The protocol follows a limited reporting strategy, since it exploits information from the whole network area but from a limited number of nodes. The nodes of each slice periodically run a small computation overhead protocol in order to elect some special nodes, the *reporters* of the slice; in particular, each node becomes a reporter independently with some appropriate probability (thus, the number of reporters is binomially distributed). The reporters act as the representatives of their slice and their task is the briefing of the control center about their criticality.

The *percentage of the nodes that will act as reporters* brings off a trade-off between the representation granularity of the network and the communication overhead on each message propagated in the network. If we set a large percentage of reporters, the control center will have a more detailed knowledge of each slice's overall criticality but the message overhead will highly increase, since each message should carry the slice reporter's current criticality. On the contrary, if we set a small percentage of reporters, the overhead will be tolerable, but the representation of a slice will be less detailed.

In order to maintain *a small set of reporters* for each slice (for communication overhead purposes), we propose that slice l that contains n_l nodes elects $\kappa_l = \frac{n_l}{N} \cdot \kappa_{total}$ reporters, with the global number of reporters being

$$\kappa_{total} = h \frac{R}{r} \log N, \text{ where } h = 1 - \frac{a}{b}$$

is a network density heterogeneity parameter. Clearly, a highly heterogeneous deployment (large b compared to a) will necessitate a higher number κ_{total} of reporters. Also, κ_{total} must be large in large networks with many nodes. Each node periodically with probability p_i becomes a reporter. In order to have an expected number of κ_l reporters in slice l, we need

$$\kappa_l = n_l \cdot p_i \Rightarrow p_i = \frac{\frac{n_l}{N} \cdot \kappa_{total}}{n_l} \Rightarrow p_i = \frac{\kappa_{total}}{N}.$$

The reporter selection is meant to occur in a local and distributed manner, that is, each node becomes a reporter with the aforementioned suitably, independently chosen probability. This random independent generation of reporters is captured by the Bernoulli trials (one per node), that is, a binomial distribution. In order to figure out possible good values for κ_{total} that maximize the limited reporting protocol (LRP) performance, we carry out a comparison operating the protocol between several reporter numbers. Figure 21.2a depicts the number of alive nodes of the network for various percentages of reporters over the total number of nodes in the network. In the particular setting, the formula for κ_{total} yields $\kappa_{total} = 5\%$. It is obvious that if the protocol defines the number of reporters to be less than 5% of the network nodes, the granularity of the slice representation is poor, resulting in reduced network lifetime. Similarly, for numbers greater than 5% of the network nodes, the lifetime is also reduced, due to the higher message exchange overhead (over total traffic) in Figure 21.2b.

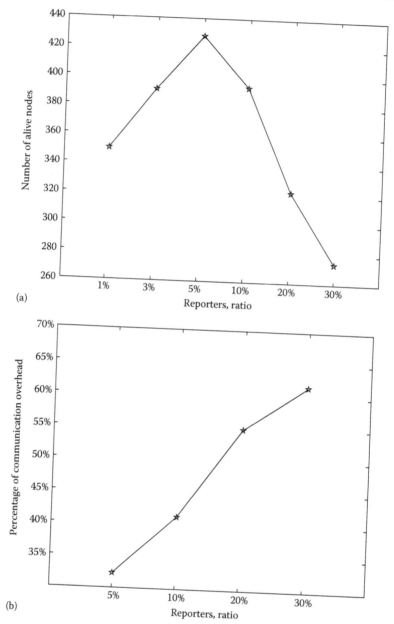

(a)

(b)

FIGURE 21.2   (a) Alive nodes and (b) communication overhead for various κ_{total} values in LRP, after 6.000 sensing events.

21.4.5   REACTIVE TRAJECTORY PROTOCOL [3]

In this protocol, a node i is propagating an *alert message* to its neighbours each time its energy drops below a set of some crucial limits. The messages are propagated for some hops and are stored at every node passed, in order for a *tree structure* rooted at i to be formed that can be *detected by the mobile charger when passing through* some tree node. Every node can root a tree and the strategy followed (toward a small tree management overhead) is the maintenance of a small tree degree with a larger tree depth.

The tree that is formed for each node is gradually growing in an analogous way to the criticality of the root node, as the gradual increase in a node's criticality is an indication of either high traffic or high energy consumption. *We use criticality as a measure of the gradual expansion of the tree*, since its value depicts both the importance of the node in the network and its energy consumption rate. We propose a strategy of message propagations which aims at covering a relatively large area of the network, while keeping energy consumption due to communication overhead low.

More specifically, each node i can alter among $\lceil \log\left(N\frac{R}{r}\right)\rceil$ alert levels that determine the characteristics of the i's rooted tree. We denote as al_i the current *alert level* of node i. The tree rooted at i is formed in a way that the *degree* $= al_i - 1$ and the *depth* $= 2^{al_i - 1} - 1$. The duration of each successive alert level is increased by a constant ratio from the previous level:

$$al_i = \begin{cases} 1 & \text{if } 0 \le c_i(t) < 0.5 \\ 2 & \text{if } 0.5 \le c_i(t) < 0.75 \\ \vdots & \vdots \\ \lceil \log\left(N\frac{R}{r}\right)\rceil & \text{if } 1 - \frac{1}{2^{\lceil \log\left(N\frac{R}{r}\right)\rceil - 1}} \le c_i(t) < 1 \end{cases}$$

$$= \left\{ \mu \mid \mu \in \left[1, 2, \ldots, \left\lceil \log\left(N\frac{R}{r}\right)\right\rceil\right] \right\}$$

with $1 - \frac{1}{2^{\mu-1}} \le c_i(t) < 1 - \frac{1}{2^{\mu}}$ where $1 - \frac{1}{2^{\mu-1}} = \sum_{k=1}^{\mu-1} \frac{1}{2^k}, 1 - \frac{1}{2^{\mu}} = \sum_{k=1}^{\mu} \frac{1}{2^k}.$

The tree management procedure aims at providing a high-level information about the local trees' state and at the same time at maintaining the node memory reservation at relatively low levels. For this reason, nodes store information solely about their parent nodes in emerging tree structures (i.e., one record per parent). Node i that is already a tree member may receive alert messages coming from nodes that belong to other trees. In this case, i stores the received alert messages from surrounding parents, the number of which is at most equal to the number of i's neighbors (since a parent node of i can only be in its transmission range). Nodes that participate in multiple trees propagate messages concerning solely the highest alert level and redirect the mobile charger (when the latter is near the current node) to follow the highest alert level tree, links. In short, each node can participate in multiple trees, reserves memory at most equal to the number of its neighbors, propagates messages about the highest priority tree, and redirects the mobile charger to it.

The mobile charger alters its state between a *patrol mode* and a *charging mode*. When in patrol mode, it follows a spiral patrol trajectory centered at the control center and does not charge any nodes until notified that the area traversed is low on energy. When so notified by a node in such an area, it pauses the patrol mode and enters the charging mode, in which it follows a different trajectory in order to accomplish the charging process in this area. If the mobile charger detects simultaneously different trees, then by a check on the *depth* of each structure, it can decide which is the most critical. After the completion of the charging process, the mobile charger resumes the patrol mode.

21.4.6 PROTOCOL EVALUATION

We perform a performance evaluation on the single-charger protocols. The overall death rate (in terms of *alive nodes over time*) of the network is vastly reduced, as shown in Figure 21.3a. The performance of both LRP and reactive trajectory protocol (RTP) approaches the performance of the GKP, powerful charger. They also outperform the local knowledge protocol (LKP) that seems to be the less adaptive.

Routing robustness is critical for ad hoc networks, as information collected needs to be sent to remote control centers. Path breakage occurs frequently due to node mobility, node failure, or channel impairments, so the maintenance of a path from each node to a control center is challenging. A way of addressing the routing robustness of a network is by considering for each node the number of its alive neighbors over time, which can be seen as an implicit measure of network connectivity. The

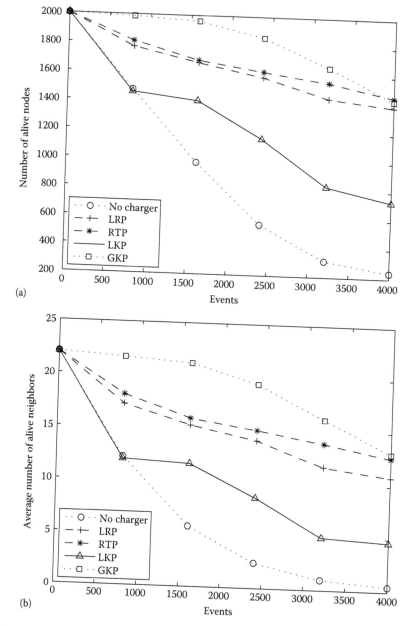

(a)

(b)

FIGURE 21.3 Lifetime and robustness of single-charger protocols. (a) Alive nodes over time and (b) average routing robustness.

average number of alive neighbors is depicted in Figure 21.3b. LRP and RTP achieve high robustness, outperforming the LKP and approaching the GKP performance.

Point coverage problem is regarding how to ensure that all points in the network are covered by enough nodes. Coverage is an important aspect in ad hoc networks (e.g., localization, target tracking). A point that is covered by k nodes is called k-covered. The coverage aging (evolution of coverage with time) of 1000 randomly selected points in the network is shown in Figure 21.4. We investigate how many points are 1-covered, 2-covered, 3-covered, and > 3-covered during an experiment of 4000 generated events. LRP and RTP maintain satisfactory levels of coverage.

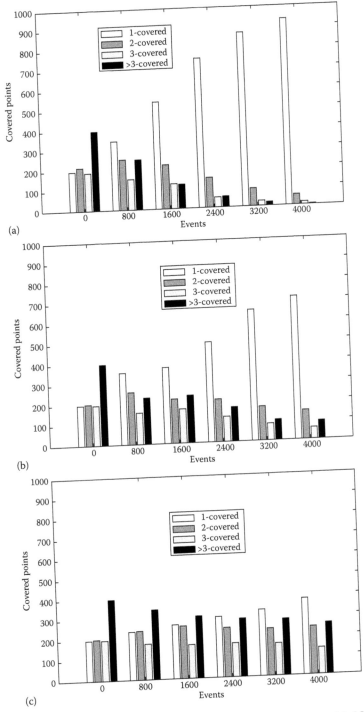

FIGURE 21.4 Coverage aging of single-charger protocols. (a) No charger; (b) LKP; and (c) LRP.
(Continued)

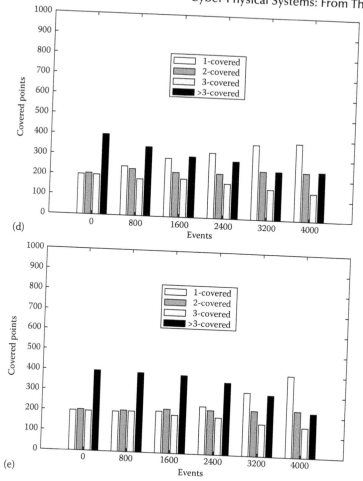

(d)

(e)

FIGURE 21.4 (*Continued*) Coverage aging of single-charger protocols. (d) RTP and (e) GKP.

21.5 PROTOCOLS USING MULTIPLE MOBILE CHARGERS

21.5.1 CENTRALIZED COORDINATION PROTOCOL [17]

The centralized coordination (CC) protocol performs CC among the chargers and assumes no knowledge on the network. In particular, the coordination process is able to use information from all mobile chargers (energy status, position, etc.) but is agnostic of the underlying network and node attributes (energy status, position, etc.). This approach virtually partitions the network elements in two completely separate levels, the mobile charger level and the node level.

Coordination phase: Each mobile charger is assigned to a network region. Since the initial charger deployment coordinates are $(x, y) = \left(\frac{R}{2}\cos(\frac{\pi}{K}(2j-1)), \frac{R}{2}\sin(\frac{\pi}{K}(2j-1))\right)$, where $j = 1, 2, \ldots, K$, we can initially assign a mobile charger to each slice. When the coordination process is initialized, the region of each charger is computed. Each charger should be assigned to a region of size analogous to its current energy level, so that the energy dissipation among the chargers is balanced. In order to compute the size of the region of charger j, it suffices to compute the central angle ϕ_j corresponding to the charger's slice. In particular

$$\phi_j = 2\pi \cdot \frac{E_j}{\sum_{j=1}^{K} E_j}, \text{ where } \sum_{j=1}^{K} \phi_j = 2\pi.$$

Charging phase: During this phase, charger j traverses the network region it is assigned to (slice defined by angle ϕ_j) and charges the corresponding nodes. The CC protocol assumes no knowledge on the network. For this reason, the path followed by the mobile charger is restricted to several naive alternatives. In our approach, we use a *blind* scanning of the region where the mobile charger starts to the control center and traverses an exhaustive path until it reaches the boundaries of the network area. The advantage of this movement is that due to its space filling attributes, the mobile charger covers the whole slice and almost every node is charged, until the energy of the mobile charger is totally depleted. On the other hand, due to lack of knowledge, this movement is not adaptive, that is, it does not take into account differences of the energy depletion rates of the network area caused by the underlying message propagation.

21.5.2 DISTRIBUTED COORDINATION PROTOCOL [17]

Coordination phase: The distributed coordination (DC) protocol performs DC among chargers and assumes no network knowledge. We again assign one slice per charger. Angle ϕ_j corresponds to the central angle of jth charger's slice. The chargers distributively define their slice limits (i.e., the two radii that define the slice), according to the size of the region each one can handle, with respect to their energy status. Each charger can shift their right and left slice limits, resulting in either a widening or a shrinkage of the region of interest. This task is performed distributively and each region limit movement is determined through a cooperation of the two adjacent mobile chargers. A limit movement of j's region is expressed as a change of ϕ_j. The coordination process uses two critical charger parameters for the definition of the region of interest, the charger's current energy level E_j, and the charger's energy consumption rate since the last coordination ζ_j. The change $\Delta\phi_j^l$ of ϕ_j for the left slice limit and the change $\Delta\phi_j^r$ of ϕ_j for the right are defined by the following computations:

if $\min\{E_j, E_{j-1}\} = E_j$ **then**
$$\Delta\phi_j^l = -\phi_j \cdot \frac{|\zeta_{j-1} - \zeta_j|}{\max\{\zeta_{j-1}, \zeta_j\}}$$
else
$$\Delta\phi_j^l = \phi_{j-1} \cdot \frac{|\zeta_{j-1} - \zeta_j|}{\max\{\zeta_{j-1}, \zeta_j\}}$$
end if

if $\min\{E_j, E_{j+1}\} = E_j$ **then**
$$\Delta\phi_j^r = -\phi_j \cdot \frac{|\zeta_j - \zeta_{j+1}|}{\max\{\zeta_j, \zeta_{j+1}\}}$$
else
$$\Delta\phi_j^r = \phi_{j+1} \cdot \frac{|\zeta_j - \zeta_{j+1}|}{\max\{\zeta_j, \zeta_{j+1}\}}$$
end if

The new angle (denoted by ϕ_j') is computed as $\phi_j' = \phi_j + \Delta\phi_j^l + \Delta\phi_j^r$. Note that between two adjacent chargers j_1 and j_2, the change in their common slice limit is $\Delta\phi_{j_1}^r = -\Delta\phi_{j_2}^l$, so that the charger with the lower energy level provides its neighbor with a portion of its region of interest. Also, it is their energy level that determines which charger should reduce its region of interest and the energy consumption rate that determines the size of the reduced area. The size of the angle change is not computed by considering the energy levels of the two chargers because energy consumption rate shows how quickly will this energy level be reduced. For example, if ζ_j is high, then j's slice is critical, causing a rapid reduction of E_j, independently of its current level.

Charging phase: Since this protocol operates under the no knowledge assumption, the charging phase follows the same pattern with the CC protocol (slice scanning).

21.5.3 CENTRALIZED COORDINATION GLOBAL KNOWLEDGE PROTOCOL [17]

The centralized coordination global knowledge (CCGK) protocol, similar to the CC protocol, performs CC. However, the assumption of global knowledge on the network further extends the mobile chargers' abilities. For this reason, it is expected to outperform all other strategies that use only local information, thus somehow representing a performance bound. The global knowledge assumption would be unrealistic for real large-scale networks, as it introduces large communication overhead (i.e., nodes and chargers have to propagate their status over large distances).

Coordination phase: Instead of using the same coordination process with the CC protocol, we integrate the global knowledge assumption in the coordination phase. As a result, the network is not partitioned in two separate levels (mobile chargers, nodes), and the mobile chargers are allowed to use network information during this phase. Each mobile charger is assigned to a network region. The region of interest of charger j is a cluster of nodes. Node i belongs to the cluster of charger

$$j' = \arg\min_{j} \left\{ \left(1 + \frac{dist_{ij}}{2R} \right) \cdot \left(2 - \frac{E_j}{E_{MC}^{max}} \right) \right\},$$

where $dist_{ij}$ is the distance between node i and charger j. In other words, a node selects which charger is close and with high amount of energy. Note that the centralized computation of the charger region in the CCGK protocol is more powerful compared to other methods, since it uses information about the distance among every charger with every node.

Charging phase: The global knowledge charging phase we suggest uses energy and distance in a ranking function. In each round, the charger moves to the node in the corresponding cluster, which minimizes the product of each node's energy times its distance from the current position of the mobile charger. More specifically, in each moving step, the charger j charges node

$$i' = \arg\min_{i \in C_j} \left\{ \left(1 + \frac{dist_{ij}}{2R} \right) \cdot \left(1 + \frac{E_i}{E_{node}^{max}} \right) \right\}.$$

In other words, this protocol prioritizes nodes with low energy and small distance to the mobile charger.

21.5.4 DISTRIBUTED COORDINATION LOCAL KNOWLEDGE PROTOCOL [17]

Coordination phase: The coordination phase follows the same pattern with the coordination phase of DC protocol (distributed ϕ_j angle computation).

Charging phase: The distributed coordination local knowledge (DCLK) protocol operates with local knowledge assumption. The slice corresponding to charger j is divided into k sectors S_{jk} of the same width. Charger j prioritizes its sectors with respect to high number of nodes with low level of residual energy.

Definition 21.2 E_{jk}^{min} is the lowest nodal residual energy level in the sector S_{jk}.

Definition 21.3 $E_{jk}^{min+\Delta}$ is an energy level close to E_{jk}^{min} :

$$E_{jk}^{min+\Delta} = E_{jk}^{min} + \delta \cdot \frac{E_{node}^{max}}{E_{jk}^{min}}, \delta \in (0, 1).$$

Definition 21.4 $N(S_{jk})$ is the number of nodes in sector S_{jk} with residual energy between E_{jk}^{min} and $E_{min+\Delta}^{jk}$:

$$N(S_{jk}) = \sum_{e=E_{jk}^{min}}^{E_{jk}^{min+\Delta}} N(e),$$

where $N(e)$ is the number of nodes with energy level e.

Charger j charges sector S_{jk} that maximizes the product

$$\max_{S_{jk}} \left\{ N(S_{jk}) \cdot (E_{node}^{max} - E_{jk}^{min}) \right\}.$$

The intuition behind this charging process is the grouping of nodes in each slice and the selection of a critical group. A critical group is a sector containing a large number of nodes that require more energy than other nodes throughout the network.

21.5.5 PROTOCOL EVALUATION

We perform a performance evaluation of the multiple-charger protocols. The application of charging protocols results in a great reduction in the overall death rate of the network (in terms of *alive nodes over time*), as shown in Figure 21.5a. The CCGK as expected outperforms all other protocols and

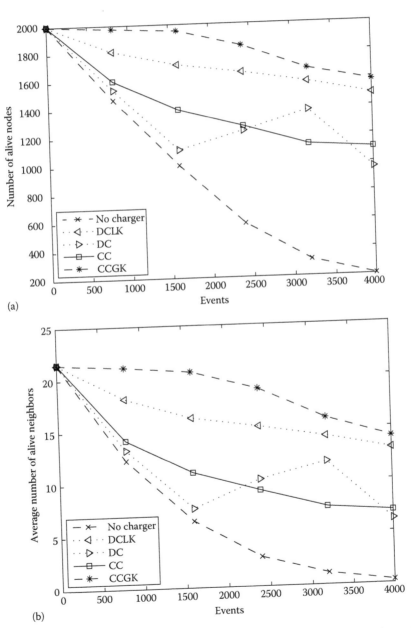

FIGURE 21.5 Lifetime and robustness of multiple-charger protocols. (a) Alive nodes over time and (b) average routing robustness.

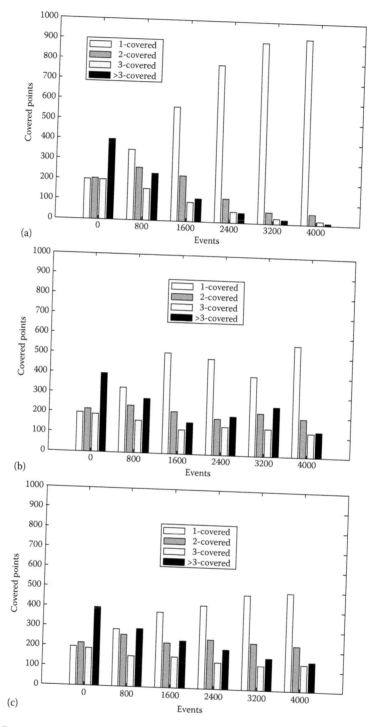

FIGURE 21.6 Coverage aging of multiple-charger protocols. (a) No charger; (b) DC; and (c) DCLK.

(*Continued*)

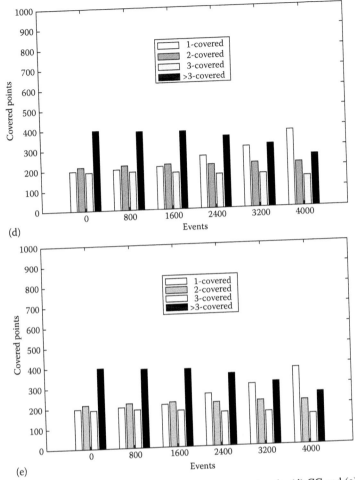

FIGURE 21.6 (*Continued*) Coverage aging of multiple-charger protocols. (d) CC and (e) CCGK.

rather serves as an upper bound of their performance. The power of CCGK comes from the great amount of knowledge it assumes and from the robust CC of the mobile chargers. The DCLK protocol, with limited network knowledge, is an efficient distributed alternative, since it manages to achieve a performance close to the upper bound. CC and DC protocols that assume no network knowledge are outperformed by other protocols. Note that the upward curve of the DC protocol after 1500 sensing events indicates a temporary starvation, after which the chargers revive some of the dead nodes when passing by areas with high amounts of dead nodes.

The CCGK protocol achieves the lowest death rate since it computes which node to charge next based on both its distance from the node and the energy status of the node. In contrast, CC and DC protocols use just the distance property for the relevant choice, so the weak nodes that are also far away from the charger may deplete their energy. DCLK protocol separates each slice into sectors and the charger chooses to charge the nodes of the most critical sector.

For our protocols, the average number of alive neighbors is depicted in Figure 21.5b. The average routing robustness follows the same pattern as the death rate of nodes in the network. This is natural, since more dead nodes in the network result in loss of neighbors for each node. CCGK and DCLK protocols achieve reliable routing robustness.

In Figure 21.6, we can see the coverage aging of 1000 randomly selected points scattered throughout the network. We examine how many points are < 2-covered, 2-covered, 3-covered, and > 3-covered for 4000 generated events. Each bar in the plot represents a number of the covered points. In the no-charger case, the number of < 2-covered points is increasing in contrast to the number of > 3-covered points that is decreasing. CCGK and DCLK protocols improve the network coverage by reducing the rate that the coverage of > 3-covered points is decreasing. The absolute difference of the number of < 2-covered and > 3-covered points, between different time instances, is not increasing quickly, compared to the no-charger case.

An important fact that comes up from the observation of the aforementioned metrics is that the CC and DC protocols are outperformed by their improved alternatives, CCGK and DCLK. The CC protocol, which uses a strong, centralized computation in order to calculate the regions of interest for the chargers, is outperformed even by the DCLK protocol that employs distributed computation (but assumes more knowledge on the network). This leads to the conclusion that the nature of the coordination procedure is less significant than the design of the charging traversal, when the latter relies on a greater amount of knowledge. Of course, when assuming the same amount of knowledge, the coordination procedure is still important (e.g., CC vs. DC).

21.6 PROTOCOLS USING HIERARCHICAL COLLABORATIVE MOBILE CHARGING

21.6.1 MODEL VARIATION

In this case, mobile chargers can either charge nodes or charge other mobile chargers. Figure 21.7 depicts the energy flow in *three different charging models*, including simple charging using multiple mobile chargers (Figure 21.7a), collaborative charging (Figure 21.7b), and hierarchical collaborative

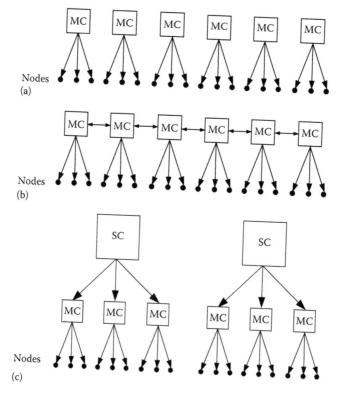

FIGURE 21.7 Energy flow models. (a) Charging model with multiple mobile chargers; (b) collaborative charging model; and (c) hierarchical, collaborative charging model.

charging model (Figure 21.7c). The arrows abstract the energy flow from chargers to other chargers/nodes. The collaborative charging model was used in Zhang, Wu, and Lu [33], for managing energy in 1D networks. In the hierarchical collaborative charging model, the *special chargers* that are the highest devices in terms of hierarchy can charge the mobile chargers and the mobile chargers can charge the nodes.

Initially,

$$E_{total} = E_{nodes} + E_{MC}(t_{init}) + E_{SC}(t_{init}),$$

where

E_{nodes} is the total amount of energy shared among the nodes
$E_{MC}(t_{init})$ is the total amount of energy shared among the mobile chargers
$E_{SC}(t_{init})$ is the total amount of energy shared among the special chargers

The maximum amount of energy that a single node, a single mobile charger, and a single special charger may store is E_{node}^{max}, E_{MC}^{max}, and E_{SC}^{max}, respectively. Energy is uniformly split among the nodes and the chargers as follows:

$$E_{node}^{max} = \frac{E_{nodes}}{N}, \quad E_{MC}^{max} = \frac{E_{MC}(t_{init})}{M}, \text{ and } E_{SC}^{max} = \frac{E_{SC}(t_{init})}{S}.$$

At first, we deploy the nodes uniformly in the circular network. Then, we divide our network into M equal-sized slices, one for each mobile charger. Thus, every mobile charger is responsible for charging nodes that belong to its slice. We denote by D_j the set of nodes that belong to slice j, that is, to the jth mobile charger's group. Finally, we divide the mobile chargers into S groups, one for each special charger. Thus, each special charger is responsible for charging the mobile chargers that belong to its group, denoted as C_k (for special charger k). Initially, these S groups are equally sized, that is, $|C_k| = \frac{M}{S}, (1 \le k \le S)$, and the mobile chargers that belong to each group are given by the following formula:

$$C_k = \left\{ j : j \in \left[(k-1)\frac{M}{S} + 1 , k\frac{M}{S} \right] \right\}, \quad (1 \le k \le S).$$

These groups may change during the protocol's coordination phase. More specifically, the special chargers communicate with each other and decide, according to their energy status, if they are still able to be in charge of the mobile chargers that belong to their group or they should delegate some of them to other special chargers.

21.6.2 NO KNOWLEDGE NO COORDINATION [18]

Coordination phase: There is no coordination between special chargers.

Trajectory: Each special charger charges the corresponding mobile chargers sequentially. When it arrives to the last mobile charger of its group, it changes direction and charges them again in a reverse order this time and so on.

Charging phase: Each special charger charges each mobile charger in its group until its battery level is E_{MC}^{max}.

21.6.3 ONE-LEVEL KNOWLEDGE DISTRIBUTED COORDINATION [18]

The one-level knowledge distributed coordination (1KDC) protocol performs a DC among special chargers, that is, every special charger can communicate with adjacent neighbors. Also, it assumes one-level network knowledge, that is, it can use information only about mobile chargers' energy status (and not about the nodes that lie one level lower).

Coordination phase: In DC, we assume that a special charger knows which are the adjacent mobile chargers on the boundaries of its region. We call next the first mobile charger that belongs to special charger $k + 1$ and previous the last mobile charger that belongs to special charger $k - 1$. More specifically,

$$n_k = \min_{j \in C_{k+1}} \{j\} : \text{next Mobile Charger (belongs to } k + 1),$$

$$p_k = \max_{j \in C_{k-1}} \{j\} : \text{previous Mobile Charger (belongs to } k - 1).$$

The special charger k, in order to coordinate with each of its neighbors ($k-1$ and $k+1$), calculates which of them has the highest energy supplies so as to charge the mobile chargers in its group and the additional mobile charger of its left or right neighbor. Thus, every special charger k estimates the residual energy in both cases (including a mobile charger of its left and right neighbor) by the following equations:

$$e_k^p = E_k - \sum_{j \in C_k} E_j^{lack} - E_{p_k}^{lack}, \quad e_k^n = E_k - \sum_{j \in C_k} E_j^{lack} - E_{n_k}^{lack},$$

where $E_j^{lack} = E_{MC}^{max} - E_j$ is the amount of energy that mobile charger j can receive until it is fully charged.

Between two adjacent special chargers, the one with the higher energy supplies takes the other's boundary mobile charger in its group. Thus, the special charger with lower energy supplies is responsible for a smaller area. In the case that their energy supplies are the same, they do not exchange any mobile chargers. More precisely, the coordination protocol is the following:

$(k, k - 1)$
if $(e_k^p > e_{k-1}^n)$ **then**
$\quad C_k = C_k \bigcup \{p_k\}$
$\quad C_{k-1} = C_{k-1} \setminus \{p_k\}$
else if $(e_k^p < e_{k-1}^n)$ **then**
$\quad C_{k-1} = C_{k-1} \bigcup \{n_{k-1}\}$
$\quad C_k = C_k \setminus \{n_{k-1}\}$
else
$\quad$ No Mobile Chargers exchange
end if

$(k, k + 1)$
if $(e_k^n > e_{k+1}^p)$ **then**
$\quad C_k = C_k \bigcup \{n_k\}$
$\quad C_{k+1} = C_{k+1} \setminus \{n_k\}$
else if $(e_k^n < e_{k+1}^p)$ **then**
$\quad C_{k+1} = C_{k+1} \bigcup \{p_{k+1}\}$
$\quad C_k = C_k \setminus \{p_{k+1}\}$
else
$\quad$ No Mobile Chargers exchange
end if

Trajectory: Special charger k should determine which mobile charger will be the next that will be charged prioritizing a mobile charger based on minimum energy and minimum distance. Considering this, k chooses to charge mobile charger m where

$$m = \arg\min_{j \in C_k} \left\{ \left(1 + \frac{E_j}{E_{MC}^{max}}\right) \cdot \left(1 + \frac{dist_{kj}}{2R}\right) \right\}.$$

Charging phase: A special charger charges a mobile charger j according to its energy consumption rate r_{MC_j}. More specifically, a mobile charger with higher consumption rate (compared to the rest mobile chargers that belong to the special charger's group) should be charged with a higher amount of energy. Motivated by that, if by m we denote the mobile charger that special charger k chose to charge, then the amount of energy that the special charger will give to it is

$$e = c_m \cdot \left(\min\left\{E_m^{lack}, E_k\right\}\right) \text{ where } c_m = \frac{r_{MC_m}}{\sum_{j \in C_k} r_{MC_j}}.$$

21.6.4 Two-Level Knowledge Distributed Coordination [18]

In contrast to the previous protocols, the two-level knowledge distributed coordination (2KDC) assumes two-level knowledge, and thus, each special charger k computes e_k^p and e_k^n using information about both the mobile chargers and the nodes as follows:

Coordination:

$$e_k^p = E_k - \sum_{j \in \mathcal{C}_k} \sum_{i \in \mathcal{D}_j} E_i^{lack} - \sum_{i \in \mathcal{D}_{P_k}} E_i^{lack}, \quad e_k^n = E_k - \sum_{j \in \mathcal{C}_k} \sum_{i \in \mathcal{D}_j} E_i^{lack} - \sum_{i \in \mathcal{D}_{n_k}} E_i^{lack}.$$

After that, the coordination protocol presented in 1KDC protocol's coordination phase is used.

Trajectory: Each mobile charger j stores a list l_j of nodes, the energy level of which is lower than a value $E_{threshold}$. Special charger k defines which mobile charger is more critical by making a query to each mobile charger in its group on the size of its list. A special charger should assign high priority to a mobile charger that has a large number of nodes of energy lower than $E_{threshold}$. Thus, special charger k selects to charge mobile charger m where

$$m = \arg\max_{j \in \mathcal{C}_k} |l_j|.$$

Charging phase: Since each special charger assumes two-level knowledge, it computes the percentage of energy to transfer, according to the lack of energy in the slice of the selected mobile charger compared to the total energy lack in all slices that this special charger is responsible for. More precisely, special charger k transfers to m an amount of energy

$$e = c_m \cdot \left(\min\left\{E_m^{lack}, E_k\right\}\right) \text{ with } c_m = \frac{\sum_{i \in \mathcal{D}_m} E_i^{lack}}{\sum_{j \in \mathcal{C}_k} \sum_{i \in \mathcal{D}_j} E_i^{lack}} \in (0, 1),$$

where $E_i^{lack} = E_{node}^{max} - E_i$ is the amount of energy that node i can receive until it is fully charged.

21.6.5 TWO-LEVEL KNOWLEDGE CENTRALIZED COORDINATION [18]

The two-level knowledge centralized coordination (2KCC) protocol performs centralized coordination and assumes two-level network knowledge, that is, it can use information both about mobile chargers' and nodes' energy status. It assigns to each special charger an amount of mobile chargers according to their residual energy. More precisely:

Coordination:

$$|C_k| = \mathcal{E}_k \cdot M, \text{ where } \mathcal{E}_k = \frac{E_k}{\sum_{t=1}^{S} E_t}, \quad (1 \le k \le S).$$

Trajectory: Since each special charger assumes two-level network knowledge, it takes into account information from both mobile chargers and nodes in order to find good trajectories. Thus, special charger k prioritizes mobile charger m where

$$m = \arg\min_{j \in \mathcal{C}_k} \left\{ \alpha \cdot \frac{E_j}{E_{MC}^{max}} + (1 - \alpha) \cdot \frac{\sum_{i \in \mathcal{D}_j} E_i}{|\mathcal{D}_j| \cdot E_{nodes}^{max}} \right\}$$

with $\alpha \in (0, 1)$, a constant allowing to select the weight of each term in the sum. We use network lifetime (one of the most indicative performance metrics) to decide which is the appropriate value of parameter α in 2KCC protocol that achieves the best performance. As shown in Figure 21.8 the most suitable value is $\alpha = 1$, which is explained by the fact that when a special charger charges a mobile charger, it should take into account its energy status only.

Charging phase: Each special charger computes the percentage of energy to transfer, according to the lack of energy in the slice of the selected mobile charger compared to the total energy lack in all

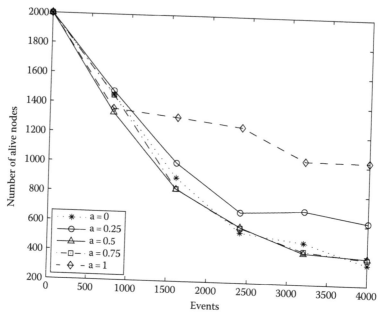

FIGURE 21.8   Alive nodes over time (varying α).

slices that this special charger is responsible for. More precisely, special charger k transfers to mobile charger m an amount of energy

$$e = c_m \cdot \left(\min\{E_m^{lack}, E_k\}\right) \text{ with } c_m = \frac{\sum_{i \in \mathcal{D}_m} E_i^{lack}}{\sum_{j \in \mathcal{C}_k} \sum_{i \in \mathcal{D}_j} E_i^{lack}} \in (0, 1),$$

where $E_i^{lack} = E_{node}^{max} - E_i$ is the amount of energy that node i can receive until it is fully charged.

21.6.6 PROTOCOL EVALUATION

We conduct a performance evaluation of the hierarchical collaborative charging protocols. In order to compare with other energy flow models, we also include measurements form a state-of-the-art noncollaborative charging protocol [23]. The protocols manage to prolong *network lifetime* (i.e., alive nodes over time) as shown in Figure 21.9a. As expected, the 2KCC outperforms the other protocols since it implements a CC process that provides the most fair partition of mobile chargers among special chargers. Despite the fact that 2KDC may not achieve the best partition since its coordination procedure takes into account only adjacent special chargers, its performance is quite close to that of 2KCC. We also observe that no knowledge no coordination (NKNC) has quite the same performance with the noncollaborative case, since it does not perform any coordination or any sophisticated trajectory procedure.

Figure 21.9b depicts the average routing robustness for our protocols. We observe that it follows the same pattern as network lifetime. This is natural since the reduction of alive nodes implies the reduction of alive neighbors. We investigate the quality of routing robustness. NKNC and noncollaborative protocols' performance is decreasing with a high rate in contrast to the 2KDC and 2KCC protocols, which achieve a better routing robustness.

We deploy 1000 random points in the network and examine how many of them are less than 2-covered, 2-covered, 3-covered, or > 3-covered over 4000 generated events. In Figure 21.10, we can

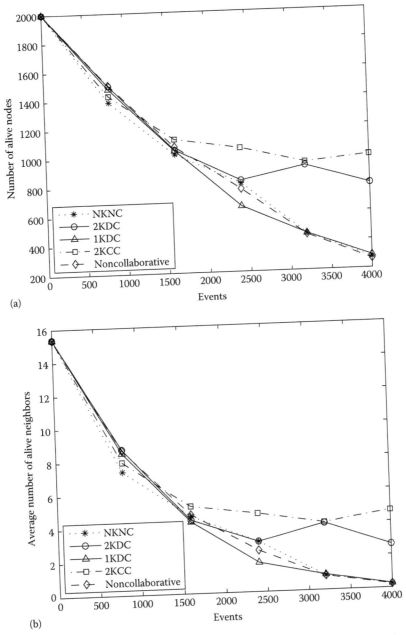

FIGURE 21.9 (a) Lifetime and (b) robustness of hierarchical collaborative charging protocols.

observe that the NKNC, noncollaborative, and 1KDC rapidly decrease the number of > 3-covered points. 2KDC and 2KCC achieve good performance, since they decrease the number of covered points in a very low rate.

By observing the performance of the aforementioned protocols, we conclude that the amount of knowledge is one of the most determinant factors. 2KDC always outperforms 1KDC and also the NKNC that has no knowledge at all. Since the coordination procedure depends on the amount of knowledge, this difference in performances indicates that the greater the amount of available knowledge the better the protocol's performance.

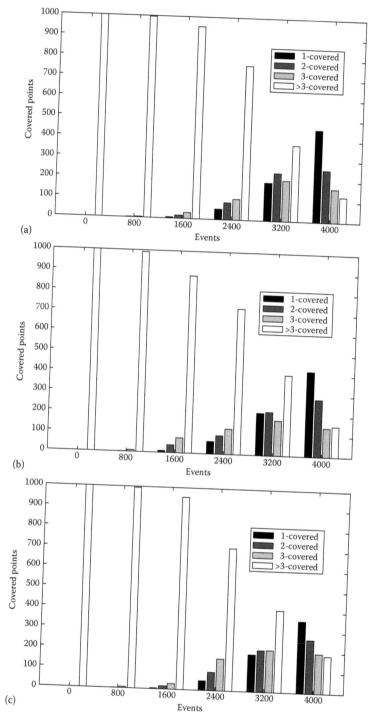

FIGURE 21.10 Coverage aging of hierarchical collaborative charging protocols. (a) Noncollaborative; (b) NKNC; and (c) 1KDC.

(*Continued*)

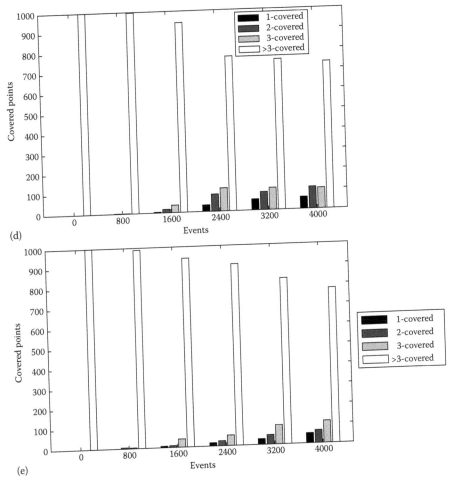

FIGURE 21.10 (*Continued*) Coverage aging of hierarchical collaborative charging protocols. (d) 2KDC, (e) 2KCC.

21.6.7 ADAPTIVITY OF THE HIERARCHICAL CONCEPT

A notable additional value of hierarchical collaborative charging is that it can be easily added on top of the noncollaborative charging protocols and further improve their performance. Figure 21.11 depicts the improvements, in terms of lifetime, of the noncollaborative protocol [23]. We transform the protocol by converting some mobile chargers to special chargers and applying hierarchy using one of our hierarchical protocols (2KCC) to achieve performance improvement. Then, we compare the initial noncollaborative protocol with our hierarchical enhanced version, as shown in Figure 21.11.

21.7 CONCLUSIONS AND FUTURE RESEARCH

Wireless ad hoc networking allows CPS nodes to communicate with each other over the wireless medium. Lack of continuous energy supply is an apparent limitation in this type of networks. Wireless power transfer using mobile chargers can be a key solution for overcoming this limitation. In this chapter, we presented some protocols designed for several critical aspects of the mobile charg-ers' configuration in order to improve energy efficiency, prolong the lifetime of the network, and also improve important network properties. The protocols are distributed and adaptive, use limited network information, can be used together with any underlying routing protocol, and are classified

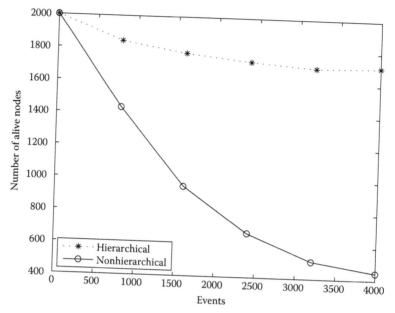

FIGURE 21.11   Hierarchical charging enhancement.

in three characteristic categories: (1) single-mobile charger protocols, (2) multiple mobile charger protocols, and (3) hierarchical collaborative mobile charging protocols. For future research, we plan to implement selected protocols in small-/medium-scale experiments, using robotic elements and wireless power transfer technology. We also plan to investigate the scheduling of mobile chargers taking into account EMR safety thresholds.

REFERENCES

1. University of California Los Angeles, Center for Embedded Networked Sensing (CENS), Los Angeles, CA. [Online]. Available: http://research.cens.ucla.edu/.
2. "The wireless power consortium." http://www.wirelesspowerconsortium.com/.
3. C. M. Angelopoulos, S. Nikoletseas, and T. P. Raptis, Wireless energy transfer in sensor networks with adaptive, limited knowledge protocols, *Computer Networks*, 70, 113–141, 2014.
4. C. M. Angelopoulos, S. Nikoletseas, T. P. Raptis, C. Raptopoulos, and F. Vasilakis, Efficient energy management in wireless rechargeable sensor networks. In *Proceedings of the 15th ACM international conference on Modeling, analysis and simulation of wireless and mobile systems (MSWiM)*, Paphos, Cyprus, 2012.
5. H. Dai, Y. Liu, G. Chen, X. Wu, and T. He, Safe charging for wireless power transfer, in *Proceedings of the 33rd Annual IEEE International Conference on Computer Communications*, ser. INFOCOM, Toronto, Canada, 2014.
6. H. Dai, Y. Liu, G. Chen, X. Wu, and T. He, SCAPE: Safe Charging with Adjustable Power. *IEEE 34th International Conference on Distributed Computing Systems (ICDCS)*, Madrid, Spain, 2014.
7. H. Dai, X. Wu, G. Chen, L. Xu, and S. Lin, Minimizing the number of mobile chargers for large-scale wireless rechargeable sensor networks, *Computer Communications*, 46, 54–65, 2014. [Online]. Available: http://www.sciencedirect.com/science/article/pii/S0140366414000875.
8. H. Dai, L. Xu, X. Wu, C. Dong, and G. Chen, Impact of mobility on energy provisioning in wireless rechargeable sensor networks, in *2013 IEEE Wireless Communications and Networking Conference (WCNC)*, Shanghai, China, April 2013, pp. 962–967.
9. M. R. Garey and D. S. Johnson, *Computers and Intractability; A Guide to the Theory of NP-Completeness*. New York: W. H. Freeman & Co., New York, 1990.
10. A. Kansal, J. Hsu, S. Zahedi, and M. B. Srivastava, Power management in energy harvesting sensor networks, *ACM Transactions on Embedded Computing Systems*, 6(4), September 2007. [Online]. Available: http://doi.acm.org/10.1145/1274858.1274870.

11. K.-D. Kim and P. Kumar, Cyber-physical systems: A perspective at the centennial, *Proceedings of the IEEE*, 100, (Special Centennial Issue), 1287–1308, May 2012.

12. E. A. Lee and S. A. Seshia, *Introduction to Embedded Systems—A Cyber-Physical Systems Approach*, 1st ed., 2010. [Online]. Available: http://chess.eecs.berkeley.edu/pubs/794.html.

13. J. Li, M. Zhao, and Y. Yang, Ower-mdg: A novel energy replenishment and data gathering mechanism in wireless rechargeable sensor networks, in *2012 IEEE Global Communications Conference (GLOBECOM)*, Anaheim, California, December 2012, pp. 5350–5355.

14. Y. Li, Z. Wang, and Y. Song, Wireless sensor network design for wildfire monitoring, in *The Sixth World Congress on, Intelligent Control and Automation, 2006. WCICA 2006*, Dalian, China, vol. 1, 2006, pp. 109–113.

15. Z. Li, Y. Peng, W. Zhang, and D. Qiao, Study of joint routing and wireless charging strategies in sensor networks, in *Proceedings of the Fifth International Conference on Wireless Algorithms, Systems, and Applications*, ser. WASA'10. Berlin, Heidelberg, Germany: Springer-Verlag, 2010, pp. 125–135. [Online]. Available: http://dl.acm.org/citation.cfm?id=1881353.1881374.

16. Z. Li, Y. Peng, W. Zhang, and D. Qiao, J-roc: A joint routing and charging scheme to prolong sensor network lifetime, in *Proceedings of the 2011 19th IEEE International Conference on Network Protocols*, ser. ICNP '11. Washington, DC: IEEE Computer Society, 2011, pp. 373–382. [Online]. Available: http://dx.doi.org/10.1109/ICNP.2011.6089076.

17. A. Madhja, S. Nikoletseas, and T. P. Raptis, Efficient, distributed coordination of multiple mobile chargers in sensor networks, in *Proceedings of the 16th ACM International Conference on Modeling, Analysis and Simulation of Wireless and Mobile Systems*, ser. MSWiM, Barcelona, Spain, 2013.

18. A. Madhja, S. Nikoletseas, and T. P. Raptis, Hierarchical, collaborative wireless charging in sensor networks, in *Proceedings of the IEEE Wireless Communications and Networking Conference*, ser. WCNC, New Orleans, Louisiana, 2015.

19. A. Mainwaring, D. Culler, J. Polastre, R. Szewczyk, and J. Anderson, Wireless sensor networks for habitat monitoring, in *Proceedings of the First ACM International Workshop on Wireless Sensor Networks and Applications*, ser. WSNA '02. New York: ACM, 2002, pp. 88–97. [Online]. Available: http://doi.acm.org/10.1145/570738.570751.

20. Y. Peng, Z. Li, W. Zhang, and D. Qiao, Prolonging sensor network lifetime through wireless charging, in *2010 IEEE 31st Real-Time Systems Symposium (RTSS)*, San Diego, California, November 2010, pp. 129–139.

21. T. Rault, A. Bouabdallah, and Y. Challal, Multi-hop wireless charging optimization in low-power networks, in *2013 IEEE Global Communications Conference (GLOBECOM)*, Atlanta, Georgia, December 2013, pp. 462–467.

22. N. Tesla, Apparatus for transmitting electrical energy, Patent US 1 119 732, December, 1914.

23. C. Wang, J. Li, F. Ye, and Y. Yang, Multi-vehicle coordination for wireless energy replenishment in sensor networks, in *2013 IEEE 27th International Symposium on Parallel Distributed Processing (IPDPS)*, Cambridge Boston, Massachusetts, May 2013, pp. 1101–1111.

24. G. Werner-Allen, K. Lorincz, M. Ruiz, O. Marcillo, J. Johnson, J. Lees, and M. Welsh, Deploying a wireless sensor network on an active volcano, *Internet Computing, IEEE*, 10(2), 18–25, March 2006.

25. F.-J. Wu, Y.-F. Kao, and Y.-C. Tseng, From wireless sensor networks towards cyber physical systems, *Pervasive and Mobile Computing*, 7(4), 397–413, 2011. [Online]. Available: http://www.sciencedirect.com/science/article/pii/S1574119211000368.

26. F. Xia, X. Kong, and Z. Xu, Cyber-physical control over wireless sensor and actuator networks with packet loss, in *Wireless Networking Based Control*, S. K. Mazumder, Ed. Springer, New York, 2011, pp. 85–102. [Online]. Available: http://dx.doi.org/10.1007/978-1-4419-7393-1_4.

27. L. Xiang, J. Luo, K. Han, and G. Shi, Fueling wireless networks perpetually: A case of multi-hop wireless power distribution, in *2013 IEEE 24th International Symposium on Personal Indoor and Mobile Radio Communications (PIMRC)*, London, UK, September 2013, pp. 1994–1999.

28. L. Xie, Y. Shi, Y. T. Hou, and H. D. Sherali, Making sensor networks immortal: An energy-renewal approach with wireless power transfer, *IEEE/ACM Transactions on Networking*, 20(6), 1748–1761, December 2012.

29. L. Xie, Y. Shi, Y. Hou, and A. Lou, Wireless power transfer and applications to sensor networks, *Wireless Communications, IEEE*, 20(4), 140–145, August 2013.

30. L. Xie, Y. Shi, Y. Hou, W. Lou, H. Sherali, and S. Midkiff, On renewable sensor networks with wireless energy transfer: The multi-node case, in *2012 9th Annual IEEE Communications Society Conference on Sensor, Mesh and Ad Hoc Communications and Networks (SECON)*, Seoul, Korea, June 2012, pp. 10–18.

31. L. Xie, Y. Shi, Y. Hou, W. Lou, H. Sherali, and S. Midkiff, Bundling mobile base station and wireless energy transfer: Modeling and optimization, in *2013 Proceedings IEEE INFOCOM*, Turin, Italy, April 2013, pp. 1636–1644.

32. W. Yao, M. Li, and M.-Y. Wu, Inductive charging with multiple charger nodes in wireless sensor networks, in *Advanced Web and Network Technologies, and Applications*, ser. Lecture Notes in Computer Science, H. Shen, J. Li, M. Li, J. Ni, and W. Wang, Eds. Springer, Berlin Heidelberg, Germany, 2006, vol. 3842, pp. 262–270. [Online]. Available: http://dx.doi.org/10.1007/11610496_34.

33. S. Zhang, J. Wu, and S. Lu, Collaborative mobile charging for sensor networks, in *2012 IEEE 9th International Conference on Mobile Adhoc and Sensor Systems (MASS)*, Las Vegas, NV, October 2012, pp. 84–92.

34. M. Zhao, J. Li, and Y. Yang, Joint mobile energy replenishment and data gathering in wireless rechargeable sensor networks, in *Proceedings of the 23rd International Teletraffic Congress*, ser. ITC '11. International Teletraffic Congress, San Francisco, CA, 2011, pp. 238–245. [Online]. Available: http://dl.acm.org/citation.cfm?id=2043468.2043506.

Index

Index